T0214168

Communications in Computer and Information Science   **1145**

Commenced Publication in 2007
Founding and Former Series Editors:
Phoebe Chen, Alfredo Cuzzocrea, Xiaoyong Du, Orhun Kara, Ting Liu,
Krishna M. Sivalingam, Dominik Ślęzak, Takashi Washio, Xiaokang Yang,
and Junsong Yuan

More information about this series at http://www.springer.com/series/7899

Milojica Jaćimović · Michael Khachay ·
Vlasta Malkova · Mikhail Posypkin (Eds.)

Optimization
and Applications

10th International Conference, OPTIMA 2019
Petrovac, Montenegro, September 30 – October 4, 2019
Revised Selected Papers

 Springer

Editors
Milojica Jaćimović 🆔
University of Montenegro
Podgorica, Montenegro

Vlasta Malkova 🆔
FRC CSC RAS
Moscow, Russia

Michael Khachay 🆔
Krasovskii Institute of Mathematics
and Mechanics
Yekaterinburg, Russia

Mikhail Posypkin 🆔
FRC CSC RAS
Moscow, Russia

ISSN 1865-0929   ISSN 1865-0937 (electronic)
Communications in Computer and Information Science
ISBN 978-3-030-38602-3   ISBN 978-3-030-38603-0 (eBook)
https://doi.org/10.1007/978-3-030-38603-0

This Springer imprint is published by the registered company Springer Nature Switzerland AG
The registered company address is: Gewerbestrasse 11, 6330 Cham, Switzerland

Preface

This volume contains the refereed proceedings of the 10th International Conference on Optimization and Applications (OPTIMA 2019)[1]. The goal of the conference is to bring together researchers and practitioners working in the field of optimization theory, methods, software, and related areas. Organized annually since 2009, the conference attracted a significant number of researchers, academics, and specialists in many fields of optimization, operations research, optimal control, game theory, and their numerous applications in practical problems of operations research, data analysis, and software development.

The broad scope of OPTIMA made it an event where researchers involved in different domains of optimization theory and numerical methods, investigating continuous and discrete extremal problems, designing heuristics and algorithms with theoretical bounds, developing optimization software and applying optimization techniques to highly relevant practical problems, can meet together and discuss their approaches and results. We strongly believe that this facilitates collaboration between researchers working in optimization theory, methods, and applications.

The conference was held during September 30 – October 4, 2019, in Petrovac, Montenegro, at the picturesque Budvanian Riviera on the azure Adriatic coast. By tradition, the main organizers of the conference were the Montenegrin Academy of Sciences and Arts, Dorodnicyn Computing Centre FRC CSC RAS, and the University of Evora. This year, the key topics of OPTIMA were grouped into five tracks:

(1) Mathematical Programming
(2) Combinatorial and Discrete Optimization
(3) Optimal Control
(4) Optimization in the Economy, Finance, and Social Sciences
(5) Applications

In the framework of the conference, a special section was held dedicated to the anniversary of Academician of the Russian Academy of Sciences, Yuri Evtushenko, a world-famous scientist in the field of computational mathematics and one of the founders of the conference.

The Program Committee (PC) and the reviewers of the conference included 113 well-known experts in continuous and discrete optimization, optimal control and game theory, data analysis, mathematical economy, and related areas from leading institutions of 25 countries including Argentina, Australia, Austria, Azerbaijan, Belgium, Finland, France, Germany, Greece, India, Israel, Italy, Kazakhstan, Lithuania, Mexico, Montenegro, Netherlands, Poland, Portugal, Russia, Serbia, Sweden, Taiwan, Ukraine, UK, and USA. This year we have received 117 submissions mostly from Russia but also from Azerbaijan, Bosnia and Herzegovina, France, Germany, India, Italy,

---

[1] http://www.agora.guru.ru/display.php?conf=optima-2019.

Kazakhstan, Montenegro, Poland, Portugal, Sweden, and Ukraine. 81 full papers were considered for review. Each submission was reviewed by at least three PC members or invited reviewers, experts in their fields, to supply detailed and helpful comments. The committee decided to accept 35 papers.

The conference featured four invited lectures, as well as several plenary and keynote talks. The invited lectures were:

- Prof. Yurii Nesterov (Center for Operations Research and Econometrics, Belgium), "Relative Smoothness: New Paradigm in Convex Optimization"
- Prof. Samir Adly (University of Limoges, France), "Quasistatic Evolution Variational Inequalities and Sweeping Process"
- Prof. Nikolai Osmolovskii (Systems Research Institute, Poland), "Necessary Conditions for an Extended Weak Minimum in Optimal Control Problems with Volterra-Type Integral Equations on a Variable Time Interval"
- Prof. Janez Povh (University of Ljubljana, Slovenia), "High-Performance Optimization"

The plenary talks were presented by:

- Prof. Anatoly Antipin (Dorodnicyn Computing Centre FRC CSC RAS, Russia) on "On methods for solving a terminal control problem with intermediate phase constraints"
- Prof. Alexey Tret'yakov (Siedlce University, Poland) on "New perspective on some basic results in optimization"
- Prof. Vladimir Krivonozhko (National University of Science and Technology MISiS, Russia) on "Three-dimensional visualization for multidimensional analysis and performance management of convex and non-convex systems"

We would like to thank all the authors for submitting their papers and the members of the PC for their efforts in providing exhaustive reviews. We would also like to express special gratitude to all the invited lectures and plenary speakers.

October 2019

Milojica Jaćimović
Michael Khachay
Vlasta Malkova
Mikhail Posypkin

Organization

Program Committee Chairs

Milojica Jaćimović	Montenegrin Academy of Sciences and Arts, Montenegro
Yuri G. Evtushenko	Dorodnicyn Computing Centre, FRC CSC RAS, Russia
Maksat Kalimoldayev	Institute of Information and Computational Technologies, Kazakhstan

Program Committee

Majid Abbasov	St. Petersburg State University, Russia
Samir Adly	University of Limoges, France
Kamil Aida-Zade	Institute of Control Systems of ANAS, Azerbaijan
Alla Albu	Dorodnicyn Computing Centre, FRC CSC RAS, Russia
Alexander P. Afanasiev	Institute for Information Transmission Problems, RAS, Russia
Yedilkhan Amirgaliyev	Suleyman Demirel University, Kazakhstan
Anatoly S. Antipin	Dorodnicyn Computing Centre, FRC CSC RAS, Russia
Sergey Astrakov	Institute of Computational Technologies, Siberian Branch RAS, Russia
Evripidis Bampis	LIP6 UPMC, France
Oleg Burdakov	Linköping University, Sweden
Olga Battaïa	ISAE-SUPAERO, France
Armen Beklaryan	National Research University Higher School of Economics, Russia
Vladimir Beresnev	Sobolev Institute of Mathematics, Russia
René Van Bevern	Novosibirsk State University, Russia
Sergiy Butenko	Texas A&M University, USA
Vladimir Bushenkov	University of Evora, Portugal
Igor A. Bykadorov	Sobolev Institute of Mathematics, Russia
Alexey Chernov	Moscow Institute of Physics and Technology, Russia
Duc-Cuong Dang	INESC TEC, Portugal
Tatjana Davidovic	Mathematical Institute of Serbian Academy of Sciences and Arts, Serbia
Stephan Dempe	TU Bergakademie Freiberg, Germany
Alexandre Dolgui	IMT Atlantique, LS2N, CNRS, France
Olga Druzhinina	FRC CSC RAS, Russia
Anton Eremeev	Omsk Branch of Sobolev Institute of Mathematics, SB RAS, Russia
Adil Erzin	Novosibirsk State University, Russia
Francisco Facchinei	University of Rome La Sapienza, Italy
Alexander V. Gasnikov	Moscow Institute of Physics and Technology, Russia

Manlio Gaudioso	Università della Calabria, Italy
Alexander I. Golikov	Dorodnicyn Computing Centre, FRC CSC RAS, Russia
Alexander Yu. Gornov	Institute System Dynamics and Control Theory, SB RAS, Russia
Edward Kh. Gimadi	Sobolev Institute of Mathematics, SB RAS, Russia
Andrei Gorchakov	Dorodnicyn Computing Centre, FRC CSC RAS, Russia
Alexander Grigoriev	Maastricht University, The Netherlands
Mikhail Gusev	N.N. Krasovskii Institute of Mathematics and Mechanics, Russia
Viktor Izhutkin	MPEI, Russia
Vladimir Jaćimović	University of Montenegro, Montenegro
Vyacheslav Kalashnikov	ITESM, Mexico
Valeriy Kalyagin	Higher School of Economics, Russia
Igor E. Kaporin	Dorodnicyn Computing Centre, FRC CSC RAS, Russia
Alexander Kazakov	Matrosov Institute for System Dynamics and Control Theory, SB RAS, Russia
Alexander V. Kelmanov	Sobolev Institute of Mathematics, Russia
Mikhail Yu. Khachay	Krasovsky Institute of Mathematics and Mechanics, Russia
Oleg V. Khamisov	L. A. Melentiev Energy Systems Institute, Russia
Andrey Kibzun	Moscow Aviation Institute, Russia
Donghyun Kim	Kennesaw State University, USA
Roman Kolpakov	Moscow State University, Russia
Igor Konnov	Kazan University, Russia
Alexander Kononov	Sobolev Institute of Mathematics, Russia
Vera Kovacevic-Vujcic	University of Belgrade, Serbia
Yury A. Kochetov	Sobolev Institute of Mathematics, Russia
Pavlo A. Krokhmal	University of Arizona, USA
Ilya Kurochkin	Institute for Information Transmission Problems, RAS, Russia
Dmitri E. Kvasov	University of Calabria, Italy
Alexander A. Lazarev	V.A. Trapeznikov Institute of Control Sciences, Russia
Vadim Levit	Ariel University, Israel
Bertrand M. T. Lin	National Chiao Tung University, Taiwan
Alexander V. Lotov	Dorodnicyn Computing Centre, FRC CSC RAS, Russia
Nikolay Lukoyanov	N.N. Krasovskii Institute of Mathematics and Mechanics, Russia
Vittorio Maniezzo	University of Bologna, Italy
Olga Masina	Bunin Yelets State University, Russia
Vladimir Mazalov	Institute of Applied Mathematical Research, Karelian Research Center, Russia
Nevena Mijajlović	University of Montenegro, Montenegro
Nenad Mladenovic	Mathematical Institute, Serbian Academy of Sciences and Arts, Serbia
Angelia Nedich	University of Illinois at Urbana Champaign, USA
Yurii Nesterov	CORE, Université Catholique de Louvain, Belgium
Yuri Nikulin	University of Turku, Finland

Evgeni Nurminski	FEFU, Russia
Nicholas N. Olenev	CEDIMES-Russie, Dorodnicyn Computing Centre, FRC CSC RAS, Russia
Panos Pardalos	University of Florida, USA
Alexander V. Pesterev	V.A. Trapeznikov Institute of Control Sciences, Russia
Alexander Petunin	Ural Federal University, Russia
Stefan Pickl	Bundeswehr University Munich, Germany
Boris T. Polyak	V.A. Trapeznikov Institute of Control Sciences, Russia
Yury S. Popkov	Institute for Systems Analysis, FRC CSC RAS, Russia
Leonid Popov	IMM UB RAS, Russia
Igor G. Pospelov	Dorodnicyn Computing Centre, FRC CSC RAS, Russia
Mikhail A. Posypkin	Dorodnicyn Computing Centre, FRC CSC RAS, Russia
Oleg Prokopyev	University of Pittsburgh, USA
Artem Pyatkin	Novosibirsk State University, Sobolev Institute of Mathematics, Russia
Ioan Bot Radu	University of Vienna, Austria
Soumyendu Raha	Indian Institute of Science, India
Andrei Raigorodskii	Moscow State University, Russia
Larisa Rybak	BSTU named after V. G. Shoukhov, Russia
Leonidas Sakalauskas	Institute of Mathematics and Informatics, Lithuania
Eugene Semenkin	Siberian State Aerospace University, Russia
Yaroslav D. Sergeyev	University of Calabria, Italy
Natalia Shakhlevich	University of Leeds, UK
Aleksandr Shananin	Moscow Institute of Physics and Technology, Russia
Angelo Sifaleras	University of Macedonia, Greece
Alexander A. Shananin	Moscow Institute of Physics and Technology, Russia
Mathias Staudigl	Maastricht University, The Netherlands
Petro Stetsyuk	V.M. Glushkov Institute of Cybernetics, Ukraine
Alexander Strekalovskiy	Matrosov Institute for System Dynamics and Control Theory, SB RAS, Russia
Vitaly Strusevich	University of Greenwich, UK
Michel Thera	University of Limoges, France
Tatiana Tchemisova	University of Aveiro, Portugal
Anna Tatarczak	Maria Curie Skłodowska University, Poland
Alexey A. Tretyakov	Dorodnicyn Computing Centre, FRC CSC RAS, Russia
Stan Uryasev	University of Florida, USA
Vladimir Voloshinov	Kharkevich Institute for Information Transmission Problems, RAS, Russia
Frank Werner	Otto-von-Guericke-Universität, Germany
Adrian Will	National Technological University, Argentina
Oleg Zaikin	Institute for System Dynamics and Control Theory, SB RAS, Russia
Vitaly G. Zhadan	Dorodnicyn Computing Centre, FRC CSC RAS, Russia
Anatoly A. Zhigljavsky	Cardiff University, UK
Julius Žilinskas	Vilnius University, Lithuania
Yakov Zinder	University of Technology, Australia

Tatiana V. Zolotova	Financial University under the Government of the Russian Federation, Russia
Vladimir I. Zubov	Dorodnicyn Computing Centre, FRC CSC RAS, Russia
Anna V. Zykina	Omsk State Technical University, Russia

Organizing Committee Chairs

Milojica Jaćimović	Montenegrin Academy of Sciences and Arts, Montenegro
Yuri G. Evtushenko	Dorodnicyn Computing Centre, FRC CSC RAS, Russia
Maksat Kalimoldayev	Institute of Information and Computational Technologies, Kazakhstan
Mikhail Posypkin	Dorodnicyn Computing Centre, FRC CSC RAS, Russia

Organizing Committee

Natalia Burova	Dorodnicyn Computing Centre, FRC CSC RAS, Russia
Alexander Golikov	Dorodnicyn Computing Centre, FRC CSC RAS, Russia
Alexander Gornov	Institute of System Dynamics and Control Theory, SB RAS, Russia
Vesna Dragović	Montenegrin Academy of Sciences and Arts, Montenegro
Vladimir Jaćimović	University of Montenegro, Montenegro
Mikhail Khachay	Krasovsky Institute of Mathematics and Mechanics, Russia
Alexander Kelmanov	Sobolev Institute of Mathematics, Russia
Yury Kochetov	Sobolev Institute of Mathematics, Russia
Vlasta Malkova	Dorodnicyn Computing Centre, FRC CSC RAS, Russia
Nevena Mijajlović	University of Montenegro, Montenegro
Oleg Obradovic	University of Montenegro, Montenegro
Nicholas Olenev	Dorodnicyn Computing Centre, FRC CSC RAS, Russia
Tatiana Tchemisova	University of Aveiro, Portugal
Yulia Trusova	Dorodnicyn Computing Centre, FRC CSC RAS, Russia
Svetlana Vladimirova	Dorodnicyn Computing Centre, FRC CSC RAS, Russia
Victor Zakharov	FRC CSC RAS, Russia
Ivetta Zonn	Dorodnicyn Computing Centre, FRC CSC RAS, Russia
Vladimir Zubov	Dorodnicyn Computing Centre, FRC CSC RAS, Russia

Additional Reviewers

Anton Anikin
Artem Baklanov
Vladimir Berikov
Pavel Borisovsky
Alexander Filatov
Stefania Funari

Konstantin Kobylkin
Stepan Kochemazov
Eloisa Macedo
Yury Morozov
Ekaterina Neznakhina
Yuri Ogorodnikov

Anna Panasenko
Alexander Plyasunov
Nikolay Pogodaev
Leonid D. Popov
Anna Romanova
Marina Sandomirskaya
Alexander Sesekin

Maxim Shishlenin
Stepan Sorokin
Maxim Staritsyn
Sergey Stepanov
Alexander Vasin
Elena Yanovskaya
Vyacheslav Zalyubovskiy

Invited Talks

Quasistatic Evolution Variational Inequalities and Sweeping Process

Samir Adly ⓘ

Université de Limoges, France
samir.adly@unilim.fr

Abstract. In this talk, we study a new variant of the Moreau's sweeping process with velocity constraint. Based on an adapted version of the Moreau's catching-up algorithm, we show the well-posedness (in the sense existence and uniqueness) of this problem in a general framework. We show the equivalence between this implicit sweeping process and a quasistatic evolution variational inequality.

It is well-known that the variational formulation of many mechanical problems with unilateral contact and friction lead to an evolution variational inequality. As an application, we reformulate the quasistatic antiplane frictional contact problem for linear elastic materials with short memory as an implicit sweeping process with velocity constraint.

The link between the implicit sweeping process and the quasistatic evolution variational inequality is possible thanks to some standard tools from convex analysis and is new in the literature.

Keyword: Moreau's sweeping process · Catching-up algorithm · Quasistatic evolution variational inequality

Relative Smoothness: New Paradigm in Convex Optimization

Yurii Nesterov ⓘ

CORE/INMA, Universite Catholique de Louvain, Belgium
Yurii.Nesterov@uclouvain.be

Abstract. Development and computational abilities of optimization methods crucially depend on the auxiliary tools provided to them by the method's designers. During the first decades of Convex Optimization, the methods were based either on the proximal setup, allowing Euclidean projections onto the basic feasible sets, or on the linear minimization framework, which assumes a possibility to minimize a linear function over the feasible set.

However, recently it was realized that any possibility of simple minimization of an auxiliary convex function leads to the efficient minimization methods for some family of more general convex functions, which are compatible with the first one. This compatibility condition, called relative smoothness, was firstly exploited for smooth convex functions (Bauschke, Bolt and Teboulle, 2016) and smooth strongly convex functions (Lu, Freund and Nesterov, 2018).

In this talk we make the final step and show how to extend this framework onto the class of nonsmooth functions. We also discuss possible consequences and applications.

Keywords: Convex Optimization · Relative smoothness · Nonsmooth functions

Necessary Conditions for an Extended Weak Minimum in Optimal Control Problems with Volterra-Type Integral Equations on a Variable Time Interval

Nikolai Osmolovskii (ID)

Systems Research Institute, Polish Academy of Sciences, Poland
osmolovski@uph.edu.pl

Abstract. We discuss an optimal control problem with Volterra-type integral equations, considered on a non-fixed time interval, subject to endpoint constraints of equality and inequality type, mixed state-control constraints of inequality and equality type, and pure state constraints of inequality type.

The main assumption is the uniform linear-positive independence of the gradients of active mixed constraints with respect to the control. We formulate first-order necessary optimality conditions for an extended weak minimum, the notion of which is a natural generalization of the notion of weak minimum with account of variations of the time. The conditions obtained generalize the corresponding ones for problems with ordinary differential equations.

This is a joint work with Andrei V. Dmitruk.

Keywords: Optimal control · Volterra-type integral equations · Extended weak minimum · Differential equations

High-Performance Optimization

Janez Povh (ID)

University of Ljubljana, Slovenia
`janez.povh@fs.uni-lj.si`

Abstract. High-Performance Computing (HPC) –with its state-of-the-art computing and storage infrastructure and the related knowledge– is an ecosystem that is essential for scientific research and industrial development. The European Commission (EC) often points out the opportunities and challenges at the interface of Big Data, High-Performance Computing and Mathematics. The recent HiPEAC Vision 2017 clearly states that Mathematics and Algorithms for extreme scale HPC systems is one out of seven current EU research priorities related to HPC. Nevertheless, the recent results of the Partnership for Advanced Computing in Europe (PRACE) reveals that the mathematical research community, including mathematical optimization, rarely decides to use these tools, although they usually do research in hard mathematical optimization problems.

In the first part of the talk we will review what exactly the strongest supercomputers within the EU offer the mathematical optimization community: what is currently the best public HPC infrastructure, how to get access to it, and how to get the necessary skills.

In the second part of the paper we will present a parallel Branch and Bound (B&B) based algorithm to solve the optimality small to medium size instances of non-convex quadratic binary problems with linear constraints. It is is available as an online solver BiqBin, running on the supercomputer owned by the University of Ljubljana, Faculty of Mechanical Engineering. This algorithm encompasses the best non-linear optimization techniques and is carefully encoded to run efficiently in parallel using state-of-the-art libraries for parallel linear algebra operations. It's online availability demonstrates new ways of how to bring high-performance scientific code closer to scientific users. We will present few implementation details and numerical results obtained by this code.

Keywords: Optimization · High-Performance Computing · Supercomputers · Branch and Bound · Parallel algorithm · BiqBin solver

Contents

Efficient Algorithms for the Routing Open Shop with Unrelated Travel Times on Cacti

Ilya Chernykh[1,2,3(✉)] and Olga Krivonogova[2]

[1] Sobolev Institute of Mathematics,
Koptyug Avenue 4, Novosibirsk 630090, Russia
idchern@math.nsc.ru
[2] Novosibirsk State University,
Pirogova Street 2, Novosibirsk 630090, Russia
krivonogova.olga@gmail.com
[3] Novosibirsk State Technical University,
Marksa Avenue 20, Novosibirsk 630073, Russia

Abstract. The object of investigation is the routing open shop problem, in which a fleet of machines have to visit all the nodes of a given transportation network to perform operations on some jobs located at those nodes. Each machine has to visit each node, to process each job and to return back to the common initial location—the depot. Operations of each job can be processed in an arbitrary sequence, any machine may perform at most one operation at a time. The goal is to construct a feasible schedule to minimize the makespan. The routing open shop problem is known to be NP-hard even in the simplest two-machine case with the transportation network consisting of just two nodes (including the depot). We consider a certain generalization of this problem in which travel times are individual for each of the two machines and the structure of the transportation network is an arbitrary cactus. We generalize an instance reduction algorithm known for the problem on a tree with identical travel times, and use it to describe new polynomially solvable cases for the problem, as well as an efficient approximation algorithm for another special case with a tight approximation ratio guarantee.

Keywords: Routing open shop · Unrelated travel times · Instance reduction · Polynomially solvable subcase · Optima localization

1 Introduction

The routing open shop problem is a natural combination of two classical discrete optimization problems: the metric traveling salesman problem (TSP) and

This research was supported by the program of fundamental scientific researches of the SB RAS No I.5.1., project No 0314-2019-0014, and by the Russian Foundation for Basic Research, projects 17-01-00170, 17-07-00513 and 18-01-00747, and by the Russian Ministry of Science and Education under the 5–100 Excellence Programme.

© Springer Nature Switzerland AG 2020
M. Jaćimović et al. (Eds.): OPTIMA 2019, CCIS 1145, pp. 1–15, 2020.
https://doi.org/10.1007/978-3-030-38603-0_1

the open shop scheduling problem. The TSP hardly needs an introduction. The open shop problem was introduced in [13] and can be described as follows. There is a set of n *jobs* $\mathcal{J}$, each of those has to be processed by each of m *machines* from a set $\mathcal{M}$ in arbitrary order. Operation of processing of job J_j by machine M_i takes a solid time interval of given length p_{ji}. Such intervals requiring the same job or the same machine cannot overlap. The goal is to minimize the *makespan* $C_{\max}$—the completion time of the last operation. Surprisingly, such a combination of machine scheduling problem and a routing problem independently appeared while considering tasks arising both in production (see, e.g. [2,3]), so in the service industry [11,25], definitely adding value to its practical significance.

According to the traditional three-field notation for scheduling problems (see [18] for example) the open shop problem with m machines is denoted by $Om||C_{\max}$. Notation $O||C_{\max}$ is used in case when the number of machines is a part of an input and is not bounded by any constant. The problem $O2||C_{\max}$ is solvable to the optimum in linear time, while for $m \geq 3$ the $Om||C_{\max}$ problem is NP-hard [13]. It is still an open question whether the $Om||C_{\max}$ problem is strongly NP-hard; a PTAS for any constant m is described in [19]. However, the $O||C_{\max}$ problem is strongly NP-hard. Moreover, there is no ρ-approximation algorithm for that problem with $\rho < \frac{5}{4}$ unless $P = NP$ [24]. On the other hand, a simple 2-approximation greedy algorithm for $O||C_{\max}$ is presented in [1].

We consider the routing open shop problem with *unrelated travel times* which generalizes both the open shop and the TSP in the following way. Jobs representing some unmovable objects are located at the nodes of a transportation network described by an edge-weighted graph $G = \langle V; E \rangle$. Machines are mobile and are initially located at some predefined node $v_0 \in V$ referred to as *the depot*. The weight $\tau(e)$ of edge $e = [v, u] \in E$ is a vector $(\tau_1(e), \ldots, \tau_m(e))$ where $\tau_i(e)$ is a travel time for machine M_i over the edge e in any direction. Machines have to travel over the network to perform operations of each job from $\mathcal{J}$. Machines are allowed to visit each node multiple times and to use the shortest paths between the nodes, therefore we may assume that travel times obey the triangle inequality.

Any number of machines can travel simultaneously over the same edge in any direction. The feasibility restrictions from the open shop problem are still in place, and a machine has to be at node v to perform operations of jobs located at v. All the machines have to return to the depot after completing all the operations. The latest *return moment* $R_{\max}$ is the makespan for this model and has to be minimized. We assume that each node with a possible exclusion of the depot contains at least one job, therefore each machine has to visit each node of G, and hence the routing open shop problem contains the metric TSP as a special case and is strongly NP-hard even for a single machine.

Following the traditional notation we denote the m-machine routing open shop problem with unrelated travel times by $ROm|Rtt|R_{\max}$. Notation Rtt stands for un**R**elated **t**ravel **t**imes and is omitted in case when travel times are identical. Additional notation $G = X$ is used if we want to specify the structure of the transportation network. In this case X is substituted with some well-known classic notation from the graph theory, like K_p for a complete graph with p nodes, *tree*, *chain*, *cycle* or similar.

The $ROm||R_{\max}$ problem was introduced in [4,5]. It was proved in [5] that the $RO2|G = K_2|R_{\max}$ problem—the simplest non-trivial case of the problem under consideration—is already NP-hard. The $RO2|G = K_2|R_{\max}$ problem is thoroughly studied in [4], where a $\frac{6}{5}$-approximation algorithm is presented. An FPTAS for the $RO2|G = K_2|R_{\max}$ problem was described later in [17], however the algorithm from [4] has one extremal property: its worst-case performance ratio guarantee is the best possible with respect to the *standard lower bound* $\bar{R}$ (see Sect. 2 for details). This means that the algorithm actually builds a schedule S with $R_{\max}(S) \leqslant \frac{6}{5}\bar{R}$, and there exists an instance for which the optimal makespan is equal to $\frac{6}{5}\bar{R}$, and thus the approximation ratio cannot be reduced in terms of $\bar{R}$. In other words, an optimal makespan for the $RO2|G = K_2|R_{\max}$ problem belongs to the *optima localization interval* $[\bar{R}, \frac{6}{5}\bar{R}]$ with tight bounds. Lately it was proved that the same optima localization interval holds for the $RO2|G = K_3|R_{\max}$ [7] and $RO2|G = tree|R_{\max}$ problems [6]. The latter result is based on an instance reduction procedure, similar to the one described in [9]. This procedure in turn uses *job aggregation* and *terminal edge contraction* operations (see Sect. 2 for details). As for the problem with individual travel times, recently it was proved that for any instance of the $RO2|Rtt, G = K_3|R_{\max}$ problem its optimum belongs to the interval $[\bar{R}, \frac{5}{4}\bar{R}]$ [8]. It was also shown that this interval is tight even for the case with $G = K_2$ and proportional travel times.

While our research focuses on the two-machine version of the problem, the progress made in the study of the m-machine problem also should be noted. A series of approximation algorithms for the $ROm||R_{\max}$ problem was developed, starting with the $\frac{m+4}{7}2$-approximation [5]. The best known algorithm up to date has the approximation ration guarantee of $O(\log m)$ [16]. A number of papers is devoted to the study of a special case with unit processing times [12,22,23].

In this paper we study the two-machine routing open shop problem with unrelated travel times. First, we generalize an instance reduction procedure for the $RO2|Rtt, G = cactus|R_{\max}$ problem by the introduction of a new instance simplification operation, namely *terminal cycle contraction* (Sect. 3). In Sect. 4 we study an optima localization interval for the special case of the $RO2|Rtt, G = cactus|R_{\max}$ problem and prove that for any instance of this problem its optimum belongs to the same interval $[\bar{R}, \frac{5}{4}\bar{R}]$ as for the problem with $G = K_3$. Section 2 contains the formal problem formulation and some preliminary results, while the last Sect. 5 contains conclusive remarks and open questions for the future investigation.

2   Problem Formulation and Preliminary Notes

2.1   Formulation and Notation

Let us give a formal description of the $RO2|Rtt|R_{\max}$ problem.

The machines M_1 and M_2 and the set of jobs $\mathcal{J} = \{J_1, \ldots, J_n\}$ are given. Each job J_j consists of two operations O_{j1} and O_{j2} to be processed by the machines M_1 and M_2 respectively. The processing of operation O_{ji} takes a solid time interval of length p_{ji}. A transportation network is represented by an edge-weighted connected graph $G = \langle V, E \rangle$. A weight of an edge $e = [u, v] \in E$ is a couple $(\tau_1(e), \tau_2(e))$ with $\tau_i(e)$ being a travel time for the machine M_i over an edge e. Jobs are distributed among the nodes of V, a set of jobs located at v is denoted by $\mathcal{J}(v)$, each node contains at least one job. The machines are mobile and are initially located at the given *depot* denoted by $v_0 \in V$. Machines have to travel over the network, perform their operations and to come back to the depot as soon as possible. Any number of machines can travel over the same edge in any direction in the same time. We assume that machines take the shortest path while traveling from v to u. Respective travel time for machine M_i is denoted by $\mathrm{dist}_i(v, u)$.

A schedule S can be described as a set of starting times for each operation's processing:

$$S = \{s(O_{ji}) | j = 1, \ldots, n, i = 1, 2\}.$$

The completion time of operation O_{ji} in some schedule under consideration is denoted by $c(O_{ji}) = s(O_{ji}) + p_{ji}$, and $P(O_{ji}) = [s(O_{ji}), c(O_{ji})]$ denotes the operation's processing interval. A schedule S is feasible if the following conditions hold:

1. If two operations O' and O'' either belong the same job or to the same machine, then intervals $P(O')$ and $P(O'')$ do not have common internal points.
2. If some machine M_i processes the operation of job $J_j \in \mathcal{J}(v)$ before the operation of job $J_k \in \mathcal{J}(u)$, then $s(O_{ki}) \geqslant c(O_{ji}) + \mathrm{dist}_i(v, u)$.
3. If an operation of job $J_j \in \mathcal{J}(v)$ is the first one processed by machine M_i, then $s(O_{ji}) \geqslant \mathrm{dist}_i(v_0, v)$.

Let $J_j \in \mathcal{J}(v)$ be the last jobs processed by the machine M_i in some schedule S. Then the *return moment* of machine M_i in S is defined as $R_i(S) = c(O_{ji}) + \mathrm{dist}_i(v, v_0)$. The *makespan* of the schedule S to be minimized is defined as

$$R_{\max}(S) = \max_i R_i.$$

In the case of $m = 2$ we use simplified notation for the operations: a_j and b_j denotes both O_{j1} and O_{j2}, respectively, and their respective processing times p_{j1} and p_{j2}.

We use the following notation for any instance I of the $RO2|Rtt|R_{\max}$ problem.

- $\ell_i = \sum_{j=1}^{n} p_{ji}$—the load of machine M_i; $\ell_{\max} = \max \ell_i$—the maximum machine load.
- $d_j = \sum_{i=1}^{m} p_{ji}$—the length of job J_j; $d_{\max}(v) = \max_{J_j \in \mathcal{J}(v)} d_j$—the maximum job length at node v.

- T_i^*—the weight of the optimal cyclic route over the graph G for machine M_i (the corresponding TSP optimum).
- $\Delta(v) = \sum_{J_j \in \mathcal{J}(v)} d_j$—the load of node v; $\Delta = \sum_{j=1}^{n} d_j$—the total load of the instance.
- $R_{\max}^*$—the optimal makespan.

In order to indicate a specific instance I, if necessary, we use notation $p_{ji}(I)$, $G(I)$, $\mathcal{J}(I;v)$, $\text{dist}_i(I;v,u)$, $\ell_i(I)$, $\ell_{\max}(I)$, $d_j(I)$, $d_{\max}(I;v)$, $T_i^*(I)$, $\Delta(I;v)$, and $R_{\max}^*(I)$.

We use the following standard lower bound from [8].

$$R_{\max}^* \geqslant \bar{R} = \max \left\{ \max_i (\ell_i + T_i^*), \max_{v \in V} (d_{\max}(v) + \text{dist}_1(v_0, v) + \text{dist}_2(v_0, v)) \right\}. \tag{1}$$

Notation $\bar{R}(I)$ is used for a specific instance I. The reference to I is omitted in case the context defines the instance clear enough.

We use the following definition, inherited from [15].

Definition 1. *A feasible schedule S is called* normal, *if $R_{\max}(S) = \bar{R}$. Instance I is* normal *if it admits constructing a normal schedule (and therefore $R_{\max}^*(I) = \bar{R}(I)$).*

As we do know from [4], not every instance of the $RO2\|R_{\max}$ problem is normal, even in case of $G = K_2$. For some instance I we define its *abnormality* as $\alpha(I) = \frac{R_{\max}^*(I)}{\bar{R}(I)}$. The abnormality of a class of instances $\mathcal{K}$ is defined as $\alpha(\mathcal{K}) = \sup_{I \in \mathcal{K}} \alpha(I)$.

Definition 2. *An interval $[\bar{R}, \alpha(\mathcal{K})\bar{R}]$ is called the* tight optima localization interval *for a class of instances $\mathcal{K}$.*

We use the following notation of instance classes from [8]: $\mathcal{I}_m^X$ for the $ROm|G = X|R_{\max}$ and $\mathcal{I}_{Rm}^X$ for the $ROm|Rtt, G = X|R_{\max}$ problems. The superscript "X" is omitted if there is no restriction on the graph structure.

Note that $G = K_1$ means we are talking about the classic open shop problem with no routing involved. From previous research we know abnormalities (and hence tight optima localization intervals) for the following classes of instances: $\alpha\left(\mathcal{I}_2^{K_1}\right) = 1$ [13], $\alpha\left(\mathcal{I}_2^{K_2}\right) = \frac{6}{5}$ [4], $\alpha\left(\mathcal{I}_2^{K_3}\right) = \frac{6}{5}$ [7], $\alpha(\mathcal{I}_2^{tree}) = \frac{6}{5}$ [6], $\alpha\left(\mathcal{I}_{R2}^{K_2}\right) = \alpha\left(\mathcal{I}_{R2}^{K_3}\right) = \frac{5}{4}$ [8], $\alpha\left(\mathcal{I}_3^{K_1}\right) = \frac{4}{3}$ [20].

The main goal of this paper is to describe new polynomially solvable subclasses with guaranteed normality of the problem under consideration, as well as the tight optima localization interval for a special case of the $RO2|Rtt, G = cactus|R_{\max}$ problem (Sect. 4).

2.2   Instance Simplification Operations

In order to determine the abnormality of some class of instances $\mathcal{K}$ one usually has to find a *critical instance* from that class, *i.e.* an instance with the greatest abnormality. Applying various *reversible simplification operations* preserving the standard lower bound $\bar{R}$ can be a very efficient way to do so. By a reversible simplification operation we understand an instance transformation $I \to I'$ with the following properties:

1. Instance I' is *simpler* than I (contains a smaller number of jobs/ machines/nodes, or a simpler structure of the transportation network).
2. Transformation is *reversible*: any feasible schedule of instance I' can be treated as a feasible schedule of instance I with the same makespan.

Definition 3. *Instance transformation $I \to I'$ is referred to as* valid *if $\bar{R}(I') = \bar{R}(I)$.*

Validity of a reversible transformation immediately implies $\alpha(I') \geqslant \alpha(I)$. In this subsection we describe known instance simplification operations—job aggregation and terminal edge contraction—and discuss their properties. A new simplification operation—terminal cycle contraction—is introduced in Sect. 3.

Job aggregation operation is based on a simple idea of replacing a number of jobs with a single *aggregated* or *composite* job with processing time equal to the total processing time of the operations combined. It was used in a series of papers, for instance, in [20] for a three-machine open shop problem. The same transformation applied to different versions of the routing open shop problem can be found in [6,7,9] (identical travel times), and in [8] (unrelated travel times).

Definition 4. *For $I \in \mathcal{I}_{R2}$, let $K \subseteq \mathcal{J}(I;v)$ for some node v. Then by* job aggregation *of set K we understand the following instance transformation $I \to I'$:*

$$G(I') = G(I), \ \mathcal{J}(I';v) = \mathcal{J}(I;v) \setminus K \cup \{J_K\}, \ p_{Ki} = \sum_{J_j \in K} p_{ji}.$$

Job aggregation is clearly a reversible transformation: any schedule of operation of a composite job can be treated as a schedule of respective operations of jobs from set K processed without any idle time in an arbitrary sequence.

Note that $\ell_i(I') = \ell_i(I)$ and $\mathrm{dist}_i(I';v,u) = \mathrm{dist}_i(I;v,u)$ for each machine M_i and each two nodes v, u. However, it is possible that $d_{\max}(I';v) = d_K > d_{\max}(I;v)$, and there is a possibility that $\bar{R}(I') > \bar{R}(I)$ and the job aggregation is not valid. Using (1) we obtain the following sufficient and necessary condition of the validity of the job aggregation transformation:

$$\bar{R}(I') = \bar{R}(I) \iff \sum_{J_j \in K} d_j \leqslant \bar{R}(I) - \mathrm{dist}_1(v_0, v) - \mathrm{dist}_2(v_0, v). \qquad (2)$$

The question is, if the job aggregation of the whole set $\mathcal{J}(I;v)$ is valid? The answer depends on the value of $\Delta(I;v)$.

Definition 5. *A node v from $G(I)$ of some problem instance I is* overloaded *if*

$$\Delta(I; v) > \bar{R}(I) - \text{dist}_1(I; v_0, v) - \text{dist}_2(I; v_0, v).$$

Otherwise the node is referred to as underloaded.

By this definition, the aggregation of $\mathcal{J}(I; v)$ is valid if and only if the node v is underloaded.

Let us describe the terminal edge contraction operation for the problem with unrelated travel times. Such an operation was described in [6,9] for identical travel times. The idea is the following: translate a single job from a terminal node v to an adjacent one u, modifying its processing times to include travel times between v and u.

Definition 6. *Let $I \in \mathcal{I}_{R2}$, $v \neq v_0$ is a terminal node in $G(I)$ and there is a single job $J_j \in \mathcal{J}(I; v)$. Let $e = [u, v]$ be an edge incident to v. Then by the* contraction of edge e *we understand the following instance transformation $I \to I'$:*

$$\mathcal{J}(I'; u) = \mathcal{J}(I; u) \cup \{J_j\}, \; G(I') = G(I) \setminus \{v\}, \; p_{ji}(I') = p_{ji}(I) + 2\tau_i(e).$$

The processing interval $P(O_{ji})$ in any feasible schedule of the transformed instance I' can be treated as a concatenation of three subintervals, representing a travel time of machine M_i from u to v, processing of the initial operation O_{ji} and a travel time back from v to u. Such a transformation is clearly reversible. Note that $\ell_i(I') = \ell_i(I) + 2\tau_i(e)$ and $T_i^*(I') = T_i^*(I) - 2\tau_i(e)$, therefore $\ell_i(I') + T_I^*(I') = \ell_i(I) + T_I^*(I)$. Moreover, $d_j(I) + \text{dist}_1(v_0, v) + \text{dist}_2(v_0, v) \leqslant \bar{R}$ due to (1). However, the transformation might not be valid since $d_j(I') = d_j(I) + 2\tau_1(e) + 2\tau_2(e)$, and $d_j(I') + \text{dist}_1(I'; v_0, u) + \text{dist}_2(I'; v_0, u) = d_j(I) + \text{dist}_1(I; v_0, v) + \text{dist}_2(I; v_0, v) + \tau_1(e) + \tau_2(e)$.

We use the following

Definition 7. *Let $I \in \mathcal{I}_{R2}$, $v \neq v_0$ is a terminal node in $G(I)$ and there is a single job $J_j \in \mathcal{J}(I; v)$. Let $e = [u, v]$ be an edge incident to v. Then edge e is* overloaded *if*

$$d_j + \text{dist}_1(v_0, u) + \text{dist}_2(v_0, u) + 2\tau_1(e) + 2\tau_2(e) > \bar{R}(I),$$

and underloaded *otherwise.*

It was proved in [7] that any instance $I \in \mathcal{I}_2$ contains at most one overloaded element (either node or edge). In Sect. 3 we introduce another overloaded element and generalize that proof for the case of unrelated travel times as well.

2.3 Superoverloaded Nodes and Unrelated Travel Times

Following [6] we use the following

Definition 8. *An instance I is called* irreducible, *if no valid job aggregation is possible for I.*

It is easy to observe that any irreducible instance contains exactly one job at any underloaded node and two or three jobs at an overloaded node (if any). The proof is based on the following inequality

$$\Delta = \ell_1 + \ell_2 \leqslant 2\bar{R} - T_1^* - T_2^*, \tag{3}$$

which follows from (1) for any instance of a two-machine problem. Moreover, any instance can be transformed into an irreducible one in linear time, see [8] for details.

Suppose that there exists a way to transform an instance $I \in \mathcal{I}_2$ with an overloaded node v by job aggregations into such an irreducible instance I' that $|\mathcal{J}(I'; v)| = 3$. In this case the node v is referred to as *superoverloaded* (see [10] for details). It turned out that the existence of a superoverloaded node is sufficient for the instance to be normal, if that node is either the depot or is adjacent to the depot in some optimal tour over $G(I)$ [10]. This idea was used in [8] for the problem with unrelated travel times on the triangular transportation network, however the authors could not guarantee normality and were only able to prove that in this case $R_{\max}^*(I) \leqslant \frac{7}{6}\bar{R}(I)$. They suggested a question for future research to investigate whether such a condition implies normality in the case of unrelated travel time. In this subsection we prove that the answer to that question is negative.

Lemma 1. *There exists an instance $I \in \mathcal{I}_{R2}^{K_2}$ with superoverloaded node such that $R_{\max}^*(I) = \frac{7}{6}\bar{R}(I)$.*

Proof. Let $G(I)$ consists of two connected nodes v_0 and v. Consider four jobs $J_0, \ldots, J_3$ with the following processing times:

$$a_0 = b_0 = b_1 = b_2 = b_3 = 0, \, a_1 = a_2 = a_3 = 2.$$

Let $\mathcal{J}(v_0) = \{J_0\}$ and $\mathcal{J}(v) = \{J_1, J_2, J_3\}$, and $\tau_1(v_0, v) = 0, \tau_2(v_0, v) = 3$. Note that

$$\bar{R}(I) = \max\{6 + 0, 0 + 6, 2 + 3\} = 6.$$

Instance I is irreducible since $d_1 + d_2 = 4 > \bar{R} - \text{dist}_1(v_0, v) - \text{dist}_2(v_0, v) = 6 - 3 = 3$, therefore the node v is superoverloaded. Consider any feasible schedule S of instance I. Without loss of generality assume that machine M_1 processes jobs from $\mathcal{J}(v)$ in order $J_1 \to J_2 \to J_3$. Let's prove that $R_{\max}(S) \geqslant 7$.

Assume otherwise, let $R_{\max}(S) < 7$. Note that all the operations of machine M_2 are processed within the interval $[3, R_{\max}(S) - 3] \subseteq [3, 4)$, as soon as $\tau_2(v_0, v) = 3$. Therefore, for any $t \in \{s(b_1), s(b_2), s(b_2)\}$ at most one of the operations a_1, a_2, a_3 is completed in S before t, so there exists such a $t \in \{s(b_1), s(b_2), s(b_2)\}$ that there are at least two of the operations a_1, a_2, a_3 which are processed after t. Therefore $R_1(S) \geqslant t + 2 + 2 \geqslant 7$. The lemma is proved by contradiction. □

We use the following extended version of Theorem 1 from [8].

Theorem 1. *Let $I \in \mathcal{I}_{R2}^{chain}$ be an irreducible instance, and $G(I)$ is a chain connecting v_0 with a superoverloaded node v (thus v contains exactly three jobs). Then one can in linear time build a feasible schedule S for I such that $R_{\max}(S) \leqslant \frac{7}{6}\bar{R}(I)$.*

Proof. Similar to the proof of [8, Theorem 1]. □

Lemma 1 implies that the bound in Theorem 1 is tight.

3   Extended Instance Reduction

Let I be an instance of the routing open shop problem and C be a cycle in graph $G(I)$. We will refer to a node $v \in C$ as to a *gate* if one of the following conditions hold:

1. $v = v_0$, or
2. the degree of v in $G(I)$ is greater than 2.

It is evident that any cyclic route over $G(I)$ enters and leaves the cycle C through the gates only. A cycle with a single gate will be referred to as a *terminal cycle*. Let us introduce the *terminal cycle contraction* operation. It is similar to the terminal edge contraction operation and can be described as follows. Let u be the only gate of the cycle C. The transformation is to replace all the edges of C and all its nodes except for u with a new job J_C located at u, with the processing times equal to the total processing times of the respective operations of jobs from C (with exception of u) plus the cyclic travel time for the respective machine over C. The processing of this new operation can be treated as traveling along the cycle while processing operations of its jobs by the way. So this simplification operation is a reversible one.

Definition 9. *Let $I \in \mathcal{I}_{R2}$, C be a terminal cycle in $G(I)$ with gate u, and each of nodes of C (with a possible exception of u) is underloaded. By the contraction of the cycle C we understand the following transformation of the instance $I \rightarrow I'$:*

$$\mathcal{J}(I'; u) = \mathcal{J}(I; u) \cup \{J_C\},\ p_{Ci}(I') = \sum_{J_j \in C\setminus\{u\}} p_{ji}(I) + T_i(C),\ G(I') = G(I)\setminus\{C\setminus\{u\}\},$$

there $T_i(C)$ is the travel time over C for machine M_i.

Accordingly, we use the following

Definition 10. *Let $I \in \mathcal{I}_{R2}$, C be a terminal cycle in $G(I)$ with a gate u, and each of nodes of C (with a possible exception of u) is underloaded. Cycle C is referred to as* overloaded *if*

$$\sum_{v \in C} \Delta(I; v) - \Delta(I; u) + T_1(C) + T_2(C) > \bar{R} - \operatorname{dist}_1(v_0, u) - \operatorname{dist}_2(v_0, u).$$

Otherwise the cycle C is underloaded.

The idea is similar to Definitions 5 and 7: the contraction of a terminal cycle C is valid if and only if the cycle is underloaded.

Theorem 2. *Let $I \in \mathcal{I}_{R2}$. Then I contains at most one overloaded element (node, edge or cycle).*

Proof. Suppose we have at least two overloaded elements. Choose any two of them. Let $l_1, \ldots, l_x$ be chosen overloaded nodes in I, $e_1, \ldots, e_y$ are overloaded edges, there $e_k = [u_k, v_k]$ with terminal node v_k with single job J_k, and $C_1, \ldots, C_z$ are terminal cycles with gates $w_1, \ldots, w_z$. (There is a possibility that some of the nodes $l_1, \ldots, l_x, u_1, \ldots, u_y, w_1, \ldots, w_z$ coincide, however all the nodes $l_1, \ldots, l_x$ are different.) We assumed that $x + y + z = 2$.

Consider the graph G' obtained from $G(I)$ by removing all the overloaded cycles (except for their respective gates) and nodes $v_1, \ldots, v_y$, and let $T_i'^*$ be the weight of an optimal cyclic route over the G' for machine M_i. As soon as the only entrance and exit point for the terminal cycles are their gates and we can only reach v_k through the edge e_k, we have

$$T_i^* = T_i'^* + \sum_{k=1}^{y} 2\tau_i(e_k) + \sum_{k=1}^{z} T_i(C_k). \tag{4}$$

Now using Definitions 5, 7 and 10 we have

$$\Delta(l_k) > \bar{R} - \mathrm{dist}_1(v_0, l_k) - \mathrm{dist}_2(v_0, l_k), \ k = 1, \ldots, x,$$

$$\Delta(v_k) = d_k > \bar{R} - \mathrm{dist}_1(v_0, u_k) - \mathrm{dist}_2(v_0, u_k) - 2\tau_1(e_k) - 2\tau_2(e_k), \ k = 1, \ldots, y,$$

$$\sum_{v \in C_k} \Delta(I; v) - \Delta(I; w_k) > \bar{R} - \mathrm{dist}_1(v_0, w_k) - \mathrm{dist}_2(v_0, w_k) - T_1(C) - T_2(C), \ k = 1, \ldots, z.$$

Note that total sum of the left-hand side of each inequality above does not exceed $\Delta(I)$. Let's denote the total sum of the right-hand parts by Q. Under the assumption $x + y + z = 2$ we have

$$\sum_{k=1}^{x}(\mathrm{dist}_1(v_0, l_k) + \mathrm{dist}_2(v_0, l_k)) + \sum_{k=1}^{y}(\mathrm{dist}_1(v_0, u_k) + \mathrm{dist}_2(v_0, u_k))$$

$$+ \sum_{k=1}^{z}(\mathrm{dist}_1(v_0, w_k) + \mathrm{dist}_2(v_0, w_k))$$

$$= \mathrm{dist}_1(v_0, \alpha) + \mathrm{dist}_2(v_0, \alpha) + \mathrm{dist}_1(v_0, \beta) + \mathrm{dist}_2(v_0, \beta) \leqslant T_1'^* + T_2'^*.$$

Hence from (4) we have

$$Q \geqslant 2\bar{R} - (T_1'^* + T_2'^*) - \sum_{k=1}^{y} 2\tau_i(e_k) - \sum_{k=1}^{z} T_i(C_k) = 2\bar{R} - T_1^* + T_2^*.$$

The theorem is proved by contradiction with (3).   □

Now consider the following procedure of instance simplification. The idea is first to obtain an irreducible instance, then apply the terminal edge and terminal cycle contractions to all underloaded terminal elements, maintaining the irreducibility after each step. For the resulting instance there will be no valid transformation (job aggregation, edge or cycle contraction) left.

Instance Simplification Procedure
INPUT: An instance $I \in \mathcal{I}_{R2}$.
OUTPUT: The reduced instance I'.
Step 1. Transform the instance to the irreducible one using valid job aggregations.
Step 2. WHILE there is an underloaded cycle or edge DO its contraction and perform valid job aggregations to obtain an irreducible instance.
Step 3. STOP.

Note that this procedure can be implemented in linear time. Indeed, Step 1 is doable in $O(n)$ time, and to enumerate the terminal edges and cycles it is sufficient to construct a block-cut tree of $G(I)$, which also can be done in linear time (see [14, 21] for instance).

The result of the Procedure applied to an instance with an arbitrary transportation network can be complicated enough. However, when applied to an instance with $G = cactus$, the Procedure might simplify the instance considerably. A cactus is a connected graph for which each block is either an edge or a cycle. Further we will consider only so-called *cycleless* instances.

Definition 11. *An instance $I \in \mathcal{I}_{R2}^{cactus}$ is referred to as* cycleless *if the Instance Simplification Procedure transforms I into an instance I' with no cycles in $G(I')$.*

Theorem 2 implies that an application of the Instance Simplification procedure to a cycleless instance I transforms it to an instance I' such that $G(I')$ is either K_1, or a chain connecting the depot to an overloaded edge, or a chain connecting the depot to an overloaded node. The first case obviously implies normality and is solvable in linear time. The same result can be proved for the second case, the proof is similar to [9, Lemma 4.1]. The third case is considered in the next section.

4   Optima Localization for the Cycleless Instances on a Cactus

The main result of this section is the following

Theorem 3. *Let $I \in \mathcal{I}_{R2}^{cactus}$ be a cycleless instance. Then one can in linear time build a feasible schedule S for I such that $R_{\max}(S) \in [\bar{R}, \frac{5}{4}\bar{R}]$.*

Proof. According to the remark at the end of the previous section, it is sufficient to prove the Theorem only for the irreducible instances on a chain, connecting v_0 with an overloaded node v. Moreover, if that node contains exactly three jobs, Theorem 3 follows from the Theorem 1.

Now consider an irreducible instance $I \in \mathcal{I}_{R2}^{chain}$ such that $G(I) = (v_0, \ldots, v_k)$, there $\mathcal{J}(v_t) = \{J_t\}$ for each $t = 0, \ldots, k-1$, and $\mathcal{J}(v_k) = \{J_\alpha, J_\beta\}$ with

$$d_\alpha + d_\beta > \bar{R} - \text{dist}_1(v_0, v_k) - \text{dist}_2(v_0, v_k), \tag{5}$$

due to the fact that v_k is overloaded.

Consider two cases.

Case 1. $\max\{d_\alpha, d_\beta\} \geqslant \frac{3}{4}\bar{R} - \text{dist}_1(v_0, v_k) - \text{dist}_2(v_0, v_k)$.

Without loss of generality assume $d_\alpha \geqslant d_\beta$. Construct an early schedule S according to the partial order of the operations from Fig. 1.

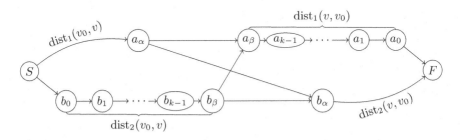

Fig. 1. Partial order of operations for Case 1.

It is easy to observe that $R_2(S) \leqslant \{\ell_2 + T_2^*, d_\alpha + \text{dist}_1(v_0, v_k) + \text{dist}_2(v_0, v_k)\} \leqslant \bar{R}$. Assume $R_1(S) > \bar{R}$, then $R_1(S) = \sum_{j=0}^{k-1} d_j + d_\beta + \text{dist}_1(v_0, v_k) + \text{dist}_2(v_0, v_k) = \Delta - d_\alpha + \text{dist}_1(v_0, v_k) + \text{dist}_2(v_0, v_k) \leqslant 2\bar{R} - T_1^* - T_2^* - \frac{3}{4}\bar{R} + T_1^* + T_2^* = \frac{5}{4}\bar{R}$ due to the Case 1 assumption and (3).

Case 2. $\max\{d_\alpha, d_\beta\} < \frac{3}{4}\bar{R} - \text{dist}_1(v_0, v_k) - \text{dist}_2(v_0, v_k)$.

Construct two early schedules, S_1 and S_2, according to the partial orders of operations from Figs. 2 and 3, respectively.

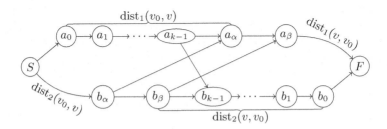

Fig. 2. Partial order for the schedule S_1.

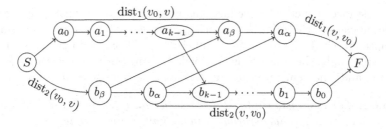

Fig. 3. Partial order for the schedule S_2.

Note that $R_2(S_1) \leqslant \bar{R}$, because due to (3) and (5) we have

$$\sum_{j=0}^{k-1} d_j + \text{dist}_1(v_0, v) + \text{dist}_1(v_0, v) = \Delta - d_\alpha - d_\beta + \text{dist}_1(v_0, v) + \text{dist}_1(v_0, v) \leqslant \bar{R}.$$

Assuming $R_{\max}(S_1) > \bar{R}$ (otherwise the Theorem is proved), we have

$$R_{\max}(S_1) = b_\alpha + \max\{a_\alpha, b_\beta\} + a_\beta + \text{dist}_1(v_0, v) + \text{dist}_2(v_0, v).$$

By similar reasoning, assuming $R_{\max}(S_2) > \bar{R}$ we obtain

$$R_{\max}(S_2) = b_\beta + \max\{b_\alpha, a_\beta\} + a_\alpha + \text{dist}_1(v_0, v) + \text{dist}_2(v_0, v).$$

Therefore, using the case assumptions and (1) we have

$$R_{\max}(S_1) + R_{\max}(S_2) = T_1^* + T_2^* + d_\alpha + d_\beta$$
$$+ \max\{a_\alpha, b_\beta\} + \max\{b_\alpha, a_\beta\} < \frac{3}{4}\bar{R} + \frac{3}{4}\bar{R} + \bar{R} \leq \frac{10}{4}\bar{R},$$

hence $\min\{R_{\max}(S_1), R_{\max}(S_2)\} < \frac{5}{4}\bar{R}$ and the Theorem is proved.   □

5   Conclusion

We have generalized the instance reduction procedure, known for identical travel times on an arbitrary tree (see [6,9]) it two directions. First, our procedure is applicable to the problem with individual travel time, and second, we introduced the terminal cycle contraction, allowing to use this procedure on a wider class of graph structures, namely *cacti*. For a special class of cycleless instances of the problem under consideration we have established the tight optima localization interval, which coincides with a known interval for a problem on a link. Note that similar result is known for the problem with identical travel times [6]. Also the terminal cycle contraction operation will simplify the investigation of a routing open shop problem on a cycle, which is an important part of research on the optima localization interval for a general case without any graph structure restriction.

The instance reduction procedure was used in [9] to establish polynomially solvable subcases of the $RO2|G = tree|R_{\max}$ problem with guaranteed normality of the optimal schedule. One of those sufficient conditions of polynomial solvability is quote simple: the depot v_0 is overloaded. Using our Instance Simplification Procedure and Theorem 2 we can now formulate a more general result.

Lemma 2. *For any instance $I \in \mathcal{I}_{R2}^{cactus}$ such that the depot v_0 is overloaded a normal (and hence optimal) schedule can be built in linear time.*

Moreover, this Lemma is also true for wider classes of instances, for which the depot *becomes* overloaded during the Instance Simplification Procedure.

A natural direction of further research is to establish tight optima localization intervals for the general two-machine problem on a cactus, both for unrelated and identical travel times. More global problems on an arbitrary transportation network are also of interest. As we have no knowledge of any instance from $\mathcal{I}_{R2}$ for which optimal makespan exceeds $\frac{5}{4}\bar{R}$, we suggest the following

Conjecture 1. For any instance $I \in \mathcal{I}_{R2}$ its optimal makespan $R_{\max}(I)$ belongs in the interval $[\bar{R}, \frac{5}{4}\bar{R}]$.

References

1. Aksyonov, V.: An approximation polynomial time algorithm for one scheduling problem. Upravlyaemye systemy **28**, 8–11 (1988). (in Russian)
2. Averbakh, I., Berman, O.: Routing two-machine flowshop problems on networks with special structure. Transp. Sci. **30**(4), 303–314 (1996). https://doi.org/10.1287/trsc.30.4.303
3. Averbakh, I., Berman, O.: A simple heuristic for m-machine flow-shop and its applications in routing-scheduling problems. Oper. Res. **47**(1), 165–170 (1999). https://doi.org/10.1287/opre.47.1.165
4. Averbakh, I., Berman, O., Chernykh, I.: A 6/5-approximation algorithm for the two-machine routing open-shop problem on a two-node network. Eur. J. Oper. Res. **166**(1), 3–24 (2005). https://doi.org/10.1016/j.ejor.2003.06.050
5. Averbakh, I., Berman, O., Chernykh, I.: The routing open-shop problem on a network: complexity and approximation. Eur. J. Oper. Res. **173**(2), 531–539 (2006). https://doi.org/10.1016/j.ejor.2005.01.034
6. Chernykh, I., Krivonogiva, O.: Optima localization for the two-machine routing open shop on a tree (2019). (in Russian), submitted to Diskretnyj Analiz i Issledovanie Operacij
7. Chernykh, I., Lgotina, E.: The 2-machine routing open shop on a triangular transportation network. In: Kochetov, Y., Khachay, M., Beresnev, V., Nurminski, E., Pardalos, P. (eds.) DOOR 2016. LNCS, vol. 9869, pp. 284–297. Springer, Cham (2016). https://doi.org/10.1007/978-3-319-44914-2_23
8. Chernykh, I., Lgotina, E.: How the difference in travel times affects the optima localization for the routing open shop. In: Khachay, M., Kochetov, Y., Pardalos, P. (eds.) MOTOR 2019. LNCS, vol. 11548, pp. 187–201. Springer, Cham (2019). https://doi.org/10.1007/978-3-030-22629-9_14

9. Chernykh, I., Lgotina, E.: Two-machine routing open shop on a tree: instance reduction and efficiently solvable subclass (2019), submitted to Optimization Methods and Software

10. Chernykh, I., Pyatkin, A.: Irreducible bin packing: complexity, solvability and application to the routing open shop. LNCS (2019, to appear)

11. Chou, S., Lin, S.: Museum visitor routing problem with the ballancing of concurrent visitors. Complex Syst. Concurr. Eng. **6**, 345–353 (2007). https://doi.org/10.1007/978-1-84628-976-7_39

12. Golovachev, M., Pyatkin, A.V.: Routing open shop with two nodes, unit processing times and equal number of jobs and machines. In: Khachay, M., Kochetov, Y., Pardalos, P. (eds.) MOTOR 2019. LNCS, vol. 11548, pp. 264–276. Springer, Cham (2019). https://doi.org/10.1007/978-3-030-22629-9_19

13. Gonzalez, T.F., Sahni, S.: Open shop scheduling to minimize finish time. J. ACM **23**(4), 665–679 (1976). https://doi.org/10.1145/321978.321985

14. Hopcroft, J., Tarjan, R.: Algorith 477: efficient algorithms for graph manipulation. Commun. ACM **16**(6), 372–378 (1973). https://doi.org/10.1145/362248.362272

15. Kononov, A., Sevastianov, S., Tchernykh, I.: When difference in machine loads leads to efficient scheduling in open shops. Ann. Oper. Res. **92**, 211–239 (1999). https://doi.org/10.1023/a:1018986731638

16. Kononov, A.: $O(\log m)$-approximation for the routing open shop problem. RAIRO-Oper. Res. **49**, 383–391 (2015). https://doi.org/10.1051/ro/2014051

17. Kononov, A.: On the routing open shop problem with two machines on a two-vertex network. J. Appl. Ind. Math. **6**(3), 318–331 (2012). https://doi.org/10.1134/s1990478912030064

18. Lawler, E.L., Lenstra, J.K., Rinnooy Kan, A.H.G., Shmoys, D.B.: Sequencing and scheduling: algorithms and complexity, Chap. 9. In: Logistics of Production and Inventory, Handbooks in Operations Research and Management Science, vol. 4, pp. 445–522. Elsevier (1993). https://doi.org/10.1016/S0927-0507(05)80189-6

19. Sevastianov, S.V., Woeginger, G.J.: Makespan minimization in open shops: a polynomial time approximation scheme. Math. Program. **82**(1–2, Ser. B), 191–198 (1998). https://doi.org/10.1007/BF01585871

20. Sevastianov, S.V., Tchernykh, I.D.: Computer-aided way to prove theorems in scheduling. In: Bilardi, G., Italiano, G.F., Pietracaprina, A., Pucci, G. (eds.) ESA 1998. LNCS, vol. 1461, pp. 502–513. Springer, Heidelberg (1998). https://doi.org/10.1007/3-540-68530-8_42

21. Tarjan, R.: Depth-first search and linear graph algorithms. SIAM J. Comput. **1**(2), 146–160 (1972). https://doi.org/10.1137/0201010

22. van Bevern, R., Pyatkin, A.V.: Completing partial schedules for open shop with unit processing times and routing. In: Kulikov, A.S., Woeginger, G.J. (eds.) CSR 2016. LNCS, vol. 9691, pp. 73–87. Springer, Cham (2016). https://doi.org/10.1007/978-3-319-34171-2_6

23. van Bevern, R., Pyatkin, A.V., Sevastyanov, S.V.: An algorithm with parameterized complexity of constructing the optimal schedule for the routing open shop problem with unit execution times. Siberian Electron. Math. Rep. **16**, 42–84 (2019). https://doi.org/10.33048/semi.2019.16.003

24. Williamson, D.P., et al.: Short shop schedules. Oper. Res. **45**(2), 288–294 (1997). https://doi.org/10.1287/opre.45.2.288

25. Yu, V.F., Lin, S., Chou, S.: The museum visitor routing problem. Appl. Math. Comput. **216**(3), 719–729 (2010). https://doi.org/10.1016/j.amc.2010.01.066

Local Strong Convexity in Hilbert Space

Maxim O. Golubev[✉][iD]

Moscow Institute of Physics and Technology,
9 Institutskiy per., Dolgoprudny, Moscow Region 141700, Russian Federation
maksimkane@mail.ru

Abstract. A metric projection from the real Hilbert space onto a subset of the boundary of a closed convex (not necessary bounded) set is considered. We show a connection between Lipschitz continuity on some subset of the Hilbert space of operator of metric projection with the Lipschitz constant strictly less than 1 and local strong convexity of the set (in terms of modulus of convexity for intersection of a set and a ball with a small radius).

Keywords: Hilbert space · Metric projection · Convex sets · Modulus of convexity

1 Introduction and Main Notations

We denote by $\mathcal{H}$ a Hilbert space over $\mathbb{R}$ with $\langle p, x \rangle$ standing for the scalar product for vectors p and x from $\mathcal{H}$. Let $\overline{\mathbb{R}} = \mathbb{R} \cup \{+\infty\}$. Let $B_r(a) = \{x \in \mathcal{H} \mid \|x - a\| \leq r\}$. For a subset $A \subset \mathcal{H}$ we denote by ∂A and $\operatorname{int} A$ the boundary and the interior of the set A, respectively. The diameter of a set A is defined by $\operatorname{diam} A = \sup_{x, y \in A} \|x - y\|$. We denote the *convex hull* of a set $A \subset \mathcal{H}$ by $\operatorname{co} A$. The *Minkowski sum* of two sets $A, B \subset \mathcal{H}$, is the set

$$A + B = \{a + b \mid a \in A, b \in B\}.$$

We denote the *normal cone* to a closed convex subset $A \subset \mathcal{H}$ at the point $a \in \partial A$ by $N(A; a)$, that is

$$N(A; a) = \{p \in \mathcal{H} \mid \langle p, a \rangle \geq \sup_{x \in A} \langle p, x \rangle\}.$$

Let a subset $A \subset \mathcal{H}$ be fixed. Then the *distance function* from the point $x \in \mathcal{H}$ to the set A is defined as follows

$$\varrho_A(x) = \inf_{a \in A} \|x - a\|,$$

and the metric projection of the point $x \in \mathcal{H}$ to the subset $A \subset \mathcal{H}$ is given by the formula

$$P_A(x) = \{a \in A \mid \|x - a\| = \varrho_A(x)\}.$$

Supported by the Russian Foundation for Basic Research, grant 18-01-00209.

M. Jaćimović et al. (Eds.): OPTIMA 2019, CCIS 1145, pp. 16–31, 2020.
https://doi.org/10.1007/978-3-030-38603-0_2

The *Hausdorff distance* between two subsets $A, B \subset \mathcal{H}$ is defined as follows

$$h(A, B) = \inf\{r > 0 \mid A \subset B + B_r(0),\ B \subset A + B_r(0)\}.$$

For the function $f: \mathcal{H} \to \overline{\mathbb{R}}$ the conjugate function $f^*: \mathcal{H} \to \overline{\mathbb{R}}$ is defined by formula $f^*(p) = \sup_{x \in \mathcal{H}}(\langle p, x \rangle - f(x))$.

Definition 1 [20]. *Let a subset $A \subset \mathcal{H}$ be convex and closed. The modulus of convexity $\delta_A: [0, \operatorname{diam} A) \to [0, +\infty)$ is the function defined by*

$$\delta_A(t) = \sup\left\{\delta \geq 0 \mid B_\delta\left(\frac{x_0 + x_1}{2}\right) \subset A,\ \forall x_0, x_1 \in A\ :\ \|x_0 - x_1\| = t\right\}.$$

For a subset $A \subset \mathcal{H}$ and a number $\varrho > 0$ we define the open ϱ-neighborhood of the set A

$$U_A(\varrho) = \{x \in \mathcal{H} \mid \varrho_A(x) < \varrho\}.$$

For a convex subset $A \subset \mathcal{H}$, a subset of its boundary $S \subset \partial A$ and a number $\varrho > 0$ we define (S, ϱ)-neighborhood of the set A as follows

$$\Phi = \Phi(A, S, \varrho) \doteq \left(\bigcup_{x \in S}\left(x + N(A; x)\right)\right) \setminus U_A(\varrho). \tag{1}$$

Definition 2 [12, 18–20]. *A nonempty subset $A \subset \mathcal{H}$ is called a strongly convex set of radius $R > 0$ if it can be represented as the intersection of closed balls of radius $R > 0$, that is there exists a subset $X \subset \mathcal{H}$ such that $A = \bigcap_{x \in X} B_R(x)$.*

It is well known, that a set $P_A(x)$ is a singleton for any closed convex subset $A \subset \mathcal{H}$ and for any point $x \in \mathcal{H}$. Moreover, for any pair of points $x_0, x_1 \in \mathcal{H}$ we have

$$\|a_0 - a_1\| \leq 1 \cdot \|x_0 - x_1\|, \tag{2}$$

where $\{a_0\} = P_A(x_0)$, $\{a_1\} = P_A(x_1)$.

The Lipschitz constant 1 in Formula (2) is the best possible in general case and it is attained, for example, on the closed affine subspace.

The aim of the present article is to characterize (in terms of modulus of convexity for intersection of a set and a ball with small radius) subset S of the boundary of a closed convex subset $A \subset \mathcal{H}$ with the following property: for any $\varrho > 0$ there exists a number $C \in (0, 1)$ such that for any pair of points $x_0, x_1 \in \Phi$ (where Φ is (S, ϱ)-neighborhood of the set A, see Formula (1)) the following inequality holds true

$$\|a_0 - a_1\| \leq C \cdot \|x_0 - x_1\|, \qquad \{a_0\} = P_A(x_0),\ \{a_1\} = P_A(x_1). \tag{3}$$

If A is closed subset in $\mathbb{R}^n$, Federer in [11] has obtained estimates for the Lipschitz constant of the metric projection operator. In papers [1, 2] Abatzoglou has considered similar question and received precise estimates for the Lipschitz constant of the metric projection operator in case of $\mathbb{R}^n$ and a Hilbert space for subsets with C^2-smooth boundary. Balashov and Golubev have proved in [6] that the property (3) (where $S = \partial A$) characterizes the class of strongly convex sets.

Alber and Notik [3,4], Björnestål [10], Penot [17] et al are also engaged in estimates of the modulus of continuity of the metric projection onto convex closed sets in Banach spaces.

The main results of the paper are Theorems 1, 2.

2 Lipschitz Property for Metric Projection on the Subset of Boundary

In this section we will prove, that if a subset $A \subset \mathcal{H}$ is locally strictly convex set (the boundary contains no nondegenerate line segments) then the metric projection onto some subset of the boundary of the set A is Lipschitz continuous and the estimate is slightly better than estimate in (2). Moreover if a subset $A \subset \mathcal{H}$ is locally strongly convex (in terms of modulus of convexity) then the metric projection onto some subset of the boundary of the set A is Lipschitz continuous with Lipschitz constant strictly less than 1.

We denote by $\angle xyz$ an angle between vectors $x - y$ and $z - y$.

Let us consider the following problem. There is a segment $[a_0, a_1] \in \mathcal{H}$, $\varphi_0, \varphi_1 \in \left[0, \dfrac{\pi}{2}\right]$. Cones K_0 and K_1 are built on the points a_0 and a_1 as a vertexes correspondingly, the segment $[a_0, a_1]$ is the axis of rotation, and for any point $x \in K_0$ the following inequality holds $\angle x a_0 a_1 \geq \pi - \varphi_0$, for any point $x \in K_1$ the next inequality holds $\angle x a_1 a_0 \geq \pi - \varphi_1$, that is the apex angle of the cone K_0 equals $2\varphi_0 < \pi$, the apex angle for the cone K_1 equals $2\varphi_1 < \pi$. We are to find

$$\min\Big\{ \|x_0 - x_1\| \mid x_0 \in K_0, x_1 \in K_1, \|x_0 - a_0\| = \varrho_0, \|x_1 - a_1\| = \varrho_1 \Big\}. \quad (4)$$

Lemma 1. *The solution of the problem (4) is*

$$\min_{\substack{x_0 \in K_0, x_1 \in K_1 \\ \|x_0 - a_0\| = \varrho_0 \\ \|x_1 - a_1\| = \varrho_1}} \|x_0 - x_1\|^2 = \varrho_0^2 + \varrho_1^2 + \|a_0 - a_1\|^2$$

$$+ 2\|a_0 - a_1\|(\varrho_0 \cos\varphi_0 + \varrho_1 \cos\varphi_1)$$
$$+ 2\varrho_0\varrho_1 \cos(\varphi_0 + \varphi_1). \quad (5)$$

Proof. Fix points $x_0 \in K_0$, $x_1 \in K_1$. Let points b_0, b_1 be orthogonal projections of the points x_0 and x_1 on the line, that contains the segment $[a_0, a_1]$, correspondingly. Let $\alpha_0 = \angle x_0 a_0 b_0$, $\alpha_1 = \angle x_1 a_1 b_1$, that is $\alpha_0 \leq \varphi_0$, $\alpha_1 \leq$. Let a point

x_1' be orthogonal projection of the point x_1 on the plane $x_0 a_0 a_1$, $\gamma = \angle x_1 b_1 x_1'$. Obviously $\gamma \leq \frac{\pi}{2}$. By Pythagorean theorem for the triangle $x_0 x_1' x_1$ we have

$$\|x_0 - x_1\|^2 = \|x_0 - x_1'\|^2 + \|x_1' - x_1\|^2$$
$$= \|b_0 - b_1\|^2 + (\|x_0 - b_0\| - \|x_1' - b_1\|)^2 + \|x_1 - x_1'\|^2. \quad (6)$$

It is clear, that $\|b_0 - b_1\| = (\varrho_0 \cos \alpha_0 + \|a_0 - a_1\| + \varrho_1 \cos \alpha_1)$, $\|x_0 - b_0\| = \varrho_0 \sin \alpha_0$, $\|x_1 - b_1\| = \varrho_1 \sin \alpha_1$, $\|x_1' - b_1\| = \varrho_1 \sin \alpha_1 \cos \gamma$, $\|x_1' - x_1\| = \varrho_1 \sin \alpha_1 \sin \gamma$. So Formula (6) takes on the following form

$$\|x_0 - x_1\|^2 = \varrho_0^2 \cos^2 \alpha_0 + \varrho_1^2 \cos^2 \alpha_1 + \|a_0 - a_1\|^2$$
$$+ 2\|a_0 - a_1\|(\varrho_0 \cos \alpha_0 + \varrho_1 \cos \alpha_1) + 2\varrho_0 \varrho_1 \cos \alpha_0 \cos \alpha_1$$
$$+ \varrho_0^2 \sin^2 \alpha_0 - 2\varrho_0 \varrho_1 \sin \alpha_0 \sin \alpha_1 \cos \gamma$$
$$+ \varrho_1^2 \sin^2 \alpha_1 \cos^2 \gamma + \varrho_1^2 \sin^2 \alpha_1 \sin^2 \gamma.$$

After transformations we get

$$\|x_0 - x_1\|^2 = \varrho_0^2 + \varrho_1^2 + \|a_0 - a_1\|^2 + 2\|a_0 - a_1\|(\varrho_0 \cos \alpha_0 + \varrho_1 \cos \alpha_1)$$
$$+ 2\varrho_0 \varrho_1 (\cos \alpha_0 \cos \alpha_1 - \sin \alpha_0 \sin \alpha_1 \cos \gamma).$$

Since $\alpha_0 < \frac{\pi}{2}$, $\alpha_1 < \frac{\pi}{2}$, $\gamma \leq \frac{\pi}{2}$, then summand $-2\varrho_0 \varrho_1 \sin \alpha_0 \sin \alpha_1 \cos \gamma < 0$ and for fixed values α_0, α_1 expression $-2\varrho_0 \varrho_1 \sin \alpha_0 \sin \alpha_1 \cos \gamma$ attains its minimum for $\cos \gamma = 1$, that is $\gamma = 0$.

Thus, for fixed angles α_0, α_1 we have

$$\min_{\substack{\gamma \in [0, \frac{\pi}{2}] \\ \alpha_0, \alpha_1 \ are fixed}} \|x_0 - x_1\|^2 = \varrho_0^2 + \varrho_1^2 + \|a_0 - a_1\|^2$$
$$+ 2\|a_0 - a_1\|(\varrho_0 \cos \alpha_0 + \varrho_1 \cos \alpha_1) + 2\varrho_0 \varrho_1 \cos(\alpha_0 + \alpha_1). \quad (7)$$

Note that the function $\cos t$ decreases for $t \in [0, \pi]$. Also notice, that in (7) for $\alpha_0 = \varphi_0$, $\alpha_1 = \varphi_1$ the arguments of all cosines are maximal, that is cosines attain their minimums.

Finally we have

$$\min_{\substack{\alpha_0 \in [0, \varphi_0], \alpha_1 \in [0, \varphi_1] \\ \gamma \in [0, \frac{\pi}{2}]}} \|x_0 - x_1\|^2 = \varrho_0^2 + \varrho_1^2 + \|a_0 - a_1\|^2$$
$$+ 2\|a_0 - a_1\|(\varrho_0 \cos \varphi_0 + \varrho_1 \cos \varphi_1) + 2\varrho_0 \varrho_1 \cos(\varphi_0 + \varphi_1).$$

Remark 1. For any number $\varrho > 0$, such that $\varrho_0 \geq \varrho$, $\varrho_1 \geq \varrho$ in terms of Problem (4) and Lemma 1 the following inequality holds

$$\min_{\substack{\alpha_0 \in [0, \varphi_0], \alpha_1 \in [0, \varphi_1] \\ \gamma \in [0, \frac{\pi}{2}]}} \|x_0 - x_1\|^2 \geq 2\varrho^2 + \|a_0 - a_1\|^2$$
$$+ 2\|a_0 - a_1\|\varrho(\cos \varphi_0 + \cos \varphi_1) + 2\varrho^2 \cos(\varphi_0 + \varphi_1). \quad (8)$$

In the case $\varrho_0 = \varrho_1 = \varrho$ the inequality (8) obviously becomes identical. Let us consider the case, when at least one of the inequalities $\varrho_i \geq \varrho$ is strict.

Since $\varphi_i < \frac{\pi}{2}$, $\varrho_i \geq \varrho$, where $i \in \{0,1\}$, then $\varrho_0 \cos\varphi_0 + \varrho_1 \cos\varphi_1 \geq \varrho(\cos\varphi_0 + \cos\varphi_1)$.

That is clear, that the following inequalities hold

$$\frac{\varrho_0^2 + \varrho_1^2 - 2\varrho^2}{2(\varrho_0\varrho_1 - \varrho^2)} \geq 1 \geq \cos(\varphi_0 + \varphi_1).$$

Since $\varrho_0\varrho_1 - \varrho^2 > 0$, then

$$\varrho_0^2 + \varrho_1^2 - 2\varrho^2 \geq 2(\varrho_0\varrho_1 - \varrho^2)\cos(\varphi_0 + \varphi_1).$$

Whence we get

$$\varrho_0^2 + \varrho_1^2 - 2\varrho_0\varrho_1\cos(\varphi_0 + \varphi_1) \geq 2\varrho^2 - 2\varrho^2\cos(\varphi_0 + \varphi_1).$$

It completes the proof of the inequality (8)

Theorem 1. *Let $A \subset \mathcal{H}$ be a closed convex set, $S \subset \partial A$, numbers $C > 0$, $p \geq 2$, $\varrho > 0$ and for any point $a \in S$ there exists number $\nu(a) > 0$, such that the subset $B = B_{\nu(a)}(a) \bigcap A$ is uniformly convex with modulus of convexity $\delta_B(t) \geq Ct^p$. The set Φ is (S, ϱ)-neighborhood of the set A (see Formula (1)). Then $\forall \varepsilon > 0$ $\forall x_0 \in \Phi$ $\exists \delta = \delta(\varepsilon, x_0) > 0$ $\forall x_1 \in \Phi : [x_0, x_1] \in \Phi$, $\|x_0 - x_1\| < \delta$, the next estimate holds*

$$\|x_0 - x_1\|^2 \geq \|a_0 - a_1\|^2 + \left(\frac{8C\varrho}{1 - 2^{1-p}} - \varepsilon\right) \|a_0 - a_1\|^p$$
$$+ \left(\frac{16C^2\varrho^2}{(1 - 2^{1-p})^2} - \varepsilon\right) \|a_0 - a_1\|^{2p-2}, \qquad (9)$$

where $\{a_0\} = P_A(x_0)$, $\{a_1\} = P_A(x_1)$

Proof. Let us fix a point $x_0 \in \Phi$ and define a point $\{a_0\} = P_A(x_0)$. We choose a point $x_1 \in \Phi$, such that for a point $\{a_1\} = P_A(x_1)$ the following conditions are fulfilled $\|a_0 - a_1\| \leq \nu(a_0)$ and $a_0 \neq a_1$ (for example, if $\|x_0 - x_1\| < \nu(a_0)$, then from convexity of the set A it follows that $\|a_0 - a_1\| < \nu(a_0)$). If $a_0 = a_1$ then Formula (9) is true. Consider the case $a_0 \neq a_1$. Let us define the set $B \doteq B_{\nu(a_0)}(a_0) \bigcap A$. Let $l_0 = \|a_0 - a_1\|$, $\delta_0 = \delta_B(l_0)$. From the definition of the function $\delta_B(t)$ it follows that $B_0 \doteq B_{\delta_0}\left(\frac{a_0+a_1}{2}\right) \subset A$. Let us choose the point y_1 the following way: $y_1 \in B_0 \bigcap \mathrm{co}\{x_0, a_0, a_1\}$, the segment $a_0 y_1$ is tangent to the ball B_0 at the point y_1. Let us construct the sequence y_k like this: for all $k \in \mathbb{N}$, $k \geq 2$, $y_k \in \mathrm{co}\{x_0, a_0, y_1\}$, $y_k \in B_{k-1} \doteq B_{\delta_B(l_{k-1})}\left(\frac{a_0+y_{k-1}}{2}\right)$ the segment $a_0 y_k$ is tangent to the ball B_{k-1} at the point y_k, where $l_k = \|a_0 - y_k\|$. Using the definition of modulus of convexity of the set B by induction for all $k \in \mathbb{N}$ we obtain inclusions $y_k \in B$. Therefore $y_k \in A$. Let $\delta_k = \delta_B(l_k)$ and $\alpha_k = \angle y_{k+1} a_0 y_k$. It is obvious, that $\alpha_k \geq \sin\alpha_k = 2\frac{\delta_k}{l_k} \geq 2Cl_k^{p-1}$, where the last inequality follows from the inequality $\delta_B(t) \geq Ct^p$.

From [8, Corollary 2.3] it follows, that there exists such a number $C_0 > 0$, that for all $t \in (0, \operatorname{diam} A)$ the inequality $\delta_A(t) \leq C_0 t^2$ holds. Let us estimate angle $\alpha = \sum\limits_{k=0}^{\infty} \alpha_k$.

$$l_1 = \sqrt{\frac{l_0^2}{4} - \delta_0^2} \geq \sqrt{\frac{l_0^2}{4} - C_0^2 l_0^4} = \frac{l_0}{2}\sqrt{1 - 4C_0^2 l_0^2}, \tag{10}$$

$$l_k = \sqrt{\frac{l_{k-1}^2}{4} - \delta_{k-1}^2} \geq \frac{l_{k-1}}{2}\sqrt{1 - 4C_0^2 l_{k-1}^2}. \tag{11}$$

Thus, for any points $a_0, a_1 \in S$ such that $l_0 < \min\left\{\frac{1}{2C_0}, \nu(a_0)\right\}$ there exists such a number $\sigma > 0$, that for any $k \in \mathbb{N}$ the following inequalities hold $l_k \geq l_{k-1}\left(\frac{1}{2} - \sigma\right)$. Hence and from estimates (10) and (11) it follows, that

$$l_k \geq l_0 \left(\frac{1}{2} - \sigma\right)^k.$$

Therefore

$$\alpha = \sum_{k=0}^{\infty} \alpha_k \geq \sum_{k=0}^{\infty} \sin \alpha_k = 2\sum_{k=0}^{\infty} \frac{\delta_k}{l_k} \geq 2C \sum_{k=0}^{\infty} l_k^{p-1}$$

$$\geq 2C l_0^{p-1} \sum_{k=0}^{\infty} \left(\frac{1}{2} - \sigma\right)^{k(p-1)} = \frac{2C l_0^{p-1}}{1 - \left(\frac{1}{2} - \sigma\right)^{(p-1)}}. \tag{12}$$

Note, that convexity of the set A implies that $\langle x_0 - a_0, a_0 \rangle \geq \langle x_0 - a_0, y_k \rangle$, that is $\langle a_0 - x_0, a_0 - y_k \rangle \leq 0$. Hence it follows, that $\angle x_0 a_0 y_k \geq \frac{\pi}{2}$, $\forall k \in \mathbb{N}$. Since $y_k \in \operatorname{co}\{x_0, a_0, a_1\}$ we have $\angle x_0 a_0 a_1 = \angle x_0 a_0 y_k + \angle y_k a_0 a_1 \geq \frac{\pi}{2} + \sum\limits_{n=0}^{k} \alpha_n$, passing the limit as $n \to \infty$ we obtain $\angle x_0 a_0 a_1 \geq \frac{\pi}{2} + \alpha$. Similarly $\angle x_1 a_1 a_0 \geq \frac{\pi}{2} + \alpha$.

Let $\varphi_0, \varphi_1 \in [0, \frac{\pi}{2}]$. Cones K_0 and K_1 are built on the points a_0 and a_1 as a vertexes correspondingly, the segment $[a_0, a_1]$ is the axis of rotation, and for any point $x \in K_0$ the following inequality holds $\angle x a_0 a_1 \geq \pi - \varphi_0$, for any point $x \in K_1$ the next inequality holds $\angle x a_1 a_0 \geq \pi - \varphi_1$. Note, that choosing $\varphi_0 = \varphi_1 = \frac{\pi}{2} - \alpha$ and taking into account inequalities $\angle x_0 a_0 a_1 \geq \frac{\pi}{2} + \alpha$ and $\angle x_1 a_1 a_0 \geq \frac{\pi}{2} + \alpha$, the inclusions $x_0 \in K_0$, $x_1 \in K_1$ are true.

Let us recall that $\|x_0 - a_0\| \geq \varrho$, $\|x_1 - a_1\| \geq \varrho$. Thus, it results from Remark 1 that the following inequality holds

$$\|x_0 - x_1\|^2 \geq 2\varrho^2 + \|a_0 - a_1\|^2 + 4\|a_0 - a_1\|\varrho \sin \alpha - 2\varrho^2 \cos 2\alpha. \tag{13}$$

Notice, that $\sin \alpha \geq \alpha - \frac{\alpha^3}{6}$, $\cos 2\alpha \leq 1 - 2\alpha^2 + \frac{2\alpha^4}{3}$. Then from estimate (12) and inequality (13) we get

$$\|x_0 - x_1\|^2 \geq \|a_0 - a_1\|^2 + \frac{8\varrho C\|a_0 - a_1\|^p}{1 - \left(\frac{1}{2} - \sigma\right)^{p-1}}$$

$$+ \frac{16\varrho^2 C^2\|a_0 - a_1\|^{2p-2}}{\left(1 - \left(\frac{1}{2} - \sigma\right)^{p-1}\right)^2} + o(\|a_0 - a_1\|^{3p-3}).$$

Since the number l_0 is small enough it follows, that for all $\varepsilon > 0$ there exists such number $\delta = \delta(\varepsilon, x_0) > 0$, that for x_1 with $\|x_0 - x_1\| < \delta$ and $[x_0, x_1] \in \Phi$ the next inequality holds

$$\|x_0 - x_1\|^2 \geq \|a_0 - a_1\|^2 + \left(\frac{8C\varrho}{1 - 2^{1-p}} - \varepsilon\right)\|a_0 - a_1\|^p$$

$$+ \left(\frac{16C^2\varrho^2}{(1 - 2^{1-p})^2} - \varepsilon\right)\|a_0 - a_1\|^{2p-2}.$$

Corollary 1. *Let the conditions of Theorem 1 be satisfied, $p = 2$, the set Φ is (S, ϱ)-neighborhood of the set A. Then for any pair of points $x_0, x_1 \in \Phi$ such that $[x_0, x_1] \subset \Phi$ the following inequality holds*

$$\|a_0 - a_1\| \leq \frac{1}{1 + 8C\varrho}\|x_0 - x_1\|, \tag{14}$$

where $\{a_i\} = P_A(x_i)$, $i \in \{0, 1\}$.

Proof. Let us fix a number $\varepsilon > 0$, $\varepsilon < \min\{16C\varrho, 64C^2\varrho^2\}$. Let $A(\varepsilon) = A(\varepsilon, C, \varrho) = 16C\varrho - \varepsilon$, $B(\varepsilon) = B(\varepsilon, C, \varrho) = 64C^2\varrho^2 - \varepsilon$. Note that for any such ε, the inequalities $A(\varepsilon) > 0$, $B(\varepsilon) > 0$ are true.

Consider a pair of points $x_0, x_1 \in \Phi$ such that $[x_0, x_1] \subset \Phi$. From the compactness of the segment $[x_0, x_1]$ it results that from its covering by open balls int $B_{\frac{\delta(\varepsilon, x)}{2}}(x)$, $x \in [x_0, x_1]$ there can be extracted finite subcovering with centers in the points $\{y_i\}_{i=0}^n$, where $y_0 = x_0$, $y_n = x_1$. Let us define points $\{b_i\} = P_A(y_i)$, $i \in \{0, n\}$. It follows from Theorem 1 that for any number $i \in \{0, n-1\}$ the next inequality holds true

$$\|y_i - y_{i+1}\| \geq \|b_i - b_{i+1}\|\sqrt{(1 + A(\varepsilon) + B(\varepsilon))}.$$

Hence and from the triangle inequality we obtain the following estimate:

$$\|x_0 - x_1\| = \sum_{i=0}^{n-1} \|y_i - y_{i+1}\|$$

$$\geq \sum_{i=0}^{n-1} \|b_i - b_{i+1}\|\sqrt{(1 + A(\varepsilon) + B(\varepsilon))} \geq \|a_0 - a_1\|\sqrt{(1 + A(\varepsilon) + B(\varepsilon))}.$$

Passing the limit as $\varepsilon \to +0$ we obtain inequality (14).

Note, that Theorem 1 and Corollary 1 are a local variant of [6, Theorem 2.2]. If $S = \partial A$ and $\delta_S(\varepsilon) = C\varepsilon^2 + o(\varepsilon^2)$, $\varepsilon \to +0$ it follows from Corollary 1 that

$$\|a_0 - a_1\| \leq \frac{1}{1 + 8C\varrho}\|x_0 - x_1\|,$$

where $x_i \in U_A(\varrho)$, $\{a_i\} = P_A(x_i)$, $i \in \{0, 1\}$. From [9, Theorem 2.1] it follows, that the set A is strongly convex with radius $R = \frac{1}{8C}$. Thus, the inequality (14) takes the form:

$$\|a_0 - a_1\| \leq \frac{R}{R + \varrho}\|x_0 - x_1\|,$$

which is exactly the same as inequality in [6, Corollary 2.1].

3   Local Strong Convexity of the Subset of Boundary with Lipschitz Property with Constant Less Than 1

In this section we will prove that if metric projection onto some subset of the boundary of the set $A \subset \mathcal{H}$ is Lipschitz continuous with Lipschitz constant strictly less than 1 then the intersection of the ball with small radius and the subset A is strongly convex.

Lemma 2. *Let* $p_i \in \mathcal{H}$, $\|p_i\| = 1$, $t_i \geq 1$, $i \in \{0, 1\}$.
 Then $\|p_0 - p_1\| \leq \|t_0 p_0 - t_1 p_1\|$

Proof. Result of Lemma 2 can be obtained easily.

Theorem 2. *Let subset* $A \subset \mathcal{H}$ *be closed and convex,* $S \subset \partial A$. *A number* $\varrho > 0$ *is fixed, the set* Φ *is* (S, ϱ)-*neighborhood of the set* A. *Suppose, that there exists a number* $C \in (0; 1)$, *such that for any points* $x_0, x_1 \in \Phi$ *the following formula holds true*

$$\|a_0 - a_1\| \leq C\|x_0 - x_1\|, \quad \{a_i\} = P_A(x_i), \quad i \in \{0, 1\}. \tag{15}$$

 Let $a \in S$ *and* $R = \frac{C\varrho}{1-C}$.
 Then for any number $\varepsilon \in (0, R)$ *with property* $B_\varepsilon(a) \bigcap \partial A \subset S$ *the subset* $A \bigcap B_\varepsilon(a)$ *is strongly convex set with radius* R.

Proof. Several auxiliary lemmas will be formulated and proved within the text of the proof of Theorem 2.

Let us fix a pair of points $x_0, x_1 \in \Phi$, such that $\varrho_A(x_0) = \varrho_A(x_1) = \varrho$. Let

$$\{a_i\} = P_A(x_i), \quad q_i = \frac{x_i - a_i}{\varrho}, \quad i \in \{0, 1\}. \tag{16}$$

Note, that $q_i \in N(A, a_i) \bigcap \partial B_1(0)$, where $i \in \{0, 1\}$. Using the condition (15) we have

$$\|a_0 - a_1\| \leq C\|x_0 - x_1\| = C\|(a_0 - a_1) + (x_0 - a_0) - (x_1 - a_1)\|$$

$$\leq C\|a_0 - a_1\| + C\varrho\left\|\frac{x_0 - a_0}{\varrho} - \frac{x_1 - a_1}{\varrho}\right\|.$$

The last inequality takes the form

$$\|a_0 - a_1\| \leq R\|q_0 - q_1\|, \tag{17}$$

where $R = \frac{C\varrho}{1-C}$.

Let us fix a number $\gamma \in (0, \varrho)$. Consider the set

$$A_\gamma = A + B_\gamma(0). \tag{18}$$

Let

$$S_\gamma = (\partial A_\gamma) \bigcap \left(\bigcup_{x \in S} \big(x + N(A; x)\big) \right). \tag{19}$$

We denote $\{b_i\} = P_{A_\gamma}(x_i)$, where $i \in \{0, 1\}$. Obviously $\|x_i - b_i\| \geq \varrho - \gamma$, $a_i + \gamma q_i \in S_\gamma$ and $\|x_i - (a_i + \gamma q_i)\| = \varrho - \gamma$. Since the metric projection on the convex set is unique then $a_i = b_i - q_i\gamma$, for $i \in \{0, 1\}$. Thus,

$$b_i = a_i + q_i\gamma, \quad i \in \{0, 1\}. \tag{20}$$

Taking it into account we can rewrite inequality (17)

$$\|b_0 - b_1\| - \gamma\|q_0 - q_1\| \leq \|(b_0 - q_0\gamma) - (b_1 - q_1\gamma)\|$$
$$= \|a_0 - a_1\| \leq R\|q_0 - q_1\|.$$

Thus,

$$\|b_0 - b_1\| \leq (R + \gamma)\|q_0 - q_1\|. \tag{21}$$

We denote

$$R_\gamma = R + \gamma. \tag{22}$$

Obviously $q_i \in N(A_\gamma, b_i) \bigcap \partial B_1(0)$, where $i \in \{0, 1\}$.

Let us prove that

$$\gamma\|q_0 - q_1\| \leq \|b_0 - b_1\|. \tag{23}$$

We will prove it by contradiction. Suppose, that the following inequality holds true

$$\gamma\|q_0 - q_1\| > \|a_0 + \gamma q_0 - (a_1 + \gamma q_1)\|.$$

Squaring the last inequality we obtain

$$0 > \|a_0 - a_1\|^2 + 2\gamma\langle a_0 - a_1, q_0 - q_1\rangle. \tag{24}$$

It follows from the inclusions $q_i \in N(A; a_i)$, $i \in \{0, 1\}$, that inequalities $\langle q_0, a_0\rangle \geq \langle q_0, a_1\rangle$ and $\langle q_1, a_1\rangle \geq \langle q_1, a_0\rangle$ hold true. Summing those inequalities we get: $\langle a_0 - a_1, q_0 - q_1\rangle \geq 0$. Hence and from the inequality (24) we obtain the following inequalities

$$0 > \|a_0 - a_1\|^2 + 2\langle a_0 - a_1, q_0 - q_1\rangle \geq 0.$$

We come to contradiction.

It follows from the inequality (23) that for any point $b \in S_\gamma$ there exists a unique vector $p_b \in \partial B_1(0) \bigcap N(A_\gamma; b) \subset \mathcal{H}$.

Fix a point $b \in S_\gamma$ and a number $\omega \in (0, \gamma)$, such that

$$(\partial A_\gamma) \bigcap B_\omega(b) \subset S_\gamma. \tag{25}$$

Let $p_b \in (\partial B_1(0)) \bigcap N(A_\gamma; b)$. Without loss of generality, we assume $b = 0$. Consider a hyperplane

$$p_b^\perp = \{z \in \mathcal{H} : \langle z, p_b \rangle = 0\}. \tag{26}$$

A closed ball with center in the point $y \in p_b^\perp$ and radius $r > 0$ in the subspace $p_b^\perp$ we will denote by $\mathcal{B}_r(y)$.

Lemma 3. *Let the conditions of Theorem 2 and Formula (25) hold true. $\hat{S}_\gamma$ - orthogonal projection of the subset $S_\gamma \bigcap B_\omega(0)$ onto hyperplane $p_b^\perp$.*

Then the function $f : \hat{S}_\gamma \to \mathbb{R}$, which is assigned by the following formula: graph $f = S_\gamma \bigcap B_\omega(0)$, where the positive direction of the ordinate axis coincides with the vector $-p_b$, is defined correctly. Moreover, if we redefine this function on the whole space, supposing that $f(z) = +\infty$ for any point $z \in p_b^\perp \setminus \hat{S}_\gamma$, then the function $f : p_b^\perp \to \overline{\mathbb{R}}$ is convex and lower semicontinuous.

Proof of Lemma 3. Let us show, that function f is defined correctly. We will prove it by contradiction. Suppose that there exist such a pair of points $b_0, b_1 \in S_\gamma \bigcap B_\omega(0)$ that $\frac{b_0 - b_1}{\|b_0 - b_1\|} = p_b$. Let $p_1 \in (\partial B_1(0)) \bigcap N(A_\gamma; b_1)$, whereby the inequalities $\langle p_1, b_0 - b_1 \rangle \leq 0$ and $\langle p_1, p_b \rangle \leq 0$ hold true. Hence $\sqrt{2} \leq \|p_1 - p_b\|$. From the inequality (23), inclusion $b_1 \in S_\gamma \bigcap B_\omega(0)$ and inclusion $\omega \in (0, \gamma)$ it follows, that

$$\sqrt{2} \leq \|p_1 - p_b\| \leq \frac{\|b_1 - b\|}{\gamma} \leq \frac{\omega}{\gamma} < 1.$$

Contradiction shows, that the function $f : \hat{S}_\gamma \to \mathbb{R}$ is defined correctly. Let us redefine the function f on the whole space $p_b^\perp$, supposing that $f(z) = +\infty$ for all $z \in p_b^\perp \setminus \hat{S}_\gamma$. Thus the function $f : p_b^\perp \to \overline{\mathbb{R}}$ is defined.

We denote by $\tilde{S}_\gamma$ orthogonal projection of the set $A_\gamma \bigcap B_\omega(0)$ on the hyperplane $p_b^\perp$. Note, that by the inclusion $S_\gamma \bigcap B_\omega(0) \subset A_\gamma \bigcap B_\omega(0)$ the inclusion $\hat{S}_\gamma \subset \tilde{S}_\gamma$ holds true. Let us show, that $\hat{S}_\gamma = \tilde{S}_\gamma$. Suppose the contrary: there exists a point $s \in \partial \tilde{S}_\gamma \setminus \hat{S}_\gamma$. Let $p_s \in N(\tilde{S}_\gamma; s) \bigcap \partial B_1(0)$. Supposition means that there exists a point $a \in \partial(A_\gamma \bigcap B_\omega(0)) \setminus S_\gamma$, which is projected into a point s, where $p_s \in N(A_\gamma \bigcap B_\omega(b); a)$. From the inclusions $(\partial A_\gamma) \bigcap B_\omega(0) \subset S_\gamma$ and $a \in \partial(A_\gamma \bigcap B_\omega(b)) \setminus S_\gamma$ it follows, that $a \in \partial B_\omega(0) \bigcap \text{int } A_\gamma$. From here and inclusion $p_s \in N(A_\gamma \bigcap B_\omega(b); a)$ we obtain $a = p_s \omega$. Hence it follows that tangent hyperplane $H_{p_b}(0) = \{x \in \mathcal{H} \mid \langle p_b, x \rangle = 0\}$ at the point $b = 0$ to the set A_γ intersects the set A_γ at the point $a \in \text{int } A_\gamma$, which contradicts the convexity of the set A_γ. Thus we prove the equality $\hat{S}_\gamma = \tilde{S}_\gamma$. This means that for any point $x \in (A_\gamma \bigcap B_\omega(0)) \setminus S_\gamma$ there exists a point $y \in S_\gamma \bigcap B_\omega(0)$ such that

$\frac{y-x}{\|y-x\|} = p_b$. This in turn means that $A_\gamma \bigcap B_\omega(0) + l_b = S_\gamma \bigcap B_\omega(0) + l_b$, where $l_b = \{-p_b t \mid t \in [0, +\infty)\}$. Note that $S_\gamma \bigcap B_\omega(0) + l_b = \text{epi } f$. Thus

$$\text{epi } f = A_\gamma \bigcap B_\omega(0) + l_b, \qquad (27)$$

where $l_b = \{-p_b t \mid t \in [0, +\infty)\}$.

The set $A_\gamma \bigcap B_\omega(0) + l_b$ is convex as a sum of two convex sets. It follows from [19, Theorem 1.13.2] that it is closed, that is epi f is closed and convex set. It follows from [19, Theorem 1.5.1] and [19, Definition 1.6.1] that the function f is convex and lower semicontinuous. **Lemma 3 is proved.**

Lemma 4. *Let the conditions of Lemma 3 hold true. Then the following inclusion holds*

$$\mathcal{B}_\tau(0) \subset \hat{S}_\gamma, \qquad (28)$$

where $\tau = \frac{\omega}{\gamma}\sqrt{\gamma^2 - \omega^2}$.

Proof of Lemma 4. Suppose the contrary. Then, using the inclusion $0 \in \hat{S}_\gamma$, we obtain that there exists a point $y_0 \in \text{int } \mathcal{B}_\tau(0) \bigcap \partial\hat{S}_\gamma$. Since $y_0 \in \partial\hat{S}_\gamma$, then there exists such a point $z_0 \in S_\gamma \bigcap \partial B_\omega(0)$, that $y_0 - z_0 = t_0 p_b$ for some $t_0 \geq 0$. Let $p_0 \in \partial B_1(0) \bigcap N(A_\gamma; z_0)$. It follows from the inequality (23), that $\|p_0 - p_b\| \leq \frac{\|z_0 - b\|}{\gamma} = \frac{\omega}{\gamma}$. Since $p_0 \in N(A_\gamma; z_0)$, then $\langle p_0, z_0 \rangle \geq \langle p_0, b \rangle = 0$. Therefore,

$$\langle p_b, z_0 \rangle = \langle p_b - p_0, z_0 \rangle + \langle p_0, z_0 \rangle \geq \langle p_b - p_0, z_0 \rangle$$

$$\geq -\|p_b - p_0\|\|z_0\| \geq -\frac{\omega^2}{\gamma}.$$

From the inclusion $y_0 \in p_b^\perp$ follows the equality $\langle p_b, y_0 \rangle = 0$. Thus,

$$t_0 = \langle p_b, y_0 - z_0 \rangle = -\langle p_b, z_0 \rangle \leq \frac{\omega^2}{\gamma}.$$

Using the equalities $y_0 - z_0 = t_0 p_b$ and $\langle p_b, y_0 \rangle = 0$, we obtain the equalities $t_0^2 + \|y_0\|^2 = \|z_0\|^2 = \omega^2$. Thus

$$\omega^2 \leq \frac{\omega^4}{\gamma^2} + \|y_0\|^2 < \frac{\omega^4}{\gamma^2} + \tau^2,$$

which contradicts the equality $\tau = \frac{\omega}{\gamma}\sqrt{\gamma^2 - \omega^2}$. So, the inclusion (28) is proved. **Lemma 4 is proved.**

Since $\text{dom } f = \hat{S}_\gamma \supset \mathcal{B}_\tau(0)$, then $\text{int dom } f \neq \emptyset$. By Lemma 2 in terms of function f the inequality (21) takes form $\forall z_0, z_1 \in \text{int dom } f \hookrightarrow$

$$\|z_0 - z_1\| \leq \|(z_0, f(z_0)) - (z_1, f(z_1))\| \leq R_\gamma\|(f'(z_0), -1) - (f'(z_1), -1)\|$$
$$\leq R_\gamma\|f'(z_0) - f'(z_1)\|, \qquad \forall z_0, z_1 \in \text{int dom } f. \qquad (29)$$

Due to [19, Theorem 1.16.4] equality $z_i = f^{*'}(p_i)$ is equivalent to equality $p_i = f'(z_i)$, $i \in \{0,1\}$. Thus the inequality (29) takes form

$$\|f^{*'}(p_0) - f^{*'}(p_1)\| \le R_\gamma \|p_0 - p_1\|, \qquad \forall p_0, p_1 \in p_b^\perp. \tag{30}$$

Note, that $\operatorname{dom} f = \hat{S}_\gamma$ is bounded set. It follows from the definition of conjugate function that $\operatorname{dom} f^* = p_b^\perp$.

From [19, Theorem 1.19.2] it follows that the function f is strongly convex with the constant $\frac{1}{R_\gamma}$.

It follows from [7, Theorem 3.1] that for any $z \in \mathcal{B}_\tau(0)$ the inequality $f(z) \ge \frac{1}{2R_\gamma}\|z\|^2 = h_\gamma(z)$ holds, whence we have the inclusion $B_\theta(0) \bigcap \operatorname{epi} f \subset B_\theta(0) \bigcap \operatorname{epi} h_\gamma$, $\forall \theta \in (0, \tau)$.

Lemma 5. *Let the conditions of Lemma 3 hold and the function $f\colon p_b^\perp \to \overline{\mathbb{R}}$ be from Lemma 3. The number ϱ is from Theorem 2, the number $\gamma \in (0, \varrho)$, and numbers ω and τ are from Lemmas 3 and 4 correspondingly, the number R_γ is defined by (22). Then for any number $\theta \in (0, \tau]$ the inclusion holds $B_\theta(0) \bigcap \operatorname{epi} f \subset B_{R_\gamma(\theta)}(-\frac{p_b}{\|p_b\|}R_\gamma(\theta))$, where $R_\gamma(\theta) = \frac{R_\gamma + \sqrt{R_\gamma^2 + \theta^2}}{2}$.*

Proof of Lemma 5. The orthogonal projection of the set $B_\theta(0) \bigcap \operatorname{epi} h_\gamma$ on the subspace $p_b^\perp$ is the ball $\mathcal{B}_{r_\theta}(0)$, where $r_\theta = R_\gamma \sqrt{2\left(\sqrt{1 + \frac{\theta^2}{R_\gamma^2}} - 1\right)}$. Let us suppose that there exists such a point z, that $\|z\| \le r_\theta$ and $\frac{1}{2R_\gamma}\|z\|^2 < R_\gamma(\theta) - \sqrt{R_\gamma(\theta)^2 - \|z\|^2}$. Hence we obtain the inequality $\sqrt{R_\gamma^2(\theta) - \|z\|^2} < R_\gamma(\theta) - \frac{1}{2R_\gamma}\|z\|^2$. By the inequality $\|z\| \le r_\theta$ and the choice of the number θ the right side of the last inequality is positive. After squaring and transformations we have $0 < 1 - \frac{R_\gamma(\theta)}{R_\gamma} + \frac{1}{4R_\gamma^2}\|z\|^2$. In other words

$$\|z\| > 2\sqrt{R_\gamma R_\gamma(\theta) - R_\gamma^2} = r_\theta.$$

Contradiction. So for any $\|z\| \le r_\theta$ we have the inequality $\frac{1}{2R_\gamma}\|z\|^2 \ge R_\gamma(\theta) - \sqrt{R_\gamma(\theta)^2 - \|z\|^2}$.

This inequality provides the inclusion $B_\theta(0) \bigcap \operatorname{epi} h_\gamma \subset B_{R_\gamma(\theta)}(-\frac{p_b}{\|p_b\|}R_\gamma(\theta))$. Hence the following inclusion holds true $B_\theta(0) \bigcap \operatorname{epi} f \subset B_{R_\gamma(\theta)}(-\frac{p_b}{\|p_b\|}R_\gamma(\theta))$. Note, that $R_\gamma(\theta) > R_\gamma$ and $R_\gamma(\theta) \to R_\gamma$, while $\theta \to 0$. **Lemma 5 is proved.**

Similarly, for any point $b \in S_\gamma$, for which exists such a number $\omega \in (0, \gamma)$, that $(\partial A_\gamma) \bigcap B_\omega(b) \subset S_\gamma$ the following inclusion holds true

$$B_\theta(b) \bigcap \operatorname{epi} f \subset B_{R_\gamma(\theta)}(b - \frac{p_b}{\|p_b\|}R_\gamma(\theta)), \tag{31}$$

where $R_\gamma(\theta) = \frac{R_\gamma + \sqrt{R_\gamma^2 + \theta^2}}{2}$. It follows from the inclusion (31), the equality (27) and conditions $\omega \in (0, \gamma)$, $\tau = \frac{\omega}{\gamma}\sqrt{\gamma^2 - \omega^2}$ and $\theta \in (0, \tau]$ that

$$B_\theta(b) \bigcap A_\gamma \subset B_{R_\gamma(\theta)}(b - \frac{p_b}{\|p_b\|}R_\gamma(\theta)), \tag{32}$$

where $R_\gamma(\theta) = \frac{R_\gamma + \sqrt{R_\gamma^2 + \theta^2}}{2}$.

Lemma 6. *Let the number R be from Theorem 2. Let $\widetilde{\varepsilon}$ and γ are fixed numbers such that $\widetilde{\varepsilon} \in (0, R)$ and $\gamma \in (0, \varrho)$. Sets A_γ and S_γ is defined by (18) and (19) correspondingly. Let $b \in S_\gamma$ and the inclusion $B_{\widetilde{\varepsilon}+2\gamma}(b) \cap \partial A_\gamma \subset S_\gamma$ holds.*
Then the set $A_\gamma \cap B_{\widetilde{\varepsilon}+\gamma}(b)$ is strongly convex set with radius R_γ.

Proof of Lemma 6. Consider a set $A_{\widetilde{\varepsilon}}(b) = A_\gamma \cap B_{\widetilde{\varepsilon}+\gamma}(b)$. For any point $x \in (\partial B_{\widetilde{\varepsilon}+\gamma}(b)) \cap A_{\widetilde{\varepsilon}}(b)$ the unit vector $p_x = \frac{b-x}{\widetilde{\varepsilon}+\gamma}$ satisfies inclusion

$$A_{\widetilde{\varepsilon}}(b) \subset B_{\widetilde{\varepsilon}+\gamma}(b) \subset B_{R_\gamma}(x - R_\gamma p_x). \tag{33}$$

Note that for any point $x \in S_\gamma \cap A_{\widetilde{\varepsilon}}(b)$ the inclusion holds true $A_\gamma \cap B_\gamma(x) \subset A_\gamma \cap B_{\widetilde{\varepsilon}+2\gamma}(b)$. Hence follows the inclusion $(\partial A_\gamma) \cap B_\gamma(x) \subset S_\gamma$. Thus for any point $x \in S_\gamma \cap A_{\widetilde{\varepsilon}}(b)$ the number $\omega = \frac{\gamma}{2}$ satisfies inclusion $(\partial A_\gamma) \cap B_\omega(x) \subset S_\gamma$ (similarly to inclusion in (25)). Taking into account that $\omega = \frac{\gamma}{2}$ we obtain that the number τ from Lemma 4 equals $\tau = \frac{\sqrt{3}}{4}$.
The inclusion (32) implies that for any point $x \in S_\gamma \cap A_{\widetilde{\varepsilon}}(b)$ and for any number $\theta \in (0, \frac{\sqrt{3}}{4}]$ the following inclusion holds

$$B_\theta(x) \cap A_{\widetilde{\varepsilon}}(b) \subset B_\theta(x) \cap A_\gamma \subset B_{R_\gamma(\theta)}(x - \frac{p_x}{\|p_x\|} R_\gamma(\theta)), \tag{34}$$

where $R_\gamma(\theta) = \frac{R_\gamma + \sqrt{R_\gamma^2 + \theta^2}}{2}$, $p_x \in N(A_\gamma, x)$.
Consider a set $\Omega = A_{\widetilde{\varepsilon}}(b)$. It follows from the inclusions (33) and (34) that for any point $x \in \partial\Omega$ there exist such a neighborhood U of a point x and a vector v, $\|v\| = 1$ that $U \cap \Omega \subset B_{R_\gamma(\theta)}(x - R_\gamma(\theta)v)$. It follows from [21, Theorem 1.2, Remark after the proof of Theorem 1.2 and Proposition 3.3] that the set $\Omega = A_{\widetilde{\varepsilon}}(b)$ is strongly convex set with radius $R_\gamma(\theta)$ for any $\theta \in (0, \frac{\sqrt{3}}{4}]$. Letting θ to zero we obtain, that the set $A_{\widetilde{\varepsilon}}(b)$ is strongly convex with radius R_γ. **Lemma 6 is proved.**

Lemma 7. *Let conditions of Theorem 2 hold true. Then for any $\widetilde{\varepsilon} \in (0, \varepsilon)$ the subset $A \cap B_{\widetilde{\varepsilon}}(a)$ is strongly convex with radius R.*

Proof of Lemma 7. Let the point $a \in S$ and the number $\varepsilon > 0$ be from Theorem 2, that is $B_\varepsilon(a) \cap \partial A \subset S$. Let $p_a \in N(A; a) \cap \partial B_1(0)$.
By analogy with sets A_γ and S_γ define the sets $A_n = A + B_{\frac{1}{n}}(0)$ and $S_n = (\partial A_n) \cap \left(\bigcup_{x \in S} (x + N(A; x)) \right)$, denote $b_n = a + \frac{1}{n} p_a$.
Let us show, that for all $n \in \mathbb{N}$ the following inclusion holds

$$B_\varepsilon(b_n) \cap \partial A_n \subset S_n. \tag{35}$$

Assume the contrary. Then there exists a point $y_n \in \partial A_n \setminus S_n$ such that $\|y_n - b_n\| \leq \varepsilon$. By the construction of the set A_n there exist a point $x_n \in \partial A \setminus S$

and vector $p_n(x) \in N(A; x) \cap \partial B_1(0)$ such that $y_n = x_n + \frac{1}{n}p_n(x)$. It follows from the inclusion $B_\varepsilon(a) \cap \partial A \subset S$ that $\|x_n - a\| > \varepsilon$. It follows from the inclusions $p_a \in N(A; a)$ and $p_n(x) \in N(A; x)$ that the inequalities $\langle p_n(x), x_n \rangle \geq \langle p_n(x), a \rangle$ and $\langle p_a, a \rangle \geq \langle p_a, x_n \rangle$ hold true. Summing the last inequalities we obtain: $\langle x_n - a, p_n(x) - p_a \rangle \geq 0$. Using the last inequality and the equalities $b_n = a + \frac{1}{n}p_a$ and $y_n = x_n + \frac{1}{n}p_n(x)$, we get

$$\|y_n - b_n\|^2 = \left\| (x_n - a) + \frac{1}{n}(p_n(x) - p_a) \right\|^2$$

$$= \|x_n - a\|^2 + \frac{2}{n}\langle x_n - a, p_n(x) - p_a \rangle + \frac{1}{n^2}\|p_n(x) - p_a\|^2$$

$$> \varepsilon^2 + \frac{1}{n^2}\|p_n(x) - p_a\|^2 \geq \varepsilon^2.$$

Which contradicts the inequality $\|y_n - b_n\| \leq \varepsilon$. Thus for all $n \in \mathbb{N}$ the inclusion $B_\varepsilon(b_n) \cap \partial A_n \subset S_n$ holds.

Fix $\widetilde{\varepsilon} \in (0, \varepsilon)$. Denote $C = A \cap B_{\widetilde{\varepsilon}}(a)$. Define the subsets

$$C_n = A_n \cap B_{\widetilde{\varepsilon} + \frac{1}{n}}(b).$$

Let $n_0 = \max\left\{ \left[\frac{2}{\varepsilon - \widetilde{\varepsilon}}\right] + 1, \left[\frac{1}{\varrho}\right] + 1 \right\}$, where $[k]$ – is the largest integer not greater than k. Then for any $n > n_0$ the inequality $\frac{1}{n} < \varrho$ and the inclusion $B_{\widetilde{\varepsilon} + \frac{2}{n}}(b_n) \cap \partial A_n \subset S_n$ hold true, that is conditions of Lemma 6 are fulfilled. Thus the subsets C_n are strongly convex with radii $R_n = R + \frac{1}{n}$.

It follows from [5, Theorem 1] that $h\big(C_n, A \cap B_{\widetilde{\varepsilon}}(a)\big) \to 0$ while $n \to \infty$. By [19, Lemma 4.3.1] for any $\widetilde{\varepsilon} \in (0, \varepsilon)$ the subset $C = A \cap B_{\widetilde{\varepsilon}}(a)$ is strongly convex with radius $\liminf_{n\to\infty} R_n = R$. **Lemma 7 is proved.**

Denote $D = A \cap B_\varepsilon(a)$. For any $n > \left[\frac{1}{\varepsilon}\right] + 1$ define the subsets

$$D_n = A \cap B_{\varepsilon - \frac{1}{n}}(a).$$

It follows from [5, Theorem 1] and [19, Lemma 4.3.1] that subset $D = A \cap B_\varepsilon(a)$ is strongly convex with radius R. The proof of Theorem 2 is completed.

Remark 2. If in Theorem 2 the number $\varepsilon > \frac{C\varrho}{1 - C}$ and the inclusion $B_\varepsilon(a) \cap \partial A \subset S$ holds, then the subset $A \cap B_\varepsilon(a)$ is strongly convex with radius ε. Let us show that such a case is possible in terms of Theorem 2. Consider the following subset

$$A = \{(x, y) \in \mathbb{R}^2 \mid x \in [-1, 1], y \geq -\sqrt{1 - x^2}\}.$$

Let

$$S = \{(x, y) \in \mathbb{R}^2 \mid x \in [-1, 1], y = -\sqrt{1 - x^2}\}.$$

Let $\varrho = 1$. It is easy to prove that for any pair points $x_0, x_1 \in \Phi(A, S, 1)$ (where a subset $\Phi(A, S, \varrho)$ is (S, ϱ)-neighborhood of the set A) the following inequality holds

$$\|a_0 - a_1\| \leq \frac{1}{2}\|x_0 - x_1\|, \quad \{a_i\} = P_A(x_i)$$

In this case the number C from Theorem 2 equals $\frac{1}{2}$ and $\frac{C\varrho}{1-C} = 1$. If we choose the point $a = (0, -1)$, then for all numbers $\varepsilon \in (0, \sqrt{2})$ inclusion $B_\varepsilon(a) \bigcap \partial A \subset S$ holds. Thus for any number $\varepsilon \in (0, 1)$ the subset $A \bigcap B_\varepsilon(a)$ is strongly convex with radius 1 and for any number $\varepsilon \in (1, \sqrt{2})$ the subset $A \bigcap B_\varepsilon(a)$ is strongly convex with radius ε.

4 Further Plans

- To consider metric projection algorithm for convex function and locally convex sets and to improve results from [13].
- To consider locally strongly convex sets in asymmetric seminormed spaces (see for example in [16]) and try to extend results of the current paper.
- To analyze is it possible to generalize results of [14] for the case of locally strongly convex sets.
- to consider differential games with locally strongly convex admissible control sets and generalize result of [15].

References

1. Abatzoglou, T.J.: The minimum norm projection on C^2-manifolds in $\mathbb{R}^n$. Trans. AMS **243**, 115–122 (1978)
2. Abatzoglou, T.J.: The Lipschitz continuity of the metric projection. J. Approx. Theory **26**, 212–218 (1979)
3. Alber, Ya.I., Notik, A.I.: On some estimates for projection operator in Banach space comm. Appl. Nonlinear Anal. **2**(1), 47–56 (1995)
4. Alber, Ya.I.: A bound for the modulus of continuity for metric projections in a uniformly convex and uniformly smooth banach space. J. Approx. Theory **85**(3), 237–249 (1996)
5. Balashov, M.V.: The continuity of the intersection of multivalued maps with strongly convex values. Electron. J. "Investig. Russia" 534–539 (2002). http://zhurnal.ape.relarn.ru/articles/2002/049.pdf. (in Russian)
6. Balashov, M.V., Golubev, M.O.: About the Lipschitz property of the metric projection in the Hilbert space. J. Math. Anal. Appl. **394**(2), 545–551 (2012)
7. Balashov, M.V., Polovinkin, E.S.: M-strongly convex subsets and their generating sets. SB MATH **191**(1), 25–60 (2000)
8. Balashov, M.V., Repovš, D.: Uniform convexity and the splitting problem for selections. J. Math. Anal. Appl. **360**(1), 307–316 (2009)
9. Balashov, M.V., Repovš, D.: Uniformly convex subsets of the Hilbert space with modulus of convexity of the second order. J. Math. Anal. Appl. **377**(2), 754–761 (2011)
10. Björnestål, B.O.: Local Lipschitz continuity of the metric projection operator. Approx. Theory Banach Center Publ. **4**, 43–54 (1979)
11. Federer, H.: Curvature measures. Trans. AMS **93**, 418–491 (1959)
12. Frankowska, H., Olech, Ch.: R-convexity of the integral of the set-valued functions. Contributions to Analysis and Geometry, pp. 117–129. John Hopkins Univ. Press, Baltimore (1981)

13. Golubev, M.O.: Gradient projection method for convex function and strongly convex set. IFAC-PapersOnLine **48**, 202–205 (2015)
14. Ivanov, G.E., Golubev, M.O.: Strong and weak convexity in nonlinear differential games. IFAC-PapersOnLine **51**(32), 13–18 (2018)
15. Ivanov, G.E., Golubev, M.O.: Alternative theorem for differential games with strongly convex admissible control sets. In: Evtushenko, Y., Jaćimović, M., Khachay, M., Kochetov, Y., Malkova, V., Posypkin, M. (eds.) OPTIMA 2018. CCIS, vol. 974, pp. 321–335. Springer, Cham (2019). https://doi.org/10.1007/978-3-030-10934-9_23
16. Ivanov, G.E., Lopushanski, M.S., Golubev, M.O.: The nearest point theorem for weakly convex sets in asymmetric seminormed spaces. In: Evtushenko, Y., Jaćimović, M., Khachay, M., Kochetov, Y., Malkova, V., Posypkin, M. (eds.) OPTIMA 2018. CCIS, vol. 974, pp. 21–34. Springer, Cham (2019). https://doi.org/10.1007/978-3-030-10934-9_2
17. Penot, J.-P.: Continuity properties of projection operators. J. Inequalities Appl. **2005**(5), 509–521 (2005)
18. Polovinkin, E.S.: Strongly convex analysis. Sbornik: Math. **187**(2), 259–286 (1996)
19. Polovinkin, E.S., Balashov, M.V.: Elements of Convex and Strongly Convex Analysis, 2nd edn. Fizmatlit (2007). (in Russian)
20. Polyak, B.T.: Existence theorems and convergence of minimizing sequences in extremum problems with restrictions. Soviet Math. **7**, 72–75 (1996)
21. Weber, A., Reißig, G.: Local characterization of strongly convex sets. J. Math. Anal. Appl. **400**(2), 743–750 (2012)

Semidefinite Relaxation and Sign-Definiteness of Quadratic Forms on the Cone

L. B. Rapoport[(✉)] [iD]

V.A. Trapeznikov Institute of Control Sciences RAS, Moscow, Russia
LBRapoport@gmail.com

Abstract. In this paper we consider the problem of the sign-definiteness of a quadratic form (QF) in the domain defined by quadratic constraints under quadratic constraints. Each constraint is determined by an inequality on a QF. Well known and widely applicable in the control theory approach consists of using so called S–procedure. The semidefinite relaxation approach investigated in this paper allows us to derive an S–procedure from duality conditions. However, the S–procedure, which gives necessary and sufficient conditions for sign-definiteness for the relaxed problem, gives only sufficient conditions for sign-definiteness for the initial problem if the number of quadratic constraints is two or more. In this paper the new approach is proposed, allowing establishment of conditional sign definiteness in some cases, when the S–procedure doesn't give an answer. The results are illustrated by an example.

Keywords: Quadratic constraints · Conditional sign definiteness · S-procedure · Semidefinite relaxation · Quadratic form · Duality conditions

1 Introduction

Let $f_i(x) = x^T F_i x$, $i = 0, \cdots, m$ be QF's in R^n satisfying the following condition

$$\{x \in R^n : f_i(x) \geq 0, \ i = 1, \cdots, m\} \neq \{0\}. \tag{1}$$

Each inequality $f_i(x) \geq 0$ determines the second order cone in R^n. The condition (1) means that all these cones intersect not at only origin. There are non-zero points in their intersection. The question often arising in the control theory can be formulated as follows: under which conditions on matrices F_i, $i = 0, \cdots, m$, inequalities $f_i(x) \geq 0$, $i = 1, \cdots, m$, and the condition $x \neq 0$ imply $f_0(x) < 0$. In other words we are interested to know when the following inclusion holds:

$$\begin{aligned} \{x \in R^n : f_i(x) \geq 0, \ i &= 1, \cdots, m, x \neq 0\} \\ &\subseteq \{x \in R^n : f_0(x) < 0\}. \end{aligned} \tag{2}$$

This work was financially supported by the Russian Foundation for Basic Research, project 18-08-00531.

M. Jaćimović et al. (Eds.): OPTIMA 2019, CCIS 1145, pp. 32–42, 2020.
https://doi.org/10.1007/978-3-030-38603-0_3

This property will be referred to as conditional sign definiteness of the QF $f_0(x)$ on the set (1) which obviously defines a cone in R^n, non-convex in general case. The commonly used technique called "the S-procedure" consists in checking if the following condition holds:

$$\exists \tau_i \geq 0 \ : \ F_0 + \sum_{i=1}^{m} \tau_i F_i \prec 0, \tag{3}$$

where symbols $\prec, \succ$ ($\preceq, \succeq$) stand for definiteness (semidefiniteness) of matrices. Thus, the S-procedure approach reduces the conditional sign definiteness problem to solving the Linear Matrix Inequality (LMI) problem (3) with respect to variables τ_i or establishing its infeasibility, see [2].

The conditional sign definiteness problem arises when applying the Lyapunov functions approach to analysis of the asymptotic stability of control systems. Particularly, application of the Lurie - Postnikov Lyapunov functions to the analysis of the absolute stability of nonlinear control systems with several stationary feedback elements subjected to sector constraints leads to the problem of the conditional sign definiteness of QF under the constraints of the special class. Each QF presented in the constraint is a product of two linear form describing margins of the sector. This particular conditional sign definiteness problem was investigated in [9, 10] where necessary and sufficient conditions were proposed. In [11] these conditions were applied to estimation of the attraction domain of the nonlinear control systems.

As for the case of the general form of quadratic forms, if $m = 1$ then (2) and (3) are equivalent (see [13]). For the case $m = 2$ the equivalence conditions of (2) and (3) are considered in [6]. A review of results related on the S-procedure is given in [5]. In general, for $m > 1$, the S-procedure gives only sufficient conditions of (2). This property is referred to as "looseness" of the S-procedure with multiple constraints. Nevertheless, even taking into account the looseness of the S-procedure, its application is attractive, since checking the condition (3) can be reduced to a convex programming problem, for which there are efficient polynomial algorithms.

Sufficient conditions of absence of conditional sign definiteness are formulated in [12] in the form of the numerical algorithm.

In this paper we consider the new conditions for conditional sign definiteness of QF under quadratic constraints for some cases for which the S-procedure does not give a result. It extends earlier published results [8] presenting new proof of the "loseness" of the S-procedure with a single constraint and considering the case of infinite number of constraints.

2   Semidefinite Relaxation and the S-Procedure

Let $N = n(n + 1)/2$ be dimension of the space of $n \times n$ real valued symmetric matrices.

Given two symmetric $n \times n$ matrices A and B, let $\langle A, B \rangle = \mathrm{tr}(AB) = \mathrm{tr}(BA)$ be the inner product, where $\mathrm{tr}(\cdot)$ is the matrix trace. Then $f_i(x) = \langle xx^T, F_i \rangle$.

Let $\mathcal{P} = \{X : X \succeq 0\}$ be the convex acute cone of positive semidefinite matrices. Then the boundary $\bar{\mathcal{P}}$ of $\mathcal{P}$ is composed of singular matrices $\{X \in \mathcal{P} : \operatorname{rank}(X) \leq n - 1\}$. The inner part $\mathcal{P}^\circ = \mathcal{P} \setminus \bar{\mathcal{P}} = \{X \in \mathcal{P} : X \succ 0\}$ is composed of strictly positive definite matrices. Let $\mathcal{P}_1 = \{xx^T : x \in R^n\} = \{X \in \bar{\mathcal{P}} : \operatorname{rank}(X) = 1\}$ be the part of the boundary, consisting of the rank one matrices. Obviously, $\mathcal{P} = \operatorname{conv}(\mathcal{P}_1)$, where $\operatorname{conv}(\cdot)$ is used to denote a convex hull.

Condition (2) can be rewritten as

$$\begin{aligned} \{X \in \mathcal{P}_1 \setminus \{0\} : \langle X, F_i \rangle \geq 0, \ i = 1, \cdots, m\} \\ \subseteq \{X \in \mathcal{P}_1 : \langle F_0, X \rangle < 0\} \end{aligned} \tag{4}$$

or, in other words,

$$\mathcal{K}_1 = \{X \in \mathcal{P}_1 : \langle X, F_i \rangle \geq 0, \ i =, 0. \cdots, m\} = \{\mathbf{0}\}. \tag{5}$$

Here and everywhere in the text the notation $\mathbf{0}$ is used for the null matrix. The condition (1) can be rewritten as

$$\{X \in \mathcal{P}_1 : \langle X, F_i \rangle \geq 0, \ i =, 1 \cdots, m\} \neq \{\mathbf{0}\}. \tag{6}$$

The set $\mathcal{P}_1 \in \mathcal{P}$ is a non-convex cone. Its substitution with a wider and convex cone $\mathcal{P}$ is commonly referred to as a "semidefinite relaxation". Applying the semidefinite relaxation to (4) we arrive at the checking of the condition

$$\begin{aligned} \{X \in \mathcal{P} \setminus \{\mathbf{0}\} : \langle X, F_i \rangle \geq 0, \ i = 1, \cdots, m\} \\ \subseteq \{X \in \mathcal{P} : \langle F_0, X \rangle < 0\} \end{aligned} \tag{7}$$

or, in other words,

$$\mathcal{K} = \{X \in \mathcal{P} : \langle X, F_i \rangle \geq 0, \ i = 0, \cdots, m\} = \{\mathbf{0}\}. \tag{8}$$

The condition (6) implies

$$\{X \in \mathcal{P} : \langle X, F_i \rangle \geq 0, \ i = 1, \cdots, m\} \neq \{\mathbf{0}\}. \tag{9}$$

It follows from the duality theorem (see [1]), that if condition (9) holds, then condition (8) is equivalent to existence of such real values $\tau_0 > 0$, $\tau_i \geq 0$, $i = 1, \cdots, m$, and matrix $Y \in \mathcal{P}^\circ$, that

$$\sum_{i=0}^{m} \tau_i F_i + Y = 0,$$

which is equivalent (after dividing it by $\tau_0 > 0$) to the condition (3).

Therefore, the S–procedure gives necessary and sufficient conditions of the conditional sign definiteness if the set $\mathcal{P}_1$ is substituted with $\mathcal{P}$. Under this condition the S–procedure becomes lossless. The idea of exploring of the semidefinite relaxation of the low rank (less than n, ideally 1) is considered in [3,4]. The main difficulty that occurs on the way of using of the incomplete rank relaxation is that it leads to non-convex numerical problems and the relaxation losses its sense.

The problem (2) and the S–procedure result can be easily extended to the case of infinite number of quadratic constraints. Really, let $f_t(x) = x^T F_t x$ be the single - parameter family of quadratic forms, the matrix F_t is supposed to be continuously dependent on $t \in [t_0, t_1]$ where $t_1 > t_0$. Along with (2) consider the problem of checking under what conditions on matrices F_t ($t \in [t_0, t_1]$) inequalities $f_t(x) \geq 0$ ($t \in [t_0, t_1]$) and $x \neq 0$ imply $f_0(x) < 0$ or, equivalently,

$$\{x \in R^n : f_t(x) \geq 0, t \in [t_0, t_1], x \neq 0\} \subseteq \{x \in R^n : f_0(x) < 0\}. \tag{10}$$

The natural generalization of the S–procedure is straightforward. It consists in checking if there exists such a continuous non-negative function $\tau_t \geq 0$ of the variable t that the following condition holds:

$$F_0 + \int_{t_0}^{t_1} \tau_t F_t dt \prec 0. \tag{11}$$

Obviously, the continuous problem setup can be extended to cases of Lebesgue integrable matrix functions F_t and necessity to consider inequalities $f_t(x) \geq 0$ that holds for almost all $t \in [t_0, t_1]$ in (10).

3   Criteria of Conditional Sign Definiteness and Absence of Conditional Sign Definiteness

If $m = 1$ then S–procedure is known to be lossless. Here we present a new proof of this fact. We start with formulation of the following lemma.

Lemma 1. *For any symmetric matrix F the following condition holds:*

$$\mathrm{conv}\{X \in \mathcal{P}_1 : \langle X, F \rangle \geq 0\} = \{X \in \mathcal{P} : \langle X, F \rangle \geq 0\}.$$

Proof. Obviously the following condition holds: $\mathrm{conv}\{X \in \mathcal{P}_1 : \langle X, F \rangle \geq 0\} \subseteq \mathcal{P} \cap \{X : \langle X, F \rangle \geq 0\}$. Let us prove the inverse inclusion. Let

$$X^0 \in \mathcal{P} \cap \{X : \langle X, F \rangle \geq 0\}. \tag{12}$$

Suppose that $X^0 \in \mathcal{P}^\circ$ i.e. $X^0 \succ 0$. We need to show that there are such vectors $x_i \in R^n$ and such values λ_i, $i = 1, \cdots, n \geq 0$, that

$$X^0 = \sum_{i=1}^n \lambda_i x_i x_i^T, \quad \sum_{i=1}^n \lambda_i = 1 \tag{13}$$

and

$$\langle x_i x_i^T, F \rangle \geq 0. \tag{14}$$

Let us choose $Y \neq 0$ satisfying conditions

$$\begin{aligned}\langle Y, F \rangle &= 0, \\ \langle Y, I \rangle &= 0, \end{aligned} \tag{15}$$

where I is an identity matrix. Since $Y \in \mathcal{S}(n)$ and the dimension of the space $\mathcal{S}(n)$ is $N = n(n+1)/2$, then in the case $n \geq 2$, there exists a matrix Y, satisfying (15).

Let $X(\alpha) = X^0 + \alpha Y$. Taking into account the first condition in (15) and (12) we have

$$\langle X(\alpha), F \rangle \geq 0.$$

Since $\langle Y, I \rangle = \mathrm{tr}(Y)$, then, due to (15), among the diagonal entries of the matrix Y there are both positive and negative values. Therefore, there are $\alpha^{1,1} > 0$ and $\alpha^{1,2} > 0$ such that the following holds:

$$X^{1,1} = X(\alpha^{1,1}) \in \bar{\mathcal{P}}, \ \ X^{1,2} = X(-\alpha^{1,2}) \in \bar{\mathcal{P}}. \tag{16}$$

Let us define

$$\lambda^1 = \frac{\alpha^{1,2}}{\alpha^{1,1} + \alpha^{1,2}}.$$

Then it is easy to verify that

$$X^0 = \lambda^1 X^{1,1} + (1 - \lambda^1) X^{1,2}, \ \langle X^{1,1}, F \rangle \geq 0, \ \langle X^{1,2}, F \rangle \geq 0, \tag{17}$$

and by virtue of (16) $\mathrm{rank}(X^{1,1}) \leq n - 1$ and $\mathrm{rank}(X^{1,2}) \leq n - 1$, which means that the matrix $X^0 \succ 0$ can be presented as a convex hull of two matrices, each belonging to the set $\bar{\mathcal{P}}$ (the boundary of the cone $\mathcal{P}$) and having rank at least one less than the matrix X^0 has. If the rank of each of these matrices is 1, then the statement (13) is proved.

Let $\mathrm{rank}(X^{1,1}) > 1$. This matrix has at least one zero eigenvalue. Let $q^{1,1}$ be the corresponding eigenvector. Let us denote $Q^{1,1} = q^{1,1}(q^{1,1})^T$. Then we have

$$\langle X^{1,1}, Q^{1,1} \rangle = 0. \tag{18}$$

Let us choose the $n \times n$ matrix $Y^{1,1} \neq 0$ satisfying conditions

$$\begin{aligned}\langle Y^{1,1}, F \rangle &= 0, \\ \langle Y^{1,1}, I \rangle &= 0, \\ \langle Y^{1,1}, Q^{1,1} \rangle &= 0. \end{aligned} \tag{19}$$

Proceeding similarly to the previous constructions, we define $X^{1,1}(\alpha) = X^{1,1} + \alpha Y^{1,1}$. By the first condition of (19) and by (17), we have

$$\langle X^{1,1}(\alpha), F \rangle \geq 0.$$

Since $\operatorname{tr}(Y^{1,1}) = 0$ (the second condition (19)), among the diagonal entries of $Y^{1,1}$ there are both positive and negative values. Therefore, there are $\alpha^{1,1,1} > 0$ and $\alpha^{1,1,2} > 0$ such that the following holds:

$$X^{1,1,1} = X^{1,1}(\alpha^{1,1,1}), \quad \operatorname{rank}(X^{1,1,1}) \leq n-2,$$
$$X^{1,1,2} = X^{1,1}(\alpha^{1,1,2}), \quad \operatorname{rank}(X^{1,1,2}) \leq n-2. \tag{20}$$

Having defined

$$\lambda^{1,1} = \frac{\alpha^{1,1,2}}{\alpha^{1,1,1} + \alpha^{1,1,2}}$$

and taking into account (17), we get

$$X^0 = \lambda^1 \lambda^{1,1} X^{1,1,1} + \lambda^1 (1-\lambda^{1,1}) X^{1,1,2}) + (1-\lambda^1) X^{1,2},$$

$$\langle X^{1,1,1}, F \rangle \geq 0, \quad \langle X^{1,1,2}, F \rangle \geq 0, \quad \langle X^{1,2}, F \rangle \geq 0.$$

Continuing induction, we come to the decomposition (13), (14), because at each step the rank of the two matrices in the expansion X^0 decreases by 1. The number of orthogonality relations in the conditions (15), (19), etc. does not exceed n while the dimension of the space of matrices Y equal to $n(n+1)/2$. So, a matrix Y at each step can be chosen.

At the beginning of the proof, we have assumed for convenience that $X^0 \succ 0$. Now suppose $X^0 \succeq 0$. This means that the matrix X^0 can lie in $\bar{\mathcal{P}}$. But this will only lead to the situation, when probably fewer number of steps are required to construct the decomposition (13), (14). This completes the proof of the Lemma 1.

Based on Lemma 1 we proceed with proof of the lossless of the S–procedure for the $m = 1$ case.

Let the condition (4) holds for $m = 1$. Then $\{X \in \mathcal{P}_1 : \langle X, F_1 \rangle \geq 0\} \subseteq \{X : \langle F_0, X \rangle < 0\} \cup \{\mathbf{0}\}$. Because the right side of the last inclusion is convex, then $\operatorname{conv}\{X \in \mathcal{P}_1 : \langle X, F_1 \rangle \geq 0\} \subseteq \{X : \langle F_0, X \rangle < 0\} \cup \{\mathbf{0}\}$. It follows from the Lemms 1 that $\{X \in \mathcal{P} : \langle X, F_1 \rangle \geq 0\} \subseteq \{X : \langle F_0, X \rangle < 0\} \cup \{\mathbf{0}\}$ and $\{X \in \mathcal{P} : \langle X, F_1 \rangle \geq 0\} \subseteq \{X \in \mathcal{P} : \langle F_0, X \rangle < 0\}$. Therefore, for $m = 1$ the condition (4) implies (7), which proves a the lossless of the S–procedure with a single constraint.

Let now $m > 1$. The following lemma holds.

Lemma 2.
$$\operatorname{conv}\{X \in \mathcal{P}_1 : \langle X, F_i \rangle \geq 0, \, i = 1, \cdots, m\}$$
$$\subseteq \{X \in \mathcal{P} : \langle X, F_i \rangle \geq 0 \, i = 1, \cdots, m\}. \tag{21}$$

In the general case, if the condition (3) doesn't hold, then the condition (4) not necessarily fails. Only relaxed condition (7) fails and the question about conditional definiteness of QF $f_0(x)$ under conditions $f_i(x) \geq 0$, $i = 1, \cdots, m$ remains open.

Suppose that the S–procedure didn't give an answer, i.e. the linear matrix inequality (3) is infeasible. Then in (8) the convex cone is not trivial: $\mathcal{K} \neq \{\mathbf{0}\}$.

However, one can not exclude that $\mathcal{K}_1 = \{\mathbf{0}\}$ and (2) holds. Let $\mathcal{K}$ doesn't touch the boundary of $\bar{\mathcal{P}} \setminus \{\mathbf{0}\}$. Then this cone doesn't contain elements of $\mathcal{P}_1 \setminus \{\mathbf{0}\}$. Therefore, $\mathcal{K} \cap \mathcal{K}_1 = \{\mathbf{0}\}$. Therefore, if $\mathcal{K}_1 \subseteq \mathcal{K}$, then $\mathcal{K}_1 = \{\mathbf{0}\}$. We just proved the following theorem:

Theorem 1. *Let* $\mathcal{K} \neq \{\mathbf{0}\}$ *and* $\mathcal{K} \subset \mathcal{P}^\circ \cup \{\mathbf{0}\}$. *Then (2) holds.*

If $n = 2$ then $\mathcal{P}_1 = \bar{\mathcal{P}}$. Therefore, we have the following

Theorem 2. *Let* $n = 2$. *If* $\mathcal{K} \neq \{\mathbf{0}\}$, *then condition* $\mathcal{K} \subset \mathcal{P}^\circ \cup \{\mathbf{0}\}$ *is necessary and sufficient for (2).*

If the cone $\mathcal{K} \subset \mathcal{P}^\circ$ is acute and can be represented as a convex hull of k rays R_j spanned on matrices $r_j \in \mathcal{P}, j = 1, \cdots, k$, i.e.

$$\mathcal{K} = \mathrm{conv}\{R_1, \cdots, R_k\}, \text{ where } R_j = \{X = \alpha r_j, \alpha \geq 0\}, \tag{22}$$

then it suffices to demand $r_j \in \mathcal{P}^\circ$ or equivalently $r_j \succ 0$.

In the more complex case of the problem (10) the cone $\mathcal{K} \subset \mathcal{P}^\circ$ can have more general representation

$$\mathcal{K} = \mathrm{conv}\{R_t : t \in \Phi\}, R_t = \{X = \alpha r_t, \alpha \geq 0\} \tag{23}$$

and the set Φ has probably infinite number of elements as is illustrated in the example described below.

Example 1. Consider the two-dimensional case $(n = 2)$ and the problem (10). Let $\varepsilon > 0$ be sufficiently small and matrices single parameter set of matrices F_t and the matrix $\bar{F}_0$ are defined as follows:

$$\bar{F}_0 = \begin{bmatrix} \varepsilon & -1 \\ -1 & \varepsilon \end{bmatrix}, \quad F_t = \begin{bmatrix} \varepsilon^2 - t & \sqrt{\varepsilon^2 - t^2} \\ \sqrt{\varepsilon^2 - t^2} & \varepsilon^2 + t \end{bmatrix}, \quad t \in [-\varepsilon, \varepsilon].$$

Here we intentionally use notation $\bar{F}_0$ instead of F_0 not to be confused with F_t for $t = 0$. To check if there exists the matrix $X^* \succ 0$ satisfying the condition $X^* \in \mathcal{K}$ consider $X^* = I$. We have $X^* \succ 0$, $\langle X^*, \bar{F}_0 \rangle = 2\varepsilon > 0$ and $\langle X^*, F_t \rangle = \langle X^*, F_2 \rangle = 2\varepsilon^2 > 0$ for all t. Therefore, $X^* \in \mathcal{K} \neq \{\mathbf{0}\}$ and S–procedure doesn't give the positive result. In the continuous problem setup we have to try to construct the (23) representation of this cone. Consider the single parameter family of matrices

$$r_t = \begin{bmatrix} 1 + t & -\sqrt{\varepsilon^2 - t^2} \\ -\sqrt{\varepsilon^2 - t^2} & 1 - t \end{bmatrix}$$

with t taking values from the segment $[-\varepsilon, \varepsilon]$ and two matrices

$$r' = \begin{bmatrix} 1 - \varepsilon & \varepsilon \\ \varepsilon & 1 + \varepsilon \end{bmatrix}, \quad r'' = \begin{bmatrix} 1 + \varepsilon & \varepsilon \\ \varepsilon & 1 - \varepsilon \end{bmatrix}.$$

Let show now that $\mathcal{K} = \text{conv}\{R_t : t \in [-\varepsilon, \varepsilon], R', R''\}$ with $R_t = \{X = \alpha r_\phi, \alpha \geq 0\}$, $R' = \{X = \alpha r', \alpha \geq 0\}$, $R'' = \{X = \alpha r'', \alpha \geq 0\}$ in the representation (23). Really, simple calculations show that:

$$\begin{aligned}
&\langle r_t, F_t \rangle = 0, \ \forall \, t \in [-\varepsilon, \varepsilon], \\
&\langle r_t, F_s \rangle \geq (t - s)^2 \geq 0, \ \forall \, s, t \in [-\varepsilon, \varepsilon], \\
&\langle r_t, \bar{F}_0 \rangle = 2\varepsilon + 2\sqrt{\varepsilon^2 - t^2} \geq 0, \ \forall \, t \in [-\varepsilon, \varepsilon], \\
&\left\langle r', \bar{F}_0 \right\rangle = 0, \ \left\langle r'', \bar{F}_0 \right\rangle = 0, \\
&\left\langle r', \bar{F}_{-\varepsilon} \right\rangle = 0, \ \left\langle r'', \bar{F}_\varepsilon \right\rangle = 0, \\
&\left\langle r'', \bar{F}_{-\varepsilon} \right\rangle \geq 0, \ \left\langle r', \bar{F}_\varepsilon \right\rangle \geq 0.
\end{aligned} \qquad (24)$$

Further, $r_t \succ 0, \forall t \in [-\varepsilon, \varepsilon]$ and $r' \succ 0, r'' \succ 0$. Therefore, Theorem 2 guarantees (10), while the S–procedure has failed.

Note that the representation (22) is impossible if $m + 1 < N$. In this case the cone $\mathcal{K}$ can contain in its inner part a linear manifold and, obviously, can not be placed inside of the acute cone $\mathcal{P}^\circ$. It means that $\mathcal{K}$ will have common points with the boundary $\bar{\mathcal{P}}$. It however doesn't mean that the cone $\mathcal{K}$ intersects with the part of the boundary $\mathcal{P}_1$ and the condition (2) doesn't hold.

Let describe the numerical algorithm allowing to find points of intersection $\mathcal{K} \cap \mathcal{P}_1 \setminus \{0\}$ and, therefore, establishing sufficient conditions for the fail of (2). This algorithm was presented in [12] but we repeat its description with more details for the sake of completeness of the presentation.

Let an $n \times n$ matrix X and a vector $x \in R^n$ be related by the matrix inequality

$$\begin{bmatrix} X & x \\ x^T & 1 \end{bmatrix} \succeq 0. \qquad (25)$$

According the Schur lemma we have

$$X \succeq xx^T \qquad (26)$$

and, therefore, $\text{tr}(X) \geq \|x\|^2$.

Consider the problem

$$\begin{aligned}
&\|x\|^2 \to \max, \\
&\text{s.t. } \text{tr}(X) = 1, \ \begin{bmatrix} X & x \\ x^T & 1 \end{bmatrix} \succeq 0, \\
&\langle X, F_i \rangle \geq 0, \ i =, 0 \cdots, m.
\end{aligned} \qquad (27)$$

If the feasible set of (27) is empty, the conditions (3) hold. Otherwise the following theorem holds.

Theorem 3. *Let $\mathcal{K} \neq \{0\}$ and X^*, x^* be a solution of the problem (27). Then $\mathcal{K} \cap \mathcal{P}_1 \neq \{0\}$ if and only if $\|x^*\|^2 = 1$. If this last condition holds, then $x^* \in \mathcal{K} \cap \mathcal{P}_1$.*

Proof. Let us prove that the equation

$$\text{tr}(X) = \|x\|^2 \tag{28}$$

implies

$$X = xx^T. \tag{29}$$

As the matter of fact, consider two cases:

(a) rank$(X) = 1$,
(b) rank$(X) > 1$.

Define $e = \frac{1}{\|x\|}x$. In the case (a), $X = yy^T$ and (28) can be rewritten as $\|y\|^2 = \|x\|^2$. Let us multiply the matrix equality (26) on the left by e^T and on the right by e. We obtain $(e^T y)^2 \geq \|x\|^2$. If $x \neq y$, then $(e^T y)^2 < \|y\|^2$ and $\|y\|^2 > \|x\|^2$. The contradiction proves (29). In the case (b) we have $e^T X e \geq \|x\|^2$. Denote by λ_{max} the maximal eigenvalue of the matrix X. Then $\lambda_{max} \geq e^T X e \geq \|x\|^2$. But if rank$(X) > 1$ then $\lambda_{max} < \text{tr}(X)$ and $\text{tr}(X) > \|x\|^2$, which contradicts to (28).

If $\|x^*\| = 1$, then $\text{tr}(x^* x^{*T}) = \text{tr}(X^*)$ and therefore $X^* = x^* x^{*T}$. Then (27) implies $X^* \in \mathcal{K} \cap \mathcal{P}_1$. The reverse is also true. If there is $X^* \in \mathcal{K} \cap \mathcal{P}_1$, then $X^* = x^* x^{*T}$ and $\text{tr}(x^* x^{*T})$ takes the maximum possible value 1.

The problem (27) is non-convex. The following algorithm allows to solve it for a local minimum, depending on the initial approximation $x = x^{(0)}$.

(1) Let $k = 0$ and choose $x^{(0)} \neq 0$,
(2) For $k = 1, 2, \cdots$ solve the following convex optimization problem with respect to p and X:

$$\begin{aligned}
&x^{(k-1)^T} p \to \max, \\
&\text{s.t. } \text{tr}(X) = 1, \\
&x^{(k-1)^T} p \geq 0, \\
&\begin{bmatrix} X & x^{(k-1)} + p \\ (x^{(k-1)} + p)^T & 1 \end{bmatrix} \succeq 0, \\
&\langle X, F_i \rangle \geq 0, \, i =, 0 \cdots, m.
\end{aligned} \tag{30}$$

Let $p^{(k)}$, $X^{(k)}$ be solution of the problem (30).

(3) Let $x^{(k)} = x^{(k-1)} + p^{(k)}$.
(4) If $\|x^{(k)}\| - \|x^{(k-1)}\| < \delta$, where $\delta > 0$ is a small constant, then the algorithm stops. Otherwise set $k := k + 1$ and go to the step (2).

Any vector satisfying conditions $X^{(0)} \geq x^{(0)} x^{(0)^T}$, $X^{(0)} \in \mathcal{K} \setminus \{0\}$ can be taken as $x^{(0)} \neq 0$. For example, arbitrary matrix from $\mathcal{K} \setminus \{0\}$ can be chosen. The eigenvector corresponding to the largest eigenvalue can be chosen as $x^{(0)}$.

The condition $\|x^{(k)}\|^2 \geq \|x^{(k-1)}\|^2$ follows from the condition $\|x^{(k)}\|^2 = \|x^{(k-1)}\|^2 + 2x^{(k-1)^T} p^{(k)} + \|p^{(k)}\|^2$ and $x^{(k-1)^T} p^{(k)} \geq 0$,

Thus, the sequence $\{\|x^{(k)}\|^2\}$ is not decreasing and upper bounded by $\|x^{(k)}\|^2 \leq 1$. Therefore, $\|x^{(k)}\|^2 \to \bar{a} \leq 1$. Moreover, the sequence of solutions of the problem (30) satisfies conditions

$$\text{tr}(X^{(k)}) = 1, \, X^{(k)} \in \mathcal{P}, \, \|x^{(k)}\|^2 \leq 1.$$

Therefore, the sequence $\{x^{(k)}\}$ belongs to the closed bounded set. There exists a converging subsequence $\{x^{(j)}\} \to \bar{x}$, $\{X^{(j)}\} \to \bar{X}$ and $\|\bar{x}\|^2 = \bar{a}$.

If the algorithm finishes with the matrix $\bar{X}$ and vector $\bar{x}$, satisfying conditions $\|\bar{x}\|^2 = 1$, then $\bar{X} = \bar{x}\bar{x}^T \in \mathcal{K} \cap \mathcal{P}_1$ and condition (2) doesn't hold. Otherwise, if $\|\bar{x}\|^2 < 1$, then the question about conditional sign definiteness (2) remains open.

Using randomization for choosing of the initial point $x^{(0)}$ improves efficiency of the locally convergent algorithm. The paper [7] describes the method of analysis of nonconvexity, sequential solution of the semidefinite relaxation problem.

Consider the following example. Let $n = 4$, $m = 2$ and matrices F_i are defined as follows

$$
F_0 = \begin{bmatrix} \varepsilon & -1 & 0 & 0 \\ -1 & \varepsilon & 0 & 0 \\ 0 & 0 & \varepsilon & -1 \\ 0 & 0 & -1 & \varepsilon \end{bmatrix}, \; F_1 = \begin{bmatrix} -1 & \varepsilon & 0 & 0 \\ \varepsilon & 1 & 0 & 0 \\ 0 & 0 & -1 & \varepsilon \\ 0 & 0 & \varepsilon & 1 \end{bmatrix}, \; F_2 = \begin{bmatrix} 1 & \varepsilon & 0 & 0 \\ \varepsilon & -1 & 0 & 0 \\ 0 & 0 & 1 & \varepsilon \\ 0 & 0 & \varepsilon & -1 \end{bmatrix},
$$

where $\varepsilon > 0$ is a positive constant. For $\varepsilon = 0.01$ the algorithm gives at the 3rd iteration the vector $\bar{x} = (0.003521, \; 0.707098, \; 0.707098, \; 0.003521)^T$, satisfying the condition $\|\bar{x}\| = 1$ and, therefore, (2) doesn't hold. Note that for $\varepsilon = 0.01$ in the first example we had the condition (2) satisfied.

4   Conclusion

In this paper, new necessary and sufficient conditions of sign-definiteness of a quadratic form under quadratic constraints are obtained. This problem arises in the stability theory and optimization theory. Sufficient conditions lead to conservative results, which, in turn, gives less strong sufficient stability conditions. Necessary and sufficient conditions of sign definiteness are known only for small number of particular cases. In this paper, we considered the case, when the S–procedure did not give a result, but, nevertheless, the conditional sign-definiteness takes place. New results are obtained for the case of finite and infinite number of constraints. Two dimensional example with infinite number of constraints confirms the obtained result.

References

1. Balakrishnan, V., Vandenberghe, L.: Semidefinite programming duality and linear time-invariant systems. IEEE Trans. Autom. Control **48**, 30–41 (2003)
2. Boyd, S., Ghaoui, L.E., Feron, E., Balakrishnan, V.: Linear Matrix Inequalities in System and Control Theory. SIAM, Philadelphia (1994)
3. Henrion, D., Meinsma, G.: Rank-one LMIs and Lyapunov's inequality. In: Proceedings of the 391h IEEE Conference on Decision and Control Sydney, Australia, pp. 1483–1488, December 2000
4. Ibaraki, I.S., Tomizuka, M.: Rank minimization approach for solving BMI problems with random search. In: Proceedings of the American Control Conference, Arlington, VA, pp. 25–27, June 2001

5. Polik, I., Terlaky, T.: A survey of the S-lemma. SIAM Rev. **49**, 371–418 (2007)
6. Polyak, B.T.: Convexity of quadratic transformations and its use in control and optimization. J. Optim. Theory App. **99**, 553–583 (1998)
7. Polyak, B., Gryazina, E.: Convexity/nonconvexity certificates for power flow analysis. In: Bertsch, V., Fichtner, W., Heuveline, V., Leibfried, T. (eds.) Advances in Energy System Optimization. TM, pp. 221–230. Springer, Cham (2017). https://doi.org/10.1007/978-3-319-51795-7_14
8. Rapoport, L.B.: Semidefinite relaxation and new conditions of sign-definiteness of quadratic forms under quadratic constraints. In: Proceedings of 14th International Conference "Stability and Oscillations of Nonlinear Control Systems" (Pyatnitskiy's Conference), Moscow, pp. 1–3, June 2018
9. Rapoport, L.B.: Sign-definiteness of a quadratic form with quadratic constraints and absolute stability of nonlinear control systems. Sov. Phys. Dokl. **33**(2), 96–98 (1988)
10. Rapoport, L.B.: Frequency criterion of absolute stability for control systems with several nonlinear stationary elements. Autom. Remote Control **50**(6), 743–750 (1989)
11. Rapoport, L.B.: Estimation of an attraction domain for multivariable Lur'e systems using looseless extension of the s-procedure. In: Proceedings American Control Conference, San Diego, pp. 2395–2396 (1999)
12. Rapoport, L.B.: Semidefinite relaxation and new conditions for sign-definiteness of the quadratic form under quadratic constraints. Autom. Remote Control **79**(11), 2073–2079 (2018)
13. Yakubovich, V.A.: The s-procedure in the nonlinear control theory. Vestnik Leningrad. Univ. **1**, 62–77 (1971). (in Russian)

Distance-Constrained Line Routing Problem

Adil Erzin[1,2]([envelope]) [ORCID] and Roman Plotnikov[1] [ORCID]

[1] Sobolev Institute of Mathematics, Novosibirsk 630090, Russia
{adilerzin,prv}@math.nsc.ru
[2] Novosibirsk State University, Novosibirsk 630090, Russia

Abstract. On the plane, the barrier is a line segment, and the mobile sensors are initially located at some points (depots). Each sensor can travel a limited-length path, starting and ending at its depot. That part of the barrier, along which sensor moved, is *covered* by this sensor. It is required to find a min-power subset of sensors covering the entire barrier. The complexity of this problem is not known. In this paper, we have found the special cases of polynomial solvability and state some necessary and sufficient conditions for the existence of the solution. An efficient (polynomial) algorithm for checking the existence of the solution is proposed. Moreover, we have developed some approximation algorithms. In particular, an efficient implementation of the dynamic programming algorithm, which in some special cases yields an *optimal* solution, is proposed.

Keywords: Barrier covering · Mobile sensors · Distance-constrained line routing

1 Introduction

The problem considered in this paper relates to the *barrier coverage* with *mobile* devices when it is necessary to efficiently monitor extended objects, such as roads, borders, pipelines, etc. The mobile device is often referred to as a *sensor* or sometimes more specifically an Unmanned Aerial Vehicle (UAV). It is assumed that the moving energy consumption of every sensor is proportional to the length of the path traveled by it. Since a new problem is being considered, for a better understanding of its place among the known barrier covering problems, below we have listed the statements and results for the related problems.

Rational use of energy allows to extend the lifetime of the mobile sensor network. One of the problems (the abbreviation MinSum is often used to denote it) arising in this context is the problem of minimizing the total length of paths traveled by sensors to cover the barrier [1,3,4,9,10,13,15]. In this case, a certain

A. Erzin thanks the Russian Foundation for Basic Research, grant 19-47-540007 (contribution: sections 1, 2, 4, 5), and R. Plonikov thanks the Russian Science Foundation, grant 18–71–00084 (contribution: section 3), for financial support.

M. Jaćimović et al. (Eds.): OPTIMA 2019, CCIS 1145, pp. 43–55, 2020.
https://doi.org/10.1007/978-3-030-38603-0_4

line is considered as a barrier, most often a segment of a straight line. The sensor
monitoring area is the sensor coverage area, which, as a rule, has the shape of
a circle, in the center of which is the sensor. A barrier point is considered to
be covered if it belongs to at least one circle of the cover. It is not necessary
to use all sensors to cover the barrier. It is enough to define a subset of sensors
involved in the covering and their final positions. In the MinSum problem on
a plane, the barrier is defined in the form of a line segment, and each sensor
is determined by the initial location and radius of the circle that it covers. It
is required to find a subset of sensors and their final positions such that after
moving the barrier will be covered, and the total length of the paths traveled by
sensors will be minimal. The MinSum problem usually requires moving sensors
(circle centers) onto the barrier [1,3,5,10,13–15]. We know only one paper in
the press in which the MinSum problem is solved without the requirement of
moving the sensors onto the barrier [16]. The MinSum problem is NP-hard in
the case of different circles [13,14]. If the circles are equal, then the complexity
status of the problem is not known. However, in two-dimensional (2D) Euclidian
space an $O(n^4)$–time algorithm, where n is the number of sensors, that builds a
$\sqrt{2}$–approximate solution, is proposed in [10], the complexity of which is reduced
to $O(n^2)$ in [15].

In the MinMax problem, the criterion is a minimization of the maximum
path length traveled by the sensors. The MinMax problem is solved optimally
with $O(n^2)$–time complexity when the sensors are initially on the line containing
the barrier (1D space) [13], and in [8] the complexity is reduced to $O(n \log n)$.
The 2D MinMax problem is considered in [14], which shows that the problem is
NP-hard in the case of different circles. In the case of identical circles in [19] an
$O(n^3 \log n)$–time algorithm is proposed that builds a solution to the problem.

In the MinNum problem, it is required to minimize the number of sensors
involved in the barrier covering. In [20] it is proved that the 2D MinNum problem
is NP-hard in general, but is polynomially solvable when the circles are equal.

In this paper, we consider a 2D MinNum problem in a slightly different
formulation. As before, a barrier in the form of a line segment and a set of
arbitrarily arranged mobile sensors are defined on a plane. But the sensor covers
only the point at which it is located (coverage disk has zero radius). This means
that in order to cover a certain part of the barrier, the sensor must move along
this part. Moreover, it is required that the sensor, after covering a part of the
barrier, must return to its depot, and the length of the path traveled by the
sensor does not exceed the specified value. Such a situation corresponds, for
example, to monitoring extended objects by UAVs having a limited supply of
energy. For a better understanding of the problem, we can assume that it is
necessary to minimize the number of trackwalkers of the railway section. Also,
each trackwalker starts and ends his journey of the limited length in the depot.
The path length of each trackwalker depends on many reasons. For example, if
the trackwalker's operating time is limited, then the speed determines the length
of the path.

Over the past few years, UAVs have become more and more popular. The
complexity of routing UAVs has not been fully investigated in the literature. In

[12] the authors provide a formal definition of the UAV Routing and Trajectory Optimization Problem. Next, they introduce a taxonomy and review recent contributions in UAV trajectory optimization, UAV routing, and articles addressing these problems. The paper [2] presents a solution for the problem of minimum time coverage of ground areas using a group of UAVs equipped with image sensors. In [6] the authors present the Drone Arc Routing Problem (DARP) and study its relation with well-known Postman Arc Routing Problems. Applications for DARPs include traffic monitoring by flying over roadways, infrastructure inspection such as by flying along power transmission lines, pipelines or fences, and surveillance along with linear features such as coastlines or territorial borders. In [7] the authors simulate UAV recognition after a possible case of diffuse damage after a seismic event in the town of Acireale (Sicily, Italy). Given a set of sites and the range of the UAV, one is able to find a number of vehicles to employ and the shortest survey path. The problem of finding the shortest survey path is an operational research problem called the vehicle routing problem which has a solution that is known to be computationally time-consuming. The authors used the simulated annealing heuristic. They also examined the distribution of the cost of the solutions varying the depot on a regular grid in order to find the best area for executing the survey. Paper [11] introduces a UAV heterogeneous fleet routing problem, dealing with vehicles limited autonomy by considering multiple charging stations and respecting operational requirements. A green routing problem is designed for overcoming difficulties that exist as a result of limited vehicle driving range. The paper [17] proposes two new distance-constrained capacitated vehicle routing problems (DCVRPs) and study potential benefits in flexibly assigning start and end depots. The first problem is an extension of the traditional symmetric DCVRP, with additional service and travel time constraints, minimization of the number of vehicles and flexible application to both symmetric and asymmetric problems. The second problem is a relaxation of DCVRP to enable the flexible assignment of start and end depots. The research in [18] focused on providing an operational UAV routing system. The authors present the statistical methodology used to devise a quick running routing heuristic that provides reasonable solutions.

However, we were unable to find results regarding the study of the MinNum problem under consideration. In particular, we do not know its complexity. In this paper, we present some properties of the solution, as well as a polynomial algorithm for checking the existence of coverage, special cases of polynomial solvability, an effective algorithm for constructing an order-preserving coverage (OPC), a strict definition of which we will give later. The main property of the order-preserving cover is that to build an optimal OPC, the dynamic programming method can be used. We present an efficient implementation of the algorithm for constructing an optimal OPC in the L_1 metric and also find special cases when this coverage is the optimal solution to the MinNum problem.

The paper is organized as follows. In the next section, we introduce the necessary definitions, formulate the problem and define some properties of the solution. In Sect. 3, we propose a polynomial algorithm for checking the existence of a solution to the problem. In Sect. 4, several algorithms are proposed for constructing a feasible solution to the problem. Cases of polynomial solvability of the problem are found. A $O(n^2)$–time dynamic programming algorithm is proposed that builds an optimal OPC. Conditions that guarantee that this solution is optimal for MinNum problem are found. In Sect. 5 we conclude the paper.

2   Statement of the Problem and Preliminary Analysis

On the plane, we introduce a Cartesian coordinate system so that the barrier, represented by a line segment, is located between the points $(0,0)$ and $(L,0)$, and the mobile sensors are at the points $p_i = (x_i, y_i)$, $i \in S$, $|S| = n$, which we call the *depots*. Without loss of generality, we assume that $y_i \geq 0$, $i \in S$. Let $Q_i > 0$ be the maximum path length that the sensor i can travel. It is assumed that the sensor covers the point where it is located and after monitoring the sensor should return to its depot.

Definition 1. *We say that the sensor i covers the segment $[a_i, b_i] \subseteq [0, L]$ if it moves from the depot p_i to the barrier at the point $(a_i, 0)$, moves along the barrier to the point $(b_i, 0)$ (we assume that $a_i < b_i$) and returns to its depot, following a path whose length is $d(p_i, (a_i, 0)) + b_i - a_i + d((b_i, 0), p_i) \leq Q_i$, where $d(p_1, p_2)$ is the distance between the points p_1 and p_2.*

Definition 2. *A cover is a subset of sensors $C \subseteq S$ and a set of segments $[a_i, b_i]$, that $\cup_{i \in C}[a_i, b_i] \supseteq [0, L]$ and $d(p_i, (a_i, 0)) + b_i - a_i + d((b_i, 0), p_i) \leq Q_i$ for each $i \in C$.*

In the problem considered in this paper, it is required to determine the min-power cover. For this problem, we use the former abbreviation MinNum, although it is formally differs from the traditional formulation when each sensor covers a circle and does not need to return to the depot. In our formulation, the MinNum problem has a lot in common with the metric Vehicle Routing Problem [7,12,17], in which the service area of each vehicle i is the segment (arc) $[a_i, b_i]$. Another similar problem is associated with optimal street cleaning, which is the arc routing problem [6]. However, in the last problem the length of the path of each device is not limited.

In the Euclidean metric, obviously, the trajectory of an arbitrary sensor i is a triangle whose perimeter does not exceed Q_i. Moreover, in this case there is such an optimal coverage in which the segments covered by different sensors do not intersect internally $((a_i, b_i) \cap (a_j, b_j) = \emptyset$, $i \neq j)$, and the perimeter of the triangle traveled by the sensor $i \in C$ is Q_i. Then, knowing a_i and Q_i, we can determine b_i. Conversely, knowing b_i and Q_i, we can find a_i.

In the MinNum problem in the L_1 metric, we can imagine that the sensor moves not along the sides of the triangle (as a matter of fact), as in the Euclidean metric, but along the sides of the rectangle. In this case, sensor i can cover a segment $[a_i, b_i]$ of length $l_i = Q_i/2 - y_i$, where $a_i \geq x_i - l_i$ and $b_i \leq x_i + l_i$.

Therefore, in this case, we can set $y_i = 0$ and assume that sensor i covers a segment of length l_i containing barrier point x_i.

Let's renumber the sensors from left to right according to their abscissas x_i, $i \in S$. Then $i < j$ only if $x_i \leq x_j$.

Definition 3. *In* order-preserving cover *(OPC) C, $b_i \leq b_j$ for all such $i, j \in C$ that $i < j$.*

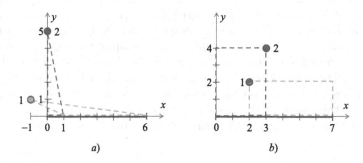

a)   *b)*

Fig. 1. Examples of non-existence of an OPC in the Euclidean metric (a) and in the metric L_1 (b).

Obviously, there is not always an optimal OPC. Figure 1 shows the covers that do not preserve order in the Euclidean metric (Fig. 1a) and in the L_1 metric (Fig. 1b). The OPC in these cases does not exist at all. First of all, we need to find out whether the problem is solvable.

3   The Existence of Solution

In the 2D Euclidian space the leftmost point of the barrier that the sensor k can cover is the point

$$\bar{a}_k = x_k - \sqrt{Q_k^2/4 - y_k^2}.$$

The rightmost barrier point that the sensor k can cover is

$$\bar{b}_k = x_k + \sqrt{Q_k^2/4 - y_k^2}.$$

The following necessary condition for the existence of a solution is obvious.

Consideration 1. *If the segments $[\bar{a}_k, \bar{b}_k]$, $k = 1, \ldots, n$, do not cover the barrier $[0, L]$, then the problem MinNum has no solution.*

Now we propose the algorithm E for checking the existence of coverage in the L_1 metric, which consists of repeating steps. It builds a cover if it exists. If the algorithm E fails to construct coverage, then the MinNum problem has no solution in the L_1 metric.

Algorithm E. Obviously, to cover the barrier (line segment), it is necessary that the left barrier endpoint is covered too. We describe the repeating Step E and illustrate its operation in Fig. 2.

Step E. Find a set of sensors S_l, each of which can cover a non-empty segment containing the left endpoint of the barrier ($S_l = \{i \in S : -l_i < x_i \leq l_i\}$). In Fig. 2a blue, red and green sensors can cover the left endpoint of the barrier. If the set S_l is empty, then the problem has no solution. Otherwise, in the set S_l, we choose a sensor k such that $x_k + l_k \leq x_i + l_i$ for all $i \in S_l$ (in Fig. 2a′ this is a blue sensor). The sensor k covers the segment $[0, \bar{l}_k]$, where $\bar{l}_k = \min\{l_k, l_k + x_k\}$. After that, we exclude the sensor k from the set S and cut the barrier on the left by $\bar{l}_k$. If the length of the remaining barrier is greater than 0 and $S \neq \emptyset$, then repeat Step E. If the length of the remaining barrier is 0, then the cover is found.

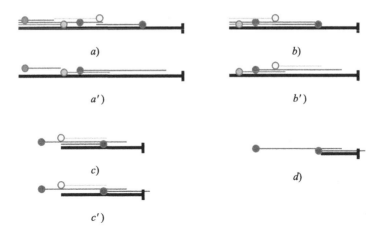

Fig. 2. Illustration of the algorithm E by example. (Color figure online)

In Fig. 2 after covering the left endpoint with a blue sensor, the segment shown in Fig. 2b remains, the left endpoint of which can cover the yellow, red and green sensors. In Fig. 2b′ we can see that in order to cover the left endpoint we need to use a green sensor. Then the barrier is truncated, and the new left endpoint is covered with yellow, red, and brown sensors (Fig. 2c). According to Step E (see Fig. 2c′), we cover the left endpoint with a yellow sensor, after which we get the situation shown in Fig. 2d. Next in the cover will be the red sensor, and then the brown one. As a result, the cover will be built.

Lemma 1. *The solution of the MinNum problem in the L_1 metric exists if and only if the algorithm E builds a barrier cover.*

Proof. If algorithm E builds a cover, then a solution exists, which proves the necessity.

Let us prove the sufficiency. Let there be a cover C of the segment $[0, L]$, in which some sensor (let it be number 1) covers the left endpoint of the segment. Let none of the sensors can be excluded from C, i.e. for each sensor in the cover C, there is a non-empty segment that is covered only by it. Suppose that among the sensors in C, the sensor $2 \in S_l$ has the following properties: (a) it can cover the left endpoint, and (b) with the maximum right shift of the coverage areas of all the sensors of the set S_l, the right endpoint of the segment, covered by sensor 2, is left than the others (i.e. $x_2 + l_2 \le x_i + l_i$, $i \in S_l$). Obviously, such a sensor exists. We will prove that there is a feasible solution in which sensor 2 covers the left endpoint of the segment. If sensor 2 coincides with sensor 1, then the cover C is the desired one. Suppose that this is not a case, i.e. sensor 2 differs from the sensor 1.

Consideration 2. *The depot of sensor 2 can be located only to the right of the left endpoint of the barrier, i.e. $x_2 > 0$.*

Proof. If sensor 2 is to the left of the segment, it can cover only the beginning of the barrier in any cover. Consequently, in the cover C, each sensor 1 and 2 covers the left endpoint of the segment, which means that the sensor, the right endpoint of the cover segment of which in the cover C to the left, can be excluded. However, in the cover C there are no sensors that can be excluded. Contradiction proves the consideration.

Consideration 3. *The depot of sensor 1 can be located only to the right of the left endpoint of the barrier, i.e. $x_1 > 0$.*

Proof. Otherwise, in solution C, the coverage area of sensor 1 is shifted as far as possible to the right. Therefore, based on a property (b) and Consideration 2, the coverage area of sensor 1 completely contains the coverage area of sensor 2. Therefore, sensor 2 can be excluded from C. However, in the solution C there are no sensors that can be excluded. Contradiction proves the consideration.

Consideration 4. *In the cover C, the depot of sensor 2 is located inside the segment covered by sensor 1.*

Proof. Since both sensors 1 and 2 can cover the barrier point 0, then $x_1 \le l_1$ and $x_2 \le l_2$. Based on Consideration 3, sensor 1 covers the segment $[0, l_1]$. Based on property (b), we have $x_2 + l_2 \le x_1 + l_1$. Then $2x_2 \le x_2 + l_2 \le x_1 + l_1 \le 2l_1$ and therefore $x_2 \le l_1$. From Consideration 2 it follows that $x_2 \ge 0$. Consequently, $x_2 \in [0, l_1]$, which is the area covered by sensor 1 in the cover C. The consideration is proved.

Corollary 1. *In the cover C, the area covered by sensors 1 and 2 is continuous and has a length of $x_2 + l_2$.*

Consideration 5. *In the cover C, we can change the coverage areas of sensors 1 and 2 so that point 0 will be covered by sensor 2, the area covered by sensors 1 and 2 will be continuous and have a length of at least $x_2 + l_2$.*

Proof. Consider two cases:

(1) $x_1 > l_2$. Let sensor 2 covers the segment $[0, l_2]$, and sensor 1 cover the segment $[l_2, l_2 + l_1]$. The covered segment is continuous and has a length of $l_1 + l_2 \geq x_2 + l_2$ (because, according to Consideration 4, $x_2 \leq l_1$).
(2) $x_1 \leq l_2$. Then let sensor 2 covers the segment $[0, l_2]$, and sensor 1 cover the segment $[x_1, x_1 + l_1]$. The covered segment is continuous and, according to property (b), has the length of $x_1 + l_1 \geq x_2 + l_2$.

The consideration is proved.

It follows from Corollary 1 and Consideration 5 that, without violating the feasibility, the cover C can be changed in such a way that the left endpoint of the segment will be covered by sensor 2. We denote the resulting cover as C_1. From the cover C_1, we exclude sensor 2 and obtain the cover C_2 of the segment $[l_2, L]$ with the remaining sensors. The same actions can be performed with cover C_2. Thus, the entire segment $[0, L]$ can be covered by successively selecting a sensor with properties (a) and (b). Then, cutting the segment to the left by the length of the interval covered by the selected sensor, and repeat this procedure for the remaining sensors. The lemma is proved.

Remark 1. If there is a solution to the problem MinNum in the L_1 metric, then there is also a solution in the Euclidean metric.

Proof. In the cover C_1 in the L_1 metric, each sensor $i \in S$ passes a path whose length is equal to the perimeter of the rectangle of height y_i and width l_i, covering the barrier segment $[a_i, b_i]$ of length $l_i = Q_i/2 - y_i$. If the length of the path is Q_i, and sensor moves from its base point p_i to the barrier point a_i, and from the barrier point b_i to the depot directly, then a path will not exceed Q_i. So, if in C_1 an arbitrary sensor i covers the segment $[a_i, b_i]$, then in the Euclidean metric it can also cover this segment.

4   Algorithms

At first, we give two simple observations.

Observation 1. *If $Q_i = Q = const$ for all $i \in S$, and all sensors are initially located in the same depot, then the problem MinNum is polynomially solvable.*

Proof. In this case, any sensor can cover a certain segment $[0, l_1]$. This sensor is excluded, and any remaining sensor covers a certain segment $[l_1, l_2]$ and is also excluded from S. Continuing the process, we will cover the entire barrier, or it will remain uncovered and not a single sensor will remain. The observation is proved.

Observation 2. *If all sensors are in the same depot, but the maximum lengths of their paths are different, then the problem MinNum is polynomially solvable in the L_1 metric. In this case, either the cover does not exist, or only one sensor participates in the covering, or the barrier is covered by two sensors.*

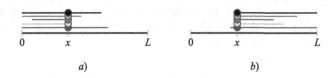

Fig. 3. Illustration to the proof of Observation 2. (Color figure online)

Proof. Indeed, the length of the segment covered by any sensor i is unchanged in the L_1 metric and equals l_i. We can assume that the depot is located on the barrier at some point x (Fig. 3). Let, as before, S_l be a set of sensors each of which can cover the left endpoint of the barrier, and S_r a set of sensors each of which can cover the right endpoint of the barrier. In Fig. 3a, the set S_l consists of black, green, yellow, and red sensors, and in Fig. 3b the set S_r consists of black and red sensors. If there is one sensor in $S_l \cap S_r$ that covers both the left and right endpoints of the barrier at the same time, then this one makes the cover. Otherwise, if there is a pair of different sensors, one of which covers the left endpoint of the barrier, and the other right endpoint, then these 2 sensors form a cover. If S_l or S_r is empty, or both of these sets consist of the same sensor, but this sensor cannot cover the entire barrier, then the cover does not exist. In Fig. 3, there is a cover of power 2, for example, a black sensor covers the segment containing the left endpoint of the barrier, and a red sensor covers the segment containing the right endpoint of the barrier. The observation is proved.

Remark 2. If there is a cover in the L_1 metric, then there is a similar cover in the Euclidean metric. The reverse is not true. Coverage may exist in the Euclidean metric, but not in the L_1 metric.

Algorithm A. It is obvious that the algorithm E after a small modification can be used to construct an approximate solution in the L_1 metric. From the example shown in Fig. 2, it is clear that at the next step, instead of the yellow sensor, we can choose a red one (Fig. 2c'). And Fig. 2d shows that the red sensor can be excluded because the brown sensor covers the rest of the barrier, i.e. brown sensor dominates red sensor.

Definition 4. *In the L_1 metric, if $x_i \leq 0$, $x_j \leq 0$ and $x_i + l_i \leq x_j + l_j$, then sensor j dominates sensor i.*

Obviously, the dominated sensors may not be used in the covering. Then the repeating step of the algorithm A, which builds an approximate cover in the L_1 metric, can be described as follows.

Step A. Find the set of sensors S_l that can cover a non-empty segment containing the left endpoint of the barrier ($S_l = \{i \in S : -l_i < x_i \le l_i\}$). If the set S_l is empty, then the problem has no solution. Otherwise, from the set S_l we delete all dominated sensors and get the set S'_l. Choose a sensor $k \in S'_l$ such that $x_k + l_k \le x_i + l_i$ for all $i \in S'_l$. Using the sensor k, we cover the segment $[0, \bar{l}_k]$, where $\bar{l}_k = \min\{l_k, l_k + x_k\}$. After that, we exclude the sensor k from the set S and cut the barrier on the left by $\bar{l}_k$. If the length of the remaining barrier is greater than 0 and $S \ne \emptyset$, then repeat Step A. If the length of the remaining barrier is 0, then the cover is built.

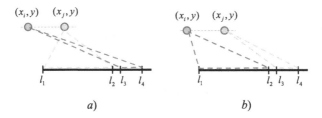

Fig. 4. Illustration to the proof of Lemma 2.

Lemma 2. *If $Q_i = Q = const$ for all $i \in S$, and the ordinates of all depots coincide (i.e. $y_i = y$, $i \in S$), then if there is a solution to the problem MinNum, then there is an optimal OPC.*

Proof. Assume that there is no optimal OPC. Then in any optimal cover C there is a pair of sensors i and j such that $i < j$, and $b_i \ge b_j$. Since the sensors are renumbered from left to right, then $x_i \le x_j$. Suppose that in C sensor j covers the segment $[l_1, l_2]$, and sensor i covers the segment $[l_3, l_4]$ and $l_2 \le l_4$ (Fig. 4a). We construct from C a cover C' in which the sensor i covers the segment $[l_1, l_2]$, and the sensor j covers the segment $[l_3, l_4]$ (Fig. 4b). Sensor i can cover the segment $[l_1, l_2]$, because the sensor j covers this segment in the cover C, and the sensor i to the left of the sensor j. The sensor j can cover the segment $[l_3, l_4]$ since the sensor i covers this segment in the cover C and the sensor j to the right of the sensor i. Consequently, the coverage C' remains feasible and its power has not changed.

Repeating such a rearrangement, we construct an OPC of the same power as the cover C, which proves the statement of the lemma.

Next, we describe the dynamic programming algorithm DP to build the optimal OPC. Since the algorithm is applicable for an arbitrary location of the depots, we will describe it in general case, although the optimality of the constructed coverage is guaranteed only in the case when $y_i = y$ and $Q_i = Q$ for all $i \in S$.

Algorithm DP. Let the function $S_k(l)$ be equal to the minimum number of sensors from the set $\{1, 2, \ldots, k\}$, $k \le n$, which cover the segment $[0, l]$, $l \in (0, L]$.

At the same time, the sensor k covers the segment $[a_k(l), b_k(l)] = [a_k(l), l]$, where in the Euclidean metric

$$a_k(l) = \frac{x_k^2 + y_k^2 - (Q_k - l - \sqrt{(l - x_k)^2 + y_k^2})^2}{2(Q_k - l - \sqrt{(l - x_k)^2 + y_k^2} + x_k)}, \tag{1}$$

when $x_k - \sqrt{4Q_k^2 - y_k^2} \le l \le x_k + \sqrt{4Q_k^2 - y_k^2}$, and in the L_1 metric

$$a_k(l) = l - Q_i/2 + y_k, \tag{2}$$

when $x_k \le l \le x_k + l_k$.

Forward Recursion. If $k = 1$, then

$$S_1(l) = \begin{cases} 1, & a_1(l) \le 0; \\ +\infty, & a_1(l) > 0. \end{cases}$$

At the step $k > 1$, it is necessary to decide whether to use the sensor k to cover the segment $[a_k(l), l]$, or just use a subset of sensors from the set $\{1, \ldots, k-1\}$ to cover the segment $[0, l]$. Therefore, for all $l \in [0, L]$ it is necessary to find

$$S_k(l) = \min\{S_{k-1}(l); \ 1 + S_{k-1}(a_k(l))\},$$

putting $S_m(l) = 0$ when $l \le 0$, and remember whether the sensor k is used in the cover: let $u_k(l) = 1$ if the sensor k is used and $u_k(l) = 0$ otherwise. Moreover, if the inequalities $x_k \le l \le x_k + l_k$ are not satisfied in the metric L_1, or the inequalities $x_k - \sqrt{4Q_k^2 - y_k^2} \le l \le x_k + \sqrt{4Q_k^2 - y_k^2}$ are not satisfied in Euclidean metric, then we set $S_k(l) = S_{k-1}(l)$.

Backward Recursion (Construction of the cover). When calculating $S_k(l)$, the values of $u_k(l)$ were found. Set $k = n$, $l = L$ and execute the next.

Step BR. If $u_k(l) = 1$, then the sensor k is included in the cover, and it covers the segment $[a_k(l), l]$. We set $k = k - 1$ and $l = a_k(l)$. If $l > 0$, then go to Step BR. Otherwise, stop, coverage is built. If $u_k(l) = 0$, then the sensor k is not included in the cover. Set $k = k - 1$ and go to Step BR.

Lemma 3. *The time complexity of the algorithm DP is $O(n^2)$.*

Proof. Obviously, $S_k(l)$, $k = 1, \ldots, n$, are piecewise constant and nondecreasing functions. Moreover, as l increases, the value of the function $S_k(l)$ changes at most n times. Between the points at which the value of the function changes (switching points) the function does not change. So, we only need to know the switching points and the function value between each pair of neighboring switching points. When calculating $S_1(l)$, we have that for l not exceeding some l_1, which is easily found from formulas (1) or (2), $S_1(l) = 1$, and for $l > l_1$ $S_1(l) = +\infty$. Then l_1 is a switching point for function $S_1(l)$. When calculating $S_2(l)$, we have the first two sensors. The segment $[0, l]$ can be covered either by one of them or by both. If $l \le l_1$, then it is enough to use one sensor 1. If $l > l_1$, then while l does not exceed some $l_2 > l_1$, which is found from formulas (1) or

(2), the segment $[0, l]$ is covered by one sensor 2 or by both sensors 1 and 2. In the first case, one sensor 2 covers the segment $[0, l]$ and $S_2(l) = 1$. In the second case, $S_2(l) = 2$ for $l \in (l_1, l_3]$, where l_3 is a such point that $a_2(l_3) = l_1$. Let $k > 2$ and the values $S_{k-1}(l)$ for all l, as well as the switching points, have already been found. Then either the sensor k covers the entire segment $[0, l]$, i.e. $a_k(l) \le 0$, or it covers the segment $[a_k(l), l]$, where $a_k(l) > 0$, and $S_{k-1}(a_k(l))$ other sensors from the set $\{1, \ldots, k-1\}$ participate in the covering of the segment $[0, a_k(l)]$. The function $S_{k-1}(l)$ was found earlier, and the values of the function change only at the switching points. Therefore, when calculating $S_k(l)$, it suffices to consider a finite number of options. If the sensor k covers a certain segment $[0, l_k]$, then $S_k(l) = 1$ for $l \le l_k$. If $l > l_k$, then other sensors participate in the coverage of the segment $[0, l]$, and we need to look between which switching points the point $a_k(l)$ falls. Until the point $a_k(l)$ becomes larger than the right boundary of this interval, $S_{k-1}(a_k(l))$ and $S_k(l)$ do not change their values. For some value of l, the point $a_k(l)$ will be in the interval between the next switching points. This is the new switching point for the function $S_k(l)$. Note that switching points can be found in advance.

In total, functions $S_k(l)$ are computed with time complexity $O(n)$ for all $k = 1, \ldots, n$. Then the time complexity of the forward recursion is $O(n^2)$. Backward recursion has less time complexity. The lemma is proved.

5 Conclusion

We considered the following problem. On the plane, a line segment $[0, L]$ is specified, as well as the depot point (x_i, y_i) for every sensors $i \in S$. Each sensor can travel a limited-length path, starting and ending at its depot. That part of the barrier, along which sensor moved, is *covered* by this sensor. It is required to find a min-power subset of sensors $C \subseteq S$ and to determine the barrier segment $[a_i, b_i]$ covered by each sensor $i \in C$ that $\cup_{i \in C}[a_i, b_i] \supseteq [0, L]$. The complexity of this problem is not known. In this paper, special cases of polynomial solvability and necessary and sufficient conditions for the existence of a solution are found. A polynomial algorithm for checking the existence of a solution is proposed. Finally, we have developed the approximation algorithms. In particular, we proposed a $O(n^2)$-time dynamic programming algorithm which in some special cases builds an *optimal* solution.

References

1. Andrews, A.M., Wang, H.: Minimizing the aggregate movements for interval coverage. Algorithmica **78**(1), 47–85 (2017)
2. Avellar, G.S.C., Pereira, G.A.S., Pimenta, L.C.A., Iscold, P.: Multi-UAV routing for area coverage and remote sensing with minimum time. Sensors **15**, 27783–27803 (2015)
3. Bar-Noy, A., Rawitz, D., Terlecky, P.: "Green" barrier coverage with mobile sensors. In: Paschos, V.T., Widmayer, P. (eds.) CIAC 2015. LNCS, vol. 9079, pp. 33–46. Springer, Cham (2015). https://doi.org/10.1007/978-3-319-18173-8_2

4. Benkoczi, R., Friggstad, Z., Gaur, D., Thom, M.: Minimizing total sensor movement for barrier coverage by non-uniform sensors on a line. In: Bose, P., Gąsieniec, L.A., Römer, K., Wattenhofer, R. (eds.) ALGOSENSORS 2015. LNCS, vol. 9536, pp. 98–111. Springer, Cham (2015). https://doi.org/10.1007/978-3-319-28472-9_8

5. Bhattacharya, B., et al.: Optimal movement of mobile sensors for barrier coverage of a planar region. Theor. Comput. Sci. **410**(52), 5515–5528 (2009)

6. Campbell, J.F., Corberan, A., Plana, I., Sanchis, J.M.: Drone arc routing problems (2018). https://www.researchgate.net/publication/326571789

7. Cannioto, M., D'Alessandro, A., Bosco, G.L., Scudero, S., Vitale, G.: Brief communication: vehicle routing problem and UAV application in the post-earthquake scenario. Nat. Hazards Earth Syst. Sci. **17**, 1939–1946 (2017)

8. Chen, D., Gu, Y., Li, J., Wang, H.: Algorithms on minimizing the maximum sensor movement for barrier coverage of a linear domain. Discret. Comput. Geom. **50**(2), 374–408 (2013)

9. Chen, D.Z., Tan, X., Wang, H., Wu, G.: Optimal point movement for covering circular regions. Algorithmica **72**(2), 379–399 (2015)

10. Cherry, A., Gudmundsson, J., Mestre, J.: Barrier coverage with uniform radii in 2D. In: Fernández Anta, A., Jurdzinski, T., Mosteiro, M.A., Zhang, Y. (eds.) ALGO-SENSORS 2017. LNCS, vol. 10718, pp. 57–69. Springer, Cham (2017). https://doi.org/10.1007/978-3-319-72751-6_5

11. Coelho, B.N., et al.: A multi-objective green UAV routing problem. Comput. Oper. Res. **88**, 306–315 (2017)

12. Coutinho, W.P., Battarra, M., Fliege, J.: The unmanned aerial vehicle routing and trajectory optimisation problem, a taxonomic review. Comput. Ind. Eng. **120**, 116–128 (2018)

13. Czyzowicz, J., et al.: On minimizing the sum of sensor movements for barrier coverage of a line segment. In: Nikolaidis, I., Wu, K. (eds.) ADHOC-NOW 2010. LNCS, vol. 6288, pp. 29–42. Springer, Heidelberg (2010). https://doi.org/10.1007/978-3-642-14785-2_3

14. Dobrev, S., et al.: Complexity of barrier coverage with relocatable sensors in the plane. Theor. Comput. Sci. **579**, 64–73 (2015)

15. Erzin, A., Lagutkina, N.: Barrier coverage problem in 2D. In: Gilbert, S., Hughes, D., Krishnamachari, B. (eds.) ALGOSENSORS 2018. LNCS, vol. 11410, pp. 118–130. Springer, Cham (2019). https://doi.org/10.1007/978-3-030-14094-6_8

16. Erzin, A., Lagutkina, N., Ioramishvili, N.: Barrier covering in 2D using mobile sensors with circular coverage areas. In: Matsatsinis, N.F., et al. (eds.) LION 2019. LNCS, vol. 11968 (2020). https://doi.org/10.1007/978-3-030-38629-0_28

17. Kek, A.G.H., Cheu, R.L., Meng, Q.: Distance-constrained capacitated vehicle routing problems with flexible assignment of start and end depots. Math. Comput. Modell. **47**, 140–152 (2008)

18. Kinney Jr., G.W., Hill, R.R., Moore, J.T.: Devising a quick-running heuristic for an unmanned aerial vehicle (UAV) routing system. J. Oper. Res. Soc. **56**(7), 776–786 (2017)

19. Li, J., Chen, J., Lai, T.: Energy-efficient intrusion detection with a barrier of probabilistic sensors. In: Proceedings IEEE INFOCOM 2012, pp. 118–126. IEEE (2012)

20. Mehrandish, M., Narayanan, L., Opatrny, J.: Minimizing the number of sensors moved on line barriers. In: 2011 IEEE Wireless Communications and Networking Conference (WCNC), pp. 653–658. IEEE (2011)

Problems of Synthesis, Analysis and Optimization of Parameters for Multidimensional Mathematical Models of Interconnected Populations Dynamics

Anastasiya Demidova[1], Olga Druzhinina[2](✉), Milojica Jaćimović[3],
Olga Masina[4], and Nevena Mijajlovic[3]

[1] Department of Applied Probability and Informatics Peoples' Friendship University
of Russia, Moscow, Russia
ademidova@sci.pfu.edu.ru

[2] Federal Research Center "Computer Science and Control" of RAS,
V.A. Trapeznikov Institute of Control Sciences of RAS, Moscow, Russia
ovdruzh@mail.ru

[3] Department of Mathematics, University of Montenegro, Podgorica, Montenegro
milojica@jacimovic.me, nevenami@ucg.ac.me

[4] Bunin Yelets State University, Yelets, Russia
olga121@inbox.ru

Abstract. The designing of multidimensional deterministic and stochastic models of populations dynamics with regard to the relations of competition and mutualism, and migration is described. The model examples in three-dimensional and four-dimensional cases are considered, qualitative and numerical investigation of models is carried out. The transition to the corresponding multidimensional nondeterministic models defined by stochastic differential equations is made. The stability analysis of stationary states is performed. The structure of multidimensional models with competition and mutualism is described with regard to the properties of the Fokker–Planck equations and the formulated rules for the transition to stochastic differential equations in the Langevin form. Numerical experiments for the studied models are carried out using the developed software package. Algorithms for generating trajectories of the Wiener process and multipoint distributions, as well as modifications of the Runge–Kutta method, are used. The comparative analysis of deterministic and stochastic models is carried out. The conditions under which stochastization has a little effect on the stability of the system are studied. Some formulations of population dynamics optimal control problems in models are proposed. The results can be applied in problems of synthesis, optimal control and stability analysis of generalized models of interconnected communities dynamics.

© Springer Nature Switzerland AG 2020
M. Jaćimović et al. (Eds.): OPTIMA 2019, CCIS 1145, pp. 56–71, 2020.
https://doi.org/10.1007/978-3-030-38603-0_5

Keywords: Multidimensional nonlinear models · Dynamics of
interconnected communities · Differential equations · Stochastic
models · Optimal control · Stability

1   Introduction

The actual problem in the study of multi-species communities dynamics is the
designing problem of multidimensional mathematical models describing various
population processes [1–3]. In [4] we considered deterministic and stochastic
multidimensional models with regard to the competition and mutualism. The
studied models are generalizations of the models considered in [5–9]. The multidimensional deterministic and stochastic models of the populations interaction
without taking into account mutualism are studied in [8]. The designing method
of self-consistent stochastic models developed in [10,11], as well as the reduction principle of the solutions stability problem of differential inclusion to the
stability problem for other types of equations are used in [7].

One of the important problems in studying the dynamics of the interconnected communities is the stability problem [1–3]. The perspective direction is
the stability analysis of nondeterministic models. The development of methods
for research of the stability of nondeterministic dynamical systems is presented
in [12–14]. In these papers the systematic approach to qualitative research is
described, which allows one to consider the stability properties of the models
described by differential equations of various types from a unified point of view.
This approach is based on the transition from the deterministic description of
the model to the stochastic one and on the principle of reduction of the stability
problem for solutions of the differential inclusion to the problem of the stability
for other types of equations.

As it is known [10], the deterministic description of the model does not
take into account the probabilistic factors that influence the behavior of the
model. The widespread method of stochastics input into the model is the additive addition of a stochastic term, which describes only the external effect, and
this method is not related to the structure of the model itself. The method for
constructing self-consistent stochastic models is developed in [10]. This method
takes into account the structure of the models and this method is based on the
idea of combinatorial methodology described in [15,16].

Some deterministic mathematical models of population dynamics, taking
into account competition and mutualism, are considered in [4,5,7] and in other
papers. Mathematical models with migration were studied in the papers of various authors (see, for example, [17,18]). In [18–22] migration flows in deterministic
population were considered. The research of distributed multidimensional population models taking into account cross-migration was carried out in papers
[23–25]. It is important to note that migration mechanisms can be described by
both linear and non-linear functions, and it lead to different effects [18]. In [19,21,
26] the issues of qualitative behavior and sustainability of population-migration
models were studied. In [7,8] a qualitative research of a three-dimensional

non-deterministic model with migration was performed. The method of constructing of stochastic self-consistent models [10] allowed us to investigate a three-dimensional model with migration.

In [26] a methodological support was developed for the analysis and synthesis of multidimensional nonlinear dynamic models describing migration flows taking into account the effects of broadband parametric and additive noise. The stability of stationary states was studied and the effects obtained for stochastic models were interpreted. The model examples show a comparison of the migration-population systems properties in deterministic and stochastic cases and the effects due to stochastic broadband perturbations were revealed.

When modeling population-migration systems are studied, various software tools are used that present wide possibilities for building computer models and carrying out computational experiments. However, many software products do not contain libraries for numerical and symbolic calculations and do not have sufficient computational complexity. In this regard, in the study of the population-migration systems models, the application of mathematical packages and general-purpose programming languages [27–29] is relevant. One of the instrumental software tools for studying population-migration models is a software package for the numerical solution of differential equations systems using modified Runge–Kutta methods. The specified software package was developed in [11, 30, 31].

The questions of optimal control in systems of population dynamics were considered in [32–35] and in other papers. In [32], some optimal control problems for distributed models of the migration populations dynamics were considered. In [33], an optimal control problem for a Volterra type distributed system was studied. In [34], the optimality criterion for auto reproduction systems was formalized and the optimal control problem was considered for analyzing evolutionally stable behavior. In [35], optimal control problems for the classical Lotka–Volterra models were formulated with allowance for phase and mixed constraints.

In this paper, we propose construction of multidimensional models, with regard to competition and mutualism, and migration flows as well. The proposed models are new and take into account several additional effects in contrast to the models studied previously. The qualitative and numerical research of models are performed. The construction of multidimensional nondeterministic models of interconnected communities dynamics based on the method of constructing stochastic self-consistent models is described. A comparative analysis of deterministic and stochastic models is carried out. As a tool for the study of the models, a software package in the Python language using the NumPy and SciPy libraries is used. The software package allows numerical experiments based on the implementation of algorithms for generating trajectories of multidimensional Wiener processes and multipoint distributions and algorithms for solving stochastic differential equations. Some formulations of optimal control problems in models of population dynamics are proposed.

2   Constructing of Deterministic Models

The multidimensional population model of dimension $n = 2k$, which takes into account competition and mutualism, is given by the system of ordinary differential equations of the following type:

$$\frac{dx_i}{dt} = a_i x_i \left(1 - \frac{x_i}{b_i} \right) - \sum_{j=1, j \neq i}^{k} \frac{p_{ij} x_i x_j}{1 + d_i x_{i+k}}, \quad i = 1, \ldots, k,$$

$$\frac{dx_i}{dt} = a_i x_i \left(1 - \frac{x_i}{u_i + \omega_i x_{i-k}} \right), \quad i = k+1, \ldots, n, \ n = 2k,$$

(1)

where x_i $(i = 1, \ldots, k)$ are the densities of populations of species-competitors, x_i $(i = k+1, \ldots, n)$ are the densities of species-mutualists populations, a_i, b_i, p_{ij}, d_i, u_i, ω_i are positive constants. The model (1) in the absence of mutualism is reduced to the multidimensional Lotka–Volterra model of competitive interaction.

The dynamic population model "competitor–competitor–mutualist" is considered in the three-dimensional case

$$\frac{dx_1}{dt} = a_1 x_1 \left(1 - \frac{x_1}{b_1} \right) - \frac{p_{12} x_1 x_2}{1 + d_1 x_3},$$

$$\frac{dx_2}{dt} = a_2 x_2 \left(1 - \frac{x_2}{b_2} \right) - p_{21} x_1 x_2,$$

(2)

$$\frac{dx_3}{dt} = a_3 x_3 \left(1 - \frac{x_3}{u + \omega x_1} \right),$$

where the following notations are used: x_i are densities of populations, respectively, of the first competitor, the second competitor and the mutualist ($i = 1, 2, 3$), $a_1, a_2, a_3, b_1, b_2, u, \omega, p_{12}, p_{21}, d_1$ are positive constants. Model (2) is the modification of the model considered in [6], and is characterized by the logistic type of the competitors populations growth. The model (2) is a classic Lotka–Volterra model of the competitive interaction in the absence of mutualism.

The dynamic population model "competitor–mutualist–competitor–mutualist" is considered in the four-dimensional case

$$\frac{dx_1}{dt} = a_1 x_1 \left(1 - \frac{x_1}{b_1} \right) - \frac{p_{12} x_1 x_2}{1 + d_1 x_3},$$

$$\frac{dx_2}{dt} = a_2 x_2 \left(1 - \frac{x_2}{b_2} \right) - \frac{p_{21} x_1 x_2}{1 + d_2 x_4},$$

$$\frac{dx_3}{dt} = a_3 x_3 \left(1 - \frac{x_3}{u_3 + \omega_3 x_1} \right),$$

$$\frac{dx_4}{dt} = a_4 x_4 \left(1 - \frac{x_4}{u_4 + \omega_4 x_2} \right),$$

(3)

where x_1, x_2 are the densities of populations of the first and the second competitors, respectively, x_3, x_4 are the densities of populations of mutualists for x_1, x_2 at any time t. The model (3) is the particular case of the model (1).

A nonlinear model with migration described by a system of ordinary differential equations of the form is considered

$$\frac{dx_1}{dt} = a_1x_1 - a_1x_1^2 - p_{13}x_1x_3 - p_{14}x_1x_4 + \beta x_2 - \gamma x_1,$$

$$\frac{dx_2}{dt} = a_2x_2 - a_2x_2^2 + \gamma x_1 - \beta x_2,$$

$$\frac{dx_3}{dt} = a_3x_3 - p_{31}x_1x_3 - a_3x_3^2 - p_{34}x_3x_4,$$

$$\frac{dx_4}{dt} = a_4x_4 - p_{41}x_1x_4 - p_{43}x_3x_4 - a_4x_4^2,$$

(4)

where x_1, x_3, x_4 are the densities of populations of competing species in the area 1, x_2 is the population density in the area 2, $p_{13}, p_{14}, p_{31}, p_{34}, p_{41}, p_{43} > 0$ are the coefficient of competition of species in the area 1, $\beta > 0$ and $\gamma > 0$ are species migration coefficients between two areas, with area 2 being a refuge. The first and the second equations describe the dynamics of the same species taking into account migration processes. The first equation defines the dynamics in the first area, the second – the dynamics in the second area. The third and fourth equations describe the dynamics of the second and third species interacting as competitors with the first species in the first range.

We propose a nonlinear model of the dynamics of interconnected communities taking into account the relations of competition and mutualism and the migration of one of the species. The model is given by a system of ordinary differential equations of the form

$$\frac{dx_1}{dt} = a_1x_1 - \frac{a_1x_1^2}{b_1} - \frac{p_{12}x_1x_2}{1 + d_1x_3},$$

$$\frac{dx_2}{dt} = a_2x_2 - \frac{a_2x_2^2}{b_2} - p_{21}x_1x_2 + \beta x_4 - \gamma x_2,$$

$$\frac{dx_3}{dt} = a_3x_3 - \frac{a_3x_3^2}{u + \omega x_1},$$

$$\frac{dx_4}{dt} = a_4x_4 - a_4x_4^2 + \gamma x_2 - \beta x_4,$$

(5)

where x_1, x_2 are the population densities of competing species (competitor 1 and competitor 2), x_3 is the population density of the mutualist interacting with the first competitor, p_{ij} are the species competition coefficients. The first three equations describe population dynamics in area 1. The fourth equation defines the dynamics of the second competitor taking into account its migration to the area 2. Using $\beta > 0$ and $\gamma > 0$ are the coefficients of migration of species between two areas, while area 2 is a refuge. In the case $p_{12} = p_{21}$ we have

$$\frac{dx_1}{dt} = a_1 x_1 - \frac{a_1 x_1^2}{b_1} - \frac{p_{12} x_1 x_2}{1 + d_1 x_3},$$

$$\frac{dx_2}{dt} = a_2 x_2 - \frac{a_2 x_2^2}{b_2} - p_{12} x_1 x_2 + \beta x_4 - \gamma x_2,$$

$$\frac{dx_3}{dt} = a_3 x_3 - \frac{a_3 x_3^2}{u + \omega x_1},$$

$$\frac{dx_4}{dt} = a_4 x_4 - a_4 x_4^2 + \gamma x_2 - \beta x_4.$$

$$(6)$$

In the case $p_{12} = p_{21}$ and migration rates are the same ($\beta = \gamma = \varepsilon$), we get a model of the form

$$\frac{dx_1}{dt} = x_1 \left(a_1 - \frac{a_1 x_1}{b_1} - \frac{p_{12} x_2}{1 + d_1 x_3} \right),$$

$$\frac{dx_2}{dt} = x_2 \left(a_2 - \frac{a_2 x_2}{b_2} - p_{12} x_1 \right) + \varepsilon(x_4 - x_2),$$

$$\frac{dx_3}{dt} = x_3 \left(a_3 - \frac{a_3 x_3}{u + \omega x_1} \right),$$

$$\frac{dx_4}{dt} = x_4(a_4 - a_4 x_4) + \varepsilon(x_2 - x_4).$$

$$(7)$$

Further, for model (6), (7) we perform qualitative and numerical analysis. In addition, we carry out the transition to the appropriate stochastic models and conduct a comparative analysis of trajectories.

3   Analysis of Model Examples in the Deterministic Case

For deterministic models "competitor–competitor–mutualist" and "competitor–mutualist–competitor–mutualist" a numerical experiment was conducted in [4] in order to conduct a comparative analysis of the dynamics of behavior and identify the impact of the second mutualist.

We consider the models (2), (3) with initial values are considered $(x_1, x_2, x_3) = (150, 165, 125)$, $(y_1, y_2, y_3, y_4) = (150, 165, 125, 125)$ and with parameter values $a_1 = 1.4$, $a_2 = 3.5$, $a_3 = 1.5$, $a_4 = 1.5$, $b_1 = 250$, $b_2 = 120$, $p_{12} = 1.95$, $p_{21} = 0.001$, $d_1 = 1.5$, $d_2 = 1.2$, $u_3 = u_4 = 200$, $\omega_3 = \omega_4 = 1.2$.

For the systems (2) and (3) the stationary states $F_1 = (182.17, 113.75, 418.60)$ and $E_1 = (197.08, 87.0, 424.49, 328.40)$ are obtained respectively. The results of the numerical experiment are shown in Fig. 1. The phase variables for the model (2) are denoted by x_i, the phase variables for the model (3) are denoted by y_i. These designations are accepted for convenience of the comparative analysis of trajectories on Fig. 1. The numerical experiment showed that the appearance of the second mutualist in the system does not qualitatively change the behavior of the system, but leads to a shift stationary state.

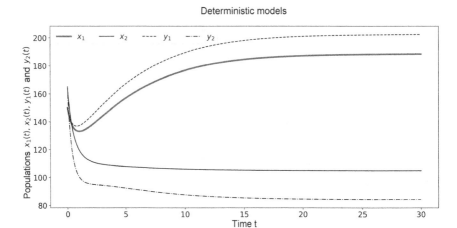

Fig. 1. The comparison of the behavior for the three-dimensional and four-dimensional models with competition and mutualism

Model (4) at $p_{13} = p_{31}$, $p_{14} = p_{41}$, $p_{34} = p_{43}$, $a_i = 1, i = 1, 2, 3, 4$, let's call it a model (4a). Model (4) at $p_{13} = p_{31}$, $p_{14} = p_{41}$, $p_{34} = p_{43}$, $a_i = 1, i = 1, 2, 3, 4$, and for $\beta = \gamma = \varepsilon$ we will call the model (4b).

Consider the models (4a) and (4b) with the same set of intraspecific and interspecific interaction coefficients. Note that in the model (4a) in general $\beta \neq \gamma$, and in the model (4b) $\beta = \gamma$. The trajectories for models (4a) and (4b) with parameter values $p_{13} = 1.2$, $p_{14} = 0.5$, $p_{34} = 1.4$ and initial values $(x_1(0), x_2(0), x_3(0), x_4(0)) = (0.5, 1, 0.8, 1)$ the time interval $[0, 25]$ is shown in Fig. 2. For the model with the same set of migration velocity values $\varepsilon = 0.3$. On the Fig. 2 a record of the form $x_i(3)$ indicates the path corresponding to the x_i phase variable for the model (4b) with the same migration rates.

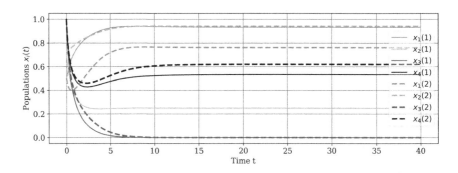

Fig. 2. Trajectories for models (4a) and (4b) with $(x_1(0), x_2(0), x_3(0), x_4(0)) = (0.5, 1, 0.8, 1)$, $p_{13} = 1.2$, $p_{14} = 0.5$, $p_{34} = 1.4$, $\beta = 1.5$, $\gamma = 0.2$, $\varepsilon = 0.3$.

Comparative analysis of trajectories showed that in a model with the same migration rates, the population density of x_4 increased insignificantly, and the density of the population of x_1 has decreased, while the population of x_3 in both models is rapidly dying out. In addition, with increasing migration rate for the model (4b) in the second area, more than favorable conditions than in the model (4a), so the population density of x_2 is significant increases.

Next, we consider the model (6) the competition, mutualism, and different migration rates. For the numerical experiment, different sets of parameters were chosen. For one of the parameters sets in the model (6) there are the following equilibrium states: P_1 (0, 0, 0, 0), P_2 (130, 0, 0, 0), P_3 (0, 0, 100, 0), P_4 (130, 0, 256, 0), P_5 (129.9, 0.00097, 255.9, 0.813), P_6 (0.761, 108.445, 100.091, 4.11). Here P_1 is an unstable node, P_2, P_3, P_4, P_6 are saddles, P_5 is the stable node. The Fig. 3 shows the trajectories of model (6) solutions at the specified initial values and the values of the parameters in the time interval [0, 10].

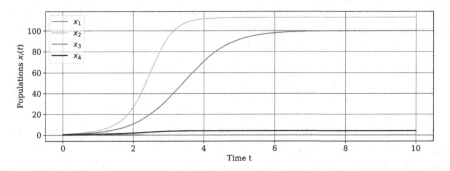

Fig. 3. Trajectories for model (6) with $x_1(0) = 0.8$, $x_2(0) = 0.9$, $x_3(0) = 0.6$, $x_4(0) = 0.3$, parameters values $a_1 = 1.4$, $a_2 = 3.5$, $a_3 = 1.5$, $a_4 = 1.6$, $b_1 = 130$, $b_2 = 120$, $p_{12} = 1.95$, $d_1 = 1.5$, $u = 100$, $\omega = 1.2$, $\beta = 0.3$, $\gamma = 0.2$

We consider the model (7) taking into account competition, mutualism and the same migration rates. For the numerical experiment of the model (7) were chosen the same sets of parameters as for model (6) except migration rate. In this case we consider $\varepsilon = 0.2$.

For one of the parameters sets in the model (7) there are the following equilibrium states: P_1 (0, 0, 0, 0), P_2 (130, 0, 0, 0), P_3 (0, 0, 100, 0), P_4 (130, 0, 256, 0), P_5 (129.9, 0.0007, 255.9, 0.875), P_6 (0.742, 108.444, 100.089, 4.145). Here P_1 is an unstable node, P_2, P_3, P_6 are saddles, P_4, P_5 is the stable node.

The results obtained for the models (6) and (7) correspond to the situation when the population of x_1 dies out, the population of x_2 and x_3 grows exponentially, and the population of x_4 is at the same level. Comparative analysis showed the similarity of the trajectories of the models (6) and (7).

The standard packages of symbolic calculations are used in the process of model calculations. Due to the rather high dimensionality of the models we are studying, serious difficulties arise in the case of alphabetic parameters.

In this regard, we conducted a series of computer experiments and considered different sets of numerical values of the parameters. In the paper we presented those results that are of the greatest interest. In the future, we plan to carry out a bifurcation analysis, which will clearly identify the corresponding types of phase portraits.

4 Stochastic Models

In this paper, stochastization of models is carried out using the method of constructing self-consistent stochastic models [10]. We proceed to the symbolic recording of all possible interactions between the elements of the system. For this, the system state operators and the system state change operator are used. Then we can get the drift and diffusion coefficients for the Fokker–Planck equation, which allows to write the equation itself and its equivalent stochastic differential equation in the Langevin form.

To obtain stochastic models corresponding to models taking into account competition, mutualism and migration, you can write a generalized scheme of interaction, which has the following form:

$$
\begin{aligned}
&X_i \xrightarrow{a_i} 2X_i, \ X_i + X_i \xrightarrow{b_i} X_i, \ i = \overline{1,n}, \\
&X_i + X_j \xrightarrow{c_{ij}} X_j, X_i \xrightarrow{d_i} X_j, \ i,j = \overline{1,n}, i \neq j.
\end{aligned}
\tag{8}
$$

In the scheme (8) n-dimensional system the first line corresponds to the natural reproduction of species in the absence of other factors, the 2nd line symbolizes the intraspecific competition, and the 3rd – interspecific competition. The latter is a description of the process of migration of the species from one area to another. The coefficient values for each of the models (2), (3), (6), (7) are given in Table 1. According to the method of constructing self-consistent stochastic models based on the interaction scheme (8) it is possible to write the Fokker–Planck equation of the form [31]:

$$
\partial_t P(x,t) = - \sum_a [A_a(x)P(x,t)] + \frac{1}{2} \sum_{a,b} \partial_a \partial_b [B_{ab}(x)P(x,t)].
$$

However, an increase in the dimension and complexity of the system under study leads to the complexity of the analytical conclusion of the necessary coefficients of the Fokker–Planck equation. As a solution to this problem, a software implementation is developed for obtaining the coefficients of the Fokker–Planck equation from interaction schemes using a symbolic computation system. This software implementation is a modification of the stochastization method for one-step processes in the computer algebra system described in [11]. This implementation is introduced as a module into a software package developed earlier for the numerical study of deterministic and stochastic models [31].

Table 1. Coefficients for the interaction scheme (8)

	Model (2)	Model (3)	Model (6)	Model (7)
a_i	$a_i = a_i,\ i = \overline{1,3}$	$a_i = a_i,\ i = \overline{1,4}$	$a_i = a_i,\ i = \overline{1,4}$	$a_i = a_i,\ i = \overline{1,4}$
b_i	$b_i = \frac{a_i}{b_i},\ i = 1,2,$ $b_3 = \frac{a_3}{u+\omega x_1}$	$b_i = \frac{a_i}{b_i},\ i = 1,2,$ $b_3 = \frac{a_3}{u+\omega_3 x_1}$ $b_4 = \frac{a_4}{u+\omega_4 x_2}$	$b_i = \frac{a_i}{b_i},\ i = 1,2,$ $b_3 = \frac{a_3}{u+\omega_3 x_1}$ $b_4 = a_4$	$b_i = \frac{a_i}{b_i},\ i = 1,2,$ $b_3 = \frac{a_3}{u+\omega x_1}, b_4 = a_4$
c_{ij}	$c_{12} = \frac{p_{12}}{1+d_1 x_3},$ $c_{21} = p_{12},$ $c_{13} = c_{31} =$ $= c_{23} = c_{32} = 0$	$c_{12} = \frac{p_{12}}{1+d_1 x_3},$ $c_{21} = \frac{p_{21}}{1+d_2 x_4},$ $c_{ij} = 0$ $i,j = \overline{1,4}, i \neq j$	$c_{12} = \frac{p_{12}}{1+d_1 x_3},$ $c_{21} = p_{31}$ $c_{ij} = 0,$ $i,j = \overline{1,4}, i \neq j$	$c_{12} = \frac{p_{12}}{1+d_1 x_3},$ $c_{21} = p_{31}$ $c_{ij} = 0,$ $i,j = \overline{1,4}, i \neq j$
d_i	$d_1 = d_2 = d_3 = 0$	$d_1 = d_2 = d_3 = d_4 = 0$	$d_2 = \beta, d_4 = a_3,$ $d_1 = d_3 = 0$	$d_2 = \varepsilon, d_4 = \varepsilon,$ $d_1 = d_3 = 0$

The following is the algorithm for obtaining the symbolic notation of a stochastic differential equation (Algorithm 1).

Data: Interaction scheme

Result: The system of differential equations in the form of Langevin

begin

> 1. Getting system state operators from the interaction scheme.
> 2. Getting the change of the system state.
> 3. Getting the transition intensities.
> 4. Record the coefficients of the Fokker–Planck equation.
> 5. Record the system of differential equations.

end

To implement the described algorithm, SymPy [27], computer computing system is used, which is a powerful symbolic computation library for the Python language. The output data obtained using the SymPy library can be transferred for numerical calculations using the NumPy [29] library and SciPy [28].

For the scheme of interaction (8) with the use of the developed software package were obtained the coefficients of the Fokker–Planck equation for the model (6) are as follows:

$$A(x) = \begin{pmatrix} x_1\left(a_1 - \frac{a_1}{b_1}x_1 - \frac{p_{12}x_2}{1+d_1 x_3}\right) \\ x_2\left(a_2 - \frac{a_2}{b_2}x_2 - p_{12}x_1\right) - \gamma x_2 + \beta x_4 \\ x_3\left(a_3 - \frac{a_3 x_3}{u+\omega x_1}\right) \\ x_4(a_4 - a_4 x_4) + \gamma x_2 - \beta x_4 \end{pmatrix},$$

$$B(x) = \begin{pmatrix} x_1\left(a_1 + \frac{a_1}{b_1}x_1 + \frac{p_{12}x_2}{1+d_1x_3}\right) & 0 & 0 & 0 \\ 0 & \begin{array}{c} x_2\left(a_2 + \frac{a_2}{b_2}x_2 + \right. \\ \left. +p_{12}x_1\right) - \\ -\gamma x_2 + \beta x_4 \end{array} & 0 & -\gamma x_2 - \beta x_4 \\ 0 & 0 & x_3\left(a_3 + \frac{a_3x_3}{u+\omega x_1}\right) & 0 \\ 0 & -\gamma x_2 - \beta x_4 & 0 & \begin{array}{c} x_4(a_4 + a_4x_4) + \\ +\gamma x_2 + \beta x_4 \end{array} \end{pmatrix}.$$

The obtained coefficients were used in the corresponding module of the software package for the numerical solution of a stochastic differential equation.

The numerical experiments were conducted for stochastic models, taking into account competition and mutualism experiments with the choice of the same parameters as for the numerical analysis of deterministic models (2) and (3). Figure 4 shows a comparison of the trajectories are three-dimensional and four-dimensional stochastic models with competition and mutualism.

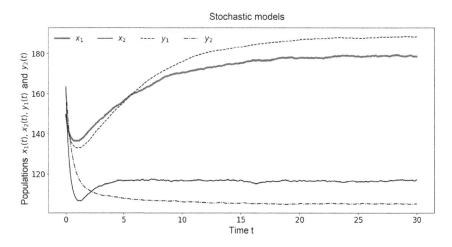

Fig. 4. The comparison of the solutions behavior for stochastic models with competition and mutualism

For the numerical experiment for the stochastic model taking into account migration, competition and mutualism, the same parameters were chosen, as for numerical analysis of deterministic models (6). Figure 5 shows a comparison of the trajectories of deterministic and stochastic models taking into account different migration rates. Solid lines in Fig. 5 trajectories of average values for 100 realizations in the time interval [0, 10] are shown.

Similar computational experiments were performed for the (7) model and the corresponding stochastic model at the same migration rates. Comparative analysis showed the proximity of the trajectories of stochastic models corresponding

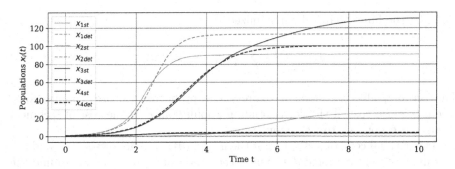

Fig. 5. The comparison of the solution behavior for model (6) and corresponding stochastic model with $x_1(0) = 0.8$, $x_2(0) = 0.9$, $x_3(0) = 0.6$, $x_4(0) = 0.3$, $a_1 = 1.4$, $a_2 = 3.5$, $a_3 = 1.5$, $a_4 = 1.6$, $b_1 = 130$, $b_2 = 120$, $p_{12} = 1.95$, $d_1 = 1.5$, $u = 100$, $\omega = 1.2$, $\beta = 0.3$, $\gamma = 0.2$

to the models (6) and (7). As follows from the obtained results, the introduction of the stochastic method of stochastization of one-step processes has led to slower migration of the second competitor from the first zone to the second. At the same time, the mutualistic relations of the first competitor and his mutualist lead to an increase in the number of both species.

5 The Problem of Optimal Control in Models of the Populations Dynamics

For generalized models of dynamics of interconnected communities, we formulate optimal control problems. The dynamics of a generalized controlled multidimensional model with competition and mutualism is described by a system of differential equations

$$\frac{dx_i}{dt} = a_i x_i \left(1 - \frac{x_i}{b_i}\right) - \sum_{j=1, j \neq i}^{k} \frac{p_{ij} x_i x_j}{1 + d_i x_{i+k}} - u_i x_i, \quad i = 1, \ldots, k,$$

$$\frac{dx_i}{dt} = a_i x_i \left(1 - \frac{x_i}{u_i + \omega_i x_{i-k}}\right) - u_i x_i, \quad i = k+1, \ldots, n, \quad n = 2k, \tag{9}$$

where $u_i(t)$ are control functions.

The constraints imposed on the control functions in (9) are as follows

$$0 \leq u_i \leq u_{i1}, \quad t \in [0, T], \quad i = 1, \ldots, n. \tag{10}$$

In addition,

$$x_i(0) = x_{i0}, \quad x_i(T) = x_{i1}, \quad i = 1, \ldots, n. \tag{11}$$

The following type of restrictions on control functions is also possible

$$0 \leq \sum_{i=1}^{k} u_i(t) \leq M, \quad u_i \geq 0, \quad i = 1, \ldots, n, \quad t \in [0, T]. \tag{12}$$

Consider the functional to be maximized:

$$J(u) = \int_0^T \sum_{i=1}^n (l_i x_i - c_i) u_i(t) dt. \tag{13}$$

The control quality criterion (13) expresses the profit from the use of populations, where l_i is the cost of the corresponding population, c_i is the cost of technical means corresponding to i-th population.

The optimal control problems for the model (9) are as follows: (1) to find the maximum of the functional $J(u)$ under the conditions (10), (11); (2) to find the maximum of the functional $J(u)$ under the conditions (11), (12).

Let us consider the special case of (9). This case is a generalization of the model (2). The dynamics of a generalized controlled three-dimensional model with competition and mutualism is described by a system of differential equations

$$\begin{aligned}
\frac{dx_1}{dt} &= a_1 x_1 \left(1 - \frac{x_1}{b_1}\right) - \frac{p_{12} x_1 x_2}{1 + d_1 x_3} - u_1 x_1, \\
\frac{dx_2}{dt} &= a_2 x_2 \left(1 - \frac{x_2}{b_2}\right) - \frac{p_{21} x_1 x_2}{1 + d_2 x_4} - u_2 x_2, \\
\frac{dx_3}{dt} &= a_3 x_3 \left(1 - \frac{x_3}{u_3 + w_3 x_1}\right) - u_3 x_3.
\end{aligned} \tag{14}$$

The constraints for the model (14) have the form:

$$\begin{aligned}
x_1(0) &= x_{10}, \; x_2(0) = x_{20}, \; x_3(0) = x_{30}, \\
x_1(T) &= x_{11}, \; x_2(T) = x_{21}, \; x_3(T) = x_{31}, \; t \in [0, T],
\end{aligned} \tag{15}$$

$$0 \le u_1 \le u_{11}, \; 0 \le u_2 \le u_{21}, \; 0 \le u_3 \le u_{31}, \; t \in [0, T]. \tag{16}$$

For (14)–(16) consider the functional to be maximized:

$$J(u) = \int_0^T \left[(l_1 x_1 - c_1) u_1(t) + (l_2 x_2 - c_2) u_2(t) + (l_3 x_3 - c_3) u_3(t)\right] dt. \tag{17}$$

The optimal control problem for the model (14) is as follows: to find the maximum of the functional (17) under the conditions (15), (16). Similar to this problem we can formulate the problem under conditions (12) modified for the model (14).

The existence and uniqueness of the maximum of the functional $J(u)$ are of theoretical interest. If all the factors of influence are known, it is possible to solve the problems on the basis of the Pontryagin's maximum principle.

In the problems of dynamics of interconnected communities, phase variables are limited in the form $x_i(t) \ge 0$, $i = 1, \ldots, n$. In addition, restrictions on the growth of i-th type $dx_i/dt \le r_i$ are often introduced, this leads to mixed constraints in optimal control problems. In this regard, it is possible to formulate optimal control problems taking into account phase and mixed constraints.

Optimal control problem statements are also considered for models (5)–(7) with migration, competition, and mutualism. Based on the method of construction of stochastic self-consistent models, we can proceed to the stochastic formulation of optimal control problems. The developed software package, taking into account the results of Sect. 4, allows to perform stochastization and further study of models with migration, competition and mutualism.

6 Conclusions

In this paper we propose a new approach to the synthesis and analysis of multidimensional models of dynamics of interconnected communities taking into account the relations of competition and mutualism, and also taking into account migration flows. Computational study of these models allowed us to obtain the results of numerical experiments for the search of trajectories and assessment parameters in case of high dimensionality of models, as well as to identify effects due to the stochasticity. For the considered models, the impact of competition, mutualism and migration on the behavior of the system is assessed. The presence of relations of mutualism and migration flows is revealed new qualitative effects in the dynamics of interacting communities. The dynamics of model trajectories in the absence of migration flows and mutualism differs significantly from classical models. Calculations for new models are shown that the presence of relationships of mutualism in a non-migratory community, taking into account the migration of another community, plays a supporting role for the non-migratory community. A number of optimal control problem are proposed for multidimensional models of population dynamics. The software package is developed in Python using the NumPy and Scipy. This software package demonstrated sufficient efficiency for computer studies of multidimensional nonlinear models with migration. Numerical experiments carried out by the aid of problem-oriented software shows the similarity of the types of trajectories to stochastic and deterministic cases. Obtained results can find application in problems of computer modeling and optimization of environmental, demographic and socio-economic systems parameters.

References

1. Aleksandrov, A.Y., Platonov, A.V., Starkov, V.N., Stepenko, N.A.: Mathematical Modeling and Research of the Stability of Biological Communities. Solo, St. Petersburg (2006)
2. Bazykin, A.D.: Nonlinear Dynamics of Interacting Populations. Institute of Computer Research, Moscow-Izhevsk (2003)
3. Svirezhev, Y.M., Logofet, D.O.: Stability of Biological Communities. Nauka, Moscow (1978)

4. Demidova, A.V., Druzhinina, O.V., Jacimovic, M., Masina, O.N., Mijajlovic, N.: Synthesis and analysis of multidimensional mathematical models of population dynamics. In: Proceedings of the Selected Papers of the 10th International Congress on Ultra Modern Telecommunications and Control Systems ICUMT, Moscow, Russia, 5–9 November 2018, IEEE Catalog Number CFP 1863G-USB, pp. 361–366. IEEE Xplore Digital Library, New York (2018)
5. Freedman, H.I., Rai, B.: Can mutualism alter competitive outcome: a mathematical analysis. Rocky Mt. **25**(1), 217–230 (1995)
6. Rai, B., Freedman, H.I., Addicott, J.F.: Analysis of three species models of mutualism in predator-prey and competitive systems. Math. Biosci. **63**, 13–50 (1983)
7. Demidova, A.V., Druzhinina, O., Jacimovic, M., Masina, O.: Construction and analysis of nondeterministic models of population dynamics. In: Vishnevskiy, V.M., Samouylov, K.E., Kozyrev, D.V. (eds.) DCCN 2016. CCIS, vol. 678, pp. 498–510. Springer, Cham (2016). https://doi.org/10.1007/978-3-319-51917-3_43
8. Demidova, A.V., Druzhinina, O.V., Masina, O.N.: Design and stability analysis of nondeterministic multidimensional populations dynamics models. In: Proceedings of the Selected Papers of the 7th International Conference "Information and Telecommunication Technologies and Mathematical Modeling of High-Tech Systems" (ITTMM 2017), Moscow, Russia, 24 April 2017, vol. 1995, pp. 14–21. CEUR (2017). http://ceur-ws.org/vol-1995/
9. Freedman, H.I., Rai, B.: Uniform persistence and global stability in models involving mutualism competitor-competitor-mutualist systems. Indian J. Math. **30**, 175–186 (1988)
10. Gevorkyan, M.N., Demidova, A.V., Egorov, A.D., Kulyabov, D.S., Korolkova, A.V., Sevastyanov, L.A.: The influence of stochastization on single-step models. Bull. Peoples' Friendship Univ. Russ. Ser. "Math. Comput. Sci. Phys." **1**, 71–85 (2014)
11. Eferina, E.G., Korolkova, A.V., Gevorkyan, M.N., Kulyabov, D.S., Sevastyanov, L.A.: One-step stochastic processes simulation software package. Bull. Peoples' Friendship Univ. Russ. Ser. "Math. Comput. Sci. Physics" **3**, 46–59 (2014)
12. Shestakov, A.A.: Generalized Direct Lyapunov Method for Systems with Distributed Parameters. URSS, Moscow (2007)
13. Merenkov, Y.N.: Stability-Like Properties of Differential Inclusions, Fuzzy and Stochastic Differential Equations. RUDN University, Moscow (2000)
14. Druzhinina, O.V., Masina, O.N.: Methods of Stability Research and Controllability of Fuzzy and Stochastic Dynamic Systems. Dorodnicyn Computing Center of RAS, Moscow (2009)
15. Van Kampen, N.G.: Stochastic Processes in Physics and Chemistry. Elsevier, Amsterdam (1992)
16. Gardiner, C.W.: Handbook of Stochastic Methods: For Physics, Chemistry and the Natural Sciences. Springer Series in Synergetics. Springer, Heidelberg (1985)
17. Tkachenko, N., Weissmann, J.D., Petersen, W.P., Lake, G., Zollikofer, C.P.E., Callegari, S.: Individual-based modelling of population growth and diffusion in discrete time. PLoS ONE **2**(4), e0176101 (2017). https://doi.org/10.1371/journal.pone.0176101
18. Tuckwell, H.C.: A study of some diffusion models of population growth. Theor. Popul. Biol. **5**(3), 345–357 (1974). https://doi.org/10.1016/0040-5809(74)90057-4
19. Zhang, X.-A., Chen, L.: The linear and nonlinear diffusion of the competitive Lotka-Volterra model. Nonlinear Anal. **66**, 2767–2776 (2007). https://doi.org/10.1016/j.na.2006.04.006

20. Svirezhev, Y.: Nonlinear Waves, Dissipative Structures and Disasters in Ecology. Science, Moscow (1987)
21. Lu, Z., Takeuchi, Y.: Global asymptotic behavior in single-species discrete diffusion systems. J. Math. Biol. **32**(1), 67–77 (1993). https://doi.org/10.1007/BF00160375
22. Cui, J., Chen, L.: The effect of diffusion on the time varying logistic population growth. Comput. Math. Appl. **36**, 1–9 (1998). https://doi.org/10.1016/S0898-1221(98)00124-2
23. Chen, L., Jungel, A.: Analysis of a multi-dimensional parabolic population model with strong cross-diffusion. SIAM J. Math. Anal. **36**, 301–322 (2004). https://doi.org/10.1137/S0036141003427798
24. Zamponi, N., Jungel, A.: Analysis of degenerate cross-diffusion population models with volume filling. Annales de l'Institut Henri Poincare C, Analyse non lineaire **34**(1), 1–29 (2017). https://doi.org/10.1016/j.anihpc.2015.08.003
25. Chen, X., Daus, E.S., Jungel, A.: Global existence analysis of cross-diffusion population systems for multiple species. Arch. Ration. Mech. Anal. **227**(2), 715–747 (2018). https://doi.org/10.1007/s00205-017-1172-6
26. Sinitsyn, I.N., Druzhinina, O.V., Masina, O.N.: Analytical modeling and stability analysis of nonlinear broadband migration flows. Nonlinear World **16**(3), 3–16 (2018)
27. Lamy, R.: Instant SymPy Starter. Packt Publishing, Birmingham (2013)
28. Oliphant, T.E.: Python for scientific computing. Comput. Sci. Eng. **9**(3), 10–20 (2007). https://doi.org/10.1109/MCSE.2007.58
29. Oliphant, T.E.: Guide to NumPy, 2nd edn. CreateSpace Independent Publishing Platform, Scotts Valley (2015)
30. Gevorkyan, M.N., Velieva, T.R., Korolkova, A.V., Kulyabov, D.S., Sevastyanov, L.A.: Stochastic Runge–Kutta software package for stochastic differential equations. In: Zamojski, W., Mazurkiewicz, J., Sugier, J., Walkowiak, T., Kacprzyk, J. (eds.) Dependability Engineering and Complex Systems. AISC, vol. 470, pp. 169–179. Springer, Cham (2016). https://doi.org/10.1007/978-3-319-39639-2_15
31. Gevorkyan, M.N., Demidova, A.V., Velieva, T.R., Korolkova, A.V., Kulyabov, D.S., Sevastyanov, L.A.: Implementing a method for stochastization of one-step processes in a computer algebra system. Program. Comput. Softw. **44**(2), 86–93 (2018). https://doi.org/10.1134/S0361768818020044
32. Moskalenko, A.I.: Methods of Nonlinear Maps in Optimal Control (Theory and Applications to Models of Natural Systems). Nauka, Novosibirsk (1983)
33. Kuzenkov, O.A.: An optimal control for a Volterra distributed system. Autom. Remote Control **67**(7), 1028–1038 (2006)
34. Kuzenkov, O.A., Kuzenkova, G.V.: Optimal control of self-reproduction systems. J. Comput. Syst. Sci. Int. **51**(4), 500–511 (2012)
35. Gusyatnikov, P.P.: Optimal control problem in the predator-prey model. Probl. Theory Saf. Stab. Syst. **7**(2), 9–12 (2005). Dorodnitsyn Computing Center of RAS, Moscow

Lipschitz Continuity of the Optimal Solution of the Infimal Convolution Problem and Subdifferential Calculus

Grigorii E. Ivanov$^{(\boxtimes)}$ (iD) and Maxim O. Golubev (iD)

Moscow Institute of Physics and Technology, 9 Institutskiy per.,
Dolgoprudny, Moscow Region 141700, Russian Federation
g.e.ivanov@mail.ru, maksimkane@mail.ru

Abstract. We consider a parametrized constrained optimization problem, which can be represented as the Moreau-type infimal convolution of the norm and some (nonconvex in general) function f. This problem arises particularly in optimal control and approximation theory. We assume that the admissible set A is weakly convex and function f is Lipschitz continuous and weakly convex on the convex hull of A. We show that the problem is Tykhonov well-posed and the solution of the problem is unique and Lipschitz continuous in some neighbourhood of A. Exact estimates for the size of the neighbourhood and for the Lipschitz constant are obtained. Based on these results we prove lower regularity of the optimal value (marginal) function of this problem in some neighbourhood of A.

Keywords: Parametrized optimization problem · Infimal convolution ·
Best approximation problem · Marginal function · Tykhonov
well-posedness · Frechet subdifferential · Limiting subdifferential

1 Introduction

Let $\mathbb{H}$ be a finite dimensional Euclidean space $\mathbb{R}^n$ or an infinite dimensional Hilbert space. The inner product of $a, b \in \mathbb{H}$ will be denoted as $\langle a, b \rangle$.

Given a subset $A \subset \mathbb{H}$ and a function $f : \mathbb{H} \to \mathbb{R}$ we consider the following constrained optimization problem

$$\mathcal{P}_{f,A} : \qquad \text{Minimize} \quad f(x) + \|x - u\| \quad \text{over } x \in A$$

with a parameter $u \in \mathbb{H}$. If $f = 0$, this is a general constrained best approximation problem

$$\text{Minimize} \quad \|x - u\| \quad \text{over } x \in A.$$

On the other hand, $\mathcal{P}_{f,A}$ is a special case of the Moreau-type infimal convolution problem

$$\text{Minimize} \quad \alpha(x) + \beta(u - x) \quad \text{over } x \in \mathbb{H}$$

Supported by the Russian Foundation for Basic Research, grant 18-01-00209.

M. Jaćimović et al. (Eds.): OPTIMA 2019, CCIS 1145, pp. 72–87, 2020.
https://doi.org/10.1007/978-3-030-38603-0_6

with $\beta(\cdot) = \|\cdot\|$ and $\alpha(\cdot) = f(\cdot) + \psi_A(\cdot)$, where $\psi_A(x) = 0$ if $x \in A$ and $\psi_A(x) = +\infty$ if $x \notin A$ is the indicator function of the set $A \subset \mathbb{H}$. The *optimal value (marginal)* function for the problem $\mathcal{P}_{f,A}$ at $u \in \mathbb{H}$ is

$$T_{f,A}(u) := \inf_{x \in A}(f(x) + \|x - u\|). \tag{1}$$

If $x \in A$ satisfies the equality $f(x) + \|x - u\| = T_{f,A}(u)$, then $x =: x_{\min}(u)$ is called a *solution* of $\mathcal{P}_{f,A}$ at $u \in \mathbb{H}$.

As shown in [1], if $\beta(\cdot)$ is a Minkowski functional of a closed bounded convex set $G \subset \mathbb{H}$, then the infimal convolution $(\alpha \square \beta)(u) := \inf_{x \in \mathbb{H}}(\alpha(x) + \beta(u - x))$ is the optimal value for an optimal control problem

$$\text{Minimize} \quad t + \alpha(\zeta(t; u)) \tag{2}$$

over all $t \geq 0$ and all solutions $\zeta(\cdot) = \zeta(\cdot; u)$ of the differential inclusion $\frac{d\zeta}{dt} \in -G$ with the initial condition $\zeta(0) = u$. Since the norm $\|\cdot\|$ is the Minkowski functional of the unit ball $\mathbb{B} = \{x \in \mathbb{H} : \|x\| \leq 1\}$ the optimal value of $\mathcal{P}_{f,A}$ coincides with the optimal value of (2) with constant dynamics $\frac{d\zeta}{dt} \in \mathbb{B}$ and $\alpha(\cdot) = f(\cdot) + \psi_A(\cdot)$.

2 Weakly Convex Sets and Functions: Motivating Examples and Definitions

Example 1. Let $\mathbb{H} = \mathbb{R}^2$, $A = \{(x_1, x_2) : x_2 \leq |x_1|\}$, $f = 0$. Then at any $u = (u_1, u_2)$ with $u_2 > |u_1| > 0$ the problem $\mathcal{P}_{f,A}$ has a unique solution $\left(\frac{|u_1| + u_2}{2} \operatorname{sign} u_1, \frac{|u_1| + u_2}{2}\right)$. This solution is locally Lipschitz continuous. If $u_1 = 0$, $u_2 > 0$, then the problem $\mathcal{P}_{f,A}$ at (u_1, u_2) has two solutions: $\left(\frac{u_2}{2}, \frac{u_2}{2}\right)$ and $\left(-\frac{u_2}{2}, \frac{u_2}{2}\right)$. In any neighborhood of $u = (0, u_2)$ with $u_2 > 0$ the solution of $\mathcal{P}_{f,A}$ is discontinuous.

The loss of uniqueness and continuity of the solution in Example 1 arises due to the fact that the set A in this example is neither convex nor smooth. Consider the class of weakly convex sets which includes both convex and smooth sets.

The distance from a point $u \in \mathbb{H}$ to a set $A \subset \mathbb{H}$ is defined as

$$\operatorname{dist}(u, A) = \inf_{x \in A} \|u - x\|.$$

Definition 1. *A closed set $A \subset \mathbb{H}$ is called* weakly convex *with radius $R > 0$ if for any $x_0, x_1 \in A$ such that $\|x_0 - x_1\| < 2R$ one has*

$$\operatorname{dist}\left(\frac{x_0 + x_1}{2}, A\right) \leq R - \sqrt{R^2 - \frac{\|x_0 - x_1\|^2}{4}}.$$

Note that any closed convex set is weakly convex with any radius $R > 0$. The set $A = \{x \in \mathbb{H} : \|x\| \geq R\}$ with $R > 0$ is an example of a non-convex weakly convex set. According to [4, Theorem 3] if the set A is smooth, namely A coincides with the closure of its interior and the unit external normal to A is Lipschitz continuous on the boundary of A with constant $C > 0$, then A is weakly convex with radius $R = \frac{1}{C}$.

We shall use $U_R(x)$ and $U_R(A)$ to define R-*neighbourhood* of $x \in \mathbb{H}$ and of $A \subset \mathbb{H}$ correspondingly:

$$U_R(x) = \{u \in \mathbb{H} : \|u - x\| < R\}, \qquad U_R(A) = \{u \in \mathbb{H} : \operatorname{dist}(u, A) < R\}.$$

The following proposition follows directly from [6, Theorem 3.1], it provides a characterization of the class of weakly convex sets in terms of existence of a unique and continuous solution of the best approximation problem, that is the problem $\mathcal{P}_{f,A}$ with $f = 0$.

Proposition 1. *A closed set $A \subset \mathbb{H}$ is weakly convex with radius $R > 0$ if the solution $x_{\min}(\cdot)$ of the problem $\mathcal{P}_{f,A}$ with $f = 0$ is unique and continuous on $U_R(A)$.*

Further we shall use the following proposition (see [5, Lemma 1.4.4]).

Proposition 2. *Let $A \subset \mathbb{H}$ be closed and weakly convex with radius $R > 0$. Then for any $\lambda \in [0,1]$ and any $x_0, x_1 \in A$ such that $\lambda(1 - \lambda)\|x_0 - x_1\|^2 \leq R^2$ there exists $\hat{x} \in A$ such that*

$$\|(1 - \lambda)x_0 + \lambda x_1 - \hat{x}\| \leq R - \sqrt{R^2 - \lambda(1 - \lambda)\|x_0 - x_1\|^2}.$$

Example 2. Let $\mathbb{H} = \mathbb{R}^2$, $A = \{(x_1, 0) : x_1 \in \mathbb{R}\}$, $f(x_1, x_2) = -L|x_1|$ with $L \in (0, 1)$. Then for any $u_2 > 0$ the problem $\mathcal{P}_{f,A}$ at the point $u = (0, u_2)$ has two solutions $x_{\min}(u) = \left(\pm\frac{L}{\sqrt{1-L^2}}u_2, \ 0\right)$. In any neighborhood of $u = (0, u_2)$ with $u_2 > 0$ the solution of $\mathcal{P}_{f,A}$ is discontinuous.

The loss of uniqueness and continuity of the solution in Example 2 arises due to the fact that the function f in this example is neither convex nor smooth. Consider the class of weakly convex functions which includes both convex and smooth functions.

Definition 2. *Let $A \subset \mathbb{H}$ be a convex set. A function $f : \mathbb{H} \to \mathbb{R}$ is called weakly convex with constant $\gamma \geq 0$ on A if the function $\varphi(x) = f(x) + \frac{\gamma}{2}\|x\|^2$ is convex on A.*

The properties of weakly convex sets and functions can be found in [2–6]. New applications of weakly and strongly convex sets in approximation theory and differential games are described in [7–9].

Further we shall prove that the solution of the problem $\mathcal{P}_{f,A}$ is Lipschitz continuous in some neighborhood of A under the following assumptions:

$$\left.\begin{array}{l} \text{the set } A \subset \mathbb{H} \text{ is closed and weakly convex with radius } R > 0; \\ \text{the function } f : \mathbb{H} \to \mathbb{R} \text{ is weakly convex with constant } \gamma \geq 0 \\ \text{and Lipschitz continuous with constant } L < 1 \text{ on conv } A, \end{array}\right\} \quad (3)$$

where conv A is the *convex hull* of A.

3 Well-Posedness of the Problem and Continuity of the Optimal Solution

The problem $\mathcal{P}_{f,A}$ is equivalent to

$$\text{Minimize} \quad h(u,x) \quad \text{over} \quad x \in \mathbb{H}$$

with

$$h(u,x) = f(x) + \psi_A(x) + \|x - u\|, \quad x \in \mathbb{H}, \ u \in \mathbb{H}. \tag{4}$$

A sequence (x_k) is called *minimizing* for $\mathcal{P}_{f,A}$ at $u \in \mathbb{H}$ if $\lim_{k \to \infty} h(u, x_k) = T_{f,A}(u)$. According to [10] the problem $\mathcal{P}_{f,A}$ is called *Tykhonov well-posed* or *strongly attained* at $u \in \mathbb{H}$ if it admits a unique solution $x_{\min}(u) \in \mathbb{H}$ and every minimizing sequence for $\mathcal{P}_{f,A}$ at u converges to $x_{\min}(u)$. The *set of ε-solutions* for $\mathcal{P}_{f,A}$ at $u \in \mathbb{H}$ and $\varepsilon \geq 0$ is

$$S_\varepsilon(u) := \{x \in \mathbb{H} : \ h(u,x) \leq T_{f,A}(u) + \varepsilon\}.$$

Recall that the *diameter* of a set $A \subset \mathbb{H}$ is

$$\operatorname{diam} A := \sup_{x,y \in A} \|x - y\|.$$

Note that the problem $\mathcal{P}_{f,A}$ is Tykhonov well-posed at $u \in \mathbb{H}$ if $\operatorname{diam} S_\varepsilon(u) \to 0$ as $\varepsilon \to +0$.

Lemma 1. *Let $A \subset \mathbb{H}$ and $f : \mathbb{H} \to \mathbb{R}$ satisfy assumptions (3). Then for any $\varepsilon > 0$, $u \in \mathbb{H}$ and $x \in S_\varepsilon(u)$ one has*

$$\|x - u\| \leq \frac{(1 + L)\operatorname{dist}(u, A) + \varepsilon}{1 - L}.$$

Proof. Fix any $a \in A$. Since $T_{f,A}(u) \leq f(a) + \|a - u\|$ it follows that $f(x) + \|x - u\| \leq f(a) + \|a - u\| + \varepsilon$. Using Lipschitz continuity of f we have

$$\|x - u\| \leq \|a - u\| + \varepsilon + L\|x - a\| \leq (1 + L)\|a - u\| + \varepsilon + L\|x - u\|.$$

Consequently,

$$(1 - L)\|x - u\| \leq (1 + L)\inf_{a \in A} \|a - u\| + \varepsilon = (1 + L)\operatorname{dist}(u, A) + \varepsilon.$$

Dividing by $(1 - L)$ we obtain the desired inequality.

Theorem 1. *Let $A \subset \mathbb{H}$ and $f : \mathbb{H} \to \mathbb{R}$ satisfy assumptions (3). Let*

$$\delta := \frac{(1 - L)^2}{\gamma + \frac{1+L}{R}}. \tag{5}$$

Then the problem $\mathcal{P}_{f,A}$ is Tykhonov well-posed at any $u \in U_\delta(A)$.

Proof. Fix any $u \in U_\delta(A)$. As diam $S_\varepsilon(u)$ is a nondecreasing function of ε, there exists the limit $\lim_{\varepsilon \to +0} \operatorname{diam} S_\varepsilon(u) =: \sigma$ and

$$\operatorname{diam} S_\varepsilon(u) \geq \sigma \qquad \forall \varepsilon > 0. \tag{6}$$

If $\sigma = 0$, then $\mathcal{P}_{f,A}$ is Tykhonov well-posed at u. Assume the contrary: $\sigma > 0$.

If $\operatorname{dist}(u, A) = 0$, then by Lemma 1 we have $\|x - u\| \leq \frac{\varepsilon}{1-L}$ for all $x \in S_\varepsilon(u)$. This contradicts (6). So, we assume that $\operatorname{dist}(u, A) > 0$.

Denote

$$d_0 := \operatorname{dist}(u, A), \quad \eta := \frac{(1-L)^2}{d_0} - \gamma - \frac{1+L}{R}, \quad \mu := \frac{1}{2} \min\left\{ \frac{\sigma^2}{4R^2}, \frac{\eta R}{1+L}, 1 \right\}.$$

Since $d_0 < \delta = \frac{(1-L)^2}{\gamma + \frac{1+L}{R}}$ we see that $\eta > 0$ and hence $\mu > 0$. Fix some $\theta \in (L, 1)$ sufficiently close to L, namely, such that

$$\frac{(1-\theta)^2}{d_0} - \gamma - \frac{1+L}{R} > \frac{\eta}{2}. \tag{7}$$

Denote

$$\varkappa := \frac{1+\theta}{1-\theta} d_0, \quad \varepsilon := \min\left\{ \frac{(\theta - L)\sigma}{2}, (1-L)\varkappa - (1+L)d_0, \frac{R^2 \mu \eta}{16} \right\}.$$

In view of $\theta \in (L, 1)$ we get $\frac{1+L}{1-L} < \frac{1+\theta}{1-\theta}$ and hence $\varepsilon > 0$. According to (6) we have diam $S_\varepsilon(u) \geq \sigma > \frac{\sigma}{2}$. Consequently, there exist $x_0, x_1 \in S_\varepsilon(u)$ such that $\|x_0 - x_1\| > \frac{\sigma}{2}$. Denote

$$\lambda := \frac{R^2 \mu}{\|x_0 - x_1\|^2}, \quad x_\lambda := (1-\lambda)x_0 + \lambda x_1.$$

Since $\mu \leq \frac{\sigma^2}{8R^2} \leq \frac{\|x_0 - x_1\|^2}{2R^2}$ and $\mu \leq \frac{1}{2}$ we obtain $\lambda \leq \frac{1}{2}$ and $R^2 - \lambda(1-\lambda)\|x_0 - x_1\|^2 \geq 0$. By Proposition 2 one can find $\hat{x}_\lambda \in A$ such that

$$\|x_\lambda - \hat{x}_\lambda\| \leq R - \sqrt{R^2 - \lambda(1-\lambda)\|x_0 - x_1\|^2}.$$

Denoting

$$t := (1-\lambda)\mu,$$

we obtain

$$\|x_\lambda - \hat{x}_\lambda\| \leq R(1 - \sqrt{1-t}). \tag{8}$$

Taking into account that $0 < t < \mu \leq \frac{1}{2}$, we deduce

$$\sqrt{1-t} \geq 1 - \frac{t}{2} - \frac{t^2}{4} \geq 1 - \frac{t}{2} - \frac{t\mu}{4} \geq 1 - \frac{t}{2} - \frac{t\eta R}{8(1+L)}$$

and by (8) we get

$$\|x_\lambda - \hat{x}_\lambda\| \leq \frac{Rt}{2}\left(1 + \frac{\eta R}{4(1+L)}\right). \tag{9}$$

Since $x_0 \in S_\varepsilon(u)$ it follows that $h(u, x_0) \leq T_{f,A}(u) + \varepsilon \leq h(u, x_1) + \varepsilon$ and, similarly, $h(u, x_1) \leq h(u, x_0) + \varepsilon$. Therefore

$$|h(u, x_0) - h(u, x_1)| \leq \varepsilon \leq \frac{(\theta - L)\sigma}{2} \leq (\theta - L)\|x_0 - x_1\|.$$

and hence by Lipschitz continuity of f

$$\left| \|x_0 - u\| - \|x_1 - u\| \right| \leq |f(x_0) - f(x_1)| + (\theta - L)\|x_0 - x_1\| \leq \theta\|x_0 - x_1\|.$$

Denoting $y_0 = x_0 - u$, $y_1 = x_1 - u$, $y_\lambda = (1 - \lambda)y_0 + \lambda y_1$, we get $\left| \|y_0\| - \|y_1\| \right| \leq \theta\|y_0 - y_1\|$ and hence

$$((1 - \lambda)\|y_0\| + \lambda\|y_1\|)^2 - \|y_\lambda\|^2 = 2\lambda(1 - \lambda)\Big(\|y_0\| \cdot \|y_1\| - \langle y_0, y_1 \rangle\Big)$$

$$= \lambda(1 - \lambda)\Big(\|y_0 - y_1\|^2 - (\|y_0\| - \|y_1\|)^2\Big) \geq \lambda(1 - \lambda)(1 - \theta^2)\|y_0 - y_1\|^2.$$

In view of $x_0 \in S_\varepsilon(u)$ Lemma 1 implies that $\|y_0\| = \|x_0 - u\| \leq \frac{(1+L)d_0 + \varepsilon}{1-L}$. Since $\varepsilon \leq (1 - L)\varkappa - (1 + L)d_0$ we have $\|x_0 - u\| \leq \varkappa$ and, similarly, $\|y_1\| \leq \varkappa$. Consequently, $(1 - \lambda)\|y_0\| + \lambda\|y_1\| + \|y_\lambda\| \leq 2$ and hence

$$2\Big((1 - \lambda)\|y_0\| + \lambda\|y_1\| - \|y_\lambda\|\Big)$$

$$\geq \Big((1 - \lambda)\|y_0\| + \lambda\|y_1\| + \|y_\lambda\|\Big) \cdot \Big((1 - \lambda)\|y_0\| + \lambda\|y_1\| - \|y_\lambda\|\Big)$$

$$= ((1 - \lambda)\|y_0\| + \lambda\|y_1\|)^2 - \|y_\lambda\|^2 \geq \lambda(1 - \lambda)(1 - \theta^2)\|y_0 - y_1\|^2.$$

It means that

$$(1 - \lambda)\|x_0 - u\| + \lambda\|x_1 - u\| - \|x_\lambda - u\|$$

$$\geq \frac{\lambda(1 - \lambda)}{2\varkappa}(1 - \theta^2)\|x_0 - x_1\|^2 = \frac{\lambda(1 - \lambda)}{2d_0}(1 - \theta)^2\|x_0 - x_1\|^2. \tag{10}$$

Since f is weakly convex with constant γ on conv A, it follows that

$$f(x_\lambda) - (1 - \lambda)f(x_0) - \lambda f(x_1)$$

$$\leq -\frac{\gamma}{2}(\|x_\lambda\|^2 - (1 - \lambda)\|x_0\|^2 - \lambda\|x_1\|^2) = \frac{\gamma\lambda(1 - \lambda)}{2}\|x_0 - x_1\|^2. \tag{11}$$

By Lipschitz continuity of f on conv A we have

$$|h(u, x_\lambda) - h(u, \hat{x}_\lambda)| \leq |f(x_\lambda) - f(\hat{x}_\lambda)| + \|x_\lambda - \hat{x}_\lambda\| \leq (1 + L)\|x_\lambda - \hat{x}_\lambda\|. \tag{12}$$

Inequalities (9), (10)–(12) imply that

$$(1 - \lambda)h(u, x_0) + \lambda h(u, x_1) - h(u, \hat{x}_\lambda)$$

$$\geq \frac{\lambda(1 - \lambda)\|x_0 - x_1\|^2}{2}\left(\frac{(1 - \theta)^2}{d_0} - \gamma\right) - (1 + L)\frac{Rt}{2}\left(1 + \frac{\eta R}{4(1 + L)}\right)$$

$$= \frac{R^2 t}{2}\left(\frac{(1 - \theta)^2}{d_0} - \gamma - \frac{1 + L}{R} - \frac{\eta}{4}\right).$$

Using (7) we obtain $(1 - \lambda)h(u, x_0) + \lambda h(u, x_1) - h(u, \hat{x}_\lambda) \geq \frac{R^2 t \eta}{8}$. Since $\lambda \leq \frac{1}{2}$ we get $t = (1 - \lambda)\mu > \frac{\mu}{2}$ and hence

$$(1 - \lambda)h(u, x_0) + \lambda h(u, x_1) - h(u, \hat{x}_\lambda) > \frac{R^2 \mu \eta}{16} \geq \varepsilon. \tag{13}$$

On the other hand, by the definition of ε-solution we have $\max\{h(u, x_0), h(u, x_0)\} \leq h(u, \hat{x}_\lambda) + \varepsilon$. So, $(1 - \lambda)h(u, x_0) + \lambda h(u, x_1) - h(u, \hat{x}_\lambda) \leq \varepsilon$. This contradicts (13) and completes the proof.

Lemma 2. *Let $A \subset \mathbb{H}$ and $f : \mathbb{H} \to \mathbb{R}$ satisfy assumptions* (3). *Then for any $u \in U_\delta(A)$ there exists a unique solution $x_{\min}(u)$ of $\mathcal{P}_{f,A}$ and this solution is continuous on $U_\delta(A)$, where δ is defined by* (5).

Proof. According to Theorem 1 the problem $\mathcal{P}_{f,A}$ is Tykhonov well-posed at any $u \in U_\delta(A)$. Consequently, for any $u \in U_\delta(A)$ the problem $\mathcal{P}_{f,A}$ admits a unique solution $x_{\min}(u)$. Fix $u_0 \in U_\delta(A)$ and any sequence of $u_k \in U_\delta(A)$ such that $u_k \to u_0$. Let us denote $x_k = x_{\min}(u_k)$ for any $k \in \mathbb{N} \cup \{0\}$. To complete the proof it suffices to show that $x_k \to x_0$. It follows from (4) that

$$h(u, x) \leq h(u', x) + \|u - u'\| \quad \forall u, u', x \in \mathbb{H}.$$

Consequently,

$$h(u_k, x_k) = \min_{x \in \mathbb{H}} h(u_k, x) \leq \min_{x \in \mathbb{H}} h(u_0, x) + \|u_k - u_0\| = h(u_0, x_0) + \|u_k - u_0\|,$$

$$h(u_0, x_k) \leq h(u_k, x_k) + \|u_k - u_0\| \leq h(u_0, x_0) + 2\|u_k - u_0\| \to h(u_0, x_0).$$

It means that (x_k) is a minimizing sequence for $\mathcal{P}_{f,A}$ at u_0. Since the problem $\mathcal{P}_{f,A}$ is Tykhonov well-posed at u_0 and x_0 is the solution of the problem it follows that $x_k \to x_0$. This completes the proof.

Remark 1. If $A \subset \mathbb{H}$ is closed and weakly convex with radius $R > 0$ and $f = 0$, then assumptions (3) are satisfied for $\gamma = 0$ and $L = 0$. In this case Lemma 2 implies that the solution $x_{\min}(u)$ of $\mathcal{P}_{f,A}$ is unique and continuous on $U_\delta(A) = U_R(A)$.

Proposition 1 implies exactness of the estimate (5) of the size of the neighborhood of A in which $x_{\min}(\cdot)$ is single-valued and continuous.

Further we shall prove that under the assumptions of Lemma 2 the solution $x_{\min}(\cdot)$ is Lipschitz continuous on $U_\delta(A)$. Lemma 2 will be essentially used in the proof of this result.

4   Lipschitz Continuity of the Optimal Solution

Given a metric space T, a vector $w \in \mathbb{H}$ is called an *increment direction* of a function $x : T \to \mathbb{H}$ at a point $t_0 \in T$ if there exists a sequence (t_k) convergent to t_0 in the space T such that the increment $x(t_k) - x(t_0)$ is nonzero for all $k \in \mathbb{N}$ and

$$w = \underset{k \to \infty}{\text{weak} \lim} \frac{x(t_k) - x(t_0)}{\|x(t_k) - x(t_0)\|}$$

is the weak limit of normalized increment $\frac{x(t_k) - x(t_0)}{\|x(t_k) - x(t_0)\|}$ of the function $x(\cdot)$. Recall that weak convergence $w = \underset{k \to \infty}{\text{weak} \lim} w_k$ means that $\langle w - w_k, y \rangle \to 0$ for any $y \in \mathbb{H}$. In case $\dim \mathbb{H} < \infty$ the weak convergence coincides with the norm convergence. We use $\mathcal{V}(x, t_0)$ to denote the set of increment directions of $x(\cdot)$ at t_0.

Lemma 3. *Let $A \subset \mathbb{H}$ and $f : \mathbb{H} \to \mathbb{R}$ satisfy assumptions (3) and $u_0 \in U_\delta(A)$, where δ is defined by (5). Let $x_0 = x_{\min}(u_0)$, where $x_{\min}(u)$ is the solution of $\mathcal{P}_{f,A}$ at u. Then*

$$|\langle x_0 - u_0, w \rangle| \le L\|x_0 - u_0\| \qquad \forall w \in \mathcal{V}(x_{\min}, u_0).$$

Proof. Let $w \in \mathcal{V}(x_{\min}, u_0)$. Then there exists a sequence $u_k \to u_0$ such that $x_k \ne x_0$ and $w = \underset{k \to \infty}{\text{weak} \lim} \frac{x_k - x_0}{\|x_k - x_0\|}$ for $x_k = x_{\min}(u_k)$. According to Lemma 2 the function $x_{\min}(\cdot)$ is continuous at u_0, consequently $x_k \to x_0$.

Since $h(u_k, x_k) \le h(u_k, x_0)$ and $h(u_0, x_0) \le h(u_0, x_k)$ by Lipschitz continuity of $f(\cdot)$ on A we get

$$\max\{\|u_k - x_k\| - \|u_k - x_0\|, \ \|u_0 - x_0\| - \|u_0 - x_k\|\} \le L\|x_k - x_0\|.$$

Hence, $|\langle x_0 - u_0, x_k - x_0 \rangle| \le L\|x_k - x_0\| \cdot \|x_0 - u_0\| + o(\|x_k - x_0\|)$. Dividing by $\|x_k - x_0\|$ and passing to the limit as $k \to \infty$, we complete the proof. $\quad\square$

Lemma 4. *For any $z_0 \in \mathbb{H} \setminus \{0\}$ one has*

$$\lim_{\substack{(z,u,v) \to (z_0,0,0) \\ u \ne 0, \ v \ne 0}} \frac{1}{\|u\| \cdot \|v\|} \left(\|z + u + v\| - \|z + u\| - \|z + v\| + \|z\| \right.$$
$$\left. + \frac{\langle z_0, u \rangle \cdot \langle z_0, v \rangle}{\|z_0\|^3} - \frac{\langle u, v \rangle}{\|z_0\|} \right) = 0 \tag{14}$$

and

$$\underset{\substack{(z,u,v) \to (z_0,0,0) \\ u \ne 0, \ v \ne 0}}{\lim \sup} \frac{\left| \|z + u + v\| - \|z + u\| - \|z + v\| + \|z\| \right|}{\|u\| \cdot \|v\|} \le \frac{1}{\|z_0\|}. \tag{15}$$

Proof. Direct calculations of the first and the second derivatives of the function $f(z) = \|z\|$ for any $z \in \mathbb{H} \setminus \{0\}$ and any $u, v \in \mathbb{H}$ give

$$f'(z) = \frac{z}{\|z\|}, \qquad f''(z)u = \frac{u}{\|z\|} - \frac{z\langle u, z \rangle}{\|z\|^3}. \tag{16}$$

In view of the relations

$$f(z + u + v) - f(z + u) - f(z + v) + f(z)$$
$$= \int_0^1 \langle f'(z + u + tv) - f'(z + tv), v \rangle \, dt = \int_0^1 d\tau \int_0^1 \langle f''(z + \tau u + tv)u, v \rangle \, dt$$

for any $u, v \in \mathbb{H} \setminus \{0\}$ and $z \in \mathbb{H}$ such that $\|u\| + \|v\| + \|z - z_0\| < \|z_0\|$ we get

$$\frac{1}{\|u\| \cdot \|v\|} \Big| f(z + u + v) - f(z + u) - f(z + v) + f(z) - \langle f''(z_0)u, v \rangle \Big|$$

$$\leq \int_0^1 d\tau \int_0^1 \|f''(z + \tau u + tv) - f''(z_0)\| \, dt \leq \sup_{\substack{t \in [0,1] \\ \tau \in [0,1]}} \|f''(z + \tau u + tv) - f''(z_0)\|.$$

Since $f''(\cdot)$ is continuous at $z_0 \neq 0$, it follows that

$$\lim_{\substack{(z,u,v) \to (z_0,0,0) \\ u \neq 0, \ v \neq 0}} \frac{1}{\|u\| \cdot \|v\|} \Big(f(z+u+v) - f(z+u) - f(z+v) + f(z) - \langle f''(z_0)u, v \rangle \Big) = 0.$$

This and (16) yield (14). The relation (15) follows directly from (14).

Given a metric space T, a function $x : T \to \mathbb{H}$ is called *Lipschitz continuous at a point* u_0 with constant L if $\|x(u) - x(u_0)\| \leq L\|u - u_0\|$ for all u in some neighbourhood of u_0.

Lemma 5. *Let $A \subset \mathbb{H}$ and $f : \mathbb{H} \to \mathbb{R}$ satisfy assumptions (3) and $u_0 \in U_\delta(A) \setminus A$, where δ is defined by (5). Then the solution $x_{\min}(\cdot)$ of $\mathcal{P}_{f,A}$ is unique and Lipschitz continuous at u_0 with any constant*

$$L_x > \frac{1}{1 - L^2 - \left(\gamma + \frac{1+L}{R}\right) \frac{1+L}{1-L} \ \text{dist}(u_0, A)}. \tag{17}$$

Proof. Let L_x satisfy (17). Assume the contrary: there exists a sequence (u_k) such that $u_k \to u_0$ and

$$\|x_{\min}(u_k) - x_{\min}(u_0)\| > L_x\|u_k - u_0\| \quad \forall k \in \mathbb{N}. \tag{18}$$

Denote for all $k \in \mathbb{N}$ and $\lambda \in (0,1)$

$$x_k = x_{\min}(u_k), \quad x_0 = x_{\min}(u_0), \quad x_k(\lambda) = (1 - \lambda)x_k + \lambda x_0.$$

According to Lemma 2 one has $x_k \to x_0$. So, without loss of generality we can assume that $\|x_k - x_0\| < 2R$ for all $k \in \mathbb{N}$. In view of Proposition 2 for any $k \in \mathbb{N}$ and $\lambda \in (0,1)$ one can find $\hat{x}_k(\lambda) \in A$ such that

$$\|\hat{x}_k(\lambda) - x_k(\lambda)\| \leq R - \sqrt{R^2 - \lambda(1 - \lambda)\|x_k - x_0\|^2}. \tag{19}$$

Since x_k is the minimizer of $h(u_k, \cdot)$ and x_0 is the minimizer of $h(u_0, \cdot)$ it follows that

$$h(u_k, x_k) \leq h(u_k, \hat{x}_k(\lambda)), \quad h(u_0, x_0) \leq h(u_0, \hat{x}_k(1 - \lambda)).$$

Hence,
$$h(u_k, x_k) + h(u_0, x_0) \leq h(u_k, \hat{x}_k(\lambda)) + h(u_0, \hat{x}_k(1 - \lambda)).$$

Taking into account that $x_k, x_0, \hat{x}_k(\lambda), \hat{x}_k(1 - \lambda) \in A$, we obtain
$$f(x_k) + \|x_k - u_k\| + f(x_0) + \|x_0 - u_0\|$$
$$\leq f(\hat{x}_k(\lambda)) + \|\hat{x}_k(\lambda) - u_k\| + f(\hat{x}_k(1 - \lambda)) + \|\hat{x}_k(1 - \lambda) - u_0\|.$$

Using Lipschitz continuity of f, we get
$$\|x_k - u_k\| + \|x_0 - u_0\| - \|x_k(\lambda) - u_k\| - \|x_k(1 - \lambda) - u_0\|$$
$$\leq f(x_k(\lambda)) + f(x_k(1 - \lambda)) - f(x_k) - f(x_0)$$
$$+ (1 + L)\Big(\|\hat{x}_k(\lambda) - x_k(\lambda)\| + \|\hat{x}_k(1 - \lambda) - x_k(1 - \lambda)\|\Big). \tag{20}$$

In view of weak convexity of the function f we see that
$$\lambda f(x_k) + (1 - \lambda)f(x_0) - f(\lambda x_k + (1 - \lambda)x_0) \geq -\frac{\lambda(1 - \lambda)\gamma}{2}\|x_k - x_0\|^2,$$
$$(1 - \lambda)f(x_k) + \lambda f(x_0) - f((1 - \lambda)x_k + \lambda x_0) \geq -\frac{\lambda(1 - \lambda)\gamma}{2}\|x_k - x_0\|^2.$$

Adding together these inequalities we get
$$f(x_k) + f(x_0) - f(x_k(\lambda)) - f(x_k(1 - \lambda)) \geq -\lambda(1 - \lambda)\gamma\|x_k - x_0\|^2.$$

This and (19), (20) yield
$$\|x_k - u_k\| + \|x_0 - u_0\| - \|x_k(\lambda) - u_k\| - \|x_k(1 - \lambda) - u_0\|$$
$$\leq \lambda(1 - \lambda)\gamma\|x_k - x_0\|^2$$
$$+ 2(1 + L)\left(R - \sqrt{R^2 - \lambda(1 - \lambda)\|x_k - x_0\|^2}\right). \tag{21}$$

By (14) we get
$$\liminf_{k \to \infty} \frac{\|x_k - u_k\| + \|x_0 - u_k\| - \|x_k(\lambda) - u_k\| - \|x_k(1 - \lambda) - u_k\|}{\|x_k - x_0\|^2}$$
$$\geq \lambda(1 - \lambda) \liminf_{k \to \infty} \frac{\|x_k - x_0\|^2 \cdot \|x_0 - u_0\|^2 - \langle x_0 - u_0, x_k - x_0\rangle^2}{\|x_k - x_0\|^2 \cdot \|x_0 - u_0\|^3}. \tag{22}$$

Lemma 3 and the Banach–Alaoglu theorem imply that
$$\limsup_{k \to \infty} \frac{|\langle x_0 - u_0, x_k - x_0\rangle|}{\|x_0 - u_0\| \cdot \|x_k - x_0\|} \leq L.$$

This and (22) yield
$$\liminf_{k \to \infty} \frac{\|x_k - u_k\| + \|x_0 - u_k\| - \|x_k(\lambda) - u_k\| - \|x_k(1 - \lambda) - u_k\|}{\|x_k - x_0\|^2}$$
$$\geq \lambda(1 - \lambda)\frac{1 - L^2}{\|x_0 - u_0\|}. \tag{23}$$

By (15) we have

$$\liminf_{k\to\infty} \frac{\|x_0 - u_0\| - \|x_0 - u_k\| + \|x_k(1-\lambda) - u_k\| - \|x_k(1-\lambda) - u_0\|}{\|x_k - x_0\| \cdot \|u_k - u_0\|}$$

$$\geq -\frac{\lambda}{\|x_0 - u_0\|}.$$

In view of (18) it follows that

$$\liminf_{k\to\infty} \frac{\|x_0 - u_0\| - \|x_0 - u_k\| + \|x_k(1-\lambda) - u_k\| - \|x_k(1-\lambda) - u_0\|}{\|x_k - x_0\|^2}$$

$$\geq -\frac{\lambda}{L_x \|x_0 - u_0\|}.$$

This and (23) imply that

$$\liminf_{k\to\infty} \frac{\|x_k - u_k\| + \|x_0 - u_0\| - \|x_k(\lambda) - u_k\| - \|x_k(1-\lambda) - u_0\|}{\|x_k - x_0\|^2}$$

$$\geq \lambda(1-\lambda)\frac{1 - L^2}{\|x_0 - u_0\|} - \frac{\lambda}{L_x \|x_0 - u_0\|}.$$

Using (21) we obtain

$$\lambda(1-\lambda)\frac{1 - L^2}{\|x_0 - u_0\|} - \frac{\lambda}{L_x \|x_0 - u_0\|}$$

$$\leq \lambda(1-\lambda)\gamma + 2(1+L)\liminf_{k\to\infty} \frac{R - \sqrt{R^2 - \lambda(1-\lambda)\|x_k - x_0\|^2}}{\|x_k - x_0\|^2}$$

$$= \lambda(1-\lambda)\left(\gamma + \frac{1+L}{R}\right).$$

Dividing by λ and then passing to the limit as $\lambda \to +0$, we have

$$\left(1 - L^2 - \frac{1}{L_x}\right)\frac{1}{\|x_0 - u_0\|} \leq \gamma + \frac{1+L}{R}.$$

It follows by Lemma 1 that $\|x_0 - u_0\| \leq \frac{1+L}{1-L}\operatorname{dist}(u_0, A)$ and hence

$$\left(1 - L^2 - \frac{1}{L_x}\right)\frac{1 - L}{(1+L)\operatorname{dist}(u_0, A)} \leq \gamma + \frac{1+L}{R}.$$

This contradicts (17) and completes the proof.

Let us show that the assumption $u_0 \notin A$ in Lemma 5 can be omitted.

Theorem 2. *Let $A \subset \mathbb{H}$ and $f : \mathbb{H} \to \mathbb{R}$ satisfy assumptions (3), $\delta > 0$ be defined by (5). Then the solution $x_{\min}(\cdot)$ of $\mathcal{P}_{f,A}$ is unique and Lipschitz continuous at any $u_0 \in U_\delta(A)$ with any constant L_x satisfying (17).*

Proof. If $u_0 \notin A$, the statement of the Theorem follows from Lemma 5. Let us assume that $u_0 \in A$. Then $\text{dist}(u_0, A) = 0$ and by (17)

$$L_x > \frac{1}{1 - L^2}.$$

Choose sufficiently small $\varepsilon > 0$ such that

$$L_x > \frac{1}{1 - L^2 - \left(\gamma + \frac{1+L}{R}\right) \frac{1+L}{1-L} \varepsilon}. \tag{24}$$

Let us fix any $u_1 \in U_\varepsilon(u_0)$. It suffices to show that

$$\|x_{\min}(u_1) - x_{\min}(u_0)\| \le L_x \|u_1 - u_0\|. \tag{25}$$

Denote $u_t := (1 - t)u_0 + tu_1$ for any $t \in [0,1]$ and $\hat{t} = \max\{t \in [0,1] : u_t \in A\}$. Maximum exists because of closedness of A. Since $u_0, u_{\hat{t}} \in A$ it follows by Lemma 1 that $x_{\min}(u_0) = u_0$ and $x_{\min}(u_{\hat{t}}) = u_{\hat{t}}$. So, if $\hat{t} = 1$, then (25) holds true. Suppose that $\hat{t} < 1$. Fix any $t \in (\hat{t}, 1)$. Let us show that

$$\|x_{\min}(u_1) - x_{\min}(u_t)\| \le L_x \|u_1 - u_t\|. \tag{26}$$

Denote $a_0 = u_t$, $b_0 = u_1$. Given $a_k, b_k \in [a_0, b_0]$, we denote $c_k = \frac{a_k + b_k}{2}$. If $\|x_{\min}(a_k) - x_{\min}(c_k)\| \ge \|x_{\min}(b_k) - x_{\min}(c_k)\|$, we define $a_{k+1} = a_k$ and $b_{k+1} = c_k$. Otherwise we define $a_{k+1} = c_k$ and $b_{k+1} = b_k$. So, we construct a sequence of nested segments $[a_k, b_k]$. For any $k \in \mathbb{N} \cup \{0\}$ we have

$$\|x_{\min}(a_k) - x_{\min}(b_k)\| \le \|x_{\min}(a_k) - x_{\min}(c_k)\| + \|x_{\min}(c_k) - x_{\min}(b_k)\|$$
$$\le 2\|x_{\min}(a_{k+1}) - x_{\min}(b_{k+1})\|.$$

Consequently,

$$\|x_{\min}(a_0) - x_{\min}(b_0)\| \le 2^k \|x_{\min}(a_k) - x_{\min}(b_k)\| \qquad \forall k \in \mathbb{N}. \tag{27}$$

According to Cantor's intersection theorem there exists $\hat{c} \in \bigcap_{k \in \mathbb{N}}[a_k, b_k]$. Observe that $\hat{c} \notin A$. Since $\hat{c} \in U_\varepsilon(u_0)$ and by (24) we have

$$L_x > \frac{1}{1 - L^2 - \left(\gamma + \frac{1+L}{R}\right) \frac{1+L}{1-L} \text{ dist}(\hat{c}, A)}$$

it follows from Lemma 5 that for sufficiently large k

$$\|x_{\min}(\hat{c}) - x_{\min}(a_k)\| \le L_x \|\hat{c} - a_k\|, \quad \|x_{\min}(\hat{c}) - x_{\min}(b_k)\| \le L_x \|\hat{c} - b_k\|.$$

Therefore,

$$\|x_{\min}(a_k) - x_{\min}(b_k)\| \le L_x \left(\|\hat{c} - a_k\| + \|\hat{c} - b_k\|\right) = L_x \|a_k - b_k\|.$$

Using (27) we get for sufficiently large k

$$\|x_{\min}(a_0) - x_{\min}(b_0)\| \le L_x 2^k \|a_k - b_k\| = L_x \|a_0 - b_0\|.$$

So, (26) is proved. Passing to the limit in (26) as $t \to \hat{t} + 0$ by continuity of $x_{\min}(\cdot)$, we obtain $\|x_{\min}(u_1) - x_{\min}(u_{\hat{t}})\| \le L_x \|u_1 - u_{\hat{t}}\|$. Since $x_{\min}(u_{\hat{t}}) = u_{\hat{t}}$ and $x_{\min}(u_0) = u_0$, we have $\|x_{\min}(u_{\hat{t}}) - x_{\min}(u_0)\| = \|u_{\hat{t}} - u_0\| \le L_x \|u_{\hat{t}} - u_0\|$. Consequently,

$$\|x_{\min}(u_1) - x_{\min}(u_0)\| \le L_x(\|u_1 - u_{\hat{t}}\| + \|u_{\hat{t}} + u_0\|) = L_x \|u_1 - u_0\|.$$

Thus, (25) is proved and the proof of the theorem is completed.

Example 3. Let $\mathbb{H} = \mathbb{R}^2$, $R > 0$, $A = \{(x_1, x_2) : x_1^2 + x_2^2 \ge R^2\}$, $f = 0$. Then assumptions (3) are satisfied for $L = 0$, $\gamma = 0$. According to (5) we have $\delta = R$. So, Theorem 2 implies that $x_{\min}(\cdot)$ is Lipschitz continuous at any $u_0 \in U_R(A)$ with any constant $L_x > \frac{R}{R - \operatorname{dist}(u_0, A)}$. One can easily see that $x_{\min}(u) = \frac{Ru}{\|u\|}$ for any $u \in U_R(A)$. The exact Lipschitz constant for $x_{\min}(\cdot)$ at $u_0 \in U_R(A)$ is $L_x^{\text{exact}} = \frac{R}{\|u_0\|} = \frac{R}{R - \operatorname{dist}(u_0, A)}$. So, the estimate of the Lipschitz constant given by Theorem 2 is exact.

Remark 2. Suppose in addition to the assumptions of Theorem 2 that the set A is convex. Then it is weakly convex with any radius $R > 0$. Passing to the limit in (5) and (17) as $R \to +\infty$, we obtain that in the case of convex A the solution $x_{\min}(\cdot)$ of $\mathcal{P}_{f,A}$ is unique and Lipschitz continuous at any $u_0 \in U_\delta(A)$ with any constant

$$L_x > \frac{1 - L}{(1 + L)\big((1 - L)^2 - \gamma \operatorname{dist}(u_0, A)\big)}, \qquad \delta = \frac{(1 - L)^2}{\gamma}.$$

It has been obtained in [11, Theorem 5.2] that if A is convex and $L \le \frac{1}{2}$, then $x_{\min}(\cdot)$ is unique and Lipschitz continuous at any $u_0 \in U_{\tilde{\delta}}(A)$ with Lipschitz constant $\tilde{L}_x = 16$ and $\tilde{\delta} = \frac{1 - L}{2(1 + L)\gamma}$. Since for any $L \in [0, \frac{1}{2}]$

$$\delta = \frac{(1 - L)^2}{\gamma} > \frac{1 - L}{2(1 + L)\gamma} = \tilde{\delta}, \qquad \frac{1 - L}{(1 + L)\big((1 - L)^2 - \gamma \tilde{\delta}\big)} = \frac{2}{1 - 2L^2} < 16$$

it follows that Theorem 2 improves the result of [11] even in the case of convex A.

5 Subdifferential Calculus and Lower Regularity of the Optimal Value Function

Differential properties of the optimal value function are very important in both theory and numerical methods. Since the optimal value function is nonsmooth in general its differential properties should be described in terms of subdifferentials. Various types of subdifferentials such as Fréchet, Clarke, limiting and proximal subdifferentials possess different properties needed for applications, in particular, in optimal conditions for optimization and optimal control problems. In regular case when the limiting and the Fréchet subdifferentials coincide, they possess the properties of each other. In this case the problem becomes much easier.

Theorems 1 and 2 imply that the problem $\mathcal{P}_{f,A}$ enjoys good stability properties in some neighborhood $U_\delta(A)$ of the set A. In particular, this problem is Lipschitz approximatively well-posed (in terms of [12]), complaint and docile (in terms of [13]) for $u \in U_\delta(A)$. These stability properties of $\mathcal{P}_{f,A}$ allow one to construct subdifferential calculus of the optimal value function $T_{f,A}(\cdot)$ using methods developed in [1,11–18]. To illustrate this we shall prove that the Fréchet subdifferential and the Mordukhovich limiting subdifferential of the optimal value function of the problem $\mathcal{P}_{f,A}$ coincide in some neighborhoods of A.

Recall that the *Fréchet subdifferential* $\partial^F f(x_0)$ of a function $f : \mathbb{H} \to \mathbb{R}$ at a point $x_0 \in \mathbb{H}$ is the set of all $\xi \in \mathbb{H}$ such that for each $\varepsilon > 0$ there exists $\delta > 0$:

$$\langle \xi, x - x_0 \rangle \le f(x) - f(x_0) + \varepsilon \|x - x_0\| \quad \forall x \in U_\delta(x_0).$$

The *Mordukhovich limiting subdifferential* $\partial^L f(x)$ of a lower semicontinuous function $f : \mathbb{H} \to \mathbb{R}$ is the set of all $\xi \in \mathbb{H}$ such that there exist sequences (x_k) and (ξ_k):

$$x_k \to x_0, \quad f(x_k) \to f(x_0), \quad \xi_k \in \partial^F f(x_k), \quad \xi = \underset{k \to \infty}{\text{weak}\lim}\, \xi_k.$$

According to [18, Corollary 2.25] one has $\partial^L f(x) \ne \emptyset$ if $f : \mathbb{H} \to \mathbb{R}$ is locally Lipschitz continuous around $x \in \mathbb{H}$. This property is very important in applications of the limiting subdifferential.

A function $f : \mathbb{H} \to \mathbb{R}$ is called *lower regular* at a point $x \in \mathbb{H}$ whenever $\partial^L f(x) = \partial^F f(x)$.

Combining Theorem 2 and [19, Theorem 4.1] we obtain the following result.

Theorem 3. *Let $A \subset \mathbb{H}$ and $f : \mathbb{H} \to \mathbb{R}$ satisfy assumptions (3) and δ be defined by (5). Then the optimal value function $T_{f,A}(\cdot)$ (see (1)) is lower regular at any $u_0 \in U_\delta(A)$.*

Theorem 3 implies that in Example 3 the optimal value function $T_{f,A}(\cdot)$ is lower regular at any $u_0 \in U_\delta(A) = \mathbb{R}^2 \setminus \{(0,0)\}$. The direct calculations show that $T_{f,A}(u_1, u_2) = \min\{R - \sqrt{u_1^2 + u_2^2}, 0\}$. Really, it is not regular at $(0,0)$. This example shows that the assumptions of Theorem 3 are essential.

It was shown in [19, Theorem 5.1] that if the set A is convex and $L \le \frac{1}{2}$, then the optimal value function $T_{f,A}(\cdot)$ is lower regular at any $u_0 \in U_{\tilde{\delta}}(A)$ with $\tilde{\delta} = \frac{1-L}{2(1+L)\gamma}$. Since $\tilde{\delta} < \delta = \frac{(1-L)^2}{\gamma}$ for any $L \in [0, \frac{1}{2}]$ Theorem 3 improves the result of [19].

6   Conclusion

Although we consider only Hilbert spaces, this approach can be applied to the study of the well-posedness and Lipschitzianness of the solution of the Moreau-type infimal convolution problem in Banach spaces.

References

1. Ivanov, G.E., Thibault, L.: Infimal convolution and optimal time control problem I: fréchet and proximal subdifferentials. Set-Valued Variational Anal. **26**(3), 581–606 (2018). https://doi.org/10.1007/s11228-016-0398-z
2. Vial, J.-P.: Strong and weak convexity of sets and functions. Math. Oper. Res. **8**, 231–259 (1983)
3. Ivanov, G.E.: Weak convexity in the senses of Vial and Efimov-Stechkin. Izv. Math. **69**(6), 1113–1135 (2005)
4. Ivanov, G.E.: Weakly convex sets and their properties. Math. Notes **79**(1), 55–78 (2006)
5. Ivanov, G.E.: Weakly Convex Sets and Functions. Theory and applications. Fizmatlit, Moscow (2006)
6. Goncharov, V.V., Ivanov, G.E.: Strong and weak convexity of closed sets in a Hilbert space. Oper. Res. Eng. Cyber Secur. **113**, 259–297 (2017)
7. Ivanov, G.E., Lopushanski, M.S., Golubev, M.O.: The nearest point theorem for weakly convex sets in asymmetric seminormed spaces. In: Evtushenko, Y., Jaćimović, M., Khachay, M., Kochetov, Y., Malkova, V., Posypkin, M. (eds.) OPTIMA 2018. CCIS, vol. 974, pp. 21–34. Springer, Cham (2019). https://doi.org/10.1007/978-3-030-10934-9_2
8. Ivanov, G.E., Golubev, M.O.: Alternative theorem for differential games with strongly convex admissible control sets. In: Evtushenko, Y., Jaćimović, M., Khachay, M., Kochetov, Y., Malkova, V., Posypkin, M. (eds.) OPTIMA 2018. CCIS, vol. 974, pp. 321–335. Springer, Cham (2019). https://doi.org/10.1007/978-3-030-10934-9_23
9. Ivanov, G.E., Golubev, M.O.: Strong and weak convexity in nonlinear differential games, IFAC PapersOnLine, 51(32), 13–18 (2018). https://www.sciencedirect.com/science/article/pii/S2405896318330428
10. Dontchev, A.L., Zolezzi, T.: Well-Posed Optimization Problems. LNM, vol. 1543. Springer, Heidelberg (1993). https://doi.org/10.1007/BFb0084195
11. Ivanov, G.E., Thibault, L.: Infimal convolution and optimal time control problem III: minimal time projection set. SIAM J. Optim. **28**(1), 30–44 (2018). https://doi.org/10.1137/16M1110212
12. Ivanov, G.E., Thibault, L.: Well-posedness and subdierentials of optimal value and infmal convolution. Set-Valued Variational Anal. (2018). https://doi.org/10.1007/s11228-018-0493-4
13. Penot, J.-P.: Differentiability properties of optimal value functions. Can. J. Math. **56**(4), 825–842 (2004)
14. Ngai, H.V., Penot, J.-P.: Subdifferentiation of regularized functions. Set-Valued Variational Anal. **24**, 167–189 (2016)
15. Penot, J.-P.: Calculus Without Derivatives. GTM, vol. 266. Springer, New York (2013). https://doi.org/10.1007/978-1-4614-4538-8
16. Nam, N.M.: Subdifferential formulas for a class of nonconvex infimal convolutions. Optimization **64**, 2213–2222 (2015)

17. Nam, N.M., Cuong, D.V.: Generalized differentiation and characterizations for differentiability of infimal convolutions. Set-Valued Variational Anal. **23**, 333–353 (2015)
18. Mordukhovich, B.S.: Variational Analysis and Generalized Differentiation I and II: Comprehensive Studies in Mathematics, vol. 330, p. 331. Springer, New York (2005). https://doi.org/10.1007/3-540-31246-3
19. Ivanov, G.E., Thibault, L.: Infimal convolution and optimal time control problem II: limiting subdifferential. Set-Valued Variational Anal. **25**(3), 517–542 (2017). https://doi.org/10.1007/s11228-017-0402-2

Polynomial-Time Solvability
of One Optimization Problem Induced
by Processing and Analyzing
Quasiperiodic ECG and PPG Signals

Alexander Kel'manov[1,2] , Sergey Khamidullin[1] , Liudmila Mikhailova[1(✉)] ,
and Pavel Ruzankin[1,2]

[1] Sobolev Institute of Mathematics, 4 Koptyug Avenue, 630090 Novosibirsk, Russia
{kelm,kham,mikh,ruzankin}@math.nsc.ru
[2] Novosibirsk State University, 2 Pirogova Street, 630090 Novosibirsk, Russia

Abstract. This paper is devoted to an unexplored discrete optimization problem, which can be interpreted as a problem of least mean squares approximation of some observed discrete-time signal (a numerical time series) by an unobservable quasiperiodic (almost periodic) pulse signal generated by a pulse with a given pattern (reference) shape. Quasiperiodicity is understood, first, in the sense of admissible fluctuations of the interval between repetitions of the reference pulse, and second, in the sense of admissible nonlinear time expansions of its reference shape. Such problems are common in biomedical applications related to monitoring and analyzing electrocardiogram (ECG), photoplethysmogram (PPG), and several other signals. In the optimization model, the number of generated (admissible or approximating) quasiperiodic pulse sequences grows exponentially with the duration of the discrete-time signal (i.e., with the number of points in the time series). The size of the admissible solutions set also grows exponentially. However, despite that exponential growth, we have constructively proved the optimization problem polynomial-time solvability. Namely, we propose an algorithm that finds an optimal solution to the problem in $\mathcal{O}(T_{\max}^3 N)$ time; where N is the duration of the observed signal (the number of points in the time series), $T_{\max} \leq N$ is a positive integer number which bounds the fluctuations of the repetition period. If $T_{\max}$ is a part of the input, then the algorithm's running time is $\mathcal{O}(N^4)$, i.e., the algorithm is polynomial. If $T_{\max}$ is a fixed parameter (that is typical for applications), then the running-time of the algorithm is $\mathcal{O}(N)$, i.e., the algorithm is linear in time. Numerical simulation examples demonstrate the robustness of the algorithm in the presence of additive noise.

Keywords: Discrete optimization problem · Polynomial-time solvability · Linear-time algorithm and Pulse train signal processing · Quasiperiodic · ECG · PPG

A. Kel'manov—Deceased 1 December 2019.

M. Jaćimović et al. (Eds.): OPTIMA 2019, CCIS 1145, pp. 88–101, 2020.
https://doi.org/10.1007/978-3-030-38603-0_7

1 Introduction

This work is devoted to studying an unexplored discrete optimization problem. The problem arises within an unconventional (alternative) approach to the applied problem of computer processing and analyzing signals forming a quasiperiodic pulse train. The applied problem is typical, in particular, for medical diagnostics based on the analysis of quasiperiodic sequences of ECG or PPG pulses. The main goal of the research is to construct an efficient algorithm with theoretical guarantees of quality (accuracy and complexity), solving this discrete optimization problem.

Traditional approaches to the problem of biomedical signal feature extraction consist in the sequential implementation of several stages: filtering out the noise, detecting certain sections of the pulse, analyzing the identified sections of the pulse, etc. Each of these steps is usually implemented via suitable well-known signal processing techniques. The main principles being employed in those algorithms are: using thresholds for the signal and its differentials, supervised machine learning algorithms such as neural networks, hidden Markov models, Bayesian approach, Fourier and wavelet transforms (e.g., see [1,2], and references therein). Heuristic threshold methods are known to suffer from certain inflexibility for significant variations of the shape of the pulse. While the classical correlation-spectral methods turn out to be ineffective for processing such quasiperiodic signals, because of fluctuations of the pulses via nonlinear expansions and contractions in time and because of fluctuations in the repetition period of the pulses in the pulse train. Those fluctuations ultimately lead to problems in interpreting the analyzed data [3].

We propose a novel approach to the problem of feature extraction, in which all the stages of signal processing are combined into one optimization process. The approach induces the corresponding discrete optimization problem. In the induced problem, the size of the admissible solutions set grows exponentially with the duration of the analyzed discrete-time signal, i.e., with the length N of the input numerical sequence. Nevertheless, we constructively prove that the problem is effectively solvable in polynomial time. In order to do it, we present a justification for the exact polynomial algorithm and show that, for fixed parameters, the running time of the algorithm is $\mathcal{O}(N)$, i.e., the algorithm is linear.

Note that this approach in a simplified form has previously demonstrated its effectiveness in solving similarly formulated applied problems of optimal processing of varying (fluctuating in time) pulse signals distorted by noise [4–11]. In an even more simplified form, this approach has been and is used at the NASA Space Research Center. In particular, it made possible to detect hundreds of new exoplanets from the highly noisy astrophysical data [12,13].

The paper has the following structure. In Sect. 2, we consider the data generation model (for quasiperiodic sequences) and formulate the applied problem of data processing and analyzing as an approximation problem. In Sect. 3, the induced discrete optimization problem is formulated. The main result is the exact polynomial algorithm presented and justified in Sect. 4. Section 5 provides several examples of numerical simulation.

2   Data Generation and Approximation Models

Pulses of certain biomedical signals are widely presented and described in the literature (e.g., see [3, 14–17]) along with annotations of characteristic waves (sections of the pulse) important for medical diagnostics. Typical pulse shapes and characteristic waves were identified by experts in medicine over the years of research. The typical shapes of the pulses and their wave markups are considered as references (patterns). It is well known that deviations of the shapes of observed pulses from typical (or reference) shapes allow diagnosing certain diseases.

On Fig. 1, the examples of the reference shapes are shown for: (a) ECG pulse and (b) PPG pulse. On the figure, characteristic waves are highlighted in color.

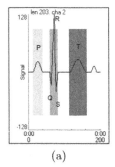 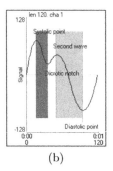

(a)   (b)

Fig. 1. Pulses having the given pattern shapes. (a): ECG-pulse, (b): PPG-pulse

In the sequel, we will assume that the elements of the input sequence are discrete-time samples of a continuous signal. Let us construct a model of generation of a quasiperiodic pulse train by some reference pulse.

The reference pulse will be given by the sequence $U = (u_1, \ldots, u_q)$, where $u_t \in \mathbb{R}$, $t = 1, \ldots, q$.

Consider the following model of admissible nonlinear time expansions of the reference pulse, i.e., of the sequence U. Denote by the natural number k_t, $t = 1, \ldots, q$, the multiplicity for repetition of the t-th element of U in the expanded sequence

$$(\underbrace{u_1, \ldots, u_1}_{k_1}, \underbrace{u_2, \ldots, u_2}_{k_2}, \ldots, \underbrace{u_q, \ldots, u_q}_{k_q}). \tag{1}$$

The sequence (1) can be treated as the reference pulse expanded in time.

Put

$$h_i = \begin{cases} u_1 & \text{if } i = 1, \ldots, k_1, \\ u_2 & \text{if } i = k_1 + 1, \ldots, k_1 + k_2, \\ \ldots \\ u_q & \text{if } i = k_1 + \ldots + k_{q-1} + 1, \ldots, k_1 + \ldots + k_q, \\ 0 & \text{if } i < 1 \text{ or } i > p, \end{cases} \tag{2}$$

where k_t, $t = 1, \ldots, q$, is the multiplicity of the element u_t in the expanded sequence, and

$$p = k_1 + \ldots + k_q \tag{3}$$

is the length of the expanded sequence.

We will assume that

$$p \leq \ell, \tag{4}$$

where ℓ is some fixed natural number which determines the upper bound for the length of an expanded reference sequence of length q. The formulas (1), (3), (4) define the set of admissible expanded sequences generated by the sequence U.

According to (2), the mapping $j(i) : \{1, \ldots, p\} \longrightarrow \{1, \ldots, q\}$ of the form

$$\begin{aligned} j(1) = 1, \; j(p) = q, \\ 0 \leq j(i+1) - j(i) \leq 1, \; i = 1, \ldots, p-1, \end{aligned} \tag{5}$$

defines the set of admissible mappings between the serial numbers of elements in the reference and expanded sequences. Besides, the following relation is valid for the multiplicity k_t of the element u_t in the expanded sequence:

$$k_t = \left| \left\{ i \,|\, j(i) = t, i \in \{1, \ldots, p\} \right\} \right|, \; t = 1, \ldots, q,$$

while the image of the set $\{1, \ldots, p\}$ under the mapping $j(i)$ can be represented as

$$\underbrace{1, \ldots, 1}_{k_1}, \underbrace{2, \ldots, 2}_{k_2}, \ldots, \underbrace{q, \ldots, q}_{k_q}.$$

Let us now construct the model for admissible fluctuations of the pulse repetition period. The top row of Fig. 2 depicts an example of a quasiperiodic sequence $X = (x_1, \ldots, x_N)$ of ECG pulses, generated by the reference pulse U (Fig. 1a). In this example, each repetition of the reference pulse is conducted by a random temporal fluctuation from the set of admissible expansions (as can be seen on the colored characteristic waves). Put $\mathcal{N} = \{1, \ldots, N\}$. Denote by $\mathcal{M}_s = \{n_1^s, n_2^s, \ldots\} \subset \mathcal{N}$ and $\mathcal{M}_e = \{n_1^e, n_2^e, \ldots\} \subset \mathcal{N}$ the collections of numbers

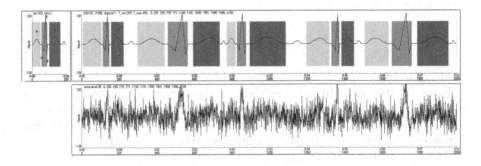

Fig. 2. ECG pulse trains

of elements of X corresponding to the starts and ends of the expanded pulses, respectively. Denote by $\mathcal{P} = \{p_1, p_2, \ldots\}$ the set of durations of the expanded pulses. We will assume that the size of the collections $\mathcal{M}_s$, $\mathcal{M}_e$, and $\mathcal{P}$ equals some unknown natural number M which stands for the number of repetitions of the reference sequence U in the generated sequence X. Evidently, we have $M \leq M_{\max} \leq N$, where $M_{\max}$ is the maximal possible number of repetitions. Put $\mathcal{M} = \{1, \ldots, M\}$. The above construction implies that

$$n^e_{m-1} < n^s_m \text{ for } m = 2, \ldots, M, \tag{6}$$

$$n^e_m - n^s_m + 1 = p_m \text{ for } m \in \mathcal{M}, \tag{7}$$

and

$$q \leq p_m \leq \ell \text{ for } m \in \mathcal{M}, \tag{8}$$

where $m \in \mathcal{M}$ is the serial number of the expanded sequence. Hereinafter we will assume that

$$q \leq n^s_m - n^s_{m-1} \leq T_{\max} \text{ for } m = 2, \ldots, M, \tag{9}$$

and

$$\ell = \lceil \alpha T_{\max} \rceil, \tag{10}$$

where $T_{\max}$ is some natural number which together with q determines the upper and lower bounds on the admissible fluctuations of the time interval between two consecutive repetitions of the pulse, and $\alpha \in (0, 1]$ is some real number. The segment of the sequence X, that contains the m-th expanded sequence with the elements' serial numbers from $n = n^s_m$ to $n = n^e_m$, can be schematically represented in the following form (the bottom line contains the serial numbers of the elements in the sequence X):

$$\ldots, 0, \overbrace{u_1, \ldots, u_1,}^{k_1^{(m)}} \ldots u_1, \overbrace{u_2, \ldots, u_2, \ldots, u_2,}^{k_2^{(m)}} \ldots, \overbrace{u_q, \ldots, u_q,}^{k_q^{(m)}} 0, \ldots, 0, u_1 \ldots \ldots$$
$$\ldots \quad , n^s_m, \ldots \ldots \ldots \ldots \ldots, \ldots \ldots \ldots \ldots \ldots \ldots \ldots, \ldots, \ldots \ldots, n^e_m, \quad \ldots \quad , n^s_{m+1} \ldots$$

The above conditions and notations allow us to represent the n-th element of the sequence X as

$$x_n = \sum_{m \in \mathcal{M}} h^{(m)}_{n - n^s_m + 1}, \quad n \in \mathcal{N}, \tag{11}$$

where $n^s_m \in \mathcal{M}_s$.

From (2) it is easy to see that the right-hand side of (11) is the sum of M expanded reference sequences that are shifted in time from the initial serial number $n = 1$ and do not overlap in discrete time. Relation (11) describes a sequence of discrete-time samples of a quasiperiodic sequence of pulses.

The constructed model for quasiperiodic repetition of pulses is suitable for many biomedical signals, in particular, for ECG and PPG signals. This model

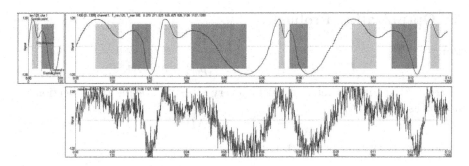

Fig. 3. PPG pulse train

reflects, first, real-life fluctuations of the pulse repetition period and, second, nonlinear temporal fluctuations of the ideal (pattern) pulse (e.g., see [3,14–17]).

In the top row of Fig. 3, an example of a quasiperiodic sequence generated by an ideal PPG-pulse (Fig. 1b) is shown. As can be seen from the above model, the reference sequence U of length q generates the set $\mathcal{X}$ of all admissible quasiperiodic sequences of length N.

We will assume that we observe a sequence $Y = (y_1, \ldots, y_N)$ which is the element-wise sum of an unobservable sequence $X \in \mathcal{X}$ and some sequence reflecting possible noise distortions. The examples of sequences Y are depicted in the bottom rows of Figs. 2 and 3.

Since the sequences Y and X can be considered as vectors in an N-dimensional space, we can state the approximation problem in the form

$$\|Y - X\|^2 \longrightarrow \min_{X \in \mathcal{X}}. \tag{12}$$

It is easy to see that the upper bound

$$|\mathcal{X}| \leq (N - q + 1) \sum_{M=1}^{M_{\max}} q^{(T_{\max} - q)M} (T_{\max} - q + 1)^{2M-1}$$

$$\leq q^{(T_{\max} - q)M_{\max}} (N - q + 1)(T_{\max} - q + 1)^{2M_{\max} - 1}$$

is valid for the size of the set $\mathcal{X}$. Evidently, except the trivial case when $T_{\max} = q$, we have

$$|\mathcal{X}| \geq 2^{\left\lfloor \frac{N-q+1}{q+1} \right\rfloor}.$$

It means that if q is fixed (which is common in applications) then we have an exponentially sized set $\mathcal{X}$ of admissible solutions. It is clear that brute-force searching through this set is hardly possible in reasonable time. Below we propose an algorithm which finds an optimal solution to the approximation problem (12) in polynomial time. Moreover, if $T_{\max}$ is fixed then the running time of the algorithm is $\mathcal{O}(N)$, i.e., the algorithm finds a solution in linear time.

3   Optimization Problem

Put

$$J^{(m)} = \{j^{(m)}(i),\ i = 1, \ldots, p_m\}, \quad m \in \mathcal{M},$$

where $j^{(m)}(i)$ satisfies relation (5) for every $m \in \mathcal{M}$. Define the collection of mappings

$$\mathcal{J} = \{J^{(m)}, m \in \mathcal{M}\}.$$

Besides, in accordance with (2), put

$$h_i^{(m)} = \begin{cases} u_{j^{(m)}(i)} & \text{if } i = 1, \ldots, p_m, \\ 0 & \text{if } i < 1 \text{ or } i > p_m, \end{cases} \tag{13}$$

where $m \in \mathcal{M}$.

Expanding the square of the norm in (12) and taking into account (11), (13), and constraints (6), (7), (8), (9), and (10), we obtain by simple calculations that

$$\|Y - X\|^2 = \sum_{n \in \mathcal{N}} y_n^2 + \sum_{m \in \mathcal{M}} \sum_{i=1}^{p_m} \{u_{j^{(m)}(i)}^2 - 2 y_{n_m^s + i - 1} u_{j^{(m)}(i)}\}.$$

Here the first term on the right-hand side is constant and, hence, problem (12) stated above is equivalent to the following problem.

Problem 1. Given: numerical sequences $Y = (y_1, \ldots, y_N)$ and $U = (u_1, \ldots, u_q)$, a natural number $T_{\max} \leq N$, and a real number $\alpha \in (0, 1]$. *Find:* collections $\mathcal{M}_s = \{n_1^s, n_2^s, \ldots n_m^s, \ldots\} \subset \mathcal{N} = \{1, \ldots, N\}$ and $\mathcal{P} = \{p_1, p_2, \ldots, p_m, \ldots\}$ of natural numbers, a collection $\mathcal{J} = \{J^{(1)}, J^{(2)}, \ldots, J^{(m)}, \ldots\}$ of mappings, and the size M of these collections, such that

$$F(\mathcal{M}_s, \mathcal{P}, \mathcal{J}) = \sum_{m \in \mathcal{M}} \sum_{i=1}^{p_m} f_{j^{(m)}(i)}(n_m^s + i - 1) \longrightarrow \min,$$

where

$$f_t(n) = u_t^2 - 2 y_n u_t, \quad t = 1, \ldots, q, \ n \in \mathcal{N}, \tag{14}$$

under the constraints

$$j^{(m)}(1) = 1, \quad j^{(m)}(p_m) = q,$$
$$0 \leq j^{(m)}(i+1) - j^{(m)}(i) \leq 1,$$
$$i = 1, \ldots, p_m - 1, \quad m = 1, \ldots, M,$$

and

$$q \leq p_m \leq \lceil \alpha T_{\max} \rceil, \quad m = 1, \ldots, M,$$
$$p_{m-1} \leq n_m^s - n_{m-1}^s \leq T_{\max}, \quad m = 2, \ldots, M,$$
$$p_M \leq N - n_M^s + 1.$$

4 Algorithm

In order to construct an algorithm solving Problem 1, we need to consider the following two auxiliary problems.

The first auxiliary problem is

Problem 2. Given: an array $\{w_{i,j},\ i = 1, \ldots, p,\ j = 1, \ldots, q\}$ of real numbers. *Find:* summing indices $j(i)$ such that

$$W = \sum_{i=1}^{p} w_{i,j(i)} \longrightarrow \min,$$

under the constraints

$$j(1) = 1,\ j(p) = q,$$
$$0 \leq j(i+1) - j(i) \leq 1,\ i = 1, \ldots, p-1.$$

The following lemma provides recurrent formulas to solve Problem 2.

Lemma 1. Let the conditions of Problem 2 hold. Then the optimal value W^* of the objective function of this problem is given by the formula

$$W^* = W_{p,q} \tag{15}$$

and the values of $W_{p,q}$ are calculated with the recurrence formulas

$$W_{s,t} = \min\{W_{s-1,t}, W_{s-1,t-1}\} + w_{s,t},\ s = 1, \ldots, p,\ t = 1, \ldots, q, \tag{16}$$

with the following initial and boundary conditions

$$W_{s,t} = \begin{cases} 0, & s = 0,\ t = 0, \\ +\infty, & s = 0,\ t = 1, \ldots, q, \\ +\infty, & s = 1, \ldots, p,\ t = 0. \end{cases} \tag{17}$$

The optimal values of the summing indices are determined by the rule

$$j^*(p) = q;$$
$$j^*(i-1) = \begin{cases} j^*(i), & \text{if } W_{i-1,j^*(i)} \leq W_{i-1,j^*(i)-1}, \\ j^*(i) - 1, & \text{if } W_{i-1,j^*(i)} > W_{i-1,j^*(i)-1}, \end{cases} \tag{18}$$
$$i = p, p-1, \ldots, 2.$$

The second auxiliary problem is

Problem 3. Given: an array $\{g_p(n),\ p = q, \ldots, \ell,\ n = 1, \ldots, N - p + 1\}$ of real numbers and a natural number $T_{\max} \geq \ell$. *Find:* a collection $\{(n_1, p_1), \ldots, (n_M, p_M)\}$, where $n_m \in \{1, \ldots, N - q - 1\}$, $p_m \in \{q, \ldots, \ell\}$, $m = 1, \ldots, M$, and the size M of this collection such that

$$G((n_1, p_1), \ldots, (n_M, p_M)) = \sum_{m=1}^{M} g_{p_m}(n_m) \longrightarrow \min,$$

under the constraints

$$p_{m-1} \leq n_m - n_{m-1} \leq T_{\max}, \quad m = 2, \ldots, M,$$

$$p_M \leq N - n_M + 1.$$

The following lemma and corollary provide recurrent formulas to solve Problem 3.

Lemma 2. Let the conditions of Problem 3 hold. Then the optimal value G^* of the objective function of that problem is given by the formula

$$G^* = \min_{p \in \{q,\ldots,\ell\}} \min_{n \in \{1,\ldots,N-p+1\}} G_p(n), \tag{19}$$

and the values of $G_p(n)$ are calculated with the recurrence formulas

$$G_p(n) = \begin{cases} g_p(n), & p = q, \ldots, \ell, \ n = 1, \ldots, q, \\ \min\left\{0, \min_{i \in \{q,\ldots,\min\{n-1,\ell\}\}} \min_{j \in \gamma_i(n)} G_i(j)\right\} + g_p(n), \\ & p = q, \ldots, \ell, \ n = q+1, \ldots, N-p+1, \end{cases} \tag{20}$$

where

$$\gamma_i(n) = \left\{k \mid \max\{n - T_{\max}, 1\} \leq k \leq n - i\right\}, \quad n = i+1, \ldots, N, \ i = q, \ldots, \ell. \tag{21}$$

In order to find the components of the optimal tuple in Problem 3, define the two functions:

$$\pi(n) = \begin{cases} 0, & \text{if } n = 1, \ldots, q, \\ 0, & \text{if } \min_{i \in \{q,\ldots,\min\{n-1,\ell\}\}} \min_{j \in \gamma_i(n)} G_i(j) \geq 0, \ n = q+1, \ldots, N-q+1, \\ \arg \min_{i \in \{q,\ldots,\min\{n-1,\ell\}\}} \left\{\min_{j \in \gamma_i(n)} G_i(j)\right\}, & \text{if} \\ & \min_{i \in \{q,\ldots,\min\{n-1,\ell\}\}} \min_{j \in \gamma_s(n)} G_i(s) < 0, \ n = q+1, \ldots, N-q+1, \end{cases} \tag{22}$$

and

$$I(n) = \begin{cases} -1, & \text{if } \pi(n) = 0, \\ \arg \min_{j \in \gamma_{\pi(n)}(n)} G_{\pi(n)}(j), & \text{if } \pi(n) > 0, \end{cases} \quad n = 1, \ldots, N-q+1. \tag{23}$$

Corollary 1. Let the conditions of Lemma 2 hold. Additionally, assume that

$$\pi_1 = \arg \min_{p \in \{q,\ldots,\ell\}} \left\{\min_{n \in \{1,\ldots,N-p+1\}} G_p(n)\right\}, \tag{24}$$

$$\nu_1 = \arg \min_{n \in \{1,\ldots,N-\pi_1+1\}} G_{\pi_1}(n), \tag{25}$$

$$\pi_m = \pi(\nu_{m-1}), \ \nu_m = I(\nu_{m-1}), \quad m = 2, \ldots, M, \tag{26}$$

where M is the minimal value of m, such that $\pi(\nu_m) = 0$. Then the tuple

$$\varphi = \{(\nu_M, \pi_M), \ldots, (\nu_1, \pi_1)\}$$

is the optimal solution to Problem 3.

In this way, we introduce the following algorithm.

Algorithm $\mathcal{A}$.

INPUT: sequences $Y = (y_1, \ldots, y_N)$ and $U = (u_1, \ldots, u_q)$, a natural number $T_{\max}$, and a real number α.

Forward pass.

STEP 1 (solving the family of Problems 2). Put $\ell = \lceil \alpha T_{\max} \rceil$.
For each $n = 1, \ldots, N - q + 1$, do:
For each $p = q, \ldots, \min\{\ell, N - n + 1\}$ do (solving Problem 2):
(1) Put

$$w_{i,j} = f_j(n + i - 1), i = 1, \ldots, p, \; j = 1, \ldots, q,$$

where $f_j(n + i - 1)$, $j = 1, \ldots, q$, $i = 1, \ldots, p$, is calculated by (14).

(2) Compute $W_{s,t}$ for all $s = 1, \ldots, p$ and $t = 1, \ldots, q$ using formulas (15), (16), and (17). Find a sequence $j(1), \ldots, j(p)$ of indices according to (18).

(3) Put $W(n, p) = W_{p,q}$, $J(n, p) = \{j(i), \; i = 1, \ldots, p\}$.

STEP 2. Put $g_p(n) = W(n, p)$ for all $p = q, \ldots, \ell$ and $n = 1, \ldots, N - p + 1$.

STEP 3. Compute $G_p(n)$ for all $p = q, \ldots, \ell$ and $n = 1, \ldots, N - p + 1$ using formulas (20) and (21), and compute G^* using formula (19). Put $F_A = G^*$.

Backward pass.

STEP 4. Compute $\pi(n)$ and $I(n)$, $n = 1, \ldots, N - q + 1$, using formulas (22) and (23).

STEP 5. Find the components of the auxiliary collections $(\pi_1, \pi_2, \ldots)$ and $(\nu_1, \nu_2, \ldots)$ and their size M by the formulas (24), (25), and (26).

STEP 6. Put $M_A = M$, $\mathcal{M}_A = \{\nu_{M_A}, \ldots, \nu_1\}$, $\mathcal{P}_A = \{\pi_{M_A}, \ldots, \pi_1\}$, $J^{(m)} = J(\nu_{M_A - m + 1}, \pi_{M_A - m + 1})$, $m = 1, \ldots, M_A$, and $\mathcal{J}_A = \{J^{(1)}, \ldots, J^{(M_A)}\}$.

OUTPUT: the natural number M_A, the collections $\mathcal{M}_A$, $\mathcal{P}_A$, $\mathcal{J}_A$, and the value F_A.

The main result of the research is the following theorem.

Theorem 1. *Algorithm $\mathcal{A}$ finds an exact solution to Problem 1 in time* $\mathcal{O}(T_{\max}^3 N)$.

The proof of the theorem is based on the following chain of equalities:

$$F^* \overset{(1)}{=} \min_{\mathcal{M}_s, \mathcal{P}, \mathcal{J}} \sum_{m=1}^{M} \sum_{i=1}^{p_m} f_{j^{(m)}(i)}(n_m + i - 1)$$

$$\overset{(2)}{=} \min_{\mathcal{M}_s, \mathcal{P}} \min_{J^{(1)}, \ldots, J^{(M)}} \sum_{m=1}^{M} \sum_{i=1}^{p_m} f_{j^{(m)}(i)}(n_m + i - 1)$$

$$\overset{(3)}{=} \min_{\mathcal{M}_s, \mathcal{P}} \sum_{m=1}^{M} \left\{ \min_{J^{(m)}} \sum_{i=1}^{p_m} f_{j^{(m)}(i)}(n_m + i - 1) \right\}$$

$$\overset{(4)}{=} \min_{\mathcal{M}_s, \mathcal{P}} \sum_{m=1}^{M} \left\{ \min_{J} \sum_{i=1}^{p_m} f_{j(i)}(n_m + i - 1) \right\} \overset{(5)}{=} \min_{\mathcal{M}_s, \mathcal{P}} \sum_{m=1}^{M} W(n_m, p_m) \overset{(6)}{=} F_A.$$

In this chain, equality (1) is a definition of optimal solution to Problem 1. Equalities (2) and (3) are valid due to the structure of the collection $\mathcal{J}$ and because each element of the inner sum in the left-hand side of (3) depends solely on $J^{(m)}$ for each $m = 1, \ldots, M$. Equality (4) is just the change of variables. For each $m = 1, \ldots, M$, the expression in the brackets in the left-hand side of (5) is the optimal value of the objective function in Problem 2 with the input data described in Step 1. Thus, equality (5) is valid by Lemma 1. Equality (6) corresponds to Step 2 of Algorithm $\mathcal{A}$ and follows from Lemma 2.

We have proved that exact solutions to Problems 2 and 3 can be found in time $\mathcal{O}(pq)$ and $\mathcal{O}(T_{\max}^3 N)$, respectively. Summing up, using the step-by-step representation of Algorithm $\mathcal{A}$ and taking into account that $q \leq p \leq \ell \leq T_{\max} \leq N$, we prove the polynomial time complexity of the algorithm.

Remark 1. If $T_{\max}$ is a part of input data, then the running time of the algorithm is $\mathcal{O}(N^4)$, since $T_{\max} \leq N$; thus, Algorithm $\mathcal{A}$ is polynomial-time.

Remark 2. If $T_{\max}$ is a fixed parameter then the running time of the algorithm is $\mathcal{O}(N)$.

5 Examples of Numerical Experiments

The figures below show examples of processing modeled (i.e., generated) signals. In the top rows of the figures, a reference sequence (pulse) U (input) and a modeled sequence (pulse train) are depicted. In those pulse trains, the repetitions of the reference sequence are conducted by random expansions, as well as fluctuations in the frequency of repetitions. The middle rows of the figures show the sequences Y (input) which are the element-wise sums of the modeled sequences and sequences of independent identically distributed Gaussian random variables (white noise). The bottom rows of the figures show the results for the algorithm, i.e., the recovered sequences X that are found by the algorithm. Those sequences are uniquely defined by the computed collections (the algorithm output) and the reference sequence.

Figure 4 shows the result of noise-robust processing of a modeled ECG signal. The example is computed for $q = 200$, $T_{\max} = 500$, $\alpha = 1$, $N = 3000$, the maximum amplitude pulse value is 120, and the noise level $\sigma = 35$.

Fig. 5 shows the result of noise-robust processing of a modeled PPG signal. The example is computed for $q = 120$, $T_{\max} = 300$, $\alpha = 1$, $N = 1000$, the maximum amplitude pulse value is 120, and the noise level $\sigma = 40$.

The numerical simulation examples show that, despite significant noise interference, the proposed algorithm perfectly finds and marks both the pulses and the characteristic waves of the pulses.

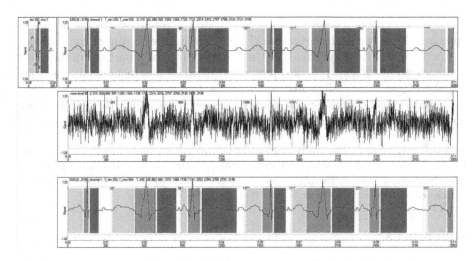

Fig. 4. Example of processing an ECG pulse train

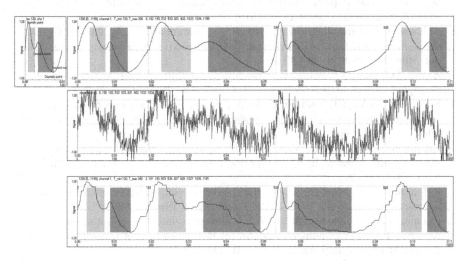

Fig. 5. Example 2 of processing a PPG pulse train

6   Conclusion

We have constructed the exact polynomial-time algorithm solving the discrete optimization problem that models the actual applied problem of noise-robust computer processing and analyzing quasiperiodic biomedical signals. For the algorithm, we employ the novel approach. The approach relies heavily on discrete optimization methods. It does not involve a breakdown of the problem into separate components (noise filtering, pulse detection, searching characteristic waves, etc.). On the contrary, within the framework of the proposed approach, all the problems are solved jointly (simultaneously) in a single process of finding an optimal solution. The examples demonstrate the robustness of the proposed approach. Thus, we have shown that the practice-induced optimization problem is exactly solvable in polynomial time and that, within the framework of the novel approach, some difficult practical problems caused by quasiperiodic fluctuations of pulsed signals are easily solvable.

In subsequent works, we are going to develop this approach to solve some applied problems of medical diagnostics.

Acknowledgments. The study was supported by the Russian Foundation for Basic Research, projects 19-07-00397 and 19-01-00308, by the Russian Academy of Science (the Program of basic research), project 0314-2019-0015, and by the Russian Ministry of Science and Education under the 5–100 Excellence Programme.

References

1. Lin, C., Mailhes, C., Tourneret, J.-Y.: P- and T-wave delineation in ECG signals using a Bayesian approach and a partially collapsed Gibbs sampler. IEEE Trans. Biomed. Eng. **57**(12), 2840–2849 (2010)
2. Akhbari, M., Shamsollahi, M.B., Sayadi, O., Armoundas, A.A., Jutten, C.: ECG segmentation and fiducial point extraction using multi hidden Markov model. Comput. Biol. Med. **79**, 21–29 (2016)
3. Zhang, D., Zuo, W., Wang, P.: Computational Pulse Signal Analysis. Springer, Singapore (2018). https://doi.org/10.1007/978-981-10-4044-3
4. Kel'manov, A.V., Khamidullin, S.A.: A posteriori detection of a given number of truncated subsequences in a quasiperiodic Sequence. Pattern Recogn. Image Anal. **10**(4), 500–513 (2000)
5. Kel'manov, A.V., Khamidullin, S.A.: Posterion detection of a given number of identical subsequences in a quasi-periodic sequence. Comp. Math. Math. Phys. **41**(5), 762–774 (2001)
6. Kel'manov, A.V., Jeon, B.: A posteriori joint detection and discrimination of pulses in a quasiperiodic pulse train. IEEE Trans. Signal Process. **52**(3), 645–656 (2004)
7. Kel'manov, A.V., Khamidullin, S.A.: A posteriori detection of a quasiperiodically recurring fragment in numerical sequences in the presence of noise and data loss. Pattern Recogn. Image Anal. **14**(3), 421–434 (2004)
8. Kel'manov, A.V., Mikhailova, L.V.: A posteriori joint detection of reference fragments in a quasi-periodic sequence. Comp. Math. Math. Phys. **48**(5), 850–865 (2008)
9. Kel'manov, A.V.: Off-line detection of a quasi-periodically recurring fragment in a numerical sequence. Proc. Steklov Inst. Math. **263**(2 Suppl.), S84–S92 (2008)
10. Kel'manov, A.V., Mikhailova, L.V.: Joint detection of a given number of reference fragments in a quasi-periodic sequence and its partition into segments containing series of identical fragments. Comp. Math. Math. Phys. **46**(1), 165–181 (2006)
11. Voskoboynikova, G., Khairetdinov, M.: Numerical modeling of posteriori algorithms for the geophysical monitoring. CCIS **549**, 190–200 (2015)
12. Carter, J.A., Agol, E., et al.: Kepler-36: a pair of planets with neighboring orbits and dissimilar densities. Science **337**(6094), 556–559 (2012)
13. Carter, J.A., Agol, E.: The quasiperiodic automated transit search algorithm. Astrophys. J. **765**(2), 132 (2013)

14. https://www.stepwards.com/?page_id=24147
15. Rajni, R., Kaur, I.: Electrocardiogram signal analysis - an overview. Int. J. Comput. Appl. **84**(7), 22–25 (2013)
16. https://www.comm.utoronto.ca/~biometrics/PPG_Dataset/index.html
17. Shelley, K., Shelley, S.: Pulse oximeter waveform: photoelectric plethysmography. In: Carol, L., Hines, R., Blitt, C. (eds.) Clinical Monitoring, pp. 420–428. W.B. Saunders Company, Philadelphia (2001)

Equity-Linked Notes Portfolio Optimization

Lev Petrov$^{(\boxtimes)}$ and Yulia Polozhishnikova

Plekhanov Russian University of Economics, Moscow, Russia
`lfp@mail.ru`, `polozhishnikova@mail.ru`

Abstract. This paper considers pricing equity-linked notes (ELN) portfolio and related portfolio optimization. ELNs are the derivative instruments which can be viewed as bonds with floating coupons. The floating coupon is represented in terms of an embedded option that depends on the behavior of a certain underlying asset or a basket of them. We provide the new optimization problem by hyperbolic absolute risk aversion (HARA) utility function approach. We obtain the solution of this problem in terms of a dynamic programming equation.

Keywords: Equity-linked note · Risk utility function · Admissible strategy · Dynamic programming equation · Hamilton-Jacobi-Bellman equations

1 Introduction

The use of exotic investment instruments became financial institutions' practice. The corresponding mathematical models were considered for example, in the papers [1–6]. This paper is addressed to the issue of pricing equity-linked notes (ELN) and the optimization of the corresponding portfolio. Notice that when ELN is traded at an exchange, it is known as ETN. ELN is a financial debt instrument designed as unsecured bond with fixed and float coupon parts. The last one depends on the behavior of the underlying asset and represents an embedded option. ELNs have a wide variety of intrinsic options because they are actively-traded and well-customized instruments with a long history. First ELNs appeared in the 1980s in U.S. and spread over the European market in the period of low interest rates [7]. Today ELN are becoming popular in developing countries: for instance, in the Russian market due to the same interest rate recession. The most actively traded ELN is iPath S&P 500 Volatility Index (VIX) Short-Term Futures ELN (VXX) issued by Barclays. VXX routinely trades $1 billion worth of shares per day or more [8]. ELNs are often developed for high-net-worth investors who use it for special purposes, so payos can be quite exotic to reflect different investment goals. Financial classification of ELNs intends to distinguish them by goals of investing. We briefly provide this one to introduce ELNs variety.

© Springer Nature Switzerland AG 2020
M. Jaćimović et al. (Eds.): OPTIMA 2019, CCIS 1145, pp. 102–114, 2020.
https://doi.org/10.1007/978-3-030-38603-0_8

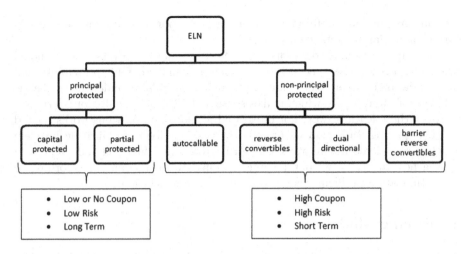

Fig. 1. Financial ELNs classification

ELNs are represented by two main classes – principle protected and non-principle protected notes (see Fig. 1) Capital protected notes return nominal value and fixed coupon together with floating coupon in case of the positive performance of underlying. Partially protected note returns slightly less than the nominal value, as a rule, 95% of it, and also floating coupon under favorable circumstances. Both types of ELNs carry low risk and have a simple payout structure. Autocallable notes are notes with an early redemption option. If the reference underlying asset is at or above the predetermined barrier on a fixed observation date, the note terminates and pays out the notional of the note multiplied by the yield of the underlying asset. Note that payoff may be less than nominal. Reverse convertibles comprise issuer's put option: if the price of the underlying asset dips below a predetermined price, the issuer returns nominal multiplied by the yield of the underlying asset. Dual directional notes contain two underlying assets. At the time of expiration, the issuer has the right to choose the yield of "worst" underlying performance and return nominal multiplied chosen yield. Barrier reverse convertibles have the same downside exposure as mentioned earlier in case of the underlying asset breaks down a predefined level.

Rational management of ELNs portfolio has become an actual issue as a result of accelerated structured market development. This boom spread on the retail sector and entailed some regulator reaction as PRIIPs (packaged retail investment and insurance products) document establishment. In 2018 Financial vehicle corporation (FVC) noticed: "For those investors seeking equity-like returns at controlled risk it is clear that over this time period structured products actually outpaced the equity market and clearly outperformed one of the most well-known funds in the recently popular absolute return fund sector." Due to an inappropriate evaluation of the portfolio of auto-callable notes, Natixis made the loss of $296 million in December 2018. Summarize it becomes obvious that precious pricing and careful management of ELNs portfolio are critical issues in current financial markets environment.

The paper is organized as follows. In Sect. 2 we obtain ELNs pricing model which comprises ELNs special features such as issuer's default probability and the correlation between underlying equity and rate. It also includes the stochastic behavior of interest rate which makes sense as ELN's are usually long-term. As we know these features were not integrated in models in previous works related to ELNs pricing. In Sect. 3 we provide the new portfolio optimization problem which is considered in two ways: though HARA utility function and Markowitz approaches. In Sect. 4 we obtain programming dynamic equation according to the optimization problem. Section 5 is devoted to conclusions.

2   Pricing Model

As noted earlier ELNs could be represented as a bond with an embedded option. In most studies, the fair values of these parts are evaluated separately. To determine the final price, the resulting estimates are summarized.

A standard approach of pricing is described in [1]. The paper considers the most common type of ELN - with an embedded European call option. A zero-coupon bond is estimated by discounting the nominal value along the risk-free Swedish Treasury rate curve. An embedded option is priced using the Black-Scholes model. This approach includes many drawbacks. For example, it does not take into account such features as the probability of default of the counterparty, the stochastic nature of interest rates, the possible correlation between rates and the underlying asset of the embedded option, the issuer's commission. In addition, the Black-Scholes model itself has significant lack, which leads to errors in determining the value of options. The assumption that stock returns have a normal distribution is contrary to empirical research: the distribution of returns is skewed and has heavy tails. Especially strong discrepancies between theoretical and real prices are observed for options with long-term maturities.

In [2] an approach for modeling the counterparty default risk and taken into account the stochastic behavior of interest rates is proposed. The authors used the classical Merton model in which the risk of default is represented as the American put option acquired by the issuer on its own assets. To describe the movement of interest rates, the authors use the model of Cox-Ingersoll-Ross. In [3] the authors used the Vasichek stochastic interest rates model and took into account the correlation between rates and the underlying asset. Analytical formula for a fair price of ETN is derived for this model. Scenario analysis is a non-standard approach for ELNs fair price estimation is represented in [4]. The approach of modelling stochastic credit spread which is added to the discount rate is developed in [9].

Further, we provide the pricing model which combines different models described above to catch up the features inherent ELNs such as a counterparty risk of default and a potential correlation between the interest rate and underlying asset. We start with the process for the underlying asset for which the

Black-Scholes model can be applied. Thus the underlying asset follows the diffusion process:

$$dS = \mu S dt + \sigma S dW_1, \tag{1}$$

where μ is the drift, σ is the volatility of the underlying and dW_1 is a random walk. Let the continuously variable spot interest rate $r(t)$, is described by the Vasicek model:

$$dr = (a - br)dt + cdW_2, \tag{2}$$

where a is the speed of the reversion to the mean, b is the long-term level of the mean, c is the rate volatility and dW_2 is a random walk. Assume that the correlation between interest rate and underlying exist and equal $\rho = < dW_1, dW_2 >$. We also assume that embedded option in ELN C is a function of underlying value, level of spot interest rate and time. If we denote the time at which the note matures by T, the valuation moment by t and the time to maturity by $\tau = T - t$, than with [5] we arrive at the PDE:

$$\frac{\partial C}{\partial t} = \frac{1}{2}\sigma^2 S^2 \frac{\partial^2 C}{\partial S^2} + \rho \sigma c S \frac{\partial^2 C}{\partial S \partial r} + \frac{1}{2}c^2 \frac{\partial^2 C}{\partial r^2} + (r - D_0)S\frac{\partial C}{\partial S} + (a - br)\frac{\partial C}{\partial r} - rC \tag{3}$$

The solution was given in [10] by Greens function:

$$C(S, r, \tau) = \int\limits_{-\infty}^{\infty} \int\limits_{0}^{\infty} G(S, \tilde{S}, r, \tilde{r}, \tau)C_0(\tilde{S}, \tilde{r})d\tilde{S}d\tilde{r}, \tag{4}$$

where D_0 is a constant dividend yield and $C_0(S, r)$ is conditional deterministic final ELN's embedded option payoff at maturity (means that it is certain in conditions that S and r are equal $\tilde{S}$ and $\tilde{r}$ respectively at maturity). The expression for G can be found in [8]. In this approach counterparty probability of default is not taken into account. Let us consider the other approach.

Now let us assume that the intensity λ_t of the Poisson process governing the default specified by the following mean reverting square-root diffusion process:

$$d\lambda = (\alpha - \beta\lambda)dt + \nu\sqrt{\lambda}dW_3, \tag{5}$$

where α and β have the same meaning as in (2) and ν is the volatility multiplier. Due to earlier designation, we denote $C_0(S, r)$ as the final ELN's payoff at maturity and now is not deterministic but stochastic. Let denote the loss given default rate as R. In the event of default ELN's holder receives just a part of the principal (recovery rate multiplied by principal) and nothing more. In the case of counterparty's survival, the holder receives the principle plus embedded option payoff at maturity. The value of ELN $P(S, r, t)$, adjusted by the probability of default, can be expressed as:

$$P(S, r, t) = E_t[C_0(S, r)]e^{-\int\limits_{t}^{t+\tau} r_u + \lambda_u du} + (1 - R)N\int\limits_{t}^{t+\tau} \lambda_u e^{-\int\limits_{t}^{v} r_v + \lambda_v dv} du. \tag{6}$$

The first term refers to the event that counterparty would not default until maturity and the second term - to the case of default. Let us assume that the stochastic processes for the spot interest rate and intensity of default are independent. Then (6) could be rewritten as:

$$P(S,r,t) = C(S,r,t)e^{-\int_t^{t+\tau} \lambda_u du} + (1-R)N \int_t^{t+\tau} \lambda_u e^{-\int_t^{v} r_v + \lambda_v dv} du, \qquad (7)$$

as the value of the embedded option $C(S,r,\tau)$ is equal to the discounted expected value of the final payoff. In [11] it is shown that Eq. (7) can be written as

$$P(S,r,t) = C(S,r,t)A(t+\tau)e^{B(t+\tau)} +$$

$$+(1-R)N \int_t^{t+\tau} e^{B(u)\lambda} Z(u)(G(u) + H(u)\lambda)du, \qquad (8)$$

where λ is the current intensity (at time t) and

$$A(t,u) = e^{\frac{\alpha(\beta+\phi)}{\sigma^2}(u-t)} \left(\frac{1-\kappa}{1-\kappa e^{\phi(u-t)}} \right)^{\frac{2\alpha}{\sigma^2}},$$

$$B(t,u) = \frac{\beta-\phi}{\sigma^2} + \frac{2\phi}{\sigma^2(1-\kappa e^{\phi(u-t)})},$$

$$Z(t,u) = E_t \left[e^{-\int_t^{u} r_v dv} \right],$$

$$G(t,u) = \frac{\alpha}{\phi} \left(e^{\phi(u-t)} - 1 \right) A(u) \left(\frac{1-\kappa}{1-\kappa e^{\phi(u-t)}} \right), \qquad (9)$$

$$H(t,u) = e^{\frac{\alpha(\beta+\phi\sigma^2)}{\sigma^2}(u-t)} \left(\frac{1-\kappa}{1-\kappa e^{\phi(u-t)}} \right)^{\frac{2\alpha}{\sigma^2}+2},$$

$$\phi = \sqrt{2\sigma^2 + \beta^2},$$

$$\kappa = \frac{\beta+\phi}{\beta-\phi}.$$

Let us combine two models given above. So our combined model comprise this system of the processes:

$$\begin{cases} dS = \mu S dt + \sigma S dW_1, \\ dr = (a-br)dt + c dW_2, \\ d\lambda = (\alpha - \beta\lambda)dt + \nu\sqrt{\lambda}dW_3, \end{cases} \qquad (10)$$

where $\rho = <dW_1, dW_2>$ and the processes in the pairs (W_1, W_3) and (W_2, W_3) are independent.

To derive the closed-form solution we can substitute the expression for $C(S,r,\tau)$ given in (4) into (8). Also we could calculate $Z(u) = E_t\left[e^{-\int_t^{u} r_v dv} \right]$

using the form of interest rate equation (Vasichek model). Note that $Z(u)$ is a price of zero-coupon bond for which analytic expression exists under Vasichek model [12] and equals to:

$$Z(t,u) = E_t\left[e^{-\int_t^u r_v dv}\right] = P(t,u)e^{-r(t)R(t,u)}, \tag{11}$$

where

$$P(t,u) = \frac{(1-e^{-b(u-t)})}{b},$$

$$R(t,u) = e^{\left(\frac{a}{b}-\frac{c^2}{2b^2}\right)(P(t,u)-u+t)-\frac{c^2}{4b}P^2(t,u)}. \tag{12}$$

Thus substituting these expression to (8) we obtain closed form-solution for the ELNs fair value:

$$P(S,r,t) = A(t+\tau)e^{B(t+\tau)} \cdot \int_{-\infty}^{\infty}\int_0^{\infty} G(S,\tilde{S},r,\tilde{r},\tau)C_0(\tilde{S},\tilde{r})d\tilde{S}d\tilde{r}+$$

$$+ (1-R)N\int_t^{t+\tau} e^{B(u)\lambda}P(t,u)e^{-r(t)R(t,u)}(G(u)+H(u)\lambda)du. \tag{13}$$

The first term in this formula corresponds to the assessment of the embedded option fair value, the second term reflects the counterparty's default risk.

3   ELN Portfolio Optimization Problem

This section is devoted ELNs portfolio optimization problem. We consider the dynamic portfolio management task, where at each point of time we rebalance the portfolio weights to maintain optimal portfolio criteria. Optimal portfolio criteria provided by the utility function approach, where different utility functions respond to special requirements over wealth. The framework of our pricing models of the ELN's returns is described by stochastic volatility processes so due to the problem of HARA utility maximization in a stochastic framework is often preferred we choose this one to construct the optimization problem. HARA utility function $U(x)$ class exhibit hyperbolic absolute risk aversion and follows:

$$U(x) = \frac{1-\gamma}{\gamma}\left(\frac{ax}{1-\gamma}+b\right)^{\gamma}, \tag{14}$$

where $a > 0$ and $\frac{ax}{1-\gamma}+b > 0$.

Our approach is related to methods for solving stochastic optimization problems in terms of viscous solutions proposed by Fleming and Sheu [12–16] for other problems. We consider the portfolio of n equity-linked notes with prices $P_1(t)$, $P_2(t)$, ..., $P_n(t)$ and risk-less bond $B(t)(d \ln B(t) = r(t)dt))$. We denote $u(t) = (u_1(t), u_2(t), ..., u_n(t))$ as the optimal time-dependent weight function such as yield of investor's wealth $P(t)$ at time t would be:

$$d \ln P(t) = u_1(t)d \ln P_1(t) + u_2(t)d \ln P_2(t) + ... + u_n(t)d \ln P_n(t)$$

$$+ (1 - \sum_{i=1}^{n} u_i(t))r(t)dt. \tag{15}$$

We assume that initial wealth $P(0) > 0$. Let us assume that the required holding period is T. According to the special case of HARA optimization problem with $a = 1$ and (b = 0) we want to maximize $U'(t) = \gamma^{-1}(1 - \gamma)^{\gamma-1}V(T)^\gamma$. As in [10] we also require the strategy $u(t)$ to be admissible which means that $u_i(t)$ is measurable via the filtration associated with portfolio prices dynamics $(\Omega, \mathfrak{F}, \mathfrak{F}_{t \geq 0})$ and the second moment of $u_i(t)$ is finite. Then the optimization problem can thus be constructed this way:

$$\max_{u(t) \in \Pi} E[U'(P(T))], \tag{16}$$

which is equivalent to:

$$\max_{u(t) \in \Pi} E[(P(T))^\gamma], \tag{17}$$

To compare HARA approach with the classical Markowitz optimization problem we provide possible mean-variance task: we balance weights to minimize the portfolio variance keeping target expected return. Now we assume static portfolio management task where portfolio vector of weights $w = (w_1, w_2, ..., w_n)$ is constant. As we operate in terms of stochastic volatility processes, we should modify classical Markowitz optimization problem by minimizing the expected portfolio variance:

$$\begin{cases} R^T w = \mu, \\ w^T \Sigma w \to min, \end{cases}$$

where R is a vector of expected returns, μ is required return and Σ is provided by:

$$\Sigma = E[Cov(\frac{dP_1(T)}{P_1(T)}, \frac{dP_2(T)}{P_2(T)}, ..., \frac{dP_n(T)}{P_n(T)})], \tag{18}$$

where Σ is the expectation of the covariance matrix.

4   Transformation the Optimization Problem to the Dynamic Programming Equation

The general theory of dynamic financial portfolio optimization problem is based on Hamilton-Jacobi-Bellman (HJB) equation in stochastic settings. Let us assume that we have stochastic process $x(t)$ determined by the stochastic differential equation:

$$dx(t) = \mu(x(t), u(t))dt + \sigma(x(t), u(t))dW(t), \tag{19}$$

where $u(t)$ is admissible strategy determined above. Let us consider the following problem:

$$\begin{cases} F(x_0) = \max_{u(t) \in U} E[\int\limits_0^{\infty} e^{-rt} f(x(t), u(t)dt], \\ x(0) = x_0. \end{cases} \tag{20}$$

Then according to HJB framework, this task is equivalent to partial derivation equation:

$$rF(x) = \max_{u(t) \in U} h(x, u) + F_x \mu(x, u) + \frac{1}{2} F_{xx}(x) \sigma^2(x, u). \tag{21}$$

The Eq. (21) could be solved numerically. In our case $P(t)$ process is considered as $x(t)$ and our first challenge is to obtain $dP(t)$ to apply HJB framework theory. We can derive $dP(t)$ based on Ito's lemma which suggested following. Let $x(t) \in R^m$ is multivariate stochastic process:

$$dx_i = \mu_i(x)dt + \sigma_i(x)dW_i. \tag{22}$$

Than for any $f(x) \in C^2$:

$$df(x) = \left(\sum_{i=1}^{n} \mu_i(x) \frac{\partial f(x)}{\partial x_i} + \frac{1}{2} \sum_{i=1}^{m} \sum_{j=1}^{n} \sigma_{ij}^2 \frac{\partial^2 f(x)}{\partial x_i x_j}\right)dt + \sum_{i=1}^{m} \sigma_i \frac{\partial f(x)}{\partial x_i} dW_i. \tag{23}$$

Let us apply the approach given above to our case. Based on Ito's lemma we obtain that:

$$dC(S, r, t) = \frac{\partial C}{\partial S} dS + \frac{\partial C}{\partial r} dr$$
$$+ \left(\frac{\partial C}{\partial t} + \frac{1}{2}\left(\frac{\partial^2 C}{\partial S^2}\sigma^2 S^2 + \frac{\partial^2 C}{\partial r^2}c^2 + 2\rho\sigma cS \frac{\partial^2 C}{\partial S\partial r}\right)\right)dt. \tag{24}$$

For both approaches introduced in Sect. 3 we should obtain the expression for $\frac{dP_i(T)}{P_i(T)}$. Based on (7):

$$P(S, r, t) = C(S, r, t)e^{-\int\limits_t^{t+\tau} \lambda_u du} + (1 - R)N \int\limits_t^{t+\tau} \lambda_u e^{-\int\limits_t^{v} r_v + \lambda_v dv} du, \tag{25}$$

and using Ito's Lemma we obtain:

$$dP_i(S,\lambda,r,t) = dC_i \cdot e^{-\int\limits_{t}^{t+\tau}\lambda_u du} - C_i e^{-\int\limits_{t}^{t+\tau}\lambda_u du}\tau d\lambda$$

$$+ (1-R)N \int\limits_{t}^{t+\tau} e^{-\int\limits_{t}^{v} r_v+\lambda_v dv} du d\lambda - (1-R)N \int\limits_{t}^{t+\tau} \lambda_u e^{-\int\limits_{t}^{v} r_v+\lambda_v dv}(u-t)du d\lambda$$

$$- (1-R)N \int\limits_{t}^{t+\tau} \lambda_u e^{-\int\limits_{t}^{v} r_v+\lambda_v dv}(u-t)du dr + C_i e^{-\int\limits_{t}^{t+\tau}\lambda_u du}(\lambda(t)-\lambda(t+\tau))dt$$

$$+ \frac{1}{2}\nu^2\lambda dt \cdot (C_i e^{-\int\limits_{t}^{t+\tau}\lambda_u du}\tau^2 - (1-R)N \int\limits_{t}^{t+\tau} \lambda_u e^{-\int\limits_{t}^{v} r_v+\lambda_v dv}(u-t)^2 du)$$

$$- \frac{1}{2}c^2(1-R)N \int\limits_{t}^{t+\tau} \lambda_u e^{-\int\limits_{t}^{v} r_v+\lambda_v dv}(u-t)^2 du dt \tag{26}$$

Substituting dC_i by (19) we obtain:

$$dP_i(S,\lambda,r,t) = Q_1 dt + Q_2 dS + Q_3 dr + Q_4 d\lambda, \tag{27}$$

where

$$Q_1 = e^{-\int\limits_{t}^{t+\tau}\lambda_u du}\frac{1}{2}(\frac{\partial^2 C_i}{\partial S^2}\sigma^2 S^2 + \frac{\partial^2 C_i}{\partial r^2}c^2 + 2\rho\sigma c S\frac{\partial^2 C_i}{\partial S\partial r})$$

$$+ C_i e^{-\int\limits_{t}^{t+\tau}\lambda_u du}(\lambda(t)-\lambda(t+\tau))$$

$$+ \frac{1}{2}\nu^2\lambda \cdot (C_i e^{-\int\limits_{t}^{t+\tau}\lambda_u du}\tau^2 - (1-R)N \int\limits_{t}^{t+\tau} e^{-\int\limits_{t}^{v} r_v+\lambda_v dv}(u-t)du$$

$$+ (1-R)N \int\limits_{t}^{t+\tau} e^{-\int\limits_{t}^{v} r_v+\lambda_v dv}(u-t)du - (1-R)N \int\limits_{t}^{t+\tau} \lambda_u e^{-\int\limits_{t}^{v} r_v+\lambda_v dv}(u-t)^2 du)$$

$$- \frac{1}{2}c^2(1-R)N \int\limits_{t}^{t+\tau} \lambda_u e^{-\int\limits_{t}^{v} r_v+\lambda_v dv}(u-t)^2 du, \tag{28}$$

$$Q_2 = \frac{\partial C}{\partial S} e^{-\int_t^{t+\tau} \lambda_u du},$$

$$Q_3 = -(1-R)N \int_t^{t+\tau} \lambda_u e^{-\int_t^v r_v + \lambda_v dv} (u-t)du,$$

$$Q_4 = -C_i e^{-\int_t^{t+\tau} \lambda_u du} \tau d\lambda + (1-R)N \int_t^{t+\tau} e^{-\int_t^v r_v + \lambda_v dv} dud\lambda$$

$$- (1-R)N \int_t^{t+\tau} \lambda_u e^{-\int_t^v r_v + \lambda_v dv} (u-t)dud\lambda. \qquad (29)$$

Substituting dS, dr, $d\lambda$ by (10) we obtain:

$$dP_i(S, \lambda, r, t) = Q_1' dt + Q_2' dW_1 + Q_3' dW_2 + Q_4' dW_3, \qquad (30)$$

where

$$Q_1' = Q_1 + \mu S Q_2 + (a - br)Q_3 + (\alpha - \beta\lambda)Q_4,$$
$$Q_2' = \sigma S Q_2,$$
$$Q_3' = c Q_3, \qquad (31)$$
$$Q_4' = \nu\sqrt{\lambda} Q_4.$$

Thus, using Ito's lemma again:

$$d\ln P_i = \frac{1}{P_i} dP_i - \frac{1}{2P_i^2}((Q_2')^2 + (Q_3')^2 + (Q_4')^2 + \rho Q_2' Q_3')dt$$

$$= (\frac{Q_1'}{P_i} - \frac{1}{2P_i^2}((Q_2')^2 + (Q_3')^2 + (Q_4')^2 + \rho Q_2' Q_3'))dt \qquad (32)$$

$$+ \frac{Q_2'}{P_i} dW_1 + \frac{Q_3'}{P_i} dW_2 + \frac{Q_4'}{P_i} dW_3.$$

Let us denote $J_{i1} = \frac{Q_1'}{P_i} - \frac{1}{2P_i^2}((Q_2')^2 + (Q_3')^2 + (Q_4')^2 + \rho Q_2' Q_3')$, $J_{i2} = \frac{Q_2'}{P_i}$, $J_{i3} = \frac{Q_3'}{P_i}$, $J_{i4} = \frac{Q_4'}{P_i}$. Substituting in (15), we derive:

$$d\ln P(t) = \sum_{i=1}^n u_i(t)(J_{i1} dt + J_{i2} dW_{i2} + J_{i3} dW_{i3} + J_{i4} dW_{i4}) + (1 - \sum_{i=1}^n u_i(t))r(t)dt$$

$$= (\sum_{i=1}^n u_i(t)J_{i1} + (1 - \sum_{i=1}^n u_i(t))r(t))dt + \sum_{i=1}^n u_i(t)(J_{i2} dW_{i2} + J_{i3} dW_{i3} + J_{i4} dW_{i4})$$

$$(33)$$

Therefore,

$$E[(P(T))^\gamma] = P(0)E[\exp \gamma (\int_0^T (\sum_{i=1}^n u_i(t)J_{i1} + (1 - \sum_{i=1}^n u_i(t))r(t))dt$$

$$+ \int_0^T \sum_{i=1}^n u_i(t)(J_{i2}dW_{i2} + J_{i3}dW_{i3} + J_{i4}dW_{i4})] \tag{34}$$

Further we fix γ to be in range $0 < \gamma < 1$. Remind that for each T we should maximize (17):

$$W(T,x) = \max_{u(t)\in\Pi} E[(P(T))^\gamma]$$

$$= P(0) \max_{u(t)\in\Pi} E[\exp (\gamma \int_0^T (\sum_{i=1}^n u_i(t)J_{i1}(t,x) + (1 - \sum_{i=1}^n u_i(t))r(t))dt \tag{35}$$

$$+ \int_0^T \sum_{i=1}^n u_i(t)(J_{i2}(t,x)dW_{i2} + J_{i3}(t,x)dW_{i3} + J_{i4}(t,x)dW_{i4})],$$

where x comprise all dependence from $S(t)$, $r(t)$, $\lambda(t)$. As in [10] we assume that $\frac{W(T,x)}{T}$ tends to a limit Λ as $T \to \infty$. Then λ is assumed as the effective long-term expected utility of wealth growth rate. As in [12] we obtain:

$$W(T,x) \sim \Lambda T + W(x), T \to \infty \tag{36}$$

Then Λ and $W(x)$ satisfy the following dynamic programming equation:

$$\Lambda = \frac{1}{2}\Delta W(x) + \frac{1}{2}|\nabla W(x)|^2$$

$$+ (\sum_{i=1}^n u_i(t)J_{i1}(t,x) + (1 - \sum_{i=1}^n u_i(t))r(t))\nabla W(x)$$

$$+ \max_{u(t)\in\Pi} [\gamma u_i(t)(J_{i2}(t,x)dW_{i2} + J_{i3}(t,x)dW_{i3} + J_{i4}(t,x)dW_{i4})\nabla W(x)$$

$$+ \gamma(\sum_{i=1}^n u_i(t)J_{i1}$$

$$+ (1 - \sum_{i=1}^n u_i(t))r(t))dt + \sum_{i=1}^n u_i(t)(J_{i2}dW_{i2} + J_{i3}dW_{i3} + J_{i4}dW_{i4})]. \tag{37}$$

The further numerical solution of dynamic programming equations are considered in [13, 15, 16].

5   Conclusion

This paper discusses the ELN portfolio optimization problem based on new ELN pricing model, which takes into account stochastic interest rates and probability of default. We suggest the portfolio optimization problem in terms of hyperbolic absolute risk aversion (HARA) utility function and obtain dynamic programming equation.The proposed approach allows constructing optimal bond portfolios with embedded options and counterparty credit risk.The proposed approach complements the existing tools for analyzing the yield and risks of financial instruments based on ELN.

In further research, we plan to take into account the impact of transaction costs on portfolio returns. This can be important when dynamically managed portfolios are rebalanced frequently. This can be realized by including the penalty function of additional costs into the optimization problem. We also plan to implement a numerical solution to the dynamic programming equations and provide a comparative analysis of the effectiveness of the proposed method and alternative algorithms for a wide range of retrospective financial data.

References

1. Frohm, D.: The Pricing of Structured Products in Sweden. Empirical Findings for Index-linked Notes Issued by Swedbank in 2005: Masters Thesis. Linkping: Linkping Institute of Technology (2007)
2. Tsun-Siou, L., Hsin-Ying, L.: The pricing of structured notes with credit risk. Investment Management and Financial Innovations (2009)
3. Mallier, R., Aglobadi, G.: Pricing equity-linked debt using the Vasichek model. Acta Math. Univ. Comenianae **71**, 211–220 (2002)
4. Hveem, M.: Portfolio management using structured products. Masters Thesis, Royal Institute of Technology, Stockholm (2011)
5. Akimkin, N., Petrov, L.: Optimization of closed-end real estate funds assets use, taking risks into account. Res., Inf., Supply Competition (RISC) **1**, 238–243 (2010). [in Russian]
6. Polozhishnikova, Y., Chernichin, A., Matveev, E.: Exchange traded notes: overview of markets and pricing methods. Financ. Risk-Manage. **4**, 276–294 (2018). [in Russian]
7. Wilkens, S., Erner, C., Roder, K.: The pricing of structured products an empirical investigation of the german market. J. Deriv. **10**(Fall), 55–68 (2003)
8. Search ETF.com Homepage, https://www.etf.com/sections/features-and-news/big-shift-volatility-etns-week?nopaging=1
9. Cserna, B., Levy, A., Wiener, Z.: Counterparty risk in Exchange Traded Notes (ETNs): theory and evidence. J. Fixed Income **23**, 76–101 (2013)
10. Mallier, R., Deakin, A.: A Greens function for a convertible bond using the Vasicek model. J. Appl. Math. **2**, 219–232 (2002)
11. Longstaf, F.A., Mithal, S., Neis, E.: Corporate yield spreads: default risk or liquidity? new evidence from the credit default swap market. J. Financ. **60**, 2213–2253 (2005)
12. Fleming, W.H., Sheu, S.J.: Risk sensitive control and an optimal investment model. Math. Financ. **10**(2), 197–213 (2000)

13. Fleming, W. H.: Optimal investment models and risk-sensitive stochatic control, IMA Mathematical Application, vol. 65, pp. 35–45, Springer, New York (1995)
14. Fleming, W.H., McEneaney, W.M.: Risk-sensitive control on an innite time horizon. SIAM J. Control Optim. **33**, 1881–1915 (1995)
15. Fleming, W.H., Rishel, R.W.: Deterministic and Stochastic Optimal Control. Springer, New York (1975). https://doi.org/10.1007/978-1-4612-6380-7
16. Fleming, W.H., Sheu, S.J.: Optimal long term growth rate of expected utility of wealth. Ann. Appl. Probab. **9**(3), 871–903 (1999)

On Optimization Problem Arising in Computer Simulation of Crystal Structures

Alla Albu[1,2], Yuri Evtushenko[1,2], and Vladimir Zubov[1,2(✉)]

[1] Dorodnicyn Computing Centre, Federal Research Center
"Computer Science and Control" of Russian Academy of Sciences, Moscow, Russia
`alla.albu@yandex.ru`, `yuri-evtushenko@yandex.ru`, `vladimir.zubov@mail.ru`
[2] Moscow Institute of Physics and Technology (National Research University),
Moscow, Russia

Abstract. Gradient optimization methods are often used to solve problems of computer simulation of the crystal structures of materials. In this case it becomes necessary to calculate the partial derivatives of the total atoms' system energy according to different parameters. Frequently the calculation of these derivatives is an extremely time-consuming and difficult problem. When describing and modeling the crystal structure of a material characterized by chemical composition, geometry, and type of chemical bond, interatomic interaction potentials are used. In this paper a special multi-step process is constructed to calculate the energy of the atoms' system in the case when the interaction of atoms is described by the Tersoff Potential. On the basis of the constructed multi-step process an algorithm for calculation the second derivatives (Hessian) of the atoms' system energy with respect to the coordinates of the atoms is presented. The above-mentioned second derivatives are provided both for the case when the material under study has a three-dimensional structure, and for the case when a two-dimensional model of a multilayer piecewise-homogeneous material is considered.

Keywords: Potentials · Energy · Gradients · Hessian

1 Introduction

When describing and modeling the crystal structure of a material characterized by chemical composition, geometry, and type of chemical bond, interatomic interaction potentials are used. The properties of crystals with a covalent connection (for example, carbon, silicon, germanium, etc.) are often described by the Tersoff Potential (see [1]). It is an example of a multiparticle potential based

---

This work was partially supported by the Russian Foundation for Basic Research (project no. 19-01-00666 A) and by Program 2 of Presidium of RAS "Mechanisms for ensuring fault tolerance in modern high-performance and highly reliable computing".

© Springer Nature Switzerland AG 2020
M. Jaćimović et al. (Eds.): OPTIMA 2019, CCIS 1145, pp. 115–126, 2020.
https://doi.org/10.1007/978-3-030-38603-0_9

on the concept of the order of connections: interaction force between two atoms is not constant, but depends on the local environment.

One of the important stages of modeling the crystal structure of the material under study is the optimization according to the coordinates of the particles, which arranges particles in positions corresponding to the minimum of the total atoms' system energy. Gradient optimization methods are often used to solve this problem (see [2–4]). At this stage it becomes necessary to calculate the partial derivatives and the Hessian of the energy of atoms' system according to the coordinates of the atoms. In the case when the energy is determined using the Tersoff Potential, the calculation of the indicated first and second derivatives is an extremely time-consuming and difficult problem.

First, as the results of the performed studies showed (see [5]), the use of the finite difference method does not allow to calculate the derivatives of the energy with acceptable accuracy. In addition, in the case of using this method it is necessary to carry out researches related to the choice of the suitable increment of atoms' coordinates at each stage of the optimization problem.

Second, the use of standard software packages for calculation of second derivatives can be associated with large restrictions on the dimension of the problem and this fact must be taken into account when a fragment of the material under study contains an enormous number of atoms.

In [6], formulas for calculating the gradient of energy of the atoms' system in the case when the interaction of atoms is described by the Tersoff Potential are given. There a comparative analysis of the calculation of above-mentioned gradient using direct differentiation formulas, using the Fast Automatic Differentiation technique (see [7]) and using the standard software package are given (see [8,9]).

In this paper a special multi-step process is constructed to calculate the energy of the atoms' system. This makes it possible to substantially simplify and increase the reliability of the calculation of the second derivatives of energy. On the basis of the constructed multi-step process, an algorithm for calculation of the exact values of the Hessian of considered cost function is proposed. The above-mentioned second derivatives are provided both for the case when the material under study has a three-dimensional structure, and for the case when a two-dimensional model of a multilayer piecewise-homogeneous material is considered.

2   Algorithm for Computing Hessian of the Cost Function

Let $\bar{r}_i = (x_{1i}, x_{2i}, x_{3i})$ are the coordinates of some lattice atom. The total interatomic energy of the atoms' system whose interaction potential is the Tersoff potential is calculated with the help of formulae $E(\bar{r}_1, \bar{r}_2, \ldots, \bar{r}_I) = \sum_{i=1}^{I} \sum_{j=1; j\neq i}^{I} V_{ij}$, where V_{ij} is the interaction potential between atoms marked i and j (i-atom and j-atom):

$$V_{ij} = f_c(r_{ij})\left(V_R(r_{ij}) - b_{ij}V_A(r_{ij})\right),$$

$$f_c(r) = \begin{cases} 1, & r < R - R_{cut}, \\ \frac{1}{2}\left(1 - \sin\left(\frac{\pi(r-R)}{2R_{cut}}\right)\right), & R - R_{cut} < r < R + R_{cut}, \\ 0, & r > R + R_{cut}, \end{cases}$$

$$V_{ij}^R = V_R(r_{ij}) = \frac{D_e}{S-1}\exp\left(-\beta\sqrt{2S}(r_{ij} - r_e)\right),$$

$$V_{ij}^A = V_A(r_{ij}) = \frac{SD_e}{S-1}\exp\left(-\beta\sqrt{\frac{2}{S}}(r_{ij} - r_e)\right),$$

$$b_{ij} = (1 + (\gamma\zeta_{ij})^\eta)^{-\frac{1}{2\eta}}, \qquad \zeta_{ij} = \sum_{k=1; k\neq i,j}^{I} f_c(r_{ik})g_{ijk}\omega_{ijk}, \qquad \omega_{ijk} = \exp(\lambda^3\tau_{ijk}),$$

$$\tau_{ijk} = (r_{ij} - r_{ik})^3, \qquad g_{ijk} = 1 + \left(\frac{c}{d}\right)^2 - \frac{c^2}{d^2 + (h - \cos\Theta_{ijk})^2}.$$

Here I is the number of atoms in the system under consideration; r_{ij} is the distance between i-atom and j-atom:

$$r_{ij} = \sqrt{(x_{1i} - x_{1j})^2 + (x_{2i} - x_{2j})^2 + (x_{3i} - x_{3j})^2};$$

Θ_{ijk} is the angle between two vectors, first vector begins at i-atom and finishes at j-atom, second vector begins at i-atom and finishes at k-atom and

$$\cos\Theta_{ijk} = q_{ijk} = \frac{r_{ij}^2 + r_{ik}^2 - r_{jk}^2}{2r_{ij}r_{ik}};$$

R and R_{cut} are known parameters, identified experimentally from geometric characteristics of substance. Tersoff Potential depends on ten parameters ($m = 10$), specific to modeled substance: $D_e, r_e, \beta, S, \eta, \gamma, \lambda, c, d, h$.

The optimization problem is to find the particles coordinates, that minimizing the summary potential energy of the considered system of atoms. In order to solve this problem by second-order methods, appears the need to determine the second derivatives of the atoms' system energy with respect to the coordinates of the atoms.

We represent the calculation of the energy of atoms' system (the interaction of atoms is described by the Tersoff Potential) in the form of a multi-step process. Let $\bar{u}$ and $\bar{z}$ be vectors having coordinates: $\bar{u}^T = [u_1, u_2, ..., u_{10}]^T, \bar{z}^T = [z_1, z_2, ..., z_{17}]^T$, where $u_1 = D_e, u_2 = r_e, u_3 = \beta, u_4 = S, u_5 = \eta, u_6 = \gamma, u_7 = \lambda, u_8 = c, u_9 = d, u_{10} = h$;

$$z_1 = \left\{ z_1^{ijk} = \sqrt{(x_{1i} - x_{1k})^2 + (x_{2i} - x_{2k})^2 + (x_{3i} - x_{3k})^2} \right\},$$

$$z_2 = \left\{ z_2^{ijk} = \sqrt{(x_{1j} - x_{1k})^2 + (x_{2j} - x_{2k})^2 + (x_{3j} - x_{3k})^2} \right\},$$

$$z_3 = \left\{ z_3^{ijk} = q_{ijk} = \frac{(z_{13}^{ij})^2 + (z_1^{ijk})^2 - (z_2^{ijk})^2}{2z_1^{ijk}z_{13}^{ij}} \right\},$$

$$z_4 = \left\{ z_4^{ijk} = f_c(z_1^{ijk}) \right\},$$

$$z_5 = \left\{ z_5^{ijk} = g_{ijk} = 1 + \left(\frac{u_8}{u_9}\right)^2 - \frac{(u_8)^2}{(u_9)^2 + (u_{10} - z_3^{ijk})^2} \right\},$$

$$z_6 = \left\{ z_6^{ijk} = \tau_{ijk} = (z_{13}^{ij} - z_1^{ijk})^3 \right\},$$

$$z_7 = \left\{ z_7^{ijk} = \omega_{ijk} = \exp((u_7)^3 z_6^{ijk}) \right\},$$

$$z_8 = \left\{ z_8^{ijk} = f_c(r_{ik})g_{ijk}\omega_{ijk} = z_4^{ijk} z_5^{ijk} z_7^{ijk} \right\},$$

$$z_9 = \left\{ z_9^{ij} = \zeta_{ij} = \sum_{k=1; k \neq i,j}^{I} z_8^{ijk} \right\},$$

$$z_{10} = \left\{ z_{10}^{ij} = \gamma\zeta_{ij} = u_6 z_9^{ij} \right\},$$

$$z_{11} = \left\{ z_{11}^{ij} = (\gamma\zeta_{ij})^{\eta} = (z_{10})^{u_5} \right\},$$

$$z_{12} = \left\{ z_{12}^{ij} = b_{ij} = (1 + z_{11}^{ij})^{-\frac{1}{2u_5}} \right\},$$

$$z_{13} = \left\{ z_{13}^{ij} = \sqrt{(x_{1i} - x_{1j})^2 + (x_{2i} - x_{2j})^2 + (x_{3i} - x_{3j})^2} \right\},$$

$$z_{14} = \left\{ z_{14}^{ij} = V_{ij}^R = \frac{u_1}{u_4 - 1} \exp\left(-u_3\sqrt{2u_4}(z_{13}^{ij} - u_2)\right) \right\},$$

$$z_{15} = \left\{ z_{15}^{ij} = V_{ij}^A = \frac{u_1 u_4}{u_4 - 1} \exp\left(-u_3\sqrt{\frac{2}{u_4}}(z_{13}^{ij} - u_2)\right) \right\},$$

$$z_{16} = \left\{ z_{16}^{ij} = f_c(z_{13}^{ij}) \right\} \equiv F(16, Z_{16}, U_{16}),$$

$$z_{17} = \left\{ z_{17}^{ij} = V_{ij} = z_{16}^{ij}(z_{14}^{ij} - z_{12}^{ij} z_{15}^{ij}) \right\},$$

$$(i = \overline{1, I}, \quad j = \overline{1, I}, \quad j \neq i, \quad k = \overline{1, I}, \quad k \neq i, j).$$

Note that each component z_l depends on a number of other components (z_l^{ij} or z_l^{ijk}). The atoms' system energy E with the help of new variables may be rewritten as follows:

$$E = E(x_{11}, x_{21}, x_{31}, ..., x_{1I}, x_{2I}, x_{3I}) = \sum_{i=1}^{I} \sum_{j=1; j \neq i}^{I} z_{17}^{ij}.$$

The derivatives with respect to the coordinates of atoms is a vector with the components:

$$\nabla E = \left(\frac{\partial E}{\partial x_{11}}, \frac{\partial E}{\partial x_{21}}, \frac{\partial E}{\partial x_{31}}, \cdots, \frac{\partial E}{\partial x_{1I}}, \frac{\partial E}{\partial x_{2I}}, \frac{\partial E}{\partial x_{3I}}\right).$$

The matrix of second derivatives has components:

$$\frac{\partial^2 E}{\partial x_{lm} \partial x_{np}}, \quad l, n = 1, 2, 3, \quad m, p = \overline{1, I}.$$

In order to determine the second derivatives of the atoms' system energy with respect to the coordinates of the atoms, there is also a need for smoothing the function $f_c(r)$. It is proposed to replace the function $f_c(r)$ as follows:

$$f_c(r) = \begin{cases} 0, & r \geq R + R_{cut}, \\ 1, & r \leq R - R_{cut}, \\ C \cdot F(\alpha r + \beta), & R - R_{cut} < r < R + R_{cut}, \end{cases}$$

where

$$F(z) = \begin{cases} f(z), & z_* \leq z \leq 0, \\ 2\, f_* - f(2\, z_* - z), & 2\, z_* \leq z \leq z_*, \end{cases}$$

$$f(z) = \begin{cases} \exp(-1/z^2), & z \neq 0, \\ 0, & z = 0, \end{cases}$$

$$C = \frac{1}{2f_*}, \quad f_* = \exp\left(-\frac{3}{2}\right), \quad z_* = -\sqrt{\frac{2}{3}}, \quad \alpha = -\frac{z_*}{R_{cut}}, \quad \beta = \frac{z_*}{R_{cut}}(R + R_{cut}).$$

For convenience of further presentation, we write the function $f_c(r)$ in the following form:

$$f_c(r) = \begin{cases} 0, & r \geq R + R_{cut}, \\ 1, & r \leq R - R_{cut}, \\ C \cdot (f_*)^{\varphi(r)}, & R \leq r < R + R_{cut}, \\ C \cdot \left(2f_* - (f_*)^{\psi(r)}\right), & R - R_{cut} < r \leq R, \end{cases}$$

where

$$\varphi(r) = \frac{R_{cut}^2}{(r - R - R_{cut})^2}, \quad \psi(r) = \frac{R_{cut}^2}{(r - R + R_{cut})^2}.$$

Derivative of function $f_c(r)$ with respect to r is calculated by the formulae:

$$\frac{\partial f_c(r)}{\partial r} = \begin{cases} 0, & r \geq R + R_{cut}, \\ 0, & r \leq R - R_{cut}, \\ C \cdot (f_*)^{\varphi(r)} \ln(f_*) \cdot \widetilde{\varphi}(r), & R \leq r < R + R_{cut}, \\ -C \cdot (f_*)^{\psi(r)} \ln(f_*) \cdot \widetilde{\psi}(r), & R - R_{cut} < r \leq R, \end{cases}$$

$$\widetilde{\varphi}(r) = \frac{-2R_{cut}^2}{(r - R - R_{cut})^3}, \quad \widetilde{\psi}(r) = \frac{-2R_{cut}^2}{(r - R + R_{cut})^3}.$$

Let us introduce the following designations: $\widetilde{z}_1, \widetilde{z}_2, ..., \widetilde{z}_{17}$ and $\widetilde{\widetilde{z}}_1, \widetilde{\widetilde{z}}_2, ..., \widetilde{\widetilde{z}}_{17}$, where

$$\widetilde{z}_s = \left\{ \widetilde{z}_s^{ijk} : \widetilde{z}_s^{ijk} = \frac{\partial z_s^{ijk}}{\partial x_{lm}} \right\}, \qquad s = \overline{1, 8},$$

$$\widetilde{z}_s = \left\{ \widetilde{z}_s^{ij} : \widetilde{z}_s^{ij} = \frac{\partial z_s^{ij}}{\partial x_{lm}} \right\}, \qquad s = \overline{9, 17},$$

$$\widetilde{\widetilde{z}}_s = \left\{ \widetilde{\widetilde{z}}_s^{ijk} : \widetilde{\widetilde{z}}_s^{ijk} = \frac{\partial^2 z_s^{ijk}}{\partial x_{lm} \partial x_{np}} \right\}, \qquad s = \overline{1, 8},$$

$$\widetilde{\widetilde{z}}_s = \left\{ \widetilde{\widetilde{z}}_s^{ij} : \widetilde{\widetilde{z}}_s^{ij} = \frac{\partial z_s^{ij}}{\partial x_{lm}\partial x_{np}} \right\}, \qquad s = \overline{9,17}.$$

Formulas for calculating first-order derivatives $\widetilde{z}_1, \widetilde{z}_2, ..., \widetilde{z}_{17}$ are given in [6].

The derivatives $\frac{\partial E}{\partial x_{lm}}$, $(l = 1,2,3, \ i = \overline{1,I})$ of the function E with respect to the coordinates of the atoms will be calculated by formulas:

$$\frac{\partial E}{\partial x_{lm}} = \sum_{i=1}^{I} \sum_{j=1; \ j \neq i}^{I} \widetilde{z}_{17}^{ij}, \qquad l = 1,2,3; \ \ m = \overline{1,I}.$$

In order to write the formulas for the second derivatives of the total atoms' system energy with respect to the coordinates of the atoms, we introduce some more notations:

$$z_s' = \left\{ z_s'^{ijk} : z_s'^{ijk} = \frac{\partial z_s^{ijk}}{\partial x_{np}} \right\}, \qquad s = \overline{1,8},$$

$$z_s' = \left\{ z_s'^{ij} : z_s'^{ij} = \frac{\partial z_s^{ij}}{\partial x_{np}} \right\}, \qquad s = \overline{9,17}.$$

These derivatives are calculated using the same formulas as the derivatives $\widetilde{z}_1, \widetilde{z}_2, ..., \widetilde{z}_{17}$, only here the index l changes to n, and the index m to p.

The formulas obtained in [6] for calculating the first derivatives $\widetilde{z}_1, \widetilde{z}_2, ..., \widetilde{z}_{17}$ with respect to the coordinates of the atoms are used as a multi-step process for obtaining the desired second derivatives. They are calculated by the formulas:

$$\widetilde{\widetilde{z}}_1^{ijk} = \frac{\partial^2 z_1^{ijk}}{\partial x_{lm}\partial x_{np}} = \begin{cases} \frac{\left(z_1^{pjk}\right)^2 - (x_{np}-x_{nk})^2}{\left(z_1^{pjk}\right)^3}, & m=i, \ n=l, \ p=m, \ m \neq k, \\[2mm] \frac{-\left(z_1^{mjp}\right)^2 - (x_{nm}-x_{np})^2}{\left(z_1^{mjp}\right)^3}, & m=i, \ n=l, \ p=k, \ m \neq k, \\[2mm] \frac{\left(z_1^{ijp}\right)^2 - (x_{np}-x_{ni})^2}{\left(z_1^{ijp}\right)^3}, & m=k, \ n=l, \ p=m, \ m \neq i, \\[2mm] \frac{-\left(z_1^{pjm}\right)^2 - (x_{np}-x_{nm})^2}{\left(z_1^{pjm}\right)^3}, & m=k, \ n=l, \ p=i, \ m \neq i, \\[2mm] \frac{(x_{lk}-x_{lm})(x_{np}-x_{nk})^2}{\left(z_1^{pjk}\right)^3}, & m=i, \ n \neq l, \ p=m, \ m \neq k, \\[2mm] \frac{(x_{lm}-x_{lk})(x_{nm}-x_{np})^2}{\left(z_1^{mjp}\right)^3}, & m=i, \ n \neq l, \ p=k, \ m \neq k, \\[2mm] \frac{(x_{li}-x_{lm})(x_{np}-x_{nm})^2}{\left(z_1^{ijp}\right)^3}, & m=k, \ n \neq l, \ p=i, \ m \neq i, \\[2mm] \frac{(x_{lm}-x_{li})(x_{ni}-x_{np})^2}{\left(z_1^{pjm}\right)^3}, & m=k, \ n \neq l, \ p=m, \ m \neq i, \\[2mm] 0, & \text{in other cases,} \end{cases}$$

$$\widetilde{\widetilde{z}}_2^{ijk} = \frac{\partial^2 z_2^{ijk}}{\partial x_{lm} \partial x_{np}} = \begin{cases} \dfrac{\left(z_2^{ipk}\right)^2 - (x_{np}-x_{nk})^2}{\left(z_2^{ipk}\right)^3}, & m=j,\ n=l,\ p=m,\ m\neq k, \\[2mm] \dfrac{-\left(z_2^{imp}\right)^2 - (x_{nm}-x_{np})^2}{\left(z_2^{imp}\right)^3}, & m=j,\ n=l,\ p=k,\ m\neq k, \\[2mm] \dfrac{\left(z_2^{ijp}\right)^2 - (x_{np}-x_{nj})^2}{\left(z_2^{ijp}\right)^3}, & m=k,\ n=l,\ p=m,\ m\neq j, \\[2mm] \dfrac{-\left(z_2^{ipm}\right)^2 - (x_{np}-x_{nm})^2}{\left(z_2^{ipm}\right)^3}, & m=k,\ n=l,\ p=j,\ m\neq j, \\[2mm] \dfrac{(x_{lk}-x_{lm})(x_{np}-x_{nk})^2}{\left(z_2^{ipk}\right)^3}, & m=j,\ n\neq l,\ p=m,\ m\neq k, \\[2mm] \dfrac{(x_{lm}-x_{lk})(x_{nm}-x_{np})^2}{\left(z_2^{imp}\right)^3}, & m=j,\ n\neq l,\ p=k,\ m\neq k, \\[2mm] \dfrac{(x_{lj}-x_{lm})(x_{np}-x_{nm})^2}{\left(z_2^{ipm}\right)^3}, & m=k,\ n\neq l,\ p=j,\ m\neq j, \\[2mm] \dfrac{(x_{lm}-x_{lj})(x_{nj}-x_{np})^2}{\left(z_2^{ijp}\right)^3}, & m=k,\ n\neq l,\ p=m,\ m\neq j, \\[2mm] 0, & \text{in other cases,} \end{cases}$$

$$\begin{aligned} \widetilde{\widetilde{z}}_3^{ijk} = &\left(z_1^{ijk} z_{13}^{ij} \cdot \left(z_1^{'ijk}(z_{13}^{ij})^2\, \widetilde{\widetilde{z}}_{13}^{ij} + 2\, z_1^{ijk}\, z_{13}^{ij}\, \widetilde{z}_{13}^{ij}\, z_{13}^{'ij} + (z_{13}^{ij})^2\, z_1^{ijk}\, \widetilde{\widetilde{z}}_{13}^{ij} \right.\right. \\ &+ (z_1^{ijk})^2\, \widetilde{z}_1^{ijk}\, z_{13}^{'ij} + 2\, z_1^{'ijk}\, z_{13}^{ij}\, \widetilde{z}_1^{ijk}\, z_1^{ijk} + \widetilde{\widetilde{z}}_1^{ijk}\, (z_1^{ijk})^2\, z_{13}^{ij} \\ &- 2\left(z_2^{ijk}\, z_1^{'ijk}\, \widetilde{z}_2^{ijk}\, z_{13}^{ij} + z_2^{ijk}\, z_1^{ijk}\, \widetilde{z}_2^{ijk}\, z_{13}^{'ij} + z_2^{'ijk}\, z_1^{ijk}\, \widetilde{z}_2^{ijk}\, z_{13}^{ij} \right. \\ &\left. + z_2^{ijk}\, z_1^{ijk}\, \widetilde{\widetilde{z}}_2^{ijk}\, z_{13}^{ij} \right) - 3\,(z_1^{ijk})^2\, \widetilde{z}_{13}^{ij}\, z_1^{'ijk} - (z_1^{ijk})^3\, \widetilde{\widetilde{z}}_{13}^{ij} + z_1^{'ijk}\,(z_2^{ijk})^2\, \widetilde{z}_{13}^{ij} \\ &+ 2\, z_1^{ijk}\, z_2^{ijk}\, z_2^{'ijk}\, \widetilde{z}_{13}^{ij} + z_1^{ijk}\,(z_2^{ijk})^2\, \widetilde{\widetilde{z}}_{13}^{ij} + \widetilde{z}_1^{ijk}\,(z_2^{ijk})^2\, z_{13}^{'ij} \\ &+ 2\, \widetilde{z}_1^{ijk}\, z_2^{ijk}\, z_2^{'ijk}\, z_{13}^{ij} + \widetilde{\widetilde{z}}_1^{ijk}\,(z_2^{ijk})^2\, z_{13}^{ij} - 3\,(z_{13}^{ij})^2\, z_{13}^{'ij}\, \widetilde{z}_1^{ijk} \\ &\left. - (z_{13}^{ij})^3\, \widetilde{\widetilde{z}}_1^{ijk} \right) - 2\left(z_1^{ijk}\, z_{13}^{'ij} + z_{13}^{ij}\, z_1^{'ijk} \right) \times \left(z_1^{ijk}\,(z_{13}^{ij})^2\, \widetilde{z}_{13}^{ij} \right. \\ &+ z_{13}^{ij}\,(z_1^{ijk})^2\, \widetilde{z}_1^{ijk} - 2\, z_1^{ijk}\, z_{13}^{ij}\, z_2^{ijk}\, \widetilde{z}_2^{ijk} - (z_1^{ijk})^3\, \widetilde{z}_{13}^{ij} + z_1^{ijk}\,(z_2^{ijk})^2\, \widetilde{z}_{13}^{ij} \\ &\left.\left. - (z_{13}^{ij})^3\, \widetilde{z}_1^{ijk} + z_{13}^{ij}\,(z_2^{ijk})^2 \right)\right) \Big/ \left(2\,(z_1^{ijk})^2\,(z_{13}^{ij})^2 \right), \end{aligned}$$

$$\widetilde{z}_4^{ijk} = \begin{cases} 0, & z_1^{ijk} \geq R^+, \\[2ex] 0, & z_1^{ijk} \leq R^-, \\[2ex] \begin{aligned} &-H \cdot C \cdot (f_*)^{\varphi(z_1^{ijk})} \times \left[-H \cdot z_1'^{ijk} \widetilde{z}_1^{ijk} / \left(z_1^{ijk} - R^+ \right)^6 \right. \\ &+ \widetilde{z}_1^{ijk} \left(z_1^{ijk} - R - R_{cut} \right) / \left(z_1^{ijk} - R^+ \right)^4 \\ &\left. -3 \ z_1'^{ijk} \ \widetilde{z}_1^{ijk} / \left(z_1^{ijk} - R^+ \right)^4 \right], \end{aligned} & R \leq z_1^{ijk} < R^+, \\[4ex] \begin{aligned} &H \cdot C \cdot (f_*)^{\psi(z_1^{ijk})} \times \left[-H \cdot z_1'^{ijk} \widetilde{z}_1^{ijk} / \left(z_1^{ijk} - R^- \right)^6 \right. \\ &+ \widetilde{z}_1^{ijk} \left(z_1^{ijk} - R + R_{cut} \right) / \left(z_1^{ijk} - R^- \right)^4 \\ &\left. -3 \ z_1'^{ijk} \ \widetilde{z}_1^{ijk} / \left(z_1^{ijk} - R^- \right)^4 \right], \end{aligned} & R \leq z_1^{ijk} < R^+, \end{cases}$$

where $H = 2 \ R_{cut}^2 \ln(f_*)$; $\quad R^- = R - R_{cut}$; $\quad R^+ = R + R_{cut}$;

$$\widetilde{z}_5^{ijk} = \frac{-2 \ (u_8)^2 \cdot \left((u_{10} - z_3^{ijk}) \ \widetilde{z}_3^{ijk} - z_3'^{ijk} \ \widetilde{z}_3^{ijk} \right) \left((u_9)^2 + (u_{10} - z_3^{ijk})^2 \right)}{\left((u_9)^2 + (u_{10} - z_3^{ijk})^2 \right)^3}$$
$$- \frac{8 \ (u_8)^2 \cdot \left(u_{10} - z_3^{ijk} \right)^2 z_3'^{ijk} \ \widetilde{z}_3^{ijk}}{\left((u_9)^2 + (u_{10} - z_3^{ijk})^2 \right)^3},$$

$$\widetilde{z}_6^{ijk} = 6 \left(z_{13}^{ij} - z_1^{ijk} \right) \left(z_{13}'^{ij} - z_1'^{ijk} \right) \left(\widetilde{z}_{13}^{ij} - \widetilde{z}_1^{ijk} \right) + 3 \left(z_{13}^{ij} - z_1^{ijk} \right)^2 \left(z_{13}'^{ij} - z_1'^{ijk} \right),$$

$$\widetilde{z}_7^{ijk} = (u_7)^3 \exp\left((u_7)^3 \ z_6^{ijk} \right) \left((u_7)^3 \ z_6'^{ijk} \ \widetilde{z}_6^{ijk} + \widetilde{z}_6^{ijk} \right),$$

$$\widetilde{z}_8^{ijk} = \widetilde{z}_4^{ijk} \ z_5^{ijk} \ z_7^{ijk} + \widetilde{z}_4^{ijk} \ z_5'^{ijk} \ z_7^{ijk} + \widetilde{z}_4^{ijk} \ z_5^{ijk} \ z_7'^{ijk} + \widetilde{z}_5^{ijk} \ z_4^{ijk} \ z_7^{ijk}$$
$$+ \widetilde{z}_5^{ijk} \ z_4'^{ijk} z_7^{ijk} + \widetilde{z}_7^{ijk} \ z_5^{ijk} \ z_4'^{ijk} + \widetilde{z}_7^{ijk} \ z_5^{ijk} z_4^{ijk} + \widetilde{z}_5^{ijk} \ z_7'^{ijk} \ z_4^{ijk} + \widetilde{z}_7^{ijk} \ z_4^{ijk} \ z_5'^{ijk},$$

$$\widetilde{z}_9^{ij} = \sum_{k=1; k \neq i,j}^{I} \widetilde{z}_8^{ijk},$$

$$\widetilde{z}_{10}^{ij} = \widetilde{z}_9^{ij} \ u_6,$$

$$\widetilde{z}_{11}^{ij} = u_5 \left(\widetilde{z}_{10}^{ij} \cdot (z_{10}^{ij})^{u_5 - 1} + (u_5 - 1) \cdot (z_{10}^{ij})^{u_5 - 2} \cdot z_{10}'^{ij} \cdot \widetilde{z}_{10}^{ij} \right),$$

$$\widetilde{z}_{12}^{ij} = -\frac{1}{2 \ u_5} \left(\widetilde{z}_{11}^{ij} \left(1 + z_{11}^{ij} \right)^{-\frac{1}{2 \ u_5} - 1} - \left(\frac{1}{2 \ u_5} + 1 \right) \left(1 + z_{11}^{ij} \right)^{-\frac{1}{2 \ u_5} - 2} \cdot z_{11}'^{ij} \widetilde{z}_{11}^{ij} \right),$$

$$\widetilde{\widetilde{z}}_{13}^{ij} = \frac{\partial^2 z_{13}^{ij}}{\partial x_{lm} \partial x_{np}} = \begin{cases} \dfrac{\left(z_{13}^{pj}\right)^2 - (x_{np} - x_{nj})^2}{\left(z_{13}^{pj}\right)^3}, & m = i, \ n = l, \ p = m, \ j \neq p, \\[3mm] \dfrac{-\left(z_{13}^{mp}\right)^2 - (x_{nm} - x_{np})^2}{\left(z_{13}^{mp}\right)^3}, & m = i, \ n = l, \ p = j, \ m \neq p, \\[3mm] \dfrac{\left(z_{13}^{ip}\right)^2 - (x_{np} - x_{ni})^2}{\left(z_{13}^{ip}\right)^3}, & m = j, \ n = l, \ p = m, \ i \neq p, \\[3mm] \dfrac{-\left(z_{13}^{pm}\right)^2 - (x_{np} - x_{nm})^2}{\left(z_{13}^{pm}\right)^3}, & m = j, \ n = l, \ p = i, \ j \neq p, \\[3mm] \dfrac{(x_{lj} - x_{lm})(x_{np} - x_{nj})^2}{\left(z_{13}^{pj}\right)^3}, & m = i, \ n \neq l, \ p = m, \ j \neq p, \\[3mm] \dfrac{(x_{lj} - x_{lm})(x_{np} - x_{nm})^2}{\left(z_{13}^{mp}\right)^3}, & m = i, \ n \neq l, \ p = j, \ m \neq p, \\[3mm] \dfrac{(x_{li} - x_{lm})(x_{np} - x_{ni})^2}{\left(z_{13}^{ip}\right)^3}, & m = j, \ n \neq l, \ p = m, \ i \neq p, \\[3mm] \dfrac{(x_{li} - x_{lm})(x_{np} - x_{nm})^2}{\left(z_{13}^{pm}\right)^3}, & m = j, \ n \neq l, \ p = i, \ j \neq p, \\[3mm] 0, & \text{in other cases,} \end{cases}$$

$$\widetilde{\widetilde{z}}_{14}^{ij} = -u_3 \sqrt{2\,u_4} \cdot \left(z_{14}^{'ij} \cdot \widetilde{z}_{13}^{ij} + z_{14}^{ij} \cdot \widetilde{\widetilde{z}}_{13}^{ij} \right),$$

$$\widetilde{\widetilde{z}}_{15}^{ij} = -u_3 \sqrt{2/u_4} \cdot \left(z_{15}^{'ij} \cdot \widetilde{z}_{13}^{ij} + z_{15}^{ij} \cdot \widetilde{\widetilde{z}}_{13}^{ij} \right),$$

$$\widetilde{\widetilde{z}}_{16}^{ij} = \begin{cases} 0, & z_{13}^{ij} \geq R^+, \\[3mm] 0, & z_{13}^{ij} \leq R^-, \\[3mm] \begin{aligned} & -H \cdot C \cdot (f_*)^{\varphi(z_{13}^{ij})} \times \Big[-H \cdot z_{13}^{'ij} \, \widetilde{z}_{13}^{ij} / \left(z_{13}^{ij} - R^+ \right)^6 \\ & + \widetilde{\widetilde{z}}_{13}^{ij} \left(z_{13}^{ij} - R^+ \right) / \left(z_{13}^{ij} - R^+ \right)^4 \\ & - 3 \, z_{13}^{'ij} \, \widetilde{z}_{13}^{ij} / \left(z_{13}^{ij} - R^+ \right)^4 \Big], \end{aligned} & R \leq z_{13}^{ij} < R^+, \\[8mm] \begin{aligned} & H \cdot C \cdot (f_*)^{\psi(z_{13}^{ij})} \times \Big[-H \cdot z_{13}^{'ij} \, \widetilde{z}_{13}^{ij} / \left(z_{13}^{ij} - R^- \right)^6 \\ & + \widetilde{\widetilde{z}}_{13}^{ij} \left(z_{13}^{ij} - R^- \right) / \left(z_{13}^{ij} - R^- \right)^4 \\ & - 3 \, z_{13}^{'ij} \, \widetilde{z}_{13}^{ij} / \left(z_{13}^{ij} - R^- \right)^4 \Big], \end{aligned} & R^- < z_{13}^{ij} \leq R, \end{cases}$$

$$\begin{aligned} \widetilde{\widetilde{z}}_{17}^{ij} = {} & \widetilde{\widetilde{z}}_{16}^{ij} \cdot z_{14}^{ij} + \widetilde{z}_{16}^{ij} \cdot z_{14}^{'ij} - \widetilde{\widetilde{z}}_{16}^{ij} \cdot z_{12}^{ij} \cdot z_{15}^{ij} + \widetilde{z}_{14}^{ij} \cdot z_{16}^{'ij} \\ & - \widetilde{z}_{16}^{ij} \cdot z_{12}^{'ij} \cdot z_{15}^{ij} - \widetilde{z}_{16}^{ij} \cdot z_{12}^{ij} \cdot z_{15}^{'ij} + \widetilde{\widetilde{z}}_{14}^{ij} \cdot z_{16}^{ij} \\ & - \widetilde{z}_{12}^{ij} \cdot z_{15}^{'ij} \cdot z_{16}^{ij} - \widetilde{z}_{12}^{ij} \cdot z_{16}^{'ij} \cdot z_{15}^{ij} - \widetilde{z}_{12}^{ij} \cdot z_{15}^{'ij} \cdot z_{16}^{ij} \\ & - \widetilde{z}_{15}^{ij} \cdot z_{16}^{'ij} \cdot z_{12}^{ij} - \widetilde{z}_{15}^{ij} \cdot z_{12}^{'ij} \cdot z_{16}^{ij} - \widetilde{\widetilde{z}}_{15}^{ij} \cdot z_{12}^{ij} \cdot z_{16}^{ij}. \end{aligned}$$

Finally, the components of the Hessian of function E are calculated by the formula:

$$\frac{\partial^2 E}{\partial x_{lm} \partial x_{np}} = \sum_{i=1}^{I} \sum_{j=1;j\neq i}^{I} \widetilde{z}_{17}^{ij}. \tag{1}$$

In the case when a two-dimensional material model is considered, the indices l and n take only the values 1 and 2.

3 Calculation the Hessian of the Energy for a Two-Dimensional Material Model with the Unloaded Condition

The two-dimensional model of a multilayer piecewise-homogeneous material proposed in [2] and [3] is considered. In this model the material is represented as a periodic piecewise homogeneous multilayer structure in which the types of atoms in different layers may be different. This model imposes the following constraints on the structure of the layers:

1. Each layer consists of identical atoms, but different layers may consist of different atoms.
2. The distances between adjacent atoms in the same level are identical, but they may be different in different layers.
3. There is a group of K parallel layers that are periodically repeated in the direction of the axis y.
4. The number of atoms in each layer and the total number of layers are potentially unbounded. Figure 1 gives an example of the model in which a group of three layers is repeated. Each layer consists of atoms of a specific type. In this model, the position of the atoms is determined by the following parameters: h_k, $k = 1, \ldots, K$ is the distance between the layer number k and the preceding layer;

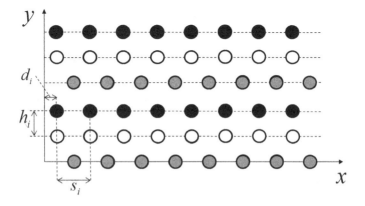

Fig. 1. Two-dimensional model of substance.

d_k, $k = 1, \ldots, K$ is the offset of the first atom in layer k with the positive abscissa relative to the zero point;

s_k, $k = 1, \ldots, K$ is the distance between the atoms in layer k.

The set of values of these parameters is called a configuration. It is required to determine the configuration corresponding to the minimum interaction energy of the atoms which enter into the simulated material fragment. The optimization problem consists in minimizing the energy $E(x)$ of the atoms' group (the potential of the interaction of atoms – the Tersoff potential) located on the K adjacent layers. The parameters of the optimization problem are the variables $x = (h_1, d_1, s_1, \ldots, h_K, d_K, s_K)$. The size of the vector x is $3K$.

Let x_i is the first coordinate of the i-th atom of a considered structure; y_i is its second coordinate. If the $i - th$ atom is an atom with an ordinal number j on the $k - th$ layer, then: $x_i = d_k + (j - 1)s_k$;

$$y_i = 0, \quad if \;\; k = 1, \qquad y_i = \sum_{m=2}^{k} h_m, \;\; if \;\; k = 2, 3, \ldots, K.$$

The formulas for calculating the second derivatives obtained in the previous section will be used here to calculate the second derivatives with respect to the parameters $h_1, d_1, s_1, \ldots, h_K, d_K, s_K$. The corresponding derivatives are calculated by the formulas:

$$\frac{\partial E}{\partial h_k} = \sum_{i=1}^{I} \frac{\partial E}{\partial y_i} \frac{\partial y_i}{\partial h_k}, \qquad \frac{\partial E}{\partial d_k} = \sum_{i=1}^{I} \frac{\partial E}{\partial x_i} \frac{\partial x_i}{\partial d_k}, \qquad \frac{\partial E}{\partial s_k} = \sum_{i=1}^{I} \frac{\partial E}{\partial x_i} \frac{\partial x_i}{\partial s_k},$$

$$\frac{\partial^2 E}{\partial h_l \partial h_k} = \sum_{i=1}^{I} \left(\sum_{p=1}^{I} \frac{\partial^2 E}{\partial y_i \partial y_p} \frac{\partial y_p}{\partial h_l} \right) \frac{\partial y_i}{\partial h_k}, \qquad \frac{\partial^2 E}{\partial d_l \partial d_k} = \sum_{i=1}^{I} \left(\sum_{p=1}^{I} \frac{\partial^2 E}{\partial x_i \partial x_p} \frac{\partial x_p}{\partial d_l} \right) \frac{\partial x_i}{\partial d_k},$$

$$\frac{\partial^2 E}{\partial s_l \partial s_k} = \sum_{i=1}^{I} \left(\sum_{p=1}^{I} \frac{\partial^2 E}{\partial x_i \partial x_p} \frac{\partial x_p}{\partial s_l} \right) \frac{\partial x_i}{\partial s_k}, \qquad \frac{\partial^2 E}{\partial d_l \partial s_k} = \sum_{i=1}^{I} \left(\sum_{p=1}^{I} \frac{\partial^2 E}{\partial x_i \partial x_p} \frac{\partial x_p}{\partial d_l} \right) \frac{\partial x_i}{\partial s_k},$$

$$\frac{\partial^2 E}{\partial h_l \partial d_k} = 0, \qquad \frac{\partial^2 E}{\partial h_l \partial s_k} = 0.$$

The second derivatives $\frac{\partial^2 E}{\partial x_i \partial x_p}$ and $\frac{\partial^2 E}{\partial y_i \partial y_p}$, which are used here, are calculated by the formula (1).

4   Conclusion

The obtained formulas are intended for solving the optimization problem which arranges particles in positions corresponding to the minimum of the total atoms' system energy using the methods of the second order. As well known the use of second-order methods lead to a significant acceleration of the optimization problem solution.

References

1. Tersoff, J.: Empirical interatomic potential for silicon with improved elastic properties. Phys. Rev. B. **38**, 9902–9905 (1988). https://doi.org/10.1103/PhysRevB.38.9902
2. Evtushenko, Y.G., Lurie, S.A., Posypkin, M.A., et al.: Application of optimization methods for finding equilibrium states of two-dimensional crystals. Comput. Math. Math. Phys. **56**, 2001–2010 (2016). https://doi.org/10.1134/S0965542516120083
3. Evtushenko, Y., Lurie, S., Posypkin, M.: New optimization problems arising in modelling of 2D-crystal lattices. AIP Conf. Proc. **1776**, 060007 (2016). https://doi.org/10.1063/1.4965341
4. Abgaryan, K.K., Posypkin, M.A.: Optimization methods as applied to parametric identification of interatomic potentials. Comput. Math. Math. Phys. **54**, 1929–1935 (2014). https://doi.org/10.1134/S0965542514120021
5. Albu, A.F.: Application of the fast automatic differentiation to the computation of the gradient of the tersoff potential. Informacionnye tekhnologii i vychislitel'nye sistemy **1**, 43–49 (2016)
6. Albu, A., Gorchakov, A., Zubov, V.: On the effectiveness of the fast automatic differentiation methodology. In: Evtushenko, Y., Jaćimović, M., Khachay, M., Kochetov, Y., Malkova, V., Posypkin, M. (eds.) OPTIMA 2018. CCIS, vol. 974, pp. 264–276. Springer, Cham (2019). https://doi.org/10.1007/978-3-030-10934-9_19
7. Evtushenko, Y.G.: Computation of exact gradients in distributed dynamic systems. Optim. Meth. Softw. **9**, 45–75 (1998). https://doi.org/10.1080/10556789808805686
8. Hogan, R.J.: Fast reverse-mode automatic differentiation using expression templates in C++. ACM Trans. Math. Softw. (TOMS) **40**(4), 26–42 (2014). https://doi.org/10.1145/2560359
9. Gorchakov, A.Y.: On software packages of fast automatic differentiation. Informacionnye tekhnologii i vychislitel'nye sistemy **1**, 30–36 (2018)

On the Complexity of Some Quadratic Euclidean Partition Problems into Balanced Clusters

Alexander Kel'manov[1,2], Vladimir Khandeev[1,2(✉)],
and Artem Pyatkin[1,2(✉)]

[1] Sobolev Institute of Mathematics, 4 Koptyug Avenue, 630090 Novosibirsk, Russia
{kelm,khan,artem}@math.nsc.ru
[2] Novosibirsk State University, 2 Pirogova Street, 630090 Novosibirsk, Russia

Abstract. We consider three problems of partitioning a finite set of N points in the d-dimensional Euclidean space into two clusters balancing the value of (1) the normalized by a cluster size sum of squared deviations from the mean, (2) the sum of squared deviations from the mean, and (3) the size-weighted sum of squared deviations from the mean. We have proved the NP-completeness of all these problems.

Keywords: Euclidean space · Balanced partition · Quadratic variance · Normalized by the cluster size · Sized-weighted · NP-completeness

1 Introduction

The subject of this study includes some discrete optimization problems. Namely, we analyze three closely related by sense problems of balanced-by-variance 2-partitioning a finite set of Euclidean points into clusters. Our aim is finding out the computational complexity status of these problems.

This research is motivated by the importance of the considered problems both for mathematical optimization theory and applications and also by the absence of any published results for them. The problems considered are related to Data analysis, Data mining and Statistics (see the next section).

The paper is organized as follows. In Sect. 2, the problems statement and motivation are given. In the same Section, some applications and interpretations of the considered problems are presented. In Sect. 3, the complexity of the problems is analyzed. Concluding remarks are given in Sect. 4.

2 Problem Statement and Related Problems

The classical mathematical statistics and a relatively new field of Data mining [1–4] have similar objectives: both are directed to the analysis of the structure of experimental data. However, there is an essential difference between these

---

A. Kel'manov—Deceased 1 December 2019.

M. Jaćimović et al. (Eds.): OPTIMA 2019, CCIS 1145, pp. 127–136, 2020.
https://doi.org/10.1007/978-3-030-38603-0_10

two subjects that is caused by the specifics of the mathematical problems solved by them. This results in different mathematical tools (methods and algorithms) created and used.

Statistics is oriented exceptionally on the analysis of homogeneous sample data (i. e. the data having the same distribution) or matching (comparing) sample data having different homogeneous distributions. Data mining sets as a goal finding out the structure of the data that, generally, have different distributions, in case when no information on the correspondence between the data and the distributions is given [1–4]. In fact, statistical methods as tools are applicable only for special cases of the various problems in Data mining.

Unfortunately, all well developed techniques of mathematical statistics turned out to be completely unusable in practice in cases when the correspondence between sample (experimental) data and distributions is absent. In such a situation, typical for Data mining, it is first necessary to find a proper (adequate) partition of the data into homogeneous groups (clusters). Only after obtaining such partition into homogeneous clusters the classical methods of mathematical statistics become correctly applicable from the mathematical point of view. The so-called exploratory search for partitions into clusters adequate to the data is one of the key problems in Data mining. The issues of clustering adequateness (i. e. the issues of correspondence of these clusterings to the experimental data) lie out of scope of this paper.

In this paper we focus on the other important problem, namely, on studying the complexity status of problems induced by the search for optimal partitions allowing at the next stage of the analysis to understand (determine, explain or interpret) a structure of experimental data. In other words, we are interested in computational complexity of the problems that should be solved for finding constructive (algorithmical) solution to the applied problems of Data mining.

There are many optimization problems of partitioning a finite set of objects (points) into clusters by various criteria. Most of the practically important Data mining problems of data clustering are NP-hard (see papers cited below). However, many of the problems have an open computation complexity status.

In the current paper we consider several such problems that are closely related to both the classical problems of statistical hypothesis verification and to some applied problems mentioned below.

Note also that all considered clustering problems are not equivalent to any of the well-known hard clustering problems such as *k-means* (or k-MSSC) [5,6], *k-median* [7,8], *k-center* [9,10], *k-Variance* [11] etc [12–15]. As far as we know, the considered problems are not also equivalent to other hard quadratic Euclidean partitioning problems studied in last years. This fact together with the facts stated above have motivated the investigation.

Throughout the paper we denoted the Euclidean norm by $\|\cdot\|$. The considered problems are stated below as decision problems.

Problem 1 (*Balanced 2-partition by the criterion of the normalized by a cluster size sum of squared deviations from the mean*). *Given*: N-element set $\mathcal{Y}$ of points in the Euclidean space of dimension d and a real number $\varepsilon > 0$.

Question: Is there a partition of the set $\mathcal{Y}$ into non-empty clusters $\mathcal{C}$ and $\mathcal{Y} \setminus \mathcal{C}$ such that

$$\left| \frac{1}{|\mathcal{C}|} \sum_{y \in \mathcal{C}} \|y - \overline{y}(\mathcal{C})\|^2 - \frac{1}{|\mathcal{Y} \setminus \mathcal{C}|} \sum_{y \in \mathcal{Y} \setminus \mathcal{C}} \|y - \overline{y}(\mathcal{Y} \setminus \mathcal{C})\|^2 \right| \le \varepsilon, \tag{1}$$

where $\overline{y}(\mathcal{C}) = \frac{1}{|\mathcal{C}|} \sum_{y \in \mathcal{C}} y$ and $\overline{y}(\mathcal{Y} \setminus \mathcal{C}) = \frac{1}{|\mathcal{Y} \setminus \mathcal{C}|} \sum_{y \in \mathcal{Y} \setminus \mathcal{C}} y$ are centroids (geometric centers) of clusters $\mathcal{C}$ and $\mathcal{Y} \setminus \mathcal{C}$ respectively?

Problem 2 (*Balanced 2-partition by the criterion of the sum of squared deviations from the mean*). *Given*: N-element set $\mathcal{Y}$ of points in the Euclidean space of dimension d and a real number $\varepsilon > 0$. *Question*: Is there a partition of the set $\mathcal{Y}$ into non-empty clusters $\mathcal{C}$ and $\mathcal{Y} \setminus \mathcal{C}$ such that

$$\left| \sum_{y \in \mathcal{C}} \|y - \overline{y}(\mathcal{C})\|^2 - \sum_{y \in \mathcal{Y} \setminus \mathcal{C}} \|y - \overline{y}(\mathcal{Y} \setminus \mathcal{C})\|^2 \right| \le \varepsilon? \tag{2}$$

Problem 3 (*Balanced 2-partition by the criterion of the size-weighted sum of squared deviations from the mean*). *Given*: N-element set $\mathcal{Y}$ of points in the Euclidean space of dimension d and a real number $\varepsilon > 0$. *Question*: Is there a partition of the set $\mathcal{Y}$ into non-empty clusters $\mathcal{C}$ and $\mathcal{Y} \setminus \mathcal{C}$ such that

$$\left| |\mathcal{C}| \sum_{y \in \mathcal{C}} \|y - \overline{y}(\mathcal{C})\|^2 - |\mathcal{Y} \setminus \mathcal{C}| \sum_{y \in \mathcal{Y} \setminus \mathcal{C}} \|y - \overline{y}(\mathcal{Y} \setminus \mathcal{C})\|^2 \right| \le \varepsilon? \tag{3}$$

In statistics there is a well-known Fisher's criterion of dispersion comparison (F-criterion) by sample data of two distributions [16]. If the clusters $\mathcal{C}$ and $\mathcal{Y} \setminus \mathcal{C}$ are considered as samples of two normal distributions with unknown means then this criterion allows to compare (check the equality) sample dispersions

$$\frac{1}{|\mathcal{C}| - 1} \sum_{y \in \mathcal{C}} \|y - \overline{y}(\mathcal{C})\|^2 \tag{4}$$

and

$$\frac{1}{|\mathcal{Y} \setminus \mathcal{C}| - 1} \sum_{y \in \mathcal{Y} \setminus \mathcal{C}} \|y - \overline{y}(\mathcal{Y} \setminus \mathcal{C})\|^2 \tag{5}$$

of these distributions by their proportion which is close to 1 in case of equality.

Formulae (4) and (5) are known in statistics as unbiased variance estimates from sample data. In Problem 1 there are biased estimates which differ from unbiased ones only by denominators. However, this difference is not essential in asymptotic sense since both estimates are asymptotically unbiased.

It is easy to see that in Problem 1 it is required to partition the input set $\mathcal{Y}$ into two clusters by the criterion of balanced sample variances. Statistical interpretation of Problem 1 is whether an unhomogeneous sample $\mathcal{Y}$ can be partitioned into two parts (subsamples) $\mathcal{C}$ and $\mathcal{Y} \setminus \mathcal{C}$ whose sample variances differ by at most some given $\varepsilon > 0$?

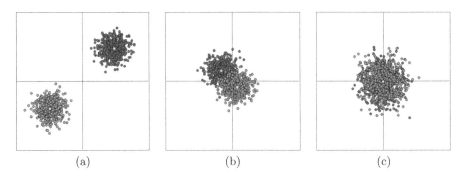

Fig. 1. Examples of two-dimensional input sets $\mathcal{Y}$ with the same spreads of the clusters: (a): centroids are far, (b): centroids are close, (c): centroids coincide.

Two-dimensional examples of input points sets are presented in Fig. 1. In these samples the input sets contain two clusters whose sum spreads and subset cardinalities are the same while the centroids are either quite far (Fig. 1a), or close (Fig. 1b), or coincide (Fig. 1c).

Problems 2 and 3 have similar sense as Problem 1, but they have different clustering criteria. In Problem 2 the sum of squared deviations from the mean (i. e. centroid) is the criterion. In Problem 3 as a criterion the size-weighted sum of squared deviations from the mean is taken. It is the same as the sum of the squared pairwise distances since for every finite set $\mathcal{Z} \subset \mathbb{R}^d$ the following easy to verify equality is true:

$$|\mathcal{Z}| \sum_{z \in \mathcal{Z}} \|z - \overline{z}(\mathcal{Z})\|^2 = \frac{1}{2} \sum_{z \in \mathcal{Z}} \sum_{x \in \mathcal{Z}} \|z - x\|^2, \tag{6}$$

where $\overline{z}(\mathcal{Z})$ is the centroid of the set $\mathcal{Z}$.

The statistical interpretations of Problems 2 and 3 as 2-partition of unhomogeneous sample $\mathcal{Y}$ are similar to the statistical interpretation of Problem 1.

Note that some other interpretations of the formulated 2-clustering problems can be given, in particular, physical or social ones. If a point of the space defines a vector (force), directed from the origin to this point, then, for example, Problem 2 can be interpreted as a search for a balanced by a sum spread 2-partition of multidirectional forces. If a coordinate of a point is treated as a numerical characteristic of the opinion on some matter of a human from some social stratum then, for instance, Problem 1 can be treated as a search for 2-partition of this stratum into groups balanced on average for the range of various opinions. Similar social or physical interpretations can be given to Problem 3 with the only difference that in this problem the desired 2-partition must be balanced both by the sum spread of the elements of the clusters and by the sizes (cardinalities) of the clusters.

Below we, in fact, prove that these easy by sense applied problems induce intractable mathematical problems. The issues of algorithmic approximability of the problems as well as the mentioned above clustering relevantness issues are out of the scopes of this paper.

3   Complexity Analysis

For finding out the complexity status of the formulated problems consider first the following problem that unifies Problems 1–3.

Problem $\Pi(g(x))$. *Given*: N-element set $\mathcal{Y}$ of points in the Euclidean space of dimension d and a real number $\varepsilon > 0$. *Question*: Is there a partition of the set $\mathcal{Y}$ into two non-empty clusters $\mathcal{C}$ and $\mathcal{Y} \setminus \mathcal{C}$ such that

$$\left| \frac{1}{2} g(|\mathcal{C}|) \sum_{y \in \mathcal{C}} \sum_{z \in \mathcal{C}} \|y - z\|^2 - \frac{1}{2} g(|\mathcal{Y} \setminus \mathcal{C}|) \sum_{y \in \mathcal{Y} \setminus \mathcal{C}} \sum_{z \in \mathcal{Y} \setminus \mathcal{C}} \|y - z\|^2 \right| \le \varepsilon? \qquad (7)$$

Taking into account (6), suppose that the weight coefficient $g(x)$ at the sums in the inequality (7) of the united problem is $1/x^2$ for Problem 1, $1/x$ for Problem 2 and 1 for Problem 3, where x is the cardinality of the corresponding cluster.

We prove the NP-completeness of Problem $\Pi(g(x))$ for each function $g(x)$ from the set $\{\frac{1}{x^2}, \frac{1}{x}, 1\}$ using the following well-known NP-complete variant of the classical problem **Bipartition** [17] that can be stated as follows.

Problem BEP (*Bipartition with Equal Parts*). *Given* a multiset of $2K$ non-negative integers $a_1, \ldots, a_{2K}$, whose sum is equal to $2W$. *Question:* is there a bipartition of this set into two multisubsets of K elements each such that the sum of the elements in each subset would be W?

The following theorem is true.

Theorem 1. *For every function* $g(x) \in \{\frac{1}{x^2}, \frac{1}{x}, 1\}$ *Problem* $\Pi(g(x))$ *is NP-complete.*

Proof. Consider an arbitrary instance of Problem **BEP**, i. e. the multiset $A = \{a_1, \ldots, a_{2K}\}$, whose elements sum up to $2W$. We may assume that $K > 3$ and $W \ge 9$ since otherwise the problem can be solved in linear time using brute force or dynamic programming, respectively.

Put $\varepsilon = 1/(3K)$ and choose rational numbers b_i, $i = 1, \ldots, 2K$, satisfying the inequalities

$$\sqrt{a_i} \le b_i \le \sqrt{a_i} + \delta, \qquad (8)$$

where

$$\delta = \frac{\varepsilon}{K(K-1)W g(K)}. \qquad (9)$$

Construct an instance of Problem $\Pi(g(x))$ by the input of Problem **BEP** in the following way. Put $N = 2K$, $d = 4K$, $\mathcal{Y} = \{y_1, \ldots, y_{2K}\}$, where for all $i = 1, \ldots, 2K$ the point y_i contains $M > 0$ (a rational parameter) in the component i and b_i in the component $2K + i$, and all other components of this point are 0.

The values of the parameter M will be determined later depending on the function $g(x) \in \{\frac{1}{x^2}, \frac{1}{x}, 1\}$.

Observe some properties of the elements from the family $\mathcal{Y}$ of points in the obtained instance of Problem $\Pi(g(x))$.

Property 1. Assume that a subset $I \subset \{1, \ldots, 2K\}$ contains the indices of points from the cluster $\mathcal{C} \subset \mathcal{Y}$. Then points from $\mathcal{C}$ satisfy the following inequality:

$$\sum_{y \in \mathcal{C}} \sum_{z \in \mathcal{C}} \|y - z\|^2 = 2|\mathcal{C}|(|\mathcal{C}| - 1)M^2 + 2(|\mathcal{C}| - 1)\sum_{i \in I} b_i^2, \tag{10}$$

that follows from the fact that in the constructed instance for every $i \neq j$, clearly,

$$\|y_i - y_j\|^2 = 2M^2 + b_i^2 + b_j^2.$$

Now note that due to (8) the bound

$$a_i \leq b_i^2 \leq a_i + \delta^2 + 2\delta\sqrt{a_i} \leq a_i + \delta(\delta + 2\sqrt{2W}) \leq a_i + W\delta \tag{11}$$

is true since $W > \delta + 2\sqrt{2W}$ when $W \geq 9$ and $\delta < \frac{1}{2}$.

Summing up inequalities (11), we obtain

Property 2. For the sum of the squares of the elements coordinates (8) the bound

$$\sum_{i=1}^{2K} b_i^2 \leq 2W + 2KW\delta = 2W(1 + K\delta) \tag{12}$$

holds.

Property 3. For every bipartition of the set $I = \{1, \ldots, 2K\}$ of indices into the subsets I_1, I_2 of the same cardinality K the inequalities

$$\left|\sum_{i \in I_1} a_i - \sum_{i \in I_2} a_i\right| - KW\delta \leq \left|\sum_{i \in I_1} b_i^2 - \sum_{i \in I_2} b_i^2\right| \leq \left|\sum_{i \in I_1} a_i - \sum_{i \in I_2} a_i\right| + KW\delta \tag{13}$$

are true, since

$$\sum_{i \in I_1} a_i \leq \sum_{i \in I_1} b_i^2 \leq \sum_{i \in I_1} a_i + KW\delta$$

and

$$\sum_{i \in I_2} a_i \leq \sum_{i \in I_2} b_i^2 \leq \sum_{i \in I_2} a_i + KW\delta$$

according to (11).

Assume now that a subset $I_1 \subset I$ contains the indices of the points from $\mathcal{C}$, and a subset $I_2 = I \setminus I_1$ contains the indices of the points from $\mathcal{Y} \setminus \mathcal{C}$ and that $|I_1| = c$. Then in Problem $\Pi(g(x))$ inequality (7) for $x = c$ due to (10) takes the following form:

$$\left|\left(c(c-1)g(c) - (2K - c)(2K - c - 1)g(2K - c)\right)M^2 \right.$$
$$\left. + (c-1)g(c)\sum_{i \in I_1} b_i^2 - (2K - c - 1)g(2K - c)\sum_{i \in I_2} b_i^2\right| \leq \varepsilon. \tag{14}$$

If a bipartition of the multiset A into two required subsets exists in Problem **BEP** then for the corresponding subsets of indices the equalities

$$c = |I_1| = |I_2| = K \tag{15}$$

and

$$\sum_{i \in I_1} a_i = \sum_{i \in I_2} a_i \tag{16}$$

hold. Then substituting (15) into (14) and taking into account (13) and (16), observe that for satisfying inequality (7) in Problem $\Pi(g(x))$ it is sufficient that

$$(K-1)g(K)\left| \sum_{i \in I_1} b_i^2 - \sum_{i \in I_2} b_i^2 \right| \le K(K-1)g(K)W\delta \le \varepsilon,$$

which is true if equality (9) holds.

So, if the required bipartition exists in Problem **BEP** then in Problem $\Pi(g(x))$ the desired clustering also exists if the parameter δ is chosen as in (9).

For proving the opposite implication, indicate the value of the parameter M for each of the three functions $g(x) \in \{\frac{1}{x^2}, \frac{1}{x}, 1\}$ in the united Problem $\Pi(g(x))$. Consider three cases.

I. If $g(x) = 1$ (which corresponds to Problem 3) then (9) implies

$$\delta = \frac{\varepsilon}{K(K-1)W}.$$

Inequality (14) in Problem $\Pi(g(x))$ turns into

$$\varepsilon \ge \left| (4Kc - 4K^2 + 2K - 2c)M^2 + (c-1)\sum_{i \in I_1} b_i^2 - (2K - c - 1)\sum_{i \in I_2} b_i^2 \right|.$$

Due to (12), estimating the sum module on the right-hand side of this inequality gives

$$\varepsilon \ge |2(K-c)(1-2K)M^2| - (c-1)\sum_{i \in I_1} b_i^2 - (2K - c - 1)\sum_{i \in I_2} b_i^2$$
$$\ge 2|K - c|(2K-1)M^2 - 4KW(K\delta + 1) \ge |K - c|M^2 - 4KW(K\delta + 1), \tag{17}$$

since $2(2K - 1) > 1$.

Choose the parameter M so that

$$M^2 > \varepsilon + 4KW(K\delta + 1).$$

Then the coefficient at M^2 in the right-hand side of (17) must be 0. Indeed, if $K \ne c$ then $|K - c| \ge 1$, and it follows from inequality (17) that

$$\varepsilon \ge M^2 - 4KW(K\delta + 1) > \varepsilon,$$

a contradiction. So, $|K - c| = 0$, and hence, $K = c$.

Further, since the elements of the multiset A are integer, if in Problem **BEP**

$$\sum_{i \in I_1} a_i \neq \sum_{i \in I_2} a_i, \tag{18}$$

then

$$\left| \sum_{i \in I_1} a_i - \sum_{i \in I_2} a_i \right| \geq 1. \tag{19}$$

Therefore, it follows from (14) in view of (13) that

$$\varepsilon \geq (K-1) \left| \sum_{i \in I_1} b_i^2 - \sum_{i \in I_2} b_i^2 \right| \geq (K-1)(1 - KW\delta) = K - 1 - \varepsilon,$$

contradicting with the condition $\varepsilon = \frac{1}{3K}$. Therefore, in Problem **BEP** equality (16) holds.

II. If $g(x) = 1/x$ (which corresponds to Problem 2), then (9) gives

$$\delta = \frac{\varepsilon}{(K-1)W} .$$

Inequality (14) becomes

$$\varepsilon \geq \left| (c - 1 - (2K - c - 1))M^2 + \frac{c-1}{c} \sum_{i \in I_1} b_i^2 - \frac{2K - c - 1}{2K - c} \sum_{i \in I_2} b_i^2 \right|.$$

Estimating the sum module on the right-hand side of this inequality using (12) gives

$$\varepsilon \geq |2(c - K)|M^2 - \sum_{i=1}^{2K} b_i^2 \geq 2|K - c|M^2 - 2W(K\delta + 1). \tag{20}$$

Let the parameter M satisfy the inequality

$$M^2 > \varepsilon + 2W(K\delta + 1).$$

Under this choice of M due to (20) we have $K = c$, as in the case I.

Similarly to the considered case I, if in Problem **BEP** there is inequality (18) then (19) holds. Therefore, (14) together with (13) gives

$$\varepsilon \geq \frac{K-1}{K} \left| \sum_{i \in I_1} b_i^2 - \sum_{i \in I_2} b_i^2 \right| \geq \frac{K-1}{K}(1 - KW\delta) = \frac{K-1}{K} - \varepsilon,$$

that contradicts the conditions $\varepsilon = \frac{1}{3K}$ and $K > 3$. Therefore, in Problem **BEP** equality (16) holds, as in the case I.

III. Finally, if $g(x) = 1/x^2$ (which corresponds to Problem 1) due to (9) we have

$$\delta = \frac{K\varepsilon}{(K-1)W} .$$

Moreover, inequality (14) has form

$$\varepsilon \geq \left| \left(\frac{c-1}{c} - \frac{2K-c-1}{2K-c} \right) M^2 + \frac{c-1}{c^2} \sum_{i \in I_1} b_i^2 - \frac{2K-c-1}{(2K-c)^2} \sum_{i \in I_2} b_i^2 \right|.$$

Due to (12), estimating the sum module on the right-hand side of this inequality gives

$$\varepsilon \geq \left| \frac{2(K-c)}{c(2K-c)} \right| M^2 - \sum_{i=1}^{2K} b_i^2 \geq \frac{|K-c|}{2K^2} M^2 - 2W(K\delta + 1). \qquad (21)$$

Introduce the parameter M so that

$$M^2 > 2K^2(\varepsilon + 2W(K\delta + 1)).$$

Then it follows from (21) that $K = c$, as in the cases I and II.

Next, as in the cases I and II above, if in Problem **BEP** inequality (18) holds, then it implies (19). Therefore from (14) in view of (13) we have

$$\varepsilon \geq \frac{K-1}{K^2} \left| \sum_{i \in I_1} b_i^2 - \sum_{i \in I_2} b_i^2 \right| \geq \frac{K-1}{K^2} (1 - KW\delta) = \frac{K-1}{K^2} - \varepsilon,$$

i. e. $K - 1 \leq 2K^2\varepsilon = 2K/3$, a contradiction with $K > 3$. Therefore, in Problem **BEP** equality (16) holds, as in the cases I and II.

So, in Problem $\Pi(g(x))$ for each function $g(x) \in \{\frac{1}{x^2}, \frac{1}{x}, 1\}$ inequality (7) yields the existence in Problem **BEP** a partition of the set A into two subsets of equal cardinality having the same sums of the elements.   □

Theorem 1 and equality (6) immediately imply the following main result of this paper:

Corollary 1. *Problems 1–3 are NP-complete.*

Clearly, the considered Problems 1–3 can be generalized into the case when the number of clusters is more than 2. In this case the question is whether the input set can be partitioned in such a way that corresponding inequalities (1), (2) and (3) would hold for all pairs of clusters. It is evident that if the number of clusters is a part of input data then these generalizations are also NP-complete. The complexity status of the parametrized case (when the number of clusters is a fixed parameter, i.e. not a part of input data) remains open for these problems.

4   Conclusion

In this paper, we have proved the NP-completeness of some quadratic Euclidean 2-partition problems of a finite set of points into balanced clusters. In addition, we have shown the close connection between these problems and some important application in Data analysis, Data mining and Statistics. Constructing algorithms with guaranteed performance for these problems is a matter of immediate prospects.

Acknowledgments. The research was supported by the Russian Foundation for Basic Research, projects 19-01-00308 and 18-31-00398, by the Russian Academy of Science (the Program of basic research), projects 0314-2019-0015 and 0314-2019-0014, and by the Russian Ministry of Science and Education under the 5–100 Excellence Programme.

References

1. Aggarwal, C.C.: Data Mining: The Textbook. Springer, Switzerland (2015). https://doi.org/10.1007/978-3-319-14142-8
2. Hastie, T., Tibshirani, R., Friedman, J.: The Elements of Statistical Learning, 2nd edn. Springer, New York (2009). https://doi.org/10.1007/978-0-387-84858-7
3. Han, J., Kamber, M., Pei, J.: Data Mining: Concepts and Techniques, 3rd edn. Morgan Kaufmann, Burlington (2012)
4. Shirkhorshidi, A.S., Aghabozorgi, S., Wah, T.Y., Herawan, T.: Big data clustering: a review. LNCS **8583**, 707–720 (2014)
5. Aloise, D., Deshpande, A., Hansen, P., Popat, P.: NP-hardness of Euclidean sum-of-squares clustering. Mach. Learn. **75**(2), 245–248 (2009)
6. Mahajan, M., Nimbhorkar, P., Varadarajan, K.: The planar k-means problem is NP-hard. Theor. Comput. Sci. **442**, 13–21 (2012)
7. Arora, S., Raghavan, P., Rao, S.: Approximation schemes for Euclidean k-medians and related problems. In: Proceedings of the 30th Annual ACM Symposium on Theory of Computing, pp. 106–113 (1998)
8. Papadimitriou, C.H.: Worst-case and probabilistic analysis of a geometric location problem. SIAM J. Comput. **10**(3), 542–557 (1981)
9. Masuyama, S., Ibaraki, T., Hasegawa, T.: The computational complexity of the m-center problems in the plane. IEEE Trans. IECE Jpn **64**(2), 57–64 (1981)
10. Hochbaum, D.S., Shmoys, D.B.: A best possible heuristic for the k-center problem. Math. Oper. Res. **10**(2), 180–184 (1985)
11. Aggarwal, H., Imai, N., Katoh, N., Suri, S.: Finding k points with minimum diameter and related problems. J. Algorithms **12**(1), 38–56 (1991)
12. Brucker, P.: On the complexity of clustering problems. Lect. Notes Econ. Math. Syst. **157**, 45–54 (1978)
13. Indyk, P.: A sublinear time approximation scheme for clustering in metric space. In: Proceedings of the 40th Annual IEEE Symposium on Foundations of Computer Science (FOCS), pp. 154–159 (1999)
14. Hansen, P., Jaumard, B., Mladenovich, N.: Minimum sum of squares clustering in a low dimensional space. J. Classification **15**, 37–55 (1998)
15. Hansen, P., Jaumard, B.: Cluster analysis and mathematical programming. Math. Programm. **79**, 191–215 (1997)
16. Snedecor, G.W., Cochran, W.G.: Statistical Methods, 8th edn. Iowa State University Press, Iowa (1989)
17. Garey, M.R., Johnson, D.S.: Computers and Intractability: A Guide to the Theory of NP-Completeness. Freeman, San Francisco (1979)

An Approximate Solution of a GNSS Satellite Selection Problem Using Semidefinite Programming

Lev Rapoport[1]([✉]) [iD] and Timofey Tormagov[1,2]([✉]) [iD]

[1] V. A. Trapeznikov Institute of Control Sciences of Russian Academy of Sciences, Moscow, Russia
`lbrapoport@gamil.com`, `tormagov@phystech.edu`

[2] Moscow Institute of Physics and Technology, Dolgoprudny, Moscow Region, Russia

Abstract. When processing multiple navigation satellite systems, including GPS, GLONASS, Galileo, Beidou, QZSS, the overall number of the pseudorange and carrier phase signals can exceed several tens. On the other hand, a much smaller number of them is usually sufficient to achieve necessary precision of positioning. Also, some parts of precise positioning algorithms, like carrier phase ambiguity resolution, are very sensitive to the problem dimension as they include the integer search. To reduce computational cost of positioning, the optimal choice of signals involved in computations should be performed. Optimization is constrained by a given number of satellite signals to be chosen for processing. This optimization problem falls into the class of binary optimization problems which are hard for precise solution. In this paper, we present approaches to an approximate solution of the optimal selection problem. After the linear relaxation of binary constraints, the relaxed problem is convex and can be transformed to semidefinite programming or second-order cone programming problems. The optimal solution of the relaxed problem can be considered as a lower bound of a combinatorial optimization problem. After rounding non-integer variables the approximate solution is obtained. As a result, two-sided bounds of the optimum are obtained. In practice, the approximate solution is very close to precise solution for most real world cases. Because the relaxed problem is convex, it can be solved efficiently.

Keywords: GNSS navigation · Second-order cone programming · Semidefinite programming · Satellite selection · GDOP

1 Introduction

Simultaneous processing of multiple Global Navigation Satellite Systems (GNSS) including multiple frequency bands leads to necessity to process a huge number of signals. The overall number of the range and carrier phase signals to be used can

This work was financially supported by the Russian Foundation for Basic Research, project 18-08-00531.

© Springer Nature Switzerland AG 2020
M. Jaćimović et al. (Eds.): OPTIMA 2019, CCIS 1145, pp. 137–149, 2020.
https://doi.org/10.1007/978-3-030-38603-0_11

exceed several tens, while a much smaller number of signals provides sufficient quality of positioning. Computational cost of positioning is limited from below by the most computationally consuming part of calculations which is either the matrix inverse, or the Cholesky factorization, or other matrix factorization, all are cubically dependent on dimensions. Integer search involved into carrier phase ambiguity resolution is part of high precision positioning. It means that the computational complexity of high precision positioning increases dramatically as the number of signal increases. Thus, the problem of choosing the optimal subset of satellites with a total number not exceeding certain value arises.

A set of satellites that is currently used for positioning often refers to as the "satellite constellation". All constellations can be evaluated by metric called GDOP (Geometric dilution of precision). GDOP reflects the influence of the geometry of satellites and satellite clocks on positioning accuracy. It depends on the number of satellites and their distribution in the sky. GDOP is defined by the formula

$$\text{GDOP} = \sqrt{\text{trace}(H^T H)^{-1}}, \tag{1}$$

where H is the matrix of directional cosines to the satellites. Let n satellites be observed and a parameter m is chosen, $m < n$. The problem is to find a constellation of no more than m satellites with minimal GDOP. Assume that all visible satellites are numbered from 1 to n and a variable x_s, $s = 1, \ldots, n$ is equal to one if the satellite with the index s is used and is equal to zero otherwise. The s-th row of matrix H consists of direction cosines to the satellite s and its time characteristics. If J is the number of GNSS systems that is used then $p = J + 3$ is the number of elements in each row of H. At least p satellites must be chosen for positioning. For one satellite system the selection problem can be formulated as follows.

Problem 1.

$$\min_{x_1, \ldots, x_n} \text{trace}(H^T \text{diag}(x_1, \ldots, x_n)H)^{-1} \tag{2}$$

subject to

$$x_s \in \{0, 1\}, \ s = 1, 2, \ldots, n, \tag{3}$$

$$p \le \sum_{s=1}^{n} x_s \le m. \tag{4}$$

The solution of the binary optimization problem 1 can be found by exhaustive search. For single satellite system (for example, for GPS only), GDOP always decreases with adding a new satellite to the constellation, see [11]. Therefore the condition (4) can be replaced by

$$\sum_{s=1}^{n} x_s = m. \tag{5}$$

In multi-constellation case, where two or more GNSS systems are used, the matrix H has different dimensions that depend on the number of GNSS-systems.

Then for two GNSS systems, for example, GPS and GLONASS, the satellite selection process can be implemented as follows.

1. Obtain the optimal solution of the *Problem* 1 for GPS and GLONASS satellites with additional conditions that at least one satellite from each systems is used.
2. Solve the problem 1 for the case of GPS satellites.
3. Solve the problem 1 for the case of GLONASS satellites.
4. Choose the solution from steps 1–3 with minimal GDOP.

In [10], it was proved that the adding the new satellite from GNSS-system that is used always decreases GDOP. But if this GNSS-system is not used then GDOP always grows. For this reason in steps 1–3 we solve the problem 1 under condition (5) that allows to reduce the computational complexity. But if the number of satellites in single systems is less than m and more than p, we must choose all satellites from this single system.

Assume that the receiver observes $n = 20$ satellites from one system (for example) and $m = 10$. Then, 184756 constellations must be tried. An effective closed-form formula for GDOP computation that is introduced in [3] allows to calculate square of GDOP for one constellation by 152 floating point operations. As result, we need to carry out more than 28 millions of floating point operations. For $n = 30$ it is more than 4 billions operations. It can be not suitable for real-time applications.

Approximate methods of the satellite selection have been considered in the recent literature. Some methods, for example, [6,7] are based on the information about elevation angles and azimuths of satellites. In [9] the linear cost function based on the directions of satellites was introduced. The method presented in [4] is based on the hypothesis that optimal subset of satellites with size k tends to share most elements with an optimal subset with size $(k - 1)$.

All known methods do not provide guaranteed lower and upper bounds of the optimal value. Some methods can be used only for definite parameter m or only for particular number J of GNSS systems. In this paper we introduce the method that use m and J as parameters of the problem. Due to linear relaxation the proposed method yields the upper bound of the approximate solution error, which is very small in practical testing.

2   Proposed Method

In proposed method we use a linear relaxation of binary constraints. Then, the relaxed problem can be transformed to semidefinite programming (SDP) or second-order cone programming problems (SOCP) and we can use primal-dual interior-point methods for the relaxed problem. After rounding the solution of the relaxed problem, we obtain an approximate solution of the satellite selection problem.

2.1   Linear Relaxation

First, *Problem* 1 is relaxed into a linear problem by replacing the binary constraint (3) by

$$x_s \in [0,1], \; s = 1, 2, \ldots, n. \tag{6}$$

Constraints (6) are weaker than (3), but (6) together with (5) (or (4)) defines the convex set.

Then we reformulate the original problem as one of the standard convex optimization problems: semidefinite programming (SDP) or second-order cone programming (SOCP). We denote by $e_j \in R^p$ the unit column vector with j-th entry being one and all other zero entries.

2.2   SDP Problem

Let $M \succeq 0$ means that the real-valued symmetric matrix M is positively semidefinite. Let $X = \mathrm{diag}(x_1, ..., x_n)$. Consider the following convex optimization problem

Problem 2.

$$\min_{x_1,\ldots,x_n,q_1,\ldots,q_p} \sum_{j=1}^{p} q_j \tag{7}$$

subject to (6), (5), and

$$\begin{bmatrix} H^T X H & e_j \\ e_i^T & q_j \end{bmatrix} \succeq 0 \; \forall j = 1, ..., p. \tag{8}$$

with $q_1, \ldots, q_p$ being auxiliary scalar variables.

Proposition 1. *Let* $(x_1^*, \ldots, x_n^*, q_1^*, \ldots, q_p^*)$ *be the optimal solution of the problem 2. Then* $(x_1^*, \ldots, x_n^*)$ *is the optimal solution of the problem (2) subject to (6) and (5).*

Proof. Note, that

$$\mathrm{trace}(H^T X H)^{-1} = \sum_{j=1}^{p} e_j^T (H^T X H)^{-1} e_j \tag{9}$$

Let auxiliary variables $q_1, \ldots, q_p$ satisfy conditions

$$q_j \geq e_j^T (H^T X H)^{-1} e_j, \; j = 1, ..., p. \tag{10}$$

Then minimization of (7) is equivalent to minimization of (2). Matrix $H^T X H$ is nonsingular because H is the matrix of directional cosines and $x_1, ..., x_n$ satisfy the conditions (6) and (5). Note that the Schur's Lemma (see [1]) guarantees that (8) is equivalent to (10).

□

2.3 SOCP Problem

Consider another convex optimization problem with w_j and t_{js} being auxiliary vectors and scalars respectively:

Problem 3.

$$\min_{x_s, w_j, t_{js}, j=1,\dots,p, s=1,\dots,n} \sum_{j=1}^{p} \sum_{s=1}^{n} t_{js} \tag{11}$$

subject to (6), (5), and

$$H^T w_j = e_j, \ w_j = (w_{j1}, \dots, w_{jn})^T, \ j = 1, \dots, p, \tag{12}$$

$$\left\| \begin{matrix} 2w_{js} \\ x_s - t_{js} \end{matrix} \right\| \le x_s + t_{js}, \ j = 1, \dots, p, \ s = 1, \dots, n. \tag{13}$$

Proposition 2. *Let $(x_1^*, \dots, x_n^*, w_1^*, \dots, w_p^*, t_{11}^*, \dots, t_{pn}^*)$ be the optimal solution of problem 3. Then $(x_1^*, \dots, x_n^*)$ is the optimal solution of problem (2) subject to (6) and (5).*

Proof. Denote $v_j = (H^T X H)^{-1} e_j$, $v_j \in R^p$, $j = 1, \dots, p$. Using (9) the problem (2) with constraints (6) and (5) can be reformulated as follows:

$$\min_{x_s, v_j, j=1,\dots,p, s=1,\dots,n} \sum_{j=1}^{p} v_j^T H^T X H v_j \tag{14}$$

subject to (6), (5), and

$$v_j = (H^T X H)^{-1} e_j, \ j = 1, \dots, p. \tag{15}$$

Define $w_j = X H v_j$, $w_j \in R^n$, $j = 1, \dots, p$. The previous problem can be transformed to

$$\min_{x_s, w_j, j=1,\dots,p, s=1,\dots,n} \sum_{j=1}^{p} w_j^T X^{-1} w_j \tag{16}$$

subject to (6), (5),

$$H^T w_j = e_j, \ j = 1, \dots, p, \tag{17}$$

$$w_j = X H v_j, \ j = 1, \dots, p. \tag{18}$$

The same way as it is done in works [1,5] we assume that if $x_s = 0$ then w_{js}^2/x_s is treated as zero if $w_{js} = 0$ and as ∞ otherwise. The conditions (18) can be omitted. To show this we fix $x_s = x_s^*$, $s = 1, \dots, n$ where x_s^* is part of optimal solution of problem (16) with conditions (6), (5), (15), and (18). The following problem

$$\min_{w_j, j=1,\dots,p} \sum_{j=1}^{p} w_j^T (X^*)^{-1} w_j \tag{19}$$

subject to
$$H^T w_j = e_j, \ j = 1, \ldots, p, \tag{20}$$

is completely determined by the solution of p problems

$$\min_{w_j} w_j^T (X^*)^{-1} w_j \tag{21}$$

subject to
$$H^T w_j = e_j, \tag{22}$$

for each j. the Karush–Kuhn–Tucker (KKT) conditions for the optimal solution $w_1^*, \ldots, w_p^*$ of the p problems (21) with (22) guarantee that there are vectors $\nu_j \in R_p$ such that

$$\frac{1}{2}(X^*)^{-1} w_j^* + H\nu_j = 0, \ j = 1, \ldots, p. \tag{23}$$

Having $v_j = -2\nu_j$ we conclude that conditions (18) are always satisfied for the optimal solution of the (14) with (6), (5), and (17). Let introduce new variables $t_{js}, \ j = 1, \ldots, p, \ s = 1, \ldots, n$ such that

$$t_{js} \geq \frac{w_{js}^2}{x_s}, \ j = 1, \ldots, p, \ s = 1, \ldots, n. \tag{24}$$

Then the problem (14) with constrains (6), (5), and (17) can be written as

$$\min_{x_s, w_j, t_{js}, j=1,\ldots,p, \, s=1,\ldots,n} \sum_{j=1}^{p} \sum_{s=1}^{n} t_{js} \tag{25}$$

subject to (6), (5) and (24). Finally, using (6) and (24) we notice that condition

$$\left\| \begin{array}{c} 2w_{j,s} \\ x_s - t_{js} \end{array} \right\| \leq x_s + t_{js}, \ j = 1, \ldots, p, \ s = 1, \ldots, n \tag{26}$$

is equivalent to

$$4w_{js}^2 + (x_s - t_{js})^2 \leq (x_s + t_{js})^2, \ j = 1, \ldots, p, \ s = 1, \ldots, n \tag{27}$$

because $t_{js} \geq 0$, $x_s \geq 0$ due to (6) and (24). Then (24) can be replaced by (26), and we arrive at the problem 3.

$\square$

2.4   Selection Algorithm

For J GNSS systems the proposed algorithm supposes calculation of GDOP for suboptimal subsets for each combination of systems. For example, in case of GPS and GLONASS ($J = 2$) we need to analyse constellations with only GPS satellites, constellations with only GLONASS satellites and GPS-GLONASS multiconstellations. In general case $2^J - 1$ subsets of GNSS systems is needed to investigate in following order.

1. Obtain the solution of problems 2 or 3 for each combination of GNSS systems with the additional condition that at least one satellite from each systems is selected.
2. Chose m satellites with maximum x_s in each case, and calculate GDOP of the solution.
3. Choose the solution with minimal GDOP.

2.5 Upper Bound of GDOP Error

Define
$$\phi(x) = \sqrt{\text{trace}(H^T X H)^{-1}}, \ X = \text{diag}(x_1, ..., x_n). \tag{28}$$

Let $\widetilde{x}$ be the value of the satellite selection vector x obtained by solving the relaxed problems 2 or 3 and $\widehat{x}$ is the value of the satellite selection vector obtained after rounding. The exact optimal solution is denoted x^*. Then $\phi(\widehat{x})$ is GDOP of the approximate solution and $\phi(x^*)$ is GDOP of the accurate solution. Due to the fact that the value of the objective function for the optimal solution of the relaxed minimization problem is always either less than or equal to such a value for the original binary optimization problem, we have

$$\phi(\widetilde{x}) \leq \phi(x^*). \tag{29}$$

The approximate method produces the solution with the same or higher GDOP value than the optimal solution. Then

$$\phi(x^*) \leq \phi(\widehat{x}). \tag{30}$$

Finally we have
$$\phi(\widetilde{x}) \leq \phi(x^*) \leq \phi(\widehat{x}). \tag{31}$$

Let Δ be the approximate solution error defined as

$$\Delta = \phi(\widehat{x}) - \phi(x^*). \tag{32}$$

Define also the gap between relaxed and approximate solutions as

$$\widetilde{\Delta} = \phi(\widehat{x}) - \phi(\widetilde{x}). \tag{33}$$

Then we have
$$\Delta \leq \widetilde{\Delta}. \tag{34}$$

3 Experimental Study

Data was processed on PC with the Intel Core i5-3230M CPU 2.60 GHz processor. The data were collected within 12 h. The number of satellites varied from 12 to 20. SCS algorithm [8] was used to solve SDP problems. ECOS algorithm [2] was applied for SOCP problems.

3.1   Results for One Epoch

The results of satellite selection for one epoch are shown in Figs. 1 and 2. The satellites constellation plots are shown in polar coordinates ρ and θ that both are measured in degrees. The value $\theta \in [0°, 360°]$ is the satellite location azimuth in projection to the local horizon. The value ρ is defined as follows: $\rho = 90° - \alpha$, where $\alpha \in [0°, 90°]$ is the satellite elevation angle. The coordinates origin corresponds to the zenith position of the satellite. The satellites with low elevation angles ($\alpha \approx 0°$) are located on a circle of the radius close to $\rho = 90°$. The plots show the position of GPS and GLONASS satellites at one epoch (time instant); 11 GPS satellites and 8 GLONASS satellites were observed. Red circles and triangles represents selected GPS and GLONASS satellites. Blue circles and triangles constitute GPS and GLONASS satellites that are not chosen for positioning.

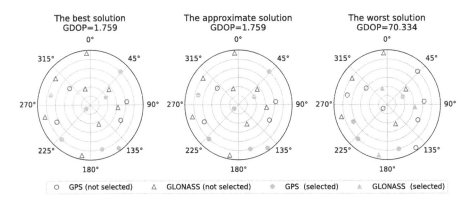

Fig. 1. The satellites constellation plots for $m = 7$. (Color figure online)

For $m = 7$ the best solution and the approximate solution are identical. In this case only GPS satellites must be chosen. Therefore the calculation for single GNSS system must be carried out. Moreover, if we use the random choice of satellites, we can obtain the worse solution with GDOP = 70.334. This constellation provides the unacceptable accuracy of positioning because GDOP > 20.

In case of $m = 8$ the exact solution and the approximate solution do not match. However the difference between GDOP of the constellations does not exceed 0.01. Then the positioning accuracy is the same. Furthermore, GDOP of the worse solution is more then 20. So random selection of 8 satellites can lead to unsatisfactory positioning results.

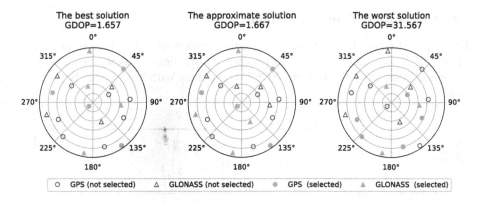

Fig. 2. The satellites constellation plots for $m = 8$. (Color figure online)

3.2   Accuracy Evaluation

The following Table 1 summarizes the accuracy results for different m taking values from 6 to 10. Second column shows the percent of cases where approximate solution coincided with the precise one. Next two columns show maximum value of Δ and its root mean square (RMS) value. Then follows two columns with the maximum and RMS values for $\widetilde{\Delta}$.

Table 1. Results of accuracy testing.

m	%	Δ_{max}	Δ_{rms}	$\widetilde{\Delta}_{max}$	$\widetilde{\Delta}_{rms}$
6	77.8	0.20	0.0364	0.37	0.0888
7	75.3	0.21	0.0355	0.29	0.0703
8	78.9	0.10	0.0117	0.15	0.0358
9	81.5	0.16	0.0150	0.19	0.0271
10	80.7	0.07	0.0070	0.09	0.0184

Results show very good accuracy performance of the proposed algorithms. The following three figures Figs. 3, 4 and 5 show plots of the Δ and $\widetilde{\Delta}$ values for three cases: $m = 6, 8, 10$.

3.3 Calculation Time Evaluation

The required computation time in seconds is presented in the Figs. 6 and 7 for the cases $m = 6$ and $m = 10$ respectively. Both figures allow a comparison of computation time of SDP and SOCP algorithms with the exhaustive search time.

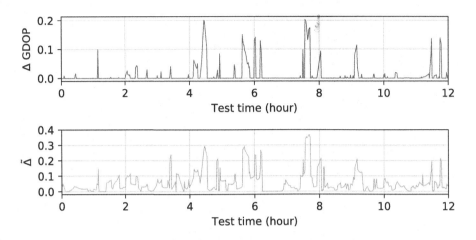

Fig. 3. The error and gap plots for $m = 6$.

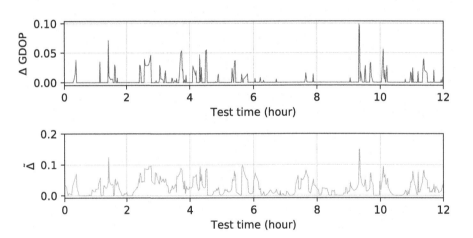

Fig. 4. The error and gap plots for $m = 8$.

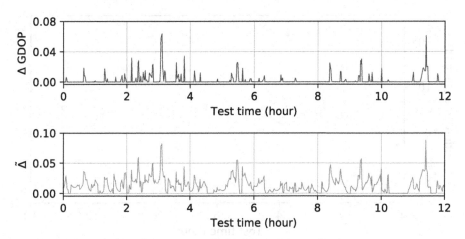

Fig. 5. The error and gap plots for $m = 10$.

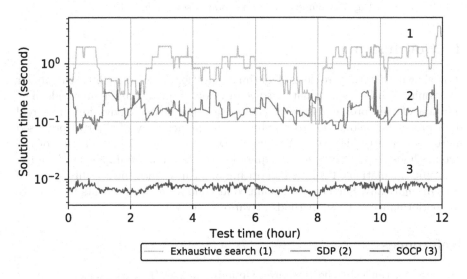

Fig. 6. Computation time comparison for $m = 6$.

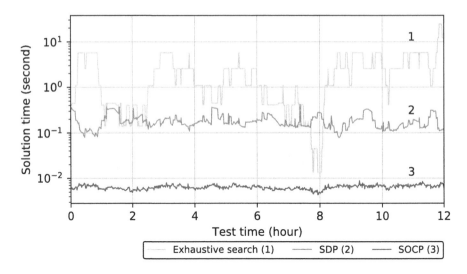

Fig. 7. Computation time comparison for $m = 10$.

4 Conclusion

It is shown that the problem of optimal selection of the satellites chosen for positioning can be approximately solved by convex programming methods. The solution of this problem is required in real-time GNSS navigation. The real data processing allows one to conclude that the accuracy of the two proposed methods is pretty high (the Δ value is low). Also, the calculation time of the two methods (especially SOCP) is quite low which allows to implement it in the real time calculations. Proposed algorithm allow us reduce the satellite selection time dramatically. Further work is aimed at conducting experiments to evaluate the performance of these methods at various receivers.

References

1. Boyd, S., Vandenberghe, L.: Convex Optimization (2004). https://doi.org/10.1017/CBO9780511804441
2. Domahidi, A., Chu, E., Boyd, S.: ECOS: an SOCP solver for embedded systems. In: 2013 European Control Conference, ECC 2013, pp. 3071–3076 (2013). https://doi.org/10.23919/ECC.2013.6669541
3. Doong, S.H.: A closed-form formula for GPS GDOP computation. GPS Solutions **13**(3), 183–190 (2009). https://doi.org/10.1007/s10291-008-0111-2
4. Liu, M., Fortin, M.A., Landry Jr, R.: A recursive quasi-optimal fast satellite selection method for GNSS receivers. In: 22nd International Technical Meeting of the Satellite Division of the Institute of Navigation 2009, ION GNSS 2009, vol. 5, pp. 3022–3032 (2009)
5. Lobo, M., Vandenberghe, L., Boyd, S., Lebret, H.: Applications of second-order cone programming. Linear Algebra Appl. **284**(1–3), 193–228 (1998). https://doi.org/10.1016/S0024-3795(98)10032-0

6. Meng, F., Wang, S., Zhu, B.: Research of fast satellite selection algorithm for multi-constellation. Chin. J. Electron. **25**(6), 1172–1178 (2016). https://doi.org/10.1049/cje.2016.10.009

7. Meng, F., Zhu, B., Wang, S.: A new fast satellite selection algorithm for BDS-GPS receivers. In: IEEE Workshop on Signal Processing Systems, SiPS: Design and Implementation, pp. 371–376 (2013). https://doi.org/10.1109/SiPS.2013.6674535

8. O'Donoghue, B., Chu, E., Parikh, N., Boyd, S.: Conic optimization via operator splitting and homogeneous self-dual embedding. J. Optim. Theory Appl. **169**(3), 1042–1068 (2016). https://doi.org/10.1007/s10957-016-0892-3

9. Park, C.W., How, J.P.: Quasi-optimal satellite selection algorithm for real-time applications. In: 14th International Technical Meeting of the Satellite Division of the Institute of Navigation (ION GPS 2001), Salt Lake City, UT, pp. 3018–3028 (2001)

10. Teng, Y., Wang, J.: New characteristics of geometric dilution of precision (GDOP) for multi-GNSS constellations. J. Navig. **67**(6), 1018–1028 (2014). https://doi.org/10.1017/S037346331400040X

11. Yarlagadda, R., Ali, I., Al-Dhahir, N., Hershey, J.: GPS GDOP metric. IEE Proc. Radar Sonar Navig. **147**(5), 259–263 (2000). https://doi.org/10.1049/ip-rsn:20000554

Dynamic Marketing Model: The Case of Piece-Wise Constant Pricing

Igor Bykadorov[1,2,3]($\boxtimes$) (iD)

[1] Sobolev Institute of Mathematics, 4 Koptyug Ave., 630090 Novosibirsk, Russia
bykadorov.igor@mail.ru
[2] Novosibirsk State University, 1 Pirogova St., 630090 Novosibirsk, Russia
[3] Novosibirsk State University of Economics and Management,
Kamenskaja Street 56, 630099 Novosibirsk, Russia

Abstract. We study a stylized vertical distribution channel where a representative manufacturer sells a single kind of good to a representative retailer. The control of the manufacturer is the price discounts, while the control of the retailer is pass-through. In the classical setting, the arising problem is quadratic with respect to wholesale price discount and pass-through. Thus, the optimal sale price is continuous. It seems elegant mathematically but not adequate economically. Therefore we assume that the controls are constant or piece-wise constant. This way, the optimal control problem reduces to the mathematical programming problem where the profit of the manufacturer is quadratic with respect to price discount level(s), while the profit of the retailer is quadratic with respect to pass-through level(s). We study the concavity property of the profits. This allows getting the optimal behavior strategies of the manufacturer and the retailer.

Keywords: Retailer · Piece-wise constant pricing · Concavity · Sale motivation

1 Introduction

To earn a reasonable profit the members of a distribution channel often adopt rather simple pricing techniques. For example, manufacturers may use cost-plus pricing, simply defining the price to be added to the desired profit margin to (variable) production costs; similarly, retailers very often use to determine shelf prices adding a fixed percentage markup to the wholesale price.

We study dynamic marketing model based on the ideas of [1–3]. The paper [4] should be recognize as the first unit work in this direction. Among many works on this subject, let us note [5–11].

In [1] we consider the concept of retailer's motivation, with the stimulation of the retailer (wholesale discount) as the manufacturer's control $\alpha(t)$. We study the case when the retailer's pass-through $\beta(t)$ is constant and maximize the

© Springer Nature Switzerland AG 2020
M. Jaćimović et al. (Eds.): OPTIMA 2019, CCIS 1145, pp. 150–163, 2020.
https://doi.org/10.1007/978-3-030-38603-0_12

manufacturer's profit with respect to $\alpha(t)$. Instead, in [2] we study the case when $\alpha(t)$ is constant and maximize the retailer's profit with respect to $\beta(t)$.

Note that in [1,2], the arising problem is quadratic with respect to wholesale price discount or pass-through. Thus, the optimal sale price is continuous. It seems elegant mathematically but not adequate economically. Indeed, it seems strange when the discounts (and therefore the prices) change continuously. In practice, the prices are piece-wise constant.

Therefore now we assume that the controls are constant or piece-wise constant[1]. This way, the optimal control problem reduces to the mathematical programming problem where the profit of the manufacturer is quadratic with respect to price discount level(s), while the profit of the retailer is quadratic with respect to pass-through level(s). In [3] we study the case of constant wholesale price discounts and constant pass-through.

The presented paper is devoted to the development of [3]. The first question is the concavity property of the profits[2]. This allows getting the optimal behavior strategies of the manufacturer and the retailer.

We prove the strict concavity of the retailer's profit with respect to pass-through levels for any fixed number of (known) switches.

2   Two Marketing Models

2.1   Pricing

As in [1] and [2], let us consider a vertical distribution channel. There are a manufacturer, retailer and consumer on the market. The firm produces and sells a single product during the time period $[t_1, t_2]$. Let p be the unit price in a situation where the firm sells the product directly to the consumer, bypassing the retailer, $p > 0$. To increase its profits, the firm uses the services of a retailer. To encourage the retailer to sell the commodity, the firm provides it with wholesale discount $\alpha(t) \in [A_1, A_2] \subset [0, 1]$. Thus, the wholesale price of the goods is $p_w(t) = (1 - \alpha(t))p$. In turn, the retailer directs pass-through, i.e., a part $\beta(t) \in [B_1, B_2] \subset [0, 1]$ of the discount $\alpha(t)$ to reduce the market price of the commodity. Therefore, the retail price of the commodity is equal to $(1 - \beta(t)\alpha(t))p$. Then the difference between retail price and wholesale price is the retailer's profit per unit from the sale and equals $\alpha(t)(1 - \beta(t))p$. Thus, the pricing process can be schematically represented as

$$p \longrightarrow p_w = (1 - \alpha)p = (1 - \beta\alpha)p - (1 - \beta)\alpha p. \tag{1}$$

---

[1] Note that we consider the case when time switches of discount levels are fixed and known. It seems realistic. Indeed, the periods of discounts usually are known. For example, Christmas sales, Winter sales. But, of course, it is possible to consider the situation with non-fixed switches of discount levels.

[2] Note that although the profits are quadratic with respect to price discount and pass-through level(s), their concavity is especially important. It is quite appropriate to recall the classic [12]: "Quadratic programming with one negative eigenvalue is NP-hard".

2.2 Profits and Motion

Let $x(t)$ be a state variable represented the accumulated sales during the period $[t_1, t]$ while c_0 be a unit production cost.

At the end of the selling period, due to (1), the total profit of the firm is

$$\Pi_m = \int_{t_1}^{t_2} (p_w(t) - c_0) \, \dot{x}(t) dt, \tag{2}$$

while the total profit of the retailer is

$$\Pi_r = p \int_{t_1}^{t_2} \dot{x}(t)\alpha(t)(1 - \beta(t)) dt. \tag{3}$$

We assume that the motivation of the retailer is determined by the state variable $M(t)$, and its dynamics is given by the differential equation[3]

$$\dot{M}(t) = \gamma \dot{x}(t) + \varepsilon \left(\alpha(t) - \overline{\alpha}\right).$$

The dynamics of the total amount of goods sold, $x(t)$, is determined by the differential equation[4]

$$\dot{x}(t) = -\theta x(t) + \delta M(t) + \eta \alpha(t)\beta(t).$$

2.3 Maximization of Manufacturer's Profit Under Constant Pass-Through

In [1] we studied the case with constant $\beta(t) = \beta$ and maximized the manufacturer's profit w.r.t. $\alpha(t)$. Let us denote $\eta_\beta = \eta\beta$. This way the problem is[5]

Manufacturer Problem:

$$\Pi_m \longrightarrow \max_\alpha$$
$$\dot{x}(t) = -\theta x(t) + \delta M(t) + \eta_\beta \alpha(t),$$
$$\dot{M}(t) = \gamma \dot{x}(t) + \varepsilon \left(\alpha(t) - \overline{\alpha}\right),$$
$$x(t_1) = 0, \ M(t_1) = \overline{M} > 0,$$
$$\alpha(t) \in [A_1, A_2] \subset [0, 1].$$

---

[3] As in [2], $\gamma > 0$ is the sales productivity in terms of motivation, $\varepsilon > 0$ is the discount productivity in terms of motivation. Parameter $\overline{\alpha} \in [A_1, A_2]$ takes into account the fact that the retailer has some expectations about the wholesale discount: the motivation is reduced if the retailer is dissatisfied with the wholesale discount, i.e., if $\alpha(t) < \overline{\alpha}$; on the contrary, the motivation increases if $\alpha(t) > \overline{\alpha}$. .

[4] $\theta > 0$ is the saturation parameter of the market, $\delta >$ is the retailer's selling skill, $\eta > 0$ is the discount productivity in terms of sales (the market sensitivity to shelf price discounts).

[5] $\overline{M} > 0$ is the initial motivation of the retailer.

2.4   Maximization of Retailer's Profit Under Constant Wholesale Discount

In [2] we studied the case with constant $\alpha(t) = \alpha$ and maximized the retailer's profit w.r.t. $\beta(t)$. Let us denote $\eta_\alpha = \eta\alpha$. This way the problem is

Retailer Problem:
$$\Pi_r \longrightarrow \max_\beta$$
$$\dot{x}(t) = -\theta x(t) + \delta M(t) + \eta_\alpha \beta(t),$$
$$\dot{M}(t) = \gamma \dot{x}(t) + \varepsilon (\alpha - \overline{\alpha}),$$
$$x(t_1) = 0, \ M(t_1) = \overline{M} > 0,$$
$$\beta(t) \in [B_1, B_2] \subset [0, 1].$$

3   Maximization of Manufacturer's and Retailer's Profits

Let us address on the *Manufacturer-Retailer Problem* defined by the objective functionals Π_m and Π_r:

Manufacturer-Retailer Problem:
$$\Pi_m \longrightarrow \max_\alpha$$
$$\Pi_r \longrightarrow \max_\beta$$
$$\dot{x}(t) = -\theta x(t) + \delta M(t) + \eta\alpha(t)\beta(t),$$
$$\dot{M}(t) = \gamma \dot{x}(t) + \varepsilon (\alpha(t) - \overline{\alpha}),$$
$$x(t_1) = 0, \ M(t_1) = \overline{M} > 0,$$
$$\alpha(t) \in [A_1, A_2] \subset [0, 1],$$
$$\beta(t) \in [B_1, B_2] \subset [0, 1].$$

3.1   The Case: Wholesale Discount and Pass-Through Are Constant

In [3] we study the *Manufacturer-Retailer Problem* in a simplified framework in which both controls must take a constant value in the whole time period $[t_1, t_2]$ and these values are decided at time t_1. In this case the solution of problems *Manufacturer Problem* and *Retailer Problem* becomes straightforward and allows to obtain some properties of *Manufacturer-Retailer Problem*.

With constant controls $\alpha(t) = \alpha \in [A_1, A_2]$ and $\beta(t) = \beta \in [B_1, B_2]$ the manufacturer's profit is (cf. (2))

$$\Pi_m = \Pi_M(\alpha, \ \beta) = (q - p\alpha)x(t_2), \tag{4}$$

where $q = p - c_0$, while the profit of retailer is (cf. (3))

$$\Pi_r = \Pi_R(\alpha, \ \beta) = p\alpha(1 - \beta)x(t_2). \tag{5}$$

The total volume of sales during $[t_1, t_2]$, $x(t_2)$, depends explicitly on α and β:

$$x(t_2) = (H\beta + L)\alpha + K, \tag{6}$$

where

$$a = \theta - \gamma\delta, \tag{7}$$

$$T = a(t_1 - t_2), \tag{8}$$

$$H = \frac{\eta}{a}\left(1 - e^T\right), \qquad L = -\frac{\delta\varepsilon}{a^2}\left(1 - e^T + T\right), \qquad K = \frac{\delta}{\eta}\overline{M}H - \overline{\alpha}L. \tag{9}$$

Under rather natural condition (the concavity of cumulative sales for constant wholesale price, see details in [1–3]), we assume

$$a > 0. \tag{10}$$

Then $T < 0, H > 0, L > 0$. Note that function (4) is quadratic and strictly concave with respect to α, while function (5) is quadratic and strictly concave with respect to β. It allows in [3] to describe and to study the equilibrium (Nash and Stackelberg)[6].

4   The Case: Wholesale Discount and Pass-Through Are Piece-Wise Constant

Let for some $t_1 = \tau_0 < \tau_1 < \ldots < \tau_n < \tau_{n+1} = t_2$

$$\alpha(t) = \begin{cases} \alpha_1, & t \in (\tau_0, \tau_1) \\ \alpha_2, & t \in (\tau_1, \tau_2) \\ \ldots \\ \alpha_{n+1}, & t \in (\tau_n, \tau_{n+1}) \end{cases} ; \qquad \beta(t) = \begin{cases} \beta_1, & t \in (\tau_0, \tau_1) \\ \beta_2, & t \in (\tau_1, \tau_2) \\ \ldots \\ \beta_{n+1}, & t \in (\tau_n, \tau_{n+1}) \end{cases} ;$$

---

[6] In particular, three Nash equilibria can be, namely $\left(\alpha^0, \beta^0\right)$, $\left(\alpha^+, \beta^+\right)$, $\left(\alpha^-, \beta^-\right)$, where

$$\alpha^0 = 0, \qquad H\beta^0 + L = \frac{pK}{q},$$

$$\alpha^+ = \frac{q(1+\Gamma)}{4p}, \qquad H\beta^+ + L = \frac{(H+L)(1+\Gamma)}{4},$$

$$\alpha^- = \frac{q(1-\Gamma)}{4p}, \qquad H\beta^- + L = \frac{(H+L)(1-\Gamma)}{4},$$

$$\Gamma = \sqrt{1 - \frac{8pK}{q(H+L)}}.$$

Besides, when the manufacturer is leader, Stackelberg equilibrium can be (α^m, β^m), where

$$\alpha^m = \frac{q(H+L) - pK}{2p(H+L)} = \frac{q\left(7 + \Gamma^2\right)}{16p}, \qquad H\beta^m + L = \frac{(H+L)\left(5 + 3\Gamma^2\right)}{2\left(7 + \Gamma^2\right)}.$$

i.e., $\alpha(t) = \alpha_i$, $\beta(t) = \beta_i$, $t \in (\tau_{i-1}, \tau_i)$, $i \in \{1, \ldots n+1\}$. Then, due to continuity of space variables,

$$x(t) = \begin{cases} x_1(t), & t \in [\tau_0, \tau_1] \\ x_2(t), & t \in [\tau_1, \tau_2] \\ \ldots \\ x_{n+1}(t), & t \in [\tau_n, \tau_{n+1}] \end{cases} ; \qquad M(t) = \begin{cases} M_1(t), & t \in [\tau_0, \tau_1] \\ M_2(t), & t \in [\tau_1, \tau_2] \\ \ldots \\ M_{n+1}(t), & t \in [\tau_n, \tau_{n+1}] \end{cases} ;$$

i.e.,

$$x(t) = x_i(t), \ M(t) = M_i(t), \ t \in [\tau_{i-1}, \tau_i], \ i \in \{1, \ldots n+1\},$$

where $x_i(t)$ and $M_i(t)$ are the solutions of the systems[7]

$$\dot{x}_i(t) = -\theta x_i(t) + \delta M_i(t) + \eta \alpha_i \beta_i,$$
$$\dot{M}_i(t) = \gamma \dot{x}_i(t) + \varepsilon(\alpha_i - \bar{\alpha}),$$
$$t \in [\tau_{i-1}, \tau_i], i \in \{1, \ldots n+1\},$$
$$x_{i+1}(\tau_i) = x_i(\tau_i), i \in \{1, \ldots n\},$$
$$M_{i+1}(\tau_i) = M_i(\tau_i), i \in \{1, \ldots n\}.$$

We get (cf. (4) and (5))

$$\Pi_m = p \sum_{i=1}^{n} (\alpha_{i+1} - \alpha_i) x_i(\tau_i) + (q - \alpha_{n+1}p) x(t_2), \tag{11}$$

$$\Pi_r = p \sum_{i=1}^{n+1} (1 - \beta_i) \alpha_i (x_i(\tau_i) - x_i(\tau_{i-1})). \tag{12}$$

Therefore, we need the expressions for $x_i(\tau_i)$. Let[8]

$$K(t) = \frac{\delta \overline{M}}{a} \cdot \left(1 - e^{a(t_1 - t)}\right) + \frac{\bar{\alpha}\delta\varepsilon}{a^2} \cdot \left(1 - e^{a(t_1-t)} + a(t_1 - t)\right),$$

$$H_i(t) = \frac{\eta}{a} \cdot \left(1 - e^{a(\tau_{i-1}-t)}\right), \ t \geq \tau_{i-1}, \ i \in \{1, \ldots n+1\},$$

$$L_i(t) = -\frac{\delta\varepsilon}{a^2} \cdot \left(1 - e^{a(\tau_{i-1}-t)} + a(\tau_{i-1} - t)\right), \ t \geq \tau_{i-1}, \ i \in \{1, \ldots n+1\},$$

(cf. (9)). The Proposition below generalizes (6).

Proposition 1. *For $t \in [\tau_{i-1}, \tau_i], \ i \in \{1, \ldots n+1\}$*

$$x_i(t) = K(t) + (H_i(t)\beta_i + L_i(t))\alpha_i$$
$$+ \sum_{j=1}^{i-1} ((H_j(t) - H_{j+1}(t))\beta_j + L_j(t) - L_{j+1}(t))\alpha_j.$$

Proof. See Sect. 5.1.

To optimize the profits, we need first to study the concavity of the profits. In this paper, we proof the strict concavity of function (12).

---

[7] Note that $x_1(\tau_0) = 0$ while $M_1(\tau_0) = \overline{M}$.
[8] Definition of a see in (7). Due to (10), these formulas are well defined.

4.1   Main Result: The Strict Concavity of Retailer's Profit

Proposition 2. *The retailer's profit Π_r is strictly concave with respect to pass-through levels β_i, $i \in \{1, \ldots n + 1\}$.*

Proof. See Sect. 5.2.

5   Proofs

5.1   Proof of Proposition 1

We use mathematical induction w.r.t. k, number of switches.

   1. **The base case.** Let $k = 0$. In $t \in [\tau_0, \tau_1] = [t_1, t_2]$, consider the system of differential equations

$$\dot{x}_1(t) = -\theta x_1(t) + \delta M_1(t) + \eta \alpha_1 \beta_1,$$

$$\dot{M}_1(t) = \gamma \dot{x}_1(t) + \varepsilon (\alpha_1 - \overline{\alpha})$$

under the initial conditions

$$x_1(t_1) = 0, \qquad M_1(t_1) = \overline{M}.$$

The solution is (cf. [3])

$$x_1(t) = -\frac{\delta \varepsilon (\alpha_1 - \overline{\alpha})}{a^2} \cdot \left(1 + a(t_1 - t) - e^{a(t_1 - t)}\right) + \frac{1}{a} \cdot \left(\delta \overline{M} + \eta \alpha_1 \beta_1\right) \left(1 - e^{a(t_1 - t)}\right),$$

$$M_1(t) = \gamma x_1(t) + \varepsilon (\alpha_1 - \overline{\alpha})(t - t_1) + \overline{M}.$$

Hence

$$x_1(t) = K(t) + (H_1(t) \beta_1 + L_1(t)) \alpha_1,$$

where

$$K(t) = \frac{\delta \overline{M}}{a} \cdot \left(1 - e^{a(t_1 - t)}\right) + \frac{\overline{\alpha} \delta \varepsilon}{a^2} \cdot \left(1 - e^{a(t_1 - t)} + a(t_1 - t)\right),$$

$$H_1(t) = \frac{\eta}{a} \cdot \left(1 - e^{a(t_1 - t)}\right), \; t \geq t_1,$$

$$L_1(t) = -\frac{\delta \varepsilon}{a^2} \cdot \left(1 - e^{a(t_1 - t)} + a(t_1 - t)\right), \; t \geq t_1.$$

   2. **Inductive step.** Assume that some unspecified value of k we get for $t \in [\tau_{k-1}, \tau_k]$:

$$x_k(t) = K(t) + (H_k(t) \beta_k + L_k(t)) \alpha_k$$
$$+ \sum_{j=1}^{k-1} ((H_j(t) - H_{j+1}(t)) \beta_j + L_j(t) - L_{j+1}(t)) \alpha_j.$$

Let us show that for $t \in [\tau_k, \tau_{k+1}]$:

$$x_{k+1}(t) = K(t) + (H_{k+1}(t) \beta_{k+1} + L_{k+1}(t)) \alpha_{k+1}$$
$$+ \sum_{j=1}^{k} ((H_j(t) - H_{j+1}(t)) \beta_j + L_j(t) - L_{j+1}(t)) \alpha_j.$$

In $[\tau_k, \tau_{k+1}]$, consider the system of differential equations

$$\dot{x}_{k+1}(t) = -\theta x_{k+1}(t) + \delta M_{k+1}(t) + \eta \alpha_{k+1}\beta_{k+1}$$

$$\dot{M}_{k+1}(t) = \gamma \dot{x}_{k+1}(t) + \varepsilon(\alpha_{k+1} - \overline{\alpha})$$

under the initial conditions

$$x_{k+1}(\tau_k) = x_k(\tau_k), \qquad M_{k+1}(\tau_k) = M_k(\tau_k).$$

Since

$$M_{k+1}(t) = \gamma(x_{k+1}(t) - x_{k+1}(\tau_k)) + \varepsilon(\alpha_{k+1} - \overline{\alpha})(t - \tau_k) + M_{k+1}(\tau_k)$$

$$= \gamma x_{k+1}(t) + \varepsilon(\alpha_{k+1} - \overline{\alpha})(t - \tau_k) + M_k(\tau_k) - \gamma x_k(\tau_k),$$

we get

$$\dot{x}_{k+1}(t) + ax_{k+1}(t)$$

$$= \delta\varepsilon(\alpha_{k+1} - \overline{\alpha})t + \delta(-\varepsilon(\alpha_{k+1} - \overline{\alpha})\tau_k + M_k(\tau_k) - \gamma x_k(\tau_k)) + \eta\alpha_{k+1}\beta_{k+1}.$$

So

$$x_{k+1}(t) = x_k(\tau_k)e^{a(\tau_k - t)} + \tfrac{\delta\varepsilon}{a}\cdot(\alpha_{k+1} - \overline{\alpha})\left(t - \tau_k e^{a(\tau_k - t)}\right)$$

$$+\tfrac{1}{a}\cdot\left(\delta\left(-\varepsilon(\alpha_{k+1} - \overline{\alpha})\left(\tau_k + \tfrac{1}{a}\right) + M_k(\tau_k) - \gamma x_k(\tau_k)\right) + \eta\alpha_{k+1}\beta_{k+1}\right)\left(1 - e^{a(\tau_k - t)}\right)$$

$$= x_k(\tau_k)e^{a(\tau_k - t)} + (H_{k+1}(t)\beta_{k+1} + L_{k+1}(t))\alpha_{k+1} - \overline{\alpha}L_{k+1}(t)$$

$$+\tfrac{\delta}{a}\cdot(M_k(\tau_k) - \gamma x_k(\tau_k))\left(1 - e^{a(\tau_k - t)}\right)$$

$$= K(t) + (H_{k+1}(t)\beta_{k+1} + L_{k+1}(t))\alpha_{k+1}$$

$$+ \sum_{j=1}^{k}((H_j(t) - H_{j+1}(t))\beta_j + L_j(t) - L_{j+1}(t))\alpha_j$$

$$+\tfrac{\delta}{a}\cdot\left(M_k(\tau_k) - \gamma x_k(\tau_k) - \overline{M} - \varepsilon\cdot(\alpha_j - \overline{\alpha})\sum_{j=1}^{k}(\tau_j - \tau_{j-1})\right)\left(1 - e^{a(\tau_k - t)}\right).$$

To finish the proof, we need to show that

$$M_k(\tau_k) - \gamma x_k(\tau_k) = \overline{M} + \varepsilon\cdot(\alpha_j - \overline{\alpha})\sum_{j=1}^{k}(\tau_j - \tau_{j-1}).$$

Indeed,

$$M_k(t) - \gamma x_k(t) = M_k(\tau_{k-1}) - \gamma x_k(\tau_{k-1}) + \varepsilon(\alpha_k - \overline{\alpha})(t - \tau_{k-1}).$$

Hence

$$M_k(\tau_k) - \gamma x_k(\tau_k) = M_{k-2}(\tau_{k-2}) - \gamma x_{k-2}(\tau_{k-2})$$

$$+\varepsilon(\alpha_{k-1} - \overline{\alpha})(\tau_{k-1} - \tau_{k-2}) + \varepsilon(\alpha_k - \overline{\alpha})(\tau_k - \tau_{k-1}) = \ldots$$

$$= M_{k-(k-1)}(\tau_1) - \gamma x_1(\tau_1) + \varepsilon\sum_{j=2}^{k}(\alpha_j - \overline{\alpha})(\tau_j - \tau_{j-1})$$

$$= M_1(t_1) - \gamma x_1(t_1) + \varepsilon\sum_{j=1}^{k}(\alpha_j - \overline{\alpha})(\tau_j - \tau_{j-1}) = \overline{M} + \varepsilon\sum_{j=1}^{k}(\alpha_j - \overline{\alpha}).$$

5.2   Proof of Proposition 2

Due to (12), we get

$$\frac{\partial^2 \Pi_r}{\partial \beta_i \partial \beta_j} = \begin{cases} -2p\alpha_i \left(\dfrac{\partial x_i(\tau_i)}{\partial \beta_i} - \dfrac{\partial x_i(\tau_{i-1})}{\partial \beta_i} \right), & i = j \\ -p\alpha_j \left(\dfrac{\partial x_j(\tau_j)}{\partial \beta_i} - \dfrac{\partial x_j(\tau_{j-1})}{\partial \beta_i} \right), & i < j \end{cases} \tag{13}$$

Now let us write the elements of matrix $(\Pi_r)''$ explicitly. Let

$$T_i = a(\tau_{i-1} - \tau_i) < 0, \; i \in \{1, \ldots, n+1\}$$

(cf. (8)) and

$$s_i = 1 - e^{T_i} = 1 - e^{a(\tau_{i-1} - \tau_i)} < 1, \; i \in \{1, \ldots, n+1\}.$$

Lemma 1. *The elements of matrix* $(\Pi_r)''$ *are*

$$\frac{\partial^2 \Pi_r}{\partial \beta_i \partial \beta_j} = \begin{cases} -\dfrac{2p\eta}{a} \cdot s_i(\alpha_i)^2, & i = j \\ \dfrac{p\eta}{a} \cdot s_i s_j \displaystyle\prod_{k=i+1}^{j-1} (1 - s_k)\,\alpha_i \alpha_j, & i < j \end{cases}$$

Proof. Due to Proposition 1,

$$-2p\alpha_i \left(\tfrac{\partial x_i(\tau_i)}{\partial \beta_i} - \tfrac{\partial x_i(\tau_{i-1})}{\partial \beta_i} \right) = -2p\,(H_i(\tau_i) - H_i(\tau_{i-1}))\,(\alpha_i)^2$$

$$= -\tfrac{2p\eta}{a} \cdot \left(1 - e^{T_i}\right)(\alpha_i)^2 = -\tfrac{2p\eta}{a} \cdot s_i(\alpha_i)^2.$$

Moreover, for $i < j$,

$$-p\alpha_j \left(\tfrac{\partial x_j(\tau_j)}{\partial \beta_i} - \tfrac{\partial x_j(\tau_{j-1})}{\partial \beta_i} \right)$$

$$= -p \cdot (H_i(\tau_j) - H_{i+1}(\tau_j) - H_i(\tau_{j-1}) - H_{i+1}(\tau_{j-1}))\,\alpha_i \alpha_j$$

$$= -\tfrac{p\eta}{a} \cdot \left(-e^{a(\tau_{i-1} - \tau_j)} + e^{a(\tau_{i-1} - \tau_{j-1})} + e^{a(\tau_i - \tau_j)} - e^{a(\tau_i - \tau_{j-1})} \right)\alpha_i \alpha_j$$

$$= -\tfrac{p\eta}{a} \cdot \left(-e^{\sum_{k=i}^{j} T_k} + e^{\sum_{k=i}^{j-1} T_k} + e^{\sum_{k=i+1}^{j} T_k} - e^{\sum_{k=i+1}^{j-1} T_k} \right)\alpha_i \alpha_j$$

$$= \tfrac{p\eta}{a} \cdot \left(e^{T_i + T_j} - e^{T_i} - e^{T_j} + 1 \right) e^{\sum_{k=i+1}^{j-1} T_k}\alpha_i \alpha_j$$

$$= \tfrac{p\eta}{a} \cdot \left(1 - e^{T_i}\right)\left(1 - e^{T_j}\right) \prod_{k=i+1}^{j-1} e^{T_k}\alpha_i \alpha_j == \tfrac{p\eta}{a} \cdot s_i s_j \prod_{k=i+1}^{j-1} (1 - s_k)\,\alpha_i \alpha_j.$$

Corollary 1. *The determinant of matrix* $(\Pi_r)''$ *is*

$$\det(\Pi_r)'' = \left(\frac{p\eta}{a}\right)^{n+1} \cdot \prod_{i=1}^{n+1} (\alpha_i)^2 \det A,$$

where

$$
A = \begin{pmatrix}
-2s_1 & s_1 s_2 & s_1 s_3 (1 - s_2) & \cdots & s_1 s_{n+1} \prod_{i=2}^{n} (1 - s_i) \\[2ex]
s_1 s_2 & -2s_2 & s_2 s_3 & \ddots & \vdots \\[2ex]
s_1 s_3 (1 - s_2) & s_2 s_3 & -2s_3 & \ddots & s_{n-1} s_{n+1} (1 - s_n) \\[2ex]
\vdots & \ddots & \ddots & \ddots & s_n s_{n+1} \\[2ex]
s_1 s_{n+1} \prod_{i=2}^{n} (1 - s_i) & \cdots & s_{n-1} s_{n+1} (1 - s_n) \, s_n s_{n+1} & & -2s_{n+1}
\end{pmatrix} .
$$

For $k \geq 2$, let us define the tridiagonal matrix

$$
B_k = \begin{pmatrix}
-2s_1 & s_1 s_2 & 0 & \cdots & & 0 \\[2ex]
s_1 s_2 & -2s_2 & (1 - s_2) s_3 & \ddots & & \vdots \\[2ex]
0 & (2 - s_2) s_3 & \dfrac{2 ((s_2 - 1) s_3 - s_2) s_3}{s_2} & \ddots & & 0 \\[2ex]
\vdots & \ddots & \ddots & \ddots & (1 - s_{k-1}) s_k & \\[2ex]
0 & \cdots & 0 & (2 - s_{k-1}) s_k & \dfrac{2 ((s_{k-1} - 1) s_k - s_{k-1}) s_k}{s_{k-1}}
\end{pmatrix} .
$$

Moreover, let

$$
C = \begin{pmatrix}
1 & 0 & 0 & \cdots & & 0 \\[2ex]
0 & 1 & \ddots & & & \vdots \\[2ex]
0 & \dfrac{(s_2 - 1) s_3}{s_2} & \ddots & \ddots & & \vdots \\[2ex]
\vdots & \ddots & \ddots & \ddots & 0 & \\[2ex]
0 & \cdots & 0 & \dfrac{(s_n - 1) s_{n+1}}{s_n} & 1
\end{pmatrix} .
$$

Then $B_{n+1} = C \cdot A \cdot C^T$, where "T" means matrix transposition.

Now, let us calculate the corners minors of matrix B_{n+1}, i.e., the determinants of matrices B_k. Define the symmetric function

$$P_{i,j} = \sum_{\substack{r_1 < \ldots < r_j \\ r_l \in \{1,\ldots,i\} \\ l = \overline{1,j}}} \prod_{l=1}^{j} s_{r_l}, \; j \in \{1, \ldots i\}.$$

Note that[9]

$$P_{i,j} = P_{i-1,j} + s_i P_{i-1,j-1}, \; j \in \{2, \ldots, i-1\}. \tag{14}$$

Lemma 2. *The corner minors of* B_{n+1} *are*

$$\det B_k = (-2)^{k-2} \left(\prod_{i=1}^{k} s_i \right) \left(4 - \sum_{j=2}^{k} (-1)^j \cdot \frac{j-1}{2^{j-2}} \cdot P_{k,j} \right).$$

Proof. Let us use the mathematical induction. Indeed,

$$\det B_2 = s_1 s_2 \cdot (4 - s_1 s_2) = (-2)^0 \, s_1 s_2 \cdot \left(4 - (-1)^2 \cdot \frac{1}{2^0} \cdot s_1 s_2 \right)$$

$$= (-2)^0 \left(\prod_{i=1}^{2} s_i \right) \left(4 - \sum_{j=2}^{2} (-1)^j \cdot \tfrac{j-1}{2^{j-2}} \cdot P_{2,j} \right).$$

Let

$$\det B_{k-1} = (-2)^{k-3} \left(\prod_{i=1}^{k-1} s_i \right) \left(4 - \sum_{j=2}^{k-1} (-1)^j \cdot \tfrac{j-1}{2^{j-2}} \cdot P_{k-1,j} \right),$$

$$\det B_k = (-2)^{k-2} \left(\prod_{i=1}^{k} s_i \right) \left(4 - \sum_{j=2}^{k} (-1)^j \cdot \tfrac{j-1}{2^{j-2}} \cdot P_{k,j} \right).$$

Then

$$\det B_{k+1} = \frac{2((s_k-1)s_{k+1}-s_k)s_{k+1}}{s_k} \cdot \det B_k - ((2 - s_k) s_{k+1})^2 \det B_{k-1}$$

$$= -\frac{((s_k-1)s_{k+1}-s_k)}{s_k} \cdot (-2)^{k-1} \left(\prod_{i=1}^{k+1} s_i \right) \left(4 - \sum_{j=2}^{k} (-1)^j \cdot \tfrac{j-1}{2^{j-2}} \cdot P_{k,j} \right)$$

$$- \frac{(2-s_k)^2}{4} \cdot \frac{s_{k+1}}{s_k} \cdot (-2)^{k-1} \left(\prod_{i=1}^{k+1} s_i \right) \left(4 - \sum_{j=2}^{k-1} (-1)^j \cdot \tfrac{j-1}{2^{j-2}} \cdot P_{k-1,j} \right).$$

---

[9] For instance,

$$P_{4,2} = s_1 s_2 + s_1 s_3 + s_1 s_4 + s_2 s_3 + s_2 s_4 + s_3 s_4$$
$$= s_1 s_2 + s_1 s_3 + s_2 s_3 + (s_1 + s_2 + s_3) s_4 = P_{3,2} + s_4 P_{3,1}.$$

So, due to (14),

$$\left((-2)^{k-1}\left(\prod_{i=1}^{k+1} s_i\right)\right)^{-1} \det B_{k+1}$$

$$= 4 - s_k s_{k+1} + \frac{((s_k-1)s_{k+1}-s_k)}{s_k} \cdot \sum_{j=2}^{k}(-1)^j \cdot \frac{j-1}{2^{j-2}} \cdot P_{k,j}$$

$$+ \frac{(2-s_k)^2}{4} \cdot \frac{s_{k+1}}{s_k} \cdot \sum_{j=2}^{k-1}(-1)^j \cdot \frac{j-1}{2^{j-2}} \cdot P_{k-1,j}$$

$$= 4 - \sum_{j=2}^{k}(-1)^j \cdot \frac{j-1}{2^{j-2}} \cdot P_{k,j} - s_k s_{k+1}$$

$$+ \frac{(s_k-1)s_{k+1}}{s_k} \cdot \sum_{j=2}^{k}(-1)^j \cdot \frac{j-1}{2^{j-2}} \cdot P_{k,j} + \frac{(2-s_k)^2}{4} \cdot \frac{s_{k+1}}{s_k} \cdot \sum_{j=2}^{k-1}(-1)^j \cdot \frac{j-1}{2^{j-2}} \cdot P_{k-1,j}$$

$$= 4 - \sum_{j=2}^{k}(-1)^j \cdot \frac{j-1}{2^{j-2}} \cdot P_{k+1,j} + s_{k+1} \cdot \sum_{j=2}^{k}(-1)^j \cdot \frac{j-1}{2^{j-2}} \cdot P_{k,j-1} - s_k \cdot s_{k+1}$$

$$+ \frac{(s_k-1)s_{k+1}}{s_k} \cdot \sum_{j=2}^{k}(-1)^j \cdot \frac{j-1}{2^{j-2}} \cdot P_{k,j} + \frac{(2-s_k)^2}{4} \cdot \frac{s_{k+1}}{s_k} \cdot \sum_{j=2}^{k-1}(-1)^j \cdot \frac{j-1}{2^{j-2}} \cdot P_{k-1,j}$$

$$= 4 - \sum_{j=2}^{k}(-1)^j \cdot \frac{j-1}{2^{j-2}} \cdot P_{k+1,j}.$$

Due to Lemma 2, to finish the proof of Proposition 2, we need only to show

Lemma 3. *The following inequality holds:*

$$\sum_{j=2}^{k}(-1)^j \cdot \frac{j-1}{2^{j-2}} \cdot P_{k,j} < 4.$$

Proof. Consider the function

$$f_k(s_1,\ldots,s_k) := \sum_{j=2}^{k}(-1)^j \cdot \frac{j-1}{2^{j-2}} \cdot P_{k,j}, \quad k \geq 2.$$

We get

$$f_{k+1}(s_1,\ldots,s_{k+1}) = \sum_{j=2}^{k+1}(-1)^j \cdot \frac{j-1}{2^{j-2}} \cdot P_{k+1,j}$$

$$= f_k(s_1,\ldots,s_k) + \left(\sum_{j=2}^{k+1}(-1)^j \cdot \frac{j-1}{2^{j-2}} \cdot P_{k,j-1}\right) s_{k+1}$$

$$= \sum_{j=2}^{k-1}(-1)^j \cdot \frac{j-1}{2^{j-2}} \cdot P_{k,j} + (-1)^k \cdot \frac{k-1}{2^{k-2}} \cdot P_{k,k}$$

$$+ \sum_{j=2}^{k+1}(-1)^j \cdot \frac{j-1}{2^{j-2}} \cdot P_{k,j-1} s_{k+1} = \ldots$$

$$= \left(\sum_{j=1}^{1}(-1)^{j+1} \cdot \frac{j \cdot P_{1,j}}{2^{j-1}}\right) s_2 + \ldots + \left(\sum_{j=1}^{k}(-1)^{j+1} \cdot \frac{j \cdot P_{k,j}}{2^{j-1}}\right) s_{k+1}$$

$$= \sum_{i=1}^{k}\left(\sum_{j=1}^{i}(-1)^{j+1} \cdot \frac{j}{2^{j-1}} \cdot P_{i,j}\right) s_{i+1}.$$

Since $s_i \in (0;1)$,

$$\max f_k(s_1,\ldots,s_k) = \lim_{k\to\infty} f_k(1,\ldots,1)$$

$$= \left(\sum_{j=1}^{1}(-1)^{j+1}\cdot\tfrac{j}{2^{j-1}}\cdot P_{1,j}\right) + \ldots + \left(\sum_{j=1}^{k}(-1)^{j+1}\cdot\tfrac{j}{2^{j-1}}\cdot P_{k,j}\right) + \ldots$$

$$= \tfrac{1}{2^0} + \left(\tfrac{1}{2^0}\cdot 2 - \tfrac{2}{2^1}\cdot 1\right) + \left(\tfrac{1}{2^0}\cdot 3 - \tfrac{2}{2^1}\cdot 3 + \tfrac{3}{2^2}\cdot 1\right)$$

$$+ \left(\tfrac{1}{2^0}\cdot 4 - \tfrac{2}{2^1}\cdot 6 + \tfrac{3}{2^2}\cdot 4 - \tfrac{4}{2^4}\cdot 1\right)$$

$$+ \left(\sum_{j=1}^{5}(-1)^{j+1}\cdot\tfrac{j}{2^{j-1}}\cdot P_{5,j}\right) + \ldots + \left(\sum_{j=1}^{k}(-1)^{j+1}\cdot\tfrac{j}{2^{j-1}}\cdot P_{k,j}\right) + \ldots$$

$$= \tfrac{1}{2^0} + \tfrac{2}{2^1} + \tfrac{3}{2^2} + \tfrac{4}{2^3} + \ldots$$

$$= \tfrac{1}{2^0} + \tfrac{1}{2^1} + \tfrac{1}{2^2} + \tfrac{1}{2^3} + \ldots + \tfrac{1}{2^1} + \tfrac{1}{2^2} + \tfrac{1}{2^3} + \tfrac{1}{2^4} + \ldots + \tfrac{1}{2^2} + \tfrac{1}{2^3} + \tfrac{1}{2^4} + \tfrac{1}{2^5} + \ldots + \ldots$$

$$= \tfrac{1}{2^0} + \tfrac{1}{2^1} + \tfrac{1}{2^2} + \tfrac{1}{2^3} + \ldots + \tfrac{1}{2^1}\cdot\left(\tfrac{1}{2^0} + \tfrac{1}{2^1} + \tfrac{1}{2^2} + \tfrac{1}{2^3} + \ldots\right)$$

$$+ \tfrac{1}{2^2}\cdot\left(\tfrac{1}{2^0} + \tfrac{1}{2^1} + \tfrac{1}{2^2} + \tfrac{1}{2^3} + \ldots\right) + \ldots + \ldots$$

$$= \left(\tfrac{1}{2^0} + \tfrac{1}{2^1} + \tfrac{1}{2^2} + \tfrac{1}{2^3} + \ldots\right)\left(\tfrac{1}{2^0} + \tfrac{1}{2^1} + \tfrac{1}{2^2} + \tfrac{1}{2^3} + \ldots\right) = \left(\cfrac{1}{1-\tfrac{1}{2}}\right)^2 = 4.$$

6   Conclusion

In this paper, we study a stylized vertical control distribution channel in the structure "manufacturer-retailer-consumer". More precisely, we consider the situation when the wholesale discount and pass-through are piece-wise constant. This case seems to be economically adequate. The arising optimization problems contain quadratic objective functions with respect to wholesale discount and pass-through level(s). In the case when the wholesale discount is constant, we prove the strict concavity of the retailer's profit with respect to pass-through levels for any fixed number of (known) switches.

As for the topics of further research, we plan to study the concavity of the manufacturer's profit (11) with respect to wholesale price discount[10]. Besides, we plan to study the equilibrium (as Nash as Stackelberg) in the structure "manufacturer-retailer-consumer". Moreover, it seems interesting to study the interaction of several manufacturers and several retailers. Finally, we can consider this kind of models in monopolistic competition framework: retailing [13], market distortion [14], investments in R&D [15–17], international trade [18].

Acknowledgments. The work was supported in part by the Russian Foundation for Basic Research, projects 18-010-00728 and 19-010-00910, by the program of fundamental scientific researches of the SB RAS, project 0314-2019-0018, and by the Russian Ministry of Science and Education under the 5–100 Excellence Programme.

---

[10] At least, it is easy to get the analogs of (13).

References

1. Bykadorov, I., Ellero, A., Moretti, E., Vianello, S.: The role of retailer's performance in optimal wholesale price discount policies. Eur. J. Oper. Res. **194**(2), 538–550 (2009)
2. Bykadorov, I.: Dynamic marketing model: optimization of retailer's role. Commun. Comput. Inf. Sci. **974**, 399–414 (2019)
3. Bykadorov, I.A., Ellero, A., Moretti, E.: Trade discount policies in the differential games framework. Int. J. Biomed. Soft Comput. Hum. Sci. **18**(1), 15–20 (2013)
4. Nerlove, M., Arrow, K.J.: Optimal advertising policy under dynamic conditions. Economica **29**(144), 129–142 (1962)
5. Bala, P.K.: A data mining model for investigating the impact of promotion in retailing. In: 2009 IEEE International Advance Computing Conference, IACC 2009, 4809092, pp. 670–674 (2009)
6. Giri, B.C., Bardhan, S.: Coordinating a two-echelon supply chain with price and inventory level dependent demand, time dependent holding cost, and partial backlogging. Int. J. Math. Oper. Res. **8**(4), 406–423 (2016)
7. Giri, B.C., Bardhan, S., Maiti, T.: Coordinating a two-echelon supply chain through different contracts under price and promotional effort-dependent demand. J. Syst. Sci. Syst. Eng. **22**(3), 295–318 (2013)
8. Printezis, A., Burnetas, A.: The effect of discounts on optimal pricing under limited capacity. Int. J. Oper. Res. **10**(2), 160–179 (2011)
9. Routroy, S., Dixit, M., Sunil Kumar, C.V.: Achieving supply chain coordination through lot size based discount. Mater. Today Proc. **2**(4–5), 2433–2442 (2015)
10. Ruteri, J.M., Xu, Q.: The new business model for SMEs food processors based on supply chain contracts. In: International Conference on Management and Service Science, MASS 2011, 5999355 (2011)
11. Sang, S.: Bargaining in a two echelon supply chain with price and retail service dependent demand. Eng. Lett. **26**(1), 181–186 (2018)
12. Pardalos, P.M., Vavasis, S.A.: Quadratic programming with one negative eigenvalue is NP-hard. J. Global Optim. **1**(1), 15–22 (1991)
13. Bykadorov, I.A., Kokovin, S.G., Zhelobodko, E.V.: Product diversity in a vertical distribution channel under monopolistic competition. Autom. Remote Control **75**(8), 1503–1524 (2014)
14. Bykadorov, I., Ellero, A., Funari, S., Kokovin, S., Pudova, M.: Chain store against manufacturers: regulation can mitigate market distortion. In: Kochetov, Y., Khachay, M., Beresnev, V., Nurminski, E., Pardalos, P. (eds.) DOOR 2016. LNCS, vol. 9869, pp. 480–493. Springer, Cham (2016). https://doi.org/10.1007/978-3-319-44914-2_38
15. Antoshchenkova, I.V., Bykadorov, I.A.: Monopolistic competition model: the impact of technological innovation on equilibrium and social optimality. Autom. Remote Control **78**(3), 537–556 (2017)
16. Bykadorov, I.: Monopolistic competition model with different technological innovation and consumer utility levels. CEUR Workshop Proc. **1987**, 108–114 (2017)
17. Bykadorov, I., Kokovin, S.: Can a larger market foster R&D under monopolistic competition with variable mark-ups? Res. Econ. **71**(4), 663–674 (2017)
18. Bykadorov, I., Gorn, A., Kokovin, S., Zhelobodko, E.: Why are losses from trade unlikely? Econ. Lett. **129**, 35–38 (2015)

Preconditioned Subspace Descent Methods for the Solution of Nonlinear Systems of Equations

Igor Kaporin[(✉)]

Dorodnicyn Computer Center of FRC CC RAS, Moscow, Russia
`igorkaporin@mail.ru`

Abstract. Nonlinear least squares type iterative solver for f(x) = 0 is considered based on successive solution of orthogonal projections of the linearized equation on a sequence of appropriately chosen low-dimensional subspaces. The bases of the latter are constructed using only the first-order derivatives of the function. The techniques based on the concept of the limiting stepsize along normalized direction (developed earlier by the author) is used to guarantee the monotone decrease of the nonlinear residual norm. The results of numerical testing are presented, including not only small-sized standard test problems, but also larger and harder examples, such as algebraic problems associated with canonical decomposition of dense and sparse 3D tensors as well as finite-difference discretizations of 2D nonlinear boundary problems for 2nd order partial differential equations.

Keywords: Nonlinear least squares · Preconditioned subspace descent · Limiting step along normalized direction · Sparse matrix methods

1 Introduction

A standard least squares scheme intended to approximate the (presumably existent) solution of a (possibly overdetermined) nonlinear equation

$$f(x) = 0, \qquad f : R^n \to R^m, \qquad m \geq n, \tag{1}$$

is to introduce the function $\varphi : R^n \to R$ of the form

$$\varphi(x) = \frac{1}{2}\|f(x)\|^2 \equiv \frac{1}{2}f^T(x)f(x), \tag{2}$$

and to find its minimum numerically. Note that even if a zero residual solution exists, any minimization method for (2) may only find a stationary point $x = x_*$

Partially supported by RFBR grant No. 19-01-00666.

of φ corresponding to a possibly nonzero residual $f(x)$. Indeed, since x_* satisfies the equation

$$g(x) \equiv \text{grad } \varphi(x) = J^T(x)f(x) = 0, \tag{3}$$

where

$$J(x) \equiv \frac{\partial f}{\partial x} \in R^{m \times n}, \tag{4}$$

the Eq. (3) shows only that the residual $f(x_*)$ lies in the nullspace of $J^T(x_*)$. The latter subspace is always nonempty when $m > n$ and therefore $f(x_*)$ may be nonzero. Even if $m \leq n$, in many applied problems one can came across with cases when $J^T(x)$ has nonempty nullspace. Further we will present a solution techniques for (1) which hopefully has a potential for the reduction of $\|f(x)\|$ and $\|g(x)\|$ in a consistent manner whenever it is possible.

2   Preconditioned Subspace Descent

Further on, we will use the notations

$$f(x) = f, \quad J(x) = J, \quad g(x) = g, \tag{5}$$

so that $g = J^T f$. Let x be the current approximation to the solution x_* and determine the next approximation as

$$\widehat{x} = x + h, \tag{6}$$

where h is the direction vector. For a sufficiently smooth f, one can linearize $f(x + h)$ near x and find an appropriate h to provide for a decrease of φ using

$$2\varphi(x + h) = \|f(x + h)\|^2 = \|f + Jh\|^2 + O(\|h\|^2).$$

Obviously, if $f^T Jh < 0$ and $\|h\| \ll 1$, then a certain decrease of the objective value $\varphi(x + h) < \varphi(x)$ can be provided. In order to proceed with a quantitative analysis, we will use the techniques based on a special scaling of the direction h first proposed in [8], modified in [9], and presented with appropriate implementation and numerical testing in [10].

2.1   General Estimate for Residual Norm Reduction

Consider the updates of the form $h = \alpha p$, that is,

$$\widehat{x} = x + \alpha p \tag{7}$$

with the stepsize satisfying $0 < \alpha < 2$ and the direction p *normalized* by the condition

$$-f^T Jp = \|Jp\|^2. \tag{8}$$

Furthermore, we assume that the stepsize α lies within $(0, \alpha_*]$, where $\alpha_* < 2$ is the so called *limiting stepsize*. The latter is defined as the maximum number (depending on f and p) such that the limiting stepsize condition

$$\|f(x + \alpha p) - f - \alpha Jp\| \leq \alpha(2 - \alpha)\frac{\|Jp\|^2}{2\|f\|} \tag{9}$$

is satisfied for all $0 < \alpha \leq \alpha_*$. For the convenience, we will also use the notation

$$\vartheta = \frac{-f^T Jp}{\|f\|\|Jp\|} \tag{10}$$

for the cosine of the acute angle between m-vectors $(-f)$ and Jp. It appears that α_* is responsible for the characterization of nonlinearity in the neighborhood of x, while ϑ determines the precision of approximate solution p of the "Newton equation" $Jp + f = 0$. Note that the latter may not (and often cannot) be solved exactly in the context of our considerations.

Theorem 1. *Let the conditions (8) and (9) hold. Then the estimate*

$$\frac{\|f(x + \alpha p)\|}{\|f\|} \leq \sqrt{1 - \tau} + \frac{\tau}{2} \tag{11}$$

is valid, where

$$\tau = \alpha(2 - \alpha)\vartheta^2 \tag{12}$$

for all $0 < \alpha \leq \alpha_*$ with ϑ determined by (10).

Proof. Indeed, using (9) one has

$$\|f(x+\alpha p)\| \leq \|f + \alpha Jp\| + \|f(x+\alpha p) - f - \alpha Jp\| \leq \|f + \alpha Jp\| + \alpha(2 - \alpha)\frac{\|Jp\|^2}{2\|f\|},$$

and then, using (8),

$$\frac{\|f(x + \alpha p)\|}{\|f\|} \leq \frac{\sqrt{\|f\|^2 - \alpha(2 - \alpha)\|Jp\|^2}}{\|f\|} + \alpha(2 - \alpha)\frac{\|Jp\|^2}{2\|f\|^2}.$$

The required result immediately follows from the relation $\vartheta = \|Jp\|/\|f\|$ obtained by combining property (8) with the definition (10).

Remark 1. Note that

$$\sqrt{1 - \tau} + \tau/2 < \sqrt{1 - \tau^2/4} \tag{13}$$

for all $0 < \tau \leq 1$ and therefore the strict decrease of the residual norm is guaranteed whenever $\tau > 0$. Here, we do not use the more restrictive definition of α_* from [10] which gives better theoretical norm reduction estimate $\sqrt{1 - \alpha\vartheta^2}$ but does not support the practical use of $\alpha > 1$.

Remark 2. The present research is directed towards complicated problems with large sparse (nearly) singular Jacobians, when the exact solution of Newton equations is not feasible. Therefore, we will concentrate on theoretical and practical results related to convergence rates not faster than linear.

2.2   Practical Method for Choosing the Stepsize

Based on the above theory, one can develop the following Armijo type procedure [1] for evaluating an appropriate stepsize α providing for a certain decrease of the residual norm. Let $\widetilde{p}$ be a direction vector satisfying

$$-f^T J\widetilde{p}/(\|f\|\|J\widetilde{p}\|) = \vartheta > 0$$

but, in general, not normalized. First we normalize it using the formula

$$p = \widetilde{p}\frac{-f^T J\widetilde{p}}{\|J\widetilde{p}\|^2};\tag{14}$$

obviously, the normalization does not change the value of ϑ. Next we check the validity of estimate (11) for a decreasing sequence of trial values of $\alpha \in (0,2)$; for instance, we have used

$$\alpha_l = 1.8^{1-l}, \qquad l = 0, 1, \ldots, l_{\max}.\tag{15}$$

However, more usual choices, such as

$$\alpha_l = 2^{-l}, \qquad l = 0, 1, \ldots, l_{\max},\tag{16}$$

may show faster convergence for certain relatively easy problems. The value $l_{\max} = 30$ was used, which approximately corresponds to $\alpha > 2 \cdot 10^{-8}$. In numerical testing, the backtracking criterion (11) was often satisfied at once for $l = 0$ with the stepsize $\alpha = \alpha_0$.

2.3   Bounding the Limiting Stepsize α_*

The following result is a generalization of similar estimates given in [9,10].

Theorem 2. *Let there exist scalar $\gamma = \gamma(x) > 0$ and $n \times n$ symmetric positive definite matrix $M = M(x) > J^T J$ such that inequality*

$$\|f(x+h) - f - Jh\| \le \frac{1}{2\|f\|}h^T(M - J^T J)h \quad \text{for all} \quad \|h\| \le \gamma \tag{17}$$

holds. Then (9) holds for all $0 < \alpha \le \widetilde{\alpha}$, where

$$\widetilde{\alpha} = \min\left(\frac{\gamma}{\|p\|}, 2\frac{p^T J^T J p}{p^T M p}\right)\tag{18}$$

and therefore, $\alpha_* \ge \widetilde{\alpha}$.

Proof. Setting $h = \alpha p$ in the right inequality (17) we find $\alpha \le \gamma/\|p\|$. On the other hand, in order for (9) be valid, it suffices to require

$$\frac{\alpha^2}{2\|f\|}p^T(M - J^T J)p = \frac{\alpha(2-\alpha)}{2}\frac{\|Jp\|^2}{\|f\|},$$

which is equivalent to $\alpha = 2p^T J^T J p / p^T M p$. The sufficiency of (18) is proved.

Remark 3. Using the above result, one can show that the number of backtracking steps for choosing the stepsize (see Sect. 2.2) is limited above as $O(\log(1/\widetilde{\alpha}))$.

Remark 4. The claim of Theorem 2 readily follows if one sets

$$M = J^T J + \Gamma \|f\| I, \tag{19}$$

in (17), where Γ is the local Lipschitz constant for the Jacobian at x, that is, $\|J(x+h) - J(x)\| \leq \Gamma\|h\|$ for all $\|h\| \leq \gamma$ (cf. [9,10]).

2.4   Reducing the Angle Between f and Jp

The direction p is constructed in the form $p = Vz$, where $V \in R^{n \times s}$ and $z \in R^s$. Here s is a small integer, typically $1 \leq s \leq 10$, and we restrict our attention to the choice of the subspace basis as

$$V = [C^{-1}J^T f \mid \widetilde{p}_{-1} \mid \ldots \mid \widetilde{p}_{-t+1}], \qquad t = \min(k,s), \tag{20}$$

where k is the (earlier omitted, see (6) above) iteration number, so that $x_{k+1} = x_k + p_k \alpha_k$, and $\widetilde{p}_{-t} = x_{k-t+1} - x_{k-t}$. Here $C \approx J^T J$ is an easily invertible preconditioning matrix, see Sect. 2.6 below. This construction is a particular case of the one proposed in [9,10]. A similar choice of the subspace basis was also discussed in [21], see also references therein. Maximizing the value of ϑ, one obtains, denoting

$$U = JV \tag{21}$$

and temporarily assuming that U has full column rank,

$$\max_{p=Vz} \vartheta^2 = \frac{1}{f^T f} \max_{p=Vz} \frac{p^T J^T f f^T J p}{p^T J^T J p} = \frac{1}{f^T f} \max_z \frac{z^T U^T f f^T U z}{z^T U^T U z}.$$

The obtained Rayleigh quotient is maximized by $z = (U^T U)^{-1} U^T f$ (up to a scalar factor) so that

$$p = V(U^T U)^{-1} U^T f, \tag{22}$$

and the optimum value of the cosine satisfies $\vartheta^2 = f^T U (U^T U)^{-1} U^T f / f^T f$. As one can see, the direction p defined by (22) is normalized in the sense of (8). Note that for any $s \geq 1$, it holds $\vartheta^2 \geq (f^T J C^{-1} J^T f)^2 \|f\|^{-2} \|J C^{-1} J^T f\|^{-2}$, which shows that it makes sense to use certain preconditioning techniques to have $C \approx J^T J$, see further Sect. 2.6.

Remark 5. Since there is no guarantee that V has full column rank, some kind of approximate pseudoinversion must be applied, for instance, $U^T U$ can be replaced by $U^T U + \delta I_s$, where $\delta \ll \|U\|^2$, or by $U^T U + \delta \mathrm{Diag}(U^T U)$, where $\delta \ll 1$. The resulting direction $\widetilde{p} = V(U^T U + \delta I_s)^{-1} U^T f$ will no longer be normalized, and therefore must be scaled according to (14).

2.5   Regularizing Subspace Projection with Account for α_*

The above presented optimization of $p = Vz$ with respect to ϑ only may not be the best choice. As follows from (11) and (12), one must rather maximize the upper bound for the product $\alpha\vartheta^2$. Therefore, using the result of Theorem 2, one finds

$$\alpha\vartheta^2 \leq 2\frac{p^T J^T J p}{p^T M p} \frac{(f^T J p)^2}{\|f\|^2 \|Jp\|^2} = \frac{2}{\|f\|^2} \frac{p^T J^T f f^T J p}{p^T M p}. \tag{23}$$

The maximum by p of the latter Rayleigh quotient is attained for

$$p = \beta M^{-1} J^T f, \quad \beta \neq 0, \tag{24}$$

and equals $2f^T J M^{-1} J^T f / f^T f$ which, along with $C \approx M$, explains our choice (20) for the first column of V. The remaining columns of V serve for further adjustment of p in order to compensate the suboptimality of C.

Noting the close relation of (24) and (19) to the Levenberg-Marquardt method (see, e.g. [3,4] and references cited therein), one can observe that the straightforward implementation of the resulting method requires the solution of linear algebraic system with the matrix M at each step. However, the involvement of general sparse linear solvers may be undesirable in certain cases. Instead, we propose to use the subspace techniques: substituting $p = Vz$ into the above Rayleigh quotient and recalling (21), one has, using (19),

$$\max_{p=Vz} p^T J^T f f^T J p / p^T M p = \max_z z^T U^T f f^T U z / z^T (U^T U + \xi V^T V)z,$$

where $\xi = \Gamma\|f\|$, so that $z = (U^T U + \xi V^T V)^{-1} U^T f$. Therefore, the corrected direction is $\widetilde{p} = V(U^T U + \xi V^T V)^{-1} U^T f$, which must further be scaled according to (14). Here, an approximate pseudoinversion arises by the construction, which allows for the use of smaller (or even zero) values for the regularization parameter δ, see Remark 5 above.

Remark 6. Assuming the validity of left inequality (17) for any γ, (e.g., with f quadratic) and using directions (24) with backtracking (16), one can prove that $\tau \geq \alpha\vartheta^2 \geq f^T J M^{-1} J^T f / f^T f = \|g\|_{M^{-1}}^2 / \|f\|^2$, where the gradient g was defined in (3). Using this with Theorem 1 and (13) gives the following estimate:

$$\|f(x_{k+1})\|^2 \leq \|f(x_k)\|^2 - \|g(x_k)\|_{M_k^{-1}}^4 / (4\|f(x_k)\|^2), \quad k = 0, 1, \ldots,$$

which shows the following convergence result for the method:

$$\min_{0 \leq j < k} \|g(x_j)\|_{M_j^{-1}} \leq \|f(x_0)\|(k/4)^{-1/4}.$$

2.6   The Use of Preconditioning

As the preconditioner

$$C = C(x) \approx J^T J \in R^{n \times n}, \tag{25}$$

we consider a symmetric positive definite preconditioning matrix, the forming of which and solving $Cz = v$ must be as cheap as possible. For instance, the simplest choice is

$$C = \text{Diag}(J^T J). \tag{26}$$

We prefer to use a more efficient $\text{SSOR}(\omega)$ preconditioning, where from the standard splitting $J^T J = D + L + L^T$ with $D = \text{Diag}(J^T J)$ diagonal and L strictly lower triangular, one obtains

$$C = C_\omega = (\omega^{-1}D + L)(I + \omega D^{-1}L^T) \equiv J^T J + (\omega^{-1} - 1)D + \omega L D^{-1} L^T.$$

Note that the condition $M > J^T J$ used in Theorem 2 holds for all $M = C_\omega$ with $0 < \omega < 1$. The choice of relaxation parameter $\omega \in (0, 2)$ is rather problem-dependent, see Sect. 3 below. The default choice for hard-to-solve problems is $\omega = 1/3$; however, the values closer to 2 are more appropriate for the solution of discretizations of nonlinear partial differential equations. Note also that one should use different values of ω for $M = C_{\omega_0}$ in Theorem 2 (and therefore in (23)) and $C = C_{\omega_1}$ for preconditioning in the definition (20) of V (because with $\omega_0 = \omega_1$ it follows $z = [1 \mid 0 \mid \dots \mid 0]^T$ at each step of the method).

It must be stressed that solving $C_\omega z = v$ for z does not require factorization or even evaluation of the product $J^T J$. The only additional feature sufficient for implementing $\text{SSOR}(\omega)$ preconditioning is the direct access to the columns of $J(x)$. Therefore, using the Column Compressed Storage format for the sparse storage of $J(x)$ one can completely avoid the need in extra workspace for the use of $\text{SSOR}(\omega)$ preconditioning.

2.7   Description of Computational Algorithm

The preconditioned subspace descent algorithm can be summarized as follows:

Algorithm 1.
Input: $J(x) \in R^{m \times n}$, $f(x) \in R^m$, $x_0 \in R^n$;
Initialization:
if (tuning $=$ 'easy') **then** $\omega = 1$, $\alpha = 1.0$; $\beta = 2.0$ **end if**;
if (tuning $=$ 'hard') **then** $\omega = 1/3$, $\alpha = 1.8$; $\beta = 1.8$ **end if**;
$\xi = 1$, $s = 5$, $\delta = 10^{-8}$, $\varepsilon = 10^{-12}$,
$\tau_{\min} = 10^{-8}$, $k_{\max} = 20000$, $l_{\max} = 30$;
$f_0 = f(x_0)$,
$\rho_0 = f_0^T f_0$;
Iterations:
for $k = 0, 1, \dots, k_{\max} - 1$:
    $A_k = J(x_k)$
    $D = \text{Diag}(A_k^T A_k)$; (**denote** $A_k^T A_k = D + L + L^T$)
    $v_1 = (\frac{1}{\omega}D + L^T)^{-1}(I + \omega L D^{-1})^{-1} A_k^T f_k$
    $t_{\max} = \min(k + 1, s)$
    for $t = 1, \dots, t_{\max}$:
        $u_t = A_k v_t$ (**denote** $V_k = [v_1 \mid \dots \mid v_{t_{\max}}]$; $U_k = [u_1 \mid \dots \mid u_{t_{\max}}]$)

end for
$$S_k = U_k^T U_k + \xi \sqrt{\rho_k} V_k^T V_k$$
$$S_k := S_k + \delta \mathrm{Diag}(S_k) + \tau \|A_k\|_F^2 I$$
$$z_k = S_k^{-1}(U_k^T f_k)$$
$$p_k = -V_k z_k$$
$$v_{2+k \bmod(s-1)} := p_k$$
$$w_k = A_k p_k$$
$$\sigma_k = f_k^T w_k / w_k^T w_k$$
$$p_k := \sigma_k p_k$$
$$\theta_k = (f_k^T w_k)^2 / (\rho_k w_k^T w_k)$$
$$\alpha^{(0)} = \alpha$$
for $l = 0, 1, \ldots, l_{\max} - 1$:
$$x_k^{(l)} = x_k + \alpha^{(l)} p_k$$
$$f_k^{(l)} = f(x_k^{(l)})$$
$$\rho_k^{(l)} = (f_k^{(l)})^T f_k^{(l)}$$
$$\tau = \alpha^{(l)}(2 - \alpha^{(l)})\theta_k$$
  if $(\tau < \tau_{\min})$ **return** x_k
  if $((\rho_k^{(l)}/\rho_k)^{1/2} > \sqrt{1-\tau} + \tau/2)$ **then**
$$\alpha^{(l+1)} = \alpha^{(l)}/\beta$$
$$x_k^{(l+1)} = x_k + \alpha^{(l+1)} p_k$$
  else
      go to NEXT
  end if
end for
NEXT: $x_{k+1} = x_k^{(l)}, \quad f_{k+1} = f_k^{(l)}, \quad \rho_{k+1} = \rho_k^{(l)};$
  if $(\rho_{k+1} < \varepsilon^2 \rho_0)$ **or** $(\rho_{k+1} \geq \rho_k)$ **return** x_{k+1}
end for

3   Test Problems and Numerical Results

Below, the test problem settings and the results obtained for Algorithm 1 are described. For the test runs, one core of Pentium(R) Dual-Core CPU E6600 3.06 GHz, 3.25 Gbytes RAM desktop PC was used. We will compare the "easy" tuning ($\alpha_l = 2^{1-l}$ and $\omega = 1$) with the "hard" one ($\alpha_l = 1.8^{2-l}$ and $\omega = 1/3$) and consider different values of subspace dimension ($s = 1, 2, 3, 5, 9$). Unless otherwise stated, the residual norm reduction was set as $\varepsilon = 10^{-14}$ with the iteration number limit $k_{max} = 100000$. In the case of nonzero residual problems, the iterations typically terminate by the condition $\tau < \tau_{\min} = 10^{-8}$. Note that the iteration number always coincides with the number of the Jacobian evaluations.

One common feature observed in the test runs is that the total number of function evaluations (as required by the proposed backtracking procedure) is only slightly larger that the iteration number. Even for Chained Rosenbrock test the ratio (#f evals.)/(#J evals.) is smaller than 3, see Table 4. Another general conclusion is that the use of SSOR preconditioning and inclusion of several previous directions in the subspace basis are often necessary for the success of the method.

3.1   Rosenbrock Function

Following [14], for $n = m = 2$ define $f(x)$ as

$$f_1 = 10(x_1^2 - x_2), \qquad f_2 = x_1 - 1.$$

The optimum value is $f^T f = 0$ at $x_* = [1,\ 1]^T$ and the starting point is $\widetilde{x} = [-1.2,\ 1]^T$. The results for 'easy' tuning are given in Table 1. As is seen from the results presented, the convergence of the method improves sharply with $s \geq 2$ compared to $s = 1$. In this case, the type of preconditioning used is not critical for the convergence. One can also notice relatively small number of backtracking steps (Table 3).

3.2   Biggs6 Test Function

Following [14], for $n = 6$ and $m = 13$ define $f(x)$ as

$$f_i = x_3 e^{-t_i x_1} - x_4 e^{-t_i x_2} + x_6 e^{-t_i x_5} - e^{-t_i} + 5e^{-10 t_i} - 3e^{-4 t_i}, \quad 1 \leq i \leq 13,$$

where $t_i = i/10$ and $e = \exp(1)$. The optimum value is $f^T f = 0$ at $x_* = [1, 10, 1, 5, 4, 3]^T$ and the starting point is set as $\widetilde{x} = [1, 2, 1, 1, 1, 1]^T$. Other exact solutions with $f = 0$ are $x_* = [4, 10, 3, 5, 1, 1]^T$ and $x_* = [10, 4, 5, 3, 1, 1]^T$. The convergence to one or another solution was observed depending on the subspace dimension s. The results are given in Table 2. This small-sized but hard enough test shows that simultaneous use of a stronger preconditioner and a larger subspace dimension s can be essential for the efficiency of the method. Test runs with Jacobi preconditioning were a complete failure and not shown in Table 2. Further, it appears that not only the choice of the initial guess, but also the size of search subspaces may lead to different solutions (if the solution is not unique).

3.3   Broyden Tridiagonal Function

Following [14], for $n = m$ and $m = 500$ define $f(x)$ as

$$f_i = (3 - 2x_i)x_i - x_{i-1} - 2x_{i+1} + 1, \quad 1 \leq i \leq 500,$$

where $x_0 = x_{n+1} = 0$. The optimum value is $f^T f = 0$ and the starting point is set as $\widetilde{x} = [-1, \ldots, -1]^T$. This test demonstrates that with SSOR preconditioning tuning the performance of Algorithm 1 can be considerably better than with Jacobi one. This test can be qualified as relatively easy due to the actual closeness of the initial guess $\widetilde{x}$ to the solution x_*.

Table 1. Performance of Algorithm 1 for Rosenbrock function with $m = n = 2$

Precond.	Tuning	s	$\#J$ evals.	$\#f$ evals.	Remark
Jacobi	Easy	1	135	146	OK
Jacobi	Easy	2	15	22	OK
Jacobi	Easy	3	15	22	OK
Jacobi	Easy	5	15	22	OK
Jacobi	Easy	9	15	22	OK
SSOR	Easy	1	709	719	OK
SSOR	Easy	2	16	23	OK
SSOR	Easy	3	15	22	OK
SSOR	Easy	5	15	22	OK
SSOR	Easy	9	15	22	OK

Table 2. Performance of Algorithm 1 for Biggs6 test with $m = 13$ and $n = 6$

Precond.	Tuning	s	$\#J$ evals.	$\#f$ evals.	Remark
SSOR	Hard	1	$\gg 100000$	n/a	Failed
SSOR	Hard	2	4209	4211	OK
SSOR	Hard	3	1880	1882	OK
SSOR	Hard	5	414	431	OK
SSOR	Hard	9	321	322	OK
SSOR	Easy	1	$\gg 100000$	n/a	Failed
SSOR	Easy	2	5000	5022	OK
SSOR	Easy	3	15994	16012	OK
SSOR	Easy	5	273	275	OK
SSOR	Easy	9	121	125	OK

Table 3. Performance of Algorithm 1 for Broyden tridiagonal test with $m = n = 500$

Precond.	Tuning	s	$\#J$ evals.	$\#f$ evals.	Remark
Jacobi	Easy	1	146	147	OK
Jacobi	Easy	2	51	52	OK
Jacobi	Easy	3	51	52	OK
Jacobi	Easy	5	46	47	OK
Jacobi	Easy	9	46	47	OK
SSOR	Easy	1	21	22	OK
SSOR	Easy	2	16	17	OK
SSOR	Easy	3	16	17	OK
SSOR	Easy	5	15	16	OK
SSOR	Easy	9	16	17	OK

Table 4. Performance of Algorithm 1 for Chained Rosenbrock test with $n = 100$

Precond.	Tuning	s	$\#J$ evals.	$\#f$ evals.	Remark
Jacobi	Easy	1	$\gg 100000$	n/a	Failed
Jacobi	Easy	2	1207	1234	OK
Jacobi	Easy	3	690	737	OK
Jacobi	Easy	5	600	655	OK
Jacobi	Easy	9	409	775	OK
SSOR	Easy	1	$\gg 100000$	N/a	Failed
SSOR	Easy	2	369	707	OK
SSOR	Easy	3	437	1098	OK
SSOR	Easy	5	350	679	OK
SSOR	Easy	9	352	697	OK

Table 5. Performance of Algorithm 1 for 3D tensor test with $m = 10^6$, $n = 1500$, and the initial guess $\tilde{x}_j = \mu(j + 2)$.

Precond.	s	$\|x_*\|_\infty$	$\#J$ evals.	$\#f$ evals.	Time, s	Remark
Jacobi	2	1.69	384	390	123.	OK
Jacobi	3	1.40	500	505	177.	OK
Jacobi	5	1.70	449	455	232.	OK
Jacobi	9	1.13	304	307	216.	OK
SSOR	2	1.59	348	354	283.	OK
SSOR	3	1.37	366	369	310.	OK
SSOR	5	2.09	591	595	549.	OK
SSOR	9	2.01	454	456	540.	OK

Table 6. Performance of Algorithm 1 with 'hard' tuning for FMM$(3,3,3;23)$ tensor test with $m = 729$, $n = 621$, and the initial guess $\tilde{x}_j = \mu(j + 29)$; "failed" means convergence to a solution with large residual norm.

Precond.	s	$\|x_*\|_\infty$	$\#J$ evals.	$\#f$ evals.	Time, s	Remark
SSOR	2	3.94	15438	15439	20.2	Failed
SSOR	3	5.68	35614	35615	49.6	Failed
SSOR	5	3.91	19873	19874	31.2	OK
SSOR	7	2.96	6059	6060	10.6	OK
SSOR	9	2.87	4088	4089	8.6	OK
SSOR	11	2.96	5582	5583	12.0	OK
SSOR	13	3.01	7123	7124	16.7	OK
SSOR	17	3.16	7895	7896	22.8	OK

Table 7. Performance of Algorithm 1 with "easy" tuning for flow in a porous medium test with $m = n = 10000$ and the initial guess $\widetilde{x}_j = 0.1$.

Precond.	s	$\|x_*\|_\infty$	#J evals.	#f evals.	Time, s	Remark
Jacobi	1	n/a	$\gg$100000	n/a	n/a	Failed
Jacobi	2	0.038	54195	54196	67.	OK
Jacobi	3	0.038	53983	53984	81.	OK
Jacobi	5	0.038	53943	53944	106.	OK
Jacobi	9	0.038	53969	53970	181.	OK
SSOR	1	n/a	$\gg$100000	n/a	n/a	Failed
SSOR	2	0.038	52506	52507	118.	OK
SSOR	3	0.038	44394	44395	110.	OK
SSOR	5	0.038	52710	52711	162.	OK
SSOR	9	0.038	52804	53805	238.	OK

3.4 Chained Rosenbrock Function

This test function was introduced in [20], and we will use its version with $m = 2n - 2$ and essentially variable coefficients:

$$f_{2i-1} = i(x_i - x_{i+1}^2), \qquad f_{2i} = 1 - x_{i+1}, \qquad i = 1, \ldots, n - 1.$$

The optimum value is $f^T f = 0$ at $x_* = [1, \ldots, 1]^T$ and the starting point is $\widetilde{x} = [-1, \ldots, -1]^T$. The results are given in Table 4 for $m = 198$ and $n = 100$. As can be seen from the convergence histories (not shown here), the convergence behavior of the method with $s \geq 2$ is rather similar to that of the linear conjugate gradient methods, though it may take more than m iterations to be observed. For this test problem, the backtracking steps were applied rather intensively (as is indicated by essential excess of the number of f evaluations over that of J evaluations). One can also notice the typical improvements of iteration number count with the use of a stronger preconditioner and a larger subspace size s.

3.5 Approximate Canonical Decomposition of Dense 3D Tensor

This problem was considered, e.g., in [11,15,17]. In general, the function is set as

$$f_{l_1+(l_2-1)m_1+(l_3-1)m_1m_2} = -t_{l_1,l_2,l_3} \tag{27}$$

$$+ \sum_{l=1}^{r} x_{(l-1)(m_1+m_2+m_3)+l_1} x_{lm_1+(l-1)(m_2+m_3)+l_2} x_{l(m_1+m_2)+(l-1)m_3+l_3},$$

$$1 \leq l_1 \leq m_1, \quad 1 \leq l_2 \leq m_2, \quad 1 \leq l_3 \leq m_3,$$

and we have $m = m_1 m_2 m_3$ and $n = (m_1 + m_2 + m_3)r$. The particular case we consider is (see, e.g., [17])

$$t_{l_1,l_2,l_3} = \left(l_1^2 + l_2^2 + l_3^2\right)^{-1/2}, \quad m_1 = m_2 = m_3 = 100, \quad r = 5.$$

The latter is rather hard-to-solve nonzero residual problem; in particular, for any x the Jacobian $J(x)$ has rank deficiency. Hence, there exist no isolated optimum solutions. Note also that $J(x)$ here is an 1000000×1500 sparse matrix containing only 15000000 nonzeroes.

It must be stressed that for such tensor decomposition problems, the choice of the initial guess is probably the most important tuning parameter. In our tests, we used a quasirandom sequence with elements in $\{-1, 0, 1\}$ given by the number-theoretic Moebius function $\mu(j)$, to form the initial guess $\tilde{x}_j = \mu(q+j)$, $j = 1, \ldots, n$, where q is an arbitrary nonnegative number. Recall that the formal definition of μ is $1 = (\sum_{k=1}^{\infty} k^{-t})(\sum_{k=1}^{\infty} \mu(k)k^{-t})$. Performance results obtained with $q = 2$ are given in Table 5. In all cases, the resulting optimal value was $\|f\| \approx 0.07388815$ while the initial value of $\|f\|$ was of the order 10^3. An important quality measure for the obtained solution x_* is its norm $\|x_*\|_\infty$ (the smaller, the better) is also shown.

For this test, only the "hard" parameter choice was used. For instance, with "easy" tuning in 6th row of Table 5 one would find numbers 7.87, 1591, 1592, 1248., i.e., the method performs 4 times slower. For the Jacobi preconditioning the difference is not so big, but the "easy" tuning is still inferior.

One can notice that the fastest test runs were performed with the use of the simplest Jacobi preconditioning.

3.6   Canonical Decomposition of Matrix Multiplication Tensor

This notoriously hard problem, also known as "Brent Equations" [6], was considered, e.g., in [11] and [19]. The problem setting is the same as in previous Subsection, but the components of the 3D tensor are specified by

$$t_{l_1,l_2,l_3} = \delta(i_2 - j_1)\delta(j_2 - k_1)\delta(k_2 - i_1), \tag{28}$$

where

$$1 \le l_1 \le n_3 n_1 = m_1, \quad l_1 = i_1 + (i_2 - 1)n_3, \quad 1 \le i_1 \le n_3, \quad 1 \le i_2 \le n_1,$$

$$1 \le l_2 \le n_1 n_2 = m_2, \quad l_2 = j_1 + (j_2 - 1)n_1, \quad 1 \le j_1 \le n_1, \quad 1 \le j_2 \le n_2,$$

$$1 \le l_3 \le n_2 n_3 = m_3, \quad l_3 = k_1 + (k_2 - 1)n_2, \quad 1 \le k_1 \le n_2, \quad 1 \le k_2 \le n_3.$$

Thus, we have $m = (n_1 n_2 n_3)^2$ and $n = (n_3 n_1 + n_1 n_2 + n_2 n_3)r$. The particular case of $n_1 = n_2 = n_3 = 2$ and $r = 7$ corresponds to a noncommutative bilinear algorithm for evaluation of the product of two 2×2 matrices using $r = 7$ multiplications, the first of which was discovered in [18], with the corresponding exact solution satisfying $x_j \in \{-1, 0, 1\}$. In general, any exact solution to the problem (27), (28) immediately yields the existence of a fast $N \times N$ matrix multiplication algorithm with arithmetic complexity $O(N^\sigma)$, where $\sigma = (3 \log r)/\log(n_1 n_2 n_3)$.

Depending on n_1, n_2, n_3, for some too small r this problem has no solutions, while for larger $r \le n_1 n_2 n_3$ it is solvable, but in all cases there exist no isolated solutions. Moreover, there are "approximate" solutions for which $\|f(x)\| \to 0$ while some components of x tend to zero, and some other ones increase to infinity.

In Table 6, the performance data obtained for the case $n_1 = n_2 = n_3 = 3$ and $r = 23$ are given (in short, $(3, 3, 3; 23)$-problem). This corresponds to the result of [12], where an exact solution with $x_j \in \{-1, 0, 1\}$ was presented, see also [7]. Of course, our approximate solutions have general real numbers as the components of x_k. The initial guess was taken as $\tilde{x}_j = \mu(j + 29)$. One can observe that with small values of s the convergence is slow. In fact, for $s = 2$ or $s = 3$ the iterates have converged to a nonzero residual solution x with $\|f(x)\|^2 = 1$. One can also notice the minimum possible number of function evaluations per one Jacobian evaluation. Similar or even better results were obtained with Algorithm 1 for another FMM tensor problems of the type $(2, 2, 2; 7)$, $(2, 3, 3; 15)$, $(2, 3, 4; 20)$, $(2, 3, 5; 25)$.

3.7 Flow in a Porous Medium 2D Problem

Next we consider the steady state case of differential equation that models the effects of capillary pressure and gravity on a fluid flow in a homogeneous medium (see [2] and references cited therein):

$$-\frac{\partial^2}{\partial x^2}(u^2) - \frac{\partial^2}{\partial y^2}(u^2) - c\frac{\partial}{\partial x}(u^3) - f(x, y) = 0, \qquad (x, y) \in \Omega = (0, 1)^2;$$

over the unit square Ω with homogeneous Dirichlet boundary conditions on $\partial\Omega$. The parameters were set as in [2]: $c = 50$ and f is a point source of magnitude 50 at the lower-left grid point $(x, y) = (h, h)$, where $h = 1/101$ is the spatial discretization step. The standard centered finite difference discretization (designed to obtain an approximation $u(ih, jh) \approx u_{i,j}$) on a uniform square grid with $n = 10000 = 100^2 = m^2$ internal nodes has the following form (with account of the scaling by h^2):

$$-u_{i,j-1}^2 - \left(u_{i-1,j}^2 - c_1 u_{i-1,j}^3\right) + 4u_{i,j}^2 - \left(u_{i+1,j}^2 + c_1 u_{i+1,j}^3\right) - u_{i,j-1}^2 - c_2\delta(i-1)\delta(j-1) = 0,$$

for $1 \leq i, j \leq m$, where $c_1 = ch/2$, $c_2 = ch^2$, and the discrete boundary conditions $u_{0,j} = u_{m+1,j} = 0$ and $u_{i,0} = u_{i,m+1} = 0$ are used to eliminate these components from the difference equations. The n-vector of unknowns was formed in a standard way as $x_{i+(j-1)m} = u_{i,j}$. The resulting nonlinear problem $f(x) = 0$ was also included in the recent nonlinear test collection [13] as Problem 3.27. With the initial guess $x_k = 0.1$, $k = 1, \ldots, n$, the solution precision $\varepsilon = 10^{-8}$, and under the "easy" tuning, the results are shown in Table 7. The best performance was obtained with $s = 2$ for the simplest Jacobi preconditioning. One can also notice the complete absence of backtrackings. On the other hand, the iteration number count of the order 50000 may seem large; nevertheless, our calculation time is several times smaller compared to that presented in [2] (345 or 545 s for this problem) corresponding to certain specialized preconditionings.

4 Concluding Remarks

In the present paper, a nonlinear least squares solver is developed which do not require second-order information, can efficiently use the sparsity of the Jacobian, and is formally applicable to all types of least squares problems. The results of extensive numerical testing (of which only a small part is included here), using not only small- and medium-sized standard test problems, but also larger and harder examples, such as finite-difference discretizations of nonlinear boundary problems for 2nd and 4th order partial differential equations as well as algebraic problems associated with matrix scaling or canonical decomposition of dense and sparse 3D tensors, have showed a promising potential of the Preconditioned Subspace Descent method in solving hard nonlinear problems.

References

1. Armijo, L.: Minimization of functions having Lipschitz continuous first partial derivatives. Pac. J. Math. **16**(1), 1–3 (1966)
2. Bellavia, S., Bertaccini, D., Morini, B.: Nonsymmetric preconditioner updates in Newton-Krylov methods for nonlinear systems. SIAM J. Sci. Comput. **33**(5), 2595–2619 (2011)
3. Bellavia, S., Gratton, S., Riccietti, E.: A Levenberg-Marquardt method for large nonlinear least-squares problems with dynamic accuracy in functions and gradients. Numer. Math. **140**(3), 791–825 (2018)
4. Bellavia, S., Morini, B.: Strong local convergence properties of adaptive regularized methods for nonlinear least squares. IMA J. Numer. Anal. **35**(2), 947–968 (2015)
5. Buckley, A., Lenir, A.: QN-like variable storage conjugate gradients. Math. Program. **27**, 155–175 (1983)
6. Brent, R.P.: Algorithms for matrix multiplication (No. STAN-CS-70-157). Stanford Univ. CA Dept. of Computer Science (1970)
7. Heule, M.J., Kauers, M., Seidl, M.: New ways to multiply 3 x 3-matrices. arXiv preprint arXiv:1905.10192 (2019)
8. Kaporin, I.E.: Estimating global convergence of inexact Newton methods via limiting step size along normalized direction. Rep 9329, Department of Mathematics, Catholic University of Nijmegen, Nijmegen, The Netherlands, 8p., July 1993
9. Kaporin, I.E.: The use of preconditioned Krylov subspaces in conjugate gradient type methods for the solution of nonlinear least square problems. (Russian) Vestnik Mosk. Univ., Ser. 15 (Computational Math. and Cybernetics), No. 3, 26–31 (1995)
10. Kaporin, I.E., Axelsson, O.: On a class of nonlinear equation solvers based on the residual norm reduction over a sequence of affine subspaces. SIAM J. Sci. Comput. **16**(1), 228–249 (1994)
11. Kazeev, V.A., Tyrtyshnikov, E.E.: Structure of the Hessian matrix and an economical implementation of Newton's method in the problem of canonical approximation of tensors. Comput. Math. Math. Phys. **50**(6), 927–945 (2010)
12. Laderman, J.D.: A noncommutative algorithm for multiplying 3 x 3 matrices using 23 multiplications. Bull. Am. Math. Soc. **82**(1), 126–128 (1976)
13. Luksan, L., Matonoha, C., Vlcek, J.: Problems for nonlinear least squares and nonlinear equations. Technical Report V-1259. ICS AS CR, Prague, 36 s (2018)
14. More, J.J., Garbow, B.S., Hillstrom, K.E.: Testing unconstrained optimization software. Argonne National Laboratory, Applied Mathematics Division, Technical Memorandum No. 324, 96 pp. July 1978
15. Oseledets, I.V., Savostyanov, D.V.: Minimization methods for approximating tensors and their comparison. Comput. Math. Math. Phys. **46**(10), 1641–1650 (2006)

16. Sterck, H.D., Winlaw, M.: A nonlinearly preconditioned conjugate gradient algorithm for rank-R canonical tensor approximation. Numer. Linear Algebra Appl. **22**(3), 410–432 (2015)
17. Sterck, H.D., Miller, K.: An adaptive algebraic multigrid algorithm for low-rank canonical tensor decomposition. SIAM J. Sci. Comput. **35**(1), B1–B24 (2013)
18. Strassen, V.: Gaussian elimination is not optimal. Numer. Math. **13**(4), 354–356 (1969)
19. Tichavsky, P., Phan, A.H., Cichocki, A.: Numerical CP decomposition of some difficult tensors. J. Comput. Appl. Math. **317**, 362–370 (2017)
20. Toint, Ph.L.: Some numerical results using a sparse matrix updating formula in unconstrained optimization. Math. Comput. **32**(143), 839–851 (1978)
21. Yuan, Y.X.: Recent advances in numerical methods for nonlinear equations and nonlinear least squares. Numer. algebra, Control Optim. **1**(1), 15–34 (2011)

Comparison of Direct and Indirect Approaches for Numerical Solution of the Optimal Control Problem by Evolutionary Methods

Askhat Diveev[1,2] and Elizaveta Shmalko[1(✉)]

[1] Federal Research Center "Computer Science and Control" of the Russian Academy of Sciences, Moscow, Russia
aidiveev@mail.ru, e.shmalko@gmail.com
[2] RUDN University, Moscow, Russia

Abstract. The optimal control problem with phase constraints is considered. A new indirect approach of synthesized optimal control is proposed as an alternative to direct methods. A comparative study of direct and indirect approaches is carried out on the problem of optimal control for a small group of mobile robots in the complex environment with phase constraints by evolutionary algorithms. With a direct approach to the numerical solution of the optimal control problem, the control function is searched in the form of piece-wise functional approximation. The indirect approach of synthesized optimal control comes from the engineering practice. Instead of reducing the optimal control problem to the problem of finite-dimensional optimization, we firstly make the object stable relative to some point in the state space by solving an additional task of synthesis of stabilizing control and then we find the coordinates of stabilization points as the desired parameters of optimal control.

Keywords: Optimal control · Phase constraints · Network operator · Evolutionary algorithms · Control synthesis

1 Introduction

Today the problem of optimal control in each specific case is numerically solved by one of two well-known approaches [1]. The first approach, which is often called direct, is to reduce the optimal control problem to a nonlinear programming problem [2]. This provides the transition from an optimization problem in an infinite-dimensional space to an optimization problem in a finite-dimensional space. Another approach, which is often called the indirect one, is to use the Pontryagin maximum principle. As a result of using this principle, the optimization problem in the infinite-dimensional space is transformed into a boundary-value

Supported by the Russian Science Foundation (project 19-11-00258).

M. Jaćimović et al. (Eds.): OPTIMA 2019, CCIS 1145, pp. 180–193, 2020.
https://doi.org/10.1007/978-3-030-38603-0_14

problem, in which it is necessary to find the initial conditions for a system of differential equations for conjugate variables.

Both of these known approaches face computational problems when there are phase constraints in the optimal control problem [3]. The usual phase constraints often lead to the loss of the unimodality property of the objective functional [4]. Then it is necessary to apply global optimization methods [5]. Moreover solving the optimal control problem based on the Pontryagin maximum principle leads to the formulation of additional conditions that the control must satisfy, along with ensuring the maximum of the Hamiltonian, that is not so easy for problems with complex phase constraints. At the same time the solution of the optimal control problem by the direct approach without using the maximum principle does not allow to establish how close the obtained solution is to the optimal one.

In this paper, we propose a new indirect approach of synthesized optimal control to solve the problem of optimal control. The method consists in solving at the first stage the problem of stabilization control synthesis ensuring the stability of the control object relative to some given point in the state space. At the second stage, the coordinates of stabilization points are searched for using finite-dimensional optimization methods. The control is performed by switching stabilization points. The switching of points is performed in a given time interval. The synthesized optimal control immediately considers the problem as finite-dimensional.

The method of synthesized optimal control can be considered as an independent computational method for solving the problem of optimal control, which can also be attributed to indirect methods.

The optimal control problem has no practical sense due to the fact that the optimal control is received from the maximum of the Hamiltonian, but it depends on the right parts of the model equations, which are never exactly known. In the synthesized optimal control, the inaccuracy of the right parts is compensated by the stability of the system relative to a point in the state space. Near a stable point, all solutions converge. Generally speaking, the proposed approach looks like some kind of parametrization, but with an essential peculiarity: parametrization is performed after solving the synthesis problem (synthesis of the stabilization system). This additional step is a key idea, it provides achievement of better results in the tasks with complex environment and noise.

This approach is inspired by practice. The developers of control systems in most cases initially make the object stable, and then look for optimal control for it. However, this approach could not be previously presented as a single computational method, since at the first stage it is necessary to solve the problem of control synthesis, which is much more complicated than the optimal control problem and has always been solved for a specific object using methods that fit specific mathematical models of objects. To solve the problem of control synthesis [6–8], we apply numerical methods of symbolic regression, which use the evolutionary algorithms to find the code of the mathematical expression of the control function as a composition of elementary functions.

The proposed indirect approach to solving the problem of optimal control through the solution of the problem of control synthesis seems to be rather complex. But it proves to be better and more stable than reduction to the problem of finite-dimensional optimization. This we illustrate in the experimental part.

2   The Problem of Optimal Control

Consider the problem of optimal control.
   A model of the control object is

$$\dot{\mathbf{x}} = \mathbf{f}(\mathbf{x}, \mathbf{u}), \tag{1}$$

where $\mathbf{x} \in \mathbb{R}^n$.
   The control is constrained

$$\mathbf{u} \in U \subseteq \mathbb{R}^m, \tag{2}$$

where U is a constrained compact set, $m \le n$.
   Given the initial conditions

$$\mathbf{x}(0) = \mathbf{x}^0. \tag{3}$$

Given the terminal conditions

$$\varphi_i(\mathbf{x}) = 0, i = \overline{1, r}. \tag{4}$$

Given the quality functional

$$J = v(\mathbf{x}(t_f)) + \int_0^{t_f} f_0(\mathbf{x})dt \rightarrow \min, \tag{5}$$

where

$$t_f = \begin{cases} t, & \text{if } t < t^+ \text{ and } \sqrt{\sum_{i=1}^r \varphi_i^2(x)} \le \varepsilon \\ t^+ & \text{otherwise} \end{cases}, \tag{6}$$

t^+ and ε are given positive values.
   It is necessary to find control as a function of time

$$\mathbf{u} = \mathbf{g}(t), \ \mathbf{g}(t) \in U \subseteq \mathbb{R}^m \ \forall t \in [0; t^+], \tag{7}$$

so that the solution

$$\mathbf{x} = \mathbf{s}(\mathbf{x}^0, t) \tag{8}$$

of the system of differential equations

$$\dot{\mathbf{x}} = \mathbf{f}(\mathbf{x}, \mathbf{g}(t)), \tag{9}$$

with initial conditions (3) within time interval $t \le t^+$ provides fulfillment with ε precision of terminal conditions (4)

$$\sqrt{\sum_{i=1}^r \varphi_i^2(\mathbf{s}(\mathbf{x}^0, t))} \le \varepsilon, \ 0 \le t \le t^+, \tag{10}$$

and minimize the quality functional

$$
\begin{aligned}
v(\mathbf{s}(\mathbf{x}^0, t_f)) &+ \int_0^{t_f} f_0(\mathbf{s}(\mathbf{x}^0, t)) dt \\
&= \min_{\forall \tilde{u} \in U} \left\{ v(\mathbf{x}(t_f)) + \int_0^{t_f} f_0(\mathbf{x}) dt \right\},
\end{aligned}
\tag{11}
$$

where $\tilde{u}$ is any admissible control ensuring the fulfillment of terminal conditions (10) for solving the system (1).

3   The Problem of Synthesized Optimal Control

Practical engineers intuitively solve the task of finding the optimal trajectory and ensuring the steady movement of objects along it in two stages: firstly, they make the object stable, and then ensure the stable movement of the object through the optimal trajectory points. The method of synthesized optimal control is close in its idea to engineering, but is mathematically formalized and based on the use of modern numerical methods of symbolic regression and evolutionary algorithms.

Initially, we ensure the stabilization of an object with respect to a certain point in the state space, solving the problem of control system synthesis.

Consider the problem statement of control synthesis.

Given a mathematical model of the control object in the form of a system of ordinary differential equations (1), and let us also be given the constraints on control (2).

The area of initial conditions is set

$$
\mathbf{x}(0) \in X_0 \subseteq \mathbb{R}^n.
\tag{12}
$$

Given the terminal state

$$
\mathbf{x}(t_f) = \mathbf{x}^f \in \mathbb{R}^n,
\tag{13}
$$

where t_f is time to reach terminal conditions.

The control quality criterion is determined by

$$
\tilde{J}_s = t_f \to \min.
\tag{14}
$$

It is necessary:

1. to find a control function in the form

$$
\tilde{\mathbf{u}} = \mathbf{h}(\mathbf{x}^* - \mathbf{x}),
\tag{15}
$$

where $\mathbf{x}^*$ is an arbitrarily given stabilization point in the state space, such that the system of differential equations

$$
\dot{x} = \mathbf{f}(\mathbf{x}, \mathbf{h}(\mathbf{x}^* - \mathbf{x}))
\tag{16}
$$

stable with respect to $\mathbf{x}^k$ for the initial state

$$||\mathbf{x}(0) - \mathbf{x}^*|| < \delta, \tag{17}$$

has a solution $\mathbf{x}(t) = \mathbf{s}(\mathbf{x}(0), t)$, that satisfies the constraints on control (2) $\mathbf{h}(\mathbf{x}^* - \mathbf{s}(\mathbf{x}(0), t)) \in U$ and for the finite specified period of time t^* verge towards a point $\mathbf{x}^*$ by the amount not exceeding the specified value ε^*

$$||\mathbf{x}^* - \mathbf{s}(\mathbf{x}(0), t^*)|| < \varepsilon^*; \tag{18}$$

2. to find stabilization points $\mathbf{x}^1, \ldots, \mathbf{x}^M$, which provide for the solution $\mathbf{s}(\mathbf{x}(0), t)$ of the system

$$\dot{\mathbf{x}} = \mathbf{f}(\mathbf{x}, \mathbf{h}(\mathbf{x}^*(t) - \mathbf{x})), \tag{19}$$

where $\mathbf{x}^*(t)$ is a piece-wise constant function defined by the relation

$$\mathbf{x}^*(t) = \begin{cases} \mathbf{x}^k, & \text{if } (k-1)\Delta t \le t < k\Delta t, \\ & k = 1, \ldots, M; \\ \mathbf{x}^f, & \text{if } \Delta t M \le t; \end{cases} \tag{20}$$

from any initial state of (12) reaching the terminal conditions (13) with the minimum value of the quality criterion (14) for the class of control, determined by the control function (15) and the values M and Δt. Here M is a given number of stabilization points, Δt is a given positive value.

Step 1. To solve the synthesis problem, we use symbolic regression methods which allow finding in an encoded form the analytic form of the mathematical expression of the control function (15).

For the numerical solution of the synthesis problem, we replace the continuous set of initial conditions (12) with the finite set of points of the initial conditions

$$\mathbf{x}(0) \in \{\mathbf{x}^{0,1}, \ldots, \mathbf{x}^{0,K}\} \subseteq \mathbb{R}^n. \tag{21}$$

Then the time to reach the terminal state depends on the initial condition, therefore we replace the functional (14) with the sum of functionals for all the initial conditions from (21)

$$\tilde{\tilde{J}}_s = \sum_{i=1}^{K} t_f(\mathbf{x}^{0,i}) \to \min. \tag{22}$$

To exclude in the numerical search such solutions that do not achieve the terminal state (13), we include the accuracy of reaching the terminal state into the functional (22), while adjusting the conditions for reaching the terminal state by (6)

$$J_s = \sum_{i=1}^{K} t_f(\mathbf{x}^{0,i}) + ||\mathbf{x}^f - \mathbf{s}(\mathbf{x}^{0,i}, t_f)|| \to \min, \tag{23}$$

where

$$t_f = \begin{cases} t, & \text{if } t < t^+ \text{ and } \sqrt{\sum_{j=1}^{n}(x_j^f - x_j)^2} < \varepsilon \\ t^+ & \text{— otherwise} \end{cases} , \tag{24}$$

t^+ and ε are given positive values.

Symbolic regression methods appeared on the basis of genetic programming [9,10]. Their main goal is to search for optimal solutions on a non-numeric space. Now there are quite a lot of methods of symbolic regression: analytical programming [11], Cartesian genetic programming [12], grammatical evolution [13], parse matrix evolution [14], network operator [15] and others. To solve the problem of control synthesis (1), (2), (13), (15), (23), we search for the control function as a code of analytic record of a mathematical expression on the space of codes of function compositions. For an effective search, we added a number of parameters as arguments to the desired mathematical expression, the values of which are also searched by criterion (23).

Step 2. We find such a sequence of stabilization points in the state space, that by switching stabilization points at fixed times, we ensure the movement of the object from the initial state to the terminal state with the optimal value of the quality criterion. Various numerical methods of optimization can be used, but from our experience the most effective methods suitable for solving the problem are evolutionary algorithms.

Further let us consider the application of the proposed method to solve the problem of optimal control for a group of robots with phase constraints and compare the results obtained with the results of solving the same problem by the known methods of finite-dimensional optimization.

4   A Comparative Example

Let us consider the problem of optimal control of a group of $N = 2$ mobile robots with phase constraints. The mathematical model of each robot is described by a system of $n = 3$ equations [16]:

$$\dot{x}_{1+(j-1)n} = 0.5(u_{1+(j-1)m} + u_{2+(j-1)m})\cos(x_{3+(j-1)n}) + B\xi,$$

$$\dot{x}_{2+(j-1)n} = 0.5(u_{1+(j-1)m} + u_{2+(j-1)m})\sin(x_{3+(j-1)n}) + B\xi, \tag{25}$$

$$\dot{x}_{3+(j-1)n} = 0.5(u_{1+(j-1)m} - u_{2+(j-1)m}) + B\xi,$$

where n is a dimension of the model of one robot, m is a dimension of the control vector for one robot, $m = 2$, $\mathbf{x} = [x_1, \ldots, x_{3+(N-1)n}]^T$ is a state vector of the whole group of robots, $\mathbf{u} = [u_1, \ldots, u_{2+(N-1)m}]^T$ is a control vector of the group of robots, $j = 1, \ldots, N$, N is the number of robots in the group. The component $B\xi$ is responsible for perturbations.

Control of each robot is similarly constrained

$$u_i^- \leq u_{i+(j-1)m} \leq u_i^+, i = \overline{1,m}, j = \overline{1,N}. \tag{26}$$

Given the initial position of each robot

$$x_{i+(j-1)n}(0) = x^0_{i+(j-1)n}, i = \overline{1,m}, j = \overline{1,N}. \tag{27}$$

Given static phase constraints

$$\beta(\mathbf{x}, j) = r^2 - (x^*_1 - x_{1+(j-1)n})^2 - (x^*_2 - x_{2+(j-1)n})^2 \le 0, j = \overline{1,N}, \tag{28}$$

where r is a given positive value, x^*_1, x^*_2 are coordinates of the center of the static phase constraints.

Given dynamic phase constraints that take into account the possibility of collision of any pair of robots between themselves

$$\begin{aligned} \delta_{k(j,i)}(\mathbf{x}(t)) = r_0^2 &- (x_{1+(j-1)n} - x_{1+(i-1)n})^2 \\ &- (x_{2+(j-1)n} - x_{1+(i-1)n})^2 \le 0, \end{aligned} \tag{29}$$

where $i \ne j$,

$$k(i,j) = i - j + (j-1)(N - 0.5j), \tag{30}$$

$j = \overline{1,N-1}$, $i = j+1, \dots, N$, r_0 is a given positive value, which determines the overall size of a single robot.

The maximum number of checks of the dynamic phase constraints is equal to the number of combinations of 2 of N. From (30) this number is obtained at $j = N - 1$ and $i = N$:

$$\begin{aligned} k(N, N-1) &= N - N + 1 + (N - 1 - 1)(N - 0.5(N-1)) \\ &= 1 + (N-2)(0.5N + 0.5) = 1 + 0.5(N^2 - N - 2) \\ &= 0.5(N^2 - N) = 0.5N(N-1). \end{aligned}$$

Terminal states are set for each robot

$$x_{i+(j-1)n} = x^f_{i+(j-1)n}, i = 1, \dots, n, j = \overline{1,N}. \tag{31}$$

Given the quality criterion of control

$$\tilde{J} = t_f \to \min, \tag{32}$$

where t_f is a time of control

$$t_f = \begin{cases} t, & \text{if } t < t^+ \text{ and} \\ & \max\{||\Delta x_{(i+j-1)n}(t)||_2 : j = \overline{1,N}\} < \varepsilon; \\ t^+ & - \text{ otherwise,} \end{cases} \tag{33}$$

$$||\Delta x_{(i+j-1)n}(t)||_2 = \sqrt{\sum_{i=1}^{n}(x_{i+(j-1)n}(t) - x^f_{i+(j-1)n})^2},$$

t^+ is the maximum possible control time, ε is a small positive value.

We include the phase constraints into the quality criterion, using the Heaviside function

$$J = t_f + \sum_{j=1}^{N} \int_0^{t_f} \vartheta(\beta(\mathbf{x}(t), j)) dt$$

$$+ \sum_{i=1}^{N-1} \sum_{j=i+1}^{N} \int_0^{t_f} \vartheta(\delta_{k(j,i)}(\mathbf{x}(t))) dt \to \min. \tag{34}$$

where $\beta(\mathbf{x}(t))$ and $\delta_k(\mathbf{x}(t))$ are determined by (28) and (29), respectively.

The optimal control by the direct and indirect approach is searched without perturbations at $B = 0$. We carry out 10 tests. The results of the algorithm were evaluated by three indicators: the best solution found, the average value of all runs and the mean square deviation.

In the computational experiment, we used the following parameter values: $n = 3$, $m = 2$, $N = 2$, $x_1^0 = 0$, $x_2^0 = 0$, $x_3^0 = 0$, $x_4^0 = 0$, $x_5^0 = 10$, $x_6^0 = 0$, $x_1^f = 10$, $x_2^f = 10$, $x_3^f = 0$, $x_4^f = 10$, $x_5^f = 10$, $x_6^f = 0$, $t^+ = 2.8$, $\varepsilon = 0.01$, $u_1^- = -10$, $u_2^- = -10$, $u_1^+ = 10$, $u_2^+ = 10$.

4.1   The Direct Approach

We reduce the infinite-dimensional optimal control problem (25)–(34) to the finite-dimensional optimization problem. Define the time interval Δt and divide the time axis into K intervals

$$K = \left\lfloor \frac{t^+}{\Delta t} \right\rfloor + 1. \tag{35}$$

In each interval, we approximate the control by a function depending on a finite number of parameters. If the approximating function goes beyond the control constraints, then replace the value of the function with the value of the violated constraint.

$$u_{i+(j-1)m}(t) = \begin{cases} u_i^+, & \text{if } \tilde{u}_{i+(j-1)m}(t) \geq u_i^+, \\ u_i^-, & \text{if } \tilde{u}_{i+(j-1)m}(t) \leq u_i^-, \\ \tilde{u}_{i+(j-1)m}(t) - \text{otherwise} \end{cases} \tag{36}$$

$$i = \overline{1, m}, j = \overline{1, N},$$

where $\tilde{u}_{i+(j-1)m}(t)$ is a value of the approximating function.

For approximation of control we use piece-wise linear function. We solved the nonlinear programming problem by a well-known algorithm particle swarm optimization (PSO) [17].

The best found solution is

$$\widetilde{q} = [19.1645 \quad 19,6240 \quad 18.3454 \quad 19.2507 \quad -19.9115 \quad 2.9243 \quad -19.8968$$
$$-14.7644 \quad -16.0936 \quad -1.9976 \quad 19.7917 \quad 10.8413 \quad 19.3851 \quad 15.0534$$
$$19.9553 \quad 19.6951 \quad -17.7232 \quad -18.0297 \quad -19.9845 \quad -17.1573 \quad -13.6002$$
$$-9.0025 \quad -13.3237 \quad -14.4927 \quad -19.8874 \quad -18.2106 \quad 6.8188 \quad -19.4085$$
$$-14.5783 \quad -16.3549 \quad -7.1894 \quad -6.3636 \quad 19.6455 \quad -14.8710 \quad -14.1214$$
$$-18.8555 \quad -10.9795 \quad -17.2831 \quad -19.6667 \quad -17.6576 \quad -18.6860 \quad -19.9358$$
$$18.2508 \quad 17.2661 \quad 8.0000]^{T}.$$

Figure 1 shows obtained trajectories of two robots on the plane. The best solution found has the functional value $J = 2.4322$, the average value of all runs is 3.088205 and the mean square deviation is 0.744130.

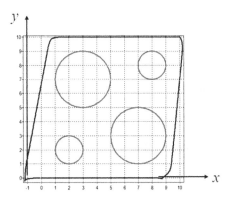

Fig. 1. Optimal trajectories of robots found by direct approach.

4.2 Synthesized Optimal Control

Now we solve the same problem by the method of synthesized optimal control. Initially, for one robot, we solve the problem of control synthesis (1), (12), (13), (15), (21), (23).

Step 1. The particular formulation of the control synthesis problem has the following description.

Given a mathematical model of the control object

$$\dot{x}_1 = 0.5(u_1 + u_2)\cos(x_3),$$

$$\dot{x}_2 = 0.5(u_1 + u_2)\sin(x_3), \tag{37}$$

$$\dot{x}_3 = 0.5(u_1 - u_2).$$

Given the constraints on control

$$-10 = u^- \le u_i \le u^+ = 10, i = 1, 2. \tag{38}$$

The terminal conditions are specified

$$x_i^f = 0, i = 1, 2, 3. \tag{39}$$

Given the set of initial conditions

$$X_0 = (\mathbf{x}^{0,1} = [-5\ -5\ -\pi/2]^T, \mathbf{x}^{0,2} = [-5\ -5\ \pi/2]^T,$$

$$\mathbf{x}^{0,3} = [-5\ 5\ -\pi/2]^T, \mathbf{x}^{0,4} = [-5\ 5\ \pi/2]^T,$$

$$\mathbf{x}^{0,5} = [5\ -5\ -\pi/2]^T, \mathbf{x}^{0,6} = [5\ -5\ \pi/2]^T, \tag{40}$$

$$\mathbf{x}^{0,7} = [5\ 5\ -\pi/2]^T, \ \mathbf{x}^{0,8} = [5\ 5\ \pi/2]^T).$$

The quality functional is given in the form (23), (24), where $K = 8$, $t^+ = 2$ s. We need to find the control in the form of (15).

To solve the problem, we use the network operator method [15]. As a result, the following control functions were obtained

$$\tilde{u}_1 = A^{-1} + \sqrt[3]{A} + \text{sgn}(q_3(x_3^f - x_3))\exp(-|q_3(x_3^f - x_3)|)$$

$$+ \text{sgn}(x_3^f - x_3) + \mu(B), \tag{41}$$

$$\tilde{u}_2 = \tilde{u}_1 + \sin(\tilde{u}_1) + \arctan(H) + \mu(B) + C - C^3, \tag{42}$$

where

$$A = \frac{1 - \exp(-D)}{1 + \exp(-D)} + \left(B + \sqrt[3]{x_1^f - x_1}\right)^3 + C + \sin(q_3(x_3^f - x_3)),$$

$$B = G + \text{sgn}(\text{sgn}(x_1^f - x_1)q_2(x_2^f - x_2)) \times \exp(-|\text{sgn}(x_1^f - x_1)q_2(x_2^f - x_2)|)$$

$$+ \sin(x_1^f - x_1) + \frac{1 - \exp(-G)}{1 + \exp(-G)} + x_1^f - x_1,$$

$$C = G + \text{sgn}(\text{sgn}(x_1^f - x_1)q_2(x_2^f - x_2))$$

$$\times \exp(-|\text{sgn}(x_1^f - x_1)q_2(x_2^f - x_2)|) + \sin(x_1^f - x),$$

$$D = H + C - C^3 + \text{sgn}(q_1(x_1^f - x_1)) + \arctan(q_1) + \vartheta(x_3^f - x_3),$$

$$G = \text{sgn}(x_1^f - x_1)q_2(x_2^f - x_2) + q_3(x_3^f - x_3) + \frac{1 - \exp(-q_1(x_1^f - x_1))}{1 + \exp(-q_1(x_1^f - x_1))},$$

$$H = \arctan(q_1(x_1^f - x_1) + \text{sgn}(W)\sqrt{|W|} + W + V + 2\text{sgn}\left(W + \frac{1 - \exp(-V)}{1 + \exp(-V)}\right)$$

$$+ \sqrt[3]{W + \frac{1 - \exp(-V)}{1 + \exp(-V)}}) + \sqrt[3]{x_1^f - x_1} + \text{sgn}(x_1^f - x_1)\sqrt{|x_1^f - x_1|}$$

$$+ \sqrt[3]{x_1^f - x_1 + \frac{1 - \exp(-V)}{1 + \exp(-V)}},$$

$$W = \text{sgn}(x_1^f - x_1) + \text{sgn}(q_2(x_2^f - x_2))\text{sgn}(x_1^f - x_1) \times \exp(-|q_2(x_2^f - x_2)(x_1^f - x_1)|) + V,$$

$$V = q_3(x_3^f - x_3) + \text{sgn}(x_1^f - x_1)q_2(x_2^f - x_2) + \frac{1 - \exp(x_1^f - x_1)}{1 + \exp(x_1^f - x_1)},$$

$$q_1 = 11.72820, q_2 = 2.02710, q_3 = 4.02222.$$

Step 2. The second step is to find the values of the terminal points as parameters of the control function:

$$q = [q_1 = x_1^{f,1} \ldots q_{nN} = x_{nN}^{f,1} \ldots \tag{43}$$

$$q_{(K-1)nN+1} = x_1^{f,K} \ldots q_{KnN} = x_{nN}^{f,K}]^T,$$

where K is the number of time intervals, N is the number of robots, n is a dimension of the mathematical model of a robot.

In this example we set the interval $\Delta t = 0.625$, with the maximum allowable control time $t^+ = 2.8$.

Constraints on the parameters are set from the possible values of the state vector:

$$q_i^- = -1 \leq q_i \leq 11 = q_i^+, i = 1, 2, 4, 5, 7, 8, 10, 11, \tag{44}$$

$$q_i^- = -1.57 \leq q_i \leq 1.57 = q_i^+, i = 3, 6, 9, 12. \tag{45}$$

To solve the problem, we use the same algorithm of PSO as for the reduced problem. We launched the algorithm 10 times.

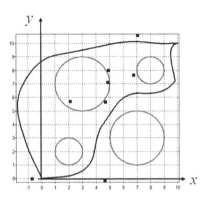

Fig. 2. Optimal trajectories of robots found by indirect approach.

The best found solution is

$$\tilde{q} = [7.1145 \quad 4.8470 \quad -0.2587 \quad 5.6976 \quad 2.1089 \quad -1.3800 \quad 7.9957$$
$$4.8950 \quad 0.4320 \quad 5.6465 \quad 4.6863 \quad -0.8672 \quad 7.6366 \quad 6.7588$$
$$0.3086 \quad -0.1725 \quad 4.6377 \quad 0.9390 \quad 10.5870 \quad 7.0599 \quad 1.4640$$
$$-0.0119 \quad -0.7168 \quad -0.3688].^T$$

Figure 2 shows obtained trajectories. The best solution found has the functional value $J = 2.5083$, the average value of all runs is 2.529371 and the mean square deviation is 0.009193.

Then, for the best solutions found by direct and synthesized methods, we simulate the system, increasing coefficient of noise B, and for different levels of B we also do 10 tests defining the best result, the average and square deviation. Table 1 contains values for direct approach, Table 2 contains values for indirect synthesized approach.

Table 1. Results for the direct approach.

Noise level	1	2	3	5
The best	4.4825	7.1111	8.9041	15.6606
Average	3.8959	1.0450	5.1924	7.6849
Mean-square deviation	0.1280	4.1890	2.1360	10.8489

Table 2. Results for the indirect approach.

Noise level	1	2	3	5
The best	2.7478	3.1479	3.9220	5.6547
Average	2.6587	2.7674	3.0777	3.6596
Mean-square deviation	0.0035	0.0394	0.1525	1.0115

To receive one solution by the direct method, the functional was calculated 404182 times, and with the synthesized approach 104182 times, that is four times less. The best solution found by the direct method, even it was better than that of the synthesized approach (2.4322 versus 2.5083) has poor square deviation (0.744 versus 0.009193). But as seen from the tables, it worsens significantly in the presence of perturbations, and for B = 5 the solution completely collapses, while that of synthesized approach preserves the quality, that can be seen in Figs. 3 and 4.

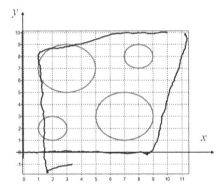

Fig. 3. Optimal trajectories of robots found by direct approach at noise level B = 5

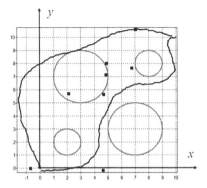

Fig. 4. Optimal trajectories of robots found by indirect approach at noise level B = 5

5   Conclusion

A comparative study of the direct and indirect approaches for solving the optimal control problem is presented on the example of solving an optimal control problem with phase constraints for a group of two mobile robots. The task was solved by reduction to the finite-dimensional optimization problem and by the method of synthesized optimal control. As a result of the computational experiment, it was shown that both approaches gave good results in the absence of perturbations, but the synthesized approach coped with the problem with a smaller number of calculations of the objective functional using the same evolutionary algorithm. And in the presence of some noise the method of synthesized optimal control allows us to find much better results than the reduction methods.

References

1. Gill, P.E., Murray, W., Wright, M.H.: Practical Optimization. Academic Press, London (1981)

2. Evtushenko, Y.G.: Optimization and fast automatic differentiation. Computing Center of RAS, Moscow (2013)
3. Diveev, A.I., Konstantinov, S.V.: Study of the practical convergence of evolutionary algorithms for the optimal program control of a wheeled robot. J. Comput. Syst. Sci. Int. **57**(4), 561–580 (2018)
4. Diveev, A., Sofronova, E., Dotsenko, A.: Violation of object functional unimodality and evolutionary algorithms for optimal control problem solution. In: IX International Conference on Optimization and Applications (OPTIMA 2018), Petrovac, Montenegro, pp. 128–140 (2018). https://doi.org/10.12783/dtcse/optim2018/27927
5. Kvasov, D.E., Sergeyev, Y.D.: Lipschitz global optimization methods in control problems. Autom. Remote Control **74**(9), 1435–1448 (2013)
6. Diveev, A.I., Shmalko, E.Yu.: Evolutionary computation for synthesis of control system for group of robots and optimum choice of trajectories for their movement. In: Proceedings of the 8th International Conference on Optimization and Applications, OPTIMA 2017, pp. 158–165 (2017)
7. Diveev A., Shmalko E.: Complete binary variational analytic programming for synthesis of control at dynamic constraints. In: ITM Web of Conferences, vol. 10 (2017)
8. Diveev, A.I., Shmalko, E.Yu.: Optimal control synthesis for group of robots by multilayer network operator. In: International Conference on Control, Decision and Information Technologies, CoDIT 2016, no. 3, pp. 77–82 (2016)
9. Koza, J.R.: Genetic Programming: On the Programming of Computers by Means of Natural Selection. MIT Press, Cambridge (1992)
10. Lee, E.B., Markus, L.: Foundations of Optimal Control Theory. Wiley, New York (1970)
11. Zelinka, I.: Analytic programming by means of soma algorithm. In: Proceedings of 8th International Conference on Soft Computing, pp. 93–101 (2002)
12. Miller, J.F., Smith, S.L.: Redundancy and computational efficiency in cartesian genetic programming. IEEE Trans. Evol. Comput. **10**(2), 167–174 (2006)
13. O'Neill, M., Ryan, C.: Grammatical evolution. IEEE Trans. Evol. Comput. **5**, 349–358 (2001)
14. Luo, C., Zhang, S.L.: Parse-matrix evolution for symbolic regression. Eng. Appl. AI **25**, 1182–93 (2012)
15. Diveev, A.I.: Numerical method for network operator for synthesis of a control system with uncertain initial values. J. Comp. Syst. Sci. Int. **51**(2), 228–243 (2012)
16. Suster, P., Jadlovska, A.: Tracking trajectory of the mobile robot Khepera II using approaches of artificial intelligence. Acta Electrotechnica et Informatica **11**(1), 38–43 (2011)
17. Kennedy, J., Eberhart, R.: Particle swarm optimization. In: Proceedings of IEEE International Conference on Neural Networks IV, pp. 1942–1948 (1995)

On PTAS for the Geometric Maximum Connected k-Factor Problem

Edward Gimadi[1,2] , Ivan Rykov[1,2], and Oxana Tsidulko[1,2($\boxtimes$)]

[1] Sobolev Institute of Mathematics, 4 Acad. Koptyug avenue, 630090 Novosibirsk, Russia
{gimadi,tsidulko}@math.nsc.ru, rykovweb@gmail.com
[2] Department of Mechanics and Mathematics, Novosibirsk State University, 1 Pirogova Street, 630090 Novosibirsk, Russia

Abstract. We consider the Connected k-factor problem (k-CFP): given a complete edge-weighted n-vertex graph, the goal is to find a connected k-regular spanning subgraph of maximum or minimum total weight. The problem is called geometric, if the vertices of a graph correspond to a set of points in a normed space $\mathbb{R}^d$ and the weight of an edge is the distance between its endpoints. The k-CFP is a natural generalization of the well-known Traveling Salesman Problem, which is equivalent to the 2-CFP. In this paper we complement the known $(1-1/k^2)$-approximation algorithm for the maximum k-CFP from [Baburin et al., 2007] with an approximation algorithm for the geometric k-CFP, that guarantees a relative error $\varepsilon = O\left((k/n)^{1/(d+2)}\right)$. Together these two algorithms form an asymptotically optimal algorithm for the geometric k-CFP with an arbitrary value of k in an arbitrary normed space of fixed dimension d. Finally, the asymptotically optimal algorithm can be easily transformed into a PTAS for the considered geometric problem.

Keywords: Asymptotically optimal algorithm · Polynomial time approximation scheme · Connected k-factor problem · Normed space · NP-hard problem

1 Introduction

In this paper we study polynomial-time approximation algorithms with performance guarantees for the following NP-hard problem.

Problem 1 (Connected k-factor problem, k-CFP).

Input: A complete undirected n-vertex simple graph $G = (V, E)$, edge weights $w : E \to \mathbb{R}_+$ and $k \in \mathbb{N}$.
Find: A connected spanning k-regular subgraph F_k (connected k-factor) of maximum (or minimum) total weight:

$$w(F_k) = \sum_{e \in F_k} w(e) \to \max \, (\min).$$

© Springer Nature Switzerland AG 2020
M. Jaćimović et al. (Eds.): OPTIMA 2019, CCIS 1145, pp. 194–205, 2020.
https://doi.org/10.1007/978-3-030-38603-0_15

If the edge weights w satisfy the triangle inequality, the problem is called *metric*. If the vertices of a graph correspond to a set of points in some normed space $(\mathbb{R}^d, || \cdot ||)$ and the weight of an edge $e = (u, v)$ is determined as $w(e) = ||u - v||$, the problem is called *geometric*.

The k-CFP is closely related to the network design, where various connectivity requirements are natural in the sense that they ensure the robustness of a network, while the regular degree requirements are also common, for example in constructing highly decentralized networks [11,24]. It is known that the problem is polynomially solvable if there is no requirement for the subgraph F_k to be connected [15,23] and NP-hard otherwise [9]. Note that the 2-CFP is the well-known NP-hard Traveling Salesman Problem (TSP), where, given a complete edge-weighted graph, it is required to find a Hamiltonian cycle of maximum or minimum total weight. The connected k-factor problem can be, in turn, considered as a generalization of the TSP. Since the problem is NP-hard even in Euclidean space [8], we are interested in approximation algorithms with performance guarantees.

In 2007 Baburin and Gimadi [5] proposed an approximation algorithm A_{BG} for the maximum k-CFP with guaranteed relative error $\varepsilon_{BG} \leq 1/k^2$ and $O(kn^3)$ running-time. Algorithm starts with computing a maximum weight k-factor using a polynomial-time algorithm by Gabow [15], and then connects the components of the obtained subgraph so that the degrees of the vertices in the subgraph remain equal to k and the total weight of the edges does not decrease much. Note that if k is a function of n that tends to infinity as n grows, then the relative error ε_{BG} of this algorithm tends to zero as n grows, and the algorithm produces the so-called *asymptotically optimal solutions*.

In this paper we supplement the result from [5], and obtain an asymptotically optimal algorithm for the geometric variant of the maximum k-CFP in case of arbitrary k and fixed dimension of a normed space $(\mathbb{R}^d, || \cdot ||)$. To this end in Sect. 2.2 we construct a polynomial-time approximation algorithm $\widetilde{A}$ for the geometric maximum k-CFP, and prove that it gives asymptotically optimal solutions in case of small or constant $k = o(n)$. This algorithm is based on the result [17] for the m-Peripatetic Salesman Problem (m-PSP), which employs geometrical ideas of Serdukov first appeared in [21] for the maximum Euclidean TSP. Later Shenmaier [22] showed that Serdukov's approach can be extended to the case of an arbitrary normed space $(\mathbb{R}^d, || \cdot ||)$.

Then in Sect. 2.3, we show how to combine algorithms A_{BG} and $\widetilde{A}$ in order to obtain an asymptotically optimal algorithm A_{geom} for the geometric maximum k-CFP for arbitrary k. Finally, from this result a PTAS follows: given an instance of the geometric k-CFP and an arbitrary fixed $\varepsilon > 0$, one should run algorithm A_{geom} if relative error $\varepsilon_{geom}(n) < \varepsilon$, and use any exact (brute-force) algorithm otherwise. The running-time of the brute-force algorithm in the latter case will be exponential in $1/\varepsilon$, but polynomial in the length of an input.

1.1   Related Work

Since the connected k-CFP contains classic TSP, without triangle inequality the minimum connected k-CFP is NP-hard to approximate within a constant factor [20]. On the contrary, the maximum k-CFP admits a polynomial-time $(1 - 1/k^2)$-approximation algorithm [5].

Usually the problem of finding connected spanning networks with certain degree constraints is studied in the metric case. The minimization variants of the problem are more investigated. In 2014 Cornelissen et al. [10] presented a polynomial-time 3-approximation algorithm for the metric minimum k-CFP in case of odd k, and a polynomial-time 2.5-approximation algorithm for the case of even k.

Later Fukunaga and Nagamochi [13] and Narayanaswamy and Rahul [19] considered the minimization version of a more general connected f-factor problem, where for each vertex v its own degree requirement $f(v)$ is given. Narayanaswamy and Rahul [19] developed a PTAS for the case of metric weights of edges and growing function of edges $f(v) \geq n/c$ for every constant c. Fukunaga and Nagamochi [13] added q-edge-connectivity constraints to the f-factor problem in a multigraph, that is the sought subgraph requires to remain connected whenever fewer than q edges are deleted from the graph and is allowed to have multiple edges between vertices. They showed a ρ-approximation algorithm for the metric q-edge-connected f-factor problem, if $f(v) \geq 2, \forall v \in V$, where $\rho = 2.5$ for even q and $\rho = 2.5 + 1.5/q$ for odd q.

In 2018 Cornelissen et al. [11] presented constant factor approximation algorithms for the metric minimum k-CFP in simple graphs with additional requirements on the edge-connectivity and vertex-connectivity of the desired k-factors. They also extended these results to the f-factor statements with non-uniform degree requirements.

Although the minimization statements are more natural, the long edges of a network also can be desirable, for example, when the neighboring elements in a network are required to be geometrically well-separated in order to avoid interferences [2].

There are fewer results on the maximization problems. Baburin and Gimadi [5] studied the maximization version of a more general f-CFP with possibly non-uniform degree requirements and proposed an approximation algorithm for the maximum f-CFP with guaranteed relative error $\varepsilon = O\left(1/k^2\right)$ and $O(kn^3)$ running-time, where $k = \min_{v \in V} f(v)$. Another work by Baburin and Gimadi [6] provides an approximation algorithm for the maximum k-CFP on random inputs with $O(n^2)$ running-time. They proved that with high probability their algorithm is asymptotically optimal in the case of i.i.d. random weights of edges with uniform distribution and $k = o(n)$.

1.2   Preliminaries

We use standard terminology and notions in graph theory. We generally study finite, undirected graphs $G = (V, E)$. We denote by $V(G) := V$ the set of vertices of G and by $E(G) := E$ the set of edges of G. We denote $n := |V|$.

Consider an edge-weighted graph $G = (V, E)$ with positive weights $w : E \to \mathbb{R}_+$. For a subgraph H of G we denote by $w(H) := \sum_{e \in E(H)} w(e)$ the total weight of H. We will consider graphs in *normed spaces* $(\mathbb{R}^d, ||\cdot||)$ where the vertices correspond to some points in $\mathbb{R}^d$ and the weights of edges are defined according to the given norm $||\cdot||$. We call such instances *geometric*. Recall that a *norm* of a normed vector space is a function $||\cdot|| : \mathbb{R}^d \to \mathbb{R}$ with the following properties for each $x, y \in \mathbb{R}^d$ and $\alpha \in \mathbb{R}$: $||x|| \geq 0$, $||x|| = 0$ iff $x = 0$, $||\alpha x|| = |\alpha| \, ||x||$, and $||x + y|| \leq ||x|| + ||y||$. Note that normed vector spaces are a subset of metric spaces.

We denote by M^* a *maximum weight matching* in G. A maximum weight matching M^* in a *complete* weighted graph with positive weights consists of $\lfloor n/2 \rfloor$ edges, since otherwise there would exist $u, v \notin V(M^*)$ and matching $(u, v) \cup M^*$ would have a larger weight. Thus, if n is even, a maximum weight matching M^* in such graph is a perfect matching, and, if n is odd, M^* covers $n - 1$ vertices of G.

A *k-factor* of a graph $G = (V, E)$ is a k-regular subgraph with vertex set V. Note that a 2-factor of G is a collection of simple vertex-disjoint cycles covering all vertices of G, that is a *cycle cover*. We denote by C^* a *maximum weight cycle cover* for G. If G is a multigraph, C^* might contain cycles of length 2.

We use standard little o notation to describe functions that grows much slower than some other functions: $f(n) = o(g(n))$ if $\lim_{n \to \infty} f(n)/g(n) = 0$.

We use the following definitions to describe the quality of approximation algorithms.

Definition 1 (ρ-approximation algorithm). *A ρ-approximation algorithm A for a given maximization problem, $0 < \rho \leq 1$, is an algorithm, such that for every instance I, the value of the approximate solution $A(I)$ and the optimum $OPT(I)$, it holds that $A(I) \geq \rho \, OPT(I)$.*

Definition 2 (PTAS). *Polynomial time approximation scheme (PTAS) for a given maximization problem is an algorithm A that obtains for any valid input I and any arbitrary fixed $\varepsilon > 0$, a solution of objective value $A(I) \geq (1 - \varepsilon)OPT(I)$ in time polynomial in the input size $|I|$.*

Definition 3 (asymptotically optimal algorithm). *Let A be an approximation algorithm for an optimization problem on a class of weighted graphs. Algorithm A is called asymptotically optimal on a class of inputs $\{I_n\}$ with n being the number of vertices of a graph in I_n, if there exist an estimation for the relative error ε_n:*

$$\left| \frac{OPT(I_n) - A(I_n)}{OPT(I_n)} \right| < \varepsilon_n \to 0, \ as \ n \to \infty.$$

2   Approximation Algorithms

In this section we show polynomial-time approximation algorithms for the maximum geometric k-CFP and prove their performance guarantees.

This section has the following structure. First, since our algorithms are based on algorithm A_{PSP} from [17] for the maximum geometric m-Peripatetic Salesman Problem, in Sect. 2.1 we briefly discuss the ideas of algorithm A_{PSP} and the general ideas behind results of this type. In Sect. 2.2 we describe algorithm $\widetilde{A}$ (based on algorithm A_{PSP}) for the maximum geometric k-CFP and prove that it is asymptotically optimal for small $k = o(n)$. Furthermore, the obtained solution will be a highly connected subgraph, therefore algorithm $\widetilde{A}$ can also give asymptotically optimal solutions for the problem with additional q-edge-connectivity requirements, $q \leq 2\lfloor k/2 \rfloor$. Finally, in Sect. 2.3 we prove that a combination of A_{BG} from [5] and $\widetilde{A}$ gives an asymptotically optimal algorithm and a PTAS for the problem in any normed space of fixed dimension and arbitrary k.

2.1 Algorithm A_{PSP} for the Geometric Maximum m-PSP

In this section we discuss an asymptotically optimal algorithm A_{PSP} from [17] for the following problem.

Problem 2 (Maximum m-Peripatetic Salesman Problem, m-PSP).

Input: A complete graph $G = (V, E)$, weights $w : E \rightarrow \mathbb{R}_+$ and $m \in \mathbb{N}$.
Find: m edge-disjoint Hamiltonian cycles $H_1, \ldots, H_m$ of maximum total weight of their edges

$$\sum_{i=1}^{m} w(H_i) = \sum_{i=1}^{m} \sum_{e \in H_i} w(e) \rightarrow \max.$$

Note that the union of m edge-disjoint Hamiltonian cycles, is a spanning $2m$-regular connected subgraph, that is, a connected $2m$-factor.

In 1987 Serdukov [21] (also discussed in English in [18, chapter 12.10]) presented a geometrical approach to obtain asymptotically optimal solutions for the maximum Euclidean TSP. Baburin and Gimadi in [7] showed how to extend this approach for the case of maximum Euclidean m-PSP. In 2014 Shenmaier [22] carried over Serdukov's approach for the maximum TSP in an arbitrary normed space $(\mathbb{R}^d, ||\cdot||)$ of fixed dimension. Finally, paper [17] gives an asymptotically optimal algorithm for the m-PSP in an arbitrary normed space of fixed dimension.

In case of arbitrary normed spaces, the approach uses definition of the so-called remote angle.

Definition 4 (remote angle). *A remote angle α between two vectors x and y in a normed space $(\mathbb{R}^d, ||\cdot||)$ is*

$$\alpha(x, y) = \begin{cases} 0, & \text{if } x = \lambda y \text{ for } \lambda \in \mathbb{R} \\ \min\{||x/||x|| - y/||y||||, ||x/||x|| + y/||y||||\}, & \text{otherwise.} \end{cases}$$

The key ideas of the approach are formulated in the following lemmas. The first lemma states that, given a graph G in a normed space and a maximum weight matching M^* of G, among a large enough number of edges of M^* there exist two almost parallel edges. Due to the second lemma, having two almost parallel edges of M^*, one can get a pair of edges $\notin M^*$ with the same endpoints and almost the same total weight. Applying these two ideas, algorithm A_{PSP} constructs Hamiltonian cycles around the maximum weight matching M^*.

Lemma 1 ([22]). *Among any t vectors in a normed space $(\mathbb{R}^d, ||\cdot||)$ there exist two vectors such that the remote angle between them doesn't exceed the value*

$$\alpha(d, t) = \frac{2d}{\lfloor (2t-1)^{1/d} \rfloor} \, .$$

Lemma 2 ([22]). *Let AB and CD be two edges (intervals) in $\mathbb{R}^d$ and α be the remote angle between them. Then*

$$\max(\|AC\| + \|BD\|, \|AD\| + \|BC\|) \geq (1 - \alpha/2)(\|AB\| + \|CD\|).$$

The general intuition of how algorithm A_{PSP} [17] for the m-PSP works is as follows. First algorithm A_{PSP} finds a maximum weight matching $M^* = \{I_1, \ldots, I_\mu\}$ ($\mu = \lfloor n/2 \rfloor$) in the given complete graph G.

At each iteration $i = 1, \ldots, m$, algorithm A_{PSP} constructs i-th Hamiltonian cycle H_i such that it has large total weight, is edge-disjoint with all $H_j, j < i$, and does not contain edges of M^*. To this end algorithm A_{PSP} reorders the edges of M^* such that the following two property hold. The first one is that there are no two successive edges $I_{i_j}, I_{i_{j+1}} \in M^*$ in this ordering such that there exist an edge $e \in \bigcup_{i' < i} H_{i'}$ adjacent to both I_{i_j} and $I_{i_{j+1}}$. This property ensures that H_i would be edge-disjoint with all the previously constructed Hamiltonian cycles. The second property is that the remote angle α between most of successive edges of M^* is small enough (like in Lemma 1). Then for each two successive edges $I_{i_j}, I_{i_{j+1}}$ of M^* the algorithm finds a pair of edges, which patches them into a maximum weight 4-cycle, and add this pair of edges to the solution H_i (see Fig. 1). Due to Lemma 2 the weight of this pair of edges will be close to the total weight of I_{i_j} and $I_{i_{j+1}}$. Thus, one gets a Hamiltonian cycle H_i of total weight close to $2w(M^*)$. See [7,17] for more details.

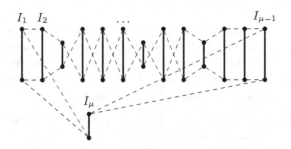

Fig. 1. Vertical lines represent the edges of M^*, dashed lines are the edges of the Hamiltonian cycle H_i constructed around M^*.

Note that since each constructed Hamiltonian cycle H_i does not contain edges of M^*, M^* can be used as a foundation in constructing further Hamiltonian cycles.

Proposition 1 ([17]). *In $O(n^3)$ time algorithm A_{PSP} finds an approximate solution for the geometric maximum m-PSP, such that it does not contain the edges of M^*. Let $w(m\text{-}PSP)$ and $w(opt\ m\text{-}PSP)$ be the weights of the solution obtained by A_{PSP} and an optimal solution, respectively. Then*

$$2mw(M^*)(1 - \varepsilon_{PSP}) \le w(m\text{-}PSP) \le w(opt\ m\text{-}PSP) \le 2mw(M^*)(1 + 1/n),$$

where

$$\varepsilon_{PSP} = 1/n + 2(d + 1)\left(\frac{m}{n}\right)^{\frac{1}{d+1}} \tag{1}$$

with d being the dimension of the normed space $(\mathbb{R}^d, ||\cdot||)$.

2.2 Asymptotically Optimal Algorithm $\widetilde{A}$ for the Geometric Maximum k-CFP in Case of Small k

Now we show how to adjust algorithm A_{PSP} for the m-PSP to obtain an approximation algorithm $\widetilde{A}$ for the maximum geometric k-CFP.

Description of algorithm $\widetilde{A}$
Input: Complete undirected graph $G = (V, E)$ in a normed space $(\mathbb{R}^d, ||\cdot||)$, edge weights $w : E \to \mathbb{R}_+$ consistent with the given norm, and $k \in \mathbb{N}$.
Output: Connected k-factor F_k.

Step 1. Let $n = |V|$, $m = \lfloor k/2 \rfloor$. Construct a maximum weight matching M^* in G. If n is even, M^* is a perfect matching of G, otherwise there is one vertex $x_0 \in V$ that is not covered by M^*. Using algorithm A_{PSP} construct an approximate solution $H_1, \ldots, H_m$ for the geometric maximum m-PSP in G, such that the solution does not contain the edges of M^*. Let $F_k := \bigcup_{i=1}^{m} H_i$.

Step 2. If $k = 2m + 1$ for some $m \in \mathbb{N} \cup \{0\}$, $F_k := F_k \cup M^*$. Note that a k-factor exists only if nk is even and $k < n$, thus, if k is odd, n is even and M^* is a perfect matching.

Return F_k.

It is clear that the algorithm constructs a feasible solution for the problem, since a union of m edge disjoint Hamiltonian cycles is a connected 2m-factor, and a union of a connected 2m-factor and an edge-disjoint perfect matching (in the case of even n) is a connected $(2m + 1)$-factor. We now show the conditions under which algorithm $\widetilde{A}$ is asymptotically optimal.

Theorem 1. *If $k = o(n)$, algorithm $\widetilde{A}$ gives asymptotically optimal solutions for the maximum k-CFP in any normed space $(\mathbb{R}^d, ||\cdot||)$ with fixed d. The running-time of the algorithm is $O(n^3)$.*

Let $m = \lfloor k/2 \rfloor$, F_k be a connected k-factor obtained by algorithm $\widetilde{A}$ and F_k^* be a maximum weight connected k-factor. To prove Theorem 1, we are going to show the lower and upper bounds on the weight $w(F_k)$ of the obtained solution.

For $m = o(n)$ using the lower bound from Proposition 1 we have:

$$2mw(M^*)(1 - \varepsilon_{PSP}) \le w(m\text{-PSP}) = w(F_{2m}) \le w(F_{2m}^*), \qquad (2)$$

if $k = 2m$, and

$$(2m + 1)w(M^*)(1 - \varepsilon_{PSP}) \le w(M^*) + 2mw(M^*)(1 - \varepsilon_{PSP})$$
$$\le w(M^*) + w(m\text{-PSP}) = w(F_{2m+1}) \le w(F_{2m+1}^*), \qquad (3)$$

if $k = 2m + 1$.

Now our goal is to prove that the weight of the optimal solution $w(F_k^*) \le kw(M^*)(1 + o(1))$. To this end, we first use the well-known Petersen's theorem on the $2m$-factor decomposition.

Proposition 2 ([1, Theorem 2.1.1]). *Any $2m$-regular (multi-)graph can be decomposed into a union of m edge-disjoint 2-regular graphs.*

According to Proposition 2, for even $k = 2m$ an optimal $2m$-factor F_{2m}^* can be decomposed into m edge-disjoint 2-factors $C_1, \ldots, C_m$. Let C^* be the maximum weight 2-factor in the initial graph, then

$$w(F_{2m}^*) = \sum_{i=1}^{m} w(C_i) \le mw(C^*). \qquad (4)$$

Consider the case of odd $k = 2m + 1$. Note that a k-factor exists only if nk is even and $k < n$. Thus, the number of vertices n in the graph is even, and there exist a perfect matching. By adding the maximum weight perfect matching M to the optimal $(2m + 1)$-factor one gets a $(2m + 2)$-factor $\widetilde{F}_{2m+2}$, which might contain double-edges. In turn, this $(2m+2)$-factor can be decomposed into $m+1$ edge-disjoint 2-factors $C_1, \ldots, C_{m+1}$, thus, we have:

$$w(F_{2m+1}^*) = w(\widetilde{F}_{2m+2}) - w(M) = \sum_{i=1}^{m+1} w(C_i) - w(M)$$
$$\le (m + 1)w(C^*) - w(M), \qquad (5)$$

where the maximum weight 2-factor C^* might contain cycles of length two. Using the following result from [16] one can estimate the weight of a 2-factor C^* in a given normed space.

Proposition 3 ([16]). *Let $G = (V, E)$ be an edge-weighted complete (multi)graph in an arbitrary normed space $(\mathbb{R}^d, \|\cdot\|)$, where the weight of an edge $e = (u, v)$ is defined as $w(e) = \|v - u\|$ for $u, v \in V \subseteq \mathbb{R}^d$. Let C^* be a maximum weight 2-factor and M^* be a maximum weight matching in G. Then: $w(C^*) \le 2w(M^*)/(1 - \varepsilon_C)$, where*

$$\varepsilon_C = \left(\frac{(d + 1)^{d+1}}{2n} \right)^{1/(d+2)}. \qquad (6)$$

Now we can finish the proof of Theorem 1.

Proof (of Theorem 1). The running-time of $\widetilde{A}$ is determined by the running-time of algorithm A_{PSP}, which is $O(n^3)$. To estimate the relative error of the algorithm consider the following.
For $k = 2m$, from the inequalities (2), (4) and Proposition 3 it follows that:

$$2mw(M^*)(1 - \varepsilon_{PSP}) \leq w(F_{2m}) \leq w(F_{2m}^*) \leq 2\,mw(M^*)/(1 - \varepsilon_C)\,.$$

For $k = 2m + 1$, from the inequalities (3), (5) and Proposition 3 it follows that:

$$
\begin{aligned}
(2m + 1)w(M^*)(1 - \varepsilon_{PSP}) &\leq w(F_{2m+1}) \leq w(F_{2m+1}^*) \\
&\leq (m + 1)w(C^*) - w(M^*) \leq (2m + 2)w(M^*)/(1 - \varepsilon_C) - w(M^*) \\
&\leq (2m + 1)w(M^*)/(1 - 2\varepsilon_C).
\end{aligned}
$$

Therefore, using definitions of ε_C (6) and ε_{PSP} (1) we get an estimation for the relative error of algorithm $\widetilde{A}$:

$$
\begin{aligned}
\varepsilon_{\widetilde{A}} = 1 - \frac{w(F_k)}{w(F_k^*)} &\leq 1 - \frac{k\,w(M^*)(1 - \varepsilon_{PSP})}{k\,w(M^*)/(1 - 2\varepsilon_C)} \leq \varepsilon_{PSP} + 2\varepsilon_C \\
&\leq 1/n + 2(d + 1)\left(\frac{k}{2n}\right)^{1/(d+1)} + 2(d + 1)\left(\frac{1}{2n}\right)^{1/(d+2)} \leq 5(d + 1)\left(\frac{k}{2n}\right)^{1/(d+2)},
\end{aligned}
$$
(7)

which tends to 0 as n grows to infinity, if $k = o(n)$ and d is a constant.   □

From this result a simple corollary follows. Recall, that a graph is called *q-edge connected*, if it remains connected whenever fewer than q edges are removed. Menger's theorem [14, Theorem 5.12] states that a graph is q-edge connected, if for every pair of vertices x and y there exists q edge disjoint paths between x and y. Since the approximate solution of the k-CFP obtained by algorithm $\widetilde{A}$ contains $\lfloor k/2 \rfloor$ edge disjoint Hamiltonian cycles, there are at least $2\lfloor k/2 \rfloor$ edge disjoint paths between any two vertices of the approximate solution F_k. Note that for odd k and $q = k$ this is not always the case.

Corollary 1. *Algorithm $\widetilde{A}$ gives asymptotically optimal solution for the problem of finding a q-edge connected k-factor in a given complete graph in any normed space of fixed dimension, if $k = o(n)$ and $q \leq 2\lfloor k/2 \rfloor$.*

2.3   PTAS and Asymptotically Optimal Algorithm in Case of Arbitrary k

Finally, in this section we show that by combining algorithm $\widetilde{A}$ for the case of small k with algorithm A_{BG} from [5] for the case of large k one can obtain an asymptotically optimal algorithm and a PTAS for the maximum geometric k-CFP for arbitrary k.

Theorem 2. *The maximum geometric k-CFP admits a polynomial-time asymptotically optimal algorithm in an arbitrary normed space $(\mathbb{R}^d, ||\cdot||)$ of fixed dimension d.*

Proof. Algorithm A_{BG} for the general maximum k-CFP from [5] runs in $O(kn^3)$ time and has relative error $\varepsilon_{BG} \leq 1/k^2$. According to Theorem 1, algorithm $\tilde{A}$ for the geometric maximum k-CFP runs in $O(n^3)$ time and has a relative error defined by (7).

Thus, applying algorithm A_{BG} if $k \geq \frac{(2n)^{1/(2d+5)}}{(5d+5)^{(d+2)/(2d+5)}}$, and algorithm $\tilde{A}$ otherwise, we get an approximation algorithm A_{geom} for the geometric maximum k-CFP with running-time $O(kn^3)$ and relative error

$$\varepsilon \leq \left(\frac{(5d+5)^{2d+4}}{4n^2}\right)^{\frac{1}{2d+5}} \leq \frac{5(d+1)}{(2n)^{2/(2d+5)}} =: \varepsilon_{geom} \xrightarrow[n\to\infty]{} 0. \qquad (8)$$

$\square$

Corollary 2. *The maximum geometric k-CFP admits a PTAS in an arbitrary normed space $(\mathbb{R}^d, ||\cdot||)$ of fixed dimension d.*

Proof. One can obtain a PTAS from the discussed asymptotically optimal algorithm as follows. Given an instance of k-CFP and $\varepsilon > 0$, apply polynomial-time asymptotically optimal algorithm A_{geom} if $\varepsilon \geq \varepsilon_{geom}$, where ε_{geom} is defined as in (8). Otherwise, from $\varepsilon < \varepsilon_{geom}$, it follows that

$$n < \frac{1}{2} \cdot \left(\frac{5d+5}{\varepsilon}\right)^{d+5/2},$$

and applying any exact (brute-force) algorithm to such instance will take time exponential in $1/\varepsilon$. Therefore, the overall time-complexity is polynomial in n (but not in $1/\varepsilon$), and we get a PTAS. $\square$

3 Conclusion

In this paper employing the extended geometrical approach by Serdukov we construct an asymptotically optimal algorithm and a PTAS for the maximum k-CFP in an arbitrary normed space of fixed dimension. Given a graph in a normed space of fixed dimension, this approach allows to build a number of edge-disjoint matchings with weight of each matching being close to the weight of the maximum weight matching. Therefore the approach is well suited for solving geometric maximization problems, where a feasible solution is a certain subgraph that can be decomposed into a union of edge-disjoint matchings.

Moreover, this approach allows to obtain structural results on the weight of the optimal solutions for such problems. For example, in an arbitrary normed space of fixed dimension from the result [22] it follows that the weight of an optimal maximum TSP solution asymptotically equals to $2w(M^*)$, the weight of a maximum weight 2-factor asymptotically equals to $2w(M^*)$ [16], the weight of a maximum m-PSP solution is asymptotically $2mw(M^*)$ for $m = o(n)$ [7,17], and, now, according to Theorem 1 and Corollary 1 the weights of a maximum weight k-factor, connected k-factor, and q-edge connected k-factor are asymptotically equal to $kw(M^*)$, if $k = o(n)$ and $q \leq 2\lfloor k/2 \rfloor$.

Note that for the *minimization* problems even in the Euclidean case and even when the vertices of an input graph are random points *uniformly* sampled from the hypercube $[0, 1]^d$, the weight of a minimum weight Hamiltonian cycle, 2-factor and the doubled weight of a minimum weight perfect matching are not asymptotically equal to each other [12].

A promising direction for the further research is studying maximization problems in metric spaces of fixed doubling dimension, which is a more general case of a metric space than the normed spaces. For the minimum TSP in metric spaces of bounded doubling dimension Bartal et al. [4] extended an approach by Arora [3] and presented a PTAS. Since such approximation algorithms are possible to construct for the minimization problems, it is worth trying to generalize Serdukov's approach for the maximization problems for these types of metrics.

Funding. The authors are supported by the program of fundamental scientific researches of the SB RAS, project 0314-2019-0014 and by the Ministry of Science and Higher Education of the Russian Federation under the 5-100 Excellence Programme.

References

1. Akiyama, J., Kano, M.: Factors and Factorizations of Graphs. Lecture Notes in Mathematics, vol. 2031, 353 p. Springer, Heidelberg (2011). https://doi.org/10.1007/978-3-642-21919-1
2. Arkin, E.M., Chiang, Y., Mitchell, J.S.B., Skiena, S.S., Yang, T.: On the maximum scatter TSP. In: Proceedings of the Eighth Annual ACM-SIAM Symposium on Discrete Algorithms, SODA 1997, pp. 211–220 (1997)
3. Arora, S.: Polynomial time approximation schemes for Euclidean traveling salesman and other geometric problems. J. ACM **45**(5), 753–782 (1998)
4. Bartal, Y., Gottlieb, L.-A., Krauthgamer, R.: The traveling salesman problem: low-dimensionality implies a polynomial time approximation scheme. In: Proceedings of the Forty-Fourth Annual ACM Symposium on Theory of Computing, STOC 2012, pp. 663–672 (2012)
5. Baburin, A.E., Gimadi, E.K.: Certain generalization of the maximum traveling salesman problem. J. Appl. Ind. Math. **1**(4), 418–423 (2007)
6. Baburin, A.E., Gimadi, E.K.: An approximation algorithm for finding a d-regular spanning connected subgraph of maximum weight in a complete graph with random weights of edges. J. Appl. Ind. Math. **2**(2), 155–166 (2008)
7. Baburin, A.E., Gimadi, E.K.: On the asymptotic optimality of an algorithm for solving the maximum m-PSP in a multidimensional Euclidean space. Proc. Steklov Inst. Math. **272**(1), 1–13 (2011)
8. Barvinok, A., Fekete, S.P., Johnson, D.S., Tamir, A., Woeginger, G.J., Woodroofe, R.: The geometric maximum traveling salesman problem. J. ACM **50**(5), 641–664 (2003)

9. Cheah, F., Corneil, D.G.: The complexity of regular subgraph recognition. Discret. Appl. Math. **27**(1–2), 59–68 (1990)

10. Cornelissen, K., Hoeksma, R., Manthey, B., Narayanaswamy, N.S., Rahul, C.S.: Approximability of connected factors. In: Kaklamanis, C., Pruhs, K. (eds.) WAOA 2013. LNCS, vol. 8447, pp. 120–131. Springer, Cham (2014). https://doi.org/10.1007/978-3-319-08001-7_11

11. Cornelissen, K., Hoeksma, R., Manthey, B., Narayanaswamy, N.S., Rahul, C.S., Waanders, M.: Approximation algorithms for connected graph factors of minimum weight. J. Theory Comput. Syst. **62**(2), 441–464 (2018)

12. Frieze, A., Pegden, W.: Separating subadditive Euclidean functionals. In: Proceedings of the Forty-Eighth Annual ACM Symposium on Theory of Computing, STOC 2016, pp. 22–35 (2016)

13. Fukunaga, T., Nagamochi, H.: Network design with edge-connectivity and degree constraints. Theory Comput. Syst. **45**(3), 512–532 (2009)

14. Harary, F.: Graph Theory. Addison-Wesley Series in Mathematics. Addison-Wesley, Reading (1969)

15. Gabow, H.N.: An efficient reduction technique for degree-constrained subgraph and bidirected network flow problems. In: 15th Annual ACM Symposium on Theory of Computing, pp. 448–456. ACM, New York (1983)

16. Gimadi, E.Kh., Tsidulko, O.Yu.: On modification of an asymptotically optimal algorithm for the maximum Euclidean traveling salesman problem. In: van der Aalst, W.M.P., et al. (eds.) AIST 2018. LNCS, vol. 11179, pp. 283–293. Springer, Cham (2018). https://doi.org/10.1007/978-3-030-11027-7_27

17. Gimadi, E.Kh., Tsidulko, O.Yu.: Asymptotically optimal algorithm for the maximum m-peripatetic salesman problem in a normed space. In: Battiti, R., Brunato, M., Kotsireas, I., Pardalos, P.M. (eds.) LION 12 2018. LNCS, vol. 11353, pp. 402–410. Springer, Cham (2019). https://doi.org/10.1007/978-3-030-05348-2_33

18. Gutin, G., Punnen, A.P. (eds.): The Traveling Salesman Problem and its Variations. Kluwer Academic Publishers, Dortrecht (2002)

19. Narayanaswamy, N.S., Rahul, C.S.: Approximation and exact algorithms for special cases of connected f-factors. In: Beklemishev, L.D., Musatov, D.V. (eds.) CSR 2015. LNCS, vol. 9139, pp. 350–363. Springer, Cham (2015). https://doi.org/10.1007/978-3-319-20297-6_23

20. Sahni, S., Gonzalez, T.: P-complete approximation problem. J. Assoc. Comput. Mach. **23**, 555–565 (1976)

21. Serdyukov, A.I.: Asymptotically exact algorithm for the travelling salesman maximum problem in Euclidean space (In Russian). Upravlyaemye Sistemy **27**, 79–87 (1987)

22. Shenmaier, V.V.: Asymptotically optimal algorithms for geometric Max TSP and Max m-PSP. Discret. Appl. Math. **163**(2), 214–219 (2014)

23. Tutte, W.T.: A short proof of the factor theorem for finite graphs. Can. J. Math. **6**, 347–352 (1954)

24. Yazicioglu, A.Y., Egerstedt, M., Shamma, J.S.: Formation of robust multi-agent networks through self-organizing random regular graphs. IEEE Trans. Netw. Sci. Eng. **2**(4), 139–151 (2015)

Generalization of Controls Bimodality Property in the Optimal Exploitation Problem for Ecological Population with Binary Structure

Alexander I. Smirnov$^{(\boxtimes)}$ (ID) and Vladimir D. Mazurov

Krasovskii Institute of Mathematics and Mechanics UB RAS, Yekaterinburg, Russia
asmi@imm.uran.ru, vldmazurov@gmail.com

Abstract. The problem of optimal exploitation of an ecological population with a binary structure is considered (there is an additional criterion for population structuring in addition to age or developmental stage). It is assumed that population state dynamics is described by a nonlinear generalization of the Leslie model. We prove a criterion for the existence of so-called quasi-preserving controls that support the sustainable population dynamics. Moreover, optimal quasi-preserving controls with a minimum number of nonzero coordinates (i.e., controls that preserve unchanged the largest number of structural units of a population) are found explicitly. The proposition about the minimum possible number of nonzero coordinates for optimal vectors is also proved. This proposition is a generalization to the case of a binary population structure of well-known bimodality property of optimal strategies for populations with one-dimensional structure.

Keywords: Rational ecosystem exploitation · Generalization of the Leslie model · Optimal quasi-preserving controls · Bimodality of optimal strategies

1 Introduction

One of the global problems, the relevance of which has been steadily increasing lately, is the preservation of sustainable exploitation regimes for the world's bioresources. The Living Planet 2018 Index [1] prepared by World Wildlife Fund (WWF) characterizes the severity of the problem, showing that between 1970 and 2014 the planet lost 60% of mammals, birds, reptiles, and fish; at least 300 mammal species have completely disappeared as a result of human impact. The world's largest purveyors of bioresources are forestry and fishing. According to WWF, the share of marine fish stocks harvested at biologically sustainable levels shows a downward trend: from 90.0% in 1974 to 66.9% in 2015. As for forestry,

This research was funded by RFBR, grant no. 19-07-01243.

M. Jaćimović et al. (Eds.): OPTIMA 2019, CCIS 1145, pp. 206–221, 2020.
https://doi.org/10.1007/978-3-030-38603-0_16

in Russia only in 2016–2017 years more than 1 million hectares of pristine forest were cut down [1].

The problem of finding the maximum sustainable level of exploitation that preserves stability of population (the Maximum Sustainable Yield, or MSY–Problem) has been studied by many authors. The pioneering works here took into account only the total biomass of the population. Early works, considering a population structure, used a linear formulation, in that the transition from the previous population state to the next one was performed by a matrix called the projection matrix. Difficulties associated with the dynamic aspects of the problem were overcome, considering only stationary states. As the latter, they used eigenvectors corresponding to the dominant eigenvalue $\lambda(A)$ of the projection matrix A. In this case, with $\lambda(A) > 1$, the removal of surplus from the age classes of the population leaves the population in a stationary state: $x_{t+1} = Ax_t - (\lambda(A) - 1)x_t = x_t$.

The first theoretical results for linear models of sustainable population exploitation were obtained in [2,3]. In [2] the concept of sustainable exploitation was formulated for a population with an age structure—the problem of finding the maximum level of allowable exploitation was formulated, and the solvability of this problem is proved. In [3] some concepts and statements adequate for the exploitation of ecosystems were formalized, and the corresponding assertions were strictly proved.

The successful attempts to generalize these results to the case of nonlinear density dependence were stimulated by [4], where the density dependence was considered only for the first age class. But this dependence was quite general; in particular, the class of functions under consideration contained concave functions with a horizontal asymptote. Further development of these studies is described in [5], where is studied density-dependent size-structured population model.

Naturally, a solution of population optimal exploitation problem essentially depends on the formulation of an objective function. In this paper, we are more interested in the properties of a feasible set; issues related to the use of various objective functions are discussed in detail in the monograph [6].

It should be noted that in the overwhelming majority of studies on optimal exploitation based on the withdrawal of resources from the system, even if the initial formulation of the problem uses additive control, as a rule, they subsequently move to proportional withdrawal, i.e., to multiplicative control that simplifies the finding and analysis of optimal strategies. A typical approach is described in [6]. Although an iterative process with additive control is considered initially, then the fractions $h_i \in (0,1)$ are determined that will be removed from structural population units; authors are turned to multiplicative stationary control in the equilibrium state. Usually, a diagonal matrix is defined with elements h_i on the main diagonal, so that $u = HAx$, and MSY–problem is considered with the primary constraint $(I - H)Ax = x$.

All of the above studies used the well-known Leslie model of the population age structure or its generalization (see review in [7]) as a biological basis.

These studies became the basis for numerous subsequent works, both theoretical and practical, mostly using similar problem statements and mathematical tools.

Despite all differences in the formulations of population optimal exploitation problem among various authors, it turned out that there is a common characteristic property of optimal solutions, consisting in the number of age classes to be exploited. This property of controls, later called the bimodality property, means existence an optimal solution having at most two such classes. Apparently, for the first time these results were obtained in papers [8,9]. Their authors established the existence of a bimodal optimal control that allows exploitation (withdrawal, partial or complete) of no more than two age classes, partial withdrawal of one age class and full—of the other (more older).

The first successful attempt of optimal exploitation problem consideration for nonlinear models of population dynamics with a general nonlinear dependence on the total biomass of a population, when this dependence is given by the properties of dynamic system step operator, is associated with the aforementioned paper [4]. For this rather general nonlinear model, the existence of bimodal optimal control strategies has also been shown. Such "two-age" strategies for models with nonlinearities were further obtained by many authors (see e.g. the review in [5]).

All the listed studies were related to a population with a one-dimensional (age or stage) structure. The paper [10] proposed a nonlinear population model with a binary structure when there is an additional criterion for population structuring other than age or developmental stage. Judging from available reviews of generalizations and modifications of the Leslie model (see e.g. [11,12]), this binary model has not been encountered previously in studies on the exploitation of ecological populations.

The goal of this paper is to obtain a generalization of the optimal controls bimodality property for non-linear Leslie model with binary structure and to find explicitly optimal controls that satisfy this generalized property.

2   Some Definitions, Notation and Preliminary Results

Some common notation is as follows: $\mathbb{R}_+^q$—the non-negative orthant of $\mathbb{R}^q$; $\overline{m,n} = \{i \in \mathbb{Z} \mid m \leq i \leq n\}$; $\overline{M}$—the closure of a set M; $|M|$—the number of elements of a finite set M; $co(M)$—the convex hull of M.

For convenience, we will sometimes use the abbreviated coordinate entry $x = (x_i)$ for a row vector $x = (x_1, x_2, \ldots, x_q)$; $I^+(x) = \{i \in \overline{1,q} \mid x_i > 0\}$, where $x = (x_i)$.   Sets of nonzero fixed points and positive fixed points of F are denoted by N_F and N_F^+, respectively.

Let us proceed to the formalization of optimal exploitation problem proposed in [13]. It is assumed that there is an ecological population consisting of $q > 1$ structural units (subdivisions), the dynamics of which in the absence of exploitation is described by an iterative process with a step operator $F \in \{\mathbb{R}_+^q \mapsto \mathbb{R}_+^q\}$. This operator, along with the trivial equilibrium state ($F(0) = 0$), also has a nontrivial one (i.e. $N_F \neq \varnothing$). For the map F that is concave on $\mathbb{R}_+^q$, we can formulate a simple sufficient condition for the existence of a nonzero fixed point. Indeed, for such map F, there are positively homogeneous maps $F_s(x) = \lim_{\alpha \to +s} \alpha^{-1} F(\alpha x)$ ($s \in \{0, \infty\}$). Using their dominant eigenvalues, a sufficient condition for the existence of a nontrivial equilibrium can be formulated as follows:

$$\lambda(F_\infty) < 1 < \lambda(F_0) \tag{1}$$

In addition to concavity, the non-burdensome requirement for an absence of identically zero components of $F(x) = (f_i(x))$, which guarantees the positivity of positive vectors images: $F(\operatorname{int} \mathbb{R}_+^q) \subseteq \operatorname{int} \mathbb{R}_+^q$, is also assumed to be fulfilled.

The exploited population is modeled by the iterative process

$$x_{t+1} = F_u(x_t), \qquad t = 0, 1, 2, \ldots, \tag{2}$$

where $F_u(x) = F(x) - u$, $x_t \geq 0$ means a population state at time $t = 0, 1, 2, \ldots$. In [13], an optimization problem was posed for system (2), the feasible set of which formalized the stability condition of population structure. The equivalence of this problem to the following mathematical programming problem was proved:

$$\max\{c(u) \mid x = F(x) - u, \ x \geq 0, \ u \geq 0\}, \tag{3}$$

where $c(u)$ is nonnegative and monotone increasing function.

We call the control $u \geq 0$ *preserving* if it preserves all structural subdivisions of the population (2), i.e. if $x_t > 0$ ($\forall t = 0, 1, 2, \ldots$). Denote by N_u and N_u^+ the sets of nonzero fixed points and positive fixed points of F_u, respectively. The set of preserving controls can be represented as follows:

$$U = \{u \in \mathbb{R}_+^q \mid N_u^+ \neq \varnothing\}. \tag{4}$$

The feasible set of the problem (3) coincides with $\overline{U} = \{u \in \mathbb{R}_+^q \mid N_u \neq \varnothing\}$.

It was shown in [13] that N_F for concave map F satisfying condition (1) is bounded, convex and contains the largest element $\bar{x}_F$. Further, the non-empty set N_u contains the largest element $\bar{x}_u$; moreover, the map $\bar{x}(u) : u \to \bar{x}_u$ is monotone decreasing and concave on $\overline{U}$.

We introduce a set

$$D = \{u \in \mathbb{R}_+^q \mid N_u \neq \varnothing, \ N_v = \varnothing \ (\forall v > u)\}$$

that forms a part of the boundary of U. The set D contains all potentially optimal vectors of the problem (3). We divide this set into two disjoint parts by criterion of the presence or absence of common points with U:

$$D' = \{u \mid N_u^+ \neq \varnothing, \ N_v = \varnothing \ (\forall v > u)\}, \tag{5}$$

$$D'' = \{u \in \mathbb{R}_+^q \mid N_u \neq \varnothing, \ N_u^+ = \varnothing\}. \tag{6}$$

As can be seen from (4), (5), although the controls from D' lie on the boundary of U, they are preserving: $D' \subset U$. The latter is not true for D''; moreover, it follows from (6) that $D'' = \overline{U}\backslash U$. But, although the controls from D'' eliminate some structural population units, in the next development cycle they are restored and continue to exist stably (until the next withdrawal). Taking this into account, controls from D'' are called *quasi-preserving*.

In [15] it was proved that for a generalization of the Leslie model [10] $D' \neq \varnothing$. This generalization is as follows: there are $m \geq 1$ intrapopulation structural units (subdivisions), each of which contains individuals of n ages (or stages) $1, 2, \ldots, n$. If at time t, the subdivisions individuals number (density, biomass), is $x_{i,j}^{(t)}$ $(t = 0, 1, 2, \ldots)$ then

$$x_{i,1}^{(t+1)} = f_i(a_t), \quad x_{i,j+1}^{(t+1)} = \alpha_{i,j}x_{i,j}^{(t)} \quad (i \in \overline{1,m}, j \in \overline{1, n-1}). \tag{7}$$

Here $\alpha_{i,j} > 0$ and $\beta_{i,j} \geq 0$ are the survival and fertility rates in the relevant subdivisions, $a_t = \sum_{i=1}^{m} \sum_{j=1}^{n} \beta_{i,j}x_{i,j}^{(t)}$—the number of newborns at time t.

We write the population state vector in the form $x = (x^{(1)}, x^{(2)}, \ldots, x^{(m)})$, where $x^{(i)} = (x_{i,1}, \ldots, x_{i,n})$ $(i \in \overline{1,m})$. The step operator $F(x) = (f_{i,j}(x))$ for the model (7) has the following form:

$$f_{i,1}(x) = f_i(a(x)), \quad f_{i,j+1}(x) = \alpha_{i,j}x_{i,j} \ (i \in \overline{1,m}, \ j \in \overline{1, n-1}). \tag{8}$$

In order for F to be concave, the following assumption must be made:

$$f_i(a) \text{ are concave on } \mathbb{R}_+ \ (\forall i \in \overline{1,m}). \tag{9}$$

Requirement $F(0) = 0$ leads to conditions $f_i(0) = 0$ $(\forall i \in \overline{1,m})$; but everywhere except zero, functions $f_i(a)$ are positive.

Denote (by convention $\prod_{\ell=r}^{s} a_\ell = 1$ for $r > s$)

$$\pi^{(i)} = \pi_n^{(i)}, \quad \pi_j^{(i)} = p_{1,j}^{(i)}, \quad p_{j,k}^{(i)} = \prod_{\ell=j}^{k-1} \alpha_{i,\ell}, (i \in \overline{1,m}, \ j,k \in \overline{1,n}), \tag{10}$$

we obtain the system of equations for finding an equilibrium of (7):

$$x_{i,j} = \pi_j^{(i)} f_i(a(x)) \ (i \in \overline{1,m}, \ j \in \overline{1,n}), \quad a(x) = \sum_{i=1}^{m}\sum_{j=1}^{n} \beta_{i,j}x_{i,j}.$$

Condition (1) for the model (7) is equivalent [10] to the following condition:

$$\sigma'(+\infty) < 1 < \sigma'(+0), \tag{11}$$

where

$$\sigma(a) = \sum_{i=1}^{m} \sigma^{(i)} f_i(a), \ \sigma^{(i)} = \sigma_n^{(i)}, \ \sigma_k^{(i)} = \sum_{s=1}^{k} \beta_{i,s} \prod_{t=1}^{s-1} \alpha_{i,t} \ (i \in \overline{1,m}, k \in \overline{0,n}). \tag{12}$$

Condition (11) leads to solvability of equation $\sigma(a) = a$. Its unique solution $\bar{a}_F$ corresponds to the unique positive fixed point $\bar{x}_F$ of the map (8): $\bar{a}_F = a(\bar{x}_F)$.

We also assume that the condition

$$\beta_{i,n} > 0 \quad (\forall i \in \overline{1,m}), \tag{13}$$

guaranteeing the local irreducibility of the map (8) is satisfied (for the corresponding definitions and properties, see [10, 14]).

3 Properties of Quasi-preserving Controls for the Generalization of the Leslie Model

The feasible set $\overline{U}$ for the model (7) in addition to the restrictions of nonnegativity of x, u is given by equations

$$x_{i,1} = f_i(a(x)) - u_{i,1}, \; x_{i,j+1} = \alpha_{i,j} x_{i,j} - u_{i,j+1} \; (i \in \overline{1,m}, \; j \in \overline{1,n-1}). \tag{14}$$

We represent the structure of the control u in accordance with the structure of state vector x in the form $u = (u^{(1)}, u^{(2)}, \ldots, u^{(m)})$, where $u^{(i)} = (u_{i,1}, \ldots, u_{i,n})$.
    We need the following notation (here $i \in \overline{1,m}, \; j,k \in \overline{1,n}$):

$$q(u) = \sum_{i=1}^{m} \sum_{j=1}^{n} q_j^{(i)} u_{i,j}, \quad q_j^{(i)} = q_{j,n}^{(i)}, \quad q_{j,k}^{(i)} = \sum_{s=j}^{k} \beta_{i,s} \prod_{t=j}^{s-1} \alpha_{i,t}, \tag{15}$$

$$p^{(i)}(u) = p_n^{(i)}(u), \quad p_j^{(i)}(u) = \sum_{k=1}^{j} p_{k,j}^{(i)} u_{i,k}, \tag{16}$$

$$\mu(a) = \sigma(a) - a, \quad \lambda_i(a) = \pi^{(i)} f_i(a), \tag{17}$$

$$\mu^* = \max_{a \geq 0} \mu(a), \quad A^* = \text{Arg max } \mu(a), \quad a_* = \min A^*, \; a^* = \max A^*. \tag{18}$$

Under assumptions (9), (11), (13), we have $0 < \mu^* < +\infty$, $A^* \neq \varnothing$, $0 < a_* \leq a^*$.
    From (14) we obtain the following explicit dependence of x on u:

$$x_{i,j} = \pi_j^{(i)} f_i(a) - p_j^{(i)}(u), \; x_{i,n} = \lambda_i(a) - p^{(i)}(u) \; (i \in \overline{1,m}, j \in \overline{1,n-1}), \tag{19}$$

where $a = a(x)$. We write the vector x with coordinates (19) as $x = x(a,u)$.
    For $u \in \overline{U}$, $\bar{x}_u = (\bar{x}_{i,j}(u))$, we define a set

$$I_0(u) = \{i \in \overline{1,m} \mid \bar{x}_{i,n}(u) = 0\}. \tag{20}$$

The set D' have the following parametric representation:

$$D' = \{u \mid p^{(i)}(u) < \lambda_i^* \; (i \in \overline{1,m}), \; q(u) = \mu^*, \; u \geq 0\}. \tag{21}$$

This set lies entirely on the hyperplane Γ given by equation $q(u) = \mu^*$ [15]:

$$D' = \Gamma \cap U, \quad D'' \cap \Gamma \subset \overline{D'}, \quad \overline{D'} = \Gamma \cap D. \tag{22}$$

A non-negative u belongs to D'' if and only if the following restrictions are met:

$$I_0(u) \neq \varnothing, \quad p^{(i)}(u) \begin{cases} = \lambda_i(a(\bar{x}_u)), i \in I_0(u), \\ < \lambda_i(a(\bar{x}_u)), i \notin I_0(u), \end{cases} \quad q(u) = \mu(a(\bar{x}_u)). \tag{23}$$

Using the notation (12), we introduce the following quantities:

$$\bar{S}_j^{(i)}(a) = \sigma_j^{(i)} f_i(a) + \sum_{k \neq i} \sigma^{(k)} f_k(a) \ (i \in \overline{1,m}, j \in \overline{0,n}). \tag{24}$$

We strive to find optimal vectors having a minimum number of nonzero coordinates. To characterize the situation when there are quasi-preserving optimal vectors with unique positive coordinate, we need the following equation:

$$\bar{S}_{j-1}^{(i)}(a) = a \quad (a \geq a^*). \tag{25}$$

The solution to this equation (if it exists) is denoted by $a_j^{(i)}$. To find conditions for solvability of (25), we introduce the following notation:

$$I_k^* = \{j \in \overline{1,n} \mid \bar{S}_{j-1}^{(k)}(a^*) < a^*\}, \quad J_k^* = \overline{1,n} \setminus I_k^* \quad (k \in \overline{1,m}), \tag{26}$$

$$I^* = \{k \in \overline{1,m} \mid I_k^* \neq \varnothing\}, \quad J^* = \{k \in \overline{1,m} \mid J_k^* \neq \varnothing\}. \tag{27}$$

Lemma 1. *Suppose that conditions (9), (11), (13) are satisfied. Then Eq. (25) has a solution (that is unique) if and only if $j \in J_i^*$.*

Proof. Suppose that the Eq. (25) is solvable. Then it follows from (9), (11) that its solution $a_j^{(i)}$ is unique. Because the function $\bar{S}_{j-1}^{(i)}(a)$ is concave, we have for $a \geq a^*$: $1 = (a_j^{(i)})^{-1} \bar{S}_{j-1}^{(i)}(a_j^{(i)}) \leq (a^*)^{-1} \bar{S}_{j-1}^{(i)}(a^*)$, i.e. $\bar{S}_{j-1}^{(i)}(a^*) \geq a^*$. But this means by (26) that $j \in J_i^*$.

Conversely, we now show that Eq. (25) is solvable for $i \in J^*$, $j \in J_i^*$. Indeed, by (26), (27) we have $\bar{S}_{j-1}^{(i)}(a^*) \geq a^*$. Condition (11) due to concavity of $\sigma(a)$ implies the existence of a point $\bar{a}$ such that $\sigma(\bar{a}) < \bar{a}$. But then, because of $S_{j-1}^{(i)}(\bar{a}) \leq \sigma(\bar{a})$ a fortiori $S_{j-1}^{(i)}(\bar{a}) < \bar{a}$. Therefore, there is a fixed point $\bar{a} \in [a^*, \bar{a}]$ of $S_{j-1}^{(i)}(a)$, i.e. the Eq. (25) is solvable. Lemma is proved.

Let us define the control $u(i,j)$ by equalities

$$|I^+(u(i,j))| = 1, \ I_i^+(u(i,j)) = \{j\}, \ u_{i,j} = \begin{cases} \mu^*(q_{j,n}^{(i)})^{-1}, & i \in I^*, j \in I_i^*, \\ \pi_j^i f_i(a_j^{(i)}), & i \in J^*, j \in J_i^*. \end{cases} \tag{28}$$

Note that $q_{j,n}^{(i)} \neq 0$ by assumption (13). The following proposition clarifies the distribution of the elements of D with a unique nonzero coordinate over D', D''.

Lemma 2. *Suppose that conditions (9), (11), (13) are satisfied and $J^* \neq \varnothing$. Then $u(i,j) \in D$; moreover,*

$$u(i,j) \in \begin{cases} D', & i \in I^*, j \in I_i^*, \\ D'', & i \in J^*, j \in J_i^*. \end{cases} \tag{29}$$

Proof. If $i \in I^*$, $j \in I_i^*$ and $u_{i,j} = \mu^*(q_{j,n}^{(i)})^{-1}$, then it follows from (15) that $q(u(i,j)) = q_{j,n}^{(i)} u_{i,j} = \mu^*$. By (21), it remains to show that $\lambda_i^* - p^{(i)}(u) = x_{i,n} > 0$. Since $x_{i,n} = p_j^{(i)} x_{i,j}$ and $p_j^{(i)} > 0$ by (10), it suffices to verify that $x_{i,j} > 0$. From (19), using (10), (12), (15), (17), (24), we find: $x_{i,j} = \pi_j^{(i)} f_i(a^*) - u_{i,j} =$

$$(q_{j,n}^{(i)})^{-1}[q_{j,n}^{(i)} \pi_j^{(i)} f_i(a^*) - (\sigma(a^*) - a^*)] = (q_{j,n}^{(i)})^{-1}[(\prod_{\ell=1}^{j-1} \alpha_{i,\ell} \cdot \sum_{s=j}^{n} \beta_{i,s} \prod_{\ell=j}^{s-1} \alpha_{i,\ell}) f_i(a^*) +$$

$$a^* - \sum_{k=1}^{m} \sigma^{(k)} f_k(a^*)] = (q_{j,n}^{(i)})^{-1}[a^* + (\sum_{s=j}^{n} \beta_{i,s} \prod_{\ell=1}^{s-1} \alpha_{i,\ell} - \sum_{s=1}^{n} \beta_{i,s} \prod_{\ell=1}^{s-1} \alpha_{i,\ell}) f_i(a^*) -$$

$$\sum_{k \neq i} \sigma^{(k)} f_k(a^*)] = (q_{j,n}^{(i)})^{-1}[a^* - (\sum_{s=1}^{j-1} \beta_{i,s} \prod_{\ell=1}^{s-1} \alpha_{i,\ell}) f_i(a^*) - \sum_{k \neq i} \sigma^{(k)} f_k(a^*)] =$$

$(q_{j,n}^{(i)})^{-1}[a^* - \bar{S}_{j-1}^{(i)}(a^*)]$. This value is positive by definitions (26), (27). According to (21), this means that $u(i,j) \in D'$.

Now let $i \in J^*$, $j \in J_i^*$. Then by Lemma 1, the Eq. (25) is solvable and has the unique solution $a_j^{(i)}$. If $u_{i,j} = \pi_j^{(i)} f_i(a_j^{(i)})$, then, by (16), $p^{(i)}(u(i,j)) = p_{j,n}^{(i)} u_{i,j} =$

$$(\prod_{k=j}^{n-1} \alpha_{i,k}) \pi_j^{(i)} f_i(a_j^{(i)}) = \prod_{k=j}^{n-1} \alpha_{i,k} (\prod_{k=1}^{j-1} \alpha_{i,k}) f_i(a_j^{(i)}) = (\prod_{k=1}^{n-1} \alpha_{i,k}) f_i(a_j^{(i)}) = \lambda_i(a_j^{(i)}).$$

This means that the first equality in (23) holds. Since $x_{k,n} > 0$ for $k \in \overline{1,m} \setminus J^*$, from (19) we obtain that the remaining restrictions on $p^{(k)}(u(i,j))$ in (23), in the form of strict inequalities, are also satisfied.

Finally, from the equality (25) with $a = a_j^{(i)}$ we get: $\mu(a_j^{(i)}) = \sigma(a_j^{(i)}) - a_j^{(i)} =$

$$\sigma(a_j^{(i)}) - \bar{S}_{j-1}^{(i)}(a_j^{(i)}) = \sum_{k=1}^{m} \sigma^{(k)} f_k(a_j^{(i)}) - \sum_{k \neq i} \sigma^{(k)} f_k(a_j^{(i)}) - (\sum_{k=1}^{j-1} \beta_{i,k} \pi_k^{(i)}) f_i(a_j^{(i)}) =$$

$$\sigma^{(i)} f_i(a_j^{(i)}) - (\sum_{k=1}^{j-1} \beta_{i,k} \pi_k^{(i)}) f_i(a_j^{(i)}) = (\sum_{k=1}^{n} \beta_{i,k} \pi_k^{(i)}) f_i(a_j^{(i)}) - (\sum_{k=1}^{j-1} \beta_{i,k} \pi_k^{(i)}) f_i(a_j^{(i)}) =$$

$$(\sum_{k=j}^{n} \beta_{i,k} \pi_k^{(i)}) f_i(a_j^{(i)}) = (\sum_{k=j}^{n} \beta_{i,k} \prod_{\ell=1}^{k-1} \alpha_{i,\ell}) f_i(a_j^{(i)}) = (\sum_{k=j}^{n} \beta_{i,k} \prod_{\ell=j}^{k-1} \alpha_{i,\ell}) \prod_{\ell=1}^{j-1} \alpha_{i,\ell}$$

$f_i(a_j^{(i)}) = q_{j,n}^{(i)} \pi_j^{(i)} f_i(a_j^{(i)}) = q_{j,n}^{(i)} u_{i,j} = q(u(i,j))$. Thus, the second equality in (23) also holds, therefore $u(i,j) \in D''$. Lemma is proved.

We now give some properties of $a = \bar{a}(u)$, which we use in the proof of the following statement. Taking in equalities (19) $u = 0$, $x_{i,n} = 0$ for $i \in I_0(u)$ and summing them with coefficients $\beta_{i,j}$, it is easy to obtain the inequality

$$\bar{S}_{n-1}^{(i)}(a) \geq a, \quad a = \bar{a}(u) \quad (\forall u \in D'', \ i \in I_0(u)). \tag{30}$$

Further, since $\mu^* = \mu(a^*)$ (see (17), (18)) it follows from (21) that $\bar{a}(u) = a^*$ $(\forall u \in D')$. Using the properties of $\mu(a)$ [15, Lemma 1], it can be shown that $\bar{a}(u) \geq a^*$ $(\forall u \in D'')$. Thus, the following property holds:

$$\bar{a}(u) = a^* \ (\forall u \in D'), \quad \bar{a}(u) \begin{cases} = a^*, \ u \in \Gamma, \\ > a^*, \ u \notin \Gamma \end{cases} (\forall u \in D''). \tag{31}$$

Further we use parametric representations of some sets. To clarify the limits of the parameter change in these representations, we introduce the following value:

$$\sup\{\bar{a}(u) \mid u \in V\} = \bar{a}(V) \quad (V \subseteq \overline{U}).$$

Note that $\bar{a}(U) = \bar{a}(\overline{U}) = \bar{a}_F$ and, by (22), $\bar{a}(D') = a^*$. The following proposition, which we give without proof, gives the upper change bound of a on D.

Lemma 3. *Suppose that conditions* (9), (11), (13) *are satisfied. If* $D'' \neq \varnothing$, *then* $\bar{a}_D = \bar{a}_{D''} = \max_{i \in J^*} a_n^{(i)}$.

Thus, by Lemma 3, the change in the value of $\bar{a}(u)$ on D'' is bounded by $[a^*, \bar{a}]$, where $\bar{a}$ is maximal among the solutions of the equations $\bar{S}_{n-1}^{(i)}(a) = a$.

The set of quasi-preserving controls D'', in contrast to D', can be empty. Let us characterize the conditions under that D'' is non-empty.

Theorem 1. *Suppose that conditions* (9), (11), (13) *are satisfied. Then* $D'' \neq \varnothing$ *if and only if* $J^* \neq \varnothing$.

Proof. Necessity. Let $D'' \neq \varnothing$, $u \in D''$, $a = \bar{a}(u)$. Then it follows from (23), (31), that $I_0(u) \neq \varnothing$ and $a \geq a^* > 0$. By (9), the function $\bar{S}_{n-1}^{(i)}(a)$ is also concave. Set $\alpha = a^{-1}a^*$, $\beta = 1-\alpha$, then $\alpha \in (0,1]$; hence $\bar{S}_{n-1}^{(i)}(a^*) = \bar{S}_{n-1}^{(i)}(\alpha a) \geq \alpha \bar{S}_{n-1}^{(i)}(a) = a^{-1}a^*\bar{S}_{n-1}^{(i)}(a)$, i.e. $(a^*)^{-1}\bar{S}_{n-1}^{(i)}(a^*) \geq a^{-1}\bar{S}_{n-1}^{(i)}(a)$ $(\forall i \in \overline{1,m})$. Therefore, by (30), a fortiori $(a^*)^{-1}\bar{S}_{n-1}^{(i)}(a^*) \geq 1$, i.e. $\bar{S}_{n-1}^{(i)}(a^*) \geq a^*$ $(\forall i \in I_0(u))$.

Further, because of monotonicity of $\bar{S}_{j-1}^{(i)}(a)$ with respect to subscript, we obtain: $J^* = \{i \in \overline{1,m} \mid J_i^* \neq \varnothing\} = \{i \in \overline{1,m} \mid I_i^* \neq \overline{1,n}\} = \{i \in \overline{1,m} \mid \exists j \in \overline{1,n}: \bar{S}_{j-1}^{(i)}(a^*) \geq a^*\} = \{i \in \overline{1,m} \mid \bar{S}_{n-1}^{(i)}(a^*) \geq a^*\}$. Due to the inequality $\bar{S}_{n-1}^{(i)}(a^*) \geq a^*$ obtained above, we have $i \in J^*$, i.e. $J^* \neq \varnothing$.

Sufficiency follows from Lemma 2. Indeed, if $J^* \neq \varnothing$ then D'' contains the elements $u(i,n)$ for $i \in J^*$. Theorem is proved.

As we saw above, the structure of D' is simple enough—D' is a part of the hyperplane. In order to study the structure of the second (in the general case nonlinear) part of D, the structure of D'', we assume that the condition

$$\exists i \in \overline{1,m}: \quad \bar{S}_{n-1}^{(i)}(a^*) \geq a^*$$

from Theorem 1 is fulfilled.

Fix $a \in [0, \bar{a}_F]$, where $\bar{a}_F = a(\bar{x}_F)$, and consider the polyhedron

$$\overline{U}(a) = \{u \mid p^{(i)}(u) \leq \lambda_i(a) \ (\forall i \in \overline{1,m}), \ q(u) = \mu(a), \ u \geq 0\}. \tag{32}$$

Clearly, $\overline{U} = \cup\{\overline{U}(a) \mid a \in [0, \bar{a}_F]\}$. We introduce the sets

$$V_{D'} = V(a^*) \cap D', \quad V_{D''}(a) = V(a) \cap D'' \quad (a \in [a^*, \bar{a}_{D''}]),$$

where $V(a)$ is the set of all vertex of polyhedron $\overline{U}(a)$. Fix $a \in [a^*, \bar{a}_{D''}]$, $i \in \overline{1, m}$, and denote $L_i(a) = M_i(a) \cap \mathbb{R}_+^{mn}$, where

$$M_i(a) = \{u \mid p^{(i)}(u) = \lambda_i(a),\ p^{(j)}(u) \leq \lambda_j(a)\ (\forall j \neq i),\ q(u) = \mu(a)\}.$$

Clearly $M_i(a)$ is an $(mn - 2)$–dimensional affine manifold $(i \in \overline{1, m})$. Denote $D(a) = \overline{U}(a) \cap D''$, we obtain the following representation for D'':

$$D'' = \cup\{D(a) \mid a \in [a^*, \bar{a}_{D''}],\quad D(a) = \overset{m}{\underset{i=1}{\cup}} L_i(a)\}. \tag{33}$$

Along with the set (20), we also define the sets

$$I_1(u) = \{i \in \overline{1, m} \mid I_i^+(u) \neq \varnothing\},\quad I_2(u) = \{i \in \overline{1, m} \mid |I_i^+(u)| = 2\}.$$

The properties of the elements of $V_{D''}(a)$ are characterized by the following

Lemma 4. *Suppose that conditions* (9), (11), (13) *are satisfied.*

(i) if $u \in V_{D'}$, then u coincides with one of the vectors $u(i,j)$ $(i \in I^*,\ j \in I_i^*)$;
(ii) if $u \in V_{D''}(a)$, then the following properties hold:

$$I_2(u) \subseteq I_0(u) \subseteq I_1(u), \tag{34}$$

$$|I_i^+(u)| \leq 2\ (\forall i \in \overline{1, m}), \tag{35}$$

$$|I_2(u)| \leq 1;\quad I_2(u) \neq \varnothing \Leftrightarrow I_0(u) = I_1(u). \tag{36}$$

Proof. The control $u \in V(a)$ is a solution of a system of linear equations with a nonsingular matrix A_u that is obtained from the matrix

$$A = \begin{bmatrix} q \\ \hline P \\ \hline I \end{bmatrix} = \left[\begin{array}{c|cccc} q^{(1)}\ q^{(2)}\ \ldots\ q^{(m-1)}\ q^{(m)} \\ \hline p^{(1)}\ \ \mathbb{O}\ \ \ldots\ \ \ \mathbb{O}\ \ \ \ \mathbb{O} \\ \mathbb{O}\ \ p^{(2)}\ \ldots\ \ \ \mathbb{O}\ \ \ \ \mathbb{O} \\ \ldots\ \ \ldots\ \ \ldots\ \ \ \ldots\ \ \ \ldots \\ \mathbb{O}\ \ \ \mathbb{O}\ \ \ldots\ p^{(m-1)}\ \ \mathbb{O} \\ \mathbb{O}\ \ \ \mathbb{O}\ \ \ldots\ \ \ \mathbb{O}\ \ \ p^{(m)} \\ \hline \multicolumn{4}{c}{I_{mn}} \end{array}\right]$$

by adding to the first line another $mn - 1$ lines from the remaining ones. Here $p^{(i)} = (p_{1,n}^{(i)}, p_{2,n}^{(i)}, \ldots, p_{n,n}^{(i)})$, $q^{(i)} = (q_1^{(i)}, q_2^{(i)}, \ldots, q_n^{(i)})$, $i \in \overline{1, m}$ (see (10), (15)), I_{mn} is the identity matrix of dimension mn.

Assertion (i) is obvious; we prove now the assertion (ii), first the property $I_0(u) \subseteq I_1(u)$ from (34). If $i \in I_0(u)$, then $p^{(i)}(u) = \lambda_i(a)$. Assuming $i \notin I_1(u)$, we get $I_i^+(u) = \varnothing$, i.e. $u_{i,j} = 0$ $(\forall j \in \overline{1, n})$. But then A_u contains simultaneously the i-st row of A and those rows of I_{mn} that have units in the columns with indices $(i - 1)n + j$ $(\forall j \in \overline{1, n})$. Expanding determinant $|A_u|$ along these unit rows, we obtain a determinant having a zero row (the part of the i-st row of

P without vector $p^{(i)}$); therefore $|A_u| = 0$. But this contradicts the way A_u is chosen. Therefore, $i \in I_1(u)$, so really $I_0(u) \subseteq I_1(u)$.

We now prove property $I_2(u) \subseteq I_0(u)$. Let $i \in I_2(u)$, so $|I_i^+(u)| = 2$. This means that two rows of E_{mn} with indices $(i-1)n + j_1$, $(i-1)n + j_2$ for some $j_1, j_2 \in \overline{1,n}$ are absent in A_u. Assuming $i \notin I_0(u)$, we also get the absence of the i-st row of P in A_u. But then columns of A_u with indices $(i-1)n+j_1$, $(i-1)n+j_2$ contain nonzero elements only in the first row. Expanding determinant $|A_u|$ along one of them, we obtain a determinant with zero column, and thus $|A_u| = 0$. This contradiction shows that $i \in I_0(u)$ and $I_2(u) \subseteq I_0(u)$.

We now prove (35). Denote $I_0 = I_0(u)$, $|I_0| = k$, $|I_\ell^+(u)| = k_\ell$, $\max_{\ell \in I_0} k_\ell = \bar{k}$. Each element from I_0 adds to A_u one row of P; each element from $I_\ell^+(u)$ removes rows from I_{mn} in amounts of k_ℓ. Because of $\sum_{\ell \in I_0} k_\ell \geq \bar{k} + (k-1)$, we have: $mn = 1 + k + (mn - \sum_{\ell \in I_0} k_\ell) \leq 1 + k + mn - \bar{k} - k + 1 = mn + 2 - \bar{k}$, so $\bar{k} \leq 2$.

We now prove the first part of (36). Denote $|I_2(u)| = k$. If A_u contains ℓ rows of P, then $\ell = |I_0(u)|$, and from $I_2(u) \subseteq I_0(u)$ we obtain $k \leq \ell$. We break up the set $\overline{1, mn}$ into disjoint parts $I_1 = \overline{1,m} \setminus I_0(u)$, $I_2 = I_0(u) \setminus I_2(u)$, $I_3 = I_2(u)$. This partition generates the corresponding partition of rows of E_{mn} belonging to A_u. We denote by $k(I)$ the number of rows of E_{mn} corresponding to I. Thus, the total number of rows of E_{mn} contained in A_u is equal to $s = k(I_1) + k(I_2) + k(I_3)$. If $i \in I_2$ (resp. $i \in I_2$), then A_u contains all rows of E_{mn} with indices $(i-1)n+j$ ($\forall j \in \overline{1,n}$), except for one (resp. two), therefore $k(I_2) = (\ell - k)(n-1)$, $k(I_3) = k(n-2)$. Since $s + l + 1 = mn$ we get $k(I_1) = mn - \ell - 1 - (\ell - k)(n-1) - k(n-2) = (m - \ell)n + k - 1$. On the other hand, $k(I_1) \leq (m - \ell)n$ since $|I_1| = m - \ell$. This implies $k \leq 1$, so that $I_2(u) \leq 1$.

Finally, we prove the second part of (36). Representing I_1 in the form $I \cup J$, where $I = I_1(u) \setminus I_0(u)$, $J = \overline{1,m} \setminus I_0(u)$, and given that A_u cannot contain the rows of E_{mn} with indices $(i-1)n+j$ ($j \in \overline{1,n}$) for $i \in I_1(u)$, we get: $k(I_1) = k(I) + k(J) = |I|(n-1) + |J|n = (|I_1(u)| - |I_0(u)|)n - |I| + (m - |I_1(u)|)n = (m - |I_0(u)|)n - |I| = (m - \ell)n - |I|$, so $k(I_1) = (m - \ell)n - |I|$.

On the other hand, $k(I_1) = (m - \ell)n + k - 1$. Hence, $|I| = 1 - k$, and, therefore, $k = 1$ (i.e., $I_2(u) \neq \varnothing$) if and only if $I = \varnothing$, i.e. $I_0(u) = I_1(u)$, as required.

All assertions of Lemma are proved.

Conclusions of Lemma 4 are interpreted as follows. Among the vertices of the polyhedron $\overline{U}(a^*)$, the vectors $u(i,j)$ ($i \in I^*$, $j \in I_i^*$) (and only they) can be preserving controls. Only one of the blocks $u^{(i)}$ of quasipreserving controls $u = (u^{(1)}, u^{(2)}, \dots, u^{(m)})$ that are the vertices of the polyhedron $\overline{U}(a)$ can have the maximum possible number of positive coordinates, equal to two. In this case, the last coordinates of blocs $x^{(k)}$ of $x = (x^{(1)}, x^{(2)}, \dots, x^{(m)})$, corresponding to not completely zero blocks $u^{(k)}$ of u, are necessarily equal to zero.

Now we can obtain an upper bound for the minimal number of nonzero coordinates for optimal controls in problem (3) with a linear non-negative function $c(u)$. Since every optimal control $\tilde{u} \in D$, we have, by (31), $\bar{a}(\tilde{u}) \geq a^*$.

If $\bar{a}(\tilde{u}) > a^*$, then $u \in D''$, and, it follows from Lemma 4 that $|I^+(\tilde{u})| \leq m+1$. Now assume that $\bar{a}(\tilde{u}) = a^*$; then $\tilde{u} \in \overline{U}(a^*)$ (see (32)). It follows from (22) that $\overline{U}(a^*) = D' \cup D(a^*)$. But $\overline{U}(a^*) = \text{co } V(a^*)$, and $V(a^*) = V_{D'} \cup V_{D''}(a^*)$, where, by Lemma 4, $V_{D'} = \{u(i,j) \mid I \in I^*, j \in I_i^*\}$ (this set may be empty). Therefore, $\overline{U}(a^*) = \mathrm{co}\,(V_{D'} \cup V_{D''}(a^*))$. As can be seen from definition (28), $|I^+(u(i,j))| = 1$. Using, as above, Lemma 4, we obtain that in case $\bar{a}(\tilde{u}) = a^*$ also $|I^+(\tilde{u})| \leq m+1$.

Thus, we get the following assertion.

Corollary 1. *Suppose that conditions* (9), (11), *and* (13) *are satisfied. Then for the problem* (3) *with a linear non-negative function* $c(u)$ *there exists an optimal vector, the number of nonzero coordinates of that does not exceed* $m + 1$.

Note that this assertion is a generalization of the above-mentioned bimodality property. Indeed, for $m = 1$, when the population structure is linear, our result indicates the existence of "two-age" optimal strategies.

Now we can get the elements of $V_{D''}(a)$ in explicit form. For this we need a generalization of the notation (24) to the case when the subscript and the superscript are sets. Henceforth, the ordered subsets of $\overline{1,m}$, $\overline{0,n}$ appear, to denote which we use the symbol of the row vector. So, the notation $I = (i_1, i_2, \ldots, i_\ell)$ indicates the order of the elements of $I = \{i_1, i_2, \ldots, i_\ell\}$.

Let $I = (i_1, i_2, \ldots, i_\ell) \subseteq \overline{1,m}$, $J = (j_1, j_2, \ldots, j_\ell) \subseteq \overline{0,n}$, and the second of these sets allow repetition of elements, in contrast to the first: $|I| = \ell$, $|J| \in \overline{1,\ell}$. We define the value of $\bar{S}_J^I(a) = \bar{S}_{j_1,\ldots,j_\ell}^{i_1,\ldots,i_\ell}(a)$ as follows:

$$\bar{S}_J^I(a) = \sum_{k=1}^\ell \sigma_{j_k}^{i_k} f_{i_k}(a) + \sum_{i \notin I} \sigma^i f_i(a). \tag{37}$$

It is easy to see that in the case of $I = \{i\}$, $J = \{j\}$, the notation (37) turns into the notation (24). We also use the following notation:

$$I = (i_1, \ldots, i_\ell),\; I' = (i_1, \ldots, i_\ell, i),\; J = (j_1, \ldots, j_\ell),\; J' = (j_1, \ldots, j_\ell, j),$$

$$K = (k_1, \ldots, k_s),\; L = (\ell_1, \ldots, \ell_s),\; L' = (\ell_1', \ldots, \ell_s'),$$

where $\ell \in \overline{1, m-1}$, $s \in \overline{1,m}$, $i_r \in \overline{1,m}$, $j_r \in \overline{0,n}$ $(r \in \overline{1,\ell})$, $k_r \in \overline{1,m}$, $\ell_r, \ell_r' \in \overline{0,n}$ $(r \in \overline{1,s})$, $j, k \in \overline{0,n}$, $i \neq i_r$ $(\forall r \in \overline{1,\ell})$,

$$\bar{\Delta}_{L,L'}^K(a) = \bar{\Delta}(k_1(\ell_1, \ell_1'), \ldots, k_s(\ell_s, \ell_s'), a) = (\bar{S}_L^K(a), \bar{S}_{L'}^K(a)],$$

$$\bar{\Delta}_{j,k}^i(I, J, a) = \bar{\Delta}_{j,k}^i(i_1(j_1), \ldots, i_\ell(j_\ell), a) = (\bar{S}_{j_1-1,\ldots,j_\ell-1,j}^{i_1,\ldots,i_\ell,i}(a), \bar{S}_{j_1-1,\ldots,j_\ell-1,k}^{i_1,\ldots,i_\ell,i}(a)],$$

$$\bar{\Delta}_j^i(I, J, a) = \bar{\Delta}_j^i(i_1(j_1), \ldots, i_\ell(j_\ell), a) = \bar{\Delta}_{j-1,j}^i(I, J, a),$$

$$\bar{\Delta}^i(I', J', a) = \bar{\Delta}^i(i_1(j_1), \ldots, i_\ell(j_\ell), i(j), a) = \bar{\Delta}_j^i(I, J, a),$$

If $|K| = |I'| = |J'| = m$ or $|I| = |J| = m - 1$, then I, J, I', J', K are omitted.

The previous statements make it possible to find the elements of $V_{D''}(a)$ in an explicit form. We present the following assertion without proof.

Theorem 2. *Suppose that conditions* (9), (11), (13) *are satisfied. Assume that* $u \in \overline{U}$, $I_0(u) \subseteq I \cup \{i\}$ *and* $I_i^+(u) \subseteq \{j, k\}$, *where* $I = (i_1, i_2, \ldots, i_\ell) \subseteq \overline{1,m} \setminus \{i\}$, $j \leq k$ $(i, j, k \in \overline{1,m})$. *Then the following statements hold:*

(i) If $I_0(u) = I$, $I_i^+(u) = \{j\}$ then $u \in V_{D''}(a)$ if and only if

$$\sum_{s=j}^{n} \beta_{i,s} > 0, \quad a \in \bar{\Delta}_{j-1,n}^{i}(I, J, a). \tag{38}$$

In this case, the positive coordinates of u are defined as follows:

$$u_{i_r,j_r} = \pi_{j_r}^{(i_r)} f_{i_r}(a) \ (r \in \overline{1,\ell}), \quad u_{i,j} = (q_{j,n}^{(i)})^{-1}(\bar{S}_{j_1-1,\ldots,j_\ell-1}^{i_1,\ldots i_\ell}(a) - a). \tag{39}$$

(ii) If $I_0(u) = I_1(u)$ then $u \in V_{D''}(a)$ if and only if

$$\sum_{s=j}^{k-1} \beta_{i,s} > 0, \quad a \in \bar{\Delta}_{j-1,k-1}^{i}(I, J, a). \tag{40}$$

In this case, the positive coordinates of u are defined as follows:

$$\left.\begin{array}{l} u_{i_r,j_r} = \pi_{j_r}^{(i_r)} f_{i_r}(a) \ (r \in \overline{1,\ell}), \\[4pt] u_{i,j} = (q_{j,k-1}^{(i)})^{-1}(\bar{S}_{j_1-1,\ldots,j_\ell-1,j-1}^{i_1,\ldots i_\ell,i}(a) - a), \\[4pt] u_{i,k} = p_{j,k}^{(i)}(q_{j,k-1}^{(i)})^{-1}(a - \bar{S}_{j_1-1,\ldots,j_\ell-1,k-1}^{i_1,\ldots i_\ell,i}(a)). \end{array}\right\} \tag{41}$$

We denote by $u_j^i(i_1(j_1), \ldots, i_\ell(j_\ell), a) = u_j^i(I, J, a)$, $u_{j,k}^i(i_1(j_1), \ldots, i_\ell(j_\ell), a) = u_{j,k}^i(I, J, a))$ the controls with coordinates (39) and (41), respectively.

Example 1. Consider a constraint system of the form (14) for non-negative vectors $x = (x_{1,1}, x_{1,2}; x_{2,1}, x_{2,2})$, $u = (u_{1,1}, u_{1,2}; u_{2,1}, u_{2,2})$, with $m = n = 2$, $\alpha_{i,1} = 1/2$, $\beta_{i,j} = 1 \, (i, j = 1, 2)$,

$$f_1(a) = \begin{cases} a, & 0 \le a < 1, \\ (a+1)/2, & a \ge 1, \end{cases} \qquad f_2(a) = \begin{cases} a, & 0 \le a < 1, \\ 1, & a \ge 1. \end{cases}$$

Using notation (12), (17), (18) we find: $\sigma_0^{(1)} = \sigma_0^{(2)} = 0$, $\sigma_1^{(1)} = \sigma_1^{(2)} = 1$, $\sigma_2^{(1)} = \sigma_2^{(2)} = \sigma^{(1)} = \sigma^{(1)} = 3/2$, $a^* = 1$, $\mu^* = 2$, $f_1(a^*) = f_2(a^*) = 1$,

$$\sigma(a) = \begin{cases} 3a, & 0 \le a < 1, \\ (3a+9)/4, & a \ge 1, \end{cases} \qquad \mu(a) = \begin{cases} 2a, & 0 \le a < 1, \\ (9-a)/4, & a \ge 1. \end{cases}$$

The functions $f_1(a)$, $f_2(a)$ are concave, all the coefficients $\beta_{i,j}$ are positive, $\sigma'(0) = 3 > 1$, $\sigma'(+\infty) = 3/4 < 1$. All conditions (9), (11), (13) are satisfied.

By (24), $\bar{S}_0^{(1)}(a) = \sigma^{(2)} f_2(a) = 3/2$, $\bar{S}_0^{(2)}(a) = \sigma^{(1)} f_1(a) = 3/2$, so we see from (26), (27) that $I^* = \varnothing$, $J^* = \{1, 2\}$. This means by Theorem 1 that $D'' \ne \varnothing$.

Solving Eq. (25), we find $a_j^{(i)}$: $a_1^{(1)} = 3/2$, $a_2^{(1)} = 4$, $a_1^{(2)} = 3$, $a_2^{(2)} = 7$, therefore (see Lemma 3) $\bar{a}_D = \bar{a}_{D''} = a_2^{(2)} = 7$.

Further, $p^{(1)}(u) = \frac{1}{2}u_{1,1} + u_{1,2}$, $p^{(2)}(u) = \frac{1}{2}u_{2,1} + u_{2,2}$. For $u \in D''$ we have, by (31), $a \ge a^*$, therefore $\mu(a) = (9-a)/4$, $\lambda_1(a) = (a+1)/4$, $\lambda_2(a) = 1/2$.

By (32), $D'' = \cup_{a \in [2,7]} D(a)$, where $D(a) = L_1(a) \cup L_2(a)$, and $L_1(a)$, $L_2(a)$ are described by systems of constraints

$$
\begin{cases}
u_{1,1} + 2u_{1,2} = (a+1)/2, \\
u_{2,1} + 2u_{2,2} \leq 1, \\
3u_{1,1} + 2u_{1,2} + 3u_{2,1} + 2u_{2,2} = \dfrac{9-a}{2}, \\
u_{1,1} \geq 0,\ u_{1,2} \geq 0,\ u_{2,1} \geq 0,\ u_{2,2} \geq 0,
\end{cases}
\qquad
\begin{cases}
u_{1,1} + 2u_{1,2} \leq (a+1)/2, \\
u_{2,1} + 2u_{2,2} = 1, \\
3u_{1,1} + 2u_{1,2} + 3u_{2,1} + 2u_{2,2} = \dfrac{9-a}{2}, \\
u_{1,1} \geq 0,\ u_{1,2} \geq 0,\ u_{2,1} \geq 0,\ u_{2,2} \geq 0,
\end{cases}
$$

respectively. Next, we find the vectors $u(i,j)$ (see (28)): $u(1,1) = (5/4, 0; 0, 0)$, $u(1,2) = (0, 5/4; 0, 0)$, $u(2,1) = (0, 0; 1, 0)$, $u(2,2) = (0, 0; 0, 1/2)$. It follows from (29) by $J^* = \{1, 2\}$ that all $u(i,j) \in D''$.

To illustrate Theorem 2, we find now the elements of $V_{D''}(a)$, for example, for $a = 2$. First of all, as seen from (38), (40), we must find all the intervals $\bar{\Delta}_j^i(I, J, a)$, $\bar{\Delta}_{j,k}^i(I, J, a)$ that contain this number: $\Delta_{0,1}^1(\varnothing, \varnothing, 2) = (\bar{S}_0^{(1)}(2), \bar{S}_1^{(1)}(2)]$, $\bar{\Delta}_2^1(\{1\}, \{2\}, 2) = (\bar{S}_{1,0}^{1,2}(2), \bar{S}_{1,1}^{1,2}(2)]$, $\bar{\Delta}_{1,2}^1(\{2\}, \{2\}, 2) = (\bar{S}_{1,0}^{2,1}(2), \bar{S}_{1,1}^{2,1}(2)(2)]$. They all coincide with $(3/2, 5/2]$, hence they contain the number $a = 2$.

Using (39), (41), we find the vectors from $V_{D''}(a)$ corresponding to these intervals: $v_1 = u_{1,2}^1(\varnothing, \varnothing, 2) = (1, 1/4, ; 0, 0)$, $v_2 = u_1^2(\{1\}, \{2\}, 2) = (0, 3/4; 2/3, 0)$, $v_3 = u_{1,2}^1(\{2\}, \{2\}, 2) = (1/2, 1/2; 0, 1/2)$. We see that in accordance with Corollary 1, the number of positive coordinates of these vectors is less than or equal to $m + 1 = 3$.

As can be seen from (22), the set D' for the model (7) always contains a part of some hyperplane (its equation is $q(u) = \mu^*$; for our example this equation has the form $3u_{1,1} + 2u_{1,2} + 3u_{2,1} + 2u_{2,2} = 4$). This example is interesting in that this property also holds for D'' (this is due to the affinity of $f_1(a)$, $f_2(a)$). Indeed, it is easy to verify that the vectors found are linearly independent. Adding to them the vector $u(1,2) = (0, 5/4; 0, 0)$ corresponding to the value of $a = 4$, we obtain four linearly independent vectors of four-dimensional space; therefore, they define a certain hyperplane. Its equation has the form $4u_{1,1} + 4u_{1,2} + 3u_{2,1} + 2u_{2,2} = 5$. Thus, the set D'' contains a part of this hyperplane.

4   Conclusion

In this article, we studied quasi-preserving controls for optimal exploitation problem of an ecological population described by the generalization of the Leslie model. This problem arose as a result of the systematic use of a general approach to formalizing the ecological populations exploitation problem [13].

We would like to emphasize once again the two basic features of our article. Firstly, additive controls are used, unlike many other works. It seems to us more adequate from the point of view of ecological interpretation, although sometimes this complicates the study of the received problem. Further, the certain novelty of our results is also determined by the transition to the population model with a binary structure. For this model, the properties of the boundary of an feasible set containing potentially optimal vectors are described in detail.

In particular, the necessary and sufficient condition for the existence of quasi-preserving controls is proved, and these controls are found explicitly (see Theorems 1, 2). Note that the explicitly found controls contain the minimum possible number of positive coordinates. It is important for populations exploitation problems, since such controls affect the minimum possible number of structural units of the population.

Further, the distribution of controls having only one positive coordinate over the sets of preserving and quasi-preserving controls is also presented (see Lemma 2). Finally, the generalization of the well-known "bimodality" property of optimal strategies is proved (see Corollary 1), previously obtained by many authors for populations with one-dimensional structure.

The significance of these main results lies, in particular, in the fact that they open the way to the construction of algorithms that take into account the features of the considered optimal exploitation problem. Due to the parametric representation (33) of the feasible set and optimal solutions (39), (41), it becomes possible to reduce this problem to the solution of a series of one-dimensional optimization problems.

We hope that this paper will show possible new directions in the development of general theory of renewable resources exploitation.

References

1. Grooten, M., Almond, R.E.A. (eds.): Living Planet Report-2018: Aiming Higher. WWF, Gland, Switzerland (2018)
2. Dunkel, G.M.: Maximum sustainable yields. SIAM J. Appl. Math. **19**(2), 367–378 (1970)
3. Doubleday, W.G.: Harvesting in matrix population model. Biometrics **31**, 189–200 (1975)
4. Reed, W.J.: Optimum age-specific harvesting in a nonlinear population model. Biometrics **36**(4), 579–593 (1980)
5. Tahvonen, O.: Optimal harvesting of size-structured biological populations. In: Moser, E., Semmler, W., Tragler, G., Veliov, V.M. (eds.) Dynamic Optimization in Environmental Economics. Dynamic Modeling and Econometrics in Economics and Finance, vol. 15, pp. 329–355. Springer, Heidelberg (2014). https://doi.org/10.1007/978-3-642-54086-8_15
6. Getz, W.M., Haight, R.G.: Population Harvesting: Demographic Models of Fish, Forest, and Animal Resources. Princeton University Press, Princeton (1989)
7. Logofet, D.O., Belova, I.N.: Nonnegative matrices as a tool to model population dynamics: classical models and contemporary expansions. J. Math. Sci. **155**(6), 894–907 (2008). https://doi.org/10.1007/s10958-008-9249-2
8. Beddington, J.R., Taylor, D.B.: Optimum age specific harvesting of a population. Biometrics **29**, 801–809 (1973)
9. Rorres, C., Fair, W.: Optimal harvesting policy for an age specific population. Math. Biosci. **24**(1/2), 31–47 (1975)
10. Smirnov, A.I.: On some nonlinear generalizations of Leslie model considering the effect of saturation. Bull. Ural Inst. Econ. Manag. Law **4**(13), 98–101 (2010). (in Russian)

11. Boucekkine, R., Hritonenko, N., Yatsenko, Yu.: Age-structured modeling: past, present, and new perspectives. In: Boucekkine, R., Hritonenko, N., Yatsenko, Yu. (eds.) Optimal Control of Age structured Populations in Economy, Demography, and the Environment. Routledge, London (2011)
12. Logofet, D.: Matrices and Graphs. Stability Problems in Mathematical Ecology. CRC Press, Boca Raton (2018)
13. Mazurov, Vl.D., Smirnov, A.I.: On the reduction of the optimal non-destructive system exploitation problem to the mathematical programming problem. In: Evtushenko, Yu.G., Khachay, M.Yu., Khamisov, O.V., et al. (eds.) Proceedings of 8th International Conference on Optimization and Applications, OPTIMA-2017, CEUR Workshop Proceedings, vol. 1987, pp. 392–398 (2017)
14. Mazurov, Vl.D., Smirnov, A.I.: Conditions of irreducibility and primitivity monotone subhomogeneous maps. Trudy Instituta Matematiki i Mekhaniki UrO RAN **22**(3), 169–177 (2016). https://doi.org/10.21538/0134-4889-2016-22-3-169-177. (in Russian)
15. Smirnov, A., Mazurov, V.: On existence of optimal non-destructive controls for ecosystem exploitation problem applied to a generalization of Leslie model. In: Evtushenko, Yu.G., Jaćimović, M., Khachay, M.Yu., et al. (eds.) Proceedings of 9th International Conference on Optimization and Applications (OPTIMA 2018) (Supplementary Volume), DEStech Transactions on Computer Science and Engineering, pp. 199–213 (2018). https://doi.org/10.12783/dtcse/optim2018/27933

On a Global Search in D.C. Optimization Problems

Alexander S. Strekalovsky[✉] [iD]

Matrosov Institute for System Dynamics and Control Theory SB RAS,
Lermontov str., 134, 664033 Irkutsk, Russia
strekal@icc.ru

Abstract. This paper addresses the nonconvex optimization problem with the cost function and equality and inequality constraints given by d.c. functions. The original problem is reduced to a problem without constraints by means of the exact penalization techniques. Furthermore, the penalized problem is presented as a d.c. minimization problem. For the latter problem, we apply the global optimality conditions (GOCs), which possess the so-called constructive (algorithmic) property. These new GOCs are generalized for the minimizing sequences, and a theoretical method is developed. Based on this theoretical foundation, a new global search scheme is designed for the auxiliary (penalized) and original problems, the convergence of which is one of the new results of the work.

Keywords: Nonconvex optimization · D.C. functions · Exact penalty · Linearized problem · Optimality conditions · Convergence

1 Introduction

A little more than 50 years ago the exact penalty method was invented by Eremin and Zangwill [7,29] independently and almost simultaneously. Furthermore, this technology has got widespread and become very popular. It is considered to be a very effective and powerful tool for solving difficult real-life problems, such as games, search for equilibria, bilevel problems, hierarchical control, etc.

On the other hand, it is not difficult to see that almost all applied problems turn out to be explicitly or implicitly nonconvex with many (and this number is often huge) local pitfalls located rather far from the set of global solutions $Sol(\mathcal{P})$.

Besides, as well-known, the classical optimization methods prove to be ineffective and even inoperative when used for finding just a global solution, and provide at best only the KKT-vectors [11–13,16,19,27,28]. Moreover, the methods of the Branch and Bound idea (and cut's approach) "suffer" the so-called "curse of dimension", when an increase in dimension of a problem under scrutiny always entails the exponential growth of computational efforts.

© Springer Nature Switzerland AG 2020
M. Jaćimović et al. (Eds.): OPTIMA 2019, CCIS 1145, pp. 222–236, 2020.
https://doi.org/10.1007/978-3-030-38603-0_17

Here we develop a different approach based on a solid theoretical foundation, in particular, on the Global Optimality Conditions (GOCs) [22,23] in nonconvex optimization problems with d.c. functions [13,19,27].

First, for Problem $(\mathcal{P})$ we introduce the penalty function $W(\cdot)$ and the corresponding auxiliary (penalized) Problem $(\mathcal{P}_\sigma)$ without equality and inequality constraints. How one can attack the similar non-convex problems in another way readers can find in [15,26].

Furthermore, we perform decomposition of the cost function $F_\sigma(\cdot)$ of Problem $(\mathcal{P}_\sigma)$ into a difference of two convex functions. Thus it becomes possible to obtain the new GOCs and study its properties.

In addition, we perform the transformation of GOCs for minimizing sequences and develop a new theoretical method for solving Problem $(\mathcal{P}_\sigma)$ and prove its convergence. On the basis of this theoretical foundation, we develop a new Global Search Scheme (GSS) along with its algorithmization.

Finally, we present the new convergence theorem for the GSS for the general d.c. optimization problem $(\mathcal{P})$.

2 Problem Statement

Consider the following problem:

$$(\mathcal{P}): \quad \left. \begin{array}{l} f_0(x) := g_0(x) - h_0(x) \downarrow \min\limits_x, \quad x \in S, \\ f_i(x) := g_i(x) - h_i(x) \le 0, \quad i \in I = \{1,\dots,m\}, \\ f_i(x) := g_i(x) - h_i(x) = 0, \quad i \in \mathcal{E} = \{m+1,\dots,l\}; \end{array} \right\}$$

where the functions $g_i(\cdot)$, $h_i(\cdot)$, $i \in \{0\} \cup I \cup \mathcal{E}$, are convex on $\mathbb{R}^n$, so that the functions $f_i(\cdot)$, $i \in \{0\} \cup I \cup \mathcal{E}$, are the d.c. functions [8,11–13,19,27]. Recall that any continuous function can be approximated by a d.c. function with any desirable accuracy. Besides, assume that the set $S \subset \mathbb{R}^n$ is convex and compact.

Next, suppose that the set $Sol(\mathcal{P})$ of global solutions to Problem $(\mathcal{P})$, $Sol(\mathcal{P}) := \{x \in \mathcal{F} \mid f_0(x) = \mathcal{V}(\mathcal{P})\}$ and the feasible set $\mathcal{F}$ of Problem $(\mathcal{P})$, $\mathcal{F} := \{x \in S \mid f_i(x) \le 0, i \in I, f_i(x) = 0, i \in \mathcal{E}\}$, are non-empty. Additionally, in what follows, the optimal value $\mathcal{V}(\mathcal{P})$ of Problem $(\mathcal{P})$ is supposed to be finite:

$$(\mathcal{H}_f): \qquad \mathcal{V}(\mathcal{P}) := \inf(f_0, \mathcal{F}) := \inf\limits_x\{f_0(x) \mid x \in \mathcal{F})\} > -\infty. \qquad (1)$$

3 The Exact Penalty

In this section, we introduce the following penalty function $W(\cdot)$ for Problem $(\mathcal{P})$

$$W(x) := \max\{0, f_1(x),\dots,f_m(x)\} + \sum\limits_{j\in\mathcal{E}} |f_j(x)|, \qquad (2)$$

and, along with Problem $(\mathcal{P})$, consider the penalized problem without the inequality and equality constraints:

$$(\mathcal{P}_\sigma): \qquad F_\sigma(x) \downarrow \min\limits_x, \quad x \in S, \qquad (3)$$

where $\sigma \geq 0$ is a penalty parameter, and the merit function

$$F_\sigma(x) := f_0(x) + \sigma W(x), \tag{4}$$

is the cost function of the auxiliary problem $(\mathcal{P}_\sigma)$–(3).

It is well-known that if $z \in Sol(\mathcal{P}_\sigma)$, and z is feasible in $(\mathcal{P})$, i.e. $z \in \mathcal{F}$, then z is a global solution to $(\mathcal{P})$: $z \in Sol(\mathcal{P})$ [1–6,10,12,15,16]. It is worth mentioning that, generally, the inverse proposition does not hold.

Hence, the key feature of the exact penalization (EP) theory is the existence of a threshold value $\sigma_* \geq 0$ of the penalty parameter $\sigma \geq 0$ for which $Sol(\mathcal{P}_\sigma) \subset Sol(\mathcal{P})$ $\forall \sigma \geq \sigma_*$. The latter means that for $\sigma \geq \sigma_*$ Problems $(\mathcal{P})$ and $(\mathcal{P}_\sigma)$ are equivalent: $Sol(\mathcal{P}) = Sol(\mathcal{P}_\sigma)$ (see [12, Chapt. VII, Lemma 1.2.1], [4]).

On the other hand, the existence of the threshold exact penalty parameter $\sigma_* \geq 0$ implies that instead of solving a sequence of unconstrained problems with $\sigma_k \to \infty$ [1,4,16] we need to solve only a single unconstrained problem.

Hence, the proof of existence of the exact penalty threshold $\sigma_* \geq 0$ is a key moment in the investigation of relations between Problems $(\mathcal{P})$ and $(\mathcal{P}_\sigma)$ [2,4–6,10,14].

Recall that under various constraint qualification (CQ) conditions (MFCQ, etc. [1,2,4–6,10,14], the error bound properties [1,2,4–6,10,14], the metric subregularity conditions, calmness of constraints systems) can help prove the existence of the exact penalty threshold $\sigma_* \geq 0$ for the local solution as well as for the global one [2,4–6,10,14].

Assume that some regularity conditions, which ensure the existence of such threshold value $\sigma_* \geq 0$ of the penalty parameter, are fulfilled.

4   Global Optimality Conditions (GOC)

Before all, we will show that the cost function $F_\sigma(\cdot)$ of Problem $(\mathcal{P}_\sigma)$ is a d.c. function, i.e. it can be represented as a difference of convex functions. Indeed, since [13,19,27].

$$|f_i(x)| = 2 \max\{g_i(x), h_i(x)\} - [g_i(x) + h_i(x)],$$

it can be readily seen that

$$F_\sigma(x) \stackrel{\triangle}{=} f_0(x) + \sigma \max\{0, f_i(x), \ i \in I\} + \sigma \sum_{i \in \mathcal{E}} |f_i(x)| = G_\sigma(x) - H_\sigma(x), \tag{5}$$

where

$$H_\sigma(x) := h_0(x) + \sigma \left[\sum_{i \in I} h_i(x) + \sum_{j \in \mathcal{E}} (g_j(x) + h_j(x)) \right], \tag{6}$$

$$G_\sigma(x) := F_\sigma(x) + H_\sigma(x) =$$
$$= g_0(x) + \sigma \max \left\{ \sum_{j \in I} h_j(x); \left[g_i(x) + \sum_{j \in I}^{j \neq i} h_j(x) \right], \ i \in I \right\} + \tag{7}$$
$$+ 2\sigma \sum_{i \in \mathcal{E}} \max\{g_i(x); h_i(x)\}.$$

It is easy to see that $G_\sigma(\cdot)$ and $H_\sigma(\cdot)$ are both convex functions [12,17,18], so that the function $F_\sigma(\cdot)$ is a d.c. function, as claimed. Besides, it is clear, that for a feasible (in $(\mathcal{P})$) point $z \in S$ we have

$$W(z) \overset{\triangle}{=} \max\{0, f_1(z), \dots, f_m(z)\} + \sum_{i \in \mathcal{E}} |f_i(z)| = 0,$$

and, therefore, for $\zeta := f_0(z)$, we obtain

$$F_\sigma(z) = f_0(z) + \sigma W(z) = f_0(z) = \zeta. \tag{8}$$

The following GOCs were proposed in [22–24].

Theorem 1. *Let a feasible point* $z \in \mathcal{F}$, $\zeta := f_0(z)$, *be a solution to Problem* $(\mathcal{P})$ *and* $\sigma \geq \sigma_* > 0$, *where* σ_* *is a threshold value of the penalty parameter, such that* $Sol(\mathcal{P}) = Sol(\mathcal{P}_\sigma)$ $\forall \sigma \geq \sigma_*$.
Then for every pair $(y, \beta) \in \mathbb{R}^n \times \mathbb{R}$, *such that*

$$H_\sigma(y) = \beta - \zeta, \tag{9}$$

the following inequality holds

$$G_\sigma(x) - \beta \geq \langle \nabla H_\sigma(y), x - y \rangle \quad \forall x \in S. \tag{10}$$

Clearly, Theorem 1 reduces the solution of the nonconvex Problem $(\mathcal{P}_\sigma)$ to a study of the family of convex (linearized) problems as follows

$$(\mathcal{P}_\sigma L(y)): \qquad \Phi_{\sigma y}(x) := G_\sigma(x) - \langle \nabla H_\sigma(y), x \rangle \downarrow \min_x, \ x \in S \tag{11}$$

depending on the pair $(y, \beta) \in \mathbb{R}^{n+1}$ fulfilling the Eq. (9).

In addition, it is worth noting that the linearization is performed with respect to the "united" nonconvexity of Problem $(\mathcal{P})$, which is accumulated by the function $H_\sigma(\cdot)$ (see $(\mathcal{P})$ and (6)).

Remark 1. Suppose, we found a triple (y, β, u), such that $(y, \beta) \in \mathbb{R}^{n+1}$, $H_\sigma(y) = \beta - \zeta$, $u \in S$, and for which the principal inequality (10) is violated, i.e.

$$0 > G_\sigma(u) - \beta - \langle \nabla H_\sigma(y), u - y \rangle.$$

Whence, using Eq. (9) and convexity of the function $H_\sigma(\cdot)$, we derive

$$0 > G_\sigma(u) - \beta - H_\sigma(u) + H_\sigma(y) = F_\sigma(u) - \zeta,$$

or

$$F_\sigma(z) > F_\sigma(u), \ z \in \mathcal{F}, \ u \in S.$$

It means that the vector z can not be a solution to $(\mathcal{P}_\sigma)$. Moreover, if, in addition, u is feasible in $(\mathcal{P})$, $z, u \in \mathcal{F}$, $W(z) = 0 = W(u)$, it yields $f_0(z) = F_\sigma(z) > F_\sigma(u) = f_0(u)$.

Hence, $z \notin Sol(\mathcal{P})$, and u is a vector better than $z \in \mathcal{F}$.

Thus, conditions (9) and (10) of Theorem 1 possess the classical constructive (algorithmic) property: once the conditions are violated, one can find a feasible vector that has a better value of the goal function than the point $(z \in \mathcal{F})$ in question.

Let now turn our attention to the corresponding properties of the minimizing sequences in Problems $(\mathcal{P})$ and $(\mathcal{P}_\sigma)$.

Consider a point $z \in S$, $\zeta := F_\sigma(z)$ and the following function

$$\varphi_\sigma(z) := \inf_{x,y,\beta}\{G_\sigma(x) - \beta - \langle \nabla H_\sigma(y), x - y \rangle \mid x \in S,$$
$$H_\sigma(y) = \beta - \zeta, \ G_\sigma(y) \leq \beta \leq \sup(G_\sigma(\cdot), S)\}. \tag{12}$$

In (12), set $y = z = x$. Then we have $\beta = \beta_0 := H_\sigma(z) + \zeta = H_\sigma(z) + F_\sigma(z) = G_\sigma(z)$, and, by Definition (12), we merely obtain the inequality

$$0 = G_\sigma(z) - \beta_0 - \langle \nabla H_\sigma(z), z - z \rangle \geq \varphi_\sigma(z),$$

which yields that

$$\varphi_\sigma(z) \leq 0 \quad \forall z \in S. \tag{13}$$

Hence, on account of (12), the optimality conditions (9) and (10) can be rewritten as: $\varphi_\sigma(z) = 0$. The language of the function $\varphi_\sigma(\cdot)$ turns out to be very appropriate to the study of properties of minimizing sequences to Problem $(\mathcal{P})$.

Definition 1. (a) A sequence $\{z^k\} \subset S$ is said to be minimizing to Problem $(\mathcal{P})$, if two following conditions hold

$$\left.\begin{array}{l} (i) \ \lim_{k\to\infty} f_0(z^k) = \mathcal{V}(\mathcal{P}) := \inf_x\{f_0(x) \mid x \in \mathcal{F}\}; \\ (ii) \ \lim_{k\to\infty} W(z^k) = 0 \ (the \ feasibility \ condition). \end{array}\right\} \tag{14}$$

(b) A sequence $\{z^k\} \subset S$ is called minimizing to Problem $(\mathcal{P}_\sigma)$, if

$$\lim_{k\to\infty} F_\sigma(z^k) = \mathcal{V}(\mathcal{P}_\sigma) := \inf_x\{F_\sigma(x) \mid x \in S\}. \tag{15}$$

Theorem 2. (i) Suppose, a sequence $\{z^k\} \subset S$ is minimizing to Problem $(\mathcal{P})$: $\{z^k\} \in \mathcal{M}(\mathcal{P})$, and $\sigma \geq \sigma_* > 0$, where σ_* is a threshold value of penalty parameter, so that $\mathcal{V}(\mathcal{P}) = \mathcal{V}(\mathcal{P}_\sigma)$. Then,

$$\lim_{k\to\infty} \varphi_\sigma(z^k) = 0. \tag{16}$$

(ii) If in addition the following assumption holds

$$(\mathcal{H}): \qquad \exists v \in S: \ \exists \gamma > 0: \ F_\sigma(v) \geq F_\sigma(z^k) + \gamma, \ k = 0,1,2,\ldots, \tag{17}$$

then condition (16) becomes sufficient for $\{z^k\}$ to be minimizing to Problem $(\mathcal{P}_\sigma)$ for any value $\sigma > 0$ of the penalty parameter.

Proof. (i) Let $\{z^k\} \subset \mathcal{M}(\mathcal{P})$. Then thanks to (14) we have $\{z^k\} \in \mathcal{M}(\mathcal{P}_\sigma)$, since

$$\lim_{k\to\infty} F_\sigma(z^k) = \lim_{k\to\infty} \left[f_0(z^k) + \sigma W(z^k)\right] = \mathcal{V}(\mathcal{P}_\sigma) = \mathcal{V}(\mathcal{P}) =$$
$$= \lim_{k\to\infty} f_0(z^k) + \sigma \lim_{k\to\infty} W(z^k). \tag{18}$$

Furthermore, due to convexity of $H_\sigma(\cdot)$, we have that

$$\forall(y, \beta) : \quad \beta - H_\sigma(y) = F_\sigma(z^k) =: \zeta_k, \; \forall x \in S,$$

the following chain takes place $(k = 0, 1, 2, \ldots)$

$$G_\sigma(x) - \beta - \langle \nabla H_\sigma(y), x - y \rangle \geq G_\sigma(x) - \beta - H_\sigma(x) + H_\sigma(y) = $$
$$= F_\sigma(x) - F_\sigma(z^k) = F_\sigma(x) - \zeta_k.$$

Thus, thanks to definition of $\{z^k\} \in \mathcal{M}(\mathcal{P})$, it yields

$$0 \geq \varphi_\sigma(z^k) \geq \inf_x (F_\sigma(\cdot), S) - F_\sigma(z^k) = \mathcal{V}(\mathcal{P}_\sigma) - F_\sigma(z^k).$$

Whence, with help of (18), we derive (16).

4.1   A Theoretical Method

Here we develop a global search theoretical scheme for the Problem

$$(\mathcal{P}_\sigma): \qquad\qquad F_\sigma(x) \overset{\triangle}{=} G_\sigma(x) - H_\sigma(x) \downarrow \min_x, \quad x \in S, \qquad (19)$$

based on GOCs (16). Theorem 2 suggests, in particular, to compute the value $\varphi_\sigma(z^k)$, in order to verify whether a current iteration $z^k \in S$, $k \in \{0, 1, 2, \ldots\}$ is a global solution.

Therefore, the next procedure, consisting of approximate and partial computation of the value $\varphi_\sigma(z^k)$ at every iteration, looks completely natural.

Let a vector $z^k \in S$ be given, and $\zeta_k := F_\sigma(z^k)$. Then the next point $z^{k+1} \in S$ is constructed to fulfill the conditions as follows

$$\left.\begin{aligned}
(\mathcal{R}1): \quad & G_\sigma(z^{k+1}) - \beta_k - \langle \nabla H_\sigma(y^k), z^{k+1} - y^k \rangle \leq \Theta_k \varphi_\sigma(z^k) + \nu_k, \\
(\mathcal{R}2): \quad & \beta_k = H_\sigma(y^k) + \zeta_k, \; \zeta_k := F_\sigma(z^k).
\end{aligned}\right\} \qquad (20)$$

$$\left.\begin{aligned}
& 0 < \Theta < \Theta_k \leq 1, \\
& \nu_k > 0, \; k = 0, 1, 2, \ldots, \sum_{k=0}^{\infty} \nu_k < +\infty.
\end{aligned}\right\} \qquad (21)$$

In addition, below, we will also use the following condition for the starting vector $z^0 \in S$

$$(\mathcal{H}_0): \qquad \exists v \in S, \quad \exists \ae > 0 : F_\sigma(z^0) \leq F_\sigma(v) - \ae - \sum_{k=0}^{\infty} \nu_k. \qquad (22)$$

It is clear that the condition (22) is neither restrictive nor overburdening according to the statement of $(\mathcal{P})$.

The cost function $f_0(x)$ of the original problem $(\mathcal{P})$ is bounded from below on the feasible set

$$\mathcal{F} = \{x \in S \mid f_i(x) \leq 0, \; i \in \mathcal{I}, \; f_j(x) = 0, \; j \in \mathcal{E}\},$$

i.e. the condition $(\mathcal{H}_f)$–(1) hold. If can be readily seen that the cost function $F_\sigma(\cdot)$ possesses the same property as $f_0(\cdot)$, thanks to the facts that $\sigma \geq 0$ and $W(x) \geq 0 \quad \forall x \in \mathbb{R}^n$, so that

$$\mathcal{V}(\mathcal{P}_\sigma) \geq \mathcal{V}(\mathcal{P}) > -\infty.$$

Theorem 3. *(i) The sequence $\{z^k\}$ constructed according to the rules (20) and (21) satisfies the optimality condition (OC) (16), i.e*

$$(\mathcal{OC}): \qquad\qquad \lim \varphi_\sigma(z^k) = 0. \qquad\qquad (16)$$

(ii) If, in addition, the assumption $(\mathcal{H}_0)$–(22) for the starting vector z^0 is fulfilled, then the sequence $\{z^k\}$ produced by the rules (20) and (21) turns out to be minimizing to Problem $(\mathcal{P}_\sigma)$.
(iii) Any limit point z of sequence $\{z^k\}$ provides infimum of the function $F_\sigma(\cdot)$ over S, and, when S is closed, this limit point turns out to be a solution to $(\mathcal{P}_\sigma)$.

Proof. (a) Since $\varphi_\sigma(z) \leq 0 \quad \forall z \in S$ (see (13)) and on account of (20) and (21) and the convexity of the function $H_\sigma(\cdot)$, we have

$$
\begin{aligned}
\nu_k &\geq \Theta_k \varphi(z^k) + \nu_k \geq G_\sigma(z^{k+1}) - \beta_k - \langle \nabla H_\sigma(y^k), z^{k+1} - y^k \rangle \geq \\
&\geq G_\sigma(z^{k+1}) - \beta_k - H_\sigma(z^{k+1}) + H_\sigma(y^k) = F_\sigma(z^{k+1}) - \zeta_k = \\
&= F_\sigma(z^{k+1}) - F_\sigma(z^k)
\end{aligned}
\qquad (23)
$$

whence it immediately follows that

$$F_\sigma(z^k) + \nu_k \geq F_\sigma(z^{k+1}). \qquad\qquad (23')$$

It means [28] that the number sequence $\{F_\sigma(z^k)\}$ is "almost" monotonously decreasing, and, therefore, there exists a finite limit

$$\lim_{k \to \infty} F_\sigma \geq \mathcal{V}(\mathcal{P}_\sigma) \geq \mathcal{V}(\mathcal{P}) > -\infty.$$

Moreover, with the help of the inequality

$$\nu_k \geq \Theta_k \varphi_\sigma(z^k) + \nu_k \geq F_\sigma(z^{k+1}) - F_\sigma(z^k),$$

we immediately obtain the condition (16).
(b) Now let us show that from the condition $(\mathcal{H}_0)$–(22) for the starting point we can merely derive the regularity condition $(\mathcal{H})$–(17) for the sequence $\{z^k\}$.

Then, in virtue of Theorem 1, the (OCs)–(16) become sufficient for the sequence $\{z^k\}$ produced by the rules $(\mathcal{R}1)$–20, $(\mathcal{R}2)$–21 to be minimizing to Problem $(\mathcal{P}_\sigma)$.

Indeed, due to (OC)–(16), (23′) and $(\mathcal{H}_0)$–22 we have

$$F_\sigma(v) - \text{\ae} - \sum_{s=0}^{\infty} \nu_s \geq F_\sigma(z^0) \geq F_\sigma(z^1) - \nu_0 \geq F_\sigma(z^k) - \sum_{s=0}^{k-1} \nu_s \geq F_\sigma(z^k) - \sum_{s=0}^{\infty} \nu_s,$$

whence it follows that

$$F_\sigma(v) - \text{\ae} \geq F_\sigma(z^k), \quad k = 0, 1, 2, \ldots$$

which coincides with $(\mathcal{H})$–(17), as was claimed.
(iii) Now, the final assertion of Theorem 3 becomes obvious.

4.2   A Global Search Scheme

Here we are interested in the same question as in Sect. 4.1, i.e. how to decide whether a feasible vector (an iterate) $z^k \in S$ is an approximate global solution to Problem $(\mathcal{P}_\sigma)$. And if not, how to construct the next iteration $z^{k+1} \in S$, improving, in a sense, the previous one z^k.

In order to do it, Theorems 2 and 3 propose to study the following auxiliary problem

$$(\mathcal{AP}_\sigma): \quad \begin{aligned} \Psi(x,y,\beta) := G_\sigma(x) - \beta - \langle \nabla H_\sigma(y), x - y \rangle \downarrow \min_{x,y,\beta}, \\ x \in S, \ (y,\beta) \in \mathbb{R}^{n+1} : \beta - H_\sigma(y) = \zeta_k := F_\sigma(z^k), \\ G_\sigma \le \beta \le \sup(G_\sigma(\cdot), S). \end{aligned} \right\} \tag{24}$$

Further, Theorems 2 and 3 state that Problem $(\mathcal{AP}_\sigma)$–(24) (at every iteration of the global search) can be solved not only approximately, but in addition, partially $(0 < \Theta < 1)$.

However, Problem $(\mathcal{AP}_\sigma)$–(24) can be assessed to be, say, of the same difficulty as $(\mathcal{P}_\sigma)$ or $(\mathcal{P})$, because there are supplementary variables and parameters in $(\mathcal{AP}_\sigma)$–(24), and, besides, even a new equation, etc.

Moreover, the problem (24) is also nonconvex. Therefore, we decided to decompose Problem (24) into several problems which are simpler and more tractable than Problem (24).

Suppose, we have a current iterate $z^k \in S$, $\zeta_k := F_\sigma(z^k)$.

(a) Let there be given a number β, such that

$$\beta_- := \inf(G_\sigma(\cdot), S) \le \beta \le \beta_+ := \sup(G_\sigma(\cdot), S). \tag{25}$$

Then, for the level surface $(\zeta_k := F_\sigma(z^k))$

$$Y_k = Y(\zeta_k, \beta) = \{ y \in \mathbb{R}^n : H_\sigma(y) = \beta - \zeta_k \}$$

of the convex function $H_\sigma(\cdot)$, we construct a finite approximation

$$\mathcal{A}(\zeta_k, \beta) = \{ y^1, ..., y^N \mid H_\sigma(y^i) = \beta - \zeta_k, \ G_\sigma(y^i) \le \beta, \ i = 1, ..., N \}.$$

(b) After that for every $y^i \in \mathcal{A}(\zeta_k, \beta) =: \mathcal{A}_k(\beta)$ we solve the linearized problem as follows

$$(\mathcal{P}_\sigma L_i): \qquad G_\sigma(x) - \langle \nabla H_\sigma(y^i), x \rangle \downarrow \min_x, \quad x \in S. \tag{26}$$

Let $\bar{u}^i \in S$ be an approximate solution to $(\mathcal{P}_\sigma L_i)$, $\bar{u}^i \in Sol(\mathcal{P}_\sigma L_i)$.

(c) By starting at the point $\bar{u}^i \in S$, we get a critical (to a Local Search Method (LSM)) vector u^i, such that u^i is an approximate solution to the linearized problem

$$(\mathcal{P}_\sigma L(u^i)): \qquad G_\sigma(x) - \langle \nabla H_\sigma(u^i), x \rangle \downarrow \min_x, \quad x \in S. \tag{27}$$

(linearized at the point u^i just). Because, as we know, effective LSMs [21,25] provide critical points (in the above sense), depending on corresponding starting points.

(d) Furthermore, we solve the level problem

$$(Lev\ \mathcal{P}_\sigma): \qquad \langle \nabla H_\sigma(v), u^i - v \rangle \uparrow \max_v, \quad H_\sigma(v) = \beta - \zeta_k. \qquad (28)$$

Let w^i be an approximate global solution to Problem $(Lev\ \mathcal{P}_\sigma)$–(28).

(e) After this, we compute the number $\eta_k(\beta) = \eta(\zeta_k, \beta) := \eta_0(\zeta, \beta) - \beta$, where

$$\eta_0(\zeta, \beta) = G_\sigma(u^j) - \langle \nabla H_\sigma(w^j), u^j - w^j \rangle :=$$
$$= \min_{1 \le i \le N} \{ G_\sigma(u^i) - \langle \nabla H_\sigma(w^i), u^i - w^i \rangle \}.$$

(f) If $\eta_k(\beta) < 0$, then the vector $u^j \in S$ will be better, than the point z^k in question, since, due to convexity of $H_\sigma(\cdot)$, we have:

$$0 > G_\sigma(u^j) - \langle H_\sigma(w^j), u^j - w^j \rangle - \beta \ge$$
$$\ge G_\sigma(u^j) - H_\sigma(u^j) + H_\sigma(w^j) - \beta = F_\sigma(u^j) - F_\sigma(z^k), \qquad (29)$$

i.e. $F_\sigma(u^j) < F_\sigma(z^k)$.

Hence, in this case we can go to the next iteration, i.e. $k := k + 1$, $z^{k+1} := u^j$.

(g) When $\eta(\zeta_k, \beta) \ge 0$, we have to change the value β for $\overline{\beta} := \beta + \Delta\beta$ with the help of one of the existing one-dimensional search methods, for instance, for minimizing the function $\Psi_1(\beta) := \eta(\zeta, \beta)$ on the interval $[\beta_-, \beta_+]$.

Remark 2. Furthermore, if the vector $z \in S$ is (an approximate) global solution to Problem $(\mathcal{P}_\sigma)$, clearly, it is impossible to improve the value $F_\sigma(z) =: \zeta$. Notwithstanding, how to understand whether the vector z is really a solution to Problem $(\mathcal{P}_\sigma)$, provided that the stages (b), (c) and (d) are implemented sufficiently well, i.e. the convex problems $(\mathcal{P}_\sigma L_i)$–(26), $(\mathcal{P}_\sigma L(u^i))$–(27) and $(Lev\ \mathcal{P}_\sigma)$–(28) (nonconvex) have been solved globally with a given accuracy.

It will be shown in the next section that from the view-point of convergence a Global Search Scheme (GSS), which we are going to propose below, to answer the question above, it is necessary to put some constraints on the stage (a), i.e. the construction of the approximation $\mathcal{A}(\zeta_k, \beta)$.

Therefore, it is clear that this approximation has to be rather representative in order to do the conclusion, whether or not the vector z is a global solution to Problem $(\mathcal{P}_\sigma)$, when $\eta(\zeta, \beta) \ge 0 \quad \forall \beta \in [\beta_-, \beta_+]$.

In the case, when the iterate z^k in question is rather far from a global solution, the number (quantity) $\eta(\zeta_k, \beta) =: \eta_k(\beta)$ has, without doubt, to be negative for some $\beta \in [\beta_-, \beta_+]$, that allows to construct a point $(u^j \in S)$ which is better than z, just as the chain (29) demonstrates.

Remark 3. It is worth noting that, when solving the convex problems $(\mathcal{P}_\sigma L_i)$ – (26), $(\mathcal{P}_\sigma L(u^i))$–(27) and $(Lev\ \mathcal{P}_\sigma)$–(28) (nonconvex), it is beneficial to use the standard methods of convex optimization [1,12,16,28], and, of course, the modern computational software, such as IBM ILOG CPLEX, XPress, Gurobi etc.

In addition, it would be useful and reasonable to apply some methods of local search, say, at the initial stage of computations to pass from a starting point to a critical vector, which allows to employ the powerful methods of convex optimization [1,12,15,16,21,25].

Now, we are able to describe a first variant of the Global Search Scheme. Let there be given a starting point $x_0 \in S$ and number sequences $\{\tau_k\}, \{\delta_k\}$, such that $\tau_k, \delta_k > 0$, $k = 0, 1, 2, \ldots$, $\tau_k \downarrow 0$, $\delta_k \downarrow 0$ $(k \to \infty)$.

Global Search Scheme 1(GSS1)

Step 0. Set $k := 0$, $x^k := x_0 \in S$.

Step 1. By starting at $x^k \in S$ and with the help of an LS method for Problem $(\mathcal{P}_\sigma)$, produce a τ_k-critical point $z^k \in S, \zeta_k := F_\sigma(z^k) \le F_\sigma(x^k)$, i.e. satisfying the following inequality:

$$G_\sigma(z^k) - \langle \nabla H_\sigma(z^k), z^k \rangle - \tau_k \le \inf_x\{G_\sigma(x) - \langle \nabla H_\sigma(z^k), x \rangle | \ x \in S\}. \quad (30)$$

Step 2. Choose a number $\beta \in [\beta_-, \beta_+]$, in particular, a one-dimensional search method can start at $\beta_- := \inf(G_\sigma(\cdot), S)$ or $\beta_1 := G_\sigma(z^k)$.

Step 3. Construct an approximation

$$\mathcal{A}_k(\beta) = \{v^1, \ldots, v^{N_k} | \ H_\sigma(v^i) = \beta - \zeta_k, i = 1, \ldots, N_k, \ N_k = N_k(\beta)\}.$$

Step 4. According to the GOCs [22–24], form a collection of indexes I_k defined as

$$I_k = I_k(\beta) = \{i \in \{1, \ldots, N_k\} | \ G_\sigma(v^i) \le \beta\}. \quad (31)$$

Step 5. For every $i \in I_k$ find a global $2\delta_k$-solution $\overline{u}^i \in S$ to the linearized convex problem $(\mathcal{P}_\sigma L_i)$–(26), so that

$$G_\sigma(\overline{u}^i) - \langle \nabla H_\sigma(v^i), \overline{u}^i \rangle - 2\delta_k \le \inf_x\{G_\sigma(x) - \langle \nabla H_\sigma(v^i), x \rangle | \ x \in S\}. \quad (32)$$

Step 6. For every $i \in I_k$, by starting at $\overline{u}^i \in S$, find a $2\tau_k$-critical vector $u^i \in S$ with the help of a LSM, so that

$$G_\sigma(u^i) - \langle \nabla H_\sigma(u^i), u^i \rangle - 2\tau_k \le \inf_x\{G_\sigma(x) - \langle \nabla H_\sigma(u^i), x \rangle | \ x \in S\}. \quad (33)$$

Step 7. For every $i \in I_k$ find a global $2\delta_k$-solution w^i: $H_\sigma(w^i) = \beta - \zeta_k$, to the Level Problem $(Lev \ \mathcal{P}_\sigma)$–(28), so that

$$\langle \nabla H_\sigma(w^i), u^i - w^i \rangle + 2\delta_k \ge \sup_v\{\langle \nabla H_\sigma(v), u^i - v \rangle | \ H_\sigma(v) = \beta - \zeta_k\}. \quad (34)$$

Step 8. Set $\eta_k(\beta) := \eta_k^0(\beta) - \beta$, where

$$\begin{aligned}\eta_k^0(\beta) := G_\sigma(u^j) - \langle \nabla H_\sigma(w^i), u^j - w^j \rangle = \\ = \min_{i \in I_k}\{G_\sigma(u^i) - \langle \nabla H_\sigma(w^i), u^i - w^i \rangle\}.\end{aligned} \quad (35)$$

Step 9. If $\eta_k(\beta) < 0$, then set $k := k + 1$, $x^{k+1} := u^j$ and loop to Step 1.

Step 10. If $\eta_k(\beta) \ge 0$, then set $\beta := \beta + \Delta\beta \in [\beta_-, \beta_+]$ and go to Step 3.

Step 11. If $\eta_k(\beta) \ge 0 \ \forall \beta \in [\beta_-, \beta_+]$ (i.e. one-dimensional search on β has been terminated), set $k := k + 1$, $x^{k+1} := z^k$ and loop to Step 1.

Remarks

(1) In order the GSS1 described above to be, to a certain extent, substantiated, introduce the following assumptions:

$(\mathcal{HL}):\quad \forall \delta > 0 \quad \forall \beta \in [\beta_-, \beta_+] \quad \forall z \in S \quad \forall v : \ H_\sigma(v) = \beta - \zeta, \ \zeta := F_\sigma(z),$
$G_\sigma(v) \le \beta$, one can find a vector $u \in S$, satisfying the following inequality

$$G_\sigma(u) - \langle \nabla H_\sigma(v), u \rangle - \delta \le \inf_x \{G_\sigma(x) - \langle \nabla H_\sigma(v), x \rangle \mid x \in S\}. \qquad (32')$$

$(Lev\ \mathcal{H}):\ \forall \delta > 0 \quad \forall \beta \in [\beta_-, \beta_+] \quad \forall z, u \in S$ one can find a vector $w :$
$H_\sigma(w) = \beta - F_\sigma(z), G_\sigma(v) \le \beta$, satisfying the following inequality

$$\langle \nabla H_\sigma(w), u - w \rangle + \delta \ge \sup_v \{\langle \nabla H_\sigma(v), u - v \rangle \mid H_\sigma(v) = \beta - F_\sigma(z)\}. \qquad (34')$$

(2) It can be readily seen that the sequence $\{z^k\}$, produced by the GSS1, is the sequence of τ_k-critical points, so that for every $k = 0, 1, 2, \ldots$ inequality (30) holds.

(3) On the Steps 1, 5, 6, 7, one can apply the well-known optimization algorithms [1,12,16] and the modern Computational Software (IBM ILOG CPLEX, XPress, Gurobi etc). The possibility to use these modern tools is, without doubts, an advantage of the proposed methodology, since it allows us to employ not only the latest advances in modern programming and computational technology, but the most prominent achievements of the optimization theory and methods.

(4) Clearly, the GSS1 described above is not yet an algorithm in the common sense, since, for example, we do not provide real algorithms of local search and algorithms for solving problems (32), (33) and (34). There are no one-dimensional methods for finding an appropriate β, either. Hence, the GSS1 can be viewed as a conceptual scheme of the Global Search.

(5) Obviously, during the real implementation of the GSS one can stop, when $\eta_k(\beta) \ge 0 \ \forall \beta \in [\beta_-, \beta_+]$, and $\tau_k, \delta_k \le \chi$, where χ is a given accuracy.

Besides, one can find necessary precisions in the next sections.

4.3   Convergence of the GSS1

In the previous section we decomposed the solution of Problem $(\mathcal{AP}_\sigma)$–(24) into several stages, in particular, (a),(b),(c),(d) etc. and a one-dimensional search on $\beta \in [\beta_-, \beta_+]$. It is clear that the choice of methods for solving problems (26), (27), and (28) and for the search of a suitable $\beta \in [\beta_-, \beta_+]$ is rather important from the view-point of the Global Search (GS).

On the other hand, the choice is merely standard, and there is a number of recognized experts in the field of Optimization Theory and Methods capable to perform this choice.

At the same time, the problem (a) of construction of a "good" approximation $\mathcal{A}_k(\beta)$ on Step 3 of the GSS1 and the choice of the parameter β on Step 2, evidently look as a new unknown procedure and an unprecedented problem without parallels in optimization.

On the other hand, as our computational experience [19–23, 25] shows, producing a relevant approximation $\mathcal{A}_k(\beta)$ with an appropriate choice of β turns out to be a crucial moment for escaping a current stationary or critical point under condition of a qualified implementation of the rest of the GSS1 or its further developments.

In order to estimate "the quality" of the approximation $\mathcal{A}_k(\beta)$ and the choice of $\beta \in [\beta_-, \beta_+]$ from the view-point of global search process, introduce the following definition. Consider an approximation

$$\mathcal{R}(\zeta, \beta) = \{v^1, \ldots, v^N \mid H_\sigma(v^i) = \beta - \zeta, \ i = 1, \ldots, N = N(\zeta, \beta)\}.$$

Let the points $u^i \in S$ and $w^i \in \mathbb{R}^n$, $H_\sigma(w^i) = \beta - \zeta$ satisfy (according to the assumptions $(\mathcal{H}L)$, $(Lev\ \mathcal{H})$ and Steps 6 and 7 of the GSS1) the inequalities (33) and (34), with β instead of β_k $(i = 1, \ldots, N)$.

Furthermore, set $(\zeta := F_\sigma(z), \ z \in S)$

$$\eta(\zeta, \beta) := G_\sigma(u^j) - \beta - \langle \nabla H_\sigma(w^j), u^j - w^j \rangle := \\ = \min_{i \in I_k} \{G_\sigma(u^i) - \beta - \langle \nabla H_\sigma(w^i), u^i - w^i \rangle\}, \tag{36}$$

where (recall) $I_k = \{i \in \{1, \ldots, N\} \mid G_\sigma(v^i) \leq \beta\}$.

Definition 2. *(i) The approximation $\mathcal{R}(\zeta, \beta)$ is said to be an $(\varepsilon, \delta, \nu, \Theta)$-resolving set for Problem $(\mathcal{P}_\sigma)$ $(\varepsilon, \delta, \nu > 0, \ 0 < \Theta < 1)$, if the inequality*

$$F_\sigma(z) > \mathcal{V}(\mathcal{P}_\sigma) + \varepsilon, \tag{37}$$

(i.e. the vector z is not an ε-solution to $(\mathcal{P}_\sigma)$) entails the two following inequalities

$$\eta(\zeta, \beta) < 0, \tag{38}$$

$$\eta(\zeta, \beta) < \Theta \varphi_\sigma(z) + \nu. \tag{39}$$

(ii) $\mathcal{R}(\zeta, \beta)$ is said to be a weakly $(\varepsilon, \delta, \nu, \Theta)$-resolving set (or collection), if the inequalities (38) and (39) are fulfilled non-strictly.

Lemma 1. *Suppose a collection $\mathcal{R}(\zeta, \beta)$ is $(\varepsilon, \delta, \nu, \Theta)$-resolving and*

$$\nu \geq \Theta \varepsilon. \tag{40}$$

Then, the inequality (38) implies the inequality (39).

It can be readily seen that Lemma 1 provides
(1) first, "the concordance condition" for the numbers ε, ν and $\Theta : \nu \geq \Theta \varepsilon$, where ε is the accuracy of solution of Problem $(\mathcal{P}_\sigma)$, ν stands for precision of the inequality (39) (recall, besides, the inequality (20)), and Θ is "a share of solution" of "the auxiliary problem" $(\mathcal{AP}_\sigma)$–(24).
(2) In addition, according to Lemma 1, during the performance of the GSS1, one can watch only the number $\eta_k := \eta(\zeta_k, \beta)$ (the inequality (38)), without paying

attention to the second inequality (39), which will take place, if $\eta_k < 0$, under condition of using a resolving set $\mathcal{R}(\zeta, \beta)$ at every iteration of the GSS1.

Therefore, to apply the concept of the GSS1 to solving Problem $\mathcal{P}_\sigma$, the next assumption looks rather natural.

$(\mathcal{HR})$: $\forall \varepsilon > 0$ $\forall \tau > 0$ and for every τ-critical (in Problem (P_σ)) vector $z \in S$, $\zeta := F_\sigma(z)$, which is not an ε-solution to (P_σ), there exists $\beta \in [\beta_-, \beta_+]$ such that

$$\forall (\delta, \nu, \Theta): \; \tau, \delta, \nu > 0, \; 0 < \Theta < 1, \; \nu \geq \Theta \varepsilon,$$

one can construct an $(\varepsilon, \delta, \nu, \Theta)$-resolving set $\mathcal{R}(\zeta, \beta)$.

In what follows, let us suppose that at every iteration of the GSS1 we construct a $(\varepsilon_k, \delta_k, \nu_k, \Theta_k)$-resolving set $\mathcal{R}(\zeta_k, \beta) := \mathcal{A}_k(\beta)$ on the step 2 and 3, and therefore the notion of the resolving set plays a crucial role in the convergence proof of GSS1.

Furthermore, introduce the following assumption for the starting vector $x^0 \in S$:

$(\mathcal{A}1)$: $\exists v \in S, \; \exists \gamma > 0: \; f_0(v) \geq f_0(x^0) + \gamma.$ (41)

In addition, let us suppose that the number sequences $\{\varepsilon_k\}$, $\{\tau_k\}$, $\{\delta_k\}$, $\{\nu_k\}$ and $\{\Theta_k\}$ satisfy the supplementary assumption as follows

$(\mathcal{A}2)$: $\left.\begin{array}{c} \varepsilon_k, \; \tau_k, \; \delta_k, \; \nu_k > 0, \; 0 < \Theta \leq \Theta_k \leq 1, k = 0, 1, 2, \dots, \\ \tau_k \downarrow 0, \; \delta_k \downarrow 0, \; \nu_k \downarrow 0 \; (k \to \infty). \end{array}\right\}$ (42)

Besides, we suppose that in the original Problem $(\mathcal{P})$ the cost function $f_0(\cdot)$ is bounded from below over the feasible set $\mathcal{F}$, which entails that the goal function $F_\sigma(\cdot)$ of the penalized Problem $(\mathcal{P}_\sigma)$ is also bounded from below on S.

Finally, let us suppose that the data of Problem $(\mathcal{P})$ is so smooth that the function $H_\sigma(\cdot)$ is differentiable on an open set Ω containing S.

Under all these assumptions we are able to prove the following result.

Theorem 4. *Suppose, that the assumptions $(\mathcal{HR})$, $(\mathcal{HL})$–$(32')$, $(Lev\ \mathcal{H})$–$(34')$, $(\mathcal{A}1)$–(41), $(\mathcal{A}2)$–(42) are fulfilled, and, besides, "the concordance condition" is also satisfied:*

$$\nu_k \geq \Theta_k \varepsilon_k, \quad k = 0, 1, 2, \dots \tag{43}$$

Then, the sequence $\{z^k\}$ produced by the Global Search $\mathcal{R}$-scheme is minimizing to Problem $(\mathcal{P}_\sigma)$: $\{z^k\} \in \mathcal{M}(\mathcal{P}_\sigma)$.

Moreover, in the case when the set S is closed, any limit point z_* of the sequence $\{z^k\}$ turns out to be a global solution to Problem $(\mathcal{P}_\sigma)$.

5   Conclusion

In this paper, we investigated the nonconvex Problem $(\mathcal{P})$ with equality and inequality constraints given by d.c. functions. The principal objectives of the paper were the new Global Optimality Conditions (GOCs) for minimizing sequences and proof of theoretical convergence of the method. However, before obtaining the desired results, we managed to use the exact penalty procedure to perform reduction of the original problem to a d.c. minimization problem without equality and inequality constraints. Only after that it became possible to develop the new GOCs which reduce the nonconvex penalized problem to a family of convex (linearized) problems. It is worth noticing that the linearization was applied to the function which accumulates all the nonconvexities of the original problem. Summarizing, we developed new mathematical tools that help not only to escape local and stationary pitfalls, but also to reach a global solution in nonconvex optimization problems with equality and inequality constraints defined by continuous functions. The numerical effectiveness of the developed approach were demonstrated in [9,19,20,25].

References

1. Bonnans, J.-F., Gilbert, J.C., Lemaréchal, C., Sagastizábal, C.A.: Numerical Optimization: Theoretical and Practical Aspects, 2nd edn. Springer, Heidelberg (2006). https://doi.org/10.1007/978-3-540-35447-5
2. Burke, J.: An exact penalization viewpoint of constrained optimization. SIAM J. Control Optim. **29**, 968–998 (1991)
3. Byrd, R., Lopez-Calva, G., Nocedal, J.: A line search exact penalty method using steering rules. Math. Program. Ser. A **133**, 39–73 (2012)
4. Demyanov, V.F.: Extremum's Conditions and Variational Calculus. High School Edition, Moscow (2005). (in Russian)
5. Di Pillo, G., Lucidi, S., Rinaldi, F.: An approach to constrained global optimization based on exact penalty functions. J. Glob. Optim. **54**, 251–260 (2012)
6. Di Pillo, G., Lucidi, S., Rinaldi, F.: A derivative-free algorithm for constrained global optimization based on exact penalty functions. J. Optim. Theory Appl. **164**, 862–882 (2015)
7. Eremin, I.: The penalty method in convex programming. Sov. Math. Dokl. **8**, 459–462 (1966)
8. Floudas, C.A., Pardalos, P.M.: Frontiers in Global Optimization. Kluwer Academic Publishers, Dordrecht (2004)
9. Gruzdeva, T.V., Strekalovskiy, A.S.: On solving the sum-of-ratios problem. Appl. Math. Comput. **318**, 260–269 (2018)
10. Han, S., Mangasarian, O.: Exact penalty functions in nonlinear programming. Math. Program. **17**, 251–269 (1979)
11. Hiriart-Urruty, J.-B.: Generalized differentiability, duality and optimization for problems dealing with difference of convex functions. In: Ponstein, J. (ed.) Convexity and Duality in Optimization. LNEM, vol. 256, pp. 37–69. Springer, Berlin (1985). https://doi.org/10.1007/978-3-642-45610-7_3
12. Hiriart-Urruty, J.-B., Lemaréchal, C.: Convex Analysis and Minimization Algorithms. Springer, Berlin (1993). https://doi.org/10.1007/978-3-662-02796-7
13. Horst, R., Tuy, H.: Global Optimization. Deterministic Approaches. Springer, Berlin (1993). https://doi.org/10.1007/978-3-662-03199-5
14. Kruger, A.: Error bounds and metric subregularity. Optimization **64**, 49–79 (2015)
15. Le Thi, H.A., Huynh, V.N., Dinh, T.P.: DC programming and DCA for general DC programs. In: van Do, T., Thi, H., Nguyen, N. (eds.) Advanced Computational Methods for Knowledge Engineering. AISC, vol. 282, pp. 15–35. Springer, Cham (2014). https://doi.org/10.1007/978-3-319-06569-4_2

16. Nocedal, J., Wright, S.J.: Numerical Optimization. Springer, New York (2006). https://doi.org/10.1007/978-0-387-40065-5
17. Rockafellar, R.T.: Convex Analysis. Princeton University Press, Princeton (1970)
18. Rockafellar, R.T., Wets, R.J.-B.: Variational Analysis. Springer, New York (1998). https://doi.org/10.1007/978-3-642-02431-3
19. Strekalovsky, A.S.: Elements of Nonconvex Optimization. Nauka, Novosibirsk (2003). (in Russian)
20. Strekalovsky, A.S.: On solving optimization problems with hidden nonconvex structures. In: Rassias, T.M., Floudas, C.A., Butenko, S. (eds.) Optimization in Science and Engineering, pp. 465–502. Springer, New York (2014). https://doi.org/10.1007/978-1-4939-0808-0_23
21. Strekalovsky, A.S.: On local search in D.C. optimization problems. Appl. Math. Comput. **255**, 73–83 (2015)
22. Strekalovsky, A.S.: Global optimality conditions in nonconvex optimization. J. Optim. Theory Appl. **173**, 770–792 (2017)
23. Strekalovsky, A.S.: Global optimality conditions and exact penalization. Optim. Lett. **13**, 597–615 (2019)
24. Strekalovsky, A.S.: New global optimality conditions in a problem with D.C. constraints. Tr. Inst. Mat. i Mekhaniki UrO RAN **25**(1), 245–261 (2019)
25. Strekalovsky, A.S., Minarchenko, I.M.: A local search method for optimization problem with D.C. inequality constraints. Appl. Math. Modell. **58**, 229–244 (2018)
26. Strongin, R.G., Sergeyev, Y.D.: Global Optimization with Non-convex Constraints: Sequential and Parallel Algorithms. Kluwer Academic Publishers, Dordrecht (2000)
27. Tuy, H.: D.C. optimization: theory, methods and algorithms. In: Horst, R., Pardalos, P.M. (eds.) Handbook of Global Optimization, pp. 149–216. Kluwer Academic Publisher, Dordrecht (1995)
28. Vasiliev, F.P.: Optimization Methods. Factorial Press, Moscow (2002). (in Russian)
29. Zangwill, W.: Non-linear programming via penalty functions. Manag. Sci. **13**, 344–358 (1967)

P-Regularity Theory and Nonlinear Optimization Problems

Yuri Evtushenko[1,2], Vlasta Malkova[1]([✉]), and Alexey Tret'yakov[1,3,4]

[1] Dorodnicyn Computing Centre, FRC CSC RAS,
Vavilov Street 40, 119333 Moscow, Russia
yuri-evtushenko@yandex.ru, vmalkova@yandex.ru
[2] Moscow Institute of Physics and Technology (National Research University),
Moscow, Russia
[3] Systems Research Institute, Polish Academy of Sciences,
Newelska 6, 01-447 Warsaw, Poland
[4] Faculty of Sciences, Siedlce University, 08-110 Siedlce, Poland
tret@ap.siedlce.pl

Abstract. The paper studies the nonlinear optimization problem in the form of optimal control problem subject to irregular constraints for which the multiplier of the objective function in Pontryagin's function may vanish. It turns out that in case of p-regularity constraints this drawback can be overcome. But for this it necessary to prove continuous dependence solution of differential equation with respect to the boundary conditions. We give a new approach to the existence of such solutions via p-regularity theory.

Keywords: Nonlinear optimization · Singular optimal control ·
p-regularity · p-factor operator · Nonlinear boundary value problems ·
Implicit function theorem

1 Non-linear Optimization Problem Formulation

We will consider the nonlinear optimal control problem in following form

$$J(x,u) = \int_{t_1}^{t_2} f(t, x(t), u(t)) \to \text{extr} \tag{1}$$

subject to

$$\dot{x} - \varphi(t, x(t), u(t)) = 0$$
$$x(t_1) = a, \qquad x(t_2) = b, \qquad u \in V, \tag{2}$$

This work was partially supported by the Russian Foundation for Basic Research (project no. 17-07-00510) and by Program 2 of Presidium of RAS "Mechanisms for ensuring fault tolerance in modern high-performance and highly reliable computing".

© Springer Nature Switzerland AG 2020
M. Jaćimović et al. (Eds.): OPTIMA 2019, CCIS 1145, pp. 237–253, 2020.
https://doi.org/10.1007/978-3-030-38603-0_18

where $f: R \times \mathbb{R}^n \times \mathbb{R}^r \rightarrow R$, $\varphi: R \times \mathbb{R}^n \times \mathbb{R}^r \rightarrow \mathbb{R}^n$. The function f and φ are assumed to be sufficiently smooth, at least, up to order $p + 1$, $x \in W^{1,1}([t_1, t_2], \mathbb{R}^n)$, $u \in L^\infty([t_1, t_2], \mathbb{R}^r)$, a.e. on $[t_1, t_2]$, and $V \subset \mathbb{R}^r$ is a fix set in $\mathbb{R}^r$. In this case, Eq. (2) and condition $u(t) \in V$ are satisfied a.e. on $[t_1, t_2]$. Below, we will consider the case $V = \mathbb{R}^r$.

System (2) can be replaced by the operator equation

$$G(x, u) = 0,$$

where $G(x, u)(\cdot) = \dot{x}(\cdot) - \varphi(\cdot, x(\cdot), u(\cdot))$, $X = \{x(\cdot) \in W^{1,1}([t_1, t_2], \mathbb{R}^n), x(t_1) = a, x(t_2) = b)\}$, $V = L^\infty([t_1, t_2], \mathbb{R}^r)$, $Y = L^1([t_1, t_2], \mathbb{R}^n)$ and $G: X \times V \rightarrow Y$.

Let $L(t, x, \dot{x}, u) = \lambda(t)(\dot{x} - \varphi(t, x, u)) + \lambda_0 F(t, x, u)$, where $\lambda(t) = (\lambda_1(t), \ldots, \lambda_n(t))^\top$. Consider Pontryagin's function $H(t, x, u, \lambda) = \lambda\varphi(t, x, u) - \lambda_0 f(t, x, u)$. Then Pontryagin's maximum principle can be formulated in the form of the following theorem.

Theorem 1. *Let (x^*, u^*) be an optimal solution of (1) and (2). Then there must be Lagrange multipliers, $\lambda_0^* \geq 0$, $\lambda^*(t): [t_1, t_2] \rightarrow \mathbb{R}^n$ that do not vanish simultaneously and such that for almost all $t \in [t_1, t_2]$, the equation*

$$\frac{d}{dt} L_{\dot{x}}(t, x^*(t), \dot{x}^*(t), u^*(t)) = L_x(t, x^*(t), \dot{x}(t), u^*(t)) \tag{3}$$

and maximum principle

$$\max_{u \in Y} \left(\lambda^*(t)\varphi(t, x^*(t), u(t)) \right) - \lambda_0^* f(t, x^*(t), u(t)) = \lambda^*(t)\varphi(t, x^*(t), u^*(t)) - \lambda_0^* f(t, x^*(t), u^*(t)). \tag{4}$$

Furthermore, if

$$\operatorname{Im} G'(x^*, u^*) = Y, \tag{5}$$

then one can set $\lambda_0^* = 1$.

In the singular (irregular, degenerate) case, when condition (5) is not fulfilled, the maximum principle (4) (as well as the Euler-Lagrange equation (3)) may be violated in the case $\lambda_0^* = 1$.

Example 1. Consider the problem

$$J(x, u) = \int_{-\frac{\pi}{2}}^{\frac{\pi}{2}} (10x_1^2 + 10x_2^2 + u^2 + u)dt \rightarrow \min$$

subject to

$$G(x, u) = \begin{pmatrix} \dot{x}_1 - x_2 + x_1^2 - \frac{1}{2}x_2^2 + u^2 \sin t \\ \dot{x}_2 + x_1 + x_1^2 - \frac{1}{2}x_2^2 + u^2 \cos t + ux_2 + ux_1 \end{pmatrix} = 0 \tag{6}$$

$$x_1\left(-\frac{\pi}{2}\right) = x_1\left(\frac{\pi}{2}\right) = 0$$

where $V = R$. The pair $x^*(t) = 0$, $u^*(t) = 0$ is a local optimum. However, it is easy to verify that λ_0^* must be equal to zero. Indeed, if $\lambda_0^* > 0$, then $-u - u^2 + \lambda_1^*(t)u^2 \sin t + \lambda_2^*(t)u^2 \cos t \leq 0 \ \forall u(t) \in V_\varepsilon = \{a \in R, \|a\| \leq \varepsilon\}$, which is not the case for small ε! In this example, it turns out that $\operatorname{Im} G'(x^*, u^*) \neq L^1([t1, t_2], \mathbb{R}^2)$, and the regularity condition (5) is not satisfied.

In turn, the mapping $G(x, u)$ is p-regular ($p = 2$) at the point (x^*, u^*), and we can formulate the p-order maximum principle, where the coefficient λ_0^* is no more equal to zero. See [1] and the main constructions of p-regularity theory you can find in [4, 9, 11].

But for proving p-order maximum principle it is necessary to prove theorem on continuous dependence of boundary value problem solutions with respect to initial conditions in p-regular case.

Let us denote

$$F(x, \mu) = \begin{pmatrix} G(x, u) \\ x(t_1) - \nu = 0 \\ x(t_2) - \rho = 0 \end{pmatrix} \tag{7}$$

where $\mu = (\nu, \rho)$ – is a parameter and $u(t)$ – some fixed feasible control. Here $F(x, u) \colon X \times M$ and $X = \{x \in W^{1,1}([t_1, t_2], \mathbb{R}^n), x(t_1) = a = \nu^*, x(t_2) = b = \rho^*\}$. Will be hold the following theorem.

Theorem 2. *Let $F(x, \mu) \in C^{p+1}(X \times M)$, $F : X \times M \to Z$, where M is finite dimensional space, X, Z are Banach spaces. Let the mappings $f_i(x, \mu)$, $i = 1, \ldots, p$ be defined by (15). Assume that $F(x^*, \mu^*) = 0$ and $\forall \bar{\mu} \in M, \|\bar{\mu}\| = 1$, $(0, \bar{\mu}) \in \bigcap_{k=1}^p \operatorname{Ker}^k f_k^{(k)}(x^*, \mu^*)$ and F is strongly p-regular with respect to M along every elements $(0, \bar{\mu})$, $\bar{\mu} \in M$, that is*

$$\|\{f_1'(x^*, \mu^*) + f_2''(x^*, \mu^*)[0, \bar{\mu}] + \ldots + f_p^{(p)}(x^*, \mu^*)[0, \bar{\mu}]^{p-1}\}^{-1}\| \leq C. \tag{8}$$

(Here $\{\cdot\}^{-1}$ means right inverse operator).

Then there exists the continuous mapping $x = x(\mu)$, $\mu \in V_\varepsilon(\mu^*)$, where $V_\varepsilon(\mu^*)$ is the neighborhood of μ^*, $x(\mu) \in C(V_\varepsilon(\mu^*))$, $\varepsilon > 0$ sufficiently small, such that $F(x(\mu), \mu) = 0$ and

$$x(\mu) = x^* + \omega(\mu), \quad \|\omega(\mu)\| = o(\|\mu - \mu^*\|), \tag{9}$$

$$\|x(\mu) - x^*\| \leq C \sum_{k=1}^p \|f_k(x^*, \mu)\|_{Z_k}^{\frac{1}{k}}, \quad \forall \mu \in V_\varepsilon(\mu^*). \tag{10}$$

In the proof of above theorem we apply the Michael selection theorem (see [8]) which we give in the some modified form:

Theorem 3. *Let X, Y - B-spaces, $A \in L(X, Y)$ and $\|A^{-1}\| \leq K$. Then there exists continuous mapping $M : Y \to X$ such that $AM(y) = y$ and $\|M(y)\| \leq c\|y\|$, where $c > 0$ is a constant independent of y.*

The illustration of our problem is the following example of the boundary value problem

$$F(x) = x'' + x + x^2 = 0, \quad x(0) = \nu, \quad x(2\pi) = \rho, \tag{11}$$

where ν and ρ are small parameters from $U(\nu^*, \rho^*) = U(0,0)$, $\nu^* = 0$, $\rho^* = 0$.

We show that for any h_ν, h_ρ such that $h_\nu \neq h_\rho$, the mapping F is 2-regular along element $H = [0, h_\nu, h_\rho]$, i.e. based on the Theorem 2 there exists a continuous solution of (11) dependent on the parameters $\mu = (\nu, \rho)$ for $h_\nu \neq h_\rho$, and for $h_\nu = h_\rho$ the mapping F is 2-regular along element $H = [\sin t, h_\nu, h_\rho]$. All of this means that based on our theorems, a continuous solution of Eq. (11) dependent on the parameter μ there exists for all μ sufficiently small.

2   Generalization of p-factor Lyusternik Theorem and p-order Implicit Function Theorem

The apparatus of p-regularity is an important tool for studying nonlinear problems. In this section we present some definitions, notations and theorems of p-regularity theory to be used in what follows (see [4–7,10,11]).

We are interested in the following nonlinear problem

$$F(x, \mu) = 0, \tag{12}$$

where the mapping $F : X \times M \to Z$ and X, M and Z are Banach spaces.

Assume that for some point $(x^*, \mu^*) \in X \times M$, $\mathrm{Im} F'(x^*, \mu^*) \neq Z$. Let

$$Z = Z_1 \oplus \ldots \oplus Z_p, \tag{13}$$

where $Z_1 = \mathrm{cl}(\mathrm{Im} F'(x^*, \mu^*))$ and $W_1 = Z$. As W_2 we use the closed complement of Z_1 in Z. Let $P_{W_2} : Z \to W_2$ be the projector onto W_2 along Z_1. By Z_2 we denote the closure of linear span of the image of the quadratic mapping $P_{W_2} F''(x^*, \mu^*)[\cdot]^2$. Then, inductively,

$$Z_i = \mathrm{cl}(\mathrm{span}\mathrm{Im} P_{W_i} F^{(i)}(x^*, \mu^*)[\cdot]^i) \subseteq W_i, i = 2, \ldots, p - 1, \tag{14}$$

where W_i is a closed complement of $Z_1 \oplus \ldots \oplus Z_{i-1}$, $i = 2, \ldots, p$ with respect to Z, and $P_{W_i} : Z \to W_i$ is a projector onto W_i along $Z_1 \oplus \ldots \oplus Z_{i-1}$, $i = 2, \ldots, p$ with respect to Z. Finally, $Z_p = W_p$. The order p is the minimal number (if it exists) for which the decomposition (13) holds.

For what follows we will denote $\varphi^{(0)} = \varphi$ for any mapping φ.

Define the following mappings

$$f_i : U \subset X \times M \to Z_i, f_i(x, \mu) = P_{Z_i} F(x, \mu), i = 1, \ldots, p, \tag{15}$$

where $P_{Z_i} : Z \to Z_i$ is the projection operator onto Z_i along $Z_1 \oplus \ldots \oplus Z_{i-1} \oplus Z_{i+1} \oplus \ldots \oplus Z_p$. Then the mapping F can be represented as

$$F(x, \mu) = f_1(x, \mu) + \ldots + f_p(x, \mu) \tag{16}$$

or

$$F(x, \mu) = (f_1(x, \mu), \ldots, f_p(x, \mu)). \tag{17}$$

Denote $h = [h_x, h_\mu]$, $h_x \in X$, $h_\mu \in M$.

Definition 1. *The linear operator* $\Psi_p(h) : X \times M \to Z$, *defined by*

$$\Psi_p(h) = f_1'(x^*, \mu^*) + f_2''(x^*, \mu^*)[h] + \ldots + f_p^{(p)}(x^*, \mu^*)[h]^{p-1} \tag{18}$$

such that

$$\Psi_p(h)[x, \mu] = f_1'(x^*, \mu^*)[x, \mu] + f_2''(x^*, \mu^*)[h][x, \mu] + \ldots + f_p^{(p)}(x^*, \mu^*)[h]^{p-1}[x, \mu] \tag{19}$$

is called p-factor operator.

Definition 2. *We say that F is completely degenerate at (x^*, μ^*) up to the order p if $F^{(i)}(x^*, \mu^*) = 0$, $i = 1, \ldots, p - 1$.*

Remark 1. *In the completely degenerate case the p-factor operator reduces to $F^{(p)}(x^*, \mu^*)[h]^{p-1}$.*

Remark 2. *For each mapping f_i, we have ([4] p. 145)*

$$f_i^{(k)}(x^*, \mu^*) = 0, \quad k = 0, 1, \ldots, i - 1, \quad \forall\, i = 1, \ldots, p. \tag{20}$$

Remark 3. *For each mapping f_i we have in completely degenerate case*

$$f_i^{(i)}(x^*, \mu^*)[h]^{i-1} = P_{Z_i} F^{(i)}(x^*, \mu^*)[h]^{i-1}, \quad i = 1, \ldots, p. \tag{21}$$

It mean that f_i are i-factor operators corresponding to completely degenerate mappings f_i up to order i. So the general degeneration of F can be reduced to the study of completely degenerated mappings f_i, $i = 1, \ldots, p$ and their compositions.

Definition 3. *The p-kernel of the operator* $\Psi_p(h)$ *is a set*

$$H_p(x^*, \mu^*) = \mathrm{Ker}^p \Psi_p(h)$$
$$= \{h \in X \times M : f_1'(x^*, \mu^*)[h] + f_2''(x^*, \mu^*)[h]^2 + \ldots + f_p^{(p)}(x^*, \mu^*)[h]^p = 0\}.$$

Note that the following relations holds:

$$\mathrm{Ker}^p \Psi_p(h) = \left\{ \bigcap_{i=1}^{p} \mathrm{Ker}^i f_i^{(i)}(x^*, \mu^*) \right\}.$$

The p-kernel of the operator $F^{(p)}(x^*, \mu^*)$ in the completely degenerate case is a set

$$\mathrm{Ker}^p F^{(p)}(x^*, \mu^*) = \{h \in X \times M : F^{(p)}(x^*, \mu^*)[h]^p = 0\}.$$

Definition 4. *A mapping F is called p-regular at (x^*, μ^*) along h $(p > 1)$ if $\mathrm{Im}\Psi_p(h) = Z$ (i.e., the operator $\Psi_p(h)$ is surjective).*

Definition 5. *A mapping F is called p-regular at (x^*, μ^*) $(p > 1)$ if either it is p-regular along every $h \in H_p(x^*, \mu^*)\backslash\{0\}$ or $H_p(x^*, \mu^*) = \{0\}$.*

Definition 6. *Let $F : X \times M \to Z = Z_1 \oplus \cdots \oplus Z_p$. The mapping $F(x, \mu)$ is called strongly p-regular at the point (x^*, μ^*) if there exist $\gamma > 0$ and $c > 0$ such that*

$$\sup_{h \in H_\gamma} \|\{\Psi_p(h)\}^{-1}\| \leq c < \infty,$$

where

$$H_\gamma = \{h = (h_x, h_\mu) \in X \times M : \|f_k^{(k)}(x^*, \mu^*)[h]^k\|_{Z_k} \leq \gamma,$$
$$\forall k = 1, \dots, p, \quad \|h\|_{X \times M} = 1\}.$$

Define *the solution set* for the mapping F as the set

$$S = S(x^*, \mu^*) = \{x \in X \times M : F(x, \mu) = F(x^*, \mu^*) = 0\} \tag{22}$$

and let $T_{(x^*, \mu^*)} S$ denote the tangent cone to the set S at the point (x^*, μ^*), i.e.,

$$T_{(x^*, \mu^*)} S = \{h \in X \times M : (x^*, \mu^*) + \varepsilon h + r(\varepsilon) \in S, \|r(\varepsilon)\| = o(\varepsilon), \varepsilon \in [0, \delta], \delta > 0\} \tag{23}$$

The following theorems describe the tangent cone to the solutions set of Eq. (12) in the p-regular case.

Theorem 4. *Let X, M and Z be the Banach spaces and let the mapping $F \in C^p(X \times M, Z)$ be p-regular at $(x^*, \mu^*) \in X \times M$ along h. Then $h \in T_{(x^*, \mu^*)} S$.*

Theorem 5 (Generalized Lyusternik Theorem, [4]). *Let X, M and Z be the Banach spaces and let the mapping $F \in C^p(X \times M, Z)$ be p-regular at $(x^*, \mu^*) \in X \times M$. Then*

$$T_{(x^*, \mu^*)} S = H_p(x^*, \mu^*). \tag{24}$$

The following lemma (see [7]) will be used in the proof of Theorem 2.

Lemma 1. *Let $F : X \times M \to Z$, where X, M, Z are Banach spaces, $z = z_1 + \dots + z_p$, $z_i \in Z_i$, $i = 1, \dots, p$, $\|h\| = 1$ and*

$$\|\{\alpha_1 f_1'(x^*) + \alpha_2 f_2'(x^*)[h] + \dots + \alpha_p f_p^{(p)}(x^*)[h]^{p-1}\}^{-1}\| = C < \infty.$$

Then

$$\|\{\alpha_1 f_1'(x^*) + \alpha_2 f_2'(x^*)[th] + \dots + \alpha_p f_p^{(p)}(x^*)[th]^{p-1}\}^{-1}(z_1 + \dots + z_p)\|$$
$$\leq C(\frac{1}{\alpha_1}\|z_1\| + \frac{1}{\alpha_2 t}\|z_2\| + \dots + \frac{1}{\alpha_p t^{p-1}}\|z_p\|),$$

where $\alpha_i \in \mathbb{R}\backslash\{0\}$, $i = 1, \dots, p$, $t \neq 0$.

The following lemma will be important in the study of surjectivity of p-factor operators in our example.

Lemma 2. *Suppose that $Y = Y_1 \oplus Y_2$, where Y_1, Y_2 are closed subspaces in Y, $A, B \in \mathcal{L}(X, Y)$, $\mathrm{Im} A = Y_1$. Let also P_2 be the projection onto Y_2 along Y_1. Then $(A + P_2 B)X = Y \Leftrightarrow (P_2 B)\mathrm{Ker} A = Y_2$.*

This lemma is a consequence of the following

Lemma 3. *Suppose that $Y = Y_1 \oplus Y_2$, where Y_1, Y_2 are closed subspaces in Y, $A_1, A_2 \in \mathcal{L}(X, Y)$, $A_1 X \subset Y_1$, $A_2 X \subset Y_2$. Then $(A_1 + A_2)X = Y$ iff $A_1 \mathrm{Ker} A_2 = Y_1$ and $A_2 \mathrm{Ker} A_1 = Y_2$.*

The proof is obvious. Lemma 2 follows from Lemma 3 if we put $A_1 = A$ and $A_2 = P_2 B$.

Some generalizations of the implicit function theorem to the p-order implicit function theorem for nonregular mappings and the p-order implicit function theorem for the nontrivial kernel, there are in [2].

The multivalued contraction mapping theorem will be used in the proof of Theorem 2. Its content is available in [3].

3   Some Generalization of Lyusternik Theorem on Tangent Cone

In this section we prove the Theorem 2 which is some analog and generalization of Lyusternik theorem on a tangent cone and says about the existence of continuous solution of the equations $F(x, \mu) = 0$.

Remark 4. The element $\mu - \mu^*$ plays the role of $\bar{\mu}$ in the Theorem 2.

Proof (of the Theorem 2). Any element $\mu \in V_\varepsilon(\mu^*)$ can be represented as $\mu^* + t\bar{\mu}$, where $t \in [0, \delta]$ for $\delta > 0$ sufficiently small. Then we are looking for a solution of equation

$$F(x^* + x(t\bar{\mu}), \mu^* + t\bar{\mu} + \tilde{\mu}(t\bar{\mu})) = 0,$$

where $\tilde{\mu} \in V_\varepsilon(0)$, $\bar{\mu} \in M$.

Consider the multimapping $\Phi : C(V_\varepsilon(0)) \times V_\varepsilon(0) \to 2^{X \times M}$, which is defined by the formula

$$\Phi(x, \tilde{\mu}) = (x, \tilde{\mu}) - \{\Psi_p(h)\}^{-1} F(x^* + x, \mu^* + t\bar{\mu} + \tilde{\mu}), \tag{25}$$

where

$$h = (0, t\bar{\mu}) \in \bigcap_{k=1}^{p} \mathrm{Ker}^k f_k^{(k)}(x^*, \mu^*)$$

and p-factor operator $\Psi_p(h) : X \times M \to Z$ has the form

$$\Psi_p(h) = f_1'(x^*, \mu^*) + f_2''(x^*, \mu^*)[h] + \ldots + f_p^{(p)}(x^*, \mu^*)[h]^{p-1}.$$

Remark that the inverse multimapping operator is the following

$$\{\Psi_p(h)\}^{-1}(z) = \{[\xi, \eta] \in X \times M : f_1'(x^*, \mu^*)[\xi, \eta] + f_2''(x^*, \mu^*)[h][\xi, \eta] + \ldots$$
$$+ f_p^{(p)}(x^*, \mu^*)[h]^{p-1}[\xi, \eta] = z\},$$

where $z = z_1 + \ldots + z_p$ or $z = (z_1, \ldots, z_p)$, $z_i \in Z_i$, $i = 1, \ldots, p$.

The "norm" of above operator is

$$\|\{\Psi_p(h)\}^{-1}\| = \sup_{\|z\|=1} \inf\{\|(x, \mu)\| : \Psi_p(h)[x, \mu] = z\}.$$

We will show that there exists an element $(x, \tilde{\mu})$, such that

$$\|(x, \tilde{\mu})\| = \|x\| + \|\tilde{\mu}\| = o(\|t\bar{\mu})\|$$

and $(x, \tilde{\mu}) \in \Phi(x, \tilde{\mu})$, i.e. $(x, \tilde{\mu})$ is a fixed point of the mapping Φ. Then

$$(0, 0) \in \{-\{\Psi_p(0, t\bar{\mu})\}^{-1} F(x^* + x(t\bar{\mu}), \mu^* + t\bar{\mu} + \tilde{\mu}(t\bar{\mu}))\}.$$

Consequently we will obtain

$$F(x^* + x(t\bar{\mu}), \mu^* + t\bar{\mu} + \tilde{\mu}(t\bar{\mu})) = 0$$

and $\|(x(t\bar{\mu}), \tilde{\mu}(t\bar{\mu}))\| = o(t)$.

In the beginning, we will prove that

$$\text{dist}((0,0), \Phi(0,0)) = \|\Phi(0,0)\| \le ct^2 = O(t^2) = o(t).$$

We have

$$\Phi(0,0) = -\{\Psi_p(0, t\bar{\mu})\}^{-1} F(x^*, \mu^* + t\bar{\mu})$$

$$\Phi(0,0) = -\{\Psi_p(0, t\bar{\mu})\}^{-1}(f_1(x^*, \mu^* + t\bar{\mu}) + f_2(x^*, \mu^* + t\bar{\mu}) + \ldots + f_p(x^*, \mu^* + t\bar{\mu}))$$

and

$$\|\Phi(0,0)\| = \| - \{\Psi_p(0, t\bar{\mu})\}^{-1}(f_1(x^*, \mu^* + t\bar{\mu}) + f_2(x^*, \mu^* + t\bar{\mu}) + \ldots + f_p(x^*, \mu^* + t\bar{\mu}))\|.$$

By Lemma 1 we obtain

$$\|\Phi(0,0)\| \le \|c(f_1(x^*, \mu^* + t\bar{\mu})\| + \frac{c}{t}\|f_2(x^*, \mu^* + t\bar{\mu})\| + \ldots + \frac{c}{t^{p-1}}\|f_p(x^*, \mu^* + t\bar{\mu})\|.$$
$$(26)$$

We apply the Taylor formula to the expressions $f_i(x^*, \mu^* + t\bar{\mu})$ for $i = 1, \ldots p$ and we have

$$\|\Phi(0,0)\| \le c\|f_1(x^*, \mu^*) + f_1'(x^*, \mu^*)[t \cdot 0, t\bar{\mu}] + O_Z(t^2)\|$$
$$+ \frac{c}{t}\|f_2(x^*, \mu^*) + f_2'(x^*, \mu^*)[t \cdot 0, t\bar{\mu}] + \frac{1}{2!}f_2''(x^*, \mu^*)[t \cdot 0, t\bar{\mu}]^2 + O_Z(t^3)\|$$
$$+ \ldots$$
$$+ \frac{c}{t^{p-1}}\|f_p(x^*, \mu^*) + f_p'(x^*, \mu^*)[t \cdot 0, t\bar{\mu}] + \ldots + \frac{1}{p!}f_p^{(p)}(x^*, \mu^*)[t \cdot 0, t\bar{\mu}]^p$$
$$+ O_Z(t^{p+1})\|.$$

Therefore based on the relations (20) we estimate

$$\|\Phi(0,0)\| \le ct^2 + \frac{c}{t}t^3 + \ldots + \frac{c}{t^{p-1}}t^{p+1} = pct^2 \tag{27}$$

and

$$\|\Phi(0,0)\| = O(t^2) = o(t). \tag{28}$$

Now we show, that for any $(x_1,\mu_1), (x_2,\mu_2) \in V_{O(t^2)}(0,0)$ the following estimation holds

$$\operatorname{dist}_H(\Phi(x_1,\mu_1), \Phi(x_2,\mu_2)) \le \theta\|(x_1,\mu_1) - (x_2,\mu_2)\|, \tag{29}$$

where $0 < \theta < 1$.

Note, at the beginning that

$$\Psi_p(th)\Phi(x_1,\mu_1) = \Psi_p(th)(x_1,\mu_1) - F((x^*,\mu^*) + th + (x_1,\mu_1)) \tag{30}$$

and

$$\Psi_p(th)\Phi(x_2,\mu_2) = \Psi_p(th)(x_2,\mu_2) - F((x^*,\mu^*) + th + (x_2,\mu_2)). \tag{31}$$

Let $(z_1,\xi_1) \in \Phi(x_1,\mu_1),\ (z_2,\xi_2) \in \Phi(x_2,\mu_2)$. Then we have

$\operatorname{dist}_H(\Phi(x_1,\mu_1), \Phi(x_2,\mu_2))$

$= \inf\{\|(z_1,\xi_1) - (z_2,\xi_2)\| : (z_i,\xi_i) \in \Phi(x_i,\mu_i), i = 1,2\}$

$= \inf\{\|(z_1,\xi_1) - (z_2,\xi_2)\| : \Psi_p(th)((z_1,\xi_1) - (z_2,\xi_2))$

$= \Psi_p(th)((x_1,\mu_1) - (x_2,\mu_2))$

$- [F((x^*,\mu^*) + th + (x_1,\mu_1)) - F((x^*,\mu^*) + th + (x_2,\mu_2))]\}$

$= \inf\{\|(z,\xi)\| : \Psi_p(th)(z,\xi)$

$= \Psi_p(th)((x_1,\mu_1) - (x_2,\mu_2))$

$- [F((x^*,\mu^*) + th + (x_1,\mu_1)) - F((x^*,\mu^*) + th + (x_2,\mu_2))]\}$

$= \inf\{\|(z,\xi)\| : \Psi_p(th)(z,\xi) = \Psi_p(th)((x_1,\mu_1) - (x_2,\mu_2))$

$- \left[\sum_{i=1}^{p}(f_i((x^*,\mu^*) + th + (x_1,\mu_1)) - f_i((x^*,\mu^*) + th + (x_2,\mu_2)))\right]\}$

$= \inf\{\|\{\Psi_p(th)\}^{-1}[f_1'(x^*,\mu^*)((x_1,\mu_1) - (x_2,\mu_2))$

$+ \sum_{i=2}^{p}f_i^{(i)}(x^*,\mu^*)[th]^{i-1}((x_1,\mu_1) - (x_2,\mu_2))$

$- \left[\sum_{i=1}^{p}(f_i((x^*,\mu^*) + th + (x_1,\mu_1)) - f_i((x^*,\mu^*) + th + (x_2,\mu_2)))\right]\|\}$

$= \inf\{\|\{\Psi_p(th)\}^{-1}$

$[[f_1'(x^*,\mu^*)((x_1,\mu_1) - (x_2,\mu_2)) - (f_1((x^*,\mu^*) + th + (x_1,\mu_1))$

$- f_1((x^*,\mu^*) + th + (x_2,\mu_2))]$

$+ \sum_{i=2}^{p}\left[f_i^{(i)}(x^*,\mu^*)[th]^{i-1}((x_1,\mu_1) - x_2,\mu_2)) - (f_i((x^*,\mu^*) + th + (x_1,\mu_1))\right.$

$- (f_i((x^*,\mu^*) + th + (x_2,\mu_2))]]\|\}.$

By Lemma 1, we obtain the following estimation

$$
\begin{aligned}
\operatorname{dist}_{\mathrm{H}}&(\Phi(x_1,\mu_1),\Phi(x_2,\mu_2))\\
\le\,& c\|f_1((x^*,\mu^*)+th+(x_1,\mu_1))-f_1((x^*,\mu^*)+th+(x_2,\mu_2))\\
&-f_1'(x^*,\mu^*)((x_1,\mu_1)-(x_2,\mu_2))\|\\
&+\frac{c}{t}\|f_2((x^*,\mu^*)+th+(x_1,\mu_1))-f_2((x^*,\mu^*)+th+(x_2,\mu_2))\\
&-f_2''[th](x^*,\mu^*)((x_1,\mu_1)-(x_2,\mu_2))\|\\
&+\cdots\\
&+\frac{c}{t^{p-1}}\|f_p((x^*,\mu^*)+th+(x_1,\mu_1))-f_p((x^*,\mu^*)+th+(x_2,\mu_2))\\
&-f_p^{(p)}[th]^{p-1}(x^*,\mu^*)((x_1,\mu_1)-(x_2,\mu_2))\|\\
=\,& A_1+A_2+\cdots+A_p,
\end{aligned}
$$

where

$$
\begin{aligned}
A_1\,&=c\|f_1((x^*,\mu^*)+th+(x_1,\mu_1))-f_1((x^*,\mu^*)+th+(x_2,\mu_2))\\
&\quad-f_1'(x^*,\mu^*)((x_1,\mu_1)-(x_2,\mu_2))\|,\\
A_2\,&=\frac{c}{t}\|f_2((x^*,\mu^*)+th+(x_1,\mu_1))-f_2((x^*,\mu^*)+th+(x_2,\mu_2))\\
&\quad-f_2''[th](x^*,\mu^*)((x_1,\mu_1)-(x_2,\mu_2))\|,\\
&\ \ \vdots\\
A_p\,&=\frac{c}{t^{p-1}}\|f_p((x^*,\mu^*)+th+(x_1,\mu_1))-f_p((x^*,\mu^*)+th+(x_2,\mu_2))\\
&\quad-f_p^{(p)}[th]^{p-1}(x^*,\mu^*)((x_1,\mu_1)-(x_2,\mu_2))\|.
\end{aligned}
$$

To the component A_1 we apply the mean value theorem and then the Taylor formula to the expression $f_1'[(x^*,\mu^*)+th+(x_2,\mu_2)+\bar\theta((x_1,\mu_1)-(x_2,\mu_2))]$. We have

$$
\begin{aligned}
A_1\,&\le c\sup_{\bar\theta\in[0,1]}\|f_1'((x^*,\mu^*)+th+(x_2,\mu_2)+\bar\theta((x_1,\mu_1)-(x_2,\mu_2)))-f_1'(x^*,\mu^*)\|\\
&\qquad\cdot\|(x_1,\mu_1)-(x_2,\mu_2)\|\\
&=c\sup_{\bar\theta\in[0,1]}\|f_1'(x^*,\mu^*)+O_Z(t)-f_1'(x^*,\mu^*)\|\|(x_1,\mu_1)-(x_2,\mu_2)\|\\
&=c\sup_{\bar\theta\in[0,1]}\|O_Z(t)\|\|(x_1,\mu_1)-(x_2,\mu_2)\|\le cc_1t\|(x_1,\mu_1)-(x_2,\mu_2)\|\\
&=k_1t\|(x_1,\mu_1)-(x_2,\mu_2)\|,
\end{aligned}
$$

where $k_1=cc_1$. Put now $\theta_1=k_1t$, where $t\in(0,\delta)$, δ is sufficiently small and then

$$
A_1\le\theta_1\|(x_1,\mu_1)-(x_2,\mu_2)\|.
$$

To the component A_2 we apply the mean value theorem and then the Taylor formula to the expression $f_2'[(x^*,\mu^*)+th+(x_2,\mu_2)+\bar\theta((x_1,\mu_1)-(x_2,\mu_2))]$. We have now

$$A_2 \le \frac{c}{t} \sup_{\bar{\theta}\in[0,1]} \|f_2'[(x^*,\mu^*) + th + (x_2,\mu_2) + \bar{\theta}((x_1,\mu_1) - (x_2,\mu_2))] - f_2''(x^*,\mu^*)[th]\|$$

$$\cdot \; \|(x_1,\mu_1) - (x_2,\mu_2)\|$$

$$= \frac{c}{t} \sup_{\bar{\theta}\in[0,1]} \|f_2'(x^*,\mu^*) + f_2''(x^*,\mu^*)[th + (x_2,\mu_2) + \bar{\theta}((x_1,\mu_1) - (x_2,\mu_2))]$$

$$+ O_Z(t^2) - f_2''(x^*)[th]\|\|(x_1,\mu_1) - (x_2,\mu_2)\|$$

$$= \frac{c}{t} \sup_{\bar{\theta}\in[0,1]} \|f_2'(x^*,\mu^*) + f_2''(x^*,\mu^*)[th]$$

$$+ f_2''(x^*,\mu^*)[(x_2,\mu_2) + \bar{\theta}((x_1,\mu_1) - (x_2,\mu_2))]$$

$$+ O_Z(t^2) - f_2''(x^*,\mu_*)[th]\|\|(x_1,\mu_1) - (x_2,\mu_2)\|.$$

Since $f_2'(x^*,\mu^*) = 0$, (see (20)) we obtain

$$A_2 \le \frac{c}{t} \sup_{\bar{\theta}\in[0,1]} \|f_2''(x^*,\mu^*)[(x_2,\mu_2) + \bar{\theta}((x_1,\mu_1) - (x_2,\mu_2))] + O_Z(t^2)\|$$

$$\cdot \; \|(x_1,\mu_1) - (x_2,\mu_2)\|,$$

therefore

$$A_2 \le \frac{c}{t} \sup_{\bar{\theta}\in[0,1]} (\|f_2''(x^*,\mu^*)[(x_2,\mu_2) + \bar{\theta}((x_1,\mu_1) - (x_2,\mu_2))]\| + \|O_Z(t^2)\|)$$

$$\cdot \; \|(x_1,\mu_1) - (x_2,\mu_2)\|,$$

Finally, using the fact that the rank of expression $f_2''(x^*,\mu^*)[(x_2,\mu_2) + \bar{\theta}((x_1,\mu_1) - (x_2,\mu_2))]$ is t^2 and the rank of $O_Z(t^2)$ is t^2 and properties of norm we conclude that

$$A_2 \le \frac{c}{t}(d_1 t^2 + d_2 t^2)\|(x_1,\mu_1) - (x_2,\mu_2)\| \le 2k_2 t\|(x_1,\mu_1) - (x_2,\mu_2)\|,$$

where $k_2 = \max\{cd_1, cd_2\}$. We can put $\theta_2 = 2k_2 t$ and then

$$A_2 \le \theta_2\|(x_1,\mu_1) - (x_2,\mu_2)\|,$$

where $t \in (0,\delta)$, $\delta > 0$ is sufficiently small.

Similarly, we will now evaluate the component A_p using also the mean value theorem and the extension of expression $f_p'[(x^*,\mu^*) + th + (x_2,\mu_2) + \bar{\theta}((x_1,\mu_1) - (x_2,\mu_2))]$ in the Taylor formula.

So let's note that

$$A_p \le \frac{c}{t^{p-1}} \sup_{\bar{\theta}\in[0,1]} \|f_p'[(x^*,\mu^*) + th + (x_2,\mu_2) + \bar{\theta}((x_1,\mu_1) - (x_2,\mu_2))]$$

$$- \frac{1}{(p-1)!} f_p^{(p)}(x^*,\mu^*)[th]^{p-1}\|\|(x_1,\mu_1) - (x_2,\mu_2)\| \tag{32}$$

By Taylor formula we have:

$$f_p'[(x^*, \mu^*) + th + (x_2, \mu_2) + \bar{\theta}((x_1, \mu_1) - (x_2, \mu_2))]$$
$$= f_p'(x^*, \mu^*) + f_p''(x^*, \mu^*)[th + (x_2, \mu_2) + \bar{\theta}((x_1, \mu_1) - (x_2, \mu_2))] + \cdots$$
$$+ \frac{1}{(p-1)!} f_p^{(p)}(x^*, \mu^*)[th + (x_2, \mu_2) + \bar{\theta}((x_1, \mu_1) - (x_2, \mu_2))]^{p-1} + O_Z(t^p).$$

According to (20) the mapping $f_p^{(i)}(x^*, \mu^*) = 0$, for $i = 1, 2, \ldots, p-1$. Then we obtain

$$A_p \leq \frac{c}{t^{p-1}} \sup_{\bar{\theta} \in [0,1]} \| \frac{1}{(p-1)!} f_p^{(p)}(x^*, \mu^*)[th + (x_2, \mu_2) + \bar{\theta}((x_1, \mu_1) - (x_2, \mu_2))]^{p-1}$$
$$+ O_Z(t^p) - \frac{1}{(p-1)!} f_p^{(p)}(x^*, \mu^*)[th]^{p-1} \| \|(x_1, \mu_1) - (x_2, \mu_2))\|$$
$$= \frac{c}{t^{p-1}} \sup_{\bar{\theta} \in [0,1]} \| \frac{1}{(p-1)!} f_p^{(p)}(x^*, \mu^*)[th]^{p-1}$$
$$+ \frac{1}{(p-1)!} f_p^{(p)}(x^*, \mu^*)[th]^{p-2}[(x_2, \mu_2) + \bar{\theta}((x_1, \mu_1) - (x_2, \mu_2))]$$
$$+ O_Z(t^p) - \frac{1}{(p-1)!} f_p^{(p)}(x^*, \mu^*)[th]^{p-1} \| \|(x_1, \mu_1) - (x_2, \mu_2))\|,$$

i.e.

$$A_p \leq \frac{c}{t^{p-1}} \sup_{\bar{\theta} \in [0,1]} \| \frac{1}{(p-1)!} f_p^{(p)}(x^*, \mu^*)[th]^{p-2}[(x_2, \mu_2) + \bar{\theta}((x_1, \mu_1) - (x_2, \mu_2))]$$
$$+ O_Z(t^p) \| \|(x_1, \mu_1) - (x_2, \mu_2))\|.$$

Then according to property of a norm we obtain

$$A_p \leq \frac{c}{t^{p-1}} \sup_{\bar{\theta} \in [0,1]} (\| \frac{1}{(p-1)!} f_p^{(p)}(x^*, \mu^*)[th]^{p-2}[(x_2, \mu_2) + \bar{\theta}((x_1, \mu_1) - (x_2, \mu_2))]\|$$
$$+ \|O_Z(t^p)\|)\|(x_1, \mu_1) - (x_2, \mu_2)\|.$$

Since $(x_1, \mu_1), (x_2, \mu_2) \in V_{O(t^2)}(0,0)$, then

$$A_p \leq \frac{c}{t^{p-1}} (\bar{d}_1 t^{p-2+2} + \bar{d}_2 t^p)\|(x_1, \mu_1) - (x_2, \mu_2)\|$$
$$\leq 2k_p t\|(x_1, \mu_1) - (x_2, \mu_2)\| = \theta_p\|(x_1, \mu_1) - (x_2, \mu_2)\|,$$

where $k_2 = \max\{c\bar{d}_1, c\bar{d}_2\}$, $t \in (0, \delta)$, $\delta > 0$ is sufficiently small and $\theta_p = 2k_p t$. Substituting $\theta = \theta_1 + \theta_2 + \cdots + \theta_p = \bar{c}t$, $\bar{c} > 0$, we obtain

$$\text{dist}_H(\Phi(x_1, \mu_1), \Phi(x_2, \mu_2)) \leq \theta\|(x_1, \mu_1) - (x_2, \mu_2)\|, \tag{33}$$

and $0 < \theta < 1$.

According to the multivalued contraction principle, we will show that

$$\varrho((0,0), \Phi(0,0)) = \|\Phi(0,0)\| < (1-\theta)\varepsilon$$

where $\theta = \bar{\bar{c}}t$, $\varepsilon = 4pct^2$, for t sufficiently small.

We can put $0 < \theta = \bar{\bar{c}}t < \frac{1}{2}$. This inequality is equivalent to $1 < 2(1 - \bar{\bar{c}}t)$. From this and inequality $\|\Phi(0,0)\| \leq pct^2$ we obtain

$$\|\Phi(0,0)\| \leq pct^2 \leq 2p(1 - \bar{\bar{c}}t)ct^2 < (1 - \bar{\bar{c}}t)4pct^2 = (1-\theta)\varepsilon,$$

which was to prove.

Therefore we proved that the mapping Φ is contraction in the set $V((0,0), ct^2)$. By multivalued contraction principle for $(z_0, \mu_0) = (0,0)$, it follows that there exists element $(x, \tilde{\mu})$, such that

$$\|(x, \tilde{\mu})\| \leq \frac{2}{1-\theta}\|\Phi(0,0)\| \leq ct^2, \tag{34}$$

i.e. $\|(x, \tilde{\mu})\| = o(t)$ and $(x, \tilde{\mu}) \in \Phi(x, \tilde{\mu})$. Therefore $(x, \tilde{\mu})$ is fixed point of the mapping Φ. Then

$$(0,0) \in \{-\{\Psi_p(0, t\bar{\mu})\}^{-1}F(x^* + x(t\bar{\mu}), \mu^* + t\bar{\mu} + \tilde{\mu}(t\bar{\mu}))\}.$$

Consequently we obtain

$$F(x^* + x(t\bar{\mu}), \mu^* + t\bar{\mu} + \tilde{\mu}(t\bar{\mu})) = 0 \tag{35}$$

and $\|(x(t\bar{\mu}), \tilde{\mu}(t\bar{\mu}))\| = o(t)$.

Summarizing we showed, that for parameter $\mu^* + t\bar{\mu}$ we have solution $(x^* + x(t\bar{\mu}), \mu^* + t\bar{\mu} + \tilde{\mu}(t\bar{\mu}))$ of equation $F(x, \mu) = 0$, i.e.

$$F(x^* + x(t\bar{\mu}), \mu^* + t\bar{\mu} + \tilde{\mu}(t\bar{\mu})) = 0.$$

Now without loss of generality, let the set M will be equal $\mathbb{R}^2$ and let us take any μ, which is a sufficiently small element from $\mathbb{R}^2$. For such μ there exists $\bar{\bar{\mu}}(\mu)$ such that

$$\|\bar{\bar{\mu}}(\mu) - \mu\| = o(\mu) \tag{36}$$

and as we showed earlier,

$$F(x^* + x(\bar{\bar{\mu}}(\mu)), \mu^* + \bar{\bar{\mu}}(\mu) + \tilde{\mu}(\bar{\bar{\mu}}(\mu))) = 0. \tag{37}$$

Now is the important moment. We are coming back to μ. From assumptions we put $\mu = \mu^* + \bar{\bar{\mu}}(\mu) + \tilde{\mu}(\bar{\bar{\mu}}(\mu))$ and marking $x^* + x(\bar{\bar{\mu}}(\mu))$ by $x(\mu)$ we obtain the equation

$$F(x(\mu), \mu) = 0. \tag{38}$$

The above equation holds, since we took any $\bar{\mu}$ from $\mathbb{R}^2$ and we proved that (35) holds for any $\bar{\mu}$. Therefore, based on the above, for any μ from $\mathbb{R}^2$ there exists

$\bar{\mu}(\mu)$ such that the contraction process is beginning in the point $(0, \mu^* + \bar{\mu}(\mu))$ and generates solutions (37) and in the end (38).

This ends the first part of the proof (we proved the existence of solutions).

Let U be a sufficiently small neighborhood of (x^*, μ^*). Let us take $\mu^* + t\bar{\mu}$, where $t > 0$ is sufficiently small and put

$$h = \frac{(0, t\bar{\mu})}{\|t\bar{\mu}\|} = (0, \bar{\mu}) \in \bigcap_{k=1}^{p} \mathrm{Ker}^k f_k^{(k)}(x^*, \mu^*),$$

where $\|\bar{\mu}\| = 1$. Then for any $k \le p$

$$f_k^{(k)}(x^*, \mu^*)[0, \bar{\mu}]^k = 0, \quad k = 1, \ldots, p.$$

We showed in Eq. (26) that

$$\|\Phi(0,0)\| \le \|c(f_1(x^*, \mu^* + t\bar{\mu})\| + \frac{c}{t}\|f_2(x^*, \mu^* + t\bar{\mu})\| + \ldots + \frac{c}{t^{p-1}}\|f_p(x^*, \mu^* + t\bar{\mu})\| \tag{39}$$

or

$$\|\Phi(0,0)\| \le c \sum_{k=1}^{p} \left\| \frac{f_k(x^*, \mu^* + t\bar{\mu})}{t^{k-1}} \right\|. \tag{40}$$

We have for $\mu^* + t\bar{\mu}$

$$\|x(\mu^* + t\bar{\mu}) - x^*\| \le \|(x(\mu^* + t\bar{\mu}) - x^*, t\bar{\mu})\| \le C\|\Phi(0,0)\|. \tag{41}$$

Let us note that the following inequality holds:

$$\|f_k(x^*, \mu^* + t\bar{\mu}) - f_k(x^*, \mu^*)\| \le C_k \|f_k(x^*, \mu^* + t\bar{\mu}) - f_k(x^*, \mu^*)\|^{\frac{1}{k}} \|(0, t\bar{\mu})\|^{k-1} \tag{42}$$

or

$$\|f_k(x^*, \mu^* + t\bar{\mu}) - f_k(x^*, \mu^*)\|^{k-1} \le C_k^k \|(0, t\bar{\mu})\|^{k(k-1)}. \tag{43}$$

This is true because by Taylor expansion and relations (20) we have

$$\|f_k(x^*, \mu^*) + f_k'(x^*, \mu^*)[0, t\bar{\mu}] + \ldots$$

$$+ \frac{1}{(k-1)!} f_k^{(k)}(x^*, \mu^*)[0, t\bar{\mu}]^k + O(t^{k+1}) - f_k(x^*, \mu^*)\|^{k-1}$$

$$= \|\frac{1}{(k-1)!} f_k^{(k)}(x^*, \mu^*)[0, t\bar{\mu}]^k + O(t^{k+1})\|^{k-1} \le C t^{k(k-1)} = C\|(0, t\bar{\mu})\|^{k(k-1)}.$$

Consequently the relation (42) is satisfied. Therefore we have

$$\|x(\mu^* + t\bar{\mu}) - x^*\| \le C\|\Phi(0,0)\| \le Cc \sum_{k=1}^{p} \left\| \frac{f_k(x^*, \mu^* + t\bar{\mu})}{t^{k-1}} \right\|$$

$$\le \bar{C} \sum_{k=1}^{p} \frac{\|f_k(x^*, \mu^* + t\bar{\mu}) - f_k(x^*, \mu^*)\|^{\frac{1}{k}}}{\|t^{k-1}\|} \|(0, t\bar{\mu})\|^{k-1}.$$

and

$$\|x(\mu^* + t\bar{\mu}) - x^*\| \leq \bar{C} \sum_{k=1}^{p} \|f_k(x^*, \mu^* + t\bar{\mu})\|^{\frac{1}{k}}, \tag{44}$$

since $f_k(x^*, \mu^*) = 0$.

Next, from (44) for any sufficiently small μ there exists $\bar{\bar{\mu}}(\mu)$, such that

$$\|x(\mu^* + \bar{\bar{\mu}}(\mu)) - x^*\| \leq \bar{C} \sum_{k=1}^{p} \|f_k(x^*, \mu^* + \bar{\bar{\mu}}(\mu))\|^{\frac{1}{k}}.$$

From this we have

$$\|x(\mu) - x^*\| \leq \bar{C} \sum_{k=1}^{p} \|f_k(x^*, \mu^* + \bar{\bar{\mu}}(\mu))\|^{\frac{1}{k}} \leq \bar{\bar{\bar{C}}} \sum_{k=1}^{p} \|f_k(x^*, \mu^* + \bar{\bar{\mu}}(\mu) + \tilde{\mu}(\mu))\|^{\frac{1}{k}}$$

$$= \bar{\bar{\bar{C}}} \sum_{k=1}^{p} \|f_k(x^*, \mu)\|^{\frac{1}{k}},$$

where

$$\mu = \mu^* + \bar{\bar{\mu}}(\mu) + \tilde{\mu}(\mu)). \tag{45}$$

We used here the following fact

$$\|f_k(x^*, \mu^* + \bar{\bar{\mu}}(\mu))\| \leq 2\|f_k(x^*, \mu^* + \bar{\bar{\mu}}(\mu) + \tilde{\mu}(\mu))\| = 2\|f_k(x^*, \mu)\|, \tag{46}$$

since

$$\|\tilde{\mu}(\mu))\| = o(\sum_{k=1}^{p} \|f_k(x^*, \mu^* + \bar{\bar{\mu}}(\mu))\|^{\frac{1}{k}})$$

where μ is sufficiently small.

Then we substantiated that (10) holds. From this immediately yields that (9) is true for $\omega(\mu) = x(\mu) - x^*$ and $\|\omega(\mu)\| = o(\mu - \mu^*)$.

The last element of the proof, i.e. the continuity of $x(\mu)$ follows from the modified form of Michael selection Theorem 3. Therefore the multimapping Φ : $C(V_\varepsilon(0)) \times V_\varepsilon(0) \to 2^{X \times M}$, which we defined in (25) by the formula

$$\Phi(x, \tilde{\mu}) = (x, \tilde{\mu}) - \{\Psi_p(h)\}^{-1} F(x^* + x, \mu^* + t\bar{\mu} + \tilde{\mu}),$$

give us the continuity selector, i.e. we can choose the continuity solutions $(x(t\bar{\mu}), \tilde{\mu}(t\bar{\mu}))$ of F. From continuity of the function $x(t\bar{\mu})$ follows continuity of the function $x(\mu)$.

This finishes the proof of the theorem.

Remark 5. If we assume that the spaces $X \times M$ and Z are finite dimensional, we can prove existence of continuous function $x(\mu)$ by consideration the following contraction process

$$(x_{k+1}, \tilde{\mu}_{k+1}) = (x_k, \tilde{\mu}_k) - \{\Psi_p(h)\}_R^{-1} F(x^* + x_k, \mu^* + t\bar{\mu} + \tilde{\mu}_k), \tag{47}$$

where $\{\Psi_p(h)\}_R^{-1} z = (x_z, \mu_z)$ is the right inverse operator and

$$\|(x_z, \mu_z)\| = \min_{\Psi_p(h)(x,\mu)=z} \|(x, \mu)\|. \tag{48}$$

Such a process will converge to the continuity mapping $x(t\bar{\mu})$.

Analogously we can prove the following two theorems.

Theorem 6 (Implicit function theorem for nontrivial kernel). *Let* $F(x,\mu) \in C^{p+1}(X \times M)$, $F : X \times M \to Z$, *where* M *is finite dimensional space,* X, Z *are Banach spaces. Assume that* $F(x^*, \mu^*) = 0$ *and* $\forall \bar{\mu} \in M, \|\bar{\mu}\| = 1$; $(0, \bar{\mu}) \in \bigcap_{k=1}^{p} \operatorname{Ker}^k f_k^{(k)}(x^*, \mu^*)$ *that is*

$$\|\{f_1'(x^*, \mu^*) + f_2''(x^*, \mu^*)[0, \bar{\mu}] + \ldots + f_p^{(p)}(x^*, \mu^*)[0, \bar{\mu}]^{p-1}\}_X^{-1}\| \leq C. \tag{49}$$

Then there exists the mapping $x = x(\mu)$, $\mu \in V_\varepsilon(\mu^*)$, $x(\mu) \in C(V_\varepsilon(\mu^*))$, $\varepsilon > 0$ *sufficiently small, such that* $F(x(\mu), \mu) = 0$ *and*

$$x(\mu) = x^* + \omega(\mu), \quad \|\omega(\mu)\| = o(\|\mu - \mu^*\|), \tag{50}$$

$$\|x(\mu) - x^*\| \leq C \sum_{k=1}^{p} \|f_k(x^*, \mu)\|_{Z_k}^{\frac{1}{k}}, \quad \forall \mu \in V_\varepsilon(\mu^*). \tag{51}$$

Theorem 7. *Let* $F(x, \mu) \in C^{p+1}(X \times M)$, $F : X \times M \to Z$, *where* M *is finite dimensional space,* X, Z *are Banach spaces. Let for* $h_\mu \neq 0$, $h_\mu \in V_\varepsilon(\mu^*)$ *there exists* $\bar{h}_x \in X$, $\|\bar{h}_x\| \leq c < \infty$, *such that* F *is p-regular along* $\bar{h} = [\bar{h}_x, \bar{h}_\mu]$, *that is*

$$\|\{f_1'(x^*, \mu^*) + f_2''(x^*, \mu^*)[\bar{h}] + \ldots + \ldots + f_p^{(p)}(x^*, \mu^*)[\bar{h}]^{p-1}\}_X^{-1}\| \leq C, \tag{52}$$

$\bar{h} \in \bigcap_{k=1}^{p} \operatorname{Ker}^k f_k^{(k)}(x^*, \mu^*)$, $\bar{h}_\mu = \frac{h_\mu}{\|h_\mu\|}$.
Then there exists the mapping $x = x(\mu)$, $\mu \in V_\varepsilon(\mu^*)$, $x(\mu) \in C(V_\varepsilon(\mu^*))$, $\varepsilon > 0$ *sufficiently small, such that* $F(x(\mu), \mu) = 0$ *and*

$$\mu = \mu^* + h_\mu, \quad x(\mu) = x^* + c(\mu)\bar{h}_x + \omega(\mu), \quad \|\omega(\mu)\| = o(\|\mu\|), \|c(\mu)\| = \|\mu\|, \tag{53}$$

$$\|x(\mu) - x^*\| \leq C \sum_{k=1}^{p} \|f_k(x^* + h_x, \mu)\|_{Z_k}^{\frac{1}{k}}. \tag{54}$$

References

1. Prisinska, A., Tret'yakov, A.: The p-order maximum principle for an irregular optimal control problem. Comput. Math. Math. Phys. **57**(9), 1453–1458 (2017)
2. Brezhneva, O.A., Tret'yakov, A.A.: Implicit function theorems for nonregular mappings in Banach spaces. Exit from singularity. In: Banach Spaces and Their Applications in Analysis, pp. 285–302. de Gruyter (2007)

3. Ioffe, A.D., Tihomirov, V.M.: Theory of Extremal Problems. North-Holland, Amsterdam (1979)
4. Izmailov, A.F., Tret'yakov, A.A.: Factor-Analysis of Nonlinear Mappings. Nauka, Moscow (1994). (in Russian)
5. Izmailov, A.F., Tret'yakov, A.A.: 2-Regular Solutions of Nonlinear Problems. Theory and Numerical Methods. Nauka, Moscow (1999). (in Russian)
6. Medak, B., Tret'yakov, A.A.: Existence of periodic solutions to nonlinear p-regular boundary value problem. Bound. Value Probl. **2015**, 91 (2015). https://doi.org/10.1186/s13661-015-0360-2
7. Medak, B., Tret'yakov, A.A.: Teoria p-regularnosti. Analiz i Prilozenia. Fizmatlit, Moskva (2017)
8. Michael, E.A.: Continuous selector. Ann. Math. **64**, 562–580 (1956)
9. Prusinska, A., Tret'yakov, A.A.: p-regularity theory. tangent cone description in the singular case. Ukrainian Math. J. **67**(8), 1236–1246 (2016)
10. Tret'yakov, A.A.: The implicit function theorem in degenerate problems. Russ. Math. Surv. **42**, 179–180 (1987)
11. Tret'yakov, A.A., Marsden, J.E.: Factor analysis of nonlinear mappings: p-regularity theory. Commun. Pure Appl. Anal. **2**(4), 425–445 (2003)

Optimization of Kernel Estimators of Probability Densities

Anton V. Voytishek[1,2(✉)] 🆔 and Tatyana E. Bulgakova[2]

[1] Institute of Computational Mathematics and Mathematical Geophysics SD RAS,
Prospect Akademika Lavrentyeva, 6, Novosibirsk, Russia
`vav@osmf.sscc.ru`
[2] Novosibirsk State University, Pirogova Street, 2, Novosibirsk, Russia
`tatyana.bulgakova@gmail.com`

Abstract. The constructive kernel algorithm for approximation of probability densities using the given sample values is proposed. This algorithm is based on the approaches of the theory of the numerical functional approximation. The critical analysis of the optimization criterion for the kernel density estimators (based on decrease of upper boundary of mean square error) is conducted. It is shown that the constructive kernel algorithm is nearly equal to the randomized projection-mesh functional numerical algorithm for approximation of the solution of the Fredholm integral equation of the second kind. In connection with this it is proposed to use the criterion of conditional optimization of functional algorithms for the kernel algorithm for approximation of probability densities. This criterion is based on minimization of the algorithm's cost for the fixed level of error. The corresponding formulae for the conditionally optimal parameters of the kernel algorithm are derived.

Keywords: Kernel estimators for approximation of probability densities · Numerical mesh approximation of functions · Optimization · Conditional optimization of randomized functional numerical algorithms · Multi-dimensional analogue of the polygon of frequencies method

1 The Constructive Kernel Algorithm for Approximation of Probability Densities

Computing Estimators of Probability Densities is an important problem is Data Sciences. In particular, many machine learning techniques including topological and dynamical data assimilation/training methods need to do this kind of estimation computationally efficiently.

This work was conducted within the framework of the budget project 0315-2019-0002 for ICMMG SB RAS.

M. Jaćimović et al. (Eds.): OPTIMA 2019, CCIS 1145, pp. 254–266, 2020.
https://doi.org/10.1007/978-3-030-38603-0_19

In the classical paper [1], the nonparametric estimator of a probability distribution density $f_{\hat{\xi}}(\mathbf{x})$, $\mathbf{x} \in X \subset \mathbf{R}^d$ of the form

$$f_{\hat{\xi}}(\mathbf{x}) \approx Z_n(\mathbf{x}) = \frac{1}{n} \sum_{j=1}^{n} \kappa^{(\mathbf{x})} \left(\hat{\xi}_j \right), \qquad (1.1)$$

using the sample values $\left\{ \hat{\xi}_1, ..., \hat{\xi}_n \right\} \subset \mathbf{R}^d$ from this distribution is considered. Here $\kappa^{(\mathbf{x})}(\mathbf{y})$ is some finite parametric, having the same shape for all values of the parameter $\mathbf{x}$ *kernel function*. The approximation (1.1) is called *the kernel estimator of the density* $f_{\hat{\xi}}(\mathbf{x})$.

For investigation of properties of the approximation (1.1), the following consequence of the large numbers law

$$Z_n(\mathbf{x}) = \frac{1}{n} \sum_{j=1}^{n} \kappa^{(\mathbf{x})} \left(\hat{\xi}_j \right) \approx \mathbf{E} \kappa^{(\mathbf{x})} \left(\hat{\xi} \right) = \int \kappa^{(\mathbf{x})}(\mathbf{y}) f_{\hat{\xi}}(\mathbf{y}) \, d\mathbf{y} \qquad (1.2)$$

for comparatively large n is used.

The evident constructive drawback of the kernel estimators theory (see, for example [1]) is related to absence of considerations about practical (firstly – numerical, computer) technique for global approximation of the function $f_{\hat{\xi}}(\mathbf{x})$. This technique can be based on the theory of mesh function approximation (see, for example, [2]). The corresponding numerical algorithm could look as follows.

Assume that the domain $X \subset \mathbf{R}^d$ is bounded and the mesh

$$X^{(M)} = \{\mathbf{x}_1, ..., \mathbf{x}_M\} \qquad (1.3)$$

is constructed in this domain, and also the mesh approximation of the form

$$f_{\hat{\xi}}(\mathbf{x}) \approx L^{(M)} f_{\hat{\xi}}(\mathbf{x}) = \sum_{i=1}^{M} w^{(i)} \left[f_{\hat{\xi}}(\mathbf{x}_1), ..., f_{\hat{\xi}}(\mathbf{x}_M) \right] \chi^{(i)}(\mathbf{x}) \qquad (1.4)$$

is considered. Here

$$\Xi^{(M)} = \left\{ \chi^{(1)}(\mathbf{x}), ..., \chi^{(M)}(\mathbf{x}) \right\}, \qquad (1.5)$$

is a set of given basic functions (as a rule, these functions depend on the mesh (1.3)). The form of functions (1.5) defines the type of the approximation (1.4). The values

$$W^{(M)} = \left\{ w^{(1)} \left[f_{\hat{\xi}}(\mathbf{x}_1), ..., f_{\hat{\xi}}(\mathbf{x}_M) \right], ..., w^{(M)} \left[f_{\hat{\xi}}(\mathbf{x}_1), ..., f_{\hat{\xi}}(\mathbf{x}_M) \right] \right\} \qquad (1.6)$$

are the coefficients which are equal to some combinations of values of the function $f_{\hat{\xi}}(\mathbf{x})$ in nodes of the mesh (1.3); more often

$$w^{(i)} \left[f_{\hat{\xi}}(\mathbf{x}_1), ..., f_{\hat{\xi}}(\mathbf{x}_M) \right] = f_{\hat{\xi}}(\mathbf{x}_i); \quad i = 1, ..., M. \qquad (1.7)$$

For simplicity of further considerations we assume that the domain X is equal to some cuboid, and the mesh (1.3) is uniform rectangular with the step h:

$$\mathbf{x}_i = \left(j_i^{(1)} h, ..., j_i^{(d)} h \right); \quad j_i^{(k)} \text{ are integer numbers; } \quad k = 1, ..., d; \quad i = 1, ..., M. \tag{1.8}$$

Algorithm 1. *Calculate the values* $\tilde{f}_{\hat{\xi}}^{(\mathbf{x}_i)}(n) = Z_n(\mathbf{x}_i)$; $i = 1, ..., M$ *with respect to the formulae of the form* (1.1) *and approximate the function* $f_{\hat{\xi}}(\mathbf{x})$ *with respect to the formula of the form* (1.4):

$$f_{\hat{\xi}}(\mathbf{x}) \approx L^{(M)} \tilde{f}_{\hat{\xi}}(\mathbf{x}) = \sum_{i=1}^{M} w^{(i)} \left[\tilde{f}_{\hat{\xi}}^{(\mathbf{x}_1)}(n), ..., \tilde{f}_{\hat{\xi}}^{(\mathbf{x}_M)}(n) \right] \chi^{(i)}(\mathbf{x}). \tag{1.9}$$

The Algorithm 1 is based on the approximate equalities

$$\int f_{\hat{\xi}}(\mathbf{y}) \kappa^{(\mathbf{x}_i)}(\mathbf{y}) \, d\mathbf{y} \approx f_{\hat{\xi}}(\mathbf{x}_i); \quad i = 1, ..., M, \tag{1.10}$$

which are in turn based on the equalities (1.1), (1.2).

2   Optimization Using the Upper Boundary of Error: Choice of the Kernel Function and the Blur Coefficient

In the paper [1], the following optimization of the approximation (1.1) is provided. It is proposed to use the kernel function $\kappa^{(\mathbf{x})}(\mathbf{y})$ in the form

$$\kappa^{(\mathbf{x})}(\mathbf{y}) = \prod_{s=1}^{d} \frac{1}{h^{(s)}(n)} \hat{\kappa}^{(s)} \left(\frac{x^{(s)} - y^{(s)}}{h^{(s)}(n)} \right). \tag{2.1}$$

Here the positive (generally speaking dependent on n) numbers $h^{(s)}(n)$; $s = 1, ..., d$ define the domain (*the blur coefficients*) of the kernel function $\kappa^{(\mathbf{x})}(\mathbf{y})$. The bounded in total even $(\hat{\kappa}^{(s)}(y) = \hat{\kappa}^{(s)}(-y))$ functions $\hat{\kappa}^{(s)}(y)$ have the unit second moment and finite m-th moments:

$$\int_{-\infty}^{+\infty} y^2 \hat{\kappa}^{(s)}(y) \, dy = 1, \quad \int_{-\infty}^{+\infty} y^m \hat{\kappa}^{(s)}(y) \, dy < \infty; \quad m > 2. \tag{2.2}$$

The informative results on optimization of the approximation (1.1) with the kernel function (2.1) are obtained only for the case

$$h^{(1)}(n) = ... = h^{(d)}(n) \equiv \hat{h}(n); \quad \hat{\kappa}^{(1)}(y) = ... = \hat{\kappa}^{(d)}(y) \equiv \hat{\kappa}(y)$$

(see, in particular, [1]); the case of various $\{h^{(s)}(n)\}$, $\{\hat{\kappa}^{(s)}(y)\}$; $s = 1, ..., d$ is not studied enough.

In the paper [1], in frames of the asymptotic (for $n \to \infty$) approach, the global mean square error

$$\left(\hat{\delta}^{(\mathbf{L_2})}\right)^2 = \frac{\int \mathbf{E}\left[f_{\hat{\xi}}(\mathbf{x}) - Z_n(\mathbf{x})\right]^2 dx}{Q}; \quad Q = \int f_{\hat{\xi}}^2(\mathbf{x})\, dx \qquad (2.3)$$

was minimized. The decomposition of the function $f_{\hat{\xi}}\left(x^{(1)} + \hat{h}(n)y^{(1)}, ..., x^{(d)} + \right.$ $\left. +\hat{h}(n)y^{(d)}\right)$ with respect to all independent variables $\mathbf{y} = \left(y^{(1)}, ..., y^{(d)}\right)$ in the point $\mathbf{x} = \left(x^{(1)}, ..., x^{(d)}\right)$ into the Taylor series was used and the following asymptotic approximation

$$\left(\hat{\delta}^{(\mathbf{L_2})}\right)^2 \sim \frac{n^{-1}\hat{h}^{-d}(n)I^d + (1/4)\hat{h}^4(n)J}{Q}; \qquad (2.4)$$

$$I = \int_{-\infty}^{+\infty} \hat{\kappa}^2(y)\, dy; \quad J = \int .. \int \left[\sum_{l=1}^{d} \frac{\partial^2 f_{\hat{\xi}}\left(x^{(1)}, .., x^{(d)}\right)}{\partial\left(x^{(l)}\right)^2}\right]^2 dx^{(1)}..dx^{(d)} \qquad (2.5)$$

was obtained.

The minimization of the obtained approximation (2.4) of the value (2.3) and the choice of the function $\hat{\kappa}(y)$, which satisfies the condition (2.2) and provides the minimal value I from (2.5), give the following optimal value of the blur coefficient and the optimal form of the function $\hat{\kappa}(y)$:

$$\hat{h}_{opt}(n) = \left(\frac{d \times I^d}{n \times J}\right)^{1/(d+4)}, \quad \hat{\kappa}_{opt}(y) = \begin{cases} \frac{3}{4\sqrt{5}} - \frac{3y^2}{20\sqrt{5}} & \text{for } |y| \leq \sqrt{5}, \\ 0 & \text{for } |y| > \sqrt{5}. \end{cases} \qquad (2.6)$$

Concerning the presented in [1] optimization of the approximation (1.1) of the density $f_{\hat{\xi}}(\mathbf{x})$, the following criticisms can be formulated. Firstly, the continuous approximation is considered, while it is more constructive, as it was mentioned above, to use the mesh approximation of the function $f_{\hat{\xi}}(\mathbf{x})$, presented in the Algorithm 1. Secondly, the formula (2.6) for $\hat{h}_{opt}(n)$ includes the indefinite constant J from (2.5). The approximation of this constant is equal to the separate (and rather difficult) problem. Thirdly, the presented optimization of the approximation (1.1) is connected to minimization of the asymptotic approximate upper boundary of error, nevertheless, more objective is the following criterion.

Criterion 1. *The best approximation of the function $f_{\hat{\xi}}(\mathbf{x})$ gives the fixed error level in minimal time (with least computer cost).*

The question about how the use of the parameters (2.6) in the Algorithm 1 (in conjunction with the choice of the step h of the mesh (1.8)), which corresponds to the Criterion 1, requires the separate detailed study.

The considerations of the Sects. 3 and 4 of this paper give the possibilities of constructive use of the Criterion 1 for optimization of the Algorithm 1.

These considerations are based on the mapping between the Algorithm 1 and the randomized projection-mesh functional algorithm (see further the Algorithm 2); for this algorithm the so called *theory of conditional optimization*, based on the Criterion 1, is elaborated.

3   The Randomized Projection-Mesh Functional Algorithm for Solving of the Fredholm Integral Equation of the Second Kind

In recent years, the theory of randomized functional algorithms is developed (especially in Novosibirsk scientific school of Monte Carlo methods); see, in particular, [3–7]. The most informative examples of these algorithms are related to approximation of the unknown solution $\varphi(\mathbf{x})$, $\mathbf{x} \in \mathbf{R}^d$ of the integral Fredholm equation of the second kind

$$\varphi(\mathbf{x}) = \int k(\mathbf{x}', \mathbf{x})\varphi(\mathbf{x}')\, d\mathbf{x}' + f(\mathbf{x}) \quad \text{or} \quad \varphi = K\varphi + f \qquad (3.1)$$

in a bounded domain $X \subset \mathbf{R}^d$; here $k(\mathbf{x}', \mathbf{x})$ (the kernel of the integral operator K) and $f(\mathbf{x})$ (the free term of the equation) are the given functions.

By analogy with considerations of the Sect. 1 of this paper (see, in particular, the formula (1.4)), for approximation of the function $\varphi(\mathbf{x})$ we use the representations of classical theory of numerical function approximation (see, for example, [2]), which have the common form

$$\varphi(\mathbf{x}) \approx L^{(M)}\varphi(\mathbf{x}) = \sum_{i=1}^{M} w^{(i)}\chi^{(i)}(\mathbf{x})$$

for some specially selected set of basic functions (1.5) and coefficients

$$\mathbf{W}^{(M)} = \left\{ w^{(1)}, ..., w^{(M)} \right\}, \qquad (3.2)$$

which are defined as functionals on the unknown approximated function $\varphi(\mathbf{x})$.

For the randomized functional algorithms, the coefficients (3.2) are calculated approximately using the Monte Carlo method with the test numbers n_i: $w^{(i)} \approx \tilde{w}^{(i)}(n_i)$ (in this paper we investigate the case $n_1 = ... = n_M \equiv n$), and the approximation

$$\varphi(\mathbf{x}) \approx L^{(M)}\tilde{\varphi}(\mathbf{x}) = \sum_{i=1}^{M} \tilde{w}^{(i)}(n)\chi^{(i)}(\mathbf{x})$$

is considered.

In the recent papers [8,9] we have proposed the new (to compare with the works [3–7]) classification of the randomized functional algorithms for approximation of the solution $\varphi(\mathbf{x})$ of the Eq. (3.1). We have distinguished *the mesh, the projection and the projection-mesh algorithms* (the type of a method is defined by the choice of the basic functions (1.5) and the coefficients (3.2)). In these papers, we also have presented the considerations why the mesh and the projection randomized functional algorithms can be non-effective (or even unrealizable) for solution of practically important problems related to solutions of integral equations of the form (3.1). In particular, for the theoretically attractive *mesh dependent test method*, the smoothness of the kernel $k(\mathbf{x}', \mathbf{x})$ of the integral operator K is needed. But the most part of kernels in applied problems has the integrable singularities (up to delta-functions) and even can not be calculated explicitly. *The mesh adjoint random walk method* is too numerically laborious because of necessity for simulating of individual set of trajectories of the corresponding applied Markov chains for every node $\mathbf{x}_i$ of the introduced mesh (1.3) in the domain X. *The projection methods* have fairly obvious numerical instability.

The projection-mesh randomized functional algorithms have no such flows. For these algorithms (as for mesh methods) the coefficients $w^{(i)} = w^{(i)}\left(\boldsymbol{\varphi}^{(M)}\right)$ from (3.2) are equal to some combinations of values

$$\boldsymbol{\varphi}^{(M)} = \{\varphi(\mathbf{x}_1), ..., \varphi(\mathbf{x}_M)\} \tag{3.3}$$

in nodes of the mesh (1.3). The "projection" part of algorithms is related to the special way of approximate calculation of the values (3.3). Choose the finite, having the same shape for all $\{\mathbf{x}_1, ..., \mathbf{x}_M\}$ functions

$$\mathbf{K}^{(M)} = \left\{\kappa^{(\mathbf{x}_1)}(\mathbf{y}), ..., \kappa^{(\mathbf{x}_M)}(\mathbf{y})\right\}; \tag{3.4}$$

essentially these functions are the versions of the kernel function $\kappa^{(\mathbf{x})}(\mathbf{y})$ from (1.1) for various values of the parameter $\mathbf{x}$.

The functions (3.4) depend on the mesh (1.3) such that

$$\int \varphi(\mathbf{y})\kappa^{(\mathbf{x}_i)}(\mathbf{y})\, d\mathbf{y} \approx \varphi(\mathbf{x}_i); \quad i = 1, ..., M \tag{3.5}$$

(these are the analogs of the approximate equalities (1.10)). Then we use the fact that the approximations (3.5) of the values (3.3) are equal to linear functionals on the solution $\varphi(\mathbf{x})$ of the Eq. (3.1). For such functionals we can construct *the main estimators* (or *the Monte Carlo collision estimates* – see, for example, the Chap. 4 of the textbook [5]) based on numerical simulation of the trajectories

$$\xi_j^{(0)}, \xi_j^{(1)}, ..., \xi_j^{(N_j)}; \quad j = 1, ..., n \tag{3.6}$$

of *the applied Markov chain*

$$\boldsymbol{\xi}^{(0)}, \boldsymbol{\xi}^{(1)}, ..., \boldsymbol{\xi}^{(N)}, \tag{3.7}$$

or homogeneous Markov chain terminated with unit probability, with the initial density $\pi(\mathbf{x})$ and the transition function $p(\mathbf{x}', \mathbf{x}) = r(\mathbf{x}', \mathbf{x}) \times \left[1 - p^{(a)}(\mathbf{x}')\right]$ (here $r(\mathbf{x}', \mathbf{x})$ is the probability transition density and $0 \leq p^{(a)}(\mathbf{x}') \leq 1$ defines the probability of a trajectory break; correspondingly, N is a random number of the break state).

Algorithm 2 [8,9]. *Simulate n trajectories (3.6) of the applied Markov chain (3.7) and get the values*

$$\tilde{\varphi}^{(\mathbf{x}_i)}(n) = \frac{1}{n} \sum_{j=1}^{n} \sum_{m=0}^{N_j} Q_j^{(m)} \kappa^{(\mathbf{x}_i)} \left(\boldsymbol{\xi}_j^{(m)} \right); \quad i = 1, ..., M;$$

here the weights $\left\{ Q_j^{(m)} \right\}$ are calculated with respect to the following recurrent formulae:

$$Q_j^{(0)} = \frac{f\left(\boldsymbol{\xi}_j^{(0)}\right)}{\pi\left(\boldsymbol{\xi}_j^{(0)}\right)}; \quad Q_j^{(m)} = Q_j^{(m-1)} \times \frac{k\left(\boldsymbol{\xi}_j^{(m-1)}, \boldsymbol{\xi}_j^{(m)}\right)}{p\left(\boldsymbol{\xi}_j^{(m-1)}, \boldsymbol{\xi}_j^{(m)}\right)}; \quad j = 1, ..., n; \quad m = 1, ..., N_j.$$

Then approximate the function $\varphi(\mathbf{x})$ with respect to the formula of the form (1.9):

$$\varphi(\mathbf{x}) \approx L^{(M)} \tilde{\varphi}(\mathbf{x}) = \sum_{i=1}^{M} w^{(i)} \left[\tilde{\varphi}^{(\mathbf{x}_1)}(n), ..., \tilde{\varphi}^{(\mathbf{x}_M)}(n) \right] \chi^{(i)}(\mathbf{x}).$$

Comparison of the Algorithms 1 and 2 gives the following important conclusion.

Remark 1. *The kernel Algorithm 1 for approximation of a probability density $f_{\hat{\boldsymbol{\xi}}}(\mathbf{x})$, based on approaches of the theory of mesh function approximation, is constructively equal to the randomized projection-mesh functional Algorithm 2 for approximation of the solution $\varphi(\mathbf{x})$ of Fredholm integral equation of the second kind (3.1).*

The difference between Algorithms 1 and 2 is defined by the distinction of forms of Monte Carlo estimators for approximate calculation of functionals (1.10) and (3.5) (which is related to the certain difference between functions $\varphi(\mathbf{x})$ and $f_{\hat{\boldsymbol{\xi}}}(\mathbf{x})$). The difference is also related to the fact that for the problem of approximation of the density $f_{\hat{\boldsymbol{\xi}}}(\mathbf{x})$, the sample $\left\{ \hat{\boldsymbol{\xi}}_1, ..., \hat{\boldsymbol{\xi}}_n \right\}$ is considered *to be given* (and the number n of sample values is fixed and cannot be increased), but for the function $\varphi(\mathbf{x})$ the number n of the simulated trajectories (3.6) of the applied Markov chain (3.7) *may vary*.

In connection with the main conclusion of the Remark 1, we can formulate the following considerations.

Remark 2. *For development of the theory of construction and conditional optimization of the randomized projection-mesh functional Algorithm 2 it is possible to use the considerations of the theory of kernel estimators of probability densities from the paper [1] (see also the Section 2 of this paper).*

The Algorithm 2 is studied in detail in the works [4, 7] only for special case, when the "absolutely stable" finite functions of *the multi-linear approximation* (or Strang – Fix approximation [10] with the basis producing function $\beta^{(1)}(u)$, which is equal to the B-spline of the first order) on a regular mesh (1.8)

$$\chi^{(i)}(\mathbf{x}) = \beta^{(1)} \left(\frac{x^{(1)}}{h} - j_i^{(1)} \right) \times ... \times \beta^{(1)} \left(\frac{x^{(d)}}{h} - j_i^{(d)} \right); \tag{3.8}$$

$$\beta^{(1)}(u) = \begin{cases} u + 1 & \text{for } -1 \leq u \leq 0; \\ -u + 1 & \text{for } 0 \leq u \leq 1; \\ 0 & \text{otherwise} \end{cases} \tag{3.9}$$

are used as basic functions (1.5). Moreover, the kernel function from the formulae (3.4), (3.5) has the simple form

$$\kappa^{(\mathbf{x})}(\mathbf{y}) = \begin{cases} \frac{1}{h^d} & \text{for } \mathbf{y} \in \Delta^{(\mathbf{x})}, \\ 0 & \text{otherwise}, \end{cases} \tag{3.10}$$

where $\Delta^{(\mathbf{x})} = \{\mathbf{y} = (y^{(1)}, ..., y^{(d)}) : x^{(s)} - h/2 \leq y^{(s)} \leq x^{(s)} + h/2; \; s = 1, ..., d; \; \mathbf{x} = (x^{(1)}, ..., x^{(d)}) \}$, thus, for this case we have

$$\hat{h}_{opt}(n) = h, \quad \hat{\kappa}_{opt}(y) = \begin{cases} 1 & \text{for } |y| \leq 1/2, \\ 0 & \text{for } |y| > 1/2. \end{cases} \tag{3.11}$$

For this case, the approximations of the coefficients (3.2) have the simplest form

$$w^{(i)} \left(\tilde{\varphi}^{(\mathbf{x}_1)}(n), ..., \tilde{\varphi}^{(\mathbf{x}_M)}(n) \right) = \tilde{\varphi}^{(\mathbf{x}_i)}(n). \tag{3.12}$$

The Algorithm 2 with functions (3.8)–(3.11) and approximation coefficients (3.12) is called in [4–7] as *the multi-dimensional analogue of the polygon of frequencies method.*

According to considerations of the Sect. 2 of this paper, it can be interesting to provide the separate investigation of the Algorithm 2 which includes the considerations of the conditional optimization theory (see [4–7], and the Sect. 4 of this paper) for the case when the kernel function (2.1) with the blur coefficient and the function $\hat{\kappa}(y)$ of the form (2.6) are used instead of (3.11) in the approximate equalities of the form (3.5).

4   Conditional Optimization of the Kernel Algorithm

In connection with the Remark 1, we can formulate the following considerations.

Remark 3. *For the kernel Algorithm 1, used for approximation of a probability density $f_{\hat{\xi}}(\mathbf{x})$, we can use considerations of the theory of conditional optimization of the randomized projection-mesh functional Algorithm 2.*

The common scheme of the conditional optimization looks as follows (see, for example, [4–7]). The problem of the coordinated choice of the parameters M and n (the number of nodes of the mesh (1.3) and the sample values from the formula (1.1)) for the investigated functional algorithm (for example, for the Algorithm 1) is stated. This choice must guarantee the given level $\gamma > 0$ of the error of approximation of the investigated function (for example, the approximation (1.9) from the Algorithm 1) together with minimal computer time costs $S(M, n)$. Thus, this approach quite consistent to the "objective" Criterion 1, which was formulated in the Sect. 2 of this paper.

Construct an upper boundary $UP^{(\mathbf{B})}(M, n)$ of the algorithm's error $\delta^{(\mathbf{B})}(M, n)$ for the used normalized functional space $\mathbf{B}(X)$, which depends on the parameters M and n:

$$\delta^{(\mathbf{B})}(M, n) = \left\| f_{\hat{\xi}} - L^{(M)} \tilde{f}_{\hat{\xi}} \right\|_{\mathbf{B}(X)} \leq UP^{(\mathbf{B})}(M, n). \qquad (4.1)$$

This two-parameter function is equated to the value γ. From the equation of the form

$$UP^{(\mathbf{B})}(M, n) = \gamma \qquad (4.2)$$

one parameter (for example, n) is presented in terms of another: $n = \psi(M)$. The last formula is substituted into the expression for the cost $S(M, n)$ (which also depends on the parameters M and n). As the result we get the function $\tilde{S}(M)$ of single independent variable M. We investigate this function for minimum using the well-known methods of mathematical or numerical analysis. The found parameters $M_{min}^{(\mathbf{B})}(\gamma) = M_{opt}^{(\mathbf{B})}(\gamma), n_{opt}^{(\mathbf{B})} = \psi\left[M_{opt}^{(\mathbf{B})}(\gamma)\right]$ are declared to be *conditionally optimal parameters* of the corresponding functional algorithm (for example, Algorithm 1).

"The conditionality" of this optimization method is related to the fact that in the left part of the Eq. (4.2) we use not the error $\delta^{(\mathbf{B})}(M, n)$ itself but the upper boundary $UP^{(\mathbf{B})}(M, n)$. By the way, the evaluation of the quality of a particular numerical algorithm by the upper boundary of error is used in the overwhelming majority of theoretical considerations of numerical mathematics (see, for example, [2]), thus, everywhere these considerations are about "the conditional optimality" of the studied numerical schemes.

When studying the error $\delta^{(\mathbf{B})}(M, n)$, it is necessary to choose both the corresponding normalized functional space $\mathbf{B}(X)$ and the probabilistic sense of the satisfying of the inequality (3.1) (after all $\delta^{(\mathbf{B})}(M, n)$ is a random variable). Following the theory of the classical numerical analysis (see, for example, [2]) we consider the spaces $\mathbf{L}_2(X)$ and $\mathbf{C}(X)$ as the normalized functional space $\mathbf{B}(X)$.

For the well developed (see, for example, [3–7]) $\mathbf{L}_2$-*approach* the convergence in mean of the error

$$\delta^{(\mathbf{L}_2)}(M, n) = \left\| f_{\hat{\xi}} - L^{(M)} \tilde{f}_{\hat{\xi}} \right\|_{\mathbf{L}_2(X)} = \left(\int_X \left[f_{\hat{\xi}}(\mathbf{x}) - L^{(M)} \tilde{f}_{\hat{\xi}}(\mathbf{x}) \right]^2 d\mathbf{x} \right)^{1/2}$$

to zero for $M, n \to \infty$ is considered, and the upper boundaries $UP^{(\mathbf{L_2})}(M, n)$ such that

$$\left[\mathbf{E}\delta^{(\mathbf{L_2})}(M, n)\right]^2 \leq UP^{(\mathbf{L_2})}(M, n)$$

are constructed.

For the **C**-*approach* [4–7] the value

$$\delta^{(\mathbf{C})}(M, n) = \left\|f_{\hat{\xi}} - L^{(M)}\tilde{f}_{\hat{\xi}}\right\|_{\mathbf{C}(X)} = \sup_{x \in X}\left|f_{\hat{\xi}}(\mathbf{x}) - L^{(M)}\tilde{f}_{\hat{\xi}}(\mathbf{x})\right|$$

is bounded above in probability:

$$\mathbf{P}\left[\delta^{(\mathbf{C})}(M, n) \leq UP^{(\mathbf{C})}(M, n)\right] > 1 - \varepsilon$$

for some comparatively small $\varepsilon > 0$.

Note that for the $\mathbf{L_2}$-approach, the comparatively "weak" integral norm $\|\cdot\|_{\mathbf{L_2}(X)}$ of the space $\mathbf{L_2}(X)$ and "strong" probabilistic convergence of error to zero (in mean) are used. In turn, in the **C**-approach for the "strict' norm $\|\cdot\|_{\mathbf{C}(X)}$, the comparatively "weak" error convergence to zero (in probability) is chosen.

Together with approximations $L^{(M)}f_{\hat{\xi}}(x)$, $L^{(M)}\tilde{f}_{\hat{\xi}}(x)$ of the function $f_{\hat{\xi}}(x)$, for which the exact and approximate values in nodes of the mesh (1.3) are used (see the formulae (1.4), (1.9)), consider the function

$$L^{(M)}\bar{f}_{\hat{\xi}}(x) = \sum_{i=1}^{M} w^{(i)}\left[\mathbf{E}\kappa^{(\mathbf{x_1})}\left(\hat{\boldsymbol{\xi}}\right), ..., \mathbf{E}\kappa^{(\mathbf{x_M})}\left(\hat{\boldsymbol{\xi}}\right)\right]\chi^{(i)}(\mathbf{x}).$$

For **C**- and $\mathbf{L_2}$-approaches we can divide the error into three components: *the deterministic* $\delta_{det}^{(\mathbf{B})}(M)$, *the stochastic* $\delta_{stoch}^{(\mathbf{B})}(M, n)$ *and the bias component* $\delta_{bias}^{(\mathbf{B})}(M)$.

In particular, for **C**-approach, using the triangle inequality, we get

$$\delta^{(\mathbf{C})}(M, n) \leq \delta_{det}^{(\mathbf{C})}(M) + \delta_{stoch}^{(\mathbf{C})}(M, n) + \delta_{bias}^{(\mathbf{C})}(M); \ \delta_{det}^{(\mathbf{C})}(M) = \left\|f_{\hat{\xi}} - L^{(M)}f_{\hat{\xi}}\right\|_{\mathbf{C}(X)},$$

$$\delta_{stoch}^{(\mathbf{C})}(M, n) = \left\|L^{(M)}\bar{f}_{\hat{\xi}} - L^{(M)}f_{\hat{\xi}}\right\|_{\mathbf{C}(X)}, \ \delta_{bias}^{(\mathbf{C})}(M) = \left\|L^{(M)}f_{\hat{\xi}} - L^{(M)}\bar{f}_{\hat{\xi}}\right\|_{\mathbf{C}(X)}.$$

For $\mathbf{L_2}$-approach, using the Cauchy – Bunyakovsky inequality and the Fubini theorem, we get

$$\left[\mathbf{E}\delta^{(\mathbf{L_2})}(M, n)\right]^2 \leq \mathbf{E}\left(\int_X\left[f_{\hat{\xi}}(\mathbf{x}) - L^{(M)}\tilde{f}_{\hat{\xi}}(\mathbf{x})\right]^2 dx\right) \times \mathbf{E}1$$

$$= \int_X \mathbf{E}\left[f_{\hat{\xi}}(\mathbf{x}) - L^{(M)}\tilde{f}_{\hat{\xi}}(\mathbf{x})\right]^2 dx.$$

Then note that

$$\mathbf{E}\left[f_{\hat{\xi}}(\mathbf{x}) - L^{(M)}\tilde{f}_{\hat{\xi}}(\mathbf{x})\right]^2 = \mathbf{E}\Big(\left[f_{\hat{\xi}}(\mathbf{x}) - L^{(M)}f_{\hat{\xi}}(\mathbf{x})\right] + \left[L^{(M)}f_{\hat{\xi}}(\mathbf{x}) - L^{(M)}\bar{f}_{\hat{\xi}}(\mathbf{x})\right]$$
$$+ \left[L^{(M)}\bar{f}_{\hat{\xi}}(\mathbf{x}) - L^{(M)}\tilde{f}_{\hat{\xi}}(\mathbf{x})\right]\Big)^2 = \Big(\left[f_{\hat{\xi}}(\mathbf{x}) - L^{(M)}f_{\hat{\xi}}(\mathbf{x})\right]$$
$$+ \left[L^{(M)}f_{\hat{\xi}}(\mathbf{x}) - L^{(M)}\bar{f}_{\hat{\xi}}(\mathbf{x})\right]\Big)^2 + \mathbf{E}\left[L^{(M)}\bar{f}_{\hat{\xi}}(\mathbf{x}) - L^{(M)}\tilde{f}_{\hat{\xi}}(\mathbf{x})\right]^2 ;$$

here we take into account that $\mathbf{E}L^{(M)}\tilde{f}_{\hat{\xi}}(\mathbf{x}) = L^{(M)}\bar{f}_{\hat{\xi}}(\mathbf{x})$. Using the evident inequality $(a+b)^2 \le 2a^2 + 2b^2$, we obtain

$$\left[\mathbf{E}\delta^{(\mathbf{L}_2)}(M,n)\right]^2 \le 2\left(\delta_{det}^{(\mathbf{L}_2)}(M)\right)^2 + \delta_{stoch}^{(\mathbf{L}_2)}(M,n) + 2\left(\delta_{bias}^{(\mathbf{L}_2)}(M)\right)^2 ;$$
$$\delta_{det}^{(\mathbf{L}_2)}(M) = \left\|f_{\hat{\xi}} - L^{(M)}f_{\hat{\xi}}\right\|_{\mathbf{L}_2(X)}, \quad \delta_{stoch}^{(\mathbf{L}_2)}(M,n) = \int_X \mathbf{D}L^{(M)}\tilde{f}_{\hat{\xi}}(\mathbf{x})\,d\mathbf{x},$$
$$\delta_{bias}^{(\mathbf{L}_2)}(M) = \left\|L^{(M)}f_{\hat{\xi}} - L^{(M)}\bar{f}_{\hat{\xi}}\right\|_{\mathbf{L}_2(X)}.$$

By analogy of the reasoning of the works [4,7], it can be shown, that if the following four conditions

(1) *the regular mesh* (1.8) *with the step h is used;*
(2) *the numerically stable basis* (1.5) *is used, which enforces the inequality*

$$\delta_{bias}^{(\mathbf{C})}(M) = \left\|L^{(M)}f_{\hat{\xi}} - L^{(M)}\bar{f}_{\hat{\xi}}\right\|_{\mathbf{C}(X)} \le H_{Leb}\max_{i=1,\dots,M}\left|f_{\hat{\xi}}(\mathbf{x}_i) - \mathbf{E}\kappa^{(\mathbf{x}_i)}\right| \quad (4.3)$$

for the Lebesgue constant H_{Leb}, which is slightly more than one;
(3) *the coefficients* (1.6) *have the simplest form* (1.7);
(4) *the blur coefficient and the function $\hat{\kappa}(y)$ of the form* (3.11) *are used (that is, the kernel function* (3.10) *is chosen)*

are fulfilled, then the upper boundaries of the bias components $\delta_{bias}^{(\mathbf{C})}(M)$ and $\delta_{bias}^{(\mathbf{L}_2)}(M)$ have the second order with respect to the mesh step $h \sim M^{-1/d}$; here the obvious inequality

$$\delta_{bias}^{(\mathbf{L}_2)}(M) \le \text{mes}X \times \delta_{bias}^{(\mathbf{C})}(M)$$

can be used.

Thus, the upper boundaries of the deterministic components $\delta_{det}^{(\mathbf{C})}(M)$ and $\delta_{det}^{(\mathbf{L}_2)}(M)$ must also have the second order with respect to h. Such upper boundaries can be provided by the basic functions (3.8), (3.9) and approximation coefficients (1.7) (here, in particular, $H_{Leb} = 1$).

Note, that violation of each of the four formulated conditions (in particular, using the blur coefficient and the function $\hat{\kappa}(y)$ of the form (2.6) instead of (3.11)) leads to the significant complication in getting upper boundaries for the bias components (in particular, for the value $\delta_{bias}^{(\mathbf{C})}(M)$ from the inequality (4.3)).

The stability properties of the form (4.3) for the basis (3.8), (3.9) and usage of the approximation coefficients (1.7) also give the possibility to construct the upper boundaries for the stochastic components $\delta_{stoch}^{(C)}(M, n)$ and $\delta_{stoch}^{(L_2)}(M, n)$.

Thus, by analogy of the reasoning of the works [4,7], for $\mathbf{C}$- and $\mathbf{L}_2$-approaches the described methodology of the conditional optimization theory can be used, and the following formulae for the conditionally optimal parameters can be obtained:

$$M_{opt}^{(\mathbf{L}_2)}(\gamma) = H_1^{(\mathbf{L}_2)} \left(\frac{d+4}{d}\right)^{d/4} \gamma^{-d/2},$$

$$n_{opt}^{(\mathbf{L}_2)}(\gamma) = H_2^{(\mathbf{L}_2)} \left(\frac{d+4}{d}\right)^{d/4} (d+4)\,\gamma^{-2-d/2}; \qquad (4.4)$$

$$M_{opt}^{(\mathbf{C})}(\gamma) = H_1^{(\mathbf{C})} \left[\frac{(2\nu+1)d+4}{(2\nu+1)d}\right]^{d/2} \gamma^{-d/2},$$

$$n_{opt}^{(\mathbf{C})}(\gamma) = H_2^{(\mathbf{C})} \frac{[(2\nu+1)d+4]^{2+d/2}}{[(2\nu+1)d]^{d/2}} (2\ln M_{opt}^{(\mathbf{C})}(\gamma) - \ln\ln M_{opt}^{(\mathbf{C})}(\gamma) + H_3^{(\mathbf{C})})\,\gamma^{-2-d/2}$$

for some positive constants $H_1^{(\mathbf{L}_2)}, H_2^{(\mathbf{L}_2)}, H_1^{(\mathbf{C})}, H_2^{(\mathbf{C})}, H_3^{(\mathbf{C})}$ and ν; the choice or approximation of these constants is equal to a separate – often complicated – problem [4].

5   Conclusion

In this paper, the modification of the kernel estimator (1.1) of the probability density $f_{\xi}(\mathbf{x})$ is proposed. This modification is based on the theory of numerical approximation of functions and leads to the constructive Algorithm 1. The critical analysis of the optimization criterion, based on minimization upper boundary of relative error for the $\mathbf{L}_2$-approach, is conducted. We have shown that the constructive kernel Algorithm 1 is equivalent to the randomized projection-mesh functional Algorithm 2 for approximation of the solution $\varphi(\mathbf{x})$ of the Fredholm integral equation of the second kind (3.1). In connection with this it is proposed to use the criterion of conditional optimization of functional algorithms, based on minimization of the computational cost $S(M, n)$ for the fixed level of error $\gamma > 0$, for the kernel Algorithm 1. We have obtained formulae for the conditionally optimal parameters (for the known $\mathbf{C}$- and $\mathbf{L}_2$-approaches – see the formulae (4.3)) for the simplest version of the Algorithm 1, for which the basic functions (3.8), (3.9) and the approximation coefficients (1.7) are used (thus, we have studied the analog of the polygon of frequencies method).

References

1. Epanechnikov, V.A.: Nonparametric estimation of a multidimensional probability density. Theory Probab. Appl. **14**(1), 153–158 (1969)
2. Bahvalov, N.S.: Numerical Methods. Nauka, Moscow (1975). (in Russian)

3. Mikhailov, G.A.: Weighted Monte Carlo Methods. SD RAS Publisher, Novosibirsk (2000). (in Russian)
4. Voytishek, A.V.: Discrete-stochastic numerical methods. Doctorial dissertation, Novosibirsk (2001). (in Russian)
5. Mikhailov, G.A., Voytishek, A.V.: Numerical Statistical Modelling. Monte Carlo Methods. Publishing House "Akademia", Moscow (2006). (in Russian)
6. Voytishek, A.V.: Functional Estimators of the Monte Carlo Method. NSU Publisher, Novosibirsk (2007). (in Russian)
7. Voytishek, A.V.: Bases of the Monte Carlo Method in Algorithms and Tasks. Part VI. Calculation of Linear Functionals on Solution of Fredholm Integral Equation of the Second Kind. Discrete-Stochastic Methods for Solving the Integral Equation of the Second Kind. NSU Publisher, Novosibirsk (2004). (in Russian)
8. Voytishek, A.V., Shipilov, N.M.: On randomized algorithms for numerical solution of applied Fredholm integral equations of the second kind. In: AIP Conference Proceedings 1907(030015) (2017)
9. Voytishek, A.V.: Development and optimization of randomized functional numerical methods for solving the practically significant Fredholm integral equations of the second kind. J. Appl. Ind. Math. 12(2), 382–394 (2018)
10. Marchuk, G.I., Agashkov, V.I.: Introduction to Projection–Mesh Methods. Nauka, Moscow (1981). (in Russian)

Golden Rule Saving Rate
for an Endogenous Production Function

Nicholas Olenev[1,2,3]($\boxtimes$) (iD)

[1] Dorodnicyn Computing Centre, FRC CSC RAS, Moscow, Russia
`nolenev@mail.ru`
[2] Peoples' Friendship University of Russia (RUDN University), Moscow, Russia
[3] Moscow Institute of Physics and Technology (MIPT), Moscow, Russia

Abstract. The paper considers the classical problem of optimal saving rate (golden rule) for an endogenous production function built on the basis of a micro-description of the dynamics of production capacity. The production capacities are distributed according to the moments of creation (vintage capacity model) and are limited by the age of their possible use. The main hypothesis of the model is that the number of workplaces on a production unit is fixed, and the capacity decreases with a constant pace. The resulting production function reflects explicitly the mechanisms for control of the production system. The average labor intensity is a short-term control, while the share of new capacities and their age limit are long-term controls. The golden rule for the Solow model is formulated in terms of capacity and labor intensity. The new endogenous production function gives new effects. The optimal level of accumulation rate does not depend on the choice of output elasticity by a production factor. The age limit of production capacity is a new production factor of the endogenous production function. It affects the value of effective labor per unit of capacity stock.

Keywords: Vintage capacity model · Endogenous production function · Russian economy model · Golden rule · Saving rate · Solow model

1 Introduction

The golden rule of capital accumulation is well known in mathematical economics and it is included in standard courses on economic growth [1,2]. The solution to this problem has been considered in many works, see, for example, [3–5]. The golden rule of capital accumulation establishes a condition under which the capital-labor ratio (capital stock per worker) maximizes average consumption (consumption per worker) in a steady state of economic growth. In a simple model without scientific and technological progress, the golden rule boils down

The publication has been prepared with the support of the "RUDN University Program 5-100".

to the fact that the marginal productivity of capital is equal to the sum of the population growth rate and the rate of capital degradation.

A clear mathematical formulation of the problem of optimal economic growth for the simplest dynamic economic model, taking into account restrictions on control, is presented, for example, in [6]. In addition, it is possible to solve the control synthesis problem, i.e. find an explicit expression of the dependence of the optimal control on the state of the system (phase coordinate) [6]. The latter allows to formulate a universal golden rule [6] for choosing the optimal level of consumption depending on the current level of capital-labor ratio, valid for all time points and for all not too large initial levels of capital-labor ratio. On the contrary, here the golden rule is not considered for the entire optimal trajectory of movement, but only for those areas where the optimal control does not take boundary values and for characteristic growth pathes. However, this problem is considered here not only for the standard Cobb-Douglas production function, but in the main for a new endogenous production function. This allows us to get new effects that have an economic applied character.

In the extreme case, if the savings are zero, all the income is consumed, which cancels the investment as well as the replacement of the capital which wears out. In the long run, when capital is fully consumed, income is reduced to almost zero. The same is true for consumption: an excessive preference for short-term consumption is to the detriment of future generations. If the savings are equal the full income, all these revenues could go to investment but consumption is zero and there is no incentive to invest. Growth is zero, too. Excessive foresight does not benefit future generations either. Between these two extremes, there is (at least) a level of savings that maximizes average growth, allowing the growth of regular and identical consumption for all generations (intergenerational solidarity). According to [4], the only way to reach this optimum is to set the real interest rate at a value equal to population growth. Indeed, if we can adjust the marginal productivity rate of capital to the population growth rate, we can also adjust the savings rate to the share of profit in the national income.

Formally writing, the nation wants to maximize intertemporal utility

$$\int_0^\infty e^{-\delta t} u(c_t) dt,$$

where $u(c_t)$ is the instant utility of consumption and δ is the subjective rate of time discount. If K is capital, L is labor and Y is output of production, then $k = K/L$ is the ratio of capital to labor, $f(k) = Y/L$ is the homogeneous of degree one production function. The evolution of consumption depends on the differential equation

$$\dot{k} = f(k) - nk - c,$$

where $c = C/L$ is the per capita consumption, $n = \dot{L}/L$ is the growth rate of the population, and the point above a variable is the derivative over time. The current value of the Hamiltonian has the form

$$H = u(c_t) + \chi \left(f(k) - nk - c \right),$$

where χ is the constant auxiliary variable. After substituting this variable under first-order conditions, we find

$$(u''(c)/u'(c))\,\dot{c} = \delta + n - f'(k).$$

Really, $\partial H/\partial c_t = \partial u/\partial c_t - \chi = 0$, $\dot{\chi} = \chi\delta - \partial H/\partial k = \chi(\delta - f'(k) + n)$, so $\chi = u'(c)$, and $\dot{\chi} = u''(c)\dot{c}$. For stationary equilibrium we have $\dot{c} = 0$ and then

$$f'(k^*) = \delta + n.$$

With this modified golden rule, the ratio of capital to labor will be smaller because of the impatience of the society represented by the discount rate of time.

Further, in the work, we will not take into account the discount rate, but instead of this we will take into account technological progress, and also move on from considering capital to considering production capacity.

In [7], on the basis of aggregating the original micro-description of production, a new class of production functions was obtained. Such a production function, along with other parameters, contains directly the growth rate of the economy, which makes the task of finding optimal accumulation more interesting, even in the absence of scientific and technological progress. The production function [7] shows the dependence of output on production factors, which are total labor and total production capacity. Production capacity is the highest possible output. For the transition from the capital to the production capacity, the capital intensity is used.

Works [8,9] give numerical representations of the endogenous production function of type [7], which, along with the growth rate of the economy and the rate of degradation of capacities, contains the maximum age of production capacity. Thus, the production capacity leaves the production process, not only due to its degradation but also due to its dismantling when exceeding the maximum age (due to its obsolescence).

In the description of the golden rule, we move from the variable capital-labor ratio to the average labor intensity of capacities.

2 Solow Model in Terms of Capacity and Labor Intensity

Here we express the Solow model of economic growth [3] in terms of the Houthakker–Johansen model [7,10,11]: $Y(t)$—total output (GDP), $M(t)$—total production capacity (maximum possible output), $L(t)$—the number of workers (it is proportional to population), $\nu(t)$—the smallest labor intensity (labor input rate per unit of product, that is the number of workers per one unit of output). The last one characterizes technical progress. In these variables, we assume that a production function sets the dependence of output (GDP) $Y(t)$ on the total production capacity (in units of output) $M(t)$ and total effective labor, taking into account its efficiency (in units of output it is the ratio of workers to the smallest labor intensity), $L(t)/\nu(t)$. Technological progress that enters in this way is known as Harrod-neutral or labor-augmenting [2]. Note, that in this formulation, the use of a homogeneous production function of the first degree does not cause problems with the units of measurement. For example, for the Cobb–Douglas production function $Y = M^\alpha (L/\nu)^\beta$ with $\alpha + \beta = 1$, we have the units of the output $[Y] = [Y]^\alpha [Y]^{1-\alpha} = [Y]$.

For a homogeneous production function $Y(t) = F(M(t), L(t)/\nu(t))$ one of two variables can be taken out in order to obtain an intensive form of production function F, function f of one variable. In our case, it is more convenient to take out the total production capacity, then

$$Y(t) = M(t)f(x), x = \frac{L(t)}{\nu(t)M(t)} \qquad (1)$$

and the production function in intensive form has the meaning of the function of loading the total capacity. If the output does not exceed the maximum possible output, $Y(t) \leq M(t)$ (the production capacity overload is not allowed), then $f(x) \leq 1$. For example, for the Cobb–Douglas production function the function of capacity loading $f(x) = x(t)^\beta$, where β is output elasticity of labor.

In a closed economy, the output $Y(t)$ is divided into the consumption $C(t)$ and the capital accumulation of $bJ(t)$:

$$Y(t) = bJ(t) + C(t). \qquad (2)$$

In (2) $b > 0$ is the coefficient of incremental capital-output ratio, it shows how much capital-forming products need to be purchased to create one unit of capacity. The value of $J(t)$ is the volume of newly created capacity.

If we use the usual dynamics of total production capacity, then

$$\dot{M}(t) \overset{\text{def}}{=} \frac{dM(t)}{dt} = J(t) - \mu M(t), \qquad (3)$$

where $\mu > 0$ is a depreciation rate of capacity.

A rate s of capital accumulation: $bJ(t) = sY(t)$. Then we have $J(t) = sf(x)M(t)/b$ and from (3) the tempo of the total production capacity is equal

$$\frac{\dot{M}(t)}{M(t)} = \frac{s}{b}f(x) - \mu. \qquad (4)$$

From the other side, if

$$\frac{\dot{L}(t)}{L(t)} = n, -\dot{\nu}(t)/\nu(t) = g, \qquad (5)$$

where n is the population growth rate, g is the rate of fall of the lowest labor intensity (rate of growth for the level of technology), then from (1) and (5) we have $M = L/(x\nu)$ and

$$\frac{\dot{M}(t)}{M(t)} = n + g - \frac{\dot{x}(t)}{x(t)}. \qquad (6)$$

The main interest of the model is the dynamics of average labor intensity of the total capacity in relative units with respect to the smallest labor intensity x, the effective labor per unit of capacity stock. In accordance of (4), (6) its behavior over time is given by the next analogy of the key equation of the Solow model in terms of x:

$$\frac{\dot{x}(t)}{x(t)} = n + g + \mu - sf(x)/b. \tag{7}$$

Stationarity condition is $\dot{x}(t) = 0$, so from here in steady-state

$$sf(x) = b(n + g + \mu). \tag{8}$$

If we denote $\dot{M}(t)/M(t) = \gamma$, then from (6) we obtain that the steady state growth rate is expressed by the relation

$$\gamma = n + g. \tag{9}$$

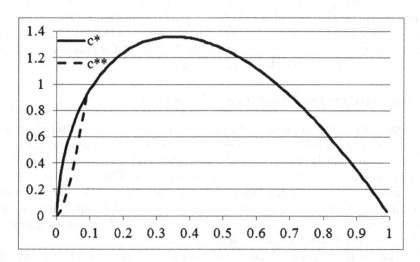

Fig. 1. The golden rule saving rate $s_{gold} = 0.35$ for the Cobb–Douglas production function $f(x) = x^\beta$. The vertical axis shows the steady-state of consumption per effective person c^* correspond to each saving rate $s \in (0, 1)$. The curve c^{**} is obtained under the condition $f(x) \le 1$.

Labor productivity in relative units $y \overset{\text{def}}{=} \nu(t)Y(t)/L(t) = f(x)/x$. Then the growth rate of labor productivity in relative units is

$$\frac{\dot{y}(t)}{y(t)} = \dot{x}(t)\left(\frac{f'(x)}{f(x)} - \frac{1}{x(t)}\right). \tag{10}$$

Consumption per effective labor has the form

$$c \overset{\text{def}}{=} \frac{\nu(t)C(t)}{L(t)} = \frac{f(x) - sf(x)}{x}. \tag{11}$$

In steady-state (8) we obtain

$$c(x) = \frac{f(x) - b(n + g + \mu)}{x}.$$ (12)

In accordance with (8) $x = x(s)$, so that $c = c(s)$, and from (12) we have

$$c'_s = (xf'(x) - f(x) + b(n + g + \mu))\frac{x'_s}{x^2}.$$ (13)

So that in terms of $x = x_{gold}$ the Solow model golden rule be

$$xf'(x) - f(x) + b(n + g + \mu) = 0.$$ (14)

For the Cobb–Douglas production function $f(x) = x^\beta$ we have from (8) $sx^\beta = b(n+g+\mu)$, and from (14) $(1-\beta)x^\beta = b(n+g+\mu)$. So that, $s = 1-\beta = \alpha = \epsilon_M$, where ϵ_M is the total output elasticity by total capacity M.

Figure 1 presents the relation between steady-state consumption per effective labor c^* and saving rate s that is implied by (12). The steady-state $x^*(s)$ is from the condition (8). For the Cobb–Douglas production function $x^*(s) = (b(n+g+\mu)/s)^{1/\beta}$. Here in Fig. 1 we use $\beta = 0.65$ and the values of the model parameters obtained on the base of statistical data 1970–2017 for the Russian economy [8]: $b = 1.1$, $n = 0.01124$, $g = 0.038115$, $\mu = 0.03155$.

3   An Exogenous Production Function

An exogenous production function for production capacities with limited age was constructed in [8]. The economic system consists of separate production units. The production unit is characterized by the technology used and the capacity—the maximum possible product output per unit of time. The production technology is fully determined by the labor intensity—the norm of the cost of human labor for the production of a unit of product. The production capabilities of such an economic system are described by the value $M(t, H)$ of the total capacity of production units whose technology is $\lambda \in H \subset R^1_+$.

The production capacity dynamics on micro-level is described by the hypothesis proposed in [7–9]. Production capacity decreases due to aging and the number of working places on it is fixed since the creation of this capacity up to reaching age limit $A(t)$.

Hypothesis 1: *The number of workplaces in the production unit remains unchanged for each time moment t from the time moment of its creation $\tau \leq t$ until the time moment of its liquidation $\theta = \tau + A(\theta) \geq t$, where $A(t)$ is the age limit of the production capacities, and the production capacity $m(t, \tau)$ decreases with a constant rate $\mu > 0$.*

If the capacities of new production units are created continuously with a speed $J(t)$ and all have the same labor input $\nu(t)$, which does not increase with time, then the measure $M(t, H)$ has a continuous density $m(t, \lambda)$, which varies according to a first-order partial differential equation [7]:

$$\frac{\partial m(t, \lambda)}{\partial t} = -2\mu m(t, \lambda) - \mu\lambda\frac{\partial m(t, \lambda)}{\partial \lambda}$$ (15)

with boundary condition

$$m(t,\nu) = \frac{J(t)}{(\mu\nu(t) - d\nu/dt)}. \tag{16}$$

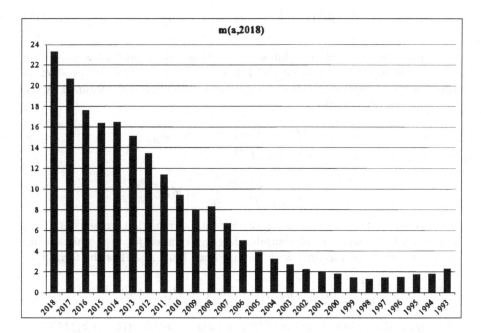

Fig. 2. Estimation of the distribution of production capacities of the Russian economy by age (vintage capacity) in 2018, at constant prices 2010. The years of investment are plotted on the horizontal axis.

Figure 2 shows the vintage production capacity in 2018. Production capacity measures the maximum output at constant prices of 2010. For the evaluation the capacity, we used the parameters of the micro-description of the model for changes in production capacity. It was assumed that the least laboriousness decreases with the rate $g(t) = -\dot{\nu}/\nu = \epsilon\sigma(t)$, and the capital intensity coefficient for new capacities also decreases due to increasing of the share of primary industries in the issue, $\dot{b}/b = -\zeta\sigma(t)$, where $\sigma = J(t)/M(t)$. Parameter identification was made by comparing the calculated and statistical time series for output and labor according to the data of 1970–2017: $b(1970) = 5.598$, $\zeta = 0.430$, $\nu(1970) = 2.512$, $\epsilon = 0.3465$, $A = 25$. Parameter estimation methodology is described in more detail in [8,9].

The expression for the production function is more convenient to derive on the basis of a micro-description of the dynamics of production capacity in the variables t, τ. The initial capacity is $m(t,t) = J(t)$. Then $m(t,\tau) = J(\tau)exp(-\mu(t - \tau))$. This reduction in capacity requires, in order to maintain the number of workplaces, an appropriate increase in labor intensity $\lambda(t,\tau) = \nu(\tau)exp(\mu(t - \tau))$, where $\nu(\tau)$ is the labor intensity on the production unit at moment of its creation τ. At every time moment t the investors choose the best technology with the smallest amount of labor $\nu(t)$. This lowest labor intensity is decreased with time due to technological progress with the rate $g(t) > 0$. The total capacity is

$$M(t) = \int_{t-A(t)}^{t} J(\tau)e^{-\mu(t-\tau)}d\tau. \tag{17}$$

Assuming the total effective labor $L(t)/\nu(t)$ is used in optimal way starting from the new production capacity with best technology $\nu(t)$ up to production capacity with age $\theta(x) \leq A$. Then the production function is determined by the next system of two equations:

$$f(x) = \int_{t-\theta(x)}^{t} \frac{M(\tau)}{M(t)}\sigma(\tau)e^{-\mu(t-\tau)}d\tau, \tag{18}$$

$$x = \int_{t-\theta(x)}^{t} \frac{M(\tau)}{M(t)}\frac{\nu(\tau)}{\nu(t)}\sigma(\tau)d\tau, \tag{19}$$

where instead of the differential Eq. (3) for total production capacity $M(t)$ in this case we have the next differential-difference equation for the total capacity $M(t)$ in the intensive variables with the fixed age limit A. It has the form [8]:

$$\frac{dM(t)}{dt} = (\sigma(t) - \mu)\,M(t) - \sigma(t-A)M(t-A)e^{-\mu A}, \tag{20}$$

where $\sigma(t)$ is the ratio of new capacities to total capacity, $\sigma(t) = J(t)/M(t)$.

If the ratio $\sigma(t)$ is constant and total production capacity grows with constant rate

$$\sigma(t) = \frac{J(t)}{M(t)} = \sigma, \frac{\dot{M}(t)}{M(t)} = \gamma, \tag{21}$$

we can find an exogenous production function.

Based on Proposition 1 from [8], taking into account our notation (1) for x and (5) for g, the following theorem can be formulated.

Theorem 1. *Let in a closed economy (2) on a balanced growth path with the rate γ,*

$$M(t) = M_0 e^{\gamma t}, Y(t) = Y_0 e^{\gamma t}, J(t) = J_0 e^{\gamma t}, C(t) = C_0 e^{\gamma t},$$

the following conditions are met:

(a) *it is true the hypothesis 1 about a fixed number of workplaces and a drop in production capacity at a rate μ up to a certain age limit A;*

(b) *it is fixed the maximum age of the production capacities, $A(t) = A = \text{const}$;*

(c) *it is fixed the ratio of incremental capital intensity, $b(t) = b = \text{const}$;*

(d) *it is reduced the least labor intensiveness due to scientific and technical progress in accordance with (5), $\dot{\nu}(t)/\nu(t) = -g$.*

Then the following statements are true:

(1) the share of new capacity is fixed: $\sigma(t) = J(t)/M(t) = \sigma = const;$
(2) the dynamics of total production capacity (20) on balanced growth path sets the relationship between the growth rate of economy (21) $\gamma = \varphi(\sigma, \mu, A) - \mu$ *and an implicit function* φ *of this parameters* σ, μ, A *by equation*

$$\varphi = \sigma(1 - e^{-\varphi A}); \tag{22}$$

(3) relation (22) gives the following expression for the production function

$$f(x) = \frac{\sigma}{\varphi} \left\{ 1 - \left[1 - \frac{(\varphi - \mu - g)}{\sigma} x \right]^{\varphi/(\varphi - \mu - g)} \right\}, \tag{23}$$

where μ *is the depreciation rate of production capacities, and* g *is the rate (5) of technological progress;*
(4) the ratio of the average labor intensity of the production capacities to the least labor intensity is constant: $x = L(t)/(\nu(t)M(t)) = const.$

Proof. It follows directly from the relations (18), (19) for the production function under the conditions (a)–(d) specified in the theorem.

Indeed, on the balanced growth path (BGP) $\sigma = J(t)/M(t) = J_0/M_0 = const$, so the statement (1) is satisfied.

Then, from $\sigma = const$, $A = const$, and condition of BGP $M(t - A)/M(t) = exp(-\gamma A)$ the Eq. (20) gives $\gamma = -\mu + \sigma(1 - exp(-(\gamma + \mu)A))$. By virtue of the notation $\varphi = \gamma + \mu$, we obtain the relation (22), so the statement (2) is satisfied.

Since on BGP $M(\tau)/M(t) = exp(-\gamma(t-\tau))$, so from (18), $\varphi = \gamma + \mu$, and $\sigma = const$ we have $f(x) = (1 - exp(-\varphi\theta(x)))\sigma/\varphi$. From (19) considering condition d), $\nu(\tau)/\nu(t) = exp(g(t-\tau))$, we have $x = (1 - exp(-(\varphi - \mu - g)\theta(x)))\sigma/(\varphi - \mu - g)$. Excluding from these relations for f and x the value $exp(-\theta(x))$ we have (23). So, the statement (3) is satisfied.

The statement (4) is satisfied because on BSP we have $f(x) = Y(t)/M(t) = M_0/Y_0 = const$, so $x = const$.

Now, by elimination the term σ from (22), (23), $\sigma = \varphi/(1 - e^{-\varphi A})$, and using (9),

$$\varphi = n + g + \mu, \tag{24}$$

we have the next form of the production function in steady-state

$$f(x, A) = \frac{1}{(1 - e^{-\varphi A})} \left\{ 1 - \left[1 - \frac{n}{\varphi}(1 - e^{-\varphi A})x \right]^{\varphi/n} \right\}. \tag{25}$$

Changes in the average labor intensity x and in the age limit of capacities A change the level of loading $f(x, A)$ of the total capacity.

4   Golden Rule for the Endogenous Production Function

In the model under consideration the total production capacity dynamics (20) differs from the usual Eq. (3), so that we should change the Eq. (4) on the next one:

$$\frac{\dot{M}(t)}{M(t)} = \frac{1}{b} s f(x, A) \left(1 - e^{-\varphi A}\right) - \mu. \tag{26}$$

In accordance of (26), (6) the effective labor per unit of capacity stock x, the key equation of our model will be

$$\frac{\dot{x}(t)}{x(t)} = \varphi - \frac{1}{b} s f(x, A) \left(1 - e^{-\varphi A}\right). \tag{27}$$

So, the dynamics of average labor intensity of the total capacity x depends of the age limit of capacities A.

Instead of (8) in steady-state ($\dot{x}(t) = 0$) we have

$$s f(x, A) = \frac{b\varphi}{\left(1 - e^{-\varphi A}\right)}. \tag{28}$$

For the consumption per effective labor (11) in steady-state (28) instead of (12) we have

$$c(x, A) = \frac{1}{x} \left(f(x, A) - \frac{b\varphi}{\left(1 - e^{-\varphi A}\right)} \right). \tag{29}$$

Then

$$c'_s = \left\{ \frac{f'_x}{x} - \frac{f(x, A) - b\varphi / \left(1 - e^{-\varphi A}\right)}{x^2} \right\} x'_s. \tag{30}$$

So, the model under consideration golden rule in terms of $x = x_{gold}$ is

$$[x f'_x(x, A) - f(x, A)] \left(1 - e^{-\varphi A}\right) + b\varphi = 0, \tag{31}$$

where due to (24) $\varphi = \gamma + \mu = n + g + \mu$.

For our production function (25) we have

$$f'_x = \left[1 - \frac{n}{\varphi} (1 - e^{-\varphi A}) x \right]^{-1 + \varphi/n}. \tag{32}$$

and from (28) and (31)

$$s_{gold} = \frac{b\varphi}{1 - [1 - zn/\varphi]^{\varphi/n}}, \tag{33}$$

where

$$z \stackrel{\text{def}}{=} (1 - e^{-\varphi A}) x_{gold} \tag{34}$$

is a root of the transcendental equation

$$1 - \frac{z}{(1 - zn/\varphi)} = (1 - b\varphi)(1 - zn/\varphi)^{\varphi/n}. \tag{35}$$

For the endogenous production function (25) from (31)–(33) we have

$$s^* = 1 - \frac{(1 - b\varphi)z^*}{b\varphi + (1 - bn)z^*}.$$ (36)

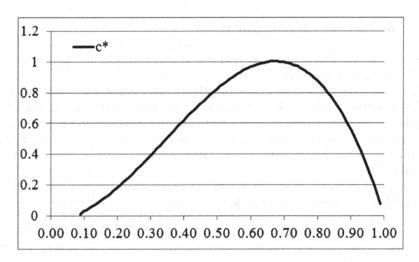

Fig. 3. The golden rule saving rate $s_{gold} = 0.67$ for the exogenous production function $f(x, A)$. The vertical axis shows the steady-state of consumption per effective person c^* correspond to each saving rate $s \in (0, 1)$.

From (28), (31) we have the next algorithm for calculating the curve $c = c(s)$ for the production function in a stable-stage:

$$z = \frac{\varphi}{n}\left[1 - \left(1 - \frac{b\varphi}{s}\right)^{n/\varphi}\right],$$ (37)

$$B \stackrel{\text{def}}{=} \left(1 - e^{-\varphi A}\right) f(x, A) = 1 - \left(1 - \frac{n}{\varphi}z\right)^{\varphi/n},$$ (38)

$$c = \frac{1}{z}(B - b\varphi).$$ (39)

The age limit A value explicitly affects the value of x_{gold} due to the formula (34). The value A is not explicitly included in the expression for the steady-state consumption per effective labor c^* due to (37)–(39):

$$c^* = \frac{nb(1/s - 1)}{1 - (1 - b\varphi/s)^{n/\varphi}}.$$ (40)

The dependence of the steady-state consumption per effective person c^* (40) on A exists since $\varphi = \varphi(\sigma, A)$ in accordance with (22).

Figure 3 presents the relation between steady-state consumption per effective labor c^* and saving rate s that is implied by (29). The steady-state $x^*(s)$ is from the condition (28). Here in Fig. 3 we use the values for parameters of the current Russian economy (in particular, see above Sect. 3): $b = 1.1$, $n = 0.01124$, $g = 0.038115$, $\mu = 0.03155$.

5   Conclusions and Implication

The introduction provides a brief reference to the literature on the problem of the golden rule saving rate.

The second section shows that the Houthakker–Johansen model notation [10,11] is suitable for describing classical problems, in particular, the problem of golden rule saving rate. At the same time, the concept of the production function in the model has a clear economic meaning of capacity utilization, and the variables have a decent dimension: either dimensionless or have dimension per unit of time.

The third section presents the endogenous production function, in which the Houthakker–Johansen model manifests itself well. Here production function is based on the distribution of production capacity for technology. In the end, the production function is presented in the stationary mode parameters.

In the fourth section, formulas for calculating the golden growth for this endogenous production function are obtained. As it turned out, the age limit of capacities does not affect the steady-state consumption per effective labor, but it affects the value of the effective labor per unit of the total capacity.

It can be concluded that the presentation of the golden rule for the exogenous production function can be used in education courses for students of classical mathematical economics. It gives food for thought about the current state of the economy.

Phelps in his Nobel lecture [12] gives an overview of works that take into account the realities of the modern economy. As an example, he points out his own models that take these realities into account, or at least remove some of the limitations of the neoclassical approach. In the model of exogenous production function presented here, an example of a departure from neoclassical principles is also given. In particular, the production function built here takes into account structural changes, and its analytical form is valid under certain conditions. In the numerical experiments with the model it is possible to use the micro-description of production sector. This saves the proposed mathematical description for the stages of the production cycle of the economy, in which there are sharp changes in the structure of production capacities.

References

1. Barro, R., Sala-i-Martin, X.: Economic Growth, 2nd edn. MIT Press, Cambridge (2004)
2. Romer, D.: Advanced Macroeconomics, 5th edn. McGraw-Hill, New York (2018)

3. Solow, R.M.: A contribution to the theory of economic growth. Q. J. Econ. **70**(1), 65–94 (1956). https://doi.org/10.2307/1884513

4. Phelps, E.: The golden rule of accumulation: a fable for growthmen. Am. Econ. Rev. **51**(4), 638–643 (1961)

5. Stoleru, L.: L'equilibre et la croissance economiques, Princioes de macroeconomie. Dunod, Paris (1969)

6. Lobanov, S.G.: On the theory of optimal economic growth (in Russian). High. Sch. Econ. Econ. J. **3**(1), 28–41 (1999)

7. Olenev, N.N., Petrov, A.A., Pospelov, I.G.: Model of change processes of production capacity and production function of industry. In: Samarsky, A.A., Moiseev, N.N., Petrov, A.A. (eds.) Mathematical Modelling: Processes in Complex Economic and Ecologic Systems (in Russian), pp. 46–60. Nauka, Moscow (1986). https://doi.org/10.13140/RG.2.1.3938.8880

8. Olenev, N.N.: Parameter identification of an endogenous production function. CEUR-WS **1987**, 428–435 (2017)

9. Olenev, N.: Identification of an aggregate production function for Polish economy. Quant. Methods Econ. **19**(4), 430–439 (2018). https://doi.org/10.22630/MIBE.2018.19.4.41

10. Houthakker, H.S.: The Pareto distribution and the Cobb-Douglas production function in activity analysis. Rev. Econ. Stud. **23**(1), 27–31 (1955). https://doi.org/10.2307/2296148

11. Johansen, L.: Production functions and the concept of capacity. Recherches recentes sur la fonction de production, collection. Economie mathematique et econometrie **2**, 49–72 (1968)

12. Phelps, E.S.: Macroeconomics for a modern economy. Am. Econ. Rev. **97**(3), 543–561 (2007). https://doi.org/10.1257/aer.97.3.543

Computational Methods for the Stable Dynamic Model

Anton Anikin[2]([✉]) [iD], Yuriy Dorn[1] [iD], and Yurii Nesterov[3] [iD]

[1] OZON.RU, Moscow, Russia
`dornyv@yandex.ru`
[2] ISDCT SB RAS, Irkutsk, Russia
`anton@anikin.xyz`
[3] CORE, Louvain-la-Neuve, Belgium
`yurii.nesterov@uclouvain.be`

Abstract. Traffic assignment problem is one of the central problems in transportation science. Various model assumptions lead to different setups corresponding to nonlinear optimization problems.

In this work, we focus on the stable dynamic model and its generalizations. We propose new equivalent representation for stable dynamic model [Nesterov and de Palma, 2003]. We use smoothing technique to derive new model, which can be interpreted as a stochastic equilibrium model.

Keywords: Traffic assignment problem · Equilibrium model · Huge-scale convex optimization · Primal-dual method · Coordinate descent method

1 Introduction

In this work, we consider transportation equilibrium models and algorithmic schemes to compute equilibrium.

It can be a coincidence, that the most popular equilibrium transportation model [3] was proposed at the same time as the Frank-Wolfe algorithm [4], the most popular algorithm for transportation equilibrium computation. The first version of Frank-Wolfe algorithm was not very practical for the traffic assignment problem, because it was applied to an optimization problem over a set of all paths in the network, which is exponential in the number of nodes. A tractable version of Frank-Wolfe method in transportation science appeared almost 20 years later [5]. Since then transportation research has motivated the development of new convex optimization methods, especially challenged for huge-scale problems.

One should note, that since its introduction Backmann's model has been almost the only one, which has been implemented as a software tool used in practice of transportation modeling. Simplicity and tractability of Backmann's

Supported by RFBR, project no. 18-29-03071 mk.

M. Jaćimović et al. (Eds.): OPTIMA 2019, CCIS 1145, pp. 280–294, 2020.
https://doi.org/10.1007/978-3-030-38603-0_21

model has led to a number of methods proposed for its computation [2,11,12]. Still Frank-Wolfe methods and their simple variants are still quite popular in engineering practice [9].

One can see that in the macroscopic branch of transportation modeling science there is a very strange situation when advances in models and algorithms do not find their place in practice. We have already mentioned that Beckmann's model is simple and easy to compute. The last feature is achieved because in this model equilibrium can be found as the solution of smooth convex optimization problem with linear constraints. That is also the reason why Frank-Wolfe algorithm works so well for this model.

On the other hand, Beckmann's model with standard choice of arc delay function (such as BPR-function) has a few drawbacks. First, the equilibrium flow can exceed arc capacity. Second, it is not clear how to model traffic jams in this model.

This is the main motivation to use the stable dynamic model [10]. In this model one can straightforward estimate traffic jams at the equilibrium point. In this model, equilibrium flow can not exceed arc capacity. In the original paper, it was shown that the equilibrium can be computed as a solution of the non-smooth convex optimization problem with linear constraints. Thus one can assume that this model must be computational tractable.

The main issue we address in this paper is that there is no computational method proposed for the Stable Dynamic model.

In our work, we show that an equilibrium in the stable dynamic model can be computed as a solution of the linear programming problem, or a non-smooth unconstrained convex optimization problem, or a saddle-point problem. All of them are equivalent reformulations of the original problem. We also propose new methods to solve those problems and provide complete complexity analysis of these methods. Finally, we describe a stochastic version of the stable dynamic model and propose a method for its computation.

Novelty and Our Contribution. In [10] it is shown that an equilibrium in the stable dynamic model can be found as a solution of the huge-scale non-smooth convex optimization problem with linear constraints. This optimization problem has a specific structure that makes most of the modern computational methods hardly inapplicable.

In this work, we propose two new representations (as a saddle point problem and as a linear programming problem) of the equilibrium problem in the stable dynamic model. Both of them allows to apply modern optimization methods for huge-scale problems such as fast gradient descent **cite**, randomized coordinate descent **cite** and others.

We also propose the smoothed version of the stable dynamic model, where the smoothness parameter is in charge of the irrationality of drivers. We prove that Beckmann's model can be considered as a special case of a smoothed version of the stable dynamic model for a number of popular cost functions.

1.1   Contents

In Sect. 2 we describe the Stable Dynamic model and its representations as a linear programming problem, a non-smooth unconstrained convex optimization problem and a saddle-point problem. In Sect. 3 we describe smoothed version of the Stable Dynamic model. In last section we provide computation experiments results, where we test Fast Gradient Methods and it's modifications on test networks.

1.2   Notation

$\Gamma(V, E)$ - weighted oriented graph Γ, loops and arc duplication are not allowed,
V - set of nodes,
$E = V \times V$ - set of arcs,
$n = |V|$ - number of nodes,
$m = |E|$ - number of arcs, we think that all arcs are enumerated,
$S \subset V$ - set of origin nodes,
$D \subset V$ - set of destination nodes,
$l = |S|$
$OD = S \times D$ - set of the origin-destination pairs (OD-pairs),
$(i, j) \in E$ - arc with the origin node i and the destination node j,
$(i, j) \in OD$ - OD-pair with origin node i and destination node j,
f_{ij} - flow on the arc (i, j),
$f = (f_{(1)}, ..., f_{(m)})^T \in R^m$ - flow (vector),
$\bar{f}_{ij}$ - capacity of the arc (i, j), constant,
$\bar{f} = (f_{(1)}, ..., f_{(m)})^T \in R^m$ - vector of capacities, constant vector
τ_{ij} - cost on the arc (i, j),
$\tau = (\tau_{(1)}, ..., \tau_{(m)})^T \in R^m$ - cost (vector),
$\bar{\tau}_{ij}$ - free flow travel time for arc (i, j), constant,
$\bar{\tau} = (\bar{\tau}_{(1)}, ..., \bar{\tau}_{(m)})^T \in R^m$ - free flow travel time vector, constant vector,
d_{ij} - demand for OD-pair (i, j),
$\{d_{ij}\}$ - OD-matrix,
$(\Gamma(V, E, \bar{f}, \bar{\tau}); \quad \{d_{ij}\})$ - the instance,
P_{ij} - set of paths for OD-pair (i, j),
$P = \cup_{(i,j) \in OD} P_{ij}$ - set of all paths,
δ_{eq}
$C_q(\tau) = \sum_{(i,j) \in E} \tau_{ij} \cdot \delta_{(i,j),q}$ - cost on path q if cost vector is τ,
$T_{ij}(\tau) = \min_{q \in P_{ij}} C_q(\tau)$ - minimal cost for OD-pair (i, j) if the cost vector is τ,
$\bigtriangledown \Phi(x)$ - gradient of the function $\Phi(x)$,
$(u)_+ = \max\{u, 0\}$,
$\triangle_l = \{x \in R^l, \ x \geq 0, \ \sum_i x_i = 1\}$,
$\langle x, u \rangle$ - scalar product of the elements x and u,
In this work we work with different primal and dual spaces equipped with corresponding norms. The primal space Ξ is endowed with a norm $\| \cdot \|_1$ and the norm for the dual space Ξ^* is defined in a standard way:

$$\|u\|^* = \max_x \{\langle x, u \rangle : \quad \|x\| = 1\}.$$

For linear operator $A : \Xi_1 \to \Xi_2^*$ we define the adjoint operator $A^* : \Xi_2 \to \Xi_1^*$ in the following way:

$$\langle Ax, u \rangle = \langle x, A^* u \rangle, \quad \forall x \in \Xi_1, \quad u \in \Xi_2.$$

The norm of such operator is defined as follows:

$$\|A\|_{1,2} = \max_{x,u}\{\langle Ax, u \rangle_2 : \quad \|x\|_1 = 1, \quad \|u\|_2 = 1\}.$$

2   Stable Dynamic Model and Its Representations

In this section we provide brief description of the Stable Dynamic model, proposed by [10]. Equilibrium in the Stable Dynamic model can be found as the solution of a huge-scale non-smooth convex optimization problem with linear constraint. The structure of this optimization problem is not clear and restricts the use of many modern computational techniques. We show that this optimization problem has several alternative representations. Each of these alternative representations has good structure and its own good properties and allows to use computational techniques for huge-scale optimization problems.

2.1   Stable Dynamic Model

Let $\Gamma(V, E)$ be a weighted oriented graph. Here V is the set of nodes, E is the set of arcs. The $\{d_{ij}\}$ is the origin-destination matrix (OD-matrix). Pair $(\Gamma(V, E), \{d_{ij}\})$ is called an instance. Each arc e is described by two parameters: free flow travel time $\bar{\tau}_e$ and arc capacity $\bar{f}_e$. Denote by P_{st} the set of all paths from node s to node t. Let $C_q(\tau) := \sum_{e \in E} \tau_e \delta_{eq}$ be the cost of a path q if the vector of costs on arcs is τ. Here δ_{eq} is equal to 1 if $e \in q$ and 0 otherwise.

At any equilibrium point (f^*, τ^*) the Wardrop's equilibrium condition holds and the following conditions are satisfied:

- If $f_e^* < \bar{f}_e$, then $\tau_e^* = \bar{\tau}_e$,
- If $f_e^* = \bar{f}_e$, then $\tau_e^* \geq \bar{\tau}_e$.

Denote by $T_{ij}(\tau) = \min_{q \in P_{st}} C_q(\tau)$ the cost on the shortest path for the OD-pair (i, j). The standard way to analyze any instance is to compute the equilibrium flow [13]. One can solve this problem by reducing equilibrium computing to the optimization problem of potential function (see, for example, [10]). Then the equilibrium in the Stable Dynamic model can be computed as a solution of the following optimization problem:

$$\sum_{(i,j) \in OD} d_{ij} T_{ij}(\tau) - \sum_{(i,j) \in E} \bar{f}_{ij}(\tau_{ij} - \bar{\tau}_{ij}) \to \max_{\tau} \tag{1}$$
$$s.t. \quad \tau \geq \bar{\tau}$$

The KKT-optimality condition for this problem ensures that in optimum for each OD-pair only shortest paths are used. This corresponds to the game-theoretic model, where a great number of infinitesimal vehicles make decisions which route to take. Each of them simply chooses the shortest path at any moment without knowledge of the global utility function itself, which gives the equilibrium conditions of the whole system. This is standard for similar perfect competition models e.g. in Economics. On the other hand, it is not natural that equilibrium can be computed as a solution of a optimization problem. Games with this property are called *potential games*. Standard traffic equilibrium problems, known in game theory as congested games, are essentially potential games. The stable dynamic model share this property.

2.2 Linear Programming Representation of the Stable Dynamic Model

The main issue about the problem (1) is that we need to compute function $T_{ij}(\tau)$. The alternative way is to expand the space of variables by including shortest path cost for every OD-pair (i, j) as independent variable T_{ij}. By definition of the shortest path the following constraints are held:

$$T_{sj} - T_{si} \leq \tau_{ij}, \quad \forall s \in S, \quad (i, j) \in E,$$
$$T_{ss} = 0, \quad \forall s \in S.$$

Then (1) can be rewritten in the following way:

$$\sum_{(i,j) \in OD} d_{ij} T_{ij} - \sum_{(i,j) \in E} \bar{f}_{ij} (\tau_{ij} - \bar{\tau}_{ij}) \to \max_{\tau, T} \tag{2}$$

$$s.t. \quad \tau \geq \bar{\tau}$$
$$T_{sj} - T_{si} \leq \tau_{ij}, \quad \forall s \in S, \quad \forall (i, j) \in E,$$
$$T_{ss} = 0, \quad \forall s \in S.$$

Equation (2) has $m + n \cdot |S|$ variables. It can also be rewritten in terms of costs from every node to every sink node and have $m + n \cdot |D|$ variables. So (1) can be rewritten as linear program with $m + n \cdot l$ variables. Because it is up to us to decide what representation (from every source to every node or from every node to every sink) to use, we can always choose representation with the minimal number of components in variable T. In the future, to be specific, we will always use "source to every node" representation. However, from the equivalence of representations we can take $|S| = l$, because if $|S| > l$ we are free to use every node to sink representation, which has the same structure, but $|D| = l$.

2.3 Node Potentials Representation

In the optimization problem (2) we can eliminate τ.
  Write (2) in the following form:

$$\sum_{(i,j) \in E} \bar{f}_{ij} (\tau_{ij} - \bar{\tau}_{ij}) - \sum_{(i,j) \in OD} d_{ij} T_{ij} \to \min_{\tau, T}$$

$$s.t. \quad \max\{\max_{s \in S}\{T_{sj} - T_{si}\}, \bar{\tau}_{ij}\} \leq \tau_{ij}, \quad \forall (i, j) \in E, \tag{3}$$

$$T_{ii} = 0, \quad \forall i \in V.$$

Note, that $\bar{f}_{ij} \geq 0$ and $\tau_{ij} - \bar{\tau}_{ij} \geq 0$ for all arcs (i,j). Thus the minimum of $\bar{f}_{ij}(\tau_{ij} - \bar{\tau}_{ij})$ is attained at

$$\tau_{ij} = \max\{\max_{s \in S}\{T_{sj} - T_{si}\}, \bar{\tau}_{ij}\}. \tag{4}$$

Thus optimization problem (3) can be reduced to the form:

$$\sum_{(i,j) \in E} \bar{f}_{ij} \cdot \max_{s \in S}(T_{sj} - T_{si} - \bar{\tau}_{ij})_+ - \sum_{(i,j) \in OD} d_{ij}(T_{ij} - T_{ii}) \to \min_T \tag{5}$$

This is the huge-scale optimization problem. The number of variables can be estimated as $|S| \times |V|$ because for each OD-pair we need to compute shortest-path tree, that include all nodes (or at least all nodes, that are used in shortest paths). For real-life networks number of variables can reach $\sim 10^7 - 10^8$.

Note that $\max_{s \in S}(T_{sj} - T_{si} - \bar{\tau}_{ij})_+$ is convex as the maximum of convex functions over T and $\sum_{(i,j)} d_{ij}(T_{ij} - T_{ii})$ is a linear function over T. Thus (5) is a non-smooth unconstrained convex optimization problem over T. We also must emphasize, that in contrast to (1), problem (5) has a good, clear structure, which allows to use modern optimization techniques for huge-scale convex optimization problems.

One can see, that if T^* is a solution of the problem (5), then every element of the set $\hat{T} = \{T \in R^{sn} | T_{ij} = T_{ij}^* + a_i, a_i \in R\}$ is also the solution of the problem (5). Thus, if T^* is the solution of the problem (5), then

$$\hat{T}_{ij} \doteq T_{ij}^* - T_{ii}^*, \quad \forall i, j \in V \tag{6}$$

is the solution of the problem (3).

Thus we loose nothing when we reduce the problem (3) to the (5).

2.4   Saddle-Point Representation

Another important type of representations of optimization problems is a saddle-point problem. For example, this representation type is required for implementing smoothing technique (see for example, [6]), described in (15). For optimization problem (1) this representation can be achieved from (5) directly.

Proposition 1. *Optimization problem (5) is equivalent to optimization problem*

$$\min_T \Phi(T) = \min_T \max_{u \in \Delta_{l+1}^m} \left\{ \sum_{(i,j) \in E} \bar{f}_{ij} \cdot \{\sum_{s \in S} u_{ij}^s \cdot (T_{sj} - T_{si} - \bar{\tau}_{ij})\} - \sum_{(i,j) \in OD} d_{ij}(T_{ij} - T_{ii}) \right\} \tag{7}$$

Proof: We use the following representation of the max:

$$\Phi(x) = \max_i \phi_i(x) = \max_u \left\{ \sum_i u_i \cdot \phi_i(x) | \quad u_i \geq 0, \sum_i u_i = 1 \right\}$$

Thus (5) can be represented as

$$\sum_{(i,j)\in E} \bar{f}_{ij} \cdot \max_{u_{ij}} \left\{ \sum_{s\in S} u_{ij}^s \cdot (T_{sj} - T_{si} - \bar{\tau}_{ij}) \middle| \ u_{ij} \in \triangle_{l+1} \right\} - \sum_{(i,j)\in OD} d_{ij}(T_{ij} - T_{ii}) \to \min_T$$

(8)

or, equivalently,

$$\min_T \Phi(T) = \min_T \max_{u\in\triangle_{l+1}^m} \left\{ \sum_{(i,j)\in E} \bar{f}_{ij} \cdot \{\sum_{s\in S} u_{ij}^s \cdot (T_{sj} - T_{si} - \bar{\tau}_{ij})\} - \sum_{(i,j)\in OD} d_{ij}(T_{ij} - T_{ii}) \right\}$$

This ends the proof.

It is clear that objective function in optimization problem (7) has the following structure:

$$\Phi(T) = \max_u \left\{ \langle u, AT \rangle + \phi(T) + \varphi(u) \middle| \ u \in \triangle_{l+1}^m \right\}, \tag{9}$$

where $\phi(T)$ and $\varphi(u)$ are linear functions and A is a $\{m \cdot (l+1)\} \times \{n \cdot (l+1)\}$ matrix.

3   Smoothed Version of the Stable Dynamic Model

In Sect. 2.4 we show, that equilibrium in the Stable Dynamic model can be computed as the solution of the saddle-point problem in the form (9). This allows to use smoothing technique from [6].

We give short introduction to this approach.

Suppose we have convex optimization problem with the following structure:

$$\Phi(x) = \phi(x) + \max_u \left\{ \langle u, Ax \rangle - \varphi(u) \middle| \ u \in Q_2 \right\} \to \min_{x\in Q_1} . \tag{10}$$

Here Q_1 is a bounded closed convex set in a finite-dimensional real vector space E_1, and Q_2 is a bounded closed convex set in a finite-dimensional real vector space E_2. Functions $\phi(x)$ and $\varphi(u)$ are assumed to be continuous and convex, and the linear operator A maps E_1 to E_2. Function $\phi(x)$ is Lipschitz continuous with the constant K.

Denote by $\text{prox}_2(u)$ prox-function for set Q_2. This means that $\text{prox}_2(u)$ is continuous and strongly convex on Q_2 with some convexity parameter σ_2.

Denote by μ a positive smoothness parameter. Now we can denote smoothed model of objective function.

$$\Phi_\mu(x) = \max_u \left\{ \langle u, Ax \rangle - \varphi(u) - \mu \cdot \text{prox}_2(u) \middle| \ u \in Q_2 \right\} \to \min_{x\in Q_1} . \tag{11}$$

Denote by $u_\mu(x)$ the optimal solution of the problem

$$\max_u \left\{ \langle u, Ax \rangle - \varphi(u) - \mu \cdot \text{prox}_2(u) \middle| \ u \in Q_2 \right\}$$

Function $\Phi_\mu(x)$ possesses two important feature:

- Function $\Phi_\mu(x)$ is well defined, convex and continuously differentiable at any $x \in E_1$. Its gradient [6]

$$\nabla \Phi_\mu(x) = A^* u_\mu(x) \tag{12}$$

  is Lipschitz continuous with constant:

$$L_\mu = K + \frac{1}{\mu \cdot \sigma_2} \|A\|_{1,2}^2 \tag{13}$$

- Function $\bar{\Phi}(x) = \phi(x) + \Phi_\mu(x)$ uniformly approximate function $\Phi(x)$:

$$\bar{\Phi}(x) \leq \Phi(x) \leq \bar{\Phi}(x) + \mu \cdot D_2, \quad \forall x \in Q_2 \tag{14}$$

  where $D_2 = \max \{\mathrm{prox}_2(u) - \min_{u \in Q_2} \mathrm{prox}_2(u)| \quad u \in Q_2\}$.

From the mathematical programming point of view the bounded rationality assumption often leads to the smoothing of the potential function.

We can model errors into two ways. First, agents estimate the shortest path cost with a stochastic error. Second, agents estimate costs of arcs with errors.

First approach leads to the following optimization problem, described in [7]:

$$\sum_{i \in S, \ j \in V} \mu \cdot d_{ij} \cdot \ln \left(\frac{1}{|P_{ij}|} \sum_{q \in P_{ij}} \exp \left(\frac{-C_q(\tau)}{\mu} \right) \right) + \sum_{e \in E} \bar{f}_e(\tau_e - \bar{\tau}_e) \to \min_\tau$$

$$s.t. \quad \tau \geq \bar{\tau} \tag{15}$$

The second approach leads to the smoothed version of the Stable Dynamic model (SVSD-model).

Consider the following function:

$$\bar{\Phi}(T) =$$

$$\mu \cdot \sum_{(i,j) \in E} \bar{\tau}_{ij} \cdot \bar{f}_{ij} \cdot \log \frac{1}{l+1} \left(\sum_{s=1}^{l} \exp \left(\frac{(T_{sj} - T_{si} - \bar{\tau}_{ij})}{\bar{\tau}_{ij} \cdot \mu} \right) + 1 \right)$$

$$- \sum_{(i,j) \in OD} d_{ij}(T_{ij} - T_{ii}) \tag{16}$$

Theorem 1. *The function $\bar{\Phi}(T)$ is a smooth uniform approximation of the function $\Phi(T)$. For every T the following inequality holds*

$$\bar{\Phi}(T) \leq \Phi(T) \leq \bar{\Phi}(T) + \mu \sum_{(i,j) \in E} \bar{\tau}_{ij} \cdot \bar{f}_{ij} \cdot \log(l+1)$$

Proof: Show that $\bar{\Phi}(T)$ can be derived via smoothing technique with appropriate choice of prox-function.

Consider the following prox-function for convex set
$\triangle_{l+1}^m = \{u \in R^{m(l+1)} | \sum_{s=1}^{(l+1)} u_{ij}^s = 1, \forall (i,j) \in E, u \geq 0\}$:

$$\text{prox}_2(u) = \sum_{(i,j) \in E} \sum_{s=1}^{l+1} \bar{\tau}_{ij} \bar{f}_{ij} \cdot \left(u_{ij}^s \cdot \log u_{ij}^s + \log(l+1)\right) \tag{17}$$

For this prox-function

$$\sigma_2 = \min_{ij} \bar{\tau}_{ij} \cdot \bar{f}_{ij}$$
$$D_2 = \sum_{(i,j) \in E} \bar{\tau}_{ij} \cdot \bar{f}_{ij} \cdot \log(l+1) \tag{18}$$

Thus smoothed version of (7) can be rewritten in the form:

$$\min_T \Phi(T) =$$
$$-\mu \sum_{(i,j) \in E} \sum_{s=1}^{l+1} \bar{\tau}_{ij} \bar{f}_{ij} \log(l+1) +$$
$$\min_T \max_{u \in \triangle_{l+1}^m} \left[\sum_{(i,j) \in E} \bar{f}_{ij} \cdot \left(\sum_{s=1}^{l} u_{ij}^s \cdot (T_{sj} - T_{si} - \bar{\tau}_{ij}) \right) \right. \tag{19}$$
$$\left. - \sum_{(i,j) \in OD} d_{ij}(T_{ij} - T_{ii}) - \mu \sum_{(i,j) \in E} \sum_{s=1}^{l+1} \bar{\tau}_{ij} \bar{f}_{ij} \cdot u_{ij}^s \cdot \log u_{ij}^s \right]$$

From KKT optimality condition on u follows:

$$\bar{f}_{ij} \cdot (T_{sj} - T_{si} - \bar{\tau}_{ij}) - \mu \cdot \bar{\tau}_{ij} \cdot \bar{f}_{ij} \cdot (1 + \log u_{ij}^s) + \lambda_{ij} - \gamma_{ij}^s = 0, \quad \forall (i,j) \in E, \quad \forall 1 \leq s \leq l \quad (s \in S) \tag{20}$$

$$-\mu \cdot \bar{\tau}_{ij} \cdot \bar{f}_{ij} \cdot (1 + \log u_{ij}^s) + \lambda_{ij} - \gamma_{ij}^s = 0, \quad \forall (i,j) \in E, \quad s = l+1 \tag{21}$$

$$\sum_{s=1}^{l+1} u_{ij}^s = 1, \quad \forall (i,j) \in E \tag{22}$$

$$\gamma_{ij}^s \cdot u_{ij}^s = 0, \quad \forall s, \quad \forall (i,j) \in E \tag{23}$$

Here λ_{ij} and γ_{ij}^s are Lagrange multipliers for $\sum_{s=1}^{l+1} u_{ij}^s = 1$ and $u_{ij}^s \geq 0$ constraints respectively. Thus, from (20) and (21) we have:

$$u_{ij}^s = \exp \left(\frac{\bar{f}_{ij} \cdot (T_{sj} - T_{si} - \bar{\tau}_{ij}) + \lambda_{ij} - \gamma_{ij}^s}{\bar{\tau}_{ij} \cdot \bar{f}_{ij} \cdot \mu} - 1 \right), \quad \forall (i,j) \in E, \quad \forall 1 \leq s \leq l$$
$$\tag{24}$$
$$u_{ij}^s = \exp \left(\frac{\lambda_{ij} - \gamma_{ij}^s}{\bar{\tau}_{ij} \cdot \bar{f}_{ij} \cdot \mu} - 1 \right), \quad \forall (i,j) \in E, \quad s = l+1$$

From (24) it is clear that $u > 0$, thus, from (23), we have $\gamma_{ij}^s = 0$ for all $(i,j) \in E$ and $1 \leq s \leq l+1$.

Hence

$$u_{ij}^s = \exp\left(\frac{\bar{f}_{ij} \cdot (T_{sj} - T_{si} - \bar{\tau}_{ij}) + \lambda_{ij}}{\bar{\tau}_{ij} \cdot \bar{f}_{ij} \cdot \mu} - 1\right), \quad \forall (i,j) \in E, \quad \forall 1 \le s \le l \quad (25)$$

$$u_{ij}^s = \exp\left(\frac{\lambda_{ij}}{\bar{\tau}_{ij} \cdot \bar{f}_{ij} \cdot \mu} - 1\right), \quad \forall (i,j) \in E, \quad s = l+1$$

Substituting (25) in (22) we get:

$$\exp\left(\frac{\lambda_{ij}}{\bar{\tau}_{ij} \cdot \bar{f}_{ij} \cdot \mu} - 1\right) \cdot \left(\sum_{s=1}^l \exp\left(\frac{\bar{f}_{ij} \cdot (T_{sj} - T_{si} - \bar{\tau}_{ij})}{\bar{\tau}_{ij} \cdot \bar{f}_{ij} \cdot \mu}\right) + 1\right) = 1$$

Finally

$$\exp\left(\frac{\lambda_{ij}}{\bar{\tau}_{ij} \cdot \bar{f}_{ij} \cdot \mu} - 1\right) = \frac{1}{\left(\sum_{s=1}^l \exp\left(\frac{(T_{sj} - T_{si} - \bar{\tau}_{ij})}{\bar{\tau}_{ij} \cdot \mu}\right) + 1\right)} \quad (26)$$

Substituting (26) in (25) we get:

$$u_{ij}^s = \frac{\exp\left(\frac{(T_{sj} - T_{si} - \bar{\tau}_{ij})}{\bar{\tau}_{ij} \cdot \mu}\right)}{\left(\sum_{s=1}^l \exp\left(\frac{(T_{sj} - T_{si} - \bar{\tau}_{ij})}{\bar{\tau}_{ij} \cdot \mu}\right) + 1\right)}, \quad \forall (i,j) \in E, \quad \forall 1 \le s \le l \quad (27)$$

$$u_{ij}^s = \frac{1}{\left(\sum_{s=1}^l \exp\left(\frac{(T_{sj} - T_{si} - \bar{\tau}_{ij})}{\bar{\tau}_{ij} \mu}\right) + 1\right)}, \quad \forall (i,j) \in E, \quad s = l+1$$

From (27) and (19) we get:

$$\min_T \sum_{(i,j) \in E} \bar{f}_{ij} \cdot \left\{ \sum_{s=1}^l \frac{\exp\left(\frac{(T_{sj} - T_{si} - \bar{\tau}_{ij})}{\bar{\tau}_{ij} \cdot \mu}\right)}{\left(\sum_{s=1}^l \exp\left(\frac{(T_{sj} - T_{si} - \bar{\tau}_{ij})}{\bar{\tau}_{ij} \cdot \mu}\right) + 1\right)} \cdot (T_{sj} - T_{si} - \bar{\tau}_{ij}) \right\}$$

$$- \sum_{(i,j) \in OD} d_{ij}(T_{ij} - T_{ii}) - \mu \cdot \sum_{(i,j) \in E} \bar{\tau}_{ij} \cdot \bar{f}_{ij} \cdot \frac{1}{\left(\sum_{s=1}^l \exp\left(\frac{(T_{sj} - T_{si} - \bar{\tau}_{ij})}{\bar{\tau}_{ij} \cdot \mu}\right) + 1\right)}$$

$$\cdot \left(-\log\left(\sum_{s=1}^l \exp\left(\frac{(T_{sj} - T_{si} - \bar{\tau}_{ij})}{\bar{\tau}_{ij} \cdot \mu}\right) + 1\right)\right) - \mu \sum_{(i,j) \in E} \sum_{s=1}^l \bar{\tau}_{ij} \cdot \bar{f}_{ij}$$

$$\cdot \frac{\exp\left(\frac{(T_{sj} - T_{si} - \bar{\tau}_{ij})}{\bar{\tau}_{ij} \cdot \mu}\right)}{\left(\sum_{s=1}^l \exp\left(\frac{(T_{sj} - T_{si} - \bar{\tau}_{ij})}{\bar{\tau}_{ij} \cdot \mu}\right) + 1\right)} \cdot \log \frac{\exp\left(\frac{(T_{sj} - T_{si} - \bar{\tau}_{ij})}{\bar{\tau}_{ij} \cdot \mu}\right)}{\left(\sum_{s=1}^l \exp\left(\frac{(T_{sj} - T_{si} - \bar{\tau}_{ij})}{\bar{\tau}_{ij} \cdot \mu}\right) + 1\right)} \quad (28)$$

which is equivalent to

$$\min_T \left\{ \bar{\Phi}(T) = \mu \cdot \sum_{(i,j) \in E} \bar{\tau}_{ij} \cdot \bar{f}_{ij} \cdot \log\left(\sum_{s=1}^l \exp\left(\frac{(T_{sj} - T_{si} - \bar{\tau}_{ij})}{\bar{\tau}_{ij} \cdot \mu}\right) + 1\right) \right.$$

$$\left. - \sum_{(i,j) \in OD} d_{ij}(T_{ij} - T_{ii}) \right\} \quad (29)$$

This ends the proof.

In both models parameter μ determines how rational the agents are [1].

Problem (15) and problem (29) are both convex with smooth objective functions. These problems have attractable structure and allow to use optimization technique for huge-scale convex optimization problems.

3.1 Optimality Conditions for SVSD-Model

Consider first order optimality conditions for (29).

$$\nabla \bar{\Phi}(T^*) = 0$$

This leads to

$$\sum_{(ji)\in E} \frac{\bar{f}_{ji} e^{\frac{1}{\bar{\tau}_{ji}\cdot\mu}(T_{si}^* - \bar{\tau}_{ji} - T_{sj}^*)}}{1 + \sum_k e^{\frac{1}{\bar{\tau}_{ji}\cdot\mu}(T_{ki}^* - \bar{\tau}_{ji} - T_{kj}^*)}} - \sum_{(ij)\in E} \frac{\bar{f}_{ij} e^{\frac{1}{\bar{\tau}_{ij}\cdot\mu}(T_{sj}^* - \bar{\tau}_{ij} - T_{si}^*)}}{1 + \sum_k e^{\frac{1}{\bar{\tau}_{ij}\cdot\mu}(T_{kj}^* - \bar{\tau}_{ij} - T_{ki}^*)}} - d_{si} = 0, \quad i \neq s \tag{30}$$

$$\sum_{(js)\in E} \frac{\bar{f}_{js} e^{\frac{1}{\bar{\tau}_{js}\cdot\mu}(T_{ss}^* - \bar{\tau}_{js} - T_{sj}^*)}}{1 + \sum_k e^{\frac{1}{\bar{\tau}_{js}\cdot\mu}(T_{ks}^* + \bar{\tau}_{js} - T_{kj}^*)}} - \sum_{(sj)\in E} \frac{\bar{f}_{sj} e^{\frac{1}{\bar{\tau}_{sj}\cdot\mu}(T_{sj}^* - \bar{\tau}_{sj} - T_{ss}^*)}}{1 + \sum_k e^{\frac{1}{\bar{\tau}_{sj}\cdot\mu}(T_{kj}^* - \bar{\tau}_{sj} - T_{ks}^*)}}$$
$$+ \sum_{j:(s,j)\in OD} d_{sj} = 0, \quad i = s \tag{31}$$

This is a simple flow balance condition, which requires that in equilibrium for every node i and for every source s the difference between incoming flow in i, generated by s, and outcoming flow from i, generated from s is equal to the demand d_{si}.

Stochastic Equilibrium Flow and Optimal Allocation Problem

From (30) it is clear that variables u_{ij}^s, which we used earlier, can be interpreted as a ratio of arc capacity $\bar{f}_{ij}$ which is used by source s.

$$f_{ij}^s(T^*) = \bar{f}_{ij} \cdot \frac{e^{\frac{1}{\bar{\tau}_{ij}\cdot\mu}(T_{sj}^* - \bar{\tau}_{ij} - T_{si}^*)}}{1 + \sum_{v\in S} e^{\frac{1}{\bar{\tau}_{ij}\cdot\mu}(T_{vj}^* - \bar{\tau}_{ij} - T_{vi}^*)}}, \tag{32}$$

Hence one can compute the stochastic equilibrium arc flow corresponding to the solution of (29) explicitly:

$$f_{ij}(T^*) = \sum_{s\in S} f_{ij}^s(T^*) = \bar{f}_{ij} \cdot \frac{\sum_{s\in S} e^{\frac{1}{\bar{\tau}_{ij}\cdot\mu}(T_{sj}^* - \bar{\tau}_{ij} - T_{si}^*)}}{1 + \sum_{v\in S} e^{\frac{1}{\bar{\tau}_{ij}\cdot\mu}(T_{vj}^* - \bar{\tau}_{ij} - T_{vi}^*)}}, \tag{33}$$

where f_{ij} is a flow from i to j, and T^* is the solution of (29).

From (32) one can see that equilibrium conditions in SVSD-model require to distribute arc capacity between source-nodes (and flows, generated by them)

according to the smoothed version of the *logit-choice* model [1]. Weighed residual
of node potentials have a role of utility:

$$U_{ij}^s = \frac{(T_{sj}^* - T_{si}^* - \bar{\tau}_{ij})}{\bar{\tau}_{ij} \cdot \mu} \tag{34}$$

The Eq. (32) can be considered as the capacity optimal allocation rule. On
the one hand this is natural, since a problem dual to the Stable Dynamic model
is a simple minimal cost flow problem, which is the problem of centralized flow
management. On the other side this is very surprising, because the Beckmann's
model and other equilibrium models are usually considered as decentralized
models.

However, this contradiction is false. If we consider Beckmann's equilibrium
model more closely, we can see that it can be represented as the problem of a
centralized optimal allocation problem.

$$\min_x \left\{ \Phi(x) = \sum_{e \in E} \int_0^{f_e} \tau(z) dz \right\}$$
$$f = \Theta x \tag{35}$$
$$x \in X = \left\{ x \in R_+^{|P|} \mid \sum_{p \in P_{ij}} x_p = d_{ij} \right\}$$

where Θ is path-arc incidence matrix.

Now we can consider a new agent, which has to find the optimal allocation
rule for external demands $\{d_{ij}\}$ on a given network with a specific arcs cost
function

$$t(f_e) = \int_0^{f_e} \tau(z) dz$$

Actually, this kind of trick can be used to every congested type game. Despite
its triviality, this allows to work with special type equilibrium models (e.g. potential games) as if they are optimal allocation problems.

4   Computational Techniques for the Stochastic Equilibrium Problems

In the original paper [10] where the Stable Dynamic model was proposed the
authors did not consider any computational technique for computing equilibrium. This issue limits the implementation of this model to the practical needs.

Important Remark:
Two different goals that can be achieved by computation of the solution of
SVSD-model. First one is that the solution of the SVSD-model is a stochastic equilibrium on the congested network, so it has its own value. If this is the
case, then the parameter μ is *fixed* and represent how irrational agents are.

The second goal is to use the solution of the SVSD-model as a stepping stone while the final aim is to compute the solution of the Stable Dynamic model. In this case the parameter μ *can vary* and should be considered as the *smoothing* parameter. This is important, because for each computational method there are two different estimations of the complexity. First corresponds to the complexity of finding ϵ-solution of the SVSD-model (with fixed μ) and the second corresponds to the complexity of finding ϵ-solution of the original Stable Dynamic model.

5   Numerical Experiments

Numerical experiments are performed on test models from well-known TNTP collection (https://github.com/bstabler/TransportationNetworks). We use FGM (accelerated gradient descent [6]), UFGM (universal FGM [8]) and AFGM (Adaptive FGM) methods. All methods were implemented with using C++ language and tested on the same machine.

We test method's "real" convergence speed for problems with different value of smoothing parameter μ. Our first tests show that decreasing of μ value leads to significant slowdown of method's convergence. This is expected result. With $\mu \leq 10^{-3}$ we can't find any "good" (with some required accuracy) solution in a reasonable time. So we limit $\mu \geq 10^{-2}$ which seems to be a compromise between model's accuracy and method's convergence speed.

The results of our tests presented in Tables 1 and 2 show that the search for high-accuracy (according to the gradient norm value) solutions requires significant computational and time costs. The FGM method showed the worst results, which can be explained by not too accurate estimation of the Lipschitz constant which is also proportional to μ^{-1}. The more complex UFGM and AFGM method work much better compared to a simple FGM, but also have some differences - AFGM works better when μ is "big" and UFGM is undisputed leader for "small" μ values.

Table 1. Solving time (sec.) for "SiouxFalls" problem

μ	Stop criteria	FGM	UFGM	AFGM
10^{-1}	$\|\nabla f(x^k)\|_\infty \leq 10^{-1}$	0.70	0.02	0.04
	$\|\nabla f(x^k)\|_\infty \leq 10^{-2}$	3.00	0.05	0.31
	$\|\nabla f(x^k)\|_\infty \leq 10^{-3}$	13.35	0.10	0.51
	$\|\nabla f(x^k)\|_\infty \leq 10^{-4}$	66.19	2.81	0.69
10^{-2}	$\|\nabla f(x^k)\|_\infty \leq 10^{-1}$	69.62	0.34	2.10
	$\|\nabla f(x^k)\|_\infty \leq 10^{-2}$	300.70	0.68	3.95
	$\|\nabla f(x^k)\|_\infty \leq 10^{-3}$	1335.30	1.41	9.25
	$\|\nabla f(x^k)\|_\infty \leq 10^{-4}$		12.41	52.70

Table 2. Solving time (sec.) for "Anaheim" problem

μ	Stop criteria	UFGM	AFGM
10^{-1}	$\|\nabla f(x^k)\|_\infty \leq 10^{-1}$	1.59	7.07
	$\|\nabla f(x^k)\|_\infty \leq 10^{-2}$	8.45	13.83
	$\|\nabla f(x^k)\|_\infty \leq 10^{-3}$	29.45	38.11
	$\|\nabla f(x^k)\|_\infty \leq 10^{-4}$	562.43	76.82
10^{-2}	$\|\nabla f(x^k)\|_\infty \leq 10^{-1}$	146.30	526.83
	$\|\nabla f(x^k)\|_\infty \leq 10^{-2}$	413.29	1251.55
	$\|\nabla f(x^k)\|_\infty \leq 10^{-3}$	1065.72	2460.64
	$\|\nabla f(x^k)\|_\infty \leq 10^{-4}$	3811.10	5710.98

It is essential to remark, that in transportation modeling we usually interested in "medium" precision. This is true because (1) we usually wish to solve macroscopic transportation equilibrium model to make long-run investment decision (i.e. where to build new road) and good decision should be robust to a small errors and (2) in our model gradients corresponds to flows (i.e. cars) in huge multi-agent system and small change in precision do not make any difference from practical point of view (you do not care if it 0.05 car per hour or 0.1 car per hour). So we are really interested in fast "medium"-precision solutions.

6   Conclusion

In this work we propose a new smoothed version of the Stable Dynamic model (29). It can be used as a stepping-stone in optimization routine to compute equilibrium in stable dynamic model or as an independent transportation model [stochastic version of stable dynamic model].

We checked that equilibrium could be efficiently computed by almost any first order optimization scheme. Probably next reasonable step is to construct reliable scheme how to choose μ. Another interesting question is to construct rule that maps choice of prox-function to a agent's stochastic decision rule [discrete choice theory].

Acknowledgment. We thank Alexander Gasnikov for comments that greatly improved the manuscript.

We are also immensely grateful to our reviewers for their important comments on an earlier version of the manuscript, although any errors are our own and should not tarnish the reputations of these esteemed persons.

References

1. Andersen, S., de Palma, A., Thisse, J.-F.: Discrete Choice Theory of Product Differentiation. MIT Press, Cambridge (1992)

2. Bar-Gera, H.: Origin-based algorithms for transportation network modeling. University of Illinois at Chicago (1999)
3. Beckmann, M., McGuire, B., Winsten, C.: Studies in the Economics of Transportation. Yale University Press, New Haven (1956)
4. Frank, M., Wolfe, P.: An algorithm for quadratic programming. Nav. Res. Logist. Q. **3**, 95–110 (1956)
5. LeBlanc, L., Morlok, E., Pierskalla, W.: An efficient approach to solving the road network equilibrium traffic assignment problem. Transp. Res. B. **9**, 309–318 (1975)
6. Nesterov, Y.: Smooth minimization of non-smooth functions. Math. Program. Ser. A **103**, 127–152 (2005)
7. Nesterov, Y.: Characteristic functions of directed graphs and applications to stochastic equilibrium problems. Optim. Eng. **8**(2), 193–214 (2007). https://doi.org/10.1007/s11081-007-9013-3
8. Nesterov, Y.: Universal gradient methods for convex optimization problems. Math. Program. **152**(1–2), 381–404 (2015)
9. Ortuzar, J., Willumsen, L.: Modelling Transport, 4th edn. Wiley, Hoboken (2011)
10. de Palma, A., Nesterov, Y.: Stationary dynamic solutions in congested transportation networks: summary and perspectives. Netw. Spat. Econ. **3**, 371–395 (2003)
11. Patriksson, M.: The Traffic Assignment Problem - Models and Methods. VSP, Utrecht, Netherlands (1994)
12. Sheffi, Y.: Urban Transportation Networks. Prentice Hall, Englewood Cliffs (1985)
13. Wardrop, J.: Some theoretical aspects of road traffic research. Proc. Inst. Civil. Eng. Part **II**(1), 325–378 (1952)

Dual Multiplicative-Barrier Methods for Linear Second-Order Cone Programming

Vitaly Zhadan[1,2]([⊠]) [ID]

[1] Dorodnicyn Computing Centre, FRC "Computer Science and Control" of RAS,
40, Vavilova str., Moscow 119333, Russia
zhadan@ccas.ru
[2] Moscow Institute of Physics and Technology (State Research University),
9 Institutskiy per., Dolgoprudny, Moscow Region 141701, Russia

Abstract. The linear second-order cone programming problem is considered. For its solution the dual multiplicative barrier methods are proposed. The methods are generalizations on the cone programming the corresponding methods for linear programming. They belong to the class of dual affine-scaling methods and can be treated as a special way for solving the optimality conditions for primal and dual problems. The local convergence of the methods with linear rate is proved.

Keywords: Second-order cone programming · Dual affine-scaling method · Local convergence

1 Introduction

The linear cone programming programs are optimization problems in which the linear objective function is minimized on the intersection of a linear manifold with a convex closed cone. In second-order cone programming (SOCP) this cone is usually a direct product of some Lorentz cones (see [1,2]). Many other optimization problems, for example, problems of robust and combinatorial optimization, can be reformulated as SOCP programs [2,3].

The numerical techniques for solving SOCP programs are obtained as generalizations of the corresponding methods for linear programming. The most popular methods from them are the primal-dual methods which are generalizations of the interior point techniques corresponding to the path-follows methods [4,5]. In the present paper, the dual methods belonging to the class of dual affine-scaling techniques for SOCP, are considered. These dual methods can be treated also as dual analogs of the primal method [6]. Moreover, the proposed methods have many common properties with the dual methods developed for linear semi-definite programming [7].

This work was partially supported by the Russian Foundation for Basic Research (project no. 17-07-00510) and by Program 2 of Presidium of RAS "Mechanisms for ensuring fault tolerance in modern high-performance and highly reliable computing".

M. Jaćimović et al. (Eds.): OPTIMA 2019, CCIS 1145, pp. 295–310, 2020.
https://doi.org/10.1007/978-3-030-38603-0_22

The paper is organized as follows. In Sect. 2, the statement of SOCP is given. In Sect. 3, the feasible and general variants of dual methods are constructed. In Sect. 4, the feasible variant of the method is considered more carefully. Finally, in Sect. 5, the local convergence of the methods is proved.

Below the symbol I_s is used for denoting the identity matrix of the order s. The symbol 0_s indicates a zero s-dimensional vector, the symbol 0_{sk} indicates a $s \times k$ zero matrix. By Diag(x) is denoted the diagonal matrix with a vector x at its diagonal. Respectively, by DIAG $(M_1, \ldots, M_k)$ is denoted a block diagonal matrix with diagonal blocks $M_1, \ldots, M_k$.

2   Primal and Dual SOCP Problems

Let $\mathcal{K} \subset \mathbb{R}^n$ denote a closed convex pointed cone with the nonempty interior $\mathcal{K}_0 = \operatorname{int} \mathcal{K}$. This cone $\mathcal{K}$ induces in $\mathbb{R}^n$ a partial order, that is $x_1 \succeq_K x_2$, if $x_1 - x_2 \in K$.

The linear cone programming problem is

$$\min \langle c, x \rangle, \quad \mathcal{A}x = b, \quad x \in \mathcal{K}, \tag{1}$$

where $\mathcal{A}$ is a $m \times n$ matrix, and $c = [c^1; \ldots; c^n] \in \mathbb{R}^n$, $b = [b^1; \ldots; b^m] \in \mathbb{R}^m$. The semicolons between vectors or components of a vector denote that these vectors or components are placed one under another. The brackets $\langle \cdot, \cdot \rangle$ denotes the usual Euclidean scalar product.

Below the special partial case of the problem (1) will be of main interest for us. Let $c_i \in \mathbb{R}^{n_i}$, $1 \leq i \leq r$. Let also matrices A_i have dimensions $m \times n_i$, $1 \leq i \leq r$. Consider the cone programming problem

$$\min \sum_{i=1}^{r} \langle c_i, x_i \rangle,$$
$$\sum_{i=1}^{r} A_i x_i = b, \quad x_1 \succeq_{K^{n_1}} 0_{n_1}, \quad \ldots, \quad x_r \succeq_{K^{n_r}} 0_{n_r}. \tag{2}$$

Here K^{n_i} is the second-order cone (the Lorentz cone) defined as:

$$K^{n_i} = \left\{ [x^0; \bar{x}] \in \mathbb{R} \times \mathbb{R}^{n_i - 1} : x^0 \geq \|\bar{x}\| \right\}, \quad 1 \leq i \leq r,$$

where $\| \cdot \|$ is the Euclidean norm. The following problem is dual to (2)

$$\max \langle b, u \rangle,$$
$$A_i^T u + y_i = c^i, \ 1 \leq i \leq r; \quad y_1 \succeq_{K^{n_1}} 0_{n_1}, \quad \ldots, \quad y_r \succeq_{K^{n_r}} 0_{n_r}, \tag{3}$$

in which $u \in \mathbb{R}^m$.

Denote $n = n_1 + \cdots + n_r$. If $c = [c_1; \ldots; c_r]$, $x = [x_1; \ldots; x_r]$, $y = [y_1; \ldots; y_r]$ and

$$\mathcal{A} = [A_1, \ldots A_r], \qquad \mathcal{K} = K^{n_1} \times \cdots \times K^{n_r},$$

then the problem (2) can be written in the form of (1). The cone $\mathcal{K}$ is self-dual, that is $\mathcal{K}^* = \mathcal{K}$. We assume that both problems (2) and (3) have solutions, and the rows of the matrix $\mathcal{A}$ are linear independent. We assume also that $r > 1$.

Let $\mathcal{F}_D = \{[u, y] \in \mathbb{R}^m \times \mathcal{K} : y = y(u)\}$ be the feasible set in problem (3). Here and in what follows: $y(u) = c - \mathcal{A}^T u$. By $\mathcal{F}_{D,u}$ we will denote the projection of the set $\mathcal{F}_D$ onto the space $\mathbb{R}^m$, i.e. the set $\mathcal{F}_{D,u} = \{u \in \mathbb{R}^m : y(u) \in \mathcal{K}\}$.

The necessary and sufficient optimality conditions for the pair of problems (2) and (3) consist of the following equalities (see [2]):

$$\langle x, y \rangle = 0, \qquad \mathcal{A}x = b, \qquad y = c - \mathcal{A}^T u, \tag{4}$$

in which $x \in \mathcal{K}$, $y \in \mathcal{K}$. Taking into account these inclusions, the equality $\langle x, y \rangle = 0$ can be replaced by n other equalities

$$x_i \circ y_i = 0_{n_i}, \quad 1 \le i \le r,$$

where the product between vectors $x_i \in \mathbb{R}^{n_i}$ and $y_i \in \mathbb{R}^{n_i}$ is defined by the following way $x_i \circ y_i = [x_i^T y_i; \ x_i^0 \bar{y} + y_i^0 \bar{x}_i]$. By introducing the matrix

$$\mathrm{Arr}\,(x_i) = \begin{bmatrix} x_i^0 & \bar{x}_i^T \\ \bar{x}_i & x_i^0 I_{n-1} \end{bmatrix},$$

the product $x_i \circ y_i$ can be represented as $x_i \circ y_i = \mathrm{Arr}\,(x_i)\,y_i = \mathrm{Arr}\,(y_i)\,x_i$.

Compose the block diagonal matrix

$$\mathcal{G}(y) = \mathrm{DIAG}\,[\mathrm{Arr}\,(y_1), \ \ldots, \ \mathrm{Arr}\,(y_r)]. \tag{5}$$

With the help of (5) equalities (4) can be rewritten as

$$\mathcal{G}(y)x = 0_n, \qquad \mathcal{A}x = b, \qquad y = c - \mathcal{A}^T u. \tag{6}$$

3   The Iterative Processes

Consider the dual method for solving problems (2) and (3). For constructing the method we multiply the second equality from (6) by the matrix $\mathcal{A}^T$ and sum it with the first equality (6). As a result, we obtain

$$\Phi(y)x = \mathcal{A}^T b, \tag{7}$$

where by $\Phi(y)$ is denoted the matrix $\Phi(y) = \mathcal{A}^T \mathcal{A} + \mathcal{G}(y)$. The matrix $\Phi(y)$ is symmetric of the order n. If $\Phi(y)$ is nonsingular, then, solving the Eq. (7), we get

$$x = x(y) = \Phi^{-1}(y)\mathcal{A}^T b. \tag{8}$$

Taking $y = y(u)$, we obtain that in this case the matrix $\Phi(y)$ depends on u.

Substituting the founded from (8) $x(y(u))$ into the second equation from (6), we derive the system of equations with respect to u, namely,

$$\left[I_m - \mathcal{A}\Phi^{-1}(y(u))\mathcal{A}^T\right] b = 0_m. \tag{9}$$

The system (9) consists of m equations. The number of unknowns is also equal to m.

Applying the fixed-point method to solve system (9), we obtain the iterative process

$$u_{k+1} = u_k + \alpha_k \left[I_m - \mathcal{A}\Phi^{-1}(y_k)\mathcal{A}^T \right] b, \qquad y_k = y(u_k), \qquad (10)$$

where $\alpha_k > 0$ is the step size. The starting point u_0 must be taken from the set $\mathcal{F}_{D,u}$.

Consider the more general iterative process. In this process both variables u and y are updated at each iteration. What is more, the equality $y_k = y(u_k)$ may not hold. For this purpose, we add to the right hand side of Eq. (7) the second equality from (6), multiplied by the preset parameter $\tau > 0$. As a result, we obtain instead of (9) the system of equations $\Phi(y)x = \mathcal{A}^T b + \tau \left(y + \mathcal{A}^T u - c \right)$. The function $x = x(u)$ from (8) is replaced by $x = x(u,y) = \Phi^{-1}(y)f(u,y)$, where $f(u,y) = \mathcal{A}^T b + \tau \left(y + \mathcal{A}^T u - c \right)$.

Substituting $x(u,y)$ in first and second equalities from (6), we obtain the system of $n + m$ equations

$$\mathcal{G}(y)\Phi^{-1}(y)f(u,y) = 0_n, \qquad b - \mathcal{A}\Phi^{-1}(y))f(u,y) = 0_m. \qquad (11)$$

Applying again the fix point method for solving (11), we derive the iterative process

$$u_{k+1} = u_k + \alpha_k \left[b - \mathcal{A}x_k \right], \quad y_{k+1} = y_k - \alpha_k \mathcal{G}(y_k)x_k, \quad x_k = \Phi^{-1}(y_k))f(u_k, y_k). \qquad (12)$$

The iterative process (12) we will call by the *dual multiplicative-barrier method*. This name is explained by introducing the matrix $\mathcal{G}(y)$ into the right-hand side of (12), which does not allow y_k to leave the cone $\mathcal{K}$. In contrast to (12) the iterative process (10) we will call by the *feasible dual multiplicative-barrier method*.

Let us give the definition of non-degeneracy of the point $[u, y] \in \mathcal{F}_D$ from [2].

Definition 1. *The point* $[u, y] \in \mathcal{F}_D$ *is called non-degenerate, if* $\mathcal{T}_{\mathcal{K}}(y) + \mathcal{R}(\mathcal{A}^T) = \mathbb{R}^n$, *where* $\mathcal{T}_{\mathcal{K}}(y)$ *is the tangent space to the cone* $\mathcal{K}$ *at the point* $y \in \mathcal{K}$ *and* $\mathcal{R}(\mathcal{A}^T)$ *is the image of the matrix* $\mathcal{A}^T$.

Let $[u, y] \in \mathcal{F}_D$, and let the vector $y \in \mathcal{K}$ be partitioned onto three blocks of components

$$y = [y_F; y_I; y_N]. \qquad (13)$$

Let, for definiteness, these blocks consist of components ordered in the following way:

$$y_F = [y_1; \dots; y_{r_F}], \quad y_I = [y_{r_F+1}; \dots; y_{r_F+r_I}], \quad y_N = [y_{r_F+r_I+1}; \dots; y_{r_F+r_B+r_N}]. \qquad (14)$$

This partition of the vector y induces the partition of the set $J^r = [1 : r]$ onto three index sets $J_F^r = J_F^r(y)$, $J_I^r = J_I^r(y)$ and $J_N^r = J_N^r(y)$, where

$$J_F^r = [1, \dots, r_F], \quad J_I^r = [r_F+1, \dots, r_F+r_I], \quad J_N^r = [r_F+r_I+1, \dots, r_F+r_I+r_N].$$

We have $r_F + r_I + r_N = r$. If $i \in J_F^r(y)$, then the component $y_i \neq 0_{n_i}$ and $y_i \in \partial \mathcal{K}^{n_i}$, where $\partial \mathcal{K}^{n_i}$ is the boundary of the cone $\mathcal{K}^{n_i}$. If $i \in J_I^r(y)$, then

$y_i = 0_{n_i}$. At last, if $i \in J_N^r(y)$, then the following inclusion $y_i \in \operatorname{int} K^{n_i}$ takes place.

In accordance with the partition of the vector y onto three blocks of component we make also partitions of the matrix $\mathcal{A}$ and the vector c:

$$\mathcal{A} = [\mathcal{A}_F, \mathcal{A}_I, \mathcal{A}_N], \qquad c = [c_F; c_I; c_N]. \tag{15}$$

For any nonzero component $y_i \in \mathbb{R}^{n_i}$, $i \in J^r$ the spectral decomposition takes place (see [2]):

$$y_i = \theta_{i,1} \mathbf{d}_{i,1} + \theta_{i,n_i} \mathbf{d}_{i,n_i}, \tag{16}$$

where $\mathbf{d}_{i,1}$ and $\mathbf{d}_{i,n_i}$ from $\mathbb{R}^{n_i}$ compose the Jordan frame of y_i. These vectors have the form

$$\mathbf{d}_{i,1} = \frac{1}{\sqrt{2}} \left[1; \frac{\bar{y}_i}{\|\bar{y}_i\|} \right], \qquad \mathbf{d}_{i,n_i} = \frac{1}{\sqrt{2}} \left[1; -\frac{\bar{y}_i}{\|\bar{y}_i\|} \right].$$

The coefficients $\theta_{i,1}$ and θ_{i,n_i} in (16) are the following:

$$\theta_{i,1} = \frac{1}{\sqrt{2}} \left(y_i^0 + \|\bar{y}_i\| \right), \qquad \theta_{i,n_i} = \frac{1}{\sqrt{2}} \left(y_i^0 - \|\bar{y}_i\| \right).$$

Both frame vectors $\mathbf{d}_{i,1}$, $\mathbf{d}_{i,n_i}$ are orthogonal each to other and their lengths equal to one.

If $y_i \in K^{n_i}$, then $\theta_{i,1} \geq 0$, $\theta_{i,n_i} \geq 0$. In the case where $y_i \neq 0_{n_i}$ and $y_i \in \partial K^{n_i}$, only the first coefficient is strictly positive, i.e. $\theta_{i,1} = \sqrt{2} y_i^0 = \sqrt{2} \|\bar{y}_i\|$.

Assume that $y_i \in K^{n_i}$ and $y_i \neq 0_{n_i}$. The matrix $\operatorname{Arr}(y_i)$ is symmetric. Let H_i be an orthogonal matrix, consisting from eigenvectors of $\operatorname{Arr}(y_i)$. The vectors $\mathbf{d}_{i,1}$ and $\mathbf{d}_{i,n_i}$ are contained in the set of eigenvectors of $\operatorname{Arr}(y_i)$, i.e. it is possible to represent the matrix H_i in the form $H_i = [\mathbf{d}_{i,1}, h_{i,2}, \ldots, h_{i,n_i-1}, \mathbf{d}_{i,n_i}]$. The eigenvectors $h_{i,2}, \ldots h_{i,n_i-1}$ may be taken arbitrary from the subspace $\mathbb{R}_0^n = \left\{ z = [z^0; \bar{z}] \in \mathbb{R}^n : z^0 = 0 \right\}$. It is important, that they have the unit length and be orthogonal each to others and to the vectors $\mathbf{d}_{i,1}$, $\mathbf{d}_{i,n_i}$.

The eigenvalues $y_i^0 + \|\bar{y}_i\|$ and $y_i^0 - \|\bar{y}_i\|$ correspond to the eigenvectors $\mathbf{d}_{i,1}$ and $\mathbf{d}_{i,n_i}$, respectively. The eigenvalue y_i^0 corresponds to all eigenvectors $h_{i,2}, \ldots h_{i,n_i-1}$. Therefore, denoting by Θ_i the diagonal matrix

$$\Theta_i = \operatorname{Diag}\left(\sqrt{2}\theta_{i,1}, y_i^0, \ldots, y_i^0, \theta_{i,n_i} \right),$$

we have $\operatorname{Arr}(y_i) = H_i^T \Theta_i H_i$.

If $i \in J_I^r(y)$, then $y_i = 0_{n_i}$. In this case the identity matrix I_{n_i} can be taken as the orthogonal matrix H_i. It is obvious, that $\Theta_i = 0_{n_i n_i}$ for such $\operatorname{Arr}(y_i)$.

Let us introduce into consideration the block diagonal matrices:

$$\mathcal{H}_F = \operatorname{DIAG}[H_1, \ldots, H_{r_F}], \quad \mathcal{H}_I = \operatorname{DIAG}[H_{r_F+1}, \ldots, H_{r_F+r_I}], \tag{17}$$

$$\mathcal{H}_N = \operatorname{DIAG}[H_{r_F+r_I+1}, \ldots, H_r], \quad \mathcal{H} = \operatorname{DIAG}[\mathcal{H}_F, \mathcal{H}_I, \mathcal{H}_N]. \tag{18}$$

All these matrices are orthogonal. In similar way we deal with matrices Θ_i, $i \in J^r$, combining them in matrices

$$\Theta_F = \text{DIAG}\,[\Theta_1, \ldots, \Theta_{r_F}], \quad \Theta_I = \text{DIAG}\,[\Theta_{r_F+1}, \ldots, \Theta_{r_F+r_I}], \qquad (19)$$

$$\Theta_N = \text{DIAG}\,[\Theta_{r_F+r_I+1}, \ldots, \Theta_r], \quad \Theta = \text{DIAG}\,[\Theta_F, \Theta_I, \Theta_N]. \qquad (20)$$

Matrices (19) and (20) are diagonal. Moreover, their diagonal elements are non-negative, if $y \in \mathcal{K}$. We have $\mathcal{G}(y) = \mathcal{H}\Theta\mathcal{H}^T$.

Proposition 1. *Let the point $[u, y] \in \mathcal{F}_D$ be non-degenerate. Then the matrix $\Phi(y)$ is nonsingular.*

Proof. Without loss of generality, we assume that all three index sets J_F^r, J_I^r and J_N^r are not empty at the point y. Then for matrices $\mathcal{H}$ and Θ representations (18) and (20) hold.

Multiplying the matrix $\Phi(y)$ from the left and the right by matrices $\mathcal{H}^T$ and $\mathcal{H}$, respectively, we obtain as a result the matrix

$$\Phi^{\mathcal{H}}(y) = \mathcal{H}^T \Phi(y) \mathcal{H} = \left(\mathcal{A}^{\mathcal{H}}\right)^T \mathcal{A}^{\mathcal{H}} + \Theta, \qquad (21)$$

where $\mathcal{A}^{\mathcal{H}} = \mathcal{A}\mathcal{H}$. Since $\mathcal{H}$ is a nonsingular matrix, $\Phi(y)$ is a nonsingular matrix if and only if the matrix $\Phi^{\mathcal{H}}(y)$ is also nonsingular.

The symmetric matrix $\Phi(y)$ is nonnegative definite. Let us show that in fact it is positive definite. For this purpose, it is sufficient to verify that the system of linear equations

$$\Phi^{\mathcal{H}}(y)z = 0_n \qquad (22)$$

has only the trivial zero solution. Indeed, after multiplying left and right parts of (22) by z^T we obtain

$$\|\mathcal{A}^{\mathcal{H}}z\|^2 + \langle z, \Theta z \rangle = 0. \qquad (23)$$

Since all diagonal elements of the matrix Θ are nonnegative, the equality (23) holds if and only if

$$\|\mathcal{A}^{\mathcal{H}}z\| = 0, \qquad \langle z, \Theta(y)z \rangle = 0. \qquad (24)$$

Partition the vector z according to the partition of the vector y onto three blocks: $z = [z_F; z_I; z_N]$. Then the right equality (24) is split onto three separate equalities:

$$\langle z_F, \Theta_F z_F \rangle = 0, \quad \langle z_I, \Theta_I z_I \rangle = 0, \quad \langle z_N, \Theta_N z_N \rangle = 0, \qquad (25)$$

and what is more, Θ_I is a zero matrix.

Since the matrix Θ_N is positive definite, it follows from (25) that $z_N = 0$. Moreover, since all diagonal elements of the matrices Θ_i, $i \in J_F^r = J_F^r(y)$, with the exception of the last element are positive, the corresponding elements of z_F are zeros. Only last components $z_i^{n_i}$, $i \in J_F^r$, may differ from zero. Let $\tilde{\mathcal{A}}_F^{\mathcal{H}}$ be the matrix $\mathcal{A}_F^{\mathcal{H}}$, from which all columns of the matrices $A_i^H = A_i H_i$, $i \in J_F^r$, are removed except for last columns. The matrix $\tilde{\mathcal{A}}_F^{\mathcal{H}}$ has the dimension $m \times r_F$.

Denote

$$A_{FI}^{\mathcal{H}} = \left[\tilde{A}_F^{\mathcal{H}}, A_I^{\mathcal{H}}\right]. \tag{26}$$

According to what has been said the first equality (24) reduces to $A_{FI}^{\mathcal{H}} z_{FI} = 0$, where $z_{FI} = \left[z_1^{n_1}; \ldots; z_{r_F}^{n_{r_F}}; z_I\right]$. But by the criterion of non-degeneracy in the dual problem the point $[u, y] \in \mathcal{F}_D$ is non-degenerate if and only if the columns of the matrix $A_{FI}^{\mathcal{H}}$ are linear independent (see [2]). Therefore, we have $z_{FI} = 0$ and $z = 0_n$. Hence the matrix $\Phi^{\mathcal{H}}(y)$ is nonsingular. Since the matrix $\mathcal{H}$ is orthogonal, $\Phi(y)$ is a nonsingular matrix. □

We call the problem (3) *non-degenerate* if all points $[u, y] \in \mathcal{F}_D$ are non-degenerate. In what follows, we assume that the problem (3) is non-degenerate. Then the right-hand sides of iterative processes (10) and (12) are defined on the feasible set $\mathcal{F}_D$. Due to continuity, it is defined also on some neighborhood of $\mathcal{F}_D$.

4   The Other Form of the Feasible Method

Define more carefully the right hand side of the iterative process (10). For this purpose, we need to evaluate the matrix $\Phi^{-1}(y)$.

Let $[u, y] \in \mathcal{F}_D$, and let for the vector y the partition (13) hold. Moreover, the blocks y_F, y_I, y_N are determined by (14) with the block y_I being zero, and the corresponding partition of the matrix $\mathcal{A}$ is given by (15). Passing from the matrix $\Phi(y)$ to matrix $\Phi^{\mathcal{H}}(y) = \mathcal{H}^T \Phi(y) \mathcal{H}$, we obtain

$$\Phi(y) = \mathcal{H} \Phi^{\mathcal{H}}(y) \mathcal{H}^T, \qquad \Phi^{-1}(y) = \mathcal{H} \left(\Phi^{\mathcal{H}}(y)\right)^{-1} \mathcal{H}^T.$$

The symmetric matrix $\Phi^{\mathcal{H}}(y)$ has the order n, and in block form it can be written as

$$\Phi^{\mathcal{H}}(y) = \begin{bmatrix} \left(A_F^{\mathcal{H}}\right)^T A_F^{\mathcal{H}} + \Theta_F & \left(A_F^{\mathcal{H}}\right)^T A_I^{\mathcal{H}} & \left(A_F^{\mathcal{H}}\right)^T A_N^{\mathcal{H}} \\ \left(A_I^{\mathcal{H}}\right)^T A_F^{\mathcal{H}} & \left(A_I^{\mathcal{H}}\right)^T A_I^{\mathcal{H}} & \left(A_I^{\mathcal{H}}\right)^T A_N^{\mathcal{H}} \\ \left(A_N^{\mathcal{H}}\right)^T A_F^{\mathcal{H}} & \left(A_N^{\mathcal{H}}\right)^T A_I^{\mathcal{H}} & \left(A_N^{\mathcal{H}}\right)^T A_N^{\mathcal{H}} + \Theta_N \end{bmatrix}.$$

If the point $[u, y]$ is non-degenerate, then by Proposition 1 the matrix $\Phi^{\mathcal{H}}(y)$ along with the matrix $\Phi(y)$ is positive definite.

Rearrange rows and columns of the matrix $\Phi^{\mathcal{H}}(y)$, adding the last column of matrices A_i^H, $i \in J_F^r(y)$, to the matrix $A_I^{\mathcal{H}}$. For definiteness, we put these columns before the matrix $A_I^{\mathcal{H}}$. As a result, we get the matrix $A_{FI}^{\mathcal{H}}$ of the form (26). The matrix $A_F^{\mathcal{H}}$, from which the last columns of the matrices A_i^H, $i \in J_F^r(y)$, are removed, we denote by $\hat{A}_F^{\mathcal{H}}$. Denote also by $\hat{\Theta}_F$ the diagonal matrix Θ_F with removing last diagonal entries of Θ_i, $i \in J_F^r(y)$. Recall, that all these removing diagonal elements are zeros. The matrix $\hat{\Theta}_F$ has the order less than the order of the matrix Θ_F at r_F. Then, taking the corresponding permutation matrix Π, we obtain that $\Phi^{\mathcal{H}}(y)$ can be written in the following form

$$\Phi^{\mathcal{H}}(y) = \Pi \begin{bmatrix} \left(\hat{A}_F^{\mathcal{H}}\right)^T \hat{A}_F^{\mathcal{H}} + \hat{\Theta}_F & \left(\hat{A}_F^{\mathcal{H}}\right)^T A_{FI}^{\mathcal{H}} & \left(\hat{A}_F^{\mathcal{H}}\right)^T A_N^{\mathcal{H}} \\ \left(A_{FI}^{\mathcal{H}}\right)^T \hat{A}_F^{\mathcal{H}} & \left(A_{FI}^{\mathcal{H}}\right)^T A_{FI}^{\mathcal{H}} & \left(A_{FI}^{\mathcal{H}}\right)^T A_N^{\mathcal{H}} \\ \left(A_N^{\mathcal{H}}\right)^T \hat{A}_F^{\mathcal{H}} & \left(A_N^{\mathcal{H}}\right)^T A_{FI}^{\mathcal{H}} & \left(A_N^{\mathcal{H}}\right)^T A_N^{\mathcal{H}} + \Theta_N \end{bmatrix} \Pi^T.$$

Set $\mathcal{W}_{22} = \Theta_N + \left(A_N^{\mathcal{H}}\right)^T A_N^{\mathcal{H}}$ and

$$\mathcal{W}_{11} = \begin{bmatrix} \left(\hat{A}_F^{\mathcal{H}}\right)^T \hat{A}_F^{\mathcal{H}} + \hat{\Theta}_F & \left(\hat{A}_F^{\mathcal{H}}\right)^T A_{FI}^{\mathcal{H}} \\ \left(A_{FI}^{\mathcal{H}}\right)^T \hat{A}_F^{\mathcal{H}} & \left(A_{FI}^{\mathcal{H}}\right)^T A_{FI}^{\mathcal{H}} \end{bmatrix}, \quad \mathcal{W}_{12} = \begin{bmatrix} \left(\hat{A}_F^{\mathcal{H}}\right)^T A_N^{\mathcal{H}} \\ \left(A_{FI}^{\mathcal{H}}\right)^T A_N^{\mathcal{H}} \end{bmatrix}$$

The matrices $\mathcal{W}_{11}$ and $\mathcal{W}_{22}$, as diagonal blocks of the positive definite matrix, are positive definite too. With the proceeding notations the matrix $\Phi^{\mathcal{H}}(y)$ and its inverse can be written as

$$\Phi^{\mathcal{H}}(y) = \Pi \begin{bmatrix} \mathcal{W}_{11} & \mathcal{W}_{12} \\ \mathcal{W}_{12}^T & \mathcal{W}_{22} \end{bmatrix} \Pi^T, \qquad \left(\Phi^{\mathcal{H}}(y)\right)^{-1} = \Pi \begin{bmatrix} \mathcal{V}_{11} & \mathcal{V}_{12} \\ \mathcal{V}_{12}^T & \mathcal{V}_{22} \end{bmatrix} \Pi^T. \tag{27}$$

Denote $\mathcal{Z} = \mathcal{W}_{22} - \mathcal{W}_{12}^T \mathcal{W}_{11}^{-1} \mathcal{W}_{12}$. Using the Frobenius formula, we derive

$$\mathcal{V}_{11} = \mathcal{W}_{11}^{-1} + \mathcal{W}_{11}^{-1} \mathcal{W}_{12} \mathcal{Z}^{-1} \mathcal{W}_{12}^T \mathcal{W}_{11}^{-1}, \qquad \mathcal{V}_{12} = -\mathcal{W}_{11}^{-1} \mathcal{W}_{12} \mathcal{Z}^{-1}, \qquad \mathcal{V}_{22} = \mathcal{Z}^{-1}. \tag{28}$$

Thus, in order to obtain the matrix $\left(\Phi^{\mathcal{H}}(y)\right)^{-1}$, we need to know $\mathcal{W}_{11}^{-1}$ and $\mathcal{Z}^{-1}$.

Determine firstly the matrix $\mathcal{W}_{11}^{-1}$. It follows from the criterion of non-degeneracy of the point $[u, y] \in \mathcal{F}_D$ (see the proof of the Proposition 1) that the right lower block $\mathcal{W}_{11}$ is a nonsingular matrix. Set

$$\mathcal{Y} = \left(\hat{A}_F^{\mathcal{H}}\right)^T \hat{A}_F^{\mathcal{H}} + \hat{\Theta}_F - \left(\hat{A}_F^{\mathcal{H}}\right)^T A_{FI}^{\mathcal{H}} \left[\left(A_{FI}^{\mathcal{H}}\right)^T A_{FI}^{\mathcal{H}}\right]^{-1} \left(A_{FI}^{\mathcal{H}}\right)^T \hat{A}_F^{\mathcal{H}}, \tag{29}$$

and denote besides $\mathcal{P} = A_{FI}^{\mathcal{H}} \left[\left(A_{FI}^{\mathcal{H}}\right)^T A_{FI}^{\mathcal{H}}\right]^{-1} \left(A_{FI}^{\mathcal{H}}\right)^T$. The matrix $\mathcal{P}$ is an orthogonal projector onto linear subspace $\mathcal{L}$, generated by columns of the matrix $A_{FI}^{\mathcal{H}}$. The matrix $\mathcal{P}_{\perp} = I_m - \mathcal{P}$ projects onto the orthogonal complement $\mathcal{L}^{\perp}$ of this subspace. By (29) the following representation $\mathcal{Y} = \hat{\Theta}_F + \left(\hat{A}_F^{\mathcal{H}}\right)^T \mathcal{P}_{\perp} \hat{A}_F^{\mathcal{H}}$ holds. Since the diagonal matrix $\hat{\Theta}_F$ is positive definite, the matrix $\mathcal{Y}$ is positive definite too.

Applying the Frobenius formula, we obtain

$$\mathcal{W}_{11}^{-1} = \begin{bmatrix} \mathcal{Y}^{-1} & -\mathcal{Y}^{-1} \left(\hat{A}_F^{\mathcal{H}}\right)^T \mathcal{T} \\ -\mathcal{T}^T \hat{A}_F^{\mathcal{H}} \mathcal{Y}^{-1} & \mathcal{E} + \mathcal{T}^T \hat{A}_F^{\mathcal{H}} \mathcal{Y}^{-1} \left(\hat{A}_F^{\mathcal{H}}\right)^T \mathcal{T} \end{bmatrix},$$

where the notations $\mathcal{E} = \left[\left(A_{FI}^{\mathcal{H}} \right)^T A_{FI}^{\mathcal{H}} \right]^{-1}$ and $\mathcal{T} = A_{FI}^{\mathcal{H}} \mathcal{E}$ are introduced. The matrix $\mathcal{P}_\perp$ is idempotent, i.e. $\mathcal{P}_\perp = \mathcal{P}_\perp^2$. Thus, by the Sherman–Morrison–Woodbury formula

$$\mathcal{Y}^{-1} = \hat{\Theta}_F^{-1} - \hat{\Theta}_F^{-1} \left(\hat{A}_F^{\mathcal{H}} \right)^T \mathcal{P}_\perp \left[I_m + \mathcal{P}_\perp \hat{A}_F^{\mathcal{H}} \hat{\Theta}_F^{-1} \left(\hat{A}_F^{\mathcal{H}} \right)^T \mathcal{P}_\perp \right]^{-1} \mathcal{P}_\perp \hat{A}_F^{\mathcal{H}} \hat{\Theta}_F^{-1}. \tag{30}$$

The matrix $\mathcal{Z}$ can be written in the form

$$\mathcal{Z} = \Theta_N + \left(A_N^{\mathcal{H}} \right)^T \left[I_m - \hat{A}_{FI}^{\mathcal{H}} \mathcal{W}_{11}^{-1} \left(\hat{A}_{FI}^{\mathcal{H}} \right)^T \right] A_N^{\mathcal{H}}, \tag{31}$$

where $\hat{A}_{FI}^{\mathcal{H}} = \left[\hat{A}_F^{\mathcal{H}}, A_{FI}^{\mathcal{H}} \right]$. The matrix $\mathcal{W}_{11}$ is positive definite, and $\mathcal{Z}$ is the Schur complement of $\mathcal{W}_{11}$ in positive definite matrix in $\Phi^{\mathcal{H}}(y)$. Therefore, $\mathcal{Z}$ is a positive definite matrix.

Proposition 2. *Let $\mathcal{S} = \hat{A}_F^{\mathcal{H}} \mathcal{Y}^{-1} \left(\hat{A}_F^{\mathcal{H}} \right)^T$, $\hat{\mathcal{S}} = \mathcal{P} + \mathcal{P}_\perp \mathcal{S} \mathcal{P}_\perp$. Then the following formula*

$$\mathcal{Z}^{-1} = \Theta_N^{-1} - \Theta_N^{-1} \left(A_N^{\mathcal{H}} \right)^T \left(I - \hat{\mathcal{S}} \right)^{1/2} \cdot$$
$$\cdot \left[I + \left(I - \hat{\mathcal{S}} \right)^{1/2} A_N^{\mathcal{H}} \Theta_N^{-1} \left(A_N^{\mathcal{H}} \right)^T \left(I - \hat{\mathcal{S}} \right)^{1/2} \right]^{-1} \left(I - \hat{\mathcal{S}} \right)^{1/2} A_N^{\mathcal{H}} \Theta_N^{-1} \tag{32}$$

holds.

Proof. Determine $\hat{A}_{FI}^{\mathcal{H}} \mathcal{W}_{11}^{-1} \left(\hat{A}_{FI}^{\mathcal{H}} \right)^T$. We have

$$\mathcal{W}_{11}^{-1} \left(\hat{A}_{FI}^{\mathcal{H}} \right)^T = \begin{bmatrix} \mathcal{Y}^{-1} & -\mathcal{Y}^{-1} \left(\hat{A}_F^{\mathcal{H}} \right)^T \mathcal{T} \\ -\mathcal{T}^T \hat{A}_F^{\mathcal{H}} \mathcal{Y}^{-1} \mathcal{E} + \mathcal{T}^T \hat{A}_F^{\mathcal{H}} \mathcal{Y}^{-1} \left(\hat{A}_F^{\mathcal{H}} \right)^T \mathcal{T} \end{bmatrix} \begin{bmatrix} \left(\hat{A}_F^{\mathcal{H}} \right)^T \\ \left(A_{FI}^{\mathcal{H}} \right)^T \end{bmatrix}.$$

Hence,

$$\mathcal{W}_{11}^{-1} \left(\hat{A}_{FI}^{\mathcal{H}} \right)^T = \begin{bmatrix} \mathcal{Y}^{-1} \left(\hat{A}_F^{\mathcal{H}} \right)^T \left[I - \mathcal{T} \left(A_{FI}^{\mathcal{H}} \right)^T \right] \\ -\mathcal{T}^T \mathcal{S} + \mathcal{T}^T + \mathcal{T}^T \mathcal{S} \mathcal{T} \left(A_{FI}^{\mathcal{H}} \right)^T \end{bmatrix} = \begin{bmatrix} \mathcal{Y}^{-1} \left(\hat{A}_F^{\mathcal{H}} \right)^T \mathcal{P}_\perp \\ \mathcal{T}^T \left(I - \mathcal{S} \mathcal{P}_\perp \right) \end{bmatrix}. \tag{33}$$

Substituting the corresponding expression (30), we amount to

$$\hat{A}_{FI}^{\mathcal{H}} \mathcal{W}_{11}^{-1} \left(\hat{A}_{FI}^{\mathcal{H}} \right)^T = \left[\hat{A}_F^{\mathcal{H}}, A_{FI}^{\mathcal{H}} \right] \begin{bmatrix} \mathcal{Y}^{-1} \left(\hat{A}_F^{\mathcal{H}} \right)^T \mathcal{P}_\perp \\ \mathcal{T}^T \left(I - \mathcal{S} \mathcal{P}_\perp \right) \end{bmatrix} = \mathcal{S} \mathcal{P}_\perp + \mathcal{P} (I - \mathcal{S} \mathcal{P}_\perp) = \hat{\mathcal{S}}. \tag{34}$$

From here and (31) we derive $\mathcal{Z} = \Theta_N + \left(A_N^{\mathcal{H}} \right)^T \left(I_m - \hat{\mathcal{S}} \right) A_N^{\mathcal{H}}$. Since Θ_N is a positive definite diagonal matrix, we obtain by the Sherman–Morrison–Woodbury formula the expression (32) for $\mathcal{Z}^{-1}$. □

Proposition 3. Let $\mathcal{U} = \mathcal{A}_N^{\mathcal{H}} \mathcal{Z}^{-1} \left(\mathcal{A}_N^{\mathcal{H}} \right)^T$. Then

$$\mathcal{A}\Phi^{-1}(y)\mathcal{A}^T = \hat{S} + (I_m - \hat{S})\mathcal{U}(I_m - \hat{S}). \tag{35}$$

Proof. First of all observe, that $\mathcal{A}\Phi^{-1}(y)\mathcal{A}^T = \mathcal{A}^{\mathcal{H}} \left(\Phi^{\mathcal{H}}(y) \right)^{-1} (\mathcal{A}^{\mathcal{H}})^T$. By (28) and (33)

$$\hat{\mathcal{A}}_{FI}^{\mathcal{H}} \mathcal{V}_{11} = \left[\mathcal{P}_\perp \hat{\mathcal{A}}_F^{\mathcal{H}} \mathcal{Y}^{-1}, \ (I_m - \mathcal{P}_\perp S)T \right] \left[I_m + \mathcal{W}_{12} \mathcal{Z}^{-1} \mathcal{W}_{12}^T \mathcal{W}_{11}^{-1} \right],$$

$$\hat{\mathcal{A}}_{FI}^{\mathcal{H}} \mathcal{V}_{12} = - \left[\mathcal{P}_\perp \hat{\mathcal{A}}_F^{\mathcal{H}} \mathcal{Y}^{-1}, \ (I_m - \mathcal{P}_\perp S)T \right] \mathcal{W}_{12} \mathcal{Z}^{-1}.$$

$$\mathcal{W}_{12} \mathcal{Z}^{-1} \mathcal{W}_{12}^T = \left(\hat{\mathcal{A}}_{FI}^{\mathcal{H}} \right)^T \mathcal{A}_N^{\mathcal{H}} \mathcal{Z}^{-1} \left(\mathcal{A}_N^{\mathcal{H}} \right)^T \hat{\mathcal{A}}_{FI}^{\mathcal{H}}.$$

Hence,

$$\hat{\mathcal{A}}_{FI}^{\mathcal{H}} \mathcal{V}_{11} \left(\hat{\mathcal{A}}_{FI}^{\mathcal{H}} \right)^T = \mathcal{P} + \mathcal{P}_\perp S \mathcal{P}_\perp$$

$$+ \left[\mathcal{P}_\perp \hat{\mathcal{A}}_F^{\mathcal{H}} \mathcal{Y}^{-1}, \ (I_m - \mathcal{P}_\perp S)T \right] \mathcal{W}_{12} \mathcal{Z}^{-1} \mathcal{W}_{12}^T \left[\mathcal{Y}^{-1} \left(\hat{\mathcal{A}}_F^{\mathcal{H}} \right)^T \mathcal{P}_\perp; \ T^T (I_m - S\mathcal{P}_\perp) \right]$$

$$= \mathcal{P} + \mathcal{P}_\perp S \mathcal{P}_\perp + \left[\mathcal{P}_\perp \hat{\mathcal{A}}_F^{\mathcal{H}} \mathcal{Y}^{-1}, \ (I_m - \mathcal{P}_\perp S)T \right] \left(\hat{\mathcal{A}}_{FI}^{\mathcal{H}} \right)^T \mathcal{U}(\mathcal{P} + \mathcal{P}_\perp S \mathcal{P}_\perp)$$

$$= \mathcal{P} + \mathcal{P}_\perp S \mathcal{P}_\perp + (\mathcal{P} + \mathcal{P}_\perp S \mathcal{P}_\perp)\mathcal{U}(\mathcal{P} + \mathcal{P}_\perp S \mathcal{P}_\perp).$$

From here we obtain

$$\hat{\mathcal{A}}_{FI}^{\mathcal{H}} \mathcal{V}_{11} \left(\hat{\mathcal{A}}_{FI}^{\mathcal{H}} \right)^T = \hat{S} + \hat{S}\mathcal{U}\hat{S}. \tag{36}$$

Because of $\hat{\mathcal{A}}_{FI}^{\mathcal{H}} \mathcal{V}_{12} = -\hat{S}\mathcal{A}_N^{\mathcal{H}} \mathcal{Z}^{-1}$ and $\hat{\mathcal{A}}_{FI}^{\mathcal{H}} \mathcal{V}_{12} \left(\mathcal{A}_N^{\mathcal{H}} \right)^T = -\hat{S}\mathcal{U}$, we have also

$$\mathcal{A}_N^{\mathcal{H}} \mathcal{V}_{12}^T \left(\hat{\mathcal{A}}_{FI}^{\mathcal{H}} \right)^T = -\mathcal{U}\hat{S}, \quad \mathcal{A}_N^{\mathcal{H}} \mathcal{V}_{22} \left(\mathcal{A}_N^{\mathcal{H}} \right)^T = \mathcal{U}. \tag{37}$$

Thus, according to (27), (36) and (37)

$$\mathcal{A}^{\mathcal{H}} \left(\Phi^{\mathcal{H}}(y) \right)^{-1} (\mathcal{A}^{\mathcal{H}})^T = \hat{S} + \hat{S}\mathcal{U}\hat{S} - \mathcal{U}\hat{S} - \hat{S}\mathcal{U} + \mathcal{U}$$

$$= \hat{S} - (I_m - \hat{S})\mathcal{U}\hat{S} + (I_m - \hat{S})\mathcal{U} = \hat{S} + (I_m - \hat{S})\mathcal{U}(I_m - \hat{S}). \tag{38}$$

Therefore, the formula (35) takes place.   □

Proposition 4. Let $\hat{\mathcal{C}} = \mathcal{P}_\perp \hat{\mathcal{A}}_F^{\mathcal{H}} \hat{\Theta}_F^{-1} \left(\hat{\mathcal{A}}_F^{\mathcal{H}} \right)^T \mathcal{P}_\perp$. Then $S = I_m - \left(I_m + \hat{\mathcal{C}} \right)^{-1}$.

Proof. Taking into account (30), we obtain

$$\mathcal{P}_\perp S \mathcal{P}_\perp = \mathcal{P}_\perp \hat{\mathcal{A}}_F^{\mathcal{H}} \mathcal{Y}^{-1} \left(\hat{\mathcal{A}}_F^{\mathcal{H}} \right)^T \mathcal{P}_\perp = \hat{\mathcal{C}} - \hat{\mathcal{C}} \left(I_m + \hat{\mathcal{C}} \right)^{-1} \hat{\mathcal{C}}.$$

After some transformations we derive from here

$$\mathcal{P}_\perp S \mathcal{P}_\perp = \hat{\mathcal{C}} - \hat{\mathcal{C}} \left(I_m - \left(I_m + \hat{\mathcal{C}} \right)^{-1} \right) = \hat{\mathcal{C}} \left(I_m + \hat{\mathcal{C}} \right)^{-1} = I_m - \left(I_m + \hat{\mathcal{C}} \right)^{-1}. \tag{39}$$

Hence, the assertion of the proposition is valid.   □

Corollary. *After inversion of the matrix $I_m + \hat{\mathcal{C}}$ by the Sherman–Morrison–Woodbury formula we get $I_m - \hat{S} = \mathcal{P}_\perp - \hat{\mathcal{Q}}$, where*

$$\hat{\mathcal{Q}} = \mathcal{P}_\perp \hat{\mathcal{A}}_F^{\mathcal{H}} \hat{\Theta}_F^{-1/2} \left[I_m + \hat{\Theta}_F^{-1/2} \left(\hat{\mathcal{A}}_F^{\mathcal{H}} \right)^T \mathcal{P}_\perp \hat{\mathcal{A}}_F^{\mathcal{H}} \hat{\Theta}_F^{-1/2} \right]^{-1} \hat{\Theta}_F^{-1/2} \left(\hat{\mathcal{A}}_F^{\mathcal{H}} \right)^T \mathcal{P}_\perp.$$

(40)

The matrix $I_m - \hat{S}$ is positive semi-definite.

Denote $\tilde{\mathcal{U}} = \left(I_m - \hat{S} \right) \mathcal{U} \left(I_m - \hat{S} \right)$ and

$$\mathcal{B} = \mathcal{A}_N^{\mathcal{H}} \Theta_N^{-1} \left(\mathcal{A}_N^{\mathcal{H}} \right)^T, \quad \tilde{\mathcal{B}} = \left(I_m - \hat{S} \right)^{1/2} \mathcal{B} \left(I_m - \hat{S} \right)^{1/2}.$$

By (32) after transformations similar to (39) we obtain

$$\tilde{\mathcal{U}} = I_m - \hat{S} - \left(I_m - \hat{S} \right)^{1/2} \left(I_m + \tilde{\mathcal{B}} \right)^{-1} \left(I_m - \hat{S} \right)^{1/2}.$$

(41)

Since by (38) $\mathcal{A}^{\mathcal{H}} \left(\Phi^{\mathcal{H}}(y) \right)^{-1} \left(\mathcal{A}^{\mathcal{H}} \right)^T = \hat{S} + \tilde{\mathcal{U}}$, the iteration formula (10) reduces to the form

$$u_{k+1} = u_k + \alpha_k \left(I_m - \hat{S} \right)^{1/2} \left(I_m + \tilde{\mathcal{B}} \right)^{-1} \left(I_m - \hat{S} \right)^{1/2} b.$$

(42)

Remind, that formula (40) is valid for the matrix $I_m - \hat{S}$. The matrices $\hat{S}$ and $\mathcal{B}$ depend on the diagonal matrices $\hat{\Theta}_F$ and Θ_N. At non-degenerate feasible points $[u_k, y_k]$, where $y_k = y(u_k)$, both diagonal matrices $\hat{\Theta}_F$ and Θ_N are positive definite.

5 The Local Convergence

Investigate the question about local convergence of the proposed methods.

Proposition 5. *Let $z_* = [u_*, y_*]$, where $y_* = y(u_*)$, be the solution of the dual problem (3). Then z_* is a stationary point of the iterative process (14). What is more, $x_* = x(z_*)$ is a solution of the primal problem (2).*

Proof. Let x^* be a solution of primal problem (2). Then $\mathcal{A}x^* = b$ and according to optimality conditions (6) the equality $\mathcal{G}(y_*)x_* = 0_n$ holds. Multiplying the equality $\mathcal{A}x_* = b$ by the matrix $\mathcal{A}^T$ and summing it with the equality $\mathcal{G}(y_*)x_*$, we obtain $\Phi(y_*)x^* = \mathcal{A}^T b$. The matrix $\Phi(y_*)$ is nonsingular in the non-degenerate point z_*. Therefore, $x^* = \Phi^{-1}(y_*)\mathcal{A}^T b$. But the point $x_* = x(z_*)$ also is determined from the system of equations $\Phi(y_*)x = \mathcal{A}^T b$. Since the solution of this system is unique, we conclude that $x(z_*) = x^*$. It follows from here that $\mathcal{A}x(z_*) = b$ or $\mathcal{A}\Phi^{-1}(y_*)\mathcal{A}^T b = b$. Moreover, $\mathcal{G}(y_*)x_* = 0_n$. Thus, z_* is a stationary point of the iterative process (14). $\square$

Denote $z = [u, y]$ and consider the mapping $\varphi(z) = \left[\varphi^{(1)}(z); \varphi^{(2)}(z)\right]$, where

$$\varphi^{(1)}(z) = \mathcal{A} \, x(z) - b, \quad \varphi^{(2)}(z) = \mathcal{G}(y) \, x(z),$$

and $x(z) = \Phi^{-1}(y) \left[A^T b + \tau \left(y + \mathcal{A}^T u - c\right)\right]$. In what follows, we need in the Jacobi matrix of the mapping $\varphi(z)$.

Lemma 1. *Let the point $z \in \mathcal{F}_D$ be non-degenerate. Then the Jacobi matrix of the mapping $\varphi(z)$ has the form*

$$\varphi_z(z) = \begin{bmatrix} \tau \mathcal{A} \Phi^{-1}(y) \mathcal{A}^T & \mathcal{A} \Phi^{-1}(y) \left[\tau I_n - \mathcal{G}\left(x(z)\right)\right] \\ \tau \mathcal{G}(y) \Phi^{-1}(y) \mathcal{A}^T & \left[I_n - \mathcal{G}(y) \Phi^{-1}(y)\right] \mathcal{G}(x(z)) + \tau \mathcal{G}(y) \Phi^{-1}(y) \end{bmatrix}. \quad (43)$$

Theorem 1. *Let x_* and $z_* = [u_*, y_*]$ be non-degenerate optimal solutions of primal and dual problems (2) and (3), respectfully. Moreover, let these solutions be strictly complementary. Then there exists $\bar{\alpha} > 0$ such that for $0 < \alpha < \bar{\alpha}$ the iterative process (12) with the step size $\alpha_k = \alpha$ locally converges to z_* with a linear rate.*

Proof. To prove the theorem, it is sufficient to show that the spectral radius of the Jacobi matrix of the mapping $F(z) = I_{m+n} - \alpha \varphi(z)$ at the point z_* is less than one.

Let η be an eigenvalue of the matrix $\varphi_z(z_*)$. Assume for definiteness that for y_* the partition onto three blocks of components $y_* = [y_{*,F}; y_{*,I}; y_{*,N}]$ holds. Then for x_* the similar partition $x_* = [x_{*,F}; x_{*,I}; x_{*,N}]$ takes place. Denote by $\bar{r}_F$, $\bar{r}_I$ and $\bar{r}_N$ the number of components in blocks for the vector x_*. It follows from strict complementarity that $\bar{r}_F = r_F$, $\bar{r}_I = r_I$ and $\bar{r}_N = r_N$. What is more, $x_{*,N}$ is a zero vector. By Proposition 5 $x(z_*) = x_*$.

According to (43) the eigenvalue η must satisfy to the following characteristic equation

$$\begin{vmatrix} \tau \mathcal{A} \Phi^{-1}(y_*) \mathcal{A}^T - \eta I_m & \mathcal{A} \Phi^{-1}(y_*) \left[\tau I_n - \mathcal{G}\left(x_*\right)\right] \\ \tau \mathcal{G}(y_*) \Phi^{-1}(y_*) \mathcal{A}^T & \left[I_n - \mathcal{G}(y_*) \Phi^{-1}(y_*)\right] \mathcal{G}(x_*) + \tau \mathcal{G}(y_*) \Phi^{-1}(y_*) - \eta I_n \end{vmatrix} = 0. \quad (44)$$

After multiplying the right column in (44) from the right by the matrix $\mathcal{A}^T$ and subtracting it from the left column the Eq. (44) is transformed to the following one

$$\begin{vmatrix} \mathcal{A} \Phi^{-1}(y_*) \mathcal{G}(x_*) \mathcal{A}^T - \eta I_m & \mathcal{A} \Phi^{-1}(y_*) \left[\tau I_n - \mathcal{G}\left(x_*\right)\right] \\ \Omega & \left[I_n - \mathcal{G}(y_*) \Phi^{-1}(y)\right] \mathcal{G}(x_*) + \tau \mathcal{G}(y_*) \Phi^{-1}(y_*) - \eta I_n \end{vmatrix} = 0, \quad (45)$$

where $\Omega = \left[\mathcal{G}(y_*) \Phi^{-1}(y) - I_n\right] \mathcal{G}(x_*) \mathcal{A}^T + \eta \mathcal{A}^T$.

Multiply further the first row in (45) from the left by the matrix $\mathcal{A}^T$ and sum it with the second row. Taking into account the equality $\left[\mathcal{A}^T \mathcal{A} + \mathcal{G}(y_*)\right] \Phi^{-1}(y_*) = I_n$, we obtain the equation

$$\begin{vmatrix} \mathcal{A} \Phi^{-1}(y_*) \mathcal{G}(x_*) \mathcal{A}^T - \eta I_m & \mathcal{A} \Phi^{-1}(y_*) \left[\tau I_n - \mathcal{G}\left(x_*\right)\right] \\ 0_{nm} & (\tau - \eta) I_n \end{vmatrix} = 0. \quad (46)$$

It follows from (46) that n eigenvalues η are equal to τ. The other eigenvalues η may be found from the characteristic equation

$$|\Psi(z_*) - \eta I_m| = 0, \quad \Psi(z_*) = \mathcal{A}\Phi^{-1}(y_*)\mathcal{G}(x_*)\mathcal{A}^T. \tag{47}$$

Let $\mathcal{G}(y_*) = \mathcal{H}\Theta_*\mathcal{H}^T$, where Θ_* is a diagonal matrix with the vector of eigenvalues of the matrix $\mathcal{G}(y_*)$ at the diagonal. Furthermore, let Λ_* be a diagonal matrix with the vector of eigenvalues of the matrix $\mathcal{G}(x_*)$, that is $\mathcal{G}(x_*) = \mathcal{H}\Lambda_*\mathcal{H}^T$. Since the matrices $\mathcal{G}(y_*)$ and $\mathcal{G}(x_*)$ commute between themselves, the orthogonal matrix $\mathcal{H}$ is the same. Due to strict complementarity, the right lower block of Λ_* is zero.

Transform the matrix $\Psi(z_*)$ to the form

$$\Psi(z_*) = \mathcal{A}\mathcal{H}\mathcal{H}^T\Phi^{-1}(y_*)\mathcal{H}\Lambda_*\mathcal{H}^T\mathcal{A}^T = \mathcal{A}^{\mathcal{H}}\left(\Phi^{\mathcal{H}}(y_*)\right)^{-1}\Lambda_*\left(\mathcal{A}^{\mathcal{H}}\right)^T.$$

From here, taking into account (27), we obtain

$$\Psi(z_*) = \mathcal{A}^{\mathcal{H}}\Pi \begin{bmatrix} \mathcal{V}_{11} & \mathcal{V}_{12} \\ \mathcal{V}_{12}^T & \mathcal{V}_{22} \end{bmatrix} \hat{\Lambda}_*\Pi^T\left(\mathcal{A}^{\mathcal{H}}\right)^T,$$

where $\hat{\Lambda}_* = \Pi^T\Lambda_*\Pi$. But, as stated above, $\mathcal{A}^{\mathcal{H}}\Pi = \left[\hat{\mathcal{A}}_{FI}^{\mathcal{H}}, \mathcal{A}_N^{\mathcal{H}}\right]$. Thus, by (27)

$$\Psi(z_*) = \left[\hat{\mathcal{A}}_{FI}^{\mathcal{H}}, \mathcal{A}_N^{\mathcal{H}}\right] \begin{bmatrix} \mathcal{V}_{11} & \mathcal{V}_{12} \\ \mathcal{V}_{12}^T & \mathcal{V}_{22} \end{bmatrix} \hat{\Lambda}_*\left[\left(\hat{\mathcal{A}}_{FI}^{\mathcal{H}}\right)^T; \left(\mathcal{A}_N^{\mathcal{H}}\right)^T\right].$$

Represent the diagonal matrix $\hat{\Lambda}_*$ as block diagonal matrix, consisting from two entries, $\hat{\Lambda}_* = \text{DIAG}\left(\hat{\Lambda}_{*,FI}, \Lambda_{*,N}\right)$. In turn, let $\hat{\Lambda}_{*,FI} = \text{DIAG}\left(\hat{\Lambda}_{*,F}, \Lambda_{*,FI}\right)$. Since the right lower block $\hat{\Lambda}_{*,N}$ is zero, then by (28)

$$\Psi(z_*) = \left[\hat{\mathcal{A}}_{FI}^{\mathcal{H}}, \mathcal{A}_N^{\mathcal{H}}\right] \begin{bmatrix} \left(\mathcal{W}_{11}^{-1} + \mathcal{W}_{11}^{-1}\mathcal{W}_{12}\mathcal{Z}^{-1}\mathcal{W}_{12}^T\mathcal{W}_{11}^{-1}\right)\hat{\Lambda}_{*,FI}\left(\hat{\mathcal{A}}_{FI}^{\mathcal{H}}\right)^T \\ -\mathcal{Z}^{-1}\mathcal{W}_{12}^T\mathcal{W}_{11}^{-1}\hat{\Lambda}_{*,FI}\left(\hat{\mathcal{A}}_{FI}^{\mathcal{H}}\right)^T \end{bmatrix}$$

or $\Psi(z_*) = \Psi_1(z_*) + \Psi_2(z_*)$, where $\Psi_1(z_*) = \hat{\mathcal{A}}_{FI}^{\mathcal{H}}\mathcal{W}_{11}^{-1}\hat{\Lambda}_{*,FI}\left(\hat{\mathcal{A}}_{FI}^{\mathcal{H}}\right)^T$ and

$$\Psi_2(z_*) = \left[\hat{\mathcal{A}}_{FI}^{\mathcal{H}}\mathcal{W}_{11}^{-1}\left(\hat{\mathcal{A}}_{FI}^{\mathcal{H}}\right)^T - I_m\right]\mathcal{A}_N^{\mathcal{H}}\mathcal{Z}^{-1}\left(\mathcal{A}_N^{\mathcal{H}}\right)^T\hat{\mathcal{A}}_{FI}^{\mathcal{H}}\mathcal{W}_{11}^{-1}\hat{\Lambda}_{*,FI}\left(\hat{\mathcal{A}}_{FI}^{\mathcal{H}}\right)^T.$$

The matrix $\Psi_2(z_*)$ by (34), (39) and with the help of introduced above notations can be rewritten as $\Psi_2(z_*) = \left[\hat{S} - I_m\right]\mathcal{U}\hat{\mathcal{A}}_{FI}^{\mathcal{H}}\mathcal{W}_{11}^{-1}\hat{\Lambda}_{*,FI}\left(\hat{\mathcal{A}}_{FI}^{\mathcal{H}}\right)^T$. Hence,

$$\Psi(z_*) = \left[I_m - \left(I_m - \hat{S}\right)\mathcal{U}\right]\hat{\mathcal{A}}_{FI}^{\mathcal{H}}\mathcal{W}_{11}^{-1}\hat{\Lambda}_{*,FI}\left(\hat{\mathcal{A}}_{FI}^{\mathcal{H}}\right)^T.$$

Let η be an eigenvalue of $\Psi(x_*, y_*)$, and let p be the corresponding eigenvector. Then η and p must satisfy the equation

$$\Psi(x_*, y_*)\, p = \eta\, p. \tag{48}$$

We first assume that $\left(\mathcal{A}_{FI}^{\mathcal{H}}\right)^T p \neq 0_{l_1}$, where $l_1 = \sum_{i \in J_I^r} n_i + r_F$. Then, multiplying (48) from the left by $\left(\mathcal{A}_{FI}^{\mathcal{H}}\right)^T$, we obtain

$$\left(\mathcal{A}_{FI}^{\mathcal{H}}\right)^T \hat{\mathcal{A}}_{FI}^{\mathcal{H}} W_{11}^{-1} \Lambda_{*,FI}^{\Pi} \left(\hat{\mathcal{A}}_{FI}^{\mathcal{H}}\right)^T p = \eta \left(\mathcal{A}_{FI}^{\mathcal{H}}\right)^T p.$$

Here we take into account that, according to corollary after Proposition 4, $\left(\mathcal{A}_{FI}^{\mathcal{H}}\right)^T \left(I_m - \hat{S}\right) = 0_{l_1}$.

Using the formula (33) and the equality $\left(\mathcal{A}_{FI}^{\mathcal{H}}\right)^T \mathcal{P}_\perp = 0_{l_1}$, we obtain

$$\left(\mathcal{A}_{FI}^{\mathcal{H}}\right)^T \hat{\mathcal{A}}_{FI}^{\mathcal{H}} W_{11}^{-1} = \left(\mathcal{A}_{FI}^{\mathcal{H}}\right)^T \left[\mathcal{P}_\perp \hat{\mathcal{A}}_F^{\mathcal{H}} \mathcal{Y}^{-1}, (I_m - \mathcal{P}_\perp S)\mathcal{T}\right] = [0_{l_1 \, l_2}, I_{l_1}],$$

where $l_2 = \sum_{i \in J_F^r} n_i - r_F$. Therefore, the Eq. (48) is reduced to the following one

$$\hat{\Lambda}_{FI} \left(\mathcal{A}_{FI}^{\mathcal{H}}\right)^T p = \eta \left(\mathcal{A}_{FI}^{\mathcal{H}}\right)^T p. \tag{49}$$

It follows from (49) that p must be orthogonal to all but one columns of $\mathcal{A}_{FI}^{\mathcal{H}}$ to be an eigenvector. The eigenvalue η is equal to the corresponding diagonal entry of $\hat{\Lambda}_{FI}$. As already noted, this entry is positive. Thus, we have l_1 eigenvectors which do not belong to the subspace $\mathcal{L}^\perp$. The corresponding eigenvalues are real and positive.

Assume now that $\left(\mathcal{A}_{FI}^{\mathcal{H}}\right)^T p = 0_{l_1}$, i.e. $p \in \mathcal{L}^\perp$. In this case, we get by (33) and (39)

$$\hat{\mathcal{A}}_{FI}^{\mathcal{H}} W_{11}^{-1} \hat{\Lambda}_{FI} \left(\hat{\mathcal{A}}_{FI}^{\mathcal{H}}\right)^T p = \mathcal{P}_\perp \hat{\mathcal{A}}_F^{\mathcal{H}} \mathcal{Y}^{-1} \hat{\Lambda}_F \left(\hat{\mathcal{A}}_F^{\mathcal{H}}\right)^T \mathcal{P}_\perp p$$
$$= \left(I + \hat{C}\right)^{-1} \mathcal{P}_\perp \hat{\mathcal{A}}_F^{\mathcal{H}} \hat{\Theta}_F^{-1} \hat{\Lambda}_F (\left(\hat{\mathcal{A}}_F^{\mathcal{H}}\right)^T \mathcal{P}_\perp p = \left(I - \hat{S}\right) \mathcal{P}_\perp \hat{\mathcal{A}}_F^{\mathcal{H}} \hat{\Theta}_F^{-1} \hat{\Lambda}_F \left(\hat{\mathcal{A}}_F^{\mathcal{H}}\right)^T \mathcal{P}_\perp p.$$

From the other hand, by (41)

$$\left[I - \left(I - \hat{S}\right)\mathcal{U}\right]\left(I - \hat{S}\right) = \left(I - \hat{S}\right) - \left(I - \hat{S}\right)\mathcal{U}\left(I - \hat{S}\right) = \left(I - \hat{S}\right)$$
$$- \left(I - \hat{S}\right)^{1/2}\left[I - \left(I + \tilde{B}\right)^{-1}\right]\left(I - \hat{S}\right)^{1/2} = \left(I - \hat{S}\right)^{1/2}\left(I + \tilde{B}\right)^{-1}\left(I - \hat{S}\right)^{1/2}.$$

Therefore, the Eq. (48) is reduced to

$$\left(I - \hat{S}\right)^{1/2}\left(I + \tilde{B}\right)^{-1}\left(I - \hat{S}\right)^{1/2} \mathcal{P}_\perp \hat{\mathcal{A}}_F^{\mathcal{H}} \hat{\Theta}_F^{-1} \hat{\Lambda}_F \left(\hat{\mathcal{A}}_F^{\mathcal{H}}\right)^T \mathcal{P}_\perp p = \eta\, p. \tag{50}$$

Denote $q = \hat{\mathcal{A}}_F^{\mathcal{H}} \hat{\Theta}_F^{-1} \hat{\Lambda}_F \left(\hat{\mathcal{A}}_F^{\mathcal{H}}\right)^T \mathcal{P}_\perp p$ and multiply both sides of (50) by the q^T. Then, taking into account that $\mathcal{P}_\perp p = p$, we get

$$\left\langle \left(I - \hat{S}\right)^{1/2} q, \left(I + \tilde{B}\right)^{-1}\left(I - \hat{S}\right)^{1/2} q \right\rangle = \eta \left\langle \left(\hat{\mathcal{A}}_F^{\mathcal{H}}\right)^T p, \hat{\Theta}_F^{-1} \hat{\Lambda}_F \left(\hat{\mathcal{A}}_F^{\mathcal{H}}\right)^T p \right\rangle.$$

Both diagonal matrices $\hat{\Theta}_F^{-1}$ and $\hat{\Lambda}_F$ are positive semi-definite. Hence, $\eta \geq 0$. But the equality $\eta = 0$ is impossible. Indeed, in this case $\hat{\Theta}_F^{-1}\hat{\Lambda}_F \left(\hat{\mathcal{A}}_F^{\mathcal{H}}\right)^T p = 0_{l_2}$. However, due to complementarity conditions, the first diagonal entries of the diagonal matrices $\hat{\Lambda}_i$, $i \in J_F^r(y_*)$, are equal to zero. Denote by $\bar{\mathcal{A}}_i^H$, $i \in J_F^r(y_*)$, the right $m \times (n_i - 1)$ sub-matrix of the matrix $\mathcal{A}_i^H$. Denote also $\bar{\mathcal{A}}_F^{\mathcal{H}} = \left[\bar{\mathcal{A}}_1^H, \ldots, \bar{\mathcal{A}}_r^H\right]$. Since $p \in \mathcal{L}_\perp$, the equality $\left(\mathcal{A}_I^{\mathcal{H}}\right)^T p = 0$ holds. Therefore, $\left(\bar{\mathcal{A}}_{FI}^{\mathcal{H}}\right)^T p = 0$, where $\bar{\mathcal{A}}_{FI}^{\mathcal{H}} = \left[\bar{\mathcal{A}}_F^{\mathcal{H}}, \mathcal{A}_I^{\mathcal{H}}\right]$. Taking into account that p is a nonzero vector, we conclude that rows of the matrix $\bar{\mathcal{A}}_{FI}^{\mathcal{H}}$ are linear dependent. This contradicts to the assumption that the point x_* is non-degenerate, because of by the criterion of non-degeneracy in the primal problem (see [2]) the point x_* is non-degenerate if and only if rows of the matrix $\bar{\mathcal{A}}_{FI}^{\mathcal{H}}$ are linear independent.

Thus, all eigenvalues of the matrix $\varphi_z(z_*)$ are strictly positive. Let η_* be the maximal eigenvalue. Taking $0 < \bar{\alpha} < 2/\eta_*$, we get that the spectral radius of the matrix $F_z(z_*)$ is less than one, if $\alpha < \bar{\alpha}$. Therefore, by the Ostrowski-Hadamard theorem, the iterative process (12) locally converges to z_* with a linear rate. $\square$

Corollary. *Let the assumptions of Theorem 1 hold. Moreover, let $u_0 \in \mathcal{F}_{D,u}$ and $y_0 = y(u_0) \in \mathcal{K}_0$. Then we can specify $\bar{\alpha} > 0$ such that the iterative process (42) for $\alpha_k = \alpha < \bar{\alpha}$ also locally converges to u_* with a linear rate.*

Conclusion

The proposed dual methods may be preferable for solving linear SOCP problems, in which the number of variables is essentially less than the dimension of the cone. Among the disadvantages of the methods there is the local convergence. It is possible to extend the convergence domain of the feasible method by using the steepest descent approach for choosing the step sizes.

References

1. Anjos, M.F., Lasserre, J.B. (eds.): Handbook of Semidefinite, Conic and Polynomial Optimization. Springer, New York (2012). https://doi.org/10.1007/978-1-4614-0769-0
2. Alizadeh, F., Goldfarb, D.: Second-order cone programming. Math. Program. Ser. B **95**, 3–51 (2003). https://doi.org/10.1016/j.ifacol.2015.11.079
3. Lobo, M.S., Vandenberghe, L., Boyd, S., Lebret, H.: Applications of second-order cone programming. Linear Algebra Appl. **284**, 193–228 (1998). https://doi.org/10.1016/S0024-3795(98)10032-0
4. Nesterov, Yu.E., Todd, M.J.: Primal-dual interior-point methods for self-scaled cones. SIAM J. Optim. **8**, 324–364 (1998). https://doi.org/10.1137/S1052623495290209
5. Monteiro, R.D.C., Tsuchiya, T.: Polynomial convergence of primal-dual algorithms for second-order cone program based on the MZ-family of directions. Math. Program. **88**(1), 61–83 (2000). https://doi.org/10.1007/S101070000137

6. Zhadan, V.G.: A variant of the affine-scaling method for a second-order cone program. Proc. Steklov Inst. Math. **303**(Suppl. 1), S231–S240 (2018). https://doi.org/10.1134/S0081543818090250
7. Zhadan, V.G., Orlov, A.A.: Dual interior point methods for linear semidefinite programming problems. Comput. Math. Math. Phys. **51**(12), 2031–2051 (2011). https://doi.org/10.1134/S0965542511120189

A Problem of Scheduling Operations at a Locomotive Maintenance Depot

A. A. Lazarev[1,2,3,4], E. G. Musatova[1], E. M. Grishin[1,2(✉)], G. V. Tarasov[1,2],
S. A. Galakhov[1,2], and N. A. Pravdivets[1]

[1] V. A. Trapeznikov Institute of Control Sciences of Russian Academy of Sciences,
Profsoyuznaya str. 65, Moscow, Russia
jobmath@mail.ru, grishin.em16@physics.msu.ru, pravdivets@ipu.ru
[2] Lomonosov Moscow State University, Leninskie Gory str. 1/2, Moscow, Russia
[3] National Research University Higher School of Economics,
Myasnitskaya str. 20, Moscow, Russia
[4] Moscow Institute of Physics and Technology (State University),
Institute lane 9, Dolgoprudny, Moscow Region, Russia
http://www.orsot.ru/index.php/en/

Abstract. In this article, we consider the problem of planning maintenance operations at a locomotive maintenance depot. There are three types of tracks at the depot: buffer tracks, access tracks and service tracks. A depot consists of up to one buffer track and a number of access tracks, each of them ending with one service track. Each of these tracks has a limited capacity measured in locomotive sections. We present a constraint programming model and a greedy algorithm for solving the problem of planning maintenance operations. Using lifelike data based on the operation of several locomotive maintenance depots in Eastern polygon of Russian Railways, we carry out numerical experiments to compare the presented approaches.

Keywords: Maintenance · Scheduling · Dynamic programming · Constraint programming · Heuristic

1 Introduction

Railway scheduling is an entangled process of managing a large number of objects, including railway infrastructure, rolling stock, etc. This process requires to take into account a lot of conditions and restrictions. Foreground goals of the planning are safety and security of the whole system and minimizing transporting delays. According to Russian Railways safety requirements, every locomotive should undergo regular maintenance. In particular, there is a kind of maintenance that is carried out every several days and includes inspection of chassis, brake system, traction motors, auxiliary equipment, transformers and

This work was supported by Russian Railways and the Russian Foundation for Basic Research, project No. 17-20-01107.

electric systems. The maintenance is carried out at special facilities, namely, Locomotive Maintenance Depots, or LMDs. Up to 100 locomotives are serviced at each LMD per day. The limitations on resources such as availability of maintenance crews, space on tracks, available special equipment, etc. result in growing downtimes. In order to reduce downtimes and thus increase efficiency in utilizing the locomotive fleet, scheduling process must be implemented.

In this paper we are going to find a solution for the described problem of scheduling locomotives maintenance. The paper is organized as follows. Section 2 includes a verbal statement of the problem, basic terms and notation, and a review of the literature on the subject. Section 3 is devoted to a constraint programming model. Section 4 proposes a heuristic algorithm. The results of numerical experiments on real data are presented and analyzed in Sect. 5.

2   Problem Statement

We consider a real-world problem of planning operation of locomotive maintenance depots of Russian Railways. There are three types of tracks at LMDs: buffer tracks, access tracks and service tracks. A typical LMD consists of up to one buffer track and a number of access tracks, each of them ending with one service track (see Fig. 1). Each of these tracks has a limited capacity measured in locomotive sections.

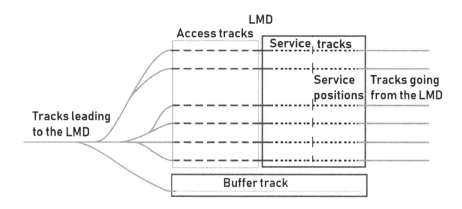

Fig. 1. Typical scheme of an LMD

Let $T = [T_0, T_1]$ be the planning horizon – the time interval during which all locomotives arrive to the LMD. We are provided a twenty-four hours plan of locomotives arriving to the LMD. All the time moments are considered to be integer. Let L be the set of all locomotives and $n = |L|$ be their number. The locomotives differ in models, number of sections and service time.

For each locomotive $l \in L$, let us define the following parameters.

- u_l is the number of sections of which l consists, $u_l \leq 4$.
- r_l is the moment of arrival of l to the LMD, $r_l \neq r_{l-1}, \forall l \in L$.
- τ_l is service time, or maintenance duration of l. Typically, the service time is from 57 to 120 min. For any locomotive, its service time is the same for any service track.
- v_l is the model number of l. Each service track can only maintain a certain set of locomotive models.

It is necessary to take into account all the limitations associated with the technical features of the locomotives and an LMD to build a feasible schedule.

Upon arriving at an LMD, each locomotive can be placed onto a buffer track. Let b be the buffer capacity, namely the number of locomotive sections that the buffer can contain at the same time. If an LMD does not include a buffer, the buffer capacity is considered to be zero. The buffer track operates according to the LIFO scheme: before a locomotive can leave the buffer, all the locomotives that came after it should leave first.

Upon leaving the buffer, each locomotive should be placed onto one of the access tracks, that are accessible from the buffer. If locomotive l was not placed into the buffer, it should be placed on an access track right at moment r_l of its arrival at the LMD. Let A be a set of all access tracks of the regarded LMD. For each access track $a \in A$, while let c_a denote the maximal number of locomotive sections that access track a can simultaneously accommodate.

Each of the access tracks ends with a single service track, and each service track can only be accessed from a single access track. The access and service tracks operate according to FIFO scheme: before a locomotive can leave, all the locomotives that came before it should leave first.

Each of the service tracks is divided into several service positions. We consider the case of 2 service positions on each service track. Each service position has its own capacity (up to 3 locomotive sections) and can only accept one locomotive at a time. A four-section locomotive takes both service positions of a service track. When several locomotives should be serviced on different positions of the same service track simultaneously, they should arrive at the track simultaneously, and leave simultaneously, too. Service sessions cannot be interrupted, and there should be a break between each pair of subsequent sessions on the same service track – typically, of at least 15 min. For simplicity, we include this time in service times of locomotives.

Upon leaving an access track, a locomotive is transported to its designated service position on the corresponding service track. Actually, any locomotive should be serviced only once and its maintenance can begin only after its arriving to the LMD.

Let P be the set of all service positions.

Let us define parameters of service position $p \in P$.

- d_p is the number of locomotive sections that p can contain at a time, $d_p \leq 3$.
- $a(p)$ is the service track that p belongs to.

- $i(p)$ is the index number of the service position on the service track (counting from the exit of the service track, starting from 1).
- R_p is the moment of time when position p becomes available. $R_p > T_0$ when, for example, position p was occupied by some locomotive at the beginning of planning horizon T.

Let $AP_{ap}, a \in A, p \in P$ be a Boolean matrix and $AP_{ap} = 1$ if it is possible to transport a locomotive from access track a to service position p, and $AP_{ap} = 0$ otherwise.

Let PM_{pm} be a Boolean matrix and $PM_{pm} = 1$ if inspection position p can be used to service locomotives with model number m, and $PM_{pm} = 0$ otherwise.

The goal is to decide for each locomotive:

- whether it should be placed onto the buffer track,
- when it should leave the buffer and head to the access tracks,
- which of the access tracks it should be placed on,
- when it should leave the access track and move onto the service track,
- which of the service positions of that track it should be serviced at.

Depending on the needs of a specific LMD, there might be different objective functions that one would need to implement. We are considering a number of different objectives to minimize, namely total idle time, total waiting time, maximum waiting time and makespan, and carrying out computational experiments.

A problem of scheduling maintenance of a large fleet of rolling stock in a single depot is investigated in [2]. The goal of this work is to schedule maintenance activities for a large number of trains arriving at the depot taking into consideration resource constraints, such as availability of service platforms, engineers, equipment, etc., while maximizing the throughput. Integer linear programming model and a heuristic algorithm are presented. However, this problem differs from ours one. It does not consider any access tracks, and a buffer track there is replaced with an infinite shunting yard with arbitrary input and output of trains.

If we could ignore the track capacity constraints and limit ourselves to considering only one service position (and, of course, only a subset of locomotives that can be serviced at that particular position), we could obtain the optimal schedule by using Smith's theorem [8]. However, the track capacity constraints are far too significant to overlook for our problem.

In [1] fifth chapter deals with the problem of optimal distribution of jobs on identical parallel machines with the same processing time. Various objective functions are considered. A polynomial algorithms is presented. This algorithm, however, does not take into account the possibility of multiple service positions existence on the same service track (i.e., some of the parallel machines would not be independent), possibility of access tracks presence (i.e. having a limited queue size for each of the machines), and various processing times.

One of the tasks of locomotive fleet management is the distribution of locomotives by location, taking into account various constraints. One of the criteria of optimal control of a transport park is matching the number of locomotive crews and locomotives. In [7], a heuristic algorithm for optimal control of a transport park is presented. The algorithm provides a high-quality solution of assembling and routing of the transport park and outperforms the IBM ILOG CPLEX [11] in terms of calculation time. However, the article does not consider maintenance of the trains.

In [3], problem of scheduling maintenance of series of machines is presented. Each unit of a series of machines (for example, a locomotive fleet) must be maintained. When maintenance of machines involves outsourcing specialists, it takes some time for specialists to arrive to the maintenance facility. Usually, the owner of a fleet has to pay not only for the work of these specialists, but he also has to cover the expenses of transporting them to a designated maintenance facility. Thus, one needs to reduce the transporting times by grouping planned maintenance events together. At the same time, the owner seeks to reduce the downtimes of the machines, while meeting necessary maintenance frequency.

In [6], the problem of the distribution a locomotive fleet between several depots for the maintenance is investigated. An algorithm for the organization of the process of operation and maintenance of locomotive fleet is presented. The algorithm allows to find the optimal distribution of the volume of maintenance of locomotives between depots and obtain a graphical solution of the unit cost of repairs. However, authors consider only the problem of distributing locomotives between LMDs, leaving the problem of operating each particular LMD aside.

Now it can be seen that in all the papers above there are significant differences from our problem, and these results cannot be directly applied to solving the problem considered in our article.

3   Constraint Programming Model

The first stage of the analysis of the problem is an attempt to find an exact solution using Constraint Programming (CP) approach. The important principle of constraint programming consists of distinguishing constraint propagation and decision-making search (see, for example, [10]). Constraint propagation is a deductive activity which consists in deducing new constraints from existing constraints. The large number of constraints in our problem contributes to high efficiency of CP methods. To use this approach, the problem needs to be formulated as a Constraint Satisfaction Problem. So, formalization of the problem in this article is made in terms of CP using the Optimization Programming Language (OPL) [12]. In the computational experiments for CP model the time moments were multiples of 15 min (arrival moments and service times were rounded off to the closest upper values).

3.1   Decision Variables, Functions and Other Denotations

To formulate in terms of constraint programming and solve the presented problem, we use the notion of interval variables. Each variable is a time interval in the horizon T. All of the intervals we use are optional, meaning that any of the processes associated with these intervals should be either carried out, according to the solution (in that case, the interval is called present), or skipped (in that case, the interval is called absent). Absent intervals are ignored by most of the constraints. Functions of interval variables can take an additional argument - a value to be returned when the interval argument of such function is absent.

We are using IBM Ilog CPLEX Optimizer 12.6.3 in Constraint Programming mode. The solver algorithm defines which intervals should be present in the solution, and what are start and end moments of those intervals that are present, given that all constraints are satisfied and the objective function is minimized (we provide the constraints and objective functions below).

Interval variables are the following.

- $intB_l$ is the interval during which locomotive l stays in the buffer $intB_l \in [r_l, T_1]$. If length of this interval is zero, or the interval is absent, the locomotive is treated as if it did not enter the buffer at all.
- $intA_{la}$ is the interval during which locomotive l occupies access track a, $intA_{la} \in [r_l, T_1]$.
- $intP_{lp}$ is the interval when locomotive l occupies service position p, $intP_{lp} \in [max(r_l, R_p), T_1]$. The length of this interval can't be less than τ_l.

Built-in OPL Functions. In order to work with the interval variables and model the constraints, we need the following built-in functions and constraints defined in the OPL framework. To make the text more laconic, we renamed built-in opl functions, see their original names in brackets.

- $s(int, dval)$ (in OPL: $startOf(int, dval)$) is a built-in function that evaluates to the start moment of interval int when it is present, and evaluates to $dval$ otherwise. If $dval = 0$, then it can be omitted.
- $e(int, dval)$ (in OPL: $endOf(int, dval)$) is a built-in function that evaluates to the end moment of interval int when it is present, and evaluates to $dval$ otherwise. If $dval = 0$, then it can be omitted.
- $p(int)$ (in OPL: $presenceOf(int)$) - a built-in function that evaluates to 1 if the interval variable int is present in the solution, and 0 if it is absent. If $dval = 0$, then it can be omitted.
- $pulse(int, h)$ is a built-in function that represents usage of a renewable resource in a process associated with interval int. h is the amount of resource used. At any time moment $s(int) \leq t < e(int)$ usage of the resource by interval int is h, and at any other time moment the resource is not used by this interval. This function can be used in constraints like $\sum_{i \in I} pulse(i, h_i) \leq a$, which means that at any time moment collective usage of some resource by intervals from set I cannot exceed a.

Built-in OPL Constraint 'startAtEnd'

- $startAtEnd(int_1, int_2)$ - built-in constraint that forces interval int_1 to start right when interval int_2 ends. If any of the intervals int_1, int_2 is absent, the constraint is automatically considered to be satisfied.

We also need to define a few cumulative functions that will represent the input of the locomotives in using different resources, such as buffer track, access tracks and service positions.

Cumulative Functions

- $BLoad = \sum_{l \in L} pulse(intB_l, u_l)$ is equal to the total number of locomotive sections in the buffer;
- $ALoad(a) = \sum_{l \in L} pulse(intA_{la}, u_l)$ is equal to the total number of sections of locomotives on access track s;
- $PLoad(p) = \sum_{l \in L} pulse(intP_{lp}, 1)$ is equal to the number of locomotives at service position p.

3.2   Objective Functions

In terms of the described decision variables, basic constraints and functions and supplementary functions, the four objective functions mentioned above take the following forms:

1. maximum waiting time $F_1(\pi) = \max\limits_{l \in L, p \in P} s(intP_{lp})$;
2. total waiting time $F_2(\pi) = \sum\limits_{l \in L, p \in P} s(intP_{lp})$;
3. total idle time $F_3(\pi) = \sum\limits_{l \in L, p \in P} e(intP_{lp}) - \tau_l - r_l$;
4. makespan $F_4(\pi) = \max\limits_{l \in L, p \in P} e(intP_{lp})$.

3.3   Constraints

The constraints of the problem will take the following form. To avoid ambiguity, here and below the symbol "$\implies$" will denote logical implication ("$A \implies B$" is equivalent to "if A, then B").

- Each locomotive, if it enters the buffer, does so right at the moment of arrival at LMD:
$$\forall l \in L : s(intB_l, r_l) = r_l.$$

- The buffer is used according to the LIFO scheme:
$$\forall l_1, l_2 \in L : p(intB_{l_1}) \wedge p(intB_{l_2}) \wedge \big(s(intB_{l_1}) \geq s(intB_{l_2})\big)$$
$$\implies \big(e(intB_{l_1}) \leq e(intB_{l_2})\big).$$

- Total count of sections of all the locomotives occupying the buffer at any time should not exceed the capacity of the buffer:
$$BLoad(b) \leq b.$$

- Each locomotive should be assigned to exactly one access track:

$$\forall l \in L : \sum_{a \in A} p(intA_{la}) = 1.$$

- Each locomotive enters its designated access track right when it leaves the buffer, or when it arrives at the LMD if it did not enter the buffer:

$$\forall l \in L, a \in A : (p(intA_{la}) \wedge p(intB_l)) \implies s(intA_{la}) == s(intB_l);$$

$$\forall l \in L, a \in A : (p(intA_{la}) \wedge !p(intB_l)) \implies s(intA_{la}) == r_l.$$

- Total count of sections of all locomotives that are occupying an access track at any given time should not exceed the capacity of the track:

$$\forall a \in A : ALoad(a) \leq c_a.$$

- The access tracks operate according to FIFO scheme:

$$\forall a \in A, l_1 \in L, l_2 \in L : \big(e(intB_{l_1}) \leq e(intB_{l_2})\big)$$
$$\implies \big(s(intA_{l_1a}) \leq s(intA_{l_2a})\big).$$

- Upon leaving an access track, the locomotive should be transported to one of the service tracks:

$$\forall l \in L, a \in A, p \in P : startAtEnd(intA_{la}, intP_{lp}).$$

- A locomotive cannot be assigned to a service position that is inaccessible from the access track which that locomotive was assigned to:

$$\forall l \in L, a \in A, p \in P : p(intA_{la}) * p(intP_{lp}) \leq AP_{ap}.$$

- At any given time, any service position cannot be occupied by more than one locomotive:

$$\forall p \in P : PLoad(p) \leq 1.$$

- A locomotive cannot be assigned to a service position that does not accept this locomotive model:

$$\forall l \in L, p \in P : p(intP_{lp}) \leq PM_{p,v_l}.$$

- Each locomotive consisting of less than 4 sections should be assigned to exactly one service position:

$$\forall l \in L, u_l < 4 : \sum_{p \in P} p(intP_{lp}) = 1.$$

- Each locomotive consisting of 4 sections should be assigned to exactly two service positions:

$$\forall l \in L, u_l = 4 : \sum_{p \in P} p(intP_{lp}) = 2.$$

These positions should be on the same service track:

$$\forall l \in L, p_1 \in P, p_2 \in P, a(p_1) \neq a(p_2) : p(intP_{lp_1}) * p(intP_{lp_2}) = 0.$$

- Total capacity of service positions occupied by a locomotive should be sufficient to accommodate the locomotive:

$$\forall l \in L : \sum_{p \in P} p(intP_{lp}) * d_p \geq u_l.$$

- When two locomotives are assigned to neighboring service positions and their service intervals begin simultaneously, these intervals should end simultaneously, too:

$$\forall l_1, l_2 \in L, p_1, p_2 \in P, a(p_1) = a(p_2) :$$
$$\left(p(intP_{l_1p_1}) * p(intP_{l_2p_2}) = 1 \right) \wedge \left(s(intP_{l_1p_1}) = s(intP_{l_2p_2}) \right)$$
$$\implies \left(e(intP_{l_1p_1}) = e(intP_{l_2p_2}) \right).$$

- The service positions operate according to FIFO scheme: when two locomotives are assigned to neighboring service positions and their inspection starts simultaneously, the locomotive that arrived at the contiguous access track earlier should be placed closer to the exit from the service track (i.e. it should be assigned to the service position with a smaller index number):

$$\forall l_1, l_2 \in L, l_1 \neq l_2, p_1, p_2 \in P, a(p_1) = a(p_2) :$$
$$\left(s(intA_{l_1a(p_1)}) < s(intA_{l_2a(p_2)}) \right) \wedge \left(p(intP_{l_1p_1}) * p(intP_{l_2p_2}) = 1 \right)$$
$$\wedge \left(s(intP_{l_1p_1}) = s(intP_{l_2p_2}) \right)$$
$$\implies \left(i(p_1) < i(p_2) \right).$$

As shown in Sect. 5, applying the exact method to solving the problem on real data takes a lot of time, which is unacceptable. Therefore, in the next section we propose a heuristic algorithm for solving the problem.

4   Greedy Algorithm

Let us number all locomotives $1, 2, \ldots, l, \ldots, n$ in the order of their arrival to the LMD. We will build a schedule step by step, where at each step l we will consider l first locomotives arrived to the LMD. Let us introduce a partial schedule $\psi(l)$, which is built for l first locomotives arrived to the LMD. Partial schedule $\psi(l)$ differs from the feasible schedule by the set of locomotives in it, and by the fact that not all locomotives must be serviced (arrived but not serviced locomotives should be buffered). At each time t for partial schedule $\psi(l)$, the following sets of locomotives can be distinguished:

- $I(\psi(l), t)$ is a set of locomotives that arrived earlier or at time t and sent to an access track immediately;
- $D(\psi(l), t)$ is a set of locomotives that arrived earlier or at time t and sent to the buffer, and then to an access track;
- $B(\psi(l), t)$ is a set of locomotives that are in the buffer at time t.

Partial schedule $\psi(l)$ provides the following information:

- a set of appropriate locomotives $\{1, 2, \ldots, l\}$;
- start and end service time of each locomotive from $I(\psi(l), t)$ and $D(\psi(l), t)$;
- service positions (and service tracks) on which each locomotive from sets $I(\psi(l), t)$ and $D(\psi(l), t)$ is serviced;
- a set of locomotives $B(\psi(l), t)$ that are buffered.

In final schedule $B(\pi, +\infty) = \emptyset$, because it is necessary to maintain all locomotives.

Let us introduce arriving time $\overline{r_e}$ of locomotive $e \in I(\psi(l), t) \cup D(\psi(l), t)$ onto an access track. For locomotives sent to an access track immediately after arriving $e \in I(\psi(l), t)$ the arriving time is equal to the time of its arrival to the LMD: $\overline{r_e} = r_e$.

Let us denote a set of all service tracks by Q. For each partial schedule $\psi(l)$ it is possible to allocate a subset $Q_1(\psi(l), t) \subset Q$ for which locomotives are standing on access tracks last and maintained on service tracks alone. We call the index of the last locomotive serviced on service track $q \in Q_1(\psi(l), t)$ under schedule $\psi(l)$ by $l_q(\psi(l), t)$.

Let us denote arriving time of locomotive e for maintenance onto a service track by $S^e(\psi(l), t), a \in I(\psi(l), t) \cup D(\psi(l), t)$; a maintenance completion time (time of exit from the LMD) by $C^e(\psi(l), t)$ in partial schedule $\psi(l)$. We introduce the concept of partial objective functions for the first l locomotives arrived to the LMD by time t:

1. maximum waiting time $F_1(\psi(l), t) = \max\limits_{i \in I(\psi(l), t) \cup D(\psi(l), t)} (S^i(\psi(l), t) - r_i)$;

2. total waiting time $F_2(\psi(l), t) = \sum\limits_{i \in I(\psi(l), t) \cup D(\psi(l), t)} S^i(\psi(l), t)$;

3. total idle time $F_3(\psi(l), t) = \sum\limits_{i \in I(\psi(l), t) \cup D(\psi(l), t)} (C^i(\psi(l), t) - \tau_i - r_i)$;

4. makespan $F_4(\psi(l), t) = \max\limits_{i \in I(\psi(l), t) \cup D(\psi(l), t)} C^i(\psi(l), t)$.

Henceforth, if it is obvious what partial schedule $\psi(l)$, what locomotive l and what time moment t we are talking about, we will omit the arguments. We associate each access track and a service track it leads to. Thus, when we point to service track $q \in Q$, we will also imply the access track and the group of service positions at the same track, taking into account all their characteristics (capacities of access tracks and service positions, and locomotive models they can service).

Let us describe procedures that are used in the heuristic algorithm. All the procedures can be performed either for locomotive e, which just arrived at the LMD, so the partial schedule will be $\psi = \psi(e - 1)$; or for last locomotive in the buffer ($e \in B(t)$) at time moment t of calling the procedure.

Choice_of_service_tracks(e, t, ψ)—a procedure for choosing a set of service tracks K^e for some locomotive e, on access track of which it can be located at time t. The output is a set of service tracks K^e (and corresponding access tracks), on which locomotive e can be serviced, in accordance with schedule ψ. The selected service tracks must correspond to the capacity of the access tracks and the type of locomotive e. If these conditions are not met at this step at time t, then locomotive e is sent to the buffer of a limited capacity (according to the LIFO scheme), and set K^e is empty.

Partial_objective_function($e, q, \overline{r_e}(q), \psi$)—a procedure for calculating the value of a partial objective function $F(q)$ when setting locomotive e to the access track leading to service track $q \in K^e$. If the number of the locomotive sections $u_e < 4$ and $q \in Q_1(\psi, r_e)$ in schedule ψ, then the algorithm considers two options. Either locomotive e is sent to an access track and will be serviced alone, or locomotive e is sent to an access track and will be serviced along with locomotive l_q on service track q. Then for locomotive e the time of its entering to service position S^e and exit from it C^e are defined as follows.

- If locomotive e is maintained at service track q alone:

$$S^e = \max\{\overline{r_e}, C^{l_q}(\psi)\}, \tag{1}$$

$$C^e = S^e + \tau_e. \tag{2}$$

- If locomotive e is maintained at service track q along with another locomotive l_q:

$$S^e = max\{\overline{r_e}, S^{l_q}(\psi)\}, \tag{3}$$

$$C^e = S^e + max\{\tau_e, \tau_{l_q}\}. \tag{4}$$

If locomotive e is a four-section one, and the set of service tracks K^e is not empty, or the number of sections of locomotive e is less than four and the set of service tracks $K^e \cap Q_1$ is empty, then only one option is considered: locomotive e is sent to the first vacant service track alone. The time of entry and exit from the service position for locomotive e is determined similarly to a non-four-section locomotive when it is serviced alone, according to formulas (1), (2).

If set K^e of service tracks is empty, or in time S^e the number of serviced locomotives is greater than the number of repair crews, then locomotive e is buffered (see *Buffering($e, \overline{r_e}(q), B$)* below), if there is a free space. Otherwise, the algorithm stops working, as it is impossible to build the schedule.

Let f^1 be a value of the partial objective function, taking into account that locomotive l is serviced on service track q alone, and f^2 is a value of the partial objective function if the locomotive is serviced together with locomotive l_q. If f^2 exists, then $F(q) = min\{f^1, f^2\}$. Otherwise, $F(q) = f^1$.

Buffering(e, t, B)—a procedure for sending locomotive e to the buffer at time t, B is the set of all locomotives in the buffer. If locomotive e is buffered, then it becomes the last locomotive in the buffer. Moreover, sets I and D remain unchanged and $B = B \cup \{e\}$. In case when there is no vacant space on the access tracks and in the buffer, the locomotive cannot get into LMD, the schedule cannot be built using this algorithm and the algorithm stops working.

Buffer_check$(r_{l-1}, r_l, \beta, \psi)$—a procedure for checking the buffer between time r_{l-1} of arriving locomotive $l-1$ and time r_l of locomotive l arriving. The procedure is performed for locomotive β, which is the last one in the buffer at the time of its call. The output information is set of service positions K^β, on which locomotive β can be maintained and possible time $\overline{r}_\beta(m)$ of sending the locomotive β on each applicable access track leading to service position $m \in K^\beta\}$. Let some locomotives have completed maintenance during the interval $(r_{l-1}, r_l]$. If a few locomotives have completed their maintenance on some service position, then we will consider only that one, which was released first. Let us denote the set of locomotives which ended their maintenance on each service position first during the interval $(r_{l-1}, r_l]$, by J, $|J| \le |M|$. Obviously, $C^j(\psi) \in (r_{l-1}, r_l]$, $j \in J$. Now call procedure *Choice_of_service_tracks(β, r_l, ψ)* for selecting a set of service tracks K^β for the locomotive β at time r_l with schedule ψ. The locomotive β can go to the access track to a service position from set K^β, as soon as maintenance on it is completed during interval $(r_{l-1}, r_l]$. For each service track $m \in K^\beta$ time moment $\overline{r}_\beta(m)$ of sending locomotive β to it will be defined as $\overline{r}_\beta(m) = C^j(\psi)$, where locomotive $j \in J$ was maintained on service track m.

In addition, if locomotive a is to be maintained on service track m along with locomotive l_m, then $S^{l_m}(\psi) = S^a(\psi), C^{l_m}(\psi) = C^a(\psi)$. For all other locomotives arrived to the LMD by time $\overline{r}_a$, the schedule remains the same.

The algorithm pseudo-code can found below, which uses the above procedures. The input information of the algorithm is: planning horizon $[T_0, T_1]$, schedule $\overline{\pi}$, built for the previous planned day and, accordingly, the set of locomotives $I(\overline{\pi}, T_0), D(\overline{\pi}, T_0), B(\overline{\pi}, T_0)$ at the start of the current planning period. The sets $I(\overline{\pi}, T_0), D(\overline{\pi}, T_0)$ determine time moments $R_p \ge T_0$ of the beginning of service positions availability. The locomotives that are buffered at the beginning of the current scheduling period (at time T_0) will be considered at the current scheduling period (for the locomotives from $I(\overline{\pi}, T_0), D(\overline{\pi}, T_0)$ the schedule remains constant). The output is schedule π of maintenance of all locomotives. If locomotive maintenance doesn't start before time T_1 in the schedule obtained using the algorithm, it will be considered in the next planning horizon $[T_1, T_2]$.

5   Results and Conclusions

The proposed approaches to solving the problem were tested on data provided by Russian Railways, which correspond to large enterprises of the Eastern Polygon. Figure 2 shows the main characteristics of the three test data sets for three LMDs. Each LMD has its own characteristics that must be considered. For example, in LMD 3 there is a shortage of repair crews, so not all service positions can work simultaneously. Four objective functions are considered: maximum waiting time (F_1), total waiting time (F_2), total idle time (F_3), makespan (F_4). Numerical experiments were carried out on a following personal computer: CPU Intel Core i7 7700 HQ 2800 MHz, 4 cores; 8 GB DDR4 RAM.

Figure 3 shows values of each objective function for each data set, obtained using the CP model and the heuristic algorithm. Figure 4 shows the average

---

Algorithm 1. Greedy algorithm

1: Input data: $T_0, T_1, I, D, B, \overline{\pi}$
2: $t' = T_0$
3: $\psi = \overline{\pi}$
4: **for all** $l = 1 \ldots n$ **do**
5:     **if** $B \neq \emptyset$ **then**
6:         **for all** $\beta \in B$ **do**
7:             $(K^\beta, [\overline{r_\beta}(m)]) \leftarrow \text{BUFFER\_CHECK}(t', r_l, \beta, \psi)$
8:             **if** $K^\beta \neq \emptyset$ **then**
9:                 **for all** $m \in K^\beta$ **do**
10:                     $F(m) \leftarrow \text{PARTIAL\_OBJECTIVE\_FUNCTION}(\beta, m, \overline{r_\beta}(m), \psi)$
11:                 **end for**
12:                 $m = \text{argmin}_{i \in K^\beta} F(i)$
13:                 $\psi \leftarrow \text{SCHEDULE\_CHANGES}(\beta, m, \overline{r_\beta}(m), \psi)$
14:             **else**
15:                 Break this cycle
16:             **end if**
17:         **end for**
18:     **end if**
19:     $K^\beta \leftarrow \text{CHOICE\_OF\_SERVICE\_TRACKS}(l, r_l, \psi)$
20:     **if** $K^l \neq \emptyset$ **then**
21:         **for all** $m \in K^l$ **do**
22:             $F(m) \leftarrow \text{PARTIAL\_OBJECTIVE\_FUNCTION}(l, m, r_l, \psi)$
23:         **end for**
24:         $m = \underset{i \in K^l}{\text{argmin}}\, F(i)$
25:         $\psi \leftarrow \text{SCHEDULE\_CHANGES}(l, m, r_l, \psi)$
26:     **else if** $K^l = \emptyset$ & $|B| \leq b$ **then**
27:         $B \leftarrow \text{BUFFERING}(l, r_l, B(r_l))$
28:         $\psi \leftarrow \text{SCHEDULE\_CHANGES}(l, buffer, r_l, \psi)$
29:     **end if**
30:     $t' = r_l$
31: **end for**
32: **if** $B \neq \emptyset$ **then**
33:     **for all** $\beta \in B$ **do**
34:         $(K^\beta, [\overline{r_\beta}(m)]) \leftarrow \text{BUFFER\_CHECK}(r_n, +\infty, \beta, \psi)$
35:         **for all** $m \in K^\beta$ **do**
36:             $F(m) \leftarrow \text{PARTIAL\_OBJECTIVE\_FUNCTION}(\beta, m, \overline{r_\beta}(m), \psi)$
37:         **end for**
38:         $m = \underset{i \in K^\beta}{\text{argmin}}\, F(i)$
39:         $\psi \leftarrow \text{SCHEDULE\_CHANGES}(\beta, m, \overline{r_\beta}(m), \psi)$
40:     **end for**
41: **end if**
42: $\pi = \psi$

	Number of locomotives	The presence of a buffer	Number of repair crews	Number of service positions
LMD 1	100	NO	28	12
LMD 2	45	YES	17	6
LMD 3	40	NO	4	8

Fig. 2. Characteristics of LMDs

deviations of the objective functions values obtained by the heuristic algorithm, relative to the values obtained by CP. The last table shows that the heuristic algorithm gives solution for the objective function F_4 (makespan) comparable to the optimizer solution. The value of the objective function obtained using heuristics is even less than that of the optimizer. This can be explained by the fact that in the CP model there is a time discretization by 15-min intervals. As you can see in Fig. 5, which represents a comparison of both methods to the current methodology of Russian Railways, a heuristic algorithm in most cases shows an advantage in comparison with the existing Russian Railways method.

	Constraint Programming				Greedy Algorithm			
Instance	F_1	F_2	F_3	F_4	F_1	F_2	F_3	F_4
LMD 1	105	1755	2415	1500	174	2051	2799	1493
LMD 2	90	375	525	1500	138	571	706	1471
LMD 3	105	165	180	1500	88	326	421	1435

Fig. 3. The values of the objective functions

F_1	F_2	F_3	F_4
33%	40%	26%	-3%

Fig. 4. The average error (CP vs GA)

In further research we plan to refine the proposed algorithms and to make new algorithms for the problem: dynamic programming and local search algorithms. A transition to a more complex problem statement is also planned. Knowing the characteristics of an LMD and the planned hourly arriving of the locomotives, it is necessary to estimate the maximum number of locomotives that can be serviced in the LMD. It is necessary to build such an autonomous model, which, having all possible combinations of locomotive arrivals, will produce a set of all possible outputs of locomotives from maintenance.

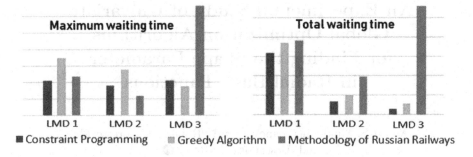

Fig. 5. Maximum Waiting Time (F_1) and Total Waiting Time (F_2)

References

1. Brucker, P.: Scheduling Algorithms. Springer, New York (2006). https://doi.org/10.1007/978-3-540-69516-5
2. Evers, R.P.: Algorithms for Scheduling of Train Maintenance, Netherlands (2010)
3. Huang, J.-Y.: New search algorithm for solving the maintenance scheduling problem for a family of machines. Optim. Methods Softw. **21**, 461–477 (2006)
4. Hansen, P., Mladenovic, N.: Variable neighborhood search: principles and application. Eur. J. Oper. Res. 335–350 (1999)
5. Hansen, P., Mladenovic, N.: Variable neighborhood decomposition search. J. Heuristics 449–467 (2001)
6. Kudayarov, M.: Selection of the optimal organization of repair of locomotives at the railway site. In: Proceedings of PTU (2012)
7. Mardanex, E., Lin, Q., Loxton, R.: A heuristic algorithm for optimal fleet composition with vehicle routing considerations. Optim. Methods Softw. **31**, 272–289 (2016)
8. Smith, W.E.: Various optimizers for single-stage production. Naval Res. Logist. Quart. **3**, 59–66 (1956)
9. Vaidyanathan, B., Ahuja, R.K., Orlin, J.B.: The locomotive routing problem. Transp. Sci. **42**, 492–507 (2008)
10. Dechter, R.: Constraint Processing. Morgan Kaufmann Publishers, San Francisco (2003)
11. IBM Ilog CPLEX. https://www.ibm.com/products/ilog-cplex-optimization-studio
12. IBM. https://www.ibm.com/analytics/data-science/prescriptive-analytics/optimization-modeling

An Experimental Study of Univariate Global Optimization Algorithms for Finding the Shape Parameter in Radial Basis Functions

Marat S. Mukhametzhanov[1,2]([✉]) [iD], Roberto Cavoretto[3] [iD],
and Alessandra De Rossi[3] [iD]

[1] University of Calabria, Via Pietro Bucci 42C, 87036 Rende, CS, Italy
m.mukhametzhanov@dimes.unical.it
[2] Lobachevsky State University, Gagarin Avenue 23, 603950 Nizhny Novgorod,
Russia
[3] Department of Mathematics "Giuseppe Peano", University of Torino,
via Carlo Alberto 10, 10123 Turin, Italy
{roberto.cavoretto,alessandra.derossi}@unito.it

Abstract. In this contribution, an interpolation problem using radial basis functions is considered. A recently proposed approach for the search of the optimal value of the shape parameter is studied. The approach consists of using global optimization algorithms to minimize the error function obtained using a leave-one-out cross validation (LOOCV) technique, which is commonly used for solving machine learning problems. In this paper, the proposed approach is studied experimentally on classes of randomly generated test problems using the GKLS-generator, which is widely used for testing global optimization algorithms. The experimental study on classes of randomly generated test problems is very important from the practical point of view, since results show the behavior of the algorithms for solving not a single test problem, but the whole class with controllable difficulty, which is the main property of the GKLS-generator. The obtained results are relevant, since the experiments have been carried out on 200 randomized test problems, and show that the algorithms are efficient for solving difficult real-life problems demonstrating a promising behavior.

Keywords: Radial basis functions · Global optimization algorithms · Shape parameter

The work of M. S. Mukhametzhanov was supported by the project "Smart Electronic Invoices Accounting" - SELINA CUP: J28C17000160006 (POR CALABRIA FESR-FSE 2014–2020) and by the INdAM-GNCS funding "Giovani Ricercatori 2018–2019". The work of R. Cavoretto and A. De Rossi was partially supported by the Department of Mathematics "Giuseppe Peano" of the University of Torino via Project 2019 "Mathematics for applications" and by the INdAM–GNCS Project 2019 "Kernel-based approximation, multiresolution and subdivision methods and related applications".

M. Jaćimović et al. (Eds.): OPTIMA 2019, CCIS 1145, pp. 326–339, 2020.
https://doi.org/10.1007/978-3-030-38603-0_24

1   Introduction

Finding an optimal value of the shape parameter is a very important problem in the Radial Basis Functions (RBF) community. As it has been shown in [5], the choice of the shape parameter influences both the accuracy and stability in interpolation using RBFs. Traditionally, there are several ways to choose the value of the shape parameter in RBF interpolation: ad hoc choices, pre-fixed constant values (see, e.g., [6,12]), using local optimization methods (see, e.g., [34]), etc. In the recent paper [4], it has been proposed to use global optimization algorithms for finding a good value of the shape parameter. A well-known Leave-One-Out Cross Validation technique has been used in order to introduce the error function, which is the objective function for the optimization problem. Well-known geometric and information univariate global optimization algorithms have been modified in order to increase their efficiency while solving this type of problems. Numerical experiments on several single benchmark test problems have shown the advantages of the proposed techniques.

In this paper, the algorithms proposed in [4] are studied experimentally on the classes of randomly generated test problems. The widely used in testing global optimization algorithms GKLS-generator of randomized test functions is used for this purpose. It can generate classes of 100 test functions with the same properties and controllable difficulty, allowing one to perform a more efficient experimental analysis of numerical algorithms. In this paper, the generator is used for constructing the objective functions for the interpolation problems, which are then used for constructing the respective optimization problems. It should be also noted that the use of classes of randomized test problems allows one not only to perform more reliable and homogeneous numerical experiments, but to visualize the results in a more clear way with respect to numerical tables, using, e.g., graphical representations or statistical notations.

The rest of the paper is organized as follows. In Sect. 2, RBF interpolation and the respective optimization problems are stated briefly. In Sect. 3, numerical algorithms used in this paper and performed experiments are described briefly. Section 4 presents the obtained results. Finally, Sect. 5 concludes the paper.

2   Problem Statement

2.1   Statement of the Interpolation Problem

Let us consider the following interpolation problem. Let the set $X_n = \{x_1, ..., x_n\}$, $x_i \in \mathbb{R}^s$, $i = 1, ..., n$, $x_i \neq x_j$, $i, j = 1, ..., n$ of n *interpolation nodes* and the corresponding set $F_n = \{f(x_1), ..., f(x_n)\}$ of *values* of the function $f : \mathbb{R}^s \to \mathbb{R}$ be given. Let $\mathcal{I}_f : \mathbb{R}^s \to \mathbb{R}$ be the radial basis function (RBF) interpolant given as the linear combination of RBFs of the form

$$\mathcal{I}_f(x) = \sum_{i=1}^{n} c_i \phi_\varepsilon(\|x - x_i\|_2), \quad x \in \mathbb{R}^s, \tag{1}$$

where c_i, $i = 1, ..., n$, are unknown real coefficients that can be found from the interpolation conditions $\mathcal{I}_f(x_i) = f(x_i)$, $i = 1, ..., n$, $||\cdot||_2$ denotes the Euclidean norm, and $\phi : \mathbb{R}_{\geq 0} \to \mathbb{R}$ is a strictly positive definite RBF depending on a *shape parameter* $\varepsilon > 0$. In this paper, the Gaussian (GA) RBF is used:

$$\phi_\varepsilon(r) = e^{-\varepsilon^2 r^2}. \tag{2}$$

The Leave-one-out cross validation technique for the search of the optimal value of the shape parameter ε can be briefly described as follows. First, for any fixed ε and each $k = 1, ..., n$, the point x_k and the respective value $f(x_k)$ are excluded from the sets X_n and F_n, respectively. Then, the partial RBF interpolant is constructed using only $n - 1$ remaining nodes and the error of interpolation at the point x_k is calculated. It has been proved in several works (see, e.g., [24]) that this error can be also calculated without solving n interpolation problems of dimension $n - 1$ as follows:

$$e_k(\varepsilon) = f(x_k) - \mathcal{I}_f^{[k]}(x_k) = \frac{c_k}{A_{kk}^{-1}}, \tag{3}$$

where c_k is the k–th coefficient of the full RBF interpolant $\mathcal{I}_f(x)$ from (1), $\mathcal{I}_f^{[k]}(x)$ is the partial RBF interpolant calculated using only $n - 1$ remaining nodes, and A_{kk}^{-1} is the inverse diagonal element of the matrix A:

$$A_{ij} = \phi_\varepsilon(||x_i - x_j||_2), i, j = 1, ..., n. \tag{4}$$

As a consequence, the value of the shape parameter ε can be fixed in order to minimize the error function $Er(\varepsilon)$ that can be defined, e.g., as follows:

$$Er(\varepsilon) = \max_{k=1,...,n} \left| \frac{c_k}{A_{kk}^{-1}} \right|. \tag{5}$$

2.2   Statement of the Optimization Problem

In this paper, the value of the shape parameter ε is fixed by solving the following optimization problem (see [4] for a detailed discussion): it is required to find the point ε^* and the corresponding value Er^* such that

$$Er^* = Er(\varepsilon^*) = \min Er(\varepsilon), \ \varepsilon \in [0, \varepsilon_{max}], \tag{6}$$

where ε_{max} is large enough (in our experiments ε_{max} was set equal to 20).

The function $Er(\varepsilon)$ can be multiextremal, non-differentiable and hard to evaluate even at one value of ε, since each its computation requires to reconstruct the interpolant (1). It is supposed that $Er(\varepsilon)$ satisfies the Lipschitz condition over the interval $[0, \varepsilon_{max}]$:

$$|Er(\varepsilon_1) - Er(\varepsilon_2)| \leq L|\varepsilon_1 - \varepsilon_2|, \varepsilon_1, \varepsilon_2 \in [0, \varepsilon_{max}], \tag{7}$$

where L, $0 < L < \infty$, is the Lipschitz constant. Since the function $Er(\varepsilon)$ can be ill-conditioned for small ε (see, e.g., [3,5,7]), then the Lipcshitz constant L can be very large.

There exist a lot of algorithms for solving global optimization problems (see, e.g., [1,2,14] for parallel optimization, [9,13] for dimensionality reduction schemes, [10,11,25] for numerical solution of real-life optimization problems, [21] for simplicial optimization methods, [15,35,36] for stochastic optimization methods, [16,17, 22,32] for univariate Lipschitz global optimization, [23] for interval branch-and-bound methods, etc. Among them, there can be distinguished two groups of algorithms: nature-inspired metaheuristic algorithms (as, for instance, genetic algorithm, firefly algorithm, particle swarm optimization, etc. (see, e.g., [31])) and deterministic mathematical programming algorithms (as, for instance, geometric or information algorithms from [20,30,33]). Even though metaheuristic algorithms are used often in practice for solving difficult multidimensional problems, it has been shown in [18,19,29] that for solving ill-conditioned univariate problems (6), (7), deterministic algorithms are more efficient. In particular, in [4], it has been shown on a class of benchmarks and two real-world problems that information global optimization algorithms can be successfully used for this purpose. In this paper, these algorithms are studied experimentally on classes of test problems generated by the GKLS-generator of test problems (see [8] for its description).

The GKLS-generator of test problems is widely used in practice for testing global optimization algorithms (see, e.g., [1,21,31]). It generates classes of 100 randomized multidimensional test problems with the controllable difficulty and a full knowledge about all local and global minimizers (including their positions and their regions of attraction). In this paper, the GKLS-generator is used to generate two-dimensional test functions $f_i(x)$, $x \in \mathbb{R}^2$, for the interpolation problem (1). Then, the generated test functions are evaluated at a uniform grid in order to generate n interpolation nodes. Two classes of test functions were used: the "simple" and "difficult"[1] two-dimensional test classes from [26], since they are used frequently for testing global optimization algorithms. In Fig. 1, an example of the GKLS-type test function is presented. The behavior of this function is typical for the functions from both the classes generated by the GKLS-generator.

3   Algorithms and Organization of Experiments

For each test problem, 128 interpolation nodes are generated on a uniform grid in the square $[-1,1] \times [-1,1]$. Hereinafter, the Information global optimization algorithm with Optimistic Local Improvement LOOCV-GOOI from [4] is used for solving (6) as one of the best global optimization algorithms. This algorithm

---

[1] Traditionally, terms "simple" and "difficult" are related to the difficulty of locating the global minimizer and are used for testing global optimization algorithms and not interpolation methods. In this paper, these terms are used only to distinguish these two classes and not to indicate the difficulty of the interpolation problem.

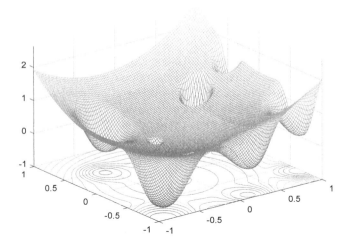

Fig. 1. An example of the GKLS-type test function.

is a locally-biased version of the information global optimization algorithm with "Maximum-Additive" local tuning and optimistic local improvement from [4]. In particular, the main advantage of this method with respect to the original information global optimization algorithm from [32] consists of the ill-conditioned region refinement and restriction of the search interval. Since it is well known that for small values of the parameter ε, the function $Er(\varepsilon)$ becomes ill-conditioned (see, e.g., [3]), then it could be reasonable to study better the region with the small values of ε. For this purpose, the algorithm is launched first on the whole search interval (the Preliminary Search step). Then, the global search is performed on the first subinterval (the Ill-Conditioned Region Refinement step). Finally, after the preliminary search and the ill-conditioned region refinement, the search interval is restricted in a neighborhood of the best obtained values of ε. After that, the search is continued over the restricted interval (the Main Search step). The standard exhaustive LOOCV method using uniform grids of the values of ε as well as the local minimization algorithm LOOCV-min implemented as MATLAB's procedure *fminbnd* from [4] are compared with the LOOCV-GOOI algorithm.

Parameters of the algorithms were set following [4]. In particular, the value $\delta = 10^{-3}$ was used for the stopping condition in LOOCV-GOOI, the reliability parameter r was set to $r = 12$ for the preliminary search, $r = 8$ for the ill-conditioned region refinement, and $r = 4$ for the main search. The local optimization algorithm LOOCV-min uses only the parameter $tolX$, which was set to 10^{-15} in our experiments in order to achieve the machine precision, for the stopping condition. Since, in [4] the LOOCV method using a uniform grid with 500 nodes has been used, but both the local and global optimization algorithms LOOCV-min and LOOCV-GOOI have not generated more than 100 trials, then it can be reasonable to study the LOOCV method with a larger stepsize.

Thus, the LOOCV method with the stepsizes $h_1 = \varepsilon_{max}/99$ and $h_2 = \varepsilon_{max}/499$ (called LOOCV-100 and LOOCV-500, respectively) are compared with the LOOCV-min and LOOCV-GOOI algorithms (the algorithm LOOCV-500 corresponds to the standard LOOCV method in [4]).

The algorithms have been coded and compiled in MATLAB (version R2016b) on a DELL Inspiron 17 5000 Series machine with 8 GB RAM and Processor Intel Core i7-8550U under the MS Windows 10 operating system.

Each algorithm has been launched on each test problem from both the classes. First, the execution times have been calculated for each algorithm on each test class as follows. The execution time T_{total}^i, $i = 1, ..., 100$, of each algorithm has been measured for each test problem of the class. Then, the average execution time T_{total}^{avg} over 100 test problems has been calculated:

$$T_{total}^{avg} = \frac{1}{100} \sum_{i=1}^{100} T_{total}^i, \tag{8}$$

as well as its standard deviation $StDev(T_{total})$:

$$StDev(T_{total}) = \sqrt{\frac{1}{99} \sum_{i=1}^{100} (T_{total}^i - T_{total}^{avg})^2}, \tag{9}$$

the smallest and the largest values T_{total}^{min} and T_{total}^{max}:

$$T_{total}^{min} = \min\{T_{total}^i, \ i = 1, ..., 100\}, \quad T_{total}^{max} = \max\{T_{total}^i, \ i = 1, ..., 100\}. \tag{10}$$

Then, for each test problem, the average execution time per trial T_{trial}^i has been calculated by division of the total execution time T_{total}^i by the number of trials N_{it}^i executed by the algorithm for finding the shape parameter: $T_{trial}^i = T_{total}^i/N_{it}^i$. Finally, the average value, the standard deviation, the smallest and the largest values have been calculated for T_{trial}^i, as well.

Then, the average value, the standard deviation, the smallest and the largest values have been calculated in the same way for the best found values ε^* and $Er(\varepsilon^*)$ from (6) and for the number of performed trials (i.e., the executed evaluations of the function $Er(\varepsilon)$ ad different values of ε) for each algorithm for each test class.

Finally, since all test problems of the same GKLS-class differ only in parameters fixed randomly (see [8,28]), then the error obtained by each algorithm on different test problems from the same class can be considered as a random variable Er^*. So, for each algorithm on each class of test problems, the linear regression model has been constructed for the best obtained error:

$$Er = \beta_0 + \beta_1 \times X, \tag{11}$$

where X is the number of the function from the class and Er is the obtained error using the value ε^* found by each algorithm for the test problem number

X. The coefficients β_0 and β_1 have been estimated using the standard Ordinary Least Squares estimator:

$$\beta_1 = \frac{\sum_{i=1}^{100}(X_i - \overline{X})(Er_i^* - \overline{Er^*})}{\sum_{i=1}^{100}(X_i - \overline{X})^2}, \tag{12}$$

$$\beta_0 = \overline{Er^*} - \beta_1\overline{X}, \tag{13}$$

where $\overline{Er^*}$ is the average error from the Tables 3 and 4:

$$\overline{Er^*} = \frac{1}{100}\sum_{i=1}^{100} Er_i^*, \text{ and } \overline{X} = \frac{1}{100}\sum_{i=1}^{100} X_i = \frac{1+100}{2} = 50.5. \tag{14}$$

4   Results of Numerical Experiments

Results of the experiments are presented in Tables 1, 2, 3 and 4. Tables 1 and 2 show the execution times for all the methods over both the classes of test problems. For each method, the average value, the standard deviation, the smallest and the largest values over 100 test problems calculated following (8)–(10) are shown for the total execution time T_{total} and for the average execution time per trial T_{trial} (i.e., the total execution time divided by the number of trials, which is equal to 100 and 500 for the methods LOOCV-100 and LOOCV-500). As it can be seen from Tables 1 and 2, the execution times for the standard LOOCV method both using 100 and 500 trials is larger, than the execution times of the local and global optimization methods LOOCV-min and LOOCV-GOOI. Moreover, the average execution time is quite similar for the LOOCV-min and LOOCV-GOOI methods.

Table 1. Execution times for "simple" class. For each method, the average, the standard deviation, the smallest and the largest values over 100 test problems are shown.

Method		Average	StDev	Min	Max
LOOCV-100	T_{total}	0.259532	0.0148	0.21462	0.29333
	T_{trial}	0.0025955	0.000148	0.00215	0.00293
LOOCV-500	T_{total}	1.6185886	0.0624	1.49634	1.77206
	T_{trial}	0.0032371	0.000124	0.00299	0.00354
LOOCV-min	T_{total}	0.0945058	0.0150	0.03967	0.15917
	T_{trial}	0.0024746	0.000234	0.00205	0.00408
LOOCV-GOOI	T_{total}	0.1330198	0.0211	0.07415	0.16174
	T_{trial}	0.0023819	0.000239	0.00181	0.00301

Table 2. Execution times for "Difficult" class. For each method, the average, the standard deviation, the smallest and the largest values over 100 test problems are shown.

Method		Average	StDev	Min	Max
LOOCV-100	T_{total}	0.2766418	0.0165	0.22933	0.33023
	T_{trial}	0.0027662	0.000165	0.00229	0.0033
LOOCV-500	T_{total}	1.7100525	0.0975	1.51752	2.13641
	T_{trial}	0.00342	0.000195	0.00304	0.00427
LOOCV-min	T_{total}	0.0995276	0.0147	0.04636	0.12692
	T_{trial}	0.002605	0.000196	0.00222	0.00331
LOOCV-GOOI	T_{total}	0.1306838	0.02652	0.07524	0.19115
	T_{trial}	0.002447	0.000287	0.00174	0.0033

Table 3. Results on the "simple" class. For each method, the average value, the standard deviation, the smallest and the largest values over 100 functions are shown.

Method		Average	StDev	Min	Max
LOOCV-100	ε^*	3.5111076	1.298	0.20202	4.84848
	Er^*	1.74588915	0.834	0.592631	4.71696
LOOCV-500	ε^*	3.0989983	1.620	0.04008	4.92986
	Er^*	1.62075683	0.781	0.488258	4.63249
LOOCV-min	N_{it}	38.32	5.412	15	43
	ε^*	3.8770097	0.473	2.79051	4.9319
	Er^*	1.84643575	0.957	0.543417	4.59887
LOOCV-GOOI	N_{it}	55.58	5.113	41	63
	ε^*	3.3596371	1.365	0.165	4.9317
	Er^*	1.66772142	0.825	0.543426	4.59973

Then, Tables 3 and 4 show the results obtained by each optimization algorithm on both the classes of test problems. In each table, the rows ε^* show the average best obtained value of the shape parameter ε by each method over all 100 test functions, its standard deviation over 100 test functions, its smallest and largest values, respectively. The rows Er^* show the average error using the best obtained values of the shape parameter ε, its standard deviation, the smallest and the largest values, respectively, over all 100 test functions. Finally, the rows N_{it} show the average number of executed trials (or evaluations of the error at different values of ε), its standard deviation, the smallest and the largest values over all 100 test functions for the methods LOOCV-min and LOOCV-GOOI.

As it can be seen from Tables 3 and 4, the best average error was obtained by the method LOOCV-500 for both the classes of test problems, while the average error obtained by the method LOOCV-GOOI is better, than the average error obtained by the methods LOOCV-100 and LOOCV-min. However, the method LOOCV-500 has executed 500 trials in order to obtain better error,

Table 4. Results on the "difficult" class. For each method, the average value, the standard deviation, the smallest and the largest values over 100 functions are shown.

Method		Average	StDev	Min	Max
LOOCV-100	ε^*	3.3010079	1.760	0.20202	8.68687
	Er^*	1.72870075	0.896	0.300373	4.71778
LOOCV-500	ε^*	2.4761517	1.891	0.08016	5.81162
	Er^*	1.46907157	0.694	0.289021	3.80709
LOOCV-min	N_{it}	38.3	5.363	16	43
	ε^*	3.9942203	0.816	2.78239	8.67908
	Er^*	1.90784202	0.954	0.286915	4.57326
LOOCV-GOOI	N_{it}	53	6.404	38	65
	ε^*	2.8477346	1.850	0.08928	6.61288
	Er^*	1.56127502	0.733	0.286961	4.17587

while the method LOOCV-GOOI has executed less than 65 trials for both the classes (see the column "Max" and the rows "N_{it}"). In average, the algorithm LOOCV-GOOI executed almost 55 trials for both the classes, which is almost 9 times smaller, than the number of trials of the method LOOCV-500. Moreover, the average and minimum values of ε for LOOCV-min are larger than those for LOOCV-100, LOOCV-500 and LOOCV-GOOI, while its standard deviation is smaller, which means that the local optimization algorithm is not able to study the ill-conditioned region with small values of ε. It stops very frequently on the locally optimal values of ε, while the smallest obtained value of ε for LOOCV-GOOI is smaller, than the value for LOOCV-100, which means that the algorithm LOOCV-GOOI studies the ill-conditioned region even better, than the "greedy" method LOOCV-100. In Fig. 2, an example of the error functions (6) is presented for the first test problem from both the "Simple" and "Difficult" classes.

Finally, Fig. 3 shows the distribution of the best obtained values $Er(\varepsilon^*)$ and the regression lines (11) for each test problem by each algorithm. As it can be seen from Fig. 3, the lowest regression line corresponds to the method LOOCV-500 for both the classes, while the regression lines of the methods LOOCV-100 and LOOCV-min are higher than the regression function of the method LOOCV-GOOI. It can be also seen from Fig. 3 that the error obtained by the LOOCV-GOOI method is the best one in several cases (see, e.g., the obtained errors for the functions number 79 and 100 of the "simple" class). The error obtained by the global optimization method LOOCV-GOOI is always not worse than the error obtained by the local optimization method, but it is better in a lot of cases.

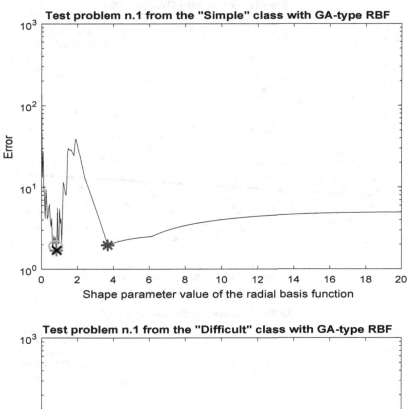

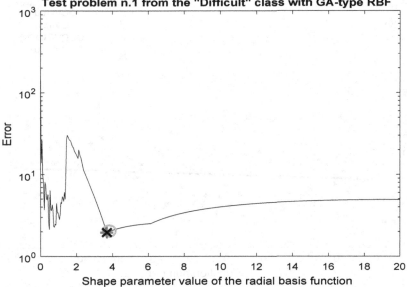

Fig. 2. The error functions for the first test problem from the "Simple" (top) and "Difficult" (bottom) classes used in the experiments. The best found values by LOOCV-100, LOOCV-500, LOOCV-min, and LOOCV-GOOI are indicated as "o", "x", "*", and "+", respectively.

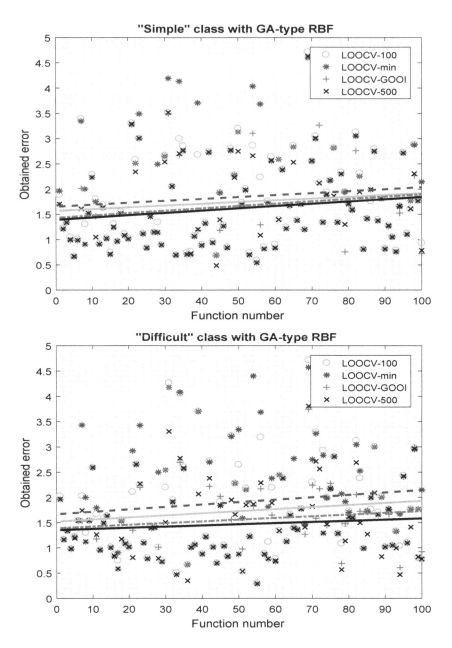

Fig. 3. Interpolation errors for test problems from the "simple" class (top) and from the "difficult" class (bottom). The regression lines for each algorithm are also indicated as blue dashed line for the method LOOCV-min, green and black solid lines - for the methods LOOCV-100 and LOOCV-500, respectively, and red dash-and-dot line for the method LOOCV-GOOI. (Color figure online)

5 Conclusion

It has been shown that global optimization methods can be successfully used for finding a good value of the shape parameter in radial basis functions. The obtained error in this case is better in average, than the error obtained by the standard LOOCV method using a uniform grid with a small size of the grid (using, e.g., 100 evaluations). It has been also shown that even a grid with 500 nodes can be not sufficient for guaranteeing the best value of ε, since in several cases the global optimization method has found a better value executing in average always no more than 55 evaluations. Finally, it has been shown that the traditionally used local optimization algorithm LOOCV-min is not able to study the ill-conditioned region with small values of the shape parameter, giving a locally optimal solution, which is worse than the globally optimal one found by the other methods. It should be also noted that the number of trials executed by the global optimization method is larger than the number of trials executed by the local optimization method, but the difference is small (almost 15 trials in average for both the classes) and not meaningful since it is related to the study of the ill-conditioned region and, in practice, the number of trials for solving ill-conditioned optimization problems is always much higher (see, e.g., [27]).

To conclude, the presented global optimization algorithm has shown a promising performance and can be successfully used in practice for finding good values of the shape parameter in radial basis functions interpolation.

References

1. Barkalov, K., Gergel, V., Lebedev, I.: Solving global optimization problems on GPU cluster. In: Simos, T.E. (ed.) AIP Conference Proceedings, vol. 1738, p. 400006 (2016)
2. Barkalov, K., Strongin, R.: Solving a set of global optimization problems by the parallel technique with uniform convergence. J. Global Optim. **71**(1), 21–36 (2018)
3. Cavoretto, R., De Rossi, A.: A trivariate interpolation algorithm using a cube-partition searching procedure. SIAM J. Sci. Comput. **37**, A1891–A1908 (2015)
4. Cavoretto, R., De Rossi, A., Mukhametzhanov, M.S., Sergeyev, Y.D.: On the search of the shape parameter in radial basis functions using univariate global optimization methods. J. Glob. Optim. (2019, in press). https://doi.org/10.1007/s10898-019-00853-3
5. Fasshauer, G.E.: Meshfree Approximation Methods with MATLAB, Interdisciplinary Mathematical Sciences, vol. 6. World Scientific, Singapore (2007)
6. Fasshauer, G.E.: Positive definite kernels: past, present and future. Dolomites Res. Notes Approx. **4**, 21–63 (2011)
7. Fornberg, B., Larsson, E., Flyer, N.: Stable computations with Gaussian radial basis functions. SIAM J. Sci. Comput. **33**, 869–892 (2011)
8. Gaviano, M., Kvasov, D.E., Lera, D., Sergeyev, Y.D.: Algorithm 829: software for generation of classes of test functions with known local and global minima for global optimization. ACM Trans. Math. Softw. **29**(4), 469–480 (2003)
9. Gergel, V.P., Grishagin, V.A., Israfilov, R.A.: Local tuning in nested scheme of global optimization. Procedia Comput. Sci. **51**, 865–874 (2015)

10. Gergel, V.P., Kuzmin, M.I., Solovyov, N.A., Grishagin, V.A.: Recognition of surface defects of cold-rolling sheets based on method of localities. Int. Rev. Autom. Control **8**(1), 51–55 (2015)
11. Gillard, J.W., Zhigljavsky, A.A.: Stochastic algorithms for solving structured low-rank matrix approximation problems. Commun. Nonlinear Sci. Numer. Simul. **21**(1–3), 70–88 (2015)
12. Golbabai, A., Mohebianfar, E., Rabiei, H.: On the new variable shape parameter strategies for radial basis functions. Comput. Appl. Math. **34**, 691–704 (2015)
13. Grishagin, V.A., Israfilov, R.A., Sergeyev, Y.D.: Convergence conditions and numerical comparison of global optimization methods based on dimensionality reduction schemes. Appl. Math. Comput. **318**, 270–280 (2018)
14. Grishagin, V.A., Sergeyev, Y.D., Strongin, R.G.: Parallel characteristic algorithms for solving problems of global optimization. J. Global Optim. **10**(2), 185–206 (1997)
15. Horst, R., Pardalos, P.M. (eds.): Handbook of Global Optimization, vol. 1. Kluwer Academic Publishers, Dordrecht (1995)
16. Khamisov, O.V., Posypkin, M.: Univariate global optimization with point-dependent Lipschitz constants. In: AIP Conference Proceedings, vol. 2070, p. 020051. AIP Publishing (2019)
17. Khamisov, O., Posypkin, M., Usov, A.: Piecewise linear bounding functions for univariate global optimization. In: Evtushenko, Y., Jaćimović, M., Khachay, M., Kochetov, Y., Malkova, V., Posypkin, M. (eds.) OPTIMA 2018. CCIS, vol. 974, pp. 170–185. Springer, Cham (2019). https://doi.org/10.1007/978-3-030-10934-9_13
18. Kvasov, D.E., Mukhametzhanov, M.S.: One-dimensional global search: nature-inspired vs. Lipschitz methods. In: Simos, T.E. (ed.) AIP Conference Proceedings, vol. 1738, p. 400012 (2016)
19. Kvasov, D.E., Mukhametzhanov, M.S.: Metaheuristic vs. deterministic global optimization algorithms: the univariate case. Appl. Math. Computat. **318**, 245–259 (2018)
20. Kvasov, D.E., Sergeyev, Y.D.: Univariate geometric Lipschitz global optimization algorithms. Numer. Algebra Control Optim. **2**(1), 69–90 (2012)
21. Paulavičius, R., Žilinskas, J.: Simplicial Global Optimization. SpringerBriefs in Optimization. Springer, New York (2014). https://doi.org/10.1007/978-1-4614-9093-7
22. Piyavskij, S.A.: An algorithm for finding the absolute extremum of a function. USSR Comput. Math. Math. Phys. **12**(4), 57–67 (1972). (in Russian) Zh. Vychisl. Mat. Mat. Fiz. **12**(4), pp. 888–896 (1972)
23. Ratz, D.: A nonsmooth global optimization technique using slopes: the one-dimensional case. J. Global Optim. **14**(4), 365–393 (1999)
24. Rippa, S.: An algorithm for selecting a good value for the parameter c in radial basis function interpolation. Adv. Comput. Math. **11**, 193–210 (1999)
25. Sergeyev, Y.D., Daponte, P., Grimaldi, D., Molinaro, A.: Two methods for solving optimization problems arising in electronic measurements and electrical engineering. SIAM J. Optim. **10**(1), 1–21 (1999)
26. Sergeyev, Y.D., Kvasov, D.E.: Global search based on efficient diagonal partitions and a set of Lipschitz constants. SIAM J. Optim. **16**(3), 910–937 (2006)
27. Sergeyev, Y.D., Kvasov, D.E., Mukhametzhanov, M.S.: On the least-squares fitting of data by sinusoids. In: Pardalos, P.M., Zhigljavsky, A., Žilinskas, J. (eds.) Advances in Stochastic and Deterministic Global Optimization. SOIA, vol. 107, pp. 209–226. Springer, Cham (2016). https://doi.org/10.1007/978-3-319-29975-4_11

28. Sergeyev, Y.D., Kvasov, D.E., Mukhametzhanov, M.S.: Emmental-type GKLS-based multiextremal smooth test problems with non-linear constraints. In: Battiti, R., Kvasov, D.E., Sergeyev, Y.D. (eds.) LION 2017. LNCS, vol. 10556, pp. 383–388. Springer, Cham (2017). https://doi.org/10.1007/978-3-319-69404-7_35

29. Sergeyev, Y.D., Kvasov, D.E., Mukhametzhanov, M.S.: Operational zones for comparing metaheuristic and deterministic one-dimensional global optimization algorithms. Math. Comput. Simul. **141**, 96–109 (2017)

30. Sergeyev, Y.D., Kvasov, D.E., Mukhametzhanov, M.S.: On strong homogeneity of a class of global optimization algorithms working with infinite and infinitesimal scales. Commun. Nonlinear Sci. Numer. Simul. **59**, 319–330 (2018)

31. Sergeyev, Y.D., Kvasov, D.E., Mukhametzhanov, M.S.: On the efficiency of nature-inspired metaheuristics in expensive global optimization with limited budget. Sci. Rep. **8**, 453 (2018)

32. Sergeyev, Y.D., Mukhametzhanov, M.S., Kvasov, D.E., Lera, D.: Derivative-free local tuning and local improvement techniques embedded in the univariate global optimization. J. Optim. Theory Appl. **171**(1), 186–208 (2016)

33. Strongin, R.G., Sergeyev, Y.D.: Global Optimization with Non-convex Constraints: Sequential and Parallel Algorithms. Kluwer Academic Publishers, Dordrecht (2000). 3rd ed. by Springer (2014)

34. Uddin, M.: On the selection of a good value of shape parameter in solving time-dependent partial differential equations using RBF approximation method. Appl. Math. Model. **38**, 135–144 (2014)

35. Zhigljavsky, A.A., Žilinskas, A.: Stochastic Global Optimization. Springer, New York (2008). https://doi.org/10.1007/978-0-387-74740-8

36. Žilinskas, A.: On similarities between two models of global optimization: statistical models and radial basis functions. J. Global Optim. **48**(1), 173–182 (2010)

Time-Optimal Control Problem
with State Constraints in a Time-Periodic
Flow Field

Roman Chertovskih$^{(\boxtimes)}$ (iD), Nathalie T. Khalil(iD), and Fernando Lobo Pereira(iD)

Research Center for Systems and Technologies (SYSTEC), Electrical and Computer
Engineering Department, Faculty of Engineering, University of Porto, Porto, Portugal
{roman,flp}@fe.up.pt, khalil.t.nathalie@gmail.com

Abstract. The following time-optimal control problem is solved numerically: compute the fastest trajectory joining two given (initial and final) points of a dynamic control system in a time-periodic flow field subject to state constraints. The considered problem mimics the real-life task of path-planning of a ship in a flow with tidal variations. The considered problem is solved using the maximum principle in Gamkrelidze's form. Under reasonable assumptions on the flow field, it is proved, that the problem is regular and the measure Lagrange multiplier, associated with the state constraint, is continuous. These properties (regularity and continuity) play a critical role in computing the field of extremals by solving the two-point boundary value problem given by the maximum principle. Some examples of time-periodic fluid flows are considered and the corresponding optimal solutions are found.

Keywords: Trajectory-planning · Optimal control · State
constraints · Indirect method · Maximum principle in Gamkrelidze
form · Two-point boundary value problems · Shooting method

1 Introduction

In this article, we consider a state-constrained time-optimal control problem in the presence of a time-periodic flow field, the so-called "navigation problem". We are interested in computing its set of extremals using an indirect method based on the necessary optimality conditions in the form of a maximum principle [1]. However, indirect methods represent significant challenges for optimal

This work was supported by FCT (Portugal): support of FCT R&D Unit SYSTEC – POCI-01-0145-FEDER-006933/SYSTEC funded by ERDF|COMPETE2020|FCT/ MEC|PT2020 extension to 2018, and Project STRIDE NORTE-01-0145-FEDER-000033, by ERDF| NORTE 2020. Results described in Sect. 5 were obtained by R. Chertovskih supported by the Russian Science Foundation (project no. 19-11-00258) in the Federal Research Center "Computer Science and Control" of the Russian Academy of Sciences. The authors are grateful to the anonymous Referees, whose comments were useful to improve the paper.

© Springer Nature Switzerland AG 2020
M. Jaćimović et al. (Eds.): OPTIMA 2019, CCIS 1145, pp. 340–354, 2020.
https://doi.org/10.1007/978-3-030-38603-0_25

control problems with state constraints. Indeed, the difficulty is due to the fact that the state constraint Lagrange multiplier appearing in the necessary optimality conditions whenever the state constraint becomes active is a mere Borel measure. Thus, in general, this multiplier is discontinuous, and this leads to serious difficulties in computing the set of extremals at the times in which the state trajectory meets the boundary of the state constraint set. Therefore, in order to overcome this difficulty, we employ a not so commonly used form of the maximum principle – the so-called Gamkrelidze form – and impose a regularity condition on the data of the problem that entails the continuity of the measure Lagrange multiplier. This is crucial to ensure the appropriate behavior of the proposed numerical procedure to find the corresponding set of extremals.

Moreover, we also present certain conditions that, once satisfied, prevent the emergence of singular control processes. These may be helpful in guiding the computational procedures, by enabling to check the absence of singular controls.

To better grasp the proposed indirect method in the framework of regular problems, we study a navigation problem in $\mathbb{R}^2$. More precisely, we consider an object moving in a closed state domain, subject to a time-dependent fluid flow vector field. The dynamics of the proposed model is affine in the control variable whose values are constrained to the unit square in $\mathbb{R}^2$, and is affected by the vector flow field action. For this model, we are interested in computing the minimum time trajectory connecting two given distinct initial and terminal points. For the problem in question, the regularity condition is satisfied under mild conditions on the vector flow field. In order to develop the proposed indirect method, we derive the corresponding necessary optimality conditions in the nondegenerate Gamkrelidze form, from which the expression of the optimal control is computed. Moreover, the regularity of the problem entails an explicit expression for the measure multiplier. The points where the extremal trajectories reach the boundary of the state constraint can be computed as a result of the continuity of the measure Lagrange multiplier. The two expressions - of the control and the measure multiplier - are functions of the state and adjoint variables, and are replaced in the associated two-point boundary value problem, solved via a shooting algorithm. From the set of all extremals, only the optimal ones, i.e., with the minimal time, are selected as solutions to the given time-optimal problem. We discuss some examples of time-periodic vector flow fields, and we plot the corresponding set of extremals.

The proposed numerical approach based indirect method was discussed in [2] for the steady flow field case. Our paper extends the analysis to time-periodic flow fields. The dependence of the flow on time is crucial for many realistic path planning problems. For example, tides play an important role in shaping water velocity fields in rivers, mainly near their mouth [3]. Moreover, in our paper we discuss, with more details, sufficient conditions for the non-occurrence of singular controls for a particular choice of control set.

The area of state-constrained optimal control problems has been widely investigated in the literature, cf. [4–14]. Questions related to the non-degeneracy of the necessary optimality conditions can be found in [15–22]. Issues on the continuity of the measure multiplier are extensively studied in [23–30]. Numerical methods for computing the set of extremals in the presence of state constraints can be found in [31–39]. For indirect methods, we refer the reader to [32,34–39], among many others.

The article is organized as follows: the formulation of the time-optimal navigation problem is described in Sect. 2 and the regularity concept is discussed. In Sect. 3, the necessary optimality conditions in the Gamkrelidze's form are given in a non-degenerate form. Section 4 is devoted to the application of the maximum principle to the problem in question when the control set is constrained to the unit square in $\mathbb{R}^2$ and to derive the explicit formulas for the corresponding measure multiplier and extremal control. Sufficient conditions for the non-occurrence of singular controls are also discussed. Numerical results and the description of the algorithm are featured in Sect. 5. Section 6 concerns a conclusion, and the Appendix contains detailed proofs of the key results.

2   Problem Formulation: Navigation Problem

We consider an object driven by a dynamical system in a two-dimensional time-space dependent flow field $v(t, x)$, and while subject to affine state constraints. The ultimate goal is to compute a control process that yields the minimum transit time between two given starting and final points A and B within the set of extremals, i.e., the set of control processes satisfying the maximum principle conditions. The corresponding problem is described as follow:

$$
\begin{aligned}
\text{Minimize} \quad & T \\
\text{subject to} \quad & \dot{x} = u + v(t, x), \\
& x(0) = A, \quad x(T) = B, \\
& |x_1| \leq 1, \\
& u \in U := \{u \ : \ \varphi(u) \leq 0\},
\end{aligned}
\tag{1}
$$

where $x = col(x_1, x_2) \in AC([0, T]; \mathbb{R}^2)$ (here $col(x_1, x_2)$ is the collection of x_1 and x_2, and AC stands for the space of absolutely continuous functions), and $u = col(u_1, u_2) \in L^1([0, T]; \mathbb{R}^2)$ are, respectively, the state and the control variables. The point A is the starting point, while B is the terminal point, and $v : [0, T] \times \mathbb{R}^2 \to \mathbb{R}^2$ is a smooth map which defines a fluid flow varying in time and space, and $\varphi : \mathbb{R}^2 \to \mathbb{R}^2$ is also smooth. Following [1], we assume that the boundary of U is regular in the sense that the vectors $\nabla \varphi_i$, $i \in I_\varphi(u)$, are linearly independent, where $I_\varphi(u)$ is the set of i's such that $\varphi_i(u) = 0$ $(i = 1, 2)$.

The state constraint set is represented by the inequality $|x_1| \leq 1$. The terminal time T is to be minimized by the optimal control process.

2.1   Regularity Condition

For the problem (1) we consider here, the function $\Gamma(t, x, u) : [0, T] \times \mathbb{R}^2 \times \mathbb{R}^2 \to \mathbb{R}$, defined by the scalar product of the gradient of the function defining the considered active inequality state constraint and the corresponding dynamics, as

$$\Gamma(t,x,u) := u_1 + v_1(t,x).$$

In the sequel of the regularity concept in [29], we state the following definition:

Definition 1. *Assume that, for all $t \in [0,T]$, $x \in \mathbb{R}^2$ and $u \in \mathbb{R}^2$, such that $|x_1| = 1$, $\varphi(u) = 0$ and $\Gamma(t,x,u) = 0$, the set of vectors $\frac{\partial \Gamma}{\partial u}$ and $\nabla \varphi_i$, for all $i \in I_\varphi(u)$ is linearly independent. Then, we say that problem (1) is regular with respect to the state constraint.*

As it will be explained in the coming sections, the regularity of the problem is, in the context of our paper, crucial for the appropriate behavior of the numerical proposed approach at points in which the trajectory meets the boundary of the state constraint set. The regularity condition might seem restrictive. However, for a large class of engineering problems it is automatically satisfied under natural assumptions. An example will be featured in Sect. 4 for a specific case of control set U.

3   Maximum Principle

In this section, we derive non-degenerate necessary optimality conditions in the Gamkrelidze's form for problem (1). We start by considering the extended time-dependent Hamilton-Pontryagin function

$$\bar{H}(t,x,u,\psi,\mu,\lambda) = \langle \psi, u + v(t,x) \rangle - \mu \Gamma(t,x,u) - \lambda,$$

where $\psi \in \mathbb{R}^2$, $\mu \in \mathbb{R}$ and $\lambda \in \mathbb{R}^+$.

In order to satisfy the notation in what follows, we denote by $f^*(y,z)$ the function $f(x,y,z)$ in which x is replaced by the reference value x^*.

Theorem 1. *We assume that problem (1) is regular in the sense of Definition 1. Then, for an optimal process (x^*, u^*, T^*), there exist a set of Lagrange multipliers: a number $\lambda \in [0,1]$, an absolutely continuous adjoint arc $\psi = (\psi_1, \psi_2) \in W_{1,\infty}([0,T^*]; \mathbb{R}^2)$, and a scalar function $\mu(\cdot)$, such that:*

(a) Adjoint equation

$$\dot{\psi}(t) = -\frac{\partial \bar{H}}{\partial x}(t, x^*(t), u^*(t), \psi(t), \mu(t), \lambda) \qquad for\ a.a.\ t \in [0,T^*];$$

(b) Maximum condition

$$u^*(t) \in \underset{\varphi(u) \leq 0}{\operatorname{argmax}} \left\{ \bar{H}(t, x^*(t), u, \psi(t), \mu(t), \lambda) \right\} \qquad for\ a.a.\ t \in [0,T^*];$$

(c) Time-transversality condition

$$h(T^*) = 0 \ \ where \ \ h(t) := \max_{\varphi(u) \leq 0} \left\{ \bar{H}^*(t,u) \right\};$$

(d) $\mu(t)$ *is constant on the time intervals where* $|x_1^*(t)| < 1$, *increasing on* $\{t \in [0, T] : x_1^*(t) = -1\}$, *and decreasing on* $\{t \in [0, T] : x_1^*(t) = 1\}$. *Moreover,* $\mu(\cdot)$ *is continuous on* $[0, T^*]$;

(e) *Non-triviality condition*

$$\lambda + |\psi_1(t) - \mu(t)| + |\psi_2(t)| > 0, \qquad for\ all\ t \in [0, T^*].$$

Remark 1.

(a) The proof of Theorem 1 follows easily from previous results, [19,40], and [21, Theorem 4.1], and, thus, is placed in the Appendix. It relies on a standard time-reparametrization technique that converts problem (1) into a fixed and time-independent one to which the maximum principle was proved in the above references.

(b) From the regularity property of the problem, the expression of the measure multiplier can be found in terms of the state and the adjoint variables. The junction points – the points at which the trajectory meets the state constraint boundary – can be computed as a result of the continuity of the measure multiplier μ (condition (d)). From these considerations, explicit formulae for the measure multiplier can be obtained. This is the core of our computational scheme proposed in this article for finding the set of extremals.

(c) The non-triviality condition (e), which asserts the non-degeneracy of the Maximum Principle, implies that

$$|\psi_1(t) - \mu(t)| + |\psi_2(t)| > 0 \quad \text{for all } t \in [0, T^*]. \tag{2}$$

4   Applications: Control Set Constrained to the Square

In this section, we consider the specific case of a control set represented as the unit square in $\mathbb{R}^2$, i.e.

$$U := \{u \in \mathbb{R}^2 \ : \ \varphi_1(u) := |u_1| \le 1 \text{ and } \varphi_2(u) := |u_2| \le 1\}. \tag{3}$$

We study how a simple assumption on the vector flow field can automatically lead to regularity in the sense of Definition 1. Thereafter, we use the necessary optimality condition derived in Sect. 3 to obtain explicit expressions of the optimal control and the measure Lagrange multiplier in terms of the state and adjoint variables. These expressions will be substituted in the associated boundary-value problem to numerically find the set of extremals (see Sect. 5).

4.1   Sufficient Condition for Regularity

In the problem considered here, the following simple assumption suffices to ensure regularity as defined above.

(H) $|v_1(t, x)| < 1$ for all $(t, x) \in \mathbb{R} \times \mathbb{R}^2$.

Indeed, if the starting and/or terminal positions are in the interior of the state constraint domain, and if the flow is much faster at the boundary of the state constraint set, $|x_1| = 1$, assumption (H) is crucial to guarantee that the moving object is able to overcome the flow field effect, and, thus, leave the boundary of the state constraint, and move across the river along the axis $0x_1$.

Proposition 1. *Assume that (H) is satisfied. Then, the problem (1), with the specific choice of the control set U, as defined in (3), is regular in the sense of Definition 1.*

Proof. For $\Gamma(t, x, u) = 0$, $|u_1| < 1$ $\mathcal{L}$-a.e., and the strict inequality $\varphi_1(u) < 0$ holds at the boundary of the state constraint set. Then, the regularity condition is satisfied if the vectors $\nabla \varphi_2$, and $\frac{\partial \Gamma}{\partial u}$ constitute a linearly independent set. For our particular problem, $\nabla \varphi_2 = col(0, 1)$, and $\frac{\partial \Gamma}{\partial u} = col(1, 0)$. Therefore, the problem (1) with the particular choice in (3) is regular.

Remark 2. Under assumption (H), the necessary conditions of optimality expressed by Theorem 1, guaranteeing the non-degeneracy of the Lagrange multipliers, and the continuity of the Borel measure μ, can be applied.

4.2   Explicit Formulas for u^* and μ

From the maximum condition (b) of Theorem 1, for a.a. $t \in [0, T^*]$,

$$\max_{|u_1| \leq 1, |u_2| \leq 1} \{(\psi_1(t) - \mu(t)) u_1 + \psi_2(t) u_2\} = (\psi_1(t) - \mu(t)) u_1^*(t) + \psi_2(t) u_2^*(t).$$

This implies that the value of the optimal control process (u_1^*, u_2^*) varies w.r.t. the sign of $\psi_1 - \mu$ and ψ_2, as follows:

$$\begin{cases} \text{if } \psi_1 - \mu \neq 0, & \text{then } u_1^* = sgn(\psi_1 - \mu) \\ \text{if } \psi_2 \neq 0, & \text{then } u_2^* = sgn(\psi_2). \end{cases} \tag{4}$$

The expressions of u^* and μ differ for points belonging to the boundary of the state constraint set or for points in its interior. Next, we discuss these two cases.

When the trajectory stays on the boundary of the state constraint set during a certain set Δ, then $u_1^*(t) = -v_1(t, x^*(t))$ $\mathcal{L}$-a.e. on Δ, and, by continuity, everywhere on Δ. Thus, under assumption (H), $|u_1^*(t)| < 1$. Therefore, from the maximum condition, we have

$$\mu(t) = \psi_1(t) \qquad \text{for all } t \text{ such that } |x^*(t)| = 1. \tag{5}$$

Moreover, as a result of (2), we have $\psi_2(t) \neq 0$ for all t such that $|x_1^*(t)| = 1$ and, thus, $u_2 = \pm 1$.

Now, let Δ be a time interval during which the trajectory lies in the interior of the state constraint set, i.e. $|x_1^*(t)| < 1$ $\forall t \in \Delta$. For any point $t \in \Delta$, the Lagrange measure multiplier μ is constant. Thus, it follows that, for some t in a nonzero Lebesgue measure subset of Δ, we have $\psi_1(t) - \mu(t) = 0$, and since $\mu(t)$ is constant on Δ, and hence, $\dot\psi_1(t) = 0$, on this subset. From the adjoint equations, we conclude that $\psi_2(t)\dfrac{\partial v_2(x)}{\partial x_1} = 0$. From the non-triviality condition, we have to have $\psi_2(t) \neq 0$ $\mathcal{L}$-a.e. on Δ, and, thus, $\dfrac{\partial v_2(x)}{\partial x_1} = 0$ $\mathcal{L}$-a.e. on Δ. In this case the maximum condition is not informative for the first component of the control. This is one case of the so-called singular control, i.e., the controls cannot be defined on a non-zero measure set. Another singular situation corresponds to the case when $\psi_2(t) = 0$ for some t in a nonzero Lebesgue measure subset of Δ. Next, we present and prove sufficient conditions that preclude the emergence of singular control by imposing additional conditions on the flow vector-field v.

Let $S_0 = \{t \in [0,T] : |x_1(t)| = 1\}$, and $S_- = \{t \in [0,T] : |x_1(t)| < 1\}$, $\Delta \subset [0,T]$ such that $\mathcal{L}$-meas$(\Delta) > 0$, and

$$S_1 = \{t \in \Delta \subset S_- : \psi_1(t) - \mu(t) = 0 \text{ and } \psi_2 \neq 0\}$$
$$S_2 = \{t \in \Delta \subset S_- : \psi_1(t) - \mu(t) \neq 0 \text{ and } \psi_2 = 0\}$$
$$S_3 = \{t \in \Delta \subset S_0 : \psi_1(t) - \mu(t) = 0 \text{ or } \psi_2 = 0\}$$

In what follows, we suppress the t-dependence of v as it does not play any role in the developments.

Proposition 2.

(a) If $\dfrac{\partial v_2(x(t))}{\partial x_1} \neq 0$ on S_1, then $\mathcal{L}$-meas$(S_1) = 0$.

(b) If $\dfrac{\partial v_1(x(t))}{\partial x_2} \neq 0$ on S_2, then $\mathcal{L}$-meas$(S_2) = 0$.

(c) On S_3 we always have $\mathcal{L}$-meas$(S_3) = 0$.

Proof. Proof of item (a). Clearly, for all $t \in S_1$, $-\dot\psi_1(t) = \psi_2(t)\dfrac{\partial v_2(x(t))}{\partial x_1}$, and, thus $-\dot\psi_1(t) \neq 0$ for all $t \in S_1$. Since $S_1 \subset S_-$, we readily conclude that $\mathcal{L}$-meas$(S_1) = 0$.

Proof of item (b). Now, for $t \in S_2$, we have $-\dot\psi_2(t) = (\psi_1(t) - \mu(t))\dfrac{\partial v_1(x(t))}{\partial x_2}$. Thus, if $\dfrac{\partial v_1(x(t))}{\partial x_2} \neq 0$, then $\dot\psi_2(t) \neq 0$, and, thus, $\mathcal{L}$-meas$(S_2) = 0$.

Proof of item (c). Consider some $t \in S_3$. The system of adjoint equations can be written as follows

$$-\dot\psi(t) = D_x^T v(x(t))\psi(t) - \mu(t)\nabla_x v_1(x(t)) \tag{6}$$

Since $\psi_1(t) - \mu(t) = 0$ on S_0, we have

$$-\dot\psi_1(t) = \dfrac{\partial v_2(x(t))}{\partial x_1}\psi_2(t) \quad \text{and} \quad -\dot\psi_2(t) = \dfrac{\partial v_2(x(t))}{\partial x_2}\psi_2(t).$$

If there exists some $\bar{t} \in \Delta$ such that $\psi_2(\bar{t}) = 0$, then $\psi_2(\bar{t}) = 0$ on a nonzero Lebesgue measure subset of Δ. From the above, also follows that $\psi_1(t)$ is constant on this subset what contradicts $\psi_1(t) - \mu(t) = 0$ on S_0. From this, and the nontriviality condition of the multipliers, we have the desired conclusion.

Remark 3. Proposition 2 represents sufficient conditions to avoid singular controls. These conditions allow the maximum condition to stay informative and to define the optimal controls on a non-zero measure set. Since the conditions are only sufficient, this means that there might be some cases in which the singular controls do not occur even if these conditions are not satisfied.

5   Numerical Results

In this section, we present and discuss some numerical results for the above stated optimal control problem by using an indirect method based on the considered Maximum Principle. The conditions of this Maximum Principle lead to the following two-point Boundary Value Problem (BVP):

$$\dot{x} = u + v, \tag{7a}$$

$$\dot{\psi} = -\psi \frac{\partial v}{\partial x} + \mu \nabla v_1, \tag{7b}$$

$$x(0) = A, \tag{7c}$$

$$x(T) = B. \tag{7d}$$

The control variables $u_1(t)$ and $u_2(t)$ are given by (4) and the measure Lagrange multiplier $\mu(t)$ is constant for trajectories not meeting the boundary and is defined by (5) along the boundary of the state constraint. Boundary conditions at 0 and T for the adjoint arc $\psi(t)$ are absent.

The BVP problem (7) is solved by a variant of the shooting method (see, e.g., [41] for a brief overview of the shooting methods). The shooting parameter is the angle θ parameterizing the initial boundary condition for ψ:

$$\psi(0) = (\cos(\theta), \sin(\theta)). \tag{8}$$

Starting from the initial conditions (7c) and (8) for a given value of θ, the Cauchy problem for the system of ordinary differential equations (7a)–(7b) is solved by the classical 4th order Runge-Kutta method with the constant time step $\tau = 10^{-4}$. The set of solutions to the BVP (7) constitute the field of extremals. By integrating the system (7a)–(7b) forward in time, the measure Lagrange multiplier $\mu(t)$ is set to zero for the trajectory in the interior of the domain (i.e. when $|x_1| < 1$). If it reaches the boundary, $|x_1| = 1$, at, say, $t = t^*$, then $\mu^* = \mu(t^*)$ is computed by (5). By using the continuity of the measure Lagrange multiplier [29], we deem, that if $|\mu^*| < 10^{-3}$, the point is a junction point of an extremal and integration of the system is continued following the domain boundary. At each time step along the boundary, the trajectories leaving it with constant values of μ are computed. This is done to find another junction point or a segment joining the boundary with the terminal point B. If at a certain time the terminal boundary condition (7d) is satisfied to the accuracy 10^{-3}, the corresponding trajectory represents an extremal. To find all extremals, the parameter θ is varied from 0 to 2π in (8) with a constant step of 10^{-2}, being the bisection method used if the required accuracy is not achieved. Once all extremals, i.e., field of extremals, are computed, the one possessing minimal travelling time among all the extremals is the optimal solution to the original control problem (1).

The first example considers the steady fluid flow

$$v(x) = (\frac{1}{4}x_1, -x_1^2), \tag{9}$$

which is represented by the black arrows in Fig. 1. We clearly notice that, for this specific field, $\frac{\partial v_1}{\partial x_2} = 0$ everywhere, while $\frac{\partial v_2}{\partial x_1} = 0$ at points such that $x_1 = 0$. However, the field of extremal can still be numerically found, supporting Remark 3. The initial and final positions are $A = (0,0)$ and $B = (-0.5, -6)$, respectively. The field of extremals is also shown in Fig. 1. The optimal extremal is the red one with travelling time 3.43 time units.

In the next example, a perturbation of the flow periodic in time (9) is considered for the same trajectory endpoints A and B. Here,

$$v(x,t) = \left(\frac{1}{4x_1} + \sin\left(\frac{\pi t}{2}\right), -x_1^2 \right)$$

and it is assumed that tidal variations affect only the component transversal to the main flow. Although Proposition 2 is violated at some points, the field of extremals is computed and it is shown in Fig. 2, displaying four extremals as in the steady case considered above. In contrast to the problem for the steady flow (9), only one extremal (shown in blue) does not meet the boundary; two extremals (black) meet the right boundary and only one (red) has the left boundary segment. The optimal extremal is again the one (shown in red, with travelling time 3.63 s) involving the left boundary segment, however, the travelling time along the extremal (black) involving the right boundary, 3.77 s, is not significantly larger.

The last example concerns the flow vector field

$$v(x,t) = \left(\frac{x_1}{4} + \frac{x_2}{10}, -x_1^2 - \frac{1}{2}\sin^2\left(\frac{\pi t}{2}\right) \right).$$

For this field, $\frac{\partial v_1}{\partial x_2} \neq 0$, while $\frac{\partial v_2}{\partial x_1} = 0$ when $x_1 = 0$. The corresponding field of extremals is shown in Fig. 3. This field contains five "inner" extremals (i.e. not meeting the state constraint boundary), one meeting the right boundary, and the optimal one meeting the left boundary and reaching the final point B in 3.19 time units.

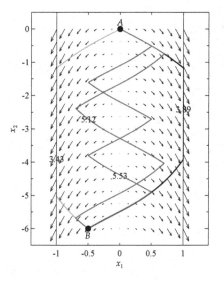

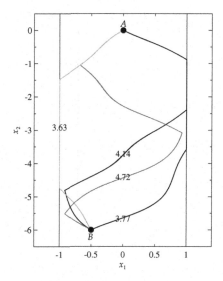

Fig. 1. Field of extremals for the steady fluid flow $v(x) = (\frac{1}{4}x_1, -x_1^2)$. Extremals not meeting the boundary are shown in blue, meeting the right and left boundaries are displayed in black and red, respectively. Inscribed numbers stand for travelling times along the corresponding extremals. (Color figure online)

Fig. 2. Field of extremals for the time-periodic fluid flow $v(x,t) = (\frac{1}{4}x_1 + \sin(\pi t/2), -x_1^2)$. Extremals not meeting the boundary are shown in blue, and meeting the right and left boundaries are displayed in black and red, respectively. Inscribed numbers stand for travelling times along the corresponding extremals. (Color figure online)

6   Conclusion

In this article, we presented an approach based on the maximum principle amenable to the numerical computation of solutions to a regular class of state-constrained optimal control navigation problems subject to a flow field effects. In order to overcome the computational difficulties due to the Borel measure associated with the state constraints, a not so common version of the maximum principle - the so-called Gamkrelidze form - was adopted and a regularity condition on the data of the problem was imposed to ensure the continuity of the Borel measure multiplier. We showed how this property plays a significant role for trajectories meeting the state constraint boundary. We also proved that this regularity assumption is not restrictive, and it is naturally satisfied by a wide class of optimal control problems. The theoretical analysis was supported by several illustrative examples (for steady and time-periodic flows mimicking real river currents) for which the corresponding fields of extremals were constructed, and optimal solutions were found.

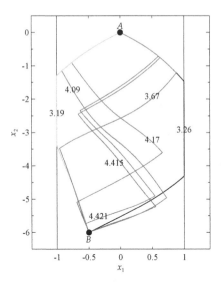

Fig. 3. Field of extremals for the fluid flow $v(x,t) = \left(\frac{1}{4}x_1 + \frac{1}{10}x_2, -x_1^2 - \frac{1}{2}\sin^2\left(\frac{1}{2}\pi t\right)\right)$. Extremals not meeting the boundary are shown in blue, meeting the right and left boundaries are displayed in black and red, respectively. Inscribed numbers stand for travelling times along the corresponding extremals. (Color figure online)

Appendix

Proof of Theorem 1. Following Remark 1 (a), by using a standard time-reparametrization we now formulate an equivalent fixed-time optimal control problem that happens to be autonomous. This is done by fixing $[0, T^*]$, and by considering time as an additional state variable. Indeed, for a minimizer (x^*, u^*, T^*) of the original problem (1), we consider the associated extended fixed, and independent-time optimal control problem. We do not relabel the state and the control variables.

$$
\begin{aligned}
\text{Minimize} \quad & x_0(T^*) \\
\text{subject to} \quad & \dot{x}_0 = u_0, \quad \dot{x} = (u + v(x_0, x))u_0, \quad \text{a.e. } t \in [0, T^*] \\
& x(0) = A, \quad x(T^*) = B, \quad x_0(0) = 0, \quad x_0(T^*) \in \mathbb{R} \\
& |x_1| \leq 1, \quad \text{for all } t \in [0, T^*] \\
& u_0 \in [1 - \alpha, 1 + \alpha] \quad \text{a.e. } t \in [0, T^*] \\
& u \in U := \{u \ : \ \varphi(u) \leq 0\},
\end{aligned} \tag{10}
$$

where x_0 is a new state variable which represents the original time variable, and u_0 the associated control taking values on $[1 - \alpha, 1 + \alpha]$, for some given $\alpha \in (0,1)$. It is straightforward to see that both problems are equivalent. If $(x_0^*(t) = t, u_0^* = 1, x^*, u^*)$ is a minimizer for (10) on $[0, T^*]$, then it follows, by the performing the inverse of the above time reparametrization that (x^*, u^*, T^*) is a minimizer for (1). The converse is easy to show by contradiction. If there is some $(\bar{x}_0, \bar{u}_0, \bar{x}, \bar{u})$ feasible for (10) with $\bar{x}_0(T^*) = \bar{T} < T^*$, then, again, by the inverse time reparametrization, we conclude that there is a control process $(\bar{x}, \bar{u}, \bar{T})$ feasible for (1), i.e., with $\bar{x}(0) = A$ and $\bar{x}(\bar{T}) = B$ as well as with the satisfaction of all other state constraints. Since $\bar{T} < T^*$ the time optimality of (x^*, u^*, T^*) is contradicted.

The corresponding time-independent Hamiltonian is denoted $\bar{H}_0$ and defined given by:

$$\bar{H}_0 := \langle \psi, [u + v(x_0, x)]u_0 \rangle + \psi_0 u_0 - \mu(u_1 + v_1(x_0, x))u_0.$$

The application of the maximum principle to (10) yields the existence of a number $\lambda \in [0, 1]$, $(\psi_0, \psi) \in W_{1,\infty}([0, T^*]; \mathbb{R}) \times W_{1,\infty}([0, T^*]; \mathbb{R}^2)$, and a scalar function $\mu(.)$, such that

(i) $(\dot{\psi}_0(t), \dot{\psi}(t)) = -\dfrac{\partial \bar{H}_0}{\partial x_0} \times \dfrac{\partial \bar{H}_0}{\partial x}(x_0^*, u_0^*, x^*, u^*, \psi_0, \psi, \mu, \lambda)$ for a.e. $t \in [0, T^*]$, and $\psi_0(T^*) = -\lambda$;

(ii) $(u_0^*(t), u^*(t)) \in \underset{u_0 \in [1-\alpha, 1+\alpha], \, \varphi(u) \leq 0}{\mathrm{argmax}} \{\bar{H}_0(x_0^*(t), u_0, x^*(t), u, \psi_0(t), \psi(t),$
$\mu(t), \lambda)\}$ for a.e. $t \in [0, T^*]$;

(iii) (Conservation law) $\underset{u_0 \in [1-\alpha, 1+\alpha], \, \varphi(u) \leq 0}{\max} \{\bar{H}_0^*(u_0, u)\} = 0$ $\forall t \in [0, T^*]$;

(iv) $\mu(t)$ is constant on the time intervals where $|x^*(t)| < 1$, increasing on $\{t \in [0, T] : x_1^*(t) = -1\}$, and decreasing on $\{t \in [0, T] : x_1^*(t) = 1\}$. Moreover, $\mu(\cdot)$ is continuous on $[0, T^*]$;

(v) $\lambda + |\psi_0(t)| + |\psi_1(t) - \mu(t)| + |\psi_2(t)| > 0$, $\forall t \in [0, T^*]$.

Remark 4. These necessary optimality conditions are results of [19,40], and [21, Theorem 4.1]. Moreover, the non-triviality condition (v) is implied by the regularity of the extended problem (10), in the sense of Definition 1. More details can be found in [2].

We explicit now these necessary conditions (i)–(v).

Condition (i) is equivalent to the following:

$$\dot{\psi}(t) = \left(-\psi(t)\frac{\partial v}{\partial x}(x_0^*(t), x^*(t)) + \mu(t)\frac{\partial \Gamma}{\partial x}(x_0^*(t), x^*(t), u^*(t)) \right) u_0^*(t)$$

$$= -\psi(t)\frac{\partial v}{\partial x}(t, x^*(t)) + \mu(t)\frac{\partial \Gamma}{\partial x}(t, x^*(t), u^*(t))$$

which proves condition (a) of Theorem 1, and

$$\dot{\psi}_0(t) = \left(-\psi(t)\frac{\partial v}{\partial x_0}(x_0^*(t), x^*(t)) + \mu(t)\frac{\partial \Gamma}{\partial x_0}(x_0^*(t), x^*(t), u^*(t)) \right) u_0^*(t)$$

$$= -\psi(t)\frac{\partial v}{\partial x_0}(t, x^*(t)) + \mu(t)\frac{\partial \Gamma}{\partial x_0}(t, x^*(t), u^*(t)). \tag{11}$$

Expliciting the maximum condition (ii) implies

$$\underset{u_0 \in [1-\alpha, 1+\alpha], \, u \in U}{\max} \{u_0 (\psi_0(t) + (\psi_1(t) - \mu(t))u_1 + \psi_2(t)u_2)\}$$

$$= \psi_0(t) + (\psi_1(t) - \mu(t))u_1^*(t) + \psi_2(t)u_2^*(t), \tag{12}$$

i.e.

$$\psi_0(t) + (\psi_1(t) - \mu(t))u_1^*(t) + \psi_2(t)u_2^*(t) \geq u_0\left(\psi_0(t) + (\psi_1(t) - \mu(t))u_1 + \psi_2(t)u_2\right)$$

for all $u \in U$ and $u_0 \in [1 - \alpha, 1 + \alpha]$, in particular for $u_0 = 1$. Therefore, for all $u \in U$

$$(\psi_1(t) - \mu(t))u_1^*(t) + \psi_2(t)u_2^*(t) \geq (\psi_1(t) - \mu(t))u_1 + \psi_2(t)u_2$$

which confirms that the maximum condition (b) of Theorem 1 holds true. Also, the fact that $u_0^*(t) = 1$ a.e. entails that $(\psi_1(t) - \mu(t))u_1^*(t) + \psi_2(t)u_2^*(t) = 0$.

Furthermore, denoting

$$h_0 := \max_{u_0 \in [1-\alpha, 1+\alpha], \ \varphi(u) \leq 0} \{\bar{H}_0^*(u_0, u)\} = \max_{u_0 \in [1-\alpha, 1+\alpha], \ \varphi(u) \leq 0} \{u_0(\bar{H}^*(u) + \lambda + \psi_0)\},$$

and since $h_0(T^*) = h_0(x_0^*(T^*)) = 0$ as a consequence of (iii), and $\psi_0(T^*) = -\lambda$ (owing to (i)), we deduce that $h(T^*) = 0$, where $h(t) := \max_{\varphi(u) \leq 0} \{\bar{H}^*(t, u)\}$, confirming the time-transversality condition (c) of Theorem 1.

The non-triviality condition (e) is a direct consequence of (2). It suffices to prove it by contradiction. Finally, condition (d) is direct from condition (iv). Therefore, Theorem 1 is proved.

References

1. Pontryagin, L.S., Boltyanskii, V.G., Gamkrelidze, R.V., Mishchenko, E.F.: The Mathematical Theory of Optimal Processes. Interscience, New York (1962)
2. Chertovskih, R., Karamzin, D., Khalil, N.T., Pereira, F.L.: An indirect method for regular state-constrained optimal control problems in flow fields. IEEE Trans. Autom. Control (2019, submitted)
3. Joseph, A.: Measuring Ocean Currents: Tools, Technologies, and Data. Elsevier Science, Amsterdam (2013)
4. Gamkrelidze, R.V.: Optimal control processes with restricted phase coordinates. Izv. Akad. Nauk SSSR Scr. Mat. **24**, 315–356 (1960)
5. Berkovitz, L.D.: On control problems with bounded state variables. J. Math. Anal. Appl. **5**, 488–498 (1962)
6. Warga, J.: Minimizing variational curves restricted to a preassigned set. Trans. Amer. Math. Soc. **112**(3), 432–455 (1964)
7. Gamkrelidze, R.V.: On some extremal problems in the theory of differential equations with applications to the theory of optimal control. SIAM J. Control **3**(1), 106–128 (1965)
8. Dubovitskii, A.Ya., Milyutin, A.A.: Extremum problems in the presence of restrictions. Zh. Vychisl. Mat. Mat. Fiz. **5**(3), 395–453 (1965). U.S.S.R. Comput. Math. Math. Phys. **5**(3), 1–80 (1965)
9. Hestenes, M.R.: Calculus of Variations and Optimal Control Theory. Wiley, New York (1966)
10. Neustadt, L.W.: An abstract variational theory with applications to a broad class of optimization problems. I: general theory. SIAM J. Control **4**(3), 505–527 (1966)

11. Neustadt, L.W.: An abstract variational theory with applications to a broad class of optimization problems. II: applications. SIAM J. Control **5**(1), 90–137 (1967)
12. Halkin, H.: A satisfactory treatment of equality and operator constraints in the Dubovitskii-Milyutin optimization formalism. J. Optim. Theory Appl. **6**(2), 138–149 (1970)
13. Russak, I.B.: On problems with bounded state variables. J. Optim. Theory Appl. **5**, 424–452 (1970)
14. Ioffe, A.D., Tikhomirov, V.M.: Theory of Extremal Problems. North-Holland, Amsterdam (1979)
15. Arutyunov, A.V., Tynyanskiy, N.T.: The maximum principle in a problem with phase constraints. Sov. J. Comput. Syst. Sci. **23**, 28–35 (1985)
16. Dubovitskii, A.Ya., Dubovitskii, V.A.: Necessary conditions for strong minimum in optimal control problems with degeneration of endpoint and phase constraints. Usp. Mat. Nauk **40**(2), 175–176 (1985)
17. Vinter, R.B., Ferreira, M.M.A.: When is the maximum principle fors tate constrained problems nondegenerate? J. Math. Anal. Appl. **187**(2), 438–467 (1994)
18. Arutyunov, A.V., Aseev, S.M.: Investigation of the degeneracy phenomenon of the maximum principle for optimal control problems with state constraints. SIAM J. Control Optim. **35**(3), 930–952 (1997)
19. Arutyunov, A.V.: Optimality Conditions: Abnormal and Degenerate Problems. Mathematics and Its Application. Kluwer Academic Publisher, Dordrecht (2000)
20. Vinter, R.B.: Optimal Control. Birkhauser, Boston (2000)
21. Arutyunov, A.V., Karamzin, D.Yu., Pereira, F.L.: The maximum principle for optimal control problems with state constraints by R.V. Gamkrelidze: revisited. J. Optim. Theory Appl. **149**(3), 474–493 (2011)
22. Bettiol, P., Khalil, N.: Non-degenerate forms of the generalized Euler-Lagrange condition for state-constrained optimal control problems. In: Variational Methods. In Imaging and Geometric Control, vol. 18. Walter de Gruyter GmbH & Co KG. (2017)
23. Hager, W.W.: Lipschitz continuity for constrained processes. SIAM J. Control Optim. **17**(3), 321–338 (1979)
24. Maurer, H.: Differential stability in optimal control problems. Appl. Math. Optim. **5**(1), 283–295 (1979)
25. Afanas'ev, A.P., Dikusar, V.V., Milyutin, A.A., Chukanov, S.A.: Necessary Condition in Optimal Control. Nauka, Moscow (1990). (in Russian)
26. Galbraith, G.N., Vinter, R.B.: Lipschitz continuity of optimal controls for state constrained problems. SIAM J. Control Optim. **42**(5), 1727–1744 (2003)
27. Bonnans, J.F., Hermant, A.: Revisiting the analysis of optimal control problems with several state constraints. Control Cybern. **38**(4), 1021–1052 (2009)
28. Arutyunov, A.V.: Properties of the Lagrange multipliers in the Pontryagin maximum principle for optimal control problems with state constraints. Differ. Eq. **48**(12), 1586–1595 (2012)
29. Arutyunov, A.V., Karamzin, D.Yu.: On some continuity properties of the measure Lagrange multiplier from the maximum principle for state constrained problems. SIAM J. Control Optim. **53**(4), 2514–2540 (2015)
30. Arutyunov, A., Karamzin, D., Pereira, F.L.: A remark on the continuity of the measure Lagrange multiplier in state constrained optimal control problems. In: IEEE Conference on Decision and Control Conference, Melbourne, Australia (2018)
31. Bryson, E.R., Ho, Y.-C.: Applied Optimal Control. Taylor & Francis, London (1969)

32. Jacobson, D., Lele, M.: A transformation technique for optimal control problems with a state variable inequality constraint. IEEE Trans. Autom. Control **14**(5), 457–464 (1969)

33. Betts, J.T., Huffman, W.P.: Path-constrained trajectory optimization using sparse sequential quadratic programming. J. Guid. Control Dyn. **16**(1), 59–68 (1993)

34. Fabien, B.C.: Numerical solution of constrained optimal control problems with parameters. Appl. Math. Comput. **80**(1), 43–62 (1996)

35. Buskens, C., Maurer, H.: SQP-methods for solving optimal control problems with control and state constraints: adjoint variables, sensitivity analysis and real-time control. J. Comput. Appl. Math. **120**, 85–108 (2000)

36. Vasiliev, F.P.: Optimization Methods. Factorial Press, Moscow (2002). (in Russian)

37. Pytlak, R.: Numerical Methods for Optimal Control Problems with State Constraints. Springer, Heidelberg (2006)

38. Haberkorn, T., Trélat, E.: Convergence results for smooth regularizations of hybrid nonlinear optimal control problems. SIAM J. Control Optim. **49**(4), 1498–1522 (2011)

39. van Keulen, T., Gillot, J., de Jager, B., Steinbuch, M.: Solution for state constrained optimal control problems applied to power split control for hybrid vehicles. Automatica **50**(1), 187–192 (2014)

40. Karamzin, D., Pereira, F.L.: On a few questions regarding the study of state-constrained problems in optimal control. J. Optim. Theory Appl. **180**(1), 235–255 (2019)

41. Press, W.H., Teukolsky, S.A., Vetterling, W.T., Flannery, B.P.: Numerical Recipes: The Art of Scientific Computing, 3rd edn. Cambridge University Press, Cambridge (2007)

The Generalized Ellipsoid Method and Its Implementation

Petro Stetsyuk[1]([✉]) [ID], Andreas Fischer[2] [ID], and Olga Khomyak[1] [ID]

[1] V.M. Glushkov Institute of Cybernetics of NAS of Ukraine, 40 Glushkov Avenue, Kyiv 03187, Ukraine
stetsyukp@gmail.com, khomiak.olha@gmail.com
[2] Faculty of Mathematics, Technische Universität Dresden, 01062 Dresden, Germany
Andreas.Fischer@tu-dresden.de

Abstract. We consider an algorithm with space dilation. For a certain choice of the dilation coefficient, this is a method of outer approximation of semi-ellipsoids by ellipsoids with monotonous decrease in their volume. It is shown that the Yudin-Nemirovski-Shor ellipsoid method is a specific case. Two forms of the algorithm are dealt with: the B-form, where the inverse space transformation matrix B is updated, and the H-form, where the symmetric matrix $H = BB^\top$ is updated. Our test results show that the B-form of the algorithm is computationally more robust to error accumulation than the H-form. The application of the algorithm for finding a minimizer of a convex function, for solving convex programming problems, and for determining a saddle point of a convex-concave function is described. Possible ways of accelerating the algorithm by deeper ellipsoid approximations are discussed as well.

Keywords: Ellipsoid method · Space transformation · Convex programming problem · Saddle point problem

1 Introduction

The classical ellipsoid method (EM) was first proposed in 1976 by Yudin and Nemirovski [1]. They derived this method from the cutting plane scheme and called it modified centered cutting method (MCCM). Independently, EM was discovered by Shor in the paper [2] from 1977. There, EM is presented as a particular case of subgradient methods with space dilation, which were proposed by Shor at the end of the sixties. The generalized ellipsoid method (GEM) [3] is an algorithm with dilation of the n-dimensional space, where the space dilation coefficient satisfies the inequality

$$\alpha + \frac{1}{\alpha} < 2\sqrt[n]{\alpha}. \tag{1}$$

Supported by Volkswagen Foundation (grant No 90 306).

M. Jaćimović et al. (Eds.): OPTIMA 2019, CCIS 1145, pp. 355–370, 2020.
https://doi.org/10.1007/978-3-030-38603-0_26

GEM is a method of outer approximation of semi-ellipsoids by ellipsoids whose volumes decrease monotonously. For $\alpha = \sqrt{\frac{n+1}{n-1}}$, GEM coincides with the Yudin-Nemirovski-Shor EM. If $\alpha = \sqrt{1 + \frac{1}{n^2}} + \frac{1}{n}$, then GEM yields the approximate ellipsoid method (AEM) in [4]. In this paper, we provide properties of two algorithmic realizations of GEM. The first algorithm is based on updating a possibly nonsymmetric matrix B, as in the Shor ellipsoid method, and the second updates a symmetric matrix $H = BB^\top$, as in the Yudin-Nemirovski ellipsoid method.

We present the **emshor** algorithm (**e**llipsoid **m**ethod of **Shor**) for computing a solution of the problem of unconstrained minimization of a convex function [5]. It updates a nonsymmetric matrix B and uses a stopping criterion that, for a convex function, guarantees to find a point at which the function value does not deviate more than a specified tolerance from the optimal function value. It is shown that the **emshor** algorithm finds sufficiently accurate approximations to the minimum point of a ravine convex function, and for functions of twenty variables it takes no longer than a few seconds on a usual PC.

The application of GEM for solving convex programming problems and for determining a saddle point of a convex-concave function is described. In these cases, we can use deeper ellipsoid approximations, i.e., minimal volume ellipsoids based on two cutting hyperplanes [6]. The anti-ravine technique, similar to that used in Shor's r-algorithm [7–10], is considered. In this case coefficients of space dilation in the direction of the difference of the normalized subgradients and in the direction of the sum of normalized subgradients are determined by the obtuse angle between subgradients.

The material is presented as follows. In Sect. 2, the H- and B-form of the GEM and their properties are described. Thereafter, Sect. 3, shows the algorithm **emshor** together with an octave-implementation and numerical results for a ravine non-smooth convex function. The latter is a function with strongly elongated level sets. In Sect. 4, we demonstrate the application of GEM for solving convex programming problems and for determining a saddle point of a convex-concave function. Possible ways of accelerating the algorithm by deeper ellipsoid approximations are discussed as well.

2   The Generalized Ellipsoid Method and Its Properties

Let a mapping $g : \mathbb{R}^n \to \mathbb{R}^n$ be given. We assume that there is $x^* \in \mathbb{R}^n$ so that $g(x)^\top (x - x^*) \geq 0$ for all $x \in \mathbb{R}^n$ and $g(x) \neq 0$ for all $x \neq x^*$. GEM is now used to approximately determine x^*.

2.1   The B-Form of the Generalized Ellipsoid Method

The B-form of GEM can be described as follows.
Step 0. Choose $x_0 \in \mathbb{R}^n$, a non-singular matrix $B_0 \in \mathbb{R}^{n \times n}$, and r_0 so that

$$\left\| B_0^{-1}(x_0 - x^*) \right\| \leq r_0.$$

Moreover, choose α according to (1) and set $k := 0$.

Step 1. Set $g_k := g(x_k)$. If $g_k = 0$, then set $x^* := x_k$ and STOP.

Step 2. Calculate

$$x_{k+1} := x_k - h_k B_k \xi_k, \quad \text{where} \quad \xi_k := \frac{B_k^\top g_k}{\|B_k^\top g_k\|}, \quad h_k := \frac{1}{2}\left(1 - \frac{1}{\alpha^2}\right) r_k.$$

Step 3. Update

$$B_{k+1} := B_k + \left(\frac{1}{\alpha} - 1\right)(B_k \xi_k)\xi_k^\top \quad \text{and} \quad r_{k+1} := \frac{1}{2}\left(\alpha + \frac{1}{\alpha}\right) r_k.$$

Step 4. Set $k := k + 1$ and go to Step 1.

Throughout, for any x_k, x_{k+1} generated by the B-form of GEM, let the ellipsoids

$$\mathcal{E}_k := \{x \mid \|B_k^{-1}(x_k - x)\| \leq r_k\}$$

be defined. Moreover, $vol(\mathcal{E})$ denotes the volume of the ellipsoid $\mathcal{E}$.

Theorem 1 ([3])**.** *Let x_k and x_{k+1} be generated by the B-form of GEM. Then,*

$$x^* \in \mathcal{E}_k \tag{2}$$

is satisfied. Moreover, the ratio of volumes of the ellipsoids $\mathcal{E}_{k+1}$ and $\mathcal{E}_k$ does not depend on k and is equal to

$$q_n(\alpha) := \frac{vol(\mathcal{E}_{k+1})}{vol(\mathcal{E}_k)} = \frac{1}{\alpha}\left(\frac{1}{2}\left(\alpha + \frac{1}{\alpha}\right)\right)^n < 1. \tag{3}$$

At each iteration of the B-form of GEM, the matrix B_k is updated. It is associated with the substitution of variables $x = B_k y$ [7,11]. Obviously, the update of the B-matrix in Step 3 requires $O(n^2)$ operations. Moreover, let us define the space dilation operator $R_\alpha(\xi) : \mathbb{R}^n \to \mathbb{R}^n$ by

$$R_\alpha(\xi) := I_n + (\alpha - 1)\xi\xi^\top,$$

where $\xi \in \mathbb{R}^n$ with $\|\xi\| = 1$ is the direction of dilation and $I_n \in \mathbb{R}^{n \times n}$ the identity matrix. The properties of this operator were studied in detail in [7]. Setting $\beta := 1/\alpha$ and denoting the inverse dilation operator by $R_\alpha^{-1}(\xi)$, we have

$$R_\alpha^{-1}(\xi) = R_\beta(\xi)$$

and

$$B_{k+1} = B_k R_\beta(\xi_k).$$

The latter shows the meaning of space dilation for the update of the B-matrices.

GEM uses an ellipsoid of smaller volume described around a half-ball of radius r in $\mathbb{R}^n$ $(n \geq 2)$. Such an ellipsoid has an oblate shape in the direction that is normal to the hyper-plane defining the half-ball. The parameters of the

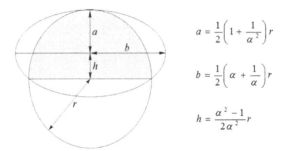

Fig. 1. The parameters of an ellipsoid containing a half-ball in $\mathbb{R}^n$.

ellipsoid are shown in Fig. 1, where a is the length of the minor semi-axis of the ellipsoid, b is the length of the major semi-axes of the ellipsoid (the number of such semi-axes is equal to $n-1$), h is the distance from the center of the ball to the center of the ellipsoid in the direction of its minor semi-axis.

The volume of this ellipsoid is $v_e = v_0 a b^{n-1}$ and the volume of the ball is $v_b = v_0 r^n$, where v_0 denotes the volume of the unit ball in $\mathbb{R}^n$. Therefore, the volume reduction factor is equal to

$$\frac{v_e}{v_b} = \left(\frac{a}{r}\right)\left(\frac{b}{r}\right)^{n-1} = \left(\frac{a}{b}\right)\left(\frac{b}{r}\right)^n = \frac{1}{\alpha}\left(\frac{1}{2}\left(\alpha + \frac{1}{\alpha}\right)\right)^n = q_n(\alpha).$$

If $\alpha = \frac{b}{a}$ satisfies condition (1), we can guarantee that $q_n(\alpha) < 1$. To transform the ellipsoid, containing a half-ball, into a new ball, it is sufficient to dilate the space of variables in the direction of the minor semi-axis with the coefficient $\alpha = \frac{b}{a}$. This can be done using the operator of space dilation $R_\alpha(\xi)$, where the direction ξ coincides with the direction of the minor semi-axis of the ellipsoid.

If $X = \mathbb{R}^n$ is the original space of variables, then in the transformed space of variables $Y = R_\alpha(\xi)X$, we get a new ball of radius b, which contains the solution of our problem. Repeating the same procedure, but for the new ball in the transformed space, we obtain GEM. Here, in Step 2, the direction of the minor semi-axis of the ellipsoid in the transformed space $Y_k = B_k^{-1}X$ is calculated and the transition to its center is performed. The calculated direction is used for the next space dilation, which is implemented in Step 3 by determining the matrix B_{k+1}. In the next transformed space $Y_{k+1} = B_{k+1}^{-1}X$, we get a ball of radius r_{k+1}.

2.2 The H-Form of the Generalized Ellipsoid Method

GEM can be written in H-form by means of a positive definite symmetric matrix H_k, which replaces $B_k B_k^\top$ in the B-form.

Step 0. Choose $x_0 \in \mathbb{R}^n$, a positive definite symmetric matrix $H_0 \in \mathbb{R}^{n \times n}$, and r_0 so that

$$(x_0 - x^*)^\top H_0^{-1}(x_0 - x^*) \le r_0^2.$$

Moreover, choose α according to (1) and set $k := 0$.

Step 1. Set $g_k := g(x_k)$. If $g_k = 0$, then set $x^* := x_k$ and STOP.

Step 2. Calculate

$$x_{k+1} := x_k - h_k \frac{H_k g_k}{\sqrt{g_k^\top H_k g_k}}, \quad \text{where } h_k := \frac{1}{2}\left(1 - \frac{1}{\alpha^2}\right) r_k.$$

Step 3. Update

$$H_{k+1} := H_k + \left(\frac{1}{\alpha^2} - 1\right)\frac{H_k g_k g_k^\top H_k}{g_k^\top H_k g_k} \quad \text{and} \quad r_{k+1} := \frac{1}{2}\left(\alpha + \frac{1}{\alpha}\right) r_k.$$

Step 4. Set $k := k + 1$ and go to Step 1.

The sequences $\{x_k\}$ and $\{\mathcal{E}_k\}$ generated by the H-form are the same as those generated by the B-form GEM if, in Step 0 of the latter, the same values for x_0, r_0, and α are chosen as for the H-form and B_0 is chosen so that $H_0 = B_0 B_0^\top$. Hence, Theorem 1 is valid for the H-form as well. To see that the H-form indeed produces the same sequences $\{x_k\}$ and $\{\mathcal{E}_k\}$, one inductively shows by simple calculations that

$$x_{k+1} = x_k - h_k B_k \xi_k = x_k - h_k \frac{H_k g_k}{\sqrt{g_k^\top H_k g_k}}$$

$$B_{k+1} B_{k+1}^\top = H_k + \left(\frac{1}{\alpha^2} - 1\right)\frac{H_k g_k g_k^\top H_k}{g_k^\top H_k g_k}$$

holds for all iterates with $k = 0, 1, 2, \ldots$ for the B-form and defines

$$H_{k+1} := B_{k+1} B_{k+1}^\top$$

for these k. Let us finally note that $\mathcal{E}_k$ has an equivalent representation by means of H_k-matrices, namely

$$\mathcal{E}_k = \{x \mid \|B_k^{-1}(x_k - x)\| \leq r_k\} = \{x \mid (x_k - x)H_k^{-1}(x_k - x) \leq r_k^2\}.$$

On the one hand, the implementation of the H-form of GEM requires just a half of operations than the B-form does. In addition, the RAM memory usage is also about half of the memory needed for the B-form if the B_k matrices are nonsymmetric. On the other hand, the H-form is computationally less stable since the matrices H_k may become unsymmetric and indefinite. Thus, the H-form requires to monitor these properties of the matrix H_k. Let us demonstrate this by means of a small example.

For $n = 2$ and $\alpha = \sqrt{\frac{n+1}{n-1}}$, we have $\alpha = \sqrt{3}$. Further, setting $H_0 := I_2$, $g_{2k} := (1, -1)^\top$ and $g_{2k+1} := (2, 1)^\top$ for $k = 0, 1, 2, \ldots$, the formula for updating H_k in Step 3 of the H-form of GEM then yields

$$H_{50} = \begin{pmatrix} 8.6163e{-}13 & 9.5855e{-}14 \\ 9.5914e{-}14 & 1.6273e{-}12 \end{pmatrix}, \quad H_{70} = \begin{pmatrix} 9.9927e{-}18 & -1.8545e{-}17 \\ 4.0653e{-}17 & -3.5467e{-}17 \end{pmatrix},$$

where H_{50} is nonsymmetric and H_{70} is neither symmetric nor positive definite.

For the B-form GEM, such problems were not observed. If we look at the same example, the matrices B_{50} and B_{70} computed by the GEM in B-form would lead to

$$B_{50}B_{50}^\top = \begin{vmatrix} 8.6162e{-}13 & 9.5889e{-}14 \\ 9.5889e{-}14 & 1.6273e{-}12 \end{vmatrix}, \quad B_{70}B_{70}^\top = \begin{vmatrix} 1.4592e{-}17 & 1.6239e{-}18 \\ 1.6239e{-}18 & 2.7559e{-}17 \end{vmatrix},$$

i.e., $H_k = B_k B_k^\top$ stay symmetric and positive definite.

2.3   Special Cases of the Generalized Ellipsoid Method

As shown in Theorem 1, the volumes of the ellipsoids containing the solution x^* converge geometrically to 0 with a rate of $q_n(\alpha) < 1$. The smallest rate is used in the classical EM of Yudin-Nemirovski-Shor. This rate corresponds to the dilation coefficient $\alpha^* = \frac{b^*}{a^*} = \sqrt{\frac{n+1}{n-1}}$ (see Fig. 2) and is reached for $q_n(\alpha^*)$, where the function q_n attains its minimum at α^*.

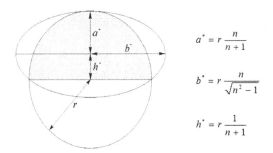

$$a^* = r\,\frac{n}{n+1}$$

$$b^* = r\,\frac{n}{\sqrt{n^2-1}}$$

$$h^* = r\,\frac{1}{n+1}$$

Fig. 2. The parameters of minimal volume ellipsoid containing a half-ball in $\mathbb{R}^n$.

For AEM [4], the dilation coefficient $\alpha^{**} = \frac{b^{**}}{a^{**}} = \sqrt{1 + \frac{1}{n^2}} + \frac{1}{n}$ (see Fig. 3) is used, where α^{**} minimizes the function Q_n, which approximates from above the function q_n according with to

$$q_n(\alpha) = \frac{1}{\alpha}\left(\frac{1}{2}\left(\alpha + \frac{1}{\alpha}\right)\right)^n$$

$$= \frac{1}{\alpha}\left(1 + \frac{1}{2}\left(\alpha + \frac{1}{\alpha} - 2\right)\right)^n$$

$$\leq \frac{1}{\alpha}\exp\left\{\frac{n}{2}\left(\alpha + \frac{1}{\alpha} - 2\right)\right\}$$

$$=: Q_n(\alpha).$$

For $n \geq 2$, the rates $q_n^* := q_n(\alpha^*)$ and $Q_n^* := Q_n(\alpha^{**})$ can be approximated from above by $q_n^* \leq 1 - \frac{1}{2n}$ and $Q_n^* \leq 1 - \frac{1}{2n} + \frac{1}{2n^2}$. Therefore, AEM can be

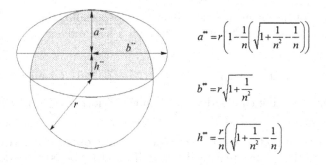

$$a^{\tilde{}} = r\left(1 - \frac{1}{n}\left(\sqrt{1 + \frac{1}{n^2} - \frac{1}{n}}\right)\right)$$

$$b^{\tilde{}} = r\sqrt{1 + \frac{1}{n^2}}$$

$$h^{\tilde{}} = \frac{r}{n}\left(\sqrt{1 + \frac{1}{n^2} - \frac{1}{n}}\right)$$

Fig. 3. The parameters of an approximating ellipsoid containing a half-ball in $\mathbb{R}^n$.

expected to have the same asymptotic volume convergence rate as EM. Indeed, it turns out that the number of iterations for AEM (almost) coincides with the one for EM, as it can be seen in the Table 1 for $n = 2, 3, \ldots, 10$. In the table, q_n^*, Q_n^*, and the ratio Q_n^*/q_n^* are given. Moreover, the last two columns k_q and k_Q show the number of iterations for EM and AEM that is sufficient to achieve a decrease of 10^{-10n} in the ellipsoid volumes, i.e., k_q and k_Q are the smallest integers with $(q_n^*)^{k_q} \leq 10^{-10n}$ and $(Q_n^*)^{k_Q} \leq 10^{-10n}$. It guarantees solution of the problem of minimization of convex function with a relative accuracy equal to 10^{-10}.

Table 1. Comparison of EM and AEM for 10^{-10n} decreasing in volume

n	q_n^*	Q_n^*	Q_n^*/q_n^*	k_q	k_Q
2	0.7698004	0.7725425	1.0035621	177	179
3	0.8437500	0.8441633	1.0004898	407	408
4	0.8813189	0.8814234	1.0001186	730	730
5	0.9042245	0.9042600	1.0000392	1144	1144
6	0.9196855	0.9197001	1.0000159	1651	1651
7	0.9308347	0.9308416	1.0000074	2249	2250
8	0.9392592	0.9392628	1.0000038	2940	2940
9	0.9458508	0.9458528	1.0000021	3723	3723
10	0.9511498	0.9511510	1.0000012	4598	4598

3 Algorithm Emshor for Minimizing Convex Functions

3.1 Algorithm Emshor

Any GEM can be used to find the (unconstrained) minimizer x^* of a convex function $f : \mathbb{R}^n \to \mathbb{R}$. The minimum value of f is denoted by $f^* := f(x^*)$. For simplicity, we assume that x^* is the only minimizer of f. To apply a GEM to the minimization problem just described let us specialize the mapping g used in Sect. 2 by $g_f : \mathbb{R}^n \to \mathbb{R}^n$ with $g_f(x)$ being a subgradient of f at x, i.e., g_f satisfies

$$(x - x^*)^\top g_f(x) \geq f(x) - f(x^*) = f(x) - f^* \geq 0 \quad \text{for all } x \in \mathbb{R}^n. \tag{4}$$

The **emshor** algorithm can be derived from the B-form of GEM from Sect. 2 to $g := g_f$ with $\alpha := \sqrt{\frac{n+1}{n-1}}$ and $B_0 := I_n$. Moreover, a more appropriate stopping criterion is used in Step 1 which guarantees that the **emshor** algorithm stops if $f(x_k) \leq f^* + \varepsilon$ for some predefined $\varepsilon > 0$. This criterion is derived as follows. For $x_k \in \mathbb{R}^n$, $B_k \in \mathbb{R}^{n \times n}$, and r_k generated by the **emshor** algorithm, we obtain

$$\begin{aligned}
f(x_k) - f^* &\leq (x_k - x^*)^\top g_f(x_k) \\
&= (B_k^{-1}(x_k - x^*))^\top B_k^\top g_f(x_k) \\
&\leq \|B_k^{-1}(x_k - x^*)\| \|B_k^\top g_f(x_k)\| \\
&\leq r_k \|B_k^\top g_f(x_k)\|,
\end{aligned} \tag{5}$$

where the first inequality follows from (4) and the last is a consequence of $x^* \in \mathcal{E}_k$ according to Theorem 1. Hence, if $r_k \|B_k^\top g_f(x_k)\| \leq \varepsilon$ is satisfied by some iterate x_k, we have $f(x_k) - f^* \leq \varepsilon$.

Step 0. Choose $x_0 \in \mathbb{R}^n$ and r_0 so that $\|x_0 - x^*\| \leq r_0$.
　　　　Moreover, choose $\varepsilon > 0$, set $B_0 := I_n$ and $k := 0$.
Step 1. If $\|B_k^\top g_f(x_k)\| r_k \leq \varepsilon$, then set $k^* := k$, $x_\varepsilon^* := x_k$ and STOP.
Step 2. Calculate

$$x_{k+1} := x_k - h_k B_k \xi_k, \quad \text{where} \quad \xi_k := \frac{B_k^\top g_f(x_k)}{\|B_k^\top g_f(x_k)\|}, \quad h_k := \frac{1}{n+1} r_k.$$

Step 3. Update

$$B_{k+1} := B_k + \left(\sqrt{\frac{n-1}{n+1}} - 1\right)(B_k \xi_k)\xi_k^\top \quad \text{and} \quad r_{k+1} := \frac{n}{\sqrt{n^2-1}} r_k.$$

Step 4. Set $k := k+1$ and go to Step 1.

The next theorem follows from Theorem 1 for $\alpha = \sqrt{\frac{n+1}{n-1}}$ and by taking into account the stopping criterion in Step 1 and its explanation above.

Theorem 2. *Let x_k and x_{k+1} be generated by the **emshor** algorithm. Then,*

$$x^* \in \mathcal{E}_k$$

is satisfied and the ratio of the volumes of the ellipsoids $\mathcal{E}_{k+1}$ and $\mathcal{E}_k$ does not depend on k and is equal to

$$q_n := \frac{vol(\mathcal{E}_{k+1})}{vol(\mathcal{E}_k)} = \frac{n}{n+1}\left(\frac{n}{\sqrt{n^2-1}}\right)^{n-1} < \exp\left\{-\frac{1}{2n}\right\} < 1.$$

*Moreover, if algorithm **emshor** stops, then $f(x_\varepsilon^*) \leq f^* + \varepsilon$ holds.*

3.2   Octave Function Emshor

The **emshor** algorithm is implemented by a program in Octave. It uses an Octave function of the form **function [f, g] = calcfg (x)**, which calculates the value $f(x)$ and a subgradient $g_f(x)$ at x. This function is provided by the user. The code of the **emshor** program and some short comments are given below.

```
# Input parameters:
# calcfg - name of the function for calculation of f and g
# x0 - starting point, x0(1:n)
# rad - the radius of the ball localizing the minimum point
# epsf, maxitn - parameters for stopping (accuracy, max. iter.)
# intp - printing interval (after each intp iterations)
# Output parameters:
# x - approximation of the minimum point, x(1:n)
# f - value of function f at the point x
# itn - number of iterations performed
# ist - exit code (1 = epsf, 4 = maxitn)
function [x,f,itn,ist] = emshor(calcfg,x0,rad,        #row01
                          epsf,maxitn,intp);
dn=double(length(x0)); beta=sqrt((dn-1.d0)/(dn+1.d0)); #row02
x=x0; radn=rad; B=eye(length(x));                     #row03
for (itn = 0:maxitn)                                  #row04
   [f, g1] = calcfg(x); g=B'*g1; dg=norm(g);          #row05
   if(radn*dg < epsf) ist = 1;   return; endif        #row06
   xi=(1.d0/dg)*g; dx = B * xi;                        #row07
   hs=radn/(dn+1.d0); x -= hs * dx;                    #row08
   B += (beta - 1) * B * xi * xi';                     #row09
   radn=radn/sqrt(1.d0-1.d0/dn)/sqrt(1.d0+1.d0/dn);    #row10
   if(mod(itn,intp)==0)                                #row11
     printf("itn %4d  f %14.6e\n",itn,f);              #row12
   endif                                               #row13
endfor                                                 #row14
ist = 4;                                               #row15
endfunction                                            #row16
```

The iterative process is executed in a **for**-loop (rows 04–14), where rows 05–06 implement Step 1, rows 07–08 implement Step 2, and rows 09–10 implement Step 3 of algorithm **emshor**. After every intp iterations in the **for**-loop intermediate results are printed (see rows 11–13). The **emshor** program is terminated if either (1) a point x_ε^* with $f(x_\varepsilon^*) \leq f^* + \varepsilon$ is found (**ist = 1** in row 06) or (2) the maximal number of iterations **maxitn** is reached (see rows 04 and 15).

3.3   Computational Experiments for Ravine Function

We will demonstrate the work of the **emshor** program on the minimization of a ravine piecewise linear convex function $f : \mathbb{R}^n \to \mathbb{R}$ with

$$f(x) = \sum_{i=1}^{n} 2^{i-1} |x_i - 1|. \tag{6}$$

Obviously, the unique minimizer of f is $x^* = (1, 1, \ldots, 1)^\top \in \mathbb{R}^n$ with optimal value $f^* = f(x^*) = 0$.

The degree of elongation of level sets of the function (6) is determined by the ratio of the maximum coefficient of $|x_i - 1|$ to the minimal coefficient. The latter is always equal to 1. For $n = 20$, the maximal coefficient equals $2^{19} \approx 5.2e + 05$.

Table 2 provides results obtained by program **emshor** for the starting values $x_0 := 0 \in \mathbb{R}^n$, $r_0 \in \{5, 500, 50000\}$, and $\varepsilon \in \{10^{-3}, 10^{-6}, 10^{-9}\}$. The calculations were performed for $n \in \{5, 10, 15, 20\}$ on a Pentium 2.5 GHz computer using GNU Octave version 3.0.0. The table shows the number of iterations (itn), the function value $f(x_{itn})$ at the last iteration, and the computing time ($time$) in

Table 2. The results for minimization function (6) by program **emshor**

$r_0 = 5$

n	$\varepsilon = 10^{-3}$			$\varepsilon = 10^{-6}$			$\varepsilon = 10^{-9}$		
	itn	$f(x_{itn})$	$time$	itn	$f(x_{itn})$	$time$	itn	$f(x_{itn})$	$time$
5	519	6.1e–06	0.18	873	1.1e–07	0.28	1201	1.2e–10	0.40
10	2484	8.7e–05	0.79	3829	7.2e–08	1.26	5246	7.9e–11	1.76
15	6561	6.5e–06	2.10	9667	6.0e–08	3.23	12786	1.5e–11	4.36
20	13101	4.8e–05	4.28	18714	3.5e–09	6.34	23416	2.0e–11	8.11

$r_0 = 500$

n	$\varepsilon = 10^{-3}$			$\varepsilon = 10^{-6}$			$\varepsilon = 10^{-9}$		
	itn	$f(x_{itn})$	$time$	itn	$f(x_{itn})$	$time$	itn	$f(x_{itn})$	$time$
5	747	1.1e–05	0.25	1080	1.6e–07	0.35	1392	1.7e–10	0.46
10	3429	9.0e–05	1.08	4810	9.3e–08	1.58	6185	6.3e–11	2.06
15	8615	5.6e–05	2.78	11704	6.5e–08	3.90	14805	2.4e–11	5.02
20	16729	1.8e–06	5.49	22404	4.4e–08	7.64	27161	1.4e–11	9.38

$r_0 = 50000$

n	$\varepsilon = 10^{-3}$			$\varepsilon = 10^{-6}$			$\varepsilon = 10^{-9}$		
	itn	$f(x_{itn})$	$time$	itn	$f(x_{itn})$	$time$	itn	$f(x_{itn})$	$time$
5	951	1.7e–04	0.32	1323	1.9e–08	0.42	1658	1.4e–10	0.55
10	4323	8.4e–05	1.38	5736	6.4e–08	1.88	7093	7.2e–11	2.36
15	10663	6.0e–05	3.45	13772	1.8e–08	4.60	16860	3.6e–11	5.70
20	20417	4.7e–05	6.75	26039	3.5e–08	8.80	30772	1.6e–11	10.54

seconds. It should be noted that all execution times can be significantly reduced when one makes use of compiled commands for matrix times vector operations.

From Table 2, it can be seen that the **emshor** program finds very precise approximations to the minimizer of the ravine convex function. The number of iterations grows slightly faster than n^2 for the given accuracy ε and the radius r_0 of the initial ball containing the minimizer x^*. This is due to the fact that, to reduce the deviation of the function value from f^* by a factor of 10, the ellipsoid method requires $\approx 4.6 n^2$ iterations (see Sect. 2.3, Table 1).

The ellipsoid method in the H-form may not approximate the minimum point of the ravine function with very high accuracy. This is confirmed by numerical experiments with the modification of the program **emshor**, in which the operators with the matrix B in lines 5, 7 and 9 are replaced by the corresponding operators with the symmetric matrix $H = BB^T$. Calculation results of minimizing function (6) with $\varepsilon = 10^{-3}$, the starting values $x_0 := 0 \in \mathbb{R}^n$, and $r_0 \in \{5, 500\}$ are given in Table 3. The table shows the number of iterations itn, the function value $f(x_{itn})$, the iteration number itr with $itr \leq itn$, where the smallest value $f(x_{itr})$ of the function f is attained during the iteration.

Table 3. Results for minimization function (6) by H-form of algorithm **emshor**

	$r_0 = 5$				$r_0 = 500$			
n	itn	$f(x_{itn})$	itr	$f(x_{itr})$	itn	$f(x_{itn})$	itr	$f(x_{itr})$
5	461	0.001310	446	0.00001033	453	0.208985	443	0.00201819
10	1664	0.030775	1467	0.00008307	1767	2.357996	1690	0.00195698
15	6541	0.000065	6528	0.00000029	8615	0.000056	7804	0.00000031
20	5627	5.700439	5356	0.00240983	5434	1142.453	4980	0.20944151

From Table 3 it is easy to see that for all $n \in \{5, 10, 15, 20\}$, the accuracies are insufficient in comparison with those from Table 2, i.e., the modified program stops long before an acceptable accuracy is achieved. This is due to the fact that the norm of the matrix H converges to zero much faster than the norm of the matrix B, and due to the accumulation of rounding errors when updating the symmetric matrix H (see Sect. 2.2).

4   Other Applications of the Generalized Ellipsoid Method

4.1   Constrained Convex Programming

Let us consider the nonlinear programming problem

$$\text{minimize}_x \quad f_0(x) \quad \text{subject to} \quad f_i(x) \leq 0 \quad i = 1, 2, \ldots, m \tag{7}$$

where, for $i = 0, 1, \ldots, m$, the functions $f_i : \mathbb{R}^n \to \mathbb{R}$ are convex and $g_i : \mathbb{R}^n \to \mathbb{R}^n$ denotes a mapping so that $g_i(x)$ is a subgradient of f_i at x. Let us assume that problem (7) has the unique solution x^* and that the Slater condition is satisfied. Moreover, let the mapping $g : \mathbb{R}^n \to \mathbb{R}^n$ be defined by

$$g(x) = g_{i(x)}(x),$$

where the mapping $x \mapsto i(x)$ satisfies the conditions

$$
\begin{aligned}
i(x) = 0, &\quad \text{if } f_1(x) \le 0, \ldots, f_m(x) \le 0, \\
i(x) \in \{i \mid f_i(x) > 0\}, &\quad \text{if } \max\{f_1(x), \ldots, f_m(x)\} > 0.
\end{aligned}
\tag{8}
$$

Then, it can be seen that $(x - x^*)^\top g(x) \ge 0$ holds for all $x \in \mathbb{R}^n$. Let us first consider the case that x satisfies $\max\{f_1(x), \ldots, f_m(x)\} \le 0$. According to (8), we have $g(x) = g_0(x)$. Moreover, by $f_0(x) \ge f_0(x^*)$ and the convexity of f, we have

$$(x - x^*)^\top g(x) = (x - x^*)^\top g_0(x) \ge f_0(x) - f_0(x^*) \ge 0.$$

If, otherwise, $\max\{f_1(x), \ldots, f_m(x)\} > 0$, then (8) implies $f_j(x) > 0$ for $j := i(x)$. By the convexity of f_j and by $f_j(x^*) \le 0$, we get

$$(x - x^*)^\top g(x) = (x - x^*)^\top g_j(x) \ge f_j(x) - f_j(x^*) \ge 0.$$

Thus, $(x - x^*)^\top g(x) \ge 0$ holds for all $x \in \mathbb{R}^n$.

Hence, to approximate x^* we can apply one of the previous algorithms. In particular, we may use the same stopping criterion as in the algorithm **emshor**, see (5) as well.

4.2   Saddle Point Problems of Convex-Concave Functions

Let $f : \mathbb{R}^n \times \mathbb{R}^m \to \mathbb{R}$ be a convex-concave function, i.e., $f(\cdot, y)$ is convex for any $y \in \mathbb{R}^m$ and $f(x, \cdot)$ is concave for any $x \in \mathbb{R}^n$. A pair $z^* := (x^*, y^*) \in \mathbb{R}^n \times \mathbb{R}^m$ is called saddle point of f if

$$f(x^*, y) \le f(x^*, y^*) \le f(x, y^*) \quad \text{for all } (x, y) \in \mathbb{R}^n \times \mathbb{R}^m.$$

Moreover, with $z := (x, y)$, let the set $G_x(z) \subset \mathbb{R}^n$ contain all partial subgradients of $f(\cdot, y)$ with y fixed, whereas $G_y(z) \subset \mathbb{R}^m$ contains all partial supergradients of $f(x, \cdot)$ with x fixed. Based on this we can define the mapping $g : \mathbb{R}^n \times \mathbb{R}^m \to \mathbb{R}^n \times \mathbb{R}^m$ by

$$g(z) := \begin{pmatrix} g_x(z) \\ -g_y(z) \end{pmatrix} \quad \text{with} \quad (g_x(z), g_y(z)) \in G_x(z) \times G_y(z).$$

By the above definition of a saddle point and by the convexity of the functions $f(\cdot, y)$ and $-f(x, \cdot)$ for arbitrarily but fixed y and y, respectively, it follows that

$$
\begin{aligned}
0 &\le f(x, y^*) - f(x^*, y) \\
&= f(x, y^*) - f(x, y) + f(x, y) - f(x^*, y) \\
&\le -g_y(z)^\top (y - y^*) + g_x(z)^\top (x - x^*) \\
&= g(z)^\top (z - z^*).
\end{aligned}
$$

Thus, $g(z)^\top (z - z^*) \geq 0$ holds for all $z \in \mathbb{R}^n \times \mathbb{R}^m$ so that GEM can be used to approximate the saddle point z^*.

GEM has also a number of other applications, for example for solving non-smooth problems of small dimensions that occur in decomposition schemes (by constraints, by variables), for special convex problems with a small number of variables with a parametrically given family of constraints. The main questions are to determine a rule for constructing cutting hyperplanes, which localize the point to find, and to develop an appropriate stopping criterion for the iterative process in GEM.

4.3   Possible Ways of Accelerating the Ellipsoid Method

As it was shown in Sect. 2.1, GEM make use of the space dilation operator $R_\alpha(\xi) = I + (\alpha - 1)\xi \xi^\top$. It transforms the ellipsoid, containing the half-ball in $\mathbb{R}^n$, into a new ball after one space dilation (1d-ellipsoid, see Fig. 1). For a 2d-ellipsoid we receive a new ball after two space dilations. The projection of the 2d-ellipsoid onto the plane is shown in Fig. 4.

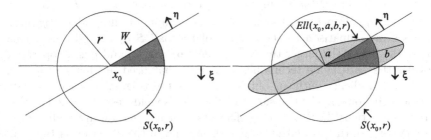

Fig. 4. Projections of the set W and the 2d-ellipsoid $Ell(x_0, a, b, r)$.

The key feature of this technique is the transformation of the 2d-ellipsoid $Ell(x_0, a, b, r)$ into a ball. The minimum volume ellipsoid $Ell(x_0, a, b, r)$ is centered at x_0 and contains the convex set $W \subset \mathbb{R}^n$ resulting from the intersection of the ball $S(x_0, r) := \{x \in \mathbb{R}^n \mid \|x - x_0\| \leq r\}$ with the two half-spaces

$$P(x_0, \xi) := \{x \in \mathbb{R}^n \mid (x - x_0)^\top \xi \leq 0\}$$
$$P(x_0, \eta) := \{x \in \mathbb{R}^n \mid (x - x_0)^\top \eta \leq 0\},$$

where $-1 < \xi^\top \eta < 0$, $\|\xi\| = 1$, $\|\eta\| = 1$. The 2d-ellipsoid has the following parameters: the length of the semi-axis in the direction $(\xi - \eta)$ is equal to $a = r\sqrt{1 + (\xi, \eta)} < r$; the length of the semi-axis in the direction $(\xi + \eta)$ is $b = r\sqrt{1 - (\xi, \eta)} > r$; in the other $(n - 2)$ directions orthogonal to ξ and η, the lengths of the semi-axes are equal to r. The ratio of the 2d-ellipsoid volume and the ball volume is equal to $q := (a/r) * (b/r) = \sqrt{1 - (\xi^\top \eta)^2} < 1$. The ratio decreases when the angle between ξ and η approaches $180°$.

Lemma 1 ([12]). *Let $B_k \in \mathbb{R}^{n \times n}$, $r > 0$, $x_k, x^* \in \mathbb{R}^n$, and $g_1, g_2 \in \mathbb{R}^n$ be given such that $\|B_k^{-1}(x_k - x^*)\| \leq r$, $(x_k - x^*)^\top g_1 \geq 0$, $(x_k - x^*)^\top g_2 \geq 0$, and $g_1^\top B_k B_k^\top g_2 < 0$ holds. Then, the updated matrix*

$$B_{k+1} := B_k R_{\beta_1}\left(\frac{\xi - \eta}{\|\xi - \eta\|}\right) R_{\beta_2}\left(\frac{\xi + \eta}{\|\xi + \eta\|}\right)$$

with

$$\beta_1 := \sqrt{1 + \xi^\top \eta}, \quad \beta_2 := \sqrt{1 - \xi^\top \eta}, \quad \xi := \frac{B_k^\top g_1}{\|B_k^\top g_1\|}, \quad \eta := \frac{B_k^\top g_2}{\|B_k^\top g_2\|}$$

has the following properties:

(i) $\|B_{k+1}^{-1}(x_k - x^*)\| \leq r$,
(ii) $\det(B_{k+1}) = \det B_k \sqrt{1 - (\xi^\top \eta)^2}$, *and*
(iii) $g_1^\top B_{k+1} B_{k+1}^\top g_2 = 0$.

Lemma 1 has the following interpretation. Property (i) means that $y^* := B_{k+1}^{-1} x^*$ belongs to the ball $S(y_k, r)$ in the transformed space $Y := B_{k+1}^{-1} X$, where $y_k := B_{k+1}^{-1} x_k$. Property (ii) shows that the volume of ellipsoid $Ell(y_k, a, b, r)$ decreases in comparison to the volume of the ball $S(y_k, r)$ and this decrease will be the bigger the larger the obtuse angle between ξ and η is. Property (iii) provides the anti-ravine technique, similar to that used in Shor's r-algorithm [7–10]. Subgradients with an obtuse angle in the original space of variables become orthogonal in the transformed space. This yields less elongated level sets of ravine functions. In this case, coefficients of space dilation in the direction of the difference of the normalized subgradients and in the direction of the sum of normalized subgradients are determined by the angle between subgradients. A more obtuse angle leads to a larger value of the coefficient of space dilation in the direction of the difference between the two normalized subgradients.

The 1d-ellipsoid (see Fig. 1) and the 2d-ellipsoid (see Fig. 4) can be used to develop accelerated variants of ellipsoid methods for solving convex programs and saddle point problems of convex-concave functions. For such accelerated methods, we can expect a convergence rate close to that of r-algorithms. This is confirmed by numerical experiments for subgradient methods with transformation of space for finding the minimizer of a convex function with a priori knowledge of the minimal function value [13]. In particular, these methods turned out to be efficient for ravine functions.

5   Conclusion

In this paper we considered the generalized method of ellipsoids and the properties of two theoretically equivalent versions. The first one is based on updating a not necessarily symmetric matrix B, as in the ellipsoid method of Shor. In the second version, a symmetric matrix $H = BB^\top$ is updated, as in the ellipsoid method of Yudin-Nemirovski. Based on the classic ellipsoid method in B-form, algorithm **emshor** (ellipsoid method of **Shor**) and its implementation in Octave for the problem of unconstrained minimization of a convex functions were developed. It is demonstrated that the algorithm **emshor** allows us to find a very accurate approximation of the minimizer of a ravine convex function in a reasonable amount of time. Therefore, algorithm **emshor** can be interesting for solving non-smooth subproblems in implementations of decomposition methods for block angular linear programming problems with a smaller number of variables or constraints.

The accelerated variants of ellipsoid methods on the basis of the 1d-ellipsoid and the 2d-ellipsoid can be used for solving a variety of problems: convex programming problems, problems of finding saddle points of convex-concave functions, and special cases of variational inequalities.

Acknowledgement. The authors would like to thank Oleksii Lykhovyd and Volodymyr Zhydkov for their help in preparing this paper. Moreover, the authors are grateful to an anonymous referee for comments that helped to improve the paper.

References

1. Yudin, D.B., Nemirovskii, A.S.: Informational complexity and efficient methods for the solution of convex extremal problems. Matekon **13**(3), 25–45 (1976)
2. Shor, N.Z.: Cut-off method with space dilation in convex programming problems. Cybernetics **13**(1), 94–96 (1977)
3. Stetsyuk, P.I., Fesiuk, O.V., Khomyak, O.N.: The generalized ellipsoid method. Cybern. Syst. Anal. **54**(4), 576–584 (2018)
4. Stetsyuk, P.I.: An approximate method of ellipsoids. Cybern. Syst. Anal. **39**(3), 435–439 (2003)
5. Stetsyuk, P.I., Fischer, A., Lyashko, V.I.: An ellipsoid method for minimization of convex function. NaUKMA Res. Pap. Comput. Sci. **2**, 16–21 (2019). https://doi.org/10.18523/2617-3808.2019.2.16-21. (in Ukranian)
6. Stetsyuk, P.I.: 2D-ellipsoid of optimal volume and its applications. In: Polyakova, L.N. (ed.) Constructive Nonsmooth Analysis and Related Topics (Dedicated to the Memory of V.F. Demyanov) (CNSA), pp. 303–306. IEEE (2017)
7. Shor, N.Z.: Minimization Methods for Non-Differentiable Functions. Springer, Heidelberg (1985). https://doi.org/10.1007/978-3-642-82118-9
8. Stetsyuk, P.I.: Shor's r-algorithms: theory and practice. In: Butenko, S., Pardalos, P.M., Shylo, V. (eds.) Optimization Methods and Applications. SOIA, vol. 130, pp. 495–520. Springer, Cham (2017). https://doi.org/10.1007/978-3-319-68640-0_24
9. Stetsyuk, P.I.: Theory and software implementations of Shor's r-algorithms. Cybern. Syst. Anal. **53**(4), 692–703 (2017)
10. Stetsyuk, P.I., Fischer, A.: Shor's r-algorithms and octave-function ralgb5a. In: International Conference "Modern Informatics: Problems, Achievements and Prospects for Development", Ukraine, Kyiv, December 2017. V.M. Glushkov Institute of Cybernetics of the NAS of Ukraine, pp. 143–146 (2017). (in Russian)
11. Sergienko, I.V., Stetsyuk, P.I.: On N.Z. Shor's three scientific ideas. Cybern. Syst. Anal. **48**(1), 2–16 (2012)

12. Stetsyuk, P.I.: r-Algorithms and ellipsoids. Cybern. Syst. Anal. **32**(1), 93–110 (1996)
13. Stetsyuk, P.I.: Methods of ellipsoids and r-algorithms, Eureka, Chisinau, Moldova (2014). (in Russian)

Investigation of the Problem of Optimal Control by a System ODE of Block Structure with Blocks Connected only by Boundary Conditions

Kamil Aida-zade[1,2(✉)] and Yegana Ashrafova[1,2]

[1] Institute of Control Systems of ANAS, B.Vahabzade 9, AZ1141 Baku, Azerbaijan
kamil_aydazade@rambler.ru
[2] Baku State University, Z.Khalilov, AZ1148 Baku, Azerbaijan
ashrafova.yegana@gmail.com
http://bsu.edu.az, http://www.isi.az

Abstract. The paper investigates the problem of optimal control by the systems of ODE of block-structure with unseparated boundary conditions between blocks. The considered complex object consists of blocks. The state of each of them is described by the system of ordinary differential equations. The blocks are interconnected in an arbitrary order only by the initial and/or final (boundary) state values. The necessary optimality conditions for the considered optimal control problem are obtained.

Note that, the obtained adjoint problem has the same specifics as the direct problem. For the numerical solution of optimal control problem, it is proposed to apply first-order optimization methods using the formulas for the functional gradient that participate in the necessary optimality conditions. To solve the direct and adjoint initial boundary-value problems of a block structure and with unseparated nonlocal boundary conditions with sparse (arbitrary filled) matrix, special schemes of the sweep method are proposed that take into account the specifics of the systems of differential equations and boundary conditions that allow the transfer of boundary conditions for each block separately.

Keywords: Optimal control · Systems of ODE · Block-structure · Gradient of the functional · Unseparated boundary conditions

1 Introduction

We investigate the problem of optimal control of an object (process), described by a system of a large number of independent subsystems of linear non-autonomous differential equations. The blocks (subsystems) of the common system are connected by nonlocal, unseparated boundary conditions [1–6].

It is assumed that most of the elements of the relation matrix are zero, nonzero elements correspond to the presence of a connection between the initial

© Springer Nature Switzerland AG 2020
M. Jaćimović et al. (Eds.): OPTIMA 2019, CCIS 1145, pp. 371–382, 2020.
https://doi.org/10.1007/978-3-030-38603-0_27

and final states of individual blocks of a complex object. The necessary optimality conditions for the considered problem are obtained, which are used to
construct a numerical method for its solution. A numerical scheme is proposed
for solving direct and adjoint initial boundary value problems of large dimension
and block structure. The method is based on the proposed special scheme of the
sweep method [1–3,5,6], which allows sweeping separately block by block. As a
result, the problem is reduced to an algebraic system of equations with a weakly
and arbitrarily filled Jacobi matrix.

The advantage of the proposed approach in comparison with the direct use
of transfer methods for the whole system [1–3,8] lies in the fact that, in the work
the transfer is carried out only with respect to those variables whose coefficients
are non-zero in the boundary conditions, while the transfer is carried out using
only the subsystem of differential equations in which the transferred variable is
involved.

2   Problem Statement

Let us consider a complex system, consisting of M arcs (blocks) arbitrarily connected by ends, which we represent in the form of oriented graph. Each arc
(edge) is an independent object (block), the state of which is described by the
system of ordinary differential equations.

We denote the set of nodes of the graph by I, and the set of arcs (edges) (k, s)
with the length $l^{(k,s)}$ with the beginning at the node $k \in I$ and with the end at
the node $s \in I$ denote by $J = \{(k, s) : k, s \in I, |I| = N, |J| = m, |I|$ indicates the
number of elements of the set I.

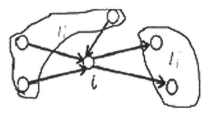

Let $J_i^+ = \{(j, i) : j \in I_i^+\}$, $J_i^- = \{(i, j) : j \in I_i^-\}$ are the sets of arcs
respectively entering and leaving the node of i, I_i^+ and I_i^- are the sets of nodes,
connected with the i-th node, which are respectively the ends and the beginnings
of the arcs from the set J_i, $J_i = J_i^+ \cup J_i^-$, $I_i = I_i^+ \cup I_i^-$. We denote by

$$\left| J_i^+ \right| = \left| I_i^+ \right| = n_i^+, \ \left| J_i^- \right| = \left| I_i^- \right| = n_i^-, \ n_i^+ + n_i^- = n_i.$$

So, it's known that

$$\sum_{i \in I} n_i^+ = \sum_{i \in I} n_i^- = N, \ \sum_{i \in I} n_i = 2m.$$

In practical applications, as a rule, take place $n_i << N, i \in I$, i.e. the number
of nodes adjacent to any other node is much less than the total number of nodes.

Let the state of each arc $((k,s) \in J$) is described by a system of χ linear non-autonomous ordinary differential equations of the following form:

$$\dot{u}^{ks}(x) = A^{ks}(x) u^{ks}(x) + B^{ks}(x) v^{ks}(x), \qquad x \in [0, l^{ks}], (k, s) \in J. \qquad (1)$$

Here $u^{ks}(x) \in R^{\chi}$ is the state of (k, s)-th arc with the length l^{ks} at the point $x \in [0, l^{ks}]$; $A^{ks}(x) \neq const$ are the known continuous $\chi-$ dimensional quadratic matrix functions, $rang\, A^{ks}(x) = \chi$; $B^{ks}(x)$ are the known continuous $\chi \times Z_{ks}-$ dimensional rectangular matrix functions; $v^{ks}(x), x \in [0, l^{ks}]$ is the control vector-function of the (k, s)-th subsystem, $v^{ks}(\cdot) \in \mho^{ks} \subset R^{z_{ks}}$, $v(x) = (v^{ks}(x) : (k, s) \in J; l^{ks} > 0$ and $\mho^{ks}$ the set of admissible values of controls $v^{ks}(x)$ are given. We assume that the sets $\mho^{ks}$ have a simple structure, for example, are parallelepipeds, that is,

$$\mho^{ks} = \{v \in R^{Z_{ks}} : \underline{v}_j^{ks} \le v_j(x) \le \overline{v}_j^{ks}, \; x \in [0, l^{ks}], j = 1, ..., Z_{ks}\}$$

Note that the system has a block structure, the total number of blocks - subsystems (1), is equal to $m-$ the number of arcs. Current states of arcs (blocks) are connected with adjacent arcs (blocks) only by unseparated boundary conditions in the form:

$$\sum_{k \in I_i^-} g_{ik}^j u^{ik}(0) + \sum_{k \in I_i^+} q_{ki}^j u^{ki}(l^{ki}) = r_i^j, \quad , j = \overline{1, \nu_i}, i \in I, \qquad (2)$$

where r_i^j is the value of j-th characteristics of external influence in the i-th node. The row-vectors $g_{ik}^j = (g_{ik}^{j1}, ..., g_{ik}^{j\chi}), k \in I_i^-$, $g_{ik}^j = (g_{ik}^{j1}, ..., g_{ik}^{j\chi}), k \in I_i^+$, $j = \overline{1, \nu_i}, i \in I$, are given. ν_i is a number of given condition at the node $i \in I$, and

$$\sum_{i=1}^N \nu_i = M.$$

The total number of differential equations in (1) as well as boundary conditions in (3) is $M = m\chi$.

External influences as a rule, are differ from zero only in the beginning and end nodes of the system. We denote the set of these nodes by I^f, $N^f = |I^f|$, and if $i \in I^f$. We denote the set of internal nodes of the graph by $I^{int} = I \backslash I^f$, $N^{int} = |I^{int}|$.

We will assume that, the coefficients involved in conditions (2) and forming sparse matrices $G = (g_{ik}^j, k \in I_i^-, j = \overline{1, \nu_i}, i \in I)$, $Q = (q_{ki}^j, k \in I_i^+, j = \overline{1, \nu_i}, i \in I,)$ such that the rangs of the extended matrices is

$$rang[G, Q] = M.$$

Note that the boundary conditions (2) have an important specificity, namely, that they are unseparated (non-local) boundary conditions, and in practical applications the matrices G, Q are sparse, while they are filled arbitrarily, which is determined by the structure of the object, namely interconnections between blocks. We will write down the expressions (2) in a more generally form:

$$Gu(0) + Qu(l) = r. \tag{3}$$

Here we used the notations:

$$u(0) = (u^{1,1}(0), \ldots, u^{1,\chi}(0), \ldots, u^{m,1}(0), \ldots u^{m,\chi})^T,$$

$$u(l) = (u^{1,1}(l^{1,1}), \ldots, u^{1,\chi}(l^{1,\chi}), \ldots, u^{m,\chi}(l^{m,\chi}))^T,$$

$$r = (r_1^1, \ldots, r_1^{\nu_1}, \ldots, r_N^1, \ldots, r_N^{\nu_N})^T.$$

We assume $u^{i,j}(0) = 0$, if the nodes i and j are not adjacent, and respectively, in matrixes G, Q dimension $2M \times M$, made of coefficients $g_{ik}^{j\nu}, q_{ki}^{j\nu}$, elements, corresponding non-existent edges are also 0. The components of the vector $r \in R^M$ corresponding to the conditions relative to the nodes in which there are no internal or external sources are equal to 0. Let's assume that the coefficients, participating in conditions (3) are such that, poorly filled augmented matrix $[G, Q]$ have the rang equal to M.

We consider the problem of minimization of the functional

$$I(v) = \sum_{(k,s) \in J} \left(\int_0^{l^{ks}} f_0^{ks} \left(u^{ks}(x), v^{ks}(x) \right) dx + \varPhi^{ks} \left(u^{ks}(0), u^{ks}(l^{ks}) \right) \right), \tag{4}$$

where the state of each of the block $u^{ks}(x)$ is described by a system of linear non-autonomous differential equations of the form (1) with non-local boundary conditions (2). Here $f_0^{ks}(u^{ks}(x), v^{ks}(x))$, $\varPhi^{ks} \left(u^{ks}(0), u^{ks}(l^{ks}) \right)$ are given continuously differentiable functions of its' arguments.

Let's note that, for example, the problem of optimal control of transient processes of fluid, gas flow in pipeline networks of complex structure, also optimal control of mechanical systems are reduced to the problem under consideration. Mathematical models of such processes are described by systems of partial differential equations, consisting of subsystems of equations of hyperbolic type, describing the process of fluid flow in each individual section. At the junction of the sections, the conditions of flow continuity and material balance are satisfied, which are determined by conditions of the form (2). The application of the method of straight lines in time or spatial variable (analogous to the use of decomposition) leads the problem of calculating the modes of fluid flow of the transport network to a problem of the form (1) and (2).

3   Necessary Optimality Conditions in Problem (1)–(4)

We formulate the necessary optimality conditions in the variational form [11] for the problem (1)–(4) as the following theorem.

Theorem 1. Let the conditions imposed on the functions and parameters involved in the problem (1)–(4) be satisfied. For optimality of the control $\widehat{v}^{ks}(x)$ it is necessary that the inequalities

$$
(-(\psi^{ks}(x))^T B^{ks}(x) + \frac{\partial f_0^{ks}(u^{ks}(x), v^{ks}(x))}{\partial v^{ks}}, \ v^{ks}(x) - \widehat{v}^{ks}(x)) \geq 0, (k, s) \in J,
$$
(5)

fulfill for all admissible values of the controls $v^{ks}(\cdot) \in \mho^{ks}$. The function $\psi^{ks}(x)$ is a solution of the system of ODE:

$$
\dot{\psi}^{ks}(x) = \frac{\partial f_0^{ks}\left(u^{ks}(x); v^{ks}(x)\right)}{\partial u^{ks}} - (A^{ks}(x))^T \psi^{ks}(x), \ x \in [0, l^{ks}], \ (k, s) \in J,
$$
(6)

with unseparated boundary conditions:

$$
\sum_{k \in I_i^+} q_{ik}^j \psi^{ik}(0) + \sum_{k \in I_i^-} g_{ik}^j \psi^{ik}(l^{ik}) = \sum_{k \in I_i^+} g_{ik}^j \frac{\partial \Phi^{ik}(u^{ik}(0), u^{ik}(l^{ik}))}{\partial u^{ik}(l^{ki})}
$$

$$
+ \sum_{k \in I_i^+} q_{ik}^j \frac{\partial \Phi^{ik}\left(u^{ik}(0), u^{ik}(l^{ik})\right)}{\partial u^{ik}(0)}, j = \overline{1, \nu_i}, i \in I.
$$
(7)

Proof: Let $\widehat{v}^{ks}(x)$ and $\widehat{u}^{ks}(x), (k, s) \in J$ are optimal solution of the problem (1)–(5). Let $v = v^{ks}(x)$ admissible control and $u^{ks}(x, v)$ corresponding solution to the boundary-value problem (1)–(2). Move the right side of the differential equation (1) to the left and multiply by yet arbitrary continuously differentiable in their arguments as $x \in (0, l^{ks})$ χ-dimensional vector functions $\psi^{ks}(x) \in R^\chi$, $(k, s) \in J$, and to the right we move the parts of constraints (2) to the left and the results are added with (4), so, we will have:

$$
\Im(v) = \sum_{(k,s) \in J} \int_0^{l^{ks}} f_0^{ks}(u^{ks}(x); v^{ks}(x))dx + \sum_{(k,s) \in J} \Phi^{ks}(u^{ks}(0), u^{ks}(l^{ks}))
$$

$$
+ \sum_{(k,s) \in J} [(\psi^{ks}(x))^T (\dot{u}^k(x) - A^{ks}(x)u^{ks}(x) - B^{ks}(x)v^{ks}(x))]
$$
(8)

Let denote $\Delta u^{ks}(x, v^{ks}) = \widehat{u}^{ks}(x, v^{ks}) - u^{ks}(x, v^{ks}), (k, s) \in J$. It is clear, as follows from (1)–(2) the function $\Delta u^{ks}(x, v)$ is a solution of the following boundary-value problem with nonlocal boundary conditions:

$$
\Delta \dot{u}^{ks}(x) = A^k(x)\Delta u^{ks}(x) + B^{ks}(x)\Delta v^{ks}(x), \quad x \in [0, l^{ks}]
$$
(9)

$$
\sum_{k \in I_i^-} g_{ik}^j \Delta u^{ik}(0) + \sum_{k \in I_i^+} q_{ki}^j \Delta u^{ki}(l^{ki}) = 0, \quad j = \overline{1, \nu_i}, \ i \in I.
$$
(10)

We assume that the coefficients involved in conditions (2) are such that the sparse augmented matrix $[G, Q]$ has a rang equal to $2M$. Without pleading for generality, for the sake of simplicity of the calculations, we will assume that $rangG = M$. Otherwise, from the considered matrix $[G, Q]$, some other submatrix with a rang equal to M would be extracted. Relations (10) in general according to (3) can be written in the matrix form:

$$G\Delta u(0) + Q\Delta u(l) = 0. \qquad (11)$$

Multiplying both sides of (11) by G^{-1} we get:

$$\Delta u(0) = -G^{-1}Q\Delta u(l). \qquad (12)$$

The increment of the functional (4), taking into account the notation adopted above, corresponding to the increment $\Delta v(x) = \Delta v^{ks}(x) = v^{ks}(x) - \hat{v}^{ks}(x)$ can be written in the following form:

$$\Delta \mathfrak{J}(v) = J(v + \Delta v) - J(v) =$$

$$\sum_{(k,s)\in J} \int_0^{l^{ks}} (f_0^{ks}(u^{ks}(x) + \Delta u^{ks}(x); v^{ks}(x) + \Delta v^{ks}(x)) - f_0^{ks}(u^{ks}(x); v^{ks}(x)))dx$$

$$\sum_{(k,s)\in J} \Phi^{ks}(u^{ks}(0) + \Delta u^{ks}(0), u^{ks}(l^{ks}) + \Delta u^{ks}(l^{ks})) - \sum_{(k,s)\in J} \Phi^{ks}(u^{ks}(0), u^{ks}(l^{ks}))$$

$$\sum_{(k,s)\in J} \int_0^{l^{ks}} [\psi^{ks}(x)^T(\dot{u}^{ks}(x) + \Delta \dot{u}^{ks}(x)) - A^{ks}(x)(u^{ks}(x) + \Delta u^{ks}(x))$$

$$- B^{ks}(x)(v^{ks}(x) + \Delta v^{ks}(x))]dx$$

$$- \sum_{(k,s)\in J} \int_0^{l^{ks}} \left[\psi^{ks}(x)^T\left(\dot{u}^{ks}(x) - A^{ks}(x)u^{ks}(x) - B^{ks}(x)v^{ks}(x)\right)\right]dx$$

$$= \sum_{(k,s)\in J} \int_0^{l^{ks}} \left[\frac{\partial f_0^{ks}\left(u^{ks}(x); v^{ks}(x)\right)^T}{\partial u^{ks}(x)}\Delta u^{ks}(x) +\right.$$

$$\left.\frac{\partial f_0^{ks}\left(u^{ks}(x); v^{ks}(x)\right)^T}{\partial v^{ks}(x)}\Delta v^{ks}(x)\right]dx$$

$$+ \sum_{i\in I}\left[\sum_{k\in I_i^-}\frac{\partial \Phi^{ks}\left(u^{ks}(0), u^{ks}(l^{ks})\right)^T}{\partial u^{ks}(0)}\Delta u^{ks}(0) +\right.$$

$$\left.\sum_{k\in I_i^+}\frac{\partial \Phi^{ks}(u^{ks}(0), u^{ks}(l^{ks}))^T}{\partial u^{ks}(l^{ks})}\Delta u^{ks}\left(l^{ks}\right)\right] +$$

$$\sum_{(k,s)\in J} \int_0^{l^{ks}} \left[\psi^{ks}(x)^T\left(\Delta \dot{u}^{ks}(x) - A^{ks}(x)\Delta u^{ks}(x) - B^{ks}(x)\Delta v^{ks}(x)\right)\right]dx + \eta,$$

$$\eta = \sum_{(k,s)\in J} \int_0^{l^{ks}} o(\|\Delta u^{ks}(x)\|_{L_2^{n^{ks}}[0, l^{ks}]} + \|\Delta v^{ks}(x)\|_{L_2^{n^{ks}}[0, l^{ks}]})dx +$$

$$+ o(\|\Delta u^{ks}(x)\|_{L_2^{n^{ks}}[0, l^{ks}]}, \|\Delta v^{ks}(x)\|_{L_2^{n^{ks}}[0, l^{ks}]}).$$

$$(13)$$

In the last term of (13) we will integrate by parts according to x:

$$\sum_{(k,s)\in J} \int_0^{l^{ks}} \left[\psi^{ks}(x)^T (\Delta \dot{u}^{ks}(x) - A^{ks}(x)\Delta u^{ks}(x) - B^{ks}(x)\Delta v^{ks}(x)) \right] dx$$

$$= \sum_{(k,s)\in J} \psi^{ks}(x)^T \Delta u^{ks}(x) \Big|_0^{l^{ks}}$$

$$- \sum_{(k,s)\in J} \int_0^{l^{ks}} \left[\left(\dot{\psi}^{ks}(x)^T - \psi^{ks}(x)^T A^{ks}(x) \right) \Delta u^{ks}(x) - \psi^{ks}(x)^T B^{ks}(x)\Delta v^{ks}(x) \right],$$

$$\sum_{(k,s)\in J} \psi^{ks}(x)^T \Delta u^{ks}(x) \Big|_0^{l^{ks}}$$

$$= \sum_{i\in I} \left[\sum_{k\in I_i^-} \psi^{ki}(l^{ki})^T \Delta u^{ks}(l^{ki}) - \sum_{k\in I_i^+} \psi^{ik}(0)^T \Delta u^{ks}(0) \right].$$

By grouping the corresponding terms in (13) and using (10), we have:

$$\Delta \mathfrak{I}(v) = \sum_{(k,s)\in J} \int_0^{l^{ks}} \left[\frac{\partial f_0^{ks}(u^{ks}(x);v^{ks}(x))^T}{\partial u^{ks}(x)} - \dot{\psi}^{ks}(x)^T - \right.$$

$$\psi^{ks}(x)^T A^{ks}(x) \Big] \Delta u^{ks}(x) dx - \sum_{(k,s)\in J} \int_0^{l^{ks}} \psi^{ks}(x)^T B^{ks}(x)\Delta v^{ks}(x)$$

$$+\sum_{i\in I}\sum_{k\in I_i^-} \left[\frac{\partial \Phi^{ik}(u^{ik}(0), u^{ik}(l^{ik}))^T}{\partial u^{ik}(0)} - \psi^{ik}(0)^T \right] \Delta u^{ik}(0)$$

$$+\sum_{i\in I}\sum_{k\in I_i^+} \left[\frac{\partial \Phi^{ki}(u^{ki}(0), u^{ki}(l^{ki}))^T}{\partial u^{ki}(l^{ki})} + \psi^{ki}(l^{ki})^T \right] \Delta u^{ki}(l^{ki}).$$

Using the arbitrariness of the functions $\psi^{ks}(x)$, we require the square bracket in the first integral to be zero. We obtain the adjoint system (6). Using assumption (12) from the third and fourth integral, we obtain conditions (7), that must be satisfied by the values of the conjugate functions $\psi^{ks}(x)$ for $x = 0$ and $x = l^{ks}$. Then we have the following formula for the functional increment:

$$\Delta \mathfrak{I}(v) = - \sum_{(k,s)\in J} \int_0^{l^{ks}} \left(\psi^{ks}(x)B^{ks}(x) + \frac{\partial f_0^{ks}(u^{ks}(x), v^{ks}(x))^T}{\partial v^{ks}} \right) \Delta v^{ks}(x)dx + \eta.$$

$$(14)$$

Here $\psi^{ks}(x), (k,s) \in J$ is the solution to the adjoint problem (6) and (7).

According to the definition, the gradient of the functional with respect to the control functions is determined by the expression for the linear part with respect to the control in the functional increment formula (14) [7,11].

Then, according to the known necessary conditions for the optimality of the functional in the variational form, conditions (5) should be satisfied, and the components of the gradient of the functional (4) are determined by the formula

$$grad_{v^{ks}} \mathfrak{I}(v) = -(\psi^{ks}(x))^T B^{ks}(x) + \frac{\partial f_0^{ks}(u^{ks}(x); v^{ks}(x))^T}{\partial v^{ks}}, (k,s) \in J. \quad (15)$$

Theorem 1 can be considered proven.

Formula (15) can be used for the numerical solution to the problem (1)–(4) applying effective first-order optimization methods. For example, one can use the gradient projection method [7, 11]:

$$\left[v^{ks}(x)\right]^{\mu+1} = P_{\mho^{ks}}[v^{ks}(x)^{\mu} - \alpha\, grad_{v^{ks}}\Im(v^{\mu}(x))], \ (k,s) \in J, \ \mu = 1, 2, ..$$

$P_{\mho^{ks}}[\bullet]$ is the projection operator of a point on an admissible set $\mho^{ks}$, α -step of one-dimensional minimization.

To calculate the components of the gradient of the functional $\Im(v)$ for the current values of the control $v(x)$ first we solve the direct boundary-value problem (1) and (2), then we solve the adjoint boundary-value problem (6) and (7). The results of the solution are substituted into the formula (15) to calculate the components of the gradient of the functional.

An approach based on the use of the operation of shifting conditions proposed in [1] is applied to solve systems of differential equations with unseparated boundary conditions. The shift operation of intermediate conditions generalizes the well-known operation of transferring boundary conditions [2, 3] to the case of unseparated boundary conditions with the participation of unknown parameters in them.

We present an approach that is analogous to the method of transferring conditions, taking into account the specifics of the system (6). We present the corresponding formulas, algorithms that do not require the simultaneous solution of all subsystems of the system (6). For simplicity of further statements, we'll rewrite (7) as follows:

$$\sum_{k\in I_i^+} q_{ik}^j \psi^{ik}(0) + \sum_{k\in I_i^-} g_{ik}^j \psi^{ik}(l^{ik}) = \overline{R}_i^j, \ j = \overline{1, \nu_i}, i \in I,$$

$$\overline{R}_i^j = \sum_{k\in I_i^+} g_{ik}^j \frac{\partial \Phi^{ik}(u^{ik}(0), u^{ik}(l^{ik}))}{\partial u^{ik}(l^{ki})} + \sum_{k\in I_i^+} q_{ik}^j \frac{\partial \Phi^{ik}(u^{ik}(0), u^{ik}(l^{ik}))}{\partial u^{ik}(0)}.$$

The proposed approach, like all methods of transferring conditions, consists in replacing the values $\psi^{ik}(0)$ (or $\psi^{ik}(l^{ks})$) in conditions (7) due to transferring to the right (left) equivalent combinations of the values $\psi^{ik}(l^{ks})$ (or $\psi^{ik}(0)$ when sweeping to the left). As a result, instead of (7), we obtain $M = m\chi$ conditions of the form:

$$\sum_{k\in I_i^-} \tilde{G}_i^j \psi^{ik}(l^{ik}) = \tilde{R}_i^j, \ i \in I, j = \overline{1, \nu_i}, \tag{16}$$

when transferring conditions (7) to the right, of the next form

$$\sum_{k\in I_i^-} \tilde{G}_i^j \psi^{ik}(0) = \tilde{R}_i^j, \ j = \overline{1, \nu_i}, i \in I, \tag{17}$$

when transferring conditions (7) to the left. Obtaining conditions of the form (16) or (17) will be carried out in stages.

Taking into account the sparseness of the matrices G, Q so as not to deal with matrix operations, each condition from (7) will be transferred separately.

In many concrete practical problems, a large number of conditions from (7) instead of a general form can be separated or, moreover, coincide with Cauchy conditions on the left or right ends. Therefore, the choice of the direction of the transfer of conditions to the left or right should be made on the basis of which end of local conditions is greater to that end and transfer the remaining conditions.

And so, consider an arbitrary j-th condition from (7), $j = 1, .., \nu_i$, given for i-th node, $i \in I$. Let transfer the j-th condition of (7) to the right end, i.e. we obtain the condition equivalent to (7):

$$\sum_{k \in I_i^-} \alpha_{ik}^j (l^{ik}) \psi^{ik}(l^{ik}) + \sum_{k \in I_i^+} g_{ik}^j \psi^{ki}(l^{ki}) = \gamma_i^j(l^{ki}), \qquad (18)$$

where $(\alpha_{ik}^j x)$ and $\gamma_i^j(x)$ are some as yet unknown χ-dimensional row vector and scalar functions; χ-dimensional vector $\psi^{ik}(l^{ik})$ unknown values of the solution of the ik-th subsystem of the system (6) at the right end.

Obtaining conditions of the form (18) we will implement in stages.

Suppose that among the vector coefficients $q_{ik}^j = (q_{ik}^{j1}, ... q_{ik}^{j\chi})$, $k \in I_i^-$ there are nonzero ones. Otherwise, it is not necessary to transfer the j-th condition to the right, since this condition involves only values of $\psi^{ik}(l^{ik})$. Let the nonzero vector coefficient be $q_{id}^j \neq 0_\chi$, $d \in I_i^-$, (0_χ is χ-dimensional vector, all components of which are equal to 0).

It should be noted that the order of choice of the non-zero coefficients is not principal. The order of transferring values of the solutions of the subsystems from the left end to the right can be carried out in an arbitrary sequence of choice of as subsystems, as well as conditions.

Definition 1. We'll say that χ-dimensional row-vector function $\alpha_{id}^j(x)$, and function $\gamma_i^j(x)$ are such that

$$\alpha_{id}^j(0) = q_{id}^j, \quad \gamma_i^j(0) = \overline{R}_i^j, \quad d \in I_i^-, \qquad (19)$$

carry out the transfer of the boundary value of the solution to (i, d)-th subsystem (6) in the j-th condition from (7) for i-th node to the right, if for an arbitrary solution $\psi^{id}(x)$ of (i, d)-th subsystem (6) the next equality holds:

$$\alpha^j{}_{id}(x)\psi^{id}(x) + \sum_{k \in I_i^-} q_{ki}^j \psi^{ki}(0) +$$
$$\sum_{k \in I_i^+} g_{ki}^j \psi^{ki}(l^{ki}) = \gamma_i^j(x), \ x \in [0, l^{id}]. \qquad (20)$$

It is clear that condition (20), taking into account (18), at $x = 0$ coincides with the j-th condition of (7) for i-th node. We'll call the functions $\alpha_{id}^j(x)$, $\gamma_i^j(x)$ are the sweep functions. Substituting the values of the functions $\alpha_{id}^j(x)$, $\gamma_i^j(x)$ for

$x = l^{id}$ in (20), we obtain an equality equivalent to the j-th condition for i-th node from (7):

$$\sum_{k \in I_i^- \setminus d} q_{ki}^j \psi^{ik}(0) + \sum_{k \in I_i^+} g_{ik}^j \psi^{ik}(l^{ki}) = \gamma_i^j(l^{id}). \tag{21}$$

The sweep functions $\alpha_{id}^j(x)$, $\gamma_i^j(x)$ used to transfer the boundary values of the solutions of the subsystems participating in boundary conditions (7) from one end to the other are not unique. In particular, their constructive formulation was proposed in the following theorem.

Theorem: Let $g_{id}^j \neq 0_\chi$ for $d \in I_i^-$ and χ -dimensional row-vector function $\alpha_{id}^j(x)$, and scalar function $\gamma_i^j(x)$ with $x \in [0, l^{id}]$ are the solution to the following Cauchy problems:

$$\dot{\alpha}_{id}^j(x) = -\alpha_{id}^j(x)(A^{id}(x))^T, \alpha_{id}^j(0) = q_{id}^j, \tag{22}$$

$$\dot{\gamma}_i^j(x) = \alpha_{id}^j(x) \frac{\partial f_0^{id}\left(u^{id}(x); v^{id}(x)\right)}{\partial u^{id}}, \quad \gamma_i^j(0) = 0 .$$

Then these functions are the sweep coefficients for transferring from left to right the boundary value of the solution to (i, d) -th subsystem (6) in the j-th condition for i-th node, for $x \in [0, l^{id}], d \in I_i^-$.

Proof: Let $\alpha_{id}^j(x)$, $\gamma_i^j(x)$ be arbitrary differentiable functions that satisfy (19) and the condition (20). We differentiate (20) and substitute in (6):

$$\dot{\alpha}_{id}^j(x)\psi^{id}(x) + \alpha_{id}^j(x)\dot{\psi}^{id}(x) = \dot{\gamma}_i^j(x), d \in I_i^- .$$

Taking into account here the (i, d)-th subsystem of the equations (6), grouping the corresponding terms, we get

$$[\dot{\alpha}_{id}^j(x) - \alpha_{id}^j(x)(A^{id}(x))^T]\psi^{id}(x)+$$
$$+\alpha_{id}^j(x)\frac{\partial f_0^{id}\left(u^{id}(x), v^{id}(x)\right)}{\partial u^{id}} = \dot{\gamma}_i^j(x), d \in I_i^- . \tag{23}$$

Taking into account the arbitrariness of the functions $\alpha_{id}^j(x)$, $\gamma_i^j(x)$ and the need to satisfy equality (23) for all solutions $\psi^{id}(x)$ of the (i, d)-th subsystem of equations (6), we require the fulfillment of equality of zero expressions in square brackets. It follows that $\alpha_{id}^j(x)$, $\gamma_i^j(x)$ are the solutions of Cauchy problems (19)–(20).

The above procedure is repeated for the next value $\psi^{ik}(0), (i, k) \in J$, for which the coefficient $q_{ik}^j \neq 0_\chi$ different from zero, until any component of the vector $\psi^{ik}(0), (i, k) \in J$ will not cease to participate in the j-th condition for i-th node. After that, it is necessary to go to the $(j + 1)$ -th condition for i-th node from (7) until these procedures are carried out for all restrictions for all nodes and a condition of the form (16) is obtained.

By solving a system of algebraic equations (16) or (17) of the M-th order, we determine the vectors $\psi^{ks}(l^{ks}), (k, s) \in J$.

To determine the desired vector functions $\psi^{ks}(x)$, $x \in [0, l^{ks}]$, $(k, s) \in J$ the components $\psi^{ks}(l^{ks})$, $(k, s) \in J$ of the found vector are used as initial values for the corresponding Cauchy problems with respect to each individual subsystem of system (6), solved in the reverse order: from $x = l^{ks}$ to $x = 0, (k, s) \in J$.

The transfer of conditions can be carried out from right to left. Obtaining auxiliary Cauchy problems with respect to the sweep coefficients, in this case, is carried out in a similar way.

4 Conclusion

In this paper, we obtain the necessary conditions for optimality in the problem of optimal control of a system of differential equations of block structure. An approach to the numerical solution of a direct and adjoint initial boundary value problems of large dimension and block structure is described. It is clear that the direct use of methods for running boundary conditions is not effective, since taking into account the block structure of conditions at the same time for the entire system as a whole, as well as for many other classes of problems, can significantly accelerate their solution. The method is based on the proposed scheme of the sweep method, which allows one to sweep separately by blocks with reducing the solution of the problem to an algebraic system of equations with a sparse Jacobi matrix.

References

1. Abdullaev, V.M., Ayda-zade, K.R.: On the numerical solution of optimal control problems with nonseparated multipoint and integral conditions. Zh. Vychisl. Math mat. Phys. **52**(12), 2163–2177 (2012). (in Russian)
2. Abramov A.A.: On the transfer of boundary conditions for systems of linear ordinary differential equations (variant of the sweep method). Zh. Vychisl. Math Math. Phys. **1**(3), 542–545 (1961). (in Russian)
3. Abdullaev, V.M.: Solution of differential equations with nonseparated multipoint and integral conditions. Sib. J. Ind. Math. **15**(3), 3–15 (2012)
4. Ashepkov, L.T.: Optimal control of a system with intermediate conditions. Prikl. **45**(2), 215–222 (1981)
5. Aida-zade, K.R., Ashrafova, Y.R.: Solving systems of differential equations of block structure with nonseparated boundary conditions. J. Appl. Ind. Math. **9**(1), 1–10 (2015)
6. Aida-zade, K.R., Ashrafova, E.R.: Calculation of the state of a system of discrete linear processes connected by unseparated boundary conditions. Sib. J. Ind. Math. **19**(68), 3–14 (2016)
7. Vasiliev, F.P.: Optimization Methods. Factorial Press, Tokyo p. 824 (2002) (in Russian)
8. Voevodin, A.F.: The sweep method for difference equations defined on a complex. Zh. Vychisl. Math Math. Phys. **13**(2), 494–497 (1973)

9. Vasilieva, O.O., Mizukami, K.: Optimality criterion for singular controllers: linear boundary conditions. J. Math. Anal. Appl. **213**(2), 620–641 (1997)
10. Dzhumabaev, D.S., Imanchiev, A.E.: The correct solvability of a linear multipoint boundary value problem. Mat. J. **5**(15), 30–38 (2005)
11. Evtushenko, Y.G.: Methods for Solving Extremal Problems and Their Application in Optimization Systems. Nauka, Moscow (1982). (in Russian)
12. Samarskii, A.A., Nikolaev, E.S.: Methods for Solving Grid Equations, p. 592. Nauka, Moscow (1978). (in Russian)

A Graph-Theoretic Approach to Multiobjective Permutation-Based Optimization

Liudmyla Koliechkina[1] , Oksana Pichugina[2(✉)] , and Sergiy Yakovlev[2]

[1] University of Lodz, Uniwersytecka Str. 3, 90-137 Lodz, Poland
liudmyla.koliechkina@wmii.uni.lodz.pl
[2] National Aerospace University "Kharkiv Aviation Institute",
17 Chkalova Street, Kharkiv 61070, Ukraine
oksanapichugina1@gmail.com, svsyak7@gmail.com

Abstract. A Generalized Coordinate Method (GCM) for linear permutation-based optimization is presented as a generalization of the Modified Coordinate Localization Method and Modified Coordinate Method is presented, and its applications to multiobjective linear optimization on permutations are outlined. The method is based on properties of linear function on a transposition graph, a decomposition of the graph, and extracting from it a multidimensional grid graph, where a directed search of an optimal solution is performed. Depending on the search parameters, GCM yields an exact or approximate solution to the original problem. An illustrative example is given for the method.

Keywords: Permutation-based optimization · Vertex located set · Convex extension · Permutation set · Permutohedron · Transposition graph · Skeleton graph · Configuration graph · Structural graph

1 Introduction

Let us consider the following decision-making problem (DP):

$$\phi_j(\pi) \to extr, \ j \in J,$$
$$\eta_i(\pi) \leq 0, i \in I, \quad (1)$$
$$\pi \in \Pi',$$

where $I, J \subset \mathbb{N}$,

$$\Pi' \text{ is a permutation set induced by a multiset set} \quad (2)$$
$$A = \{a_1, .., a_n\} \text{ with a ground set } S(A) = \{a'_1, .., a'_k\}.$$

First, consider the problem (1) assuming that Π' is a combinatorial set with a given topology. Then (1) is a combinatorial DP, which is: (a) an ordinary single-criterion combinatorial optimization problem (*a combinatorial optimization problem, COP*), if $|J| = 1$; (b) a multi-criterion COP (*a multiobjective COP, MCOP*), if $|J| > 1$; (c) an unconstrained COP/MCOP, if $I = \emptyset$

© Springer Nature Switzerland AG 2020
M. Jaćimović et al. (Eds.): OPTIMA 2019, CCIS 1145, pp. 383–400, 2020.
https://doi.org/10.1007/978-3-030-38603-0_28

(*UCOP/UMCOP*); (d) a problem of finding a feasible point of E (*a feasibility problem, FP*), if $J = \emptyset$. On Π', a topology can be defined differently. It assumes defining of a neighborhood system for any $\pi \in \Pi'$ making it possible usage of concepts of local optimizer and optimum. It allows using local search techniques to solve COP started from a feasible $\pi^0 \in \Pi'$ as soon as the corresponding FP has been solved yielding π^0. An issue is that solving this FP can be complicated. Also, there is no guarantee on obtaining in such a way an exact solution of COP or the one close to it. Adding the constraint (2) to (1) specifies a combinatorial type of the problem, emphasizing that it is a *permutation-based problem* (*PBP*) [23, 24]. Respectively, this typology can be extended to PBP, and such classes as permutation-based COP (*PB-OP*), permutation-based MCOP (*PB-MOP*), etc. can be singled out.

The PBP-class covers a variety of real-world DPs, where decisions are associated with an ordering of a certain set or a multiset. Among them, PB-OP and PB-MOP are most important [17, 23, 24, 26]. Indeed, it is known that scheduling, assignment, routing problems, as well as many others, can be modeled as optimization problems on permutations [11, 17–19, 23–25]. From a theoretical point of view, PB-OP is also of interest. Here, $\Pi' = \Pi_{nk}(A)$ is a set of permutation configurations [28] consisting of all ordered n-samples from A such that $\forall \pi = \langle \pi_1, ..., \pi_n \rangle \in \Pi' \{\pi_1, ..., \pi_n\} = A$. A neighborhood of $\pi \in \Pi'$ can be defined, for instance, as permutations differed from π by a single transposition. Another interesting peculiarity of $\Pi_{nk}(A)$ is a possibility to consider it as *a finite point configuration* (*FPC*) [12] in $\mathbb{R}^n$ in case if A is numerical. Indeed, if $A \subset \mathbb{R}^1$, then $\forall \pi \in \Pi' \langle \pi_1, ..., \pi_n \rangle$ is a numerical tuple, in other words, it is a vector in Euclidean space – $\pi = (\pi_1, ..., \pi_n) \in \mathbb{R}^n$. After such a mapping into Euclidean space, it is possible to formulate a COP as a discrete optimization problem, in particular, consider a linear COP on $\Pi' = \Pi_{nk}(A)$ (further referred to as a *PB-LOP*).

The class PB-LOP is an object of our study with its application to solving a *multiobjective PB-LOP* (*PB-MLOP*). Among exact methods to solve PB-LOP are Branch and Bound techniques [11, 17, 24], a combinatorial cutting method [38], and other cutting methods [11, 24, 36]. A tightening constraints method is an example of approximate techniques [31] to solve such problems. Finally, [18, 19, 24] offer heuristic methods for a case of specific additional constraints in PB-LOP. Another group of methods is graph-theoretic approaches. They explore an idea that $\Pi_{nk}(G)$, embedded in Euclidean space, is a node-set of a skeleton graph of its convex hull [5, 6, 13–16, 30]. Thus the search can be restricted to this node-set. Among such approaches are: (a) horizontal methods [16] intended to solve PB-LOP and other *linear COP* (*LCOP*) on sets allowing solving linear programs on Π' effectively; (b) coordinate localization methods for solving FP with a linear equality-constraint (*FP1*) [5, 6, 15]; (c) coordinate methods for solving the general PB-LOP and LCOP on other sets [13, 14, 30]. Some generalizations of the methods to solve MCOP and some nonlinear COP are offered in [13–15, 30]. The paper [15] presents a concept of a structural graph of the following PB-LOP, further referred to as *PB-LOP0*: $c^T x \to$ max, $c^T x \leq b$, $x \in E$, where E is a permutation set induced by a numerical multiset. The graph utilizes a transposition graph [4,10]. Common features of the graph-theoretic methods are applying a partition of graphs under consideration into subgraphs and single outing two-dimensional grid graphs [1,15] from them.

In this paper, the concept of a structural graph is generalized to a structural graph of the general COP and then is specified for PB-COP; the partition into subgraphs uses single-outing multidimensional grid graphs from this graph; a new graph-theoretic approach to solving PB-LOP titled a Generalized Coordinate Method *GCM* is presented. GCM generalizes *the Modified Coordinate Localization Method (MCLM)* presented in [15] and the Modified Coordinate Method offered in [14]. In the scope of the paper, we restrict our consideration to optimization on a set of permutation without repetitions only in order to use specific properties of this combinatorial set and its image in Euclidean space. It will be shown that GLM is flexible enough and, depending on input, allows solving PB-LOP approximately in polynomial time or exactly in exponential time.

This paper is organized as follows. Section 2 describes the advantages of embedding combinatorial sets into Euclidean space when COP is solved. Special attention is paid to geometric graphs emerged as a result of this embedding. A concept of a structural graph (*SG*) of the general Euclidean setting of COP is presented. Section 3 describes the properties of permutation sets embedded into Euclidean space underlying GCM. In Sect. 4, a concept of a grid graph of PB-COP is presented, and its properties are studied. Also, a description of GCM is provided. An illustrative example of GCM is given in Sect. 5. Section 6 outlines applications of GCM to multiobjective linear permutation-based optimization. Finally, Sect. 7 presents our conclusions.

2 Embedding into Euclidean Space and Graph-Theoretic Approaches to Solving COP

PBP and some other COP allow reformulating in terms of Cartesian variables regardless of whether the induced set of Π' is numerical or not. The process of this reformulation is called an *embedding* such COP into Euclidean space [20,27,31], and the resulting settings of the problems as discrete optimization ones are called their Euclidean settings (Euclidean COP, *ECOP*) [27,28,31]. Similarly, MCOP on finite point configurations (*FPCs*) in $\mathbb{R}^n$ is a class of *Euclidean MCOP (MECOP)* [15,16].

A general ECOP is:

$$f(x) \to \max; \tag{3}$$
$$h_i(x) \leq 0, \ i \in J_m = \{1, ..., m\}, \tag{4}$$
$$x \in E' \subset \mathbb{R}^n, \tag{5}$$
$$1 < |E'| < \infty, \tag{6}$$

while MECOP is

$$f_l(x) \to \max, l \in J_L, \tag{7}$$

where $L > 1$, subject to constraints (4)–(6).

Here,

$$E = \{x \in E' : h_i(x) \leq 0, \ i \in J_m\} \tag{8}$$

is an image of a feasible set $\Pi = \{\pi \in \Pi' : \eta_i(\pi) \leq 0, i \in I\}$ in Euclidean space. Thus, there exists a mapping $\varphi : \Pi \to \mathbb{R}^n$ such that $E = \varphi(\Pi)$. If, in addition, $\Pi = \varphi^{-1}(E)$, i.e., the mapping is bijective, such sets Π, E are called *an e-set* [31] and *a C-set* [28], respectively. Further, we will assume that $\varphi : \Pi' \to \mathbb{R}^n$ is bijective, i.e., $E' = \varphi(\Pi')$ and $\Pi' = \varphi^{-1}(E')$, and will map the whole Π' into Euclidean space. Embedding Π' into a continuous space, such as $\mathbb{R}^n$, allows utilizing various features of the space in optimization. For instance, $\mathbb{R}^n$ is equipped with a topology, limits, continuity, completeness, a metric, an inner product, and so on [20,22]. Power analytic tools like derivatives, convexity, and optimization algorithms are built on their basis. So, while elements in combinatorial sets are collections of objects without any relations between them and able to be even different by nature, their images in $\mathbb{R}^n$ are FPCs. On the one hand, the embedding preserves the combinatorial structure of preimage of E'. It allows using in optimization all the listed tools, as well as geometric properties of E' and other geometric structures induced by it, such as its convex hull – a combinatorial polytope $\mathbf{P}' = conv \, E'$ associated with E', its faces such as facets, edges $\mathbf{E}' = edges \, \mathbf{P}'$, and vertices $\mathbf{V}' = vert \, \mathbf{P}'$. Together, $\mathbf{V}'$ and $\mathbf{E}'$ form a graph

$$\mathcal{G}' = \langle \mathbf{V}', \mathbf{E}' \rangle, \tag{9}$$

which a skeleton graph of $\mathbf{P}'$ [15,39].

Suppose, the general COP needs to be solved. Effectiveness of the embedding strategy for solving COP has proven to be powerful in many cases, e.g., a linear program over an FPC is equivalent to solving a polyhedral relaxation of this ECOP, where (5) is replaced by a condition $x \in \mathbf{P}'$. It gave rise to Polyhedral Combinatorics [17,29] studying algebraic topological and topological metric properties of combinatorial sets embedded in $\mathbb{R}^n$ along with the corresponding combinatorial polytopes. The same concerns Euclidean Combinatorial Optimization [27,31,33] focused on studying e-sets by C-sets and solving complicated real-world problems employing nonlinear programming, discrete or continuous. In this research field, combinatorial structures presented in the problems are derived and represented as an e-set Π', and then they are embedded into $\mathbb{R}^n$. After that, the corresponding ECOP on Π'-image is formed followed by exploring the image' features and extreme properties of functions involved in the mathematical model. Then the original COP is solved by optimization on the image.

Another relaxation of ECOP, we will refer to as *a graph-relaxation*, is applicable to those problems, where E' is *a vertex-located set* (*VLS*) [33]:

$$E' = vert \, \mathbf{P}'. \tag{10}$$

When moving from ECOP on a VLS to its graph-relaxation, the constraint (5) is replaced by a condition [5, 6, 21]:

$$x \in \mathcal{G}'. \tag{11}$$

Respectively, a search domain is reduced to $\{\mathbf{V}', \mathbf{E}'\} \subset \mathbf{P}'$. Now, moving from one legal point of E' to another is performed based on adjacency criteria for vertices of $\mathbf{P}'$. If E' is not a VLS, replacing (11) by (5), generally, does not yield a relaxation of ECOP since $E' \nsubseteq vert\ \mathbf{P}'$. However, the problem (3), (4), (11) can be used as a base for heuristics yielding a vertex of $\mathbf{P}'$ as an approximate solution to the original ECOP.

If we aim to solve exactly a COP on E', which is not a VLS, by exploring skeleton graphs, any technique of transforming the problem into one on VLSs, such as presented in [28, 34], can be applied first. Thus, without loss of generality, it can be assumed that the condition (10) holds, and (9) has a form:

$$\mathcal{G}' = \langle E', \mathbf{E}' \rangle, \tag{12}$$

A graph-theoretic way to solve ECOP/MECOP is not restricted to exploring a skeleton graph $\mathcal{G}'$ of $\mathbf{P}'$ as a search domain. In accordance to [21], an undirected *configuration graph* $\mathcal{G}_c$ (*CG*) can be associated with $E' \to \mathcal{G}_c = \langle E', \mathbf{E}_c \rangle$, where $\mathbf{E}_c$ is chosen such that $\mathcal{G}_c$ is connected. A choice of $\mathbf{E}_c = \mathbf{E}'$ corresponds to selecting $\mathcal{G}_c = \mathcal{G}'$, which is the only one graph from a wide family of connected graphs induced by E'. By construction, such $\mathcal{G}_c$ has a path between any its nodes. Another option is to use a complete graph with E' as a node-set. The third one is a spanning tree of $\mathcal{G}'$. The goal of constructing and utilizing $\mathcal{G}_c$ instead of $\mathcal{G}'$ is organizing an effective search on E'. In order to do this, a size $|\mathbf{E}_c|$ of $\mathcal{G}_c$ is chosen depending on its diameter that we aim to achieve.

In application to combinatorial optimization techniques based on utilizing a configuration graph $\mathcal{G}_c$, a directed configuration graph G^* of a particular ECOP, also called a *structural graph of the ECOP* is known in literature [15]. Unlike $\mathcal{G}_c$, the graph G^* is oriented, and directions of its arcs are chosen toward nondecreasing objective function. Another difference is that vertex set of G^* coincides with E, not with $E' \supseteq E$. Therefore, to be able to get from a source, which is a local minimizer of $f(x)$ on E, to any other its node along a legal directed path, it can be insufficient to form a structural graph $G_c = \overrightarrow{\mathcal{G}}_c = (E', \overrightarrow{\mathbf{E}_c})^1$ of Euclidean UCOP (3), (5), (6) first, then taking its subgraph induced by E. To overcome this drawback, constructing the configuration graph $\mathcal{G}_c$ needs taking into account the transition from its subgraph $\mathcal{G}^* = (E, \mathbf{E}^*)$ induced by E to a consideration of the directed graph $\overrightarrow{\mathcal{G}}^*$.

Definition 1. *A directed graph* $G^* = (V^*, \overrightarrow{\mathbf{E}}^*)$ *is called a structural graph (SG) of a ECOP with objective function* f, *if it satisfies the following conditions:*

---

1 For a graph $\mathbf{G} = (\mathbf{V}, \mathbf{E})$ and a function $f : \mathbf{V} \to \mathbb{R}^1$, $\overrightarrow{\mathbf{G}} = (\mathbf{V}, \overrightarrow{\mathbf{E}})$ is a digraph, where $\overrightarrow{\mathbf{E}}$ is formed as follows – $\forall u, v \in \mathbf{V}$: (a) if $f(u) < f(v)$, then $(u, v) \in \overrightarrow{\mathbf{E}}$; (a) if $f(u) > f(v)$, then $(v, u) \in \overrightarrow{\mathbf{E}}$; (c) if $f(u) = f(v)$, then $(u, v), (v, u) \in \overrightarrow{\mathbf{E}}$.

(a) $V^* = E$; (b) $\overrightarrow{\mathbf{E}}^* \subseteq \overrightarrow{\mathbf{E}_c}$; (c) $\forall x, y \in E$, when and edge $\{x, y\} \in \mathbf{E}_c$, then: if $f(x) \leq f(y)$, then a directed edge $(x, y) \in \overrightarrow{\mathbf{E}}^*$; if $f(x) \geq f(y)$, then $(y, x) \in \overrightarrow{\mathbf{E}}^*$; (d) there is a directed path from any local minimizer of f on G^* to its global maximizer.

Thus, $G^* = \overrightarrow{\mathcal{G}}^*$ satisfying (d). Definition 1 is a direct generalization of Definition 1 presented in [15] from PB-LOP0 to the general ECOP. The new condition (d) is added to the definition to ensure a possibility of obtaining a global solution by a directed search on a relevant SG towards non-decreasing f.

Remark 1. For LCOP, the condition (d) becomes "there is a directed path from global minimizer of f on G^* to its global maximizer".

3   Permutation-Based Optimization Problems

Let optimization be conducted on Π' satisfying (2), i.e., PB-OP or PB-MOP needs to be solved. The corresponding ECOP or MECOP are permutation-based optimization problems as well (further referred to them as *PB-EOP* and *PB-EMOP*, respectively). This means that it can be assumed that FPC E' in (6) is the generalized permutation *basic $\mathcal{C}$-set ($\mathcal{C}_b$-set)* [28] (*the multipermutation $\mathcal{C}_b$-set* [22,28]) $E_{nk}(G)$ induced by n-element numerical multiset $G = \{g_1, ..., g_n\}$, $g_1 \leq ... \leq g_n$, containing k different elements $S(G) = \{g'_1, ..., g'_k\}$, $g'_1 < ... < g'_k$. The bijective mapping $\varphi : \Pi' = \Pi_{nk}(A) \to E' = E_{nk}(G)$ can be defined easily though a bijection mapping $\psi : A \to G: g_i = \psi(a_i)$, $i \in J_k$. Most common special classes of are $E_{nk}(G)$: (a) $E_n(G)$ is the permutation $\mathcal{C}_b$-set (without repetitions) induced by a set $G: g_1 < ... < g_n$; (b) $E_n = E_n(J_n)$; (c) $B_n(m) = E_{n2}(G)$ is the Boolean permutation $\mathcal{C}_b$-set, where $G = \{0^{n-m}, 1^m\}$, $m \in J_{n-1}$ [27,28,34,35,37].

In this paper, we focus on solving PB-LOP on

$$E' = E_n(G). \tag{13}$$

The permutation set $E_n(G)$ and its Boolean invariant – the permutation matrices set – are well studied [2,3,33,37,39]. For instance, for them, it is known H-representations of convex hulls, main combinatorial characteristics, solutions of linear optimization and projection problems, etc. However, real problems require fulfilling many conditions on permutations. In mathematical models, they are represented by additional constraints, typically spoiled a nice combinatorial structure of permutation set, therefore, requiring developing specific techniques for solving the corresponding optimization and feasibility problems. From a theoretical point of view, the permutation set, as well as some of its subsets are of interest. Among such subsets are even, odd, alternating, distance, cyclic, circular, complete permutations [28,32,37]. It is a separate task to construct additional constraints to single out any of these subsets, which is considered in a continuous functional representation research area [7,27,28]. When it is solved, we come to a constrained PB-EOP again.

If PB-EOP is solved on the set (13) the following properties of the set are used in optimization [27,28,31,33–37,39]: (a) $E' = E_n(G)$ is a VLS; (b) E' is inscribed into a hypersphere; (c) an E'-convex hull $P_n(G) = conv\, E_n(G)$ is the permutohedron; (d) a skeleton graph $\mathcal{G}' = \mathcal{G}_n(G)$ is a graph of the permutohedron, whose adjacent vertices are differed by an adjacent transposition of their coordinates.

An idea of solving a permutation-based FP1 on finding $x \in E_n(G)$ such that $a^T x = a_0$ is presented in [5]. It is based on considering a transposition graph $\mathcal{G}_n^T(G)$ [4,10] and its partition by 2-dimensional 2×3 grid graphs [1], as well as on directed search on the graphs. These ideas were exploited in different ways in [6,14,15,30]. In particular, in the last paper [15], a concept of a structural graph of a specific PB-LOP as a directed graph formed based on $\mathcal{G}_n^T(G)$, a way to extract 2-dimensional $(Q-1) \times Q$ grid graphs from $\mathcal{G}_n^T(G)$, where $Q \geq 3$, and their usage in optimization were presented.

The current paper continues the research [15] on approaches to solving PB-LOP on $E_n(G)$ based on exploring structural graphs of COPs.

Consider the general PB-LOP:

$$f(x) = c^T x \to \max; \tag{14}$$

$$h_i(x) = a^{iT} x - b^i \leq 0,\ i \in J_m; \tag{15}$$

$$x \in E_n(G); \tag{16}$$

$$\text{where } c = (c_j)_{j \in J_n},\, c_1 \geq c_2 \geq ... \geq c_n. \tag{17}$$

Let $\langle x^*, f^* \rangle$ be a solution of the PB-LOP, $z^{min} = c^T x^{min} = \min\limits_{x \in E_n(G)} c^T x$, $z^{max} = c^T x^{max} = \max\limits_{x \in E_n(G)} c^T x$. We will focus on a version of PB-LOP, where

$$x^{min} \in E. \tag{18}$$

To this new problem we will refer to as *PB-LOP1*.

By (8), PB-LOP1 can be represented in a form: (14), (17), (18),

$$x \in E, \tag{19}$$

where

$$E = \{x \in E_n(G) : a^{iT} x - b^i \leq 0,\ i \in J_m\}. \tag{20}$$

By (17) and according to properties of a linear function on $E_n(G)$ [33], formula (18) can be rewritten as follows:

$$x^{min} = (g_1, ..., g_n) \in E. \tag{21}$$

Any PB-LOP is solved to optimality, if

$$x^{max} = (g_n, ..., g_1) \in E \tag{22}$$

yielding $x^* = x^{max}$. Let us assume that (22) does not hold, then a directed search of x^* will be performed either on a SG G^* of PB-LOP or on a SG $G_c = \overrightarrow{\mathcal{G}_c}$ of an *unconstrained PB-LOP* (14), (16), (17) (*PB-ULOP*). It is desirable to search on G^* only, that is possible for special cases of PB-LOP such as PB-LOP1 subject to a constraint $a_{i1} \geq a_{i2} \geq \ldots \geq a_{in}$, $i \in J_m$ (further referred to as *PB-LOP2*). This peculiarity of PB-COP2 underlies MCLM to its solution [15]. Since we aim to solve PB-COP1, which is a generalization of PB-COP2, we will examine a graph G_c instead of G^*. Respectively, $E' = E_n(G)$ will be a search domain instead of E. Moreover, a transposition graph will be used as $\mathcal{G}_c$, hence G_c will be its directed version – $G_c = \vec{\mathcal{G}}_n^T(G)$. GLM will be presented in terms of the CG $\mathcal{G}_c$ and its partition into transposition graphs of lower dimensions. However, within the subgraphs, a directed search towards non-decreasing objective function will be held, which is equivalent to exploring the SG $\mathcal{G}_c$. In such a way, a connection of GLM-results with a directed search on G_c is established.

To a certain $Q \leq n - 3$ and each ordered sample

$$\Lambda = (\lambda_q)_{q \in J_Q} \text{ such that } \{\Lambda\} = \{\lambda_q\}_{q \in J_Q} \subset J_n, \tag{23}$$

an auxiliary CG $\mathcal{G}_c(\Lambda)$ is associated, which is a subgraph of $\mathcal{G}_c$ induced by

$$E'(\Lambda) = \{x \in E' : x_{n-q+1} = g_{\lambda_q}, \, q \in J_Q\}. \tag{24}$$

A family $\Psi = \{\Lambda : \Lambda \text{ satisfies } (23)\}$ induces a partition of $\mathcal{G}_c$ into A_n^Q subgraphs $\Omega = \{\mathcal{G}_c(\Lambda)\}_{\Lambda \in \Psi}$ of an order $N_Q = (n - Q)!$ isomorphic to $\mathcal{G}_{n-Q}^T(J_{n-Q})$, where A_n^Q is the number of Q-permutations from n. For any $\Lambda \in \Psi$, $\mathcal{G}_c(\Lambda)$ is a CG of a PB-LOP (14), (16), (17),

$$x_{n-q+1} = g_{\lambda_q}, \, q \in J_Q \tag{25}$$

(further referred to as *PB-LOP(Λ)*). Which is a configuration graph $\mathcal{G}_c(\Lambda)$ of the subset $E'(\Lambda)$ of $E' = E_n(G)$ combinatorially isomorphic to $E_{n-Q}(J_{n-Q})$. It is clear that the original PB-LOP is reduced to a directed search on these subgraphs.

Note that PB-LOP is a special case of PB-LOP(Λ), namely, PB-LOP= =PB-LOP($\emptyset$). In addition, for any PB-LOP(Λ), formulas (21) and (22) are generalized in the following way:

$$x^{min}(\Lambda) = arg \min_{x \in E'(\Lambda)} f(x) = (g_{\mu_1}, \ldots, g_{\mu_{n-Q}}, g_{\lambda_Q}, \ldots, g_{\lambda_2}, g_{\lambda_1}) \in E;$$

$$x^{max}(\Lambda) = arg \max_{x \in E'(\Lambda)} f(x) = (g_{\mu_{n-Q}}, \ldots, g_1, g_{\lambda_Q}, \ldots, g_{\lambda_2}, g_{\lambda_1}) \in E, \tag{26}$$

where $\{\mu_i\}_{i \in J_{n-Q}} = J_n \backslash \Lambda$, $g_{\mu_1} < \ldots < g_{\mu_{n-Q}}$.

Remark 2. If Q does not depend on n, then Ω is a $\mathcal{G}_c$-partition into a polynomial number of subgraphs. If a search of x^* is organized in such a way that, for examining elements Ω, it is sufficient considering a polynomial number of nodes, the original PB-LOP1 will be polynomially solvable as well. We will achieve this by a fixation one more parameter $R \leq n - Q$, and examining only those nodes of $\mathcal{G}_c$, where the first $n - Q - R$ their coordinates are ordered increasingly.

4   Generalized Coordinate Method (GCM) for PB-LOP1

Let $S, R, Q \in \mathbb{Z}_n^+$, Λ be as follows – $S + R + Q = n$, $R \geq 3$, $\Lambda \in \Psi$, and a PB-LOP1 needs to be solved. Denote $E'(Gr_R(\Lambda, G)) = E' \cap Gr_R(\Lambda, G)$, $\mathbf{I} = J_{S+1}, ..., j \in J_{n-Q}$.

Definition 2. *For $E' = E_n(G)$, $\Lambda \in \Omega$, a graph $Gr_R(\Lambda, G)$ of PB-LOP1 is a $(S+1) \times ... \times (n-Q)$ grid graph of the dimension R with the following properties:*

- *it is a subgraph of $\mathcal{G}_c(\Lambda)$;*
- *its node-set $E'(Gr_R(\Lambda, G)) = \{p_{i,...,j}\}_{(i,...,j) \in \mathbf{I}}$;*
- *its source is $p_{1,...,1} = x^{min}(\Lambda)$;*
- *last Q of its nodes' coordinates satisfy (25);*
- *towards the last direction D_R, $n - Q$-th coordinate of the nodes decreases gradually from the value $g_{\mu_{n-Q}}$ to g_{μ_1}, while first $n - Q - 1$ coordinates are ordered increasingly;*
- *towards last but one direction D_{R-1}, an $n - Q$-th coordinate of the nodes is already fixed, and its $n - Q - 1$-th coordinate decreases gradually from its original value $g_{\mu_{n-Q-1}}$ to g_{μ_1}, while the rest first $n - Q - 2$ coordinates are ordered increasingly, etc;*
- *towards the first direction D_1, an $S + 2$-th coordinate of the nodes is already fixed, and their $S + 1$-th coordinate decreases gradually from its original value $g_{\mu_{S+1}}$ to g_{μ_1}, and the rest first S coordinates are ordered increasingly, etc.*

The order of $Gr_R(\Lambda, G)$ is

$$N(S, Q) = (S + 1) \cdot ... \cdot (n - Q). \tag{27}$$

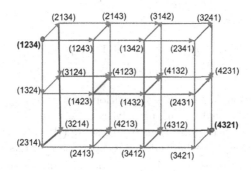

Fig. 1. The grid digraph $Gr_3(J_4)$

As a result, a sink $p_{S+1,...,n-Q}$ will have entries $\{g_{\mu_i}\}_{i \in J_{n-Q}}$ ordered decreasingly on positions $S + 1, ..., n - Q$, while elements on the first S positions will be ordered increasingly. It is seen on Figs. 1 and 2, where a projection of a digraph $\overrightarrow{Gr_3}(J_4)$ formed from $Gr_3(J_4) = Gr(\emptyset, J_4)$ onto $\mathbb{R}^3$ and $Gr_3((7), J_7)$ are

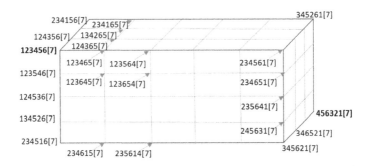

Fig. 2. The grid graph $Gr_3((7), J_7)$

depicted, respectively. Definition 2 generalizes a definition of a two-dimensional grid graph $Gr_2(\Lambda)$ given in [15] to higher dimensions. Namely, in our notations, $Gr_2(\Lambda) = Gr_2(\Lambda, G)$. Let us also generalize a concept a lexicographic order on the grid's nodes and Theorem 1 offered in [15], in the following way:

$$\forall\, (i, ..., j), (i', ..., j') \in \mathbf{I} \quad p_{i,...,j} \preceq p_{i',...,j'} \Leftrightarrow f(p_{i,...,j}) \leq f(p_{i',...,j'}).$$

Two types of transpositions of coordinates of $x \in E'$ will be distinguished from all possible: (a) a positive transposition $f.tr^+$ of x-coordinates, which is a transposition leading to increasing a linear function f; (b) a negative transposition $f.tr^-$ of x-coordinates, which is their transposition leading to decreasing the value. Without loss of generality, it can be assumed that (17) holds, wherefrom $f.tr^+$ is a transposition of a lower value located to the left with a larger value located to the right.

Theorem 1. *If $\Lambda \in \Psi$, then for $Gr(\Lambda)$ it is true:*

$$\forall (i_1, ..., i_{j^*}, ..., i_R), (i_1, ..., i_{j^*} + 1, ..., i_R) \in \mathbf{I}$$
$$y = p_{i_1,...,i_{j^*}+1,...,i_R} \text{ is formed from } x = p_{i_1,...,i_{j^*},...,i_R} \text{ by } f.tr^+.$$

Proof. By construction, last $n - S - j^*$ positions of $x = (x_i)_{i \in J_n}$ and $y = (y_i)_{i \in J_n}$ are identical and pairwise different, namely, $x_i = y_i = g_{j_i}$, $i \in J_n \backslash J_{j^*}$. Thus, the rest positions are $j^* + S$-permutations from a set $G'(\Lambda) = G(\Lambda) \backslash \{g_{i_j}\}_{j \in J_R \backslash J_{j^*}} = \{g'_j\}_{j \in J_{j^*}} : g'_1 < ... < g'_{R-j^*}$, where $G(\Lambda) = G \backslash \{g_{\lambda_q}\}_{q \in J_Q}$. x has an entry $g'_{j^*-i_{j^*}+1}$ at the j^*-th position, while $y_{j^*} = g'_{j^*-i_{j^*}} < x_{j^*}$. This means that $\exists j \in J_{k-1}$ such that $x_j = y_{j^*}$, $x_{j-1} < x_j < x_{j+1}$. If we make $x_{j^*} \leftrightarrow y_{j^*}$-transposition of x-coordinates, we obtain exactly y, since the last its position is $g'_{j^*-i_{j^*}+1}$. Forming y from x is equivalent to replacing x_{j^*} by y_{j^*} and ordering of the rest left coordinates. Since $y_{j-1} = x_{j-1} < y_j = g'_{j^*-i_{j^*}+1} << y_{j+1} = x_{j+1}$, the ordering holds. For x, this $x_{j^*} \leftrightarrow y_{j^*}$-transposition is $f.tr^+$, since $x_{j^*} > y_{j^*}$ and $j < j^*$.

Corollary 1. *If $\Lambda \in \Psi$ then for $Gr_R(\Lambda, G)$ it is true:*

$$\forall(i, ..., k, ..., j), (i, ..., k+1, ..., j) \in \mathbf{I} \quad p_{i,...,k,...,j} \preceq p_{i,...,k+1,...,j}.$$

Corollary 2. *In a sink of the grid graph $Gr_R(\Lambda, G)$, it is attained a maximum of f on its nodes.*

Theorem 1 and Corollary 1 state that it is reasonable to conduct a directed search of x^* examining $E'(Gr_R(\Lambda, G))$ from the source to the sink, thus moving from a node of $Gr_R(\Lambda, G)$ to those its adjacent ones, where the objective function does not decrease. In other words, it is worth to perform the search on $\overrightarrow{Gr}_R(\Lambda, G)$. Finally, Corollary 2 says that for examining $Gr_R(\Lambda, G)$, it is sufficient to check the only sink, if it belongs to E.

Let us establish a connection between $x^{min}(Gr_R(\Lambda, G)) = arg \min\limits_{x \in E'(Gr_R(\Lambda,G))} f(x)$, $x^{max}(Gr_R(\Lambda, G)) = arg \max\limits_{x \in E'(Gr_R(\Lambda,G))} f(x)$ and x^{min}, x^{max}. By (18), By construction and (18),

$$x^{min}(Gr_R(\emptyset, G)) = x^{min} = (g_1, ..., g_n) \in E;$$
$$x^{min}(Gr_R(\Lambda, G)) = x^{min}(\Lambda) = (g_{\mu_1}, ..., g_{\mu_{n-Q}}, g_{\lambda_Q}, ..., g_{\lambda_2}, g_{\lambda_1});$$
$$x^{max}(\Lambda) = (g_{\mu_{n-Q}}, ..., g_{\mu_1}, g_{\lambda_Q}, ..., g_{\lambda_2}, g_{\lambda_1});$$
$$x^{max}(Gr_R(\Lambda, G)) = (g_{\mu_{R+1}}, ..., g_{\mu_{n-Q}}, g_{\mu_R}, ..., g_{\mu_1}, g_{\lambda_Q}, ..., g_{\lambda_2}, g_{\lambda_1}).$$

Thus, $z^{min} \leq z^{min}(\Lambda) = z^{min}(Gr_R(\Lambda, G)) \leq z^{max}(Gr_R(\Lambda, G)) \leq z^{min}(\Lambda) \leq\leq z^{min}$, where $z^{[.]}([..]) = f(x^{[.]}([..]))$.

Algorithm of GCM

Input: PB-LOP1, $S, R, Q \in \mathbb{Z}_n^+$: $S + R + Q = n$, $R \geq 3$.

Output: if $S = 0 - \langle x^*, z^* \rangle$, otherwise, an approximate solution of PB-LOP1 $\langle x^{**}, z^{**} \rangle$.

GLM-algorithm outline:

- **Step 0.** Set $x^{**} = g$, $z^{**} = c^T g$. Find $\langle x^{max}, z^{max} \rangle$.
- **Step 1.** If (22) holds, then $\langle x^*, z^* \rangle = \langle x^{max}, z^{max} \rangle$. Finish.
- **Step 2.** Form the family Ψ.
- **Step 3.** For each $\Lambda \in \Psi$, verify:
  1. if $z^{max}(\Lambda) \leq z^{**}$, go to Step 3;
  2. if
     (a) $x^{max}(\Lambda)$ satisfy (26), then $x^{max}(\Lambda) = arg \max\limits_{x \in E(\Lambda)} f(x)$; if $z^{**} < z^{max}(\Lambda)$, then $\langle x^{**}, z^{**} \rangle = \langle x^{max}(\Lambda), z^{max}(\Lambda) \rangle$.
     (b) otherwise, i.e., if

$$x^{max}(\Lambda) \notin E, \tag{28}$$

then apply a search on a grid graph $Gr_R(\Lambda, G)$ in accordance to Algorithm 1.

Step 4. if $S = 0$, then

$$\langle x^*, z^* \rangle = \langle x^{**}, z^{**} \rangle. \tag{29}$$

Algorithm 1
Algorithm 1 intends for examining $Gr_R(\Lambda, G)$ in order to improve z^{**}.

- Step A.1. If $z^{max}(Gr_R(\Lambda, G)) \leq z^{**}$, then finish.
  If $x^{max}(Gr_R(\Lambda, G)) \in E$ and $z^{**} < z^{max}(Gr_R(\Lambda, G))$, then $\langle x^{**}, z^{**} \rangle =$
  $= \langle x^{max}(Gr_R(\Lambda, G)), z^{max}(Gr_R(\Lambda, G)) \rangle$. Finish.
- Step A.2. Examine $E'(Gr_R(\Lambda, G))$ moving from a source to a sink, improving
  z^{**}, and updating x^{**} as soon as a node in E with grater f-value is found.

Remark 3. For each examined point $x \in E$, a statistics will be given in a form
$stat(x) = [f(x), \Delta_1(x), ..., \Delta_m(x)]$, where $\Delta_i(x) = -h_i(x)$ is a residual of the
i-th constraint (15) at x, $i \in J_m$.
$x \in E'$ is feasible, i.e., it satisfies (19), iff

$$\Delta_i(x) \geq 0, i \in J_m. \tag{30}$$

Remark 4. The fact of obtaining an approximate solution $\langle x^{**}, z^{**} \rangle$ of PB-COP
is provided by an additional condition (18). If $S = 0$, (27) becomes $N(0, Q) =$
$(n - Q)!$ is an order of each of A_n^Q grid graphs $Gr_R(\Lambda, G)$, $\Lambda \in \Psi$, which are
pairwise disjoint. Together, their vertex sets form a partition of $E_n(G)$ as $|A_n^Q| \cdot$
$N = \frac{n!}{(n-Q)!}(n - Q)! = n! = |E_n(G)|$. Thus, (29) holds.

5   GLCM Illustration

Along with $E'(Gr_R(\Lambda, G))$, the following denotation will be used:

$$E(Gr_R(\Lambda, G)) = E \cap Gr_R(\Lambda, G).$$

Let us consider the following example: solve PB-LOP for $n = 7$, $m = 2$,
$G = \{g_1, g_2, g_3, g_4, g_5, g_6, g_7\} = \{1, 2, 4, 7, 8, 9, 11\}$, $c = (8, 7, 7, 6, 4, 2, 2)$, $a^1 =$
$(5, 7, 4, 5, 2, 2, 4)$, $a^2 = (2, 1, 4, 3, 2, 3, 1)$, $b^1 = 174$, $b^2 = 96$.
$x^{min} = (1, 2, 4, 7, 8, 9, 11)$, $stat(x^{min}) = [z^{min}, \Delta_1(x^{min}), \Delta_2(x^{min})] =$
$[164, 26, 1]$. The constraint (30) holds, hence, (18) is satisfied, thus, we deal with
PB-LOP1 and GLM is applicable to it.
Let $S = 3$, $R = 3$, $Q = 1$ be chosen.

Step 0. $\langle x^{**}, z^{**} \rangle = \langle x^{min}, z^{min} \rangle = \langle (1, 2, 4, 7, 8, 9, 11), 164 \rangle$.
Step 1. $x^{max} = (11, 9, 8, 7, 4, 2, 1)$, $stat(x^{max}) = [z^{max}, \Delta_1(x^{max}), \Delta_2(x^{max})]$
$= [271, -27, -3]$. No one of the constraints (18) holds, hence, $x^{max} \notin E$.
Step 2. $\Psi = \{7, 6, 5, 4, 3, 2, 1\}$.
Step 3. Explore Ψ sequentially from $j = 7 \in \Psi$ to $j = 1$.
Step 3.1. $j = 7$, $\Lambda = (j) = (7)$.

$x^{\max}(\Lambda) = x^{\max}(7) = (g_6, g_5, g_4, g_3, g_2, g_1, g_7) = (9, 8, 7, 4, 2, 1, 11), stat(x^{\max}$
$(7)) = [233, -25, 12]$. $z^{\max}(7) = 244 > z^{**} = 164$ and $x^{\max}(\Lambda) \notin E$. Continue.
(28) holds, go to Algorithm 1.

Step A.1. $x^{\max}(Gr_3(\Lambda, G)) = x^{\max}(Gr_3(7, G)) = (g_3, g_5, g_6, g_3, g_2, g_1, g_7) =$
$= (7, 8, 9, 4, 2, 1, 11), stat(x^{\max}(Gr_3(7, G))) = [231, -23, 8]$. $z_3^{\max}(Gr_3(7, G)) =$
$231 > 264 = z^{**}, x_3^{\max}(Gr_3(7, G)) \notin E$.

Step A.2. Examine $E'(Gr_3(7, G))$ starting at a source $x^{\min}(Gr_3(7, G)) =$
$= (g_1, g_2, g_3, g_4, g_5, g_6, g_7) = (1, 2, 4, 7, 8, 9, 11) = x^{\min} \in E, stat(x_3^{\min}(7)) =$
$stat(x^{\min}) = [164, 26, 1]$ to the sink $x^{\max}(Gr_3(7, G))$.

By (27), $|E'(Gr_3(7, G))| = 6 \cdot 5 \cdot 4 = 120$ nodes are examined, and results are
presented in Tables 2, 3, 4 and 5, where $stat(x)$ for $x \in Gr_3(7, G)$ are collected,
and the three-dimensional grid is presented as a decomposition into $n - Q +$
$R - 1 = 4$ two-dimensional grid graphs $Gr_3^l(7)$, $l \in J_4$, lying in parallel planes
of $Gr_3(7, G)$ (see Fig. 2). In particular, $Gr_3^1(7)$ is a front-face-grid containing
the source $x^{min}(\Lambda)$, which statistics are given in Table 2; $Gr_3^2(7)$ is next parallel
(see Table 3) to it, etc. Finally, $Gr_3^4(7)$ is a back-face-grid with the sink $x^{max}(\Lambda)$
(see Table 5).

Examine them consecutively, updating the current record 164.

(a) $Gr_3^1(7)$: from Table 2, in addition to $x^{\min}(Gr_3(7, G)) = x_3^{\min}(7, G)$, whose
statistics is in $(1, 1)$-cell, there are 10 more feasible shadowed grid-nodes.
Among them, the maximum of $f(x)$ is 203 attained at $x^{max}(E(Gr_3^1(7))) =$
$(2, 4, 7, 9, 8, 1, 11)$ associated with the cell $(2, 6)$ of the table with an entry
$[203, 1, 3]$. Since $z^{max}(E(Gr_3^1(7))) = 203 > 164$, update $\langle x^{**}, z^{**} \rangle =$
$< x^{max}(E(Gr_3^1(7))), z^{max}(E(Gr_3((7), 1)) >= \langle (2, 4, 7, 9, 8, 1, 11), 203 \rangle$.

(b) $Gr_3^2(7)$: from Table 3, there are 5 feasible shadowed grid-nodes. Maximum
of $f(x)$ at these nodes is 207, which is attained at $(2, 4, 9, 8, 7, 1, 11)$ corresponding the cell $(3, 6)$ with an entry $[207, 0, 0]$. $z^{max}(E(Gr_3^2(7))) = 207 >>$
$203 = z^{**}$. Update: $\langle x^{**}, z^{**} \rangle = \langle (2, 4, 9, 8, 7, 1, 11), 207 \rangle$.

(c) $Gr_3^3(7)$: from Table 4, exactly half of the nodes are feasible. However,
$z^{max}(E(Gr_3^3(7))) = 202 < z^{**}$. A new record was not found.

(d) $Gr_3^4(7)$: from Table 5, it is seen that 12 feasible shadowed grid-nodes are
found. $z^{max}(E(Gr_3^4(7))) = 208 > 207 = z^{**}$, thus a record is improved:
$\langle x^{**}, z^{**} \rangle = \langle x^{max}(Gr_3^4(7))), z^{max}(E(Gr_3^4(7))) \rangle = \langle (4, 7, 9, 1, 8, 2, 11), 208 \rangle$.
Finish. Go to Step 3.

$j = j - 1 = 6$. Similarly examine $Gr_3(6, G), ..., Gr_3(1, G)$. Results on examining $Gr_3(j, G)$, $j \in J_6$ are collected in Table 1, namely, values $z^{max}(\Lambda)$,
$z^{max}(Gr_3(\Lambda, G))$, $\Lambda \in \Psi$ increasing as Λ decreases; a number of feasible points
in $E(Gr_3(\Lambda, G))$; the consecutively improved bound z^{**} and a node, where it is
attained.

A result of applying GML is the following: $\langle x^{**}, z^{**} \rangle = \langle (7, 8, 9, 1, 11, 2,$
$4), 237 \rangle$. Since $S > 0$, we can only say that the solution to ECOP1 is approxi-
mate. At the same time, an exact solution is $\langle x^*, z^* \rangle = \langle ((9, 8, 7, 2, 11, 4, 1), 243 \rangle$,
thus a relative error of the solution x^{**} is $\delta^* = |\frac{z^* - z^{**}}{z^{**}}| = 2, 5\%$.

Remark 5. From Table 1, it is seen that exploring these grid graphs from $Gr_3(1, G)$ to $Gr_3(7, G)$ allows reduce the search by applying the Step 3.1-condition for $\Lambda = (1)$, because $233 = z^{max}(1) \leq z^{**} = 237$.

Remark 6. This example shows that the requirement (18) can be relaxed and replaced with the following: it is known $x^0 \in E$. If this FP has been solved, x^0 can be used as x^{**}. In turn, such an FP, in some cases, can be solved easily based on the meaning of the optimization problem under consideration. In general case, we recommend solving an auxiliary continuous optimization problem obtained from PB-LOP by replacing the constraint (16) by a functional representation of $E_n(G)$ [27,28] and applying global optimization approaches to its solution. This way is more preferable, in this case, the infeasibility of the original PB-LOP will be proven or x^0 will be a local minimizer of the auxiliary problem, whose use as a starting point can reduce a search domain significantly (Fig.3).

Table 1. Results on analysis of $E(Gr_3((j), G))$, $j \in J_7$

Λ	$z^{max}(\Lambda)$	# of feasible points	$z^{max}(Gr_3(\Lambda))$	x^{**}	z^{**}
7	233	47	208	(4,7,9,1,8,2,11)	208
6	245	24	218	(4,7,11,1,8,2,9)	218
5	250	24	220	(4,7,11,1,9,2,8)	220
4	255	9	223	(4,8,9,2,11,1,7)	223
3	267	1	237	(7,8,9,1,11,2,4)	237
2	271	0	–	(7,8,9,1,11,2,4)	237
1	271	0	–	(7,8,9,1,11,2,4)	237

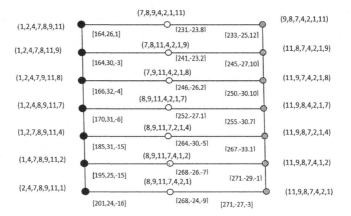

Fig. 3. The grid-graph $Gr_3((7), G)$

6   GLM for Permutation-Based Multiobjective Linear Optimization

If PB-MLOP $f_j(x) = c^{jT}x \to \max$, $j \in J_L$, $L > 1$ needs to be solved, and a prior information on importance of criteria $f_j(x)$, $j \in J_L$ is known and presented by their relative weights $\lambda_j \in (0,1)$, $\in J_L$, GCM can be used directly, if priori methods of multiobjective optimization, such as method of scalarizing the multiobjective problem, the lexicographic method, the method of successive concessions, etc. [8,9,26]. If the first method is applied, GCM is used after a preliminary convolution into a supercriterion $F(x) = \lambda_1 f_1(x) + \lambda_L f_L(x)$. According to Remark 5, in this case, an initial feasible point $x^{**} \in E$ is required, which can be found by applying the continuous functional representations of $E_n(G)$. If the last two methods are utilized, GCM is applied on each step of the corresponding

Table 2. $Gr_3((7),1)$

$j_2 \backslash j_1$	1	2	3	4	5	6
1	[164,26,1]	[166,26,2]	[170,23,2]	[185,17,-1]	[195,7,3]	[201,4,4]
2	[166,23,0]	[170,20,0]	[172,20,1]	[187,14,-2]	[197,4,2]	[203,1,3]
3	[175,17,-6]	[179,14,-6]	[184,12,-7]	[190,12,-4]	[200,2,0]	[206,-1,1]
4	[181,7,-4]	[185,4,-4]	[190,2,-5]	[205,-13,1]	[209,-13,3]	[215,-16,4]
5	[185,4,-4]	[189,1,-4]	[194,-1,-5]	[209,-16,1]	[221,-22,3]	[223,-22,4]

Table 3. $Gr_3((7),2)$

$j_2 \backslash j_1$	1	2	3	4	5	6
1	[167,29,-2]	[169,29,-1]	[174,27,-2]	[186,18,-2]	[196,8,2]	[202,5,3]
2	[170,27,-4]	[175,25,-5]	[177,25,-4]	[189,16,-4]	[199,6,0]	[205,3,1]
3	[176,18,-7]	[181,16,-8]	[185,13,-8]	[191,13,-5]	[201,3,-1]	[207,0,0]
4	[182,8,-5]	[187,6,-6]	[191,3,-6]	[206,-12,0]	[210,-12,2]	[216,-15,3]
5	[186,5,-5]	[191,3,-6]	[195,0,-6]	[210,-15,0]	[222,-21,2]	[224,-21,3]

Table 4. $Gr_3((7),3)$

$j_2 \backslash j_1$	1	2	3	4	5	6
1	[169,25,2]	[171,25,3]	[176,23,2]	[191,8,8]	[199,2,8]	[205,-1,9]
2	[172,23,0]	[177,21,-1]	[179,21,0]	[194,6,6]	[202,0,6]	[208,-3,7]
3	[181,8,3]	[186,6,2]	[191,1,4]	[197,1,7]	[205,-5,7]	[211,-8,8]
4	[185,2,1]	[190,0,0]	[195,-5,2]	[207,-14,2]	[211,-14,4]	[217,-17,5]
5	[189,-1,1]	[194,-3,0]	[199,-8,2]	[211,-17,2]	[223,-23,4]	[225,-23,5]

Table 5. $Gr_3((7),4)$

$j_2 \backslash j_1$	1	2	3	4	5	6
1	[171,25,3]	[173,25,4]	[178,23,3]	[193,8,9]	[205,2,11]	[209,-1,11]
2	[174,23,1]	[179,21,0]	[181,21,1]	[196,6,7]	[208,0,9]	[212,-3,9]
3	[183,8,4]	[188,6,3]	[193,1,5]	[199,1,8]	[211,-5,10]	[215,-8,10]
4	[191,2,4]	[196,0,3]	[201,-5,5]	[219,-14,8]	[223,-14,10]	[227,-17,10]
5	[193,-1,3]	[198,-3,2]	[203,-8,4]	[221,-17,7]	[229,-23,7]	[231,-23,8]

iterative procedure, where a single criterion is maximized. During the process, the additional constraints (15) are complemented by new ones. Note that the feasibility problem is solved only on an initial iteration. For the rest iterations, already found solutions of auxiliary PB-LOPs can be used as x^{**} and z^{**}.

7 Conclusion

A Generalized Coordinate Method (GCM) for linear constraint optimization on permutations is offered. It generalizes the Modified Coordinate Method and Modified Coordinate Localization Method of permutation-based linear optimization and can be used directly to solve multiobjective permutation-based problems after a convolution of criteria or as an auxiliary problems solved on each step of iterative methods of multiobjective optimization where a single-objective optimization problem is solved on each stage GCM uses properties of linear function on a transposition graph, a decomposition of the graph into isomorphic transposition graphs of lower dimension, as well as single outing grid graphs from it, which define a search domain in GLM. Depending on the search parameters, GLM yields an exact or approximate solution to the linear permutation-based problem. An illustration example of GLM-algorithm is given.

References

1. Brouwer, A.E., Cohen, A.M., Neumaier, A.: Distance-Regular Graphs. Springer, Heidelberg (1989). https://doi.org/10.1007/978-3-642-74341-2
2. Burkard, R.E.: Quadratic assignment problems. In: Pardalos, P.M., Du, D.-Z., Graham, R.L. (eds.) Handbook of Combinatorial Optimization, pp. 2741–2814. Springer, New York (2013). https://doi.org/10.1007/978-1-4419-7997-1_22
3. Burkard, R., Dell'Amico, M., Martello, S.: Assignment Problems. Society for Industrial and Applied Mathematics (2012). https://doi.org/10.1137/1.9781611972238
4. Chase, P.: Transposition graphs. SIAM J. Comput. **2**, 128–133 (1973)
5. Donets, G.A., Kolechkina, L.N.: Method of ordering the values of a linear function on a set of permutations. Cybern. Syst. Anal. **45**, 204–213 (2009)
6. Donec, G.A., Kolechkina, L.M.: Construction of Hamiltonian paths in graphs of permutation polyhedra. Cybernet. Syst. Anal. **46**, 7–13 (2010)
7. Farzad, B., Pichugina, O., Koliechkina, L.: Multi-layer community detection. In: 2018 International Conference on Control, Artificial Intelligence, Robotics Optimization (ICCAIRO), pp. 133–140 (2018)
8. Ehrgott, M.: Multicriteria Optimization. Springer, New York (2005). https://doi.org/10.1007/3-540-27659-9
9. Ehrgott, M., Gandibleux, X.: Multiobjective combinatorial optimization - theory, methodology, and applications. In: Ehrgott, M., Gandibleux, X. (eds.) Multiple Criteria Optimization: State of the Art Annotated Bibliographic Surveys, pp. 369–444. Springer, Boston (2003). https://doi.org/10.1007/0-306-48107-3_8
10. Ganesan, A.: Automorphism group of the complete transposition graph. J. Algebr. Comb. **42**, 793–801 (2015). https://doi.org/10.1007/s10801-015-0602-5
11. Gimadi, E., Khachay, M.: Extremal Problems on Sets of Permutations (2016). (in Russian)

12. Grande, F.: On k-level matroids: geometry and combinatorics (2015). http://www. diss.fu-berlin.de/diss/receive/FUDISS_thesis_000000100434

13. Koliechkina, L.M., Dvirna, O.A.: Solving extremum problems with linear fractional objective functions on the combinatorial configuration of permutations under multicriteriality. Cybern. Syst. Anal. 53, 590–599 (2017)

14. Koliechkina, L.N., Dvernaya, O.A., Nagornaya, A.N.: Modified coordinate method to solve multicriteria optimization problems on combinatorial configurations. Cybern. Syst. Anal. 50, 620–626 (2014). https://doi.org/10.1007/s10559-014-9650-4

15. Koliechkina, L., Pichugina, O.: Multiobjective optimization on permutations with applications. DEStech Trans. Comput. Sci. Eng., 61–75 (2018). https://doi.org/ 10.12783/dtcse/optim2018/27922

16. Koliechkina, L., Pichugina, O.: A horizontal method of localizing values of a linear function in permutation-based optimization. In: Le Thi, H.A., Le, H.M., Pham Dinh, T. (eds.) WCGO 2019. AISC, vol. 991, pp. 355–364. Springer, Cham (2020). https://doi.org/10.1007/978-3-030-21803-4_36

17. Korte, B., Vygen, J.: Combinatorial Optimization: Theory and Algorithms. Springer, Heidelberg (2012). https://doi.org/10.1007/3-540-29297-7

18. Kozin, I.V., Kryvtsun, O.V., Pinchuk, V.P.: Evolutionary-fragmentary model of the routing problem. Cybern. Syst. Anal. 51, 432–437 (2015)

19. Kozin, I.V., Maksyshko, N.K., Perepelitsa, V.A.: Fragmentary structures in discrete optimization problems. Cybern. Syst. Anal. 53, 931–936 (2017)

20. Lane, T., Yackley, B., Plis, S., McCracken, S., Anderson, B.: Geometric embedding for learning combinatorial structures 3. http://citeseerx.ist.psu.edu/viewdoc/ summary?doi.org/10.1.1.387.7534

21. Lengauer, T.: Operations research and statistics. In: Combinatorial Algorithms for Integrated Circuit Layout. pp. 137–217. Vieweg+Teubner Verlag (1990)

22. Liu, X., Draper, S.C.: LP-decodable multipermutation codes. IEEE Trans. Inf. Theory 62, 1631–1648 (2016). https://doi.org/10.1109/TIT.2016.2526655

23. Mehdi, M.: Parallel hybrid optimization methods for permutation based problems (2011). https://tel.archives-ouvertes.fr/tel-00841962/document

24. Onwubolu, G.C., Davendra, D. (eds.): Differential Evolution: A Handbook for Global Permutation-Based Combinatorial Optimization. Springer, Heidelberg (2009). https://doi.org/10.1007/978-3-540-92151-6

25. Pardalos, P.M., Du, D., Graham, R.L.: Handbook of Combinatorial Optimization. Springer, New York (2005). https://doi.org/10.1007/978-1-4419-7997-1

26. Pardalos, P.M., Žilinskas, A., Žilinskas, J.: Non-Convex Multi-Objective Optimization. Springer, Heidelberg (2017). https://doi.org/10.1007/978-3-319-61007-8

27. Pichugina, O.S., Yakovlev, S.V.: Continuous representations and functional extensions in combinatorial optimization. Cybern. Syst. Anal. 52(6), 921–930 (2016). https://doi.org/10.1007/s10559-016-9894-2

28. Pichugina, O., Yakovlev, S.: Euclidean combinatorial configurations: continuous representations and convex extensions. In: Lytvynenko, V., Babichev, S., Wójcik, W., Vynokurova, O., Vyshemyrskaya, S., Radetskaya, S. (eds.) ISDMCI 2019. AISC, vol. 1020, pp. 65–80. Springer, Cham (2020). https://doi.org/10.1007/978-3-030-26474-1_5

29. Schrijver, A.: Combinatorial Optimization: Polyhedra and Efficiency. Springer Science & Business Media, Heidelberg (2002)

30. Semenova, N.V., Kolechkina, L.N.: A polyhedral approach to solving multicriterion combinatorial optimization problems over sets of polyarrangements. Cybern. Syst. Anal. 45, 438–445 (2009). https://doi.org/10.1007/s10559-009-9110-8

31. Stoyan, Y.G., Yemets', O.: Theory and methods of Euclidean combinatorial optimization. ISSE, Kiev (1993). (in Ukrainian)
32. Weisstein, E.W.: CRC Concise Encyclopedia of Mathematics, 2nd edn. Chapman and Hall/CRC, Boca Raton (2002)
33. Yakovlev, S.V.: The theory of convex continuations of functions on vertices of convex polyhedra. Comp. Math. Math. Phys. **34**, 1112–1119 (1994)
34. Yakovlev, S., Pichugina, O., Yarovaya, O.: On optimization problems on the polyhedral-spherical configurations with their properties. In: 2018 IEEE First International Conference on System Analysis Intelligent Computing (SAIC), pp. 94–100 (2018). https://doi.org/10.1109/SAIC.2018.8516801
35. Yakovlev, S., Pichugina, O., Yarovaya, O.: Polyhedral-spherical configurations in discrete optimization problems. J. Autom. Inf. Sci. **51**, 26–40 (2019). https://doi.org/10.1615/JAutomatInfScien.v51.i1.30
36. Yakovlev, S.V., Valuiskaya, O.A.: Optimization of linear functions at the vertices of a permutation polyhedron with additional linear constraints. Ukr. Math. J. **53**, 1535–1545 (2001). https://doi.org/10.1023/A:1014374926840
37. Yemelicher, V.A., Kovalëv, M.M., Dravtsov, M.K., Lawden, G.: Polytopes, Graphs and Optimisation. Cambridge University Press, Cambridge (1984)
38. Yemets, O.A., Yemets, Y.M., Chilikina, T.V.: Combinatorial cutting while solving optimization nonlinear conditional problems of the vertex located sets. JAI(S) **42**, 21–29 (2010). https://doi.org/10.1615/JAutomatInfScien.v42.i5.30
39. Ziegler, G.M.: Lectures on Polytopes. Springer, New York (1995). https://doi.org/10.1007/978-1-4613-8431-1

A Smoothing Lagrange Multiplier Method for Solving the Quasi-variational Signorini's Inequality

Robert Namm[1]([✉]) [ID], Georgiy Tsoy[1] [ID], and Ellina Vikhtenko[2] [ID]

[1] Computing Center of Far Eastern Branch Russian Academy of Sciences,
65 Kim Yu Chen Street, Khabarovsk 680000, Russia
rnamm@yandex.ru, tsoy.dv@mail.ru
[2] Pacific National University, 136 Tikhookeanskaya Street,
Khabarovsk 680035, Russia
vikht.el@gmail.com
http://ccfebras.ru

Abstract. For semicoercive contact elastic problem with friction the smooth duality scheme is investigated, which allows on each step of successive approximation method to define simultaneously the displacement vector of elastic body points and normal contact stress defining the friction force on the next step of successive approximation method.

Keywords: Contact problem · Modified Lagrangian functional · Saddle point · Uzawa method · Sensitivity functional

1 Introduction

The modern statements of the contact problem in elasticity contain the unilateral nonpenetration condition for normal displacement in contact zone with a rigid surface and friction condition for tangential displacement. We note also that friction force according to Coulomb law depends on the desired solution. With regard to these indicated factors we obtain the quasi-variational Signorini's inequality [1–3]. The well-known a successive approximation method for solving quasi-variational Signorini's inequality is based on the iterative process in which on each step one must solve the auxiliary problem with given friction. It is convenient to use duality schemes for solving an auxiliary problem as duality schemes allow finding the solution of initial and dual problems simultaneously and the solution of dual problem defines the friction force on the next step of successive approximation method. It is known that duality schemes based on classical Lagrangian functional can be used in coercive problem only. But the convergence of the duality method in coercive problems takes place if there is a dependence of step to the dual variable on the constant of strong convexity of

This study was supported by the Russian Foundation for Basic Research (Project 17-01-00682 A).

minimizing functional. To overcome difficulties we investigated duality schemes based on modified Lagrangian functional. It was shown that modified duality methods converge to saddle point under suitable regularity of auxiliary problem solution [4–6]. In this paper we continue the further investigations of modified duality methods.

2   Semicoercive Contact Problem in Elasticity with Friction. Quasi-Variational Inequality

We consider a two-dimensional contact problem between an elastic body Ω and absolutely rigid support (Fig. 1). The boundary Γ of domain Ω is equal to $\bar{\Gamma}_0 \bigcup \bar{\Gamma}_K \bigcup \bar{\Gamma}_P$, where Γ_0, Γ_K and Γ_P are open pairwise disjoint subsets of Γ such that $\mathrm{mes}(\Gamma_0)$ and $\mathrm{mes}(\Gamma_K)$ are positive.

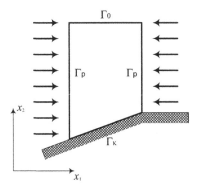

Fig. 1. The contact body

For the displacement vector $u = (u_1, u_2)$ we define the strain tensor

$$\varepsilon_{ij}(u) = \frac{1}{2}\left(\frac{\partial u_i}{\partial x_j} + \frac{\partial u_j}{\partial x_i}\right), \quad i, j = 1, 2,$$

and the stress tensor $\sigma_{ij}(u) = c_{ijpm}\varepsilon_{pm}(u)$, where i, j, p, $m = 1, 2$; $c_{ijpm} = c_{jipm} = c_{pmij}$; summation is implied over repeated indexes.

We denote $n = (n_1, n_2)$ is a unit outward normal vector on Γ; $u_n = u \cdot n$; $u_t = u - u_n n$; $\sigma_i(u) = \sigma_{ij}(u)\, n_j$ for $i = 1, 2$; $\sigma(u) = (\sigma_1(u), \sigma_2(u))$; $\sigma_n(u) = \sigma_{ij}(u)\, n_i\, n_j$; $\sigma_t(u) = \sigma(u) - \sigma_n(u)\, n$; $\sigma_{ij,j}(u) = \partial\sigma_{ij}(u)/\partial x_j$, $i, j = 1, 2$; $\mathcal{F}$ is the frictional coefficient with $\mathcal{F} \geq 0$ on Γ_K.

For given functions $F = (f_1, f_2)$ and $T = (T_1, T_2)$, consider the boundary value problem [1, 2]

$$-\sigma_{ij,j}(u) = f_i \quad \text{in } \Omega, \ i = 1, 2;$$
$$u_n = 0, \ \sigma_t(u) = 0 \text{ on } \Gamma_0;$$
$$\sigma_{ij}(u)\, n_j = T_i \quad \text{on } \Gamma_P, \ i = 1, 2.$$

The following conditions are set on the interface Γ_K between the elastic body and the absolutely rigid support

$$u_n \leq 0, \ \sigma_n(u) \leq 0, \ u_n\sigma_n(u) = 0 \ \text{(unilateral Signorini conditions)};$$
$$|\sigma_t(u)| \leq \mathcal{F}|\sigma_n(u)|, \ \sigma_t(u)u_t + \mathcal{F}|\sigma_n(u)||u_t| = 0 \ \text{(friction conditions).}$$
(1)

The basic difficulty in the study of this problem is that the frictional force $\mathcal{F}|\sigma_n(u)|$ is a function of the desired solution u. Together with the boundary value problem, we consider the corresponding quasi-variational inequality.

Define the set (see Fig. 1)

$$\mathcal{K} = \left\{ v \in [H^1(\Omega)]^2 : \ v = v_n = 0 \text{ on } \Gamma_0, \ v_n \leq 0 \text{ on } \Gamma_K \right\}.$$

Assume that $c_{ijpm} \in L_\infty(\Omega)$, $i, j, p, m = 1, 2$; $F \in [L_2(\Omega)]^2$ and $T \in [L_2(\Gamma_P)]^2$. Suppose that the boundary value problem with friction has a solution $u \in [H^2(\Omega)]^2$. Then it can be shown that u satisfies the quasi-variational inequality

$$a(u, v - u) + \int\limits_{\Gamma_K} \mathcal{F}|\sigma_n(u)|(|v_t| - |u_t|)d\Gamma \geq$$
$$\geq \int\limits_{\Omega} f_i(v_i - u_i)d\Omega + \int\limits_{\Gamma_P} T_i(v_i - u_i)d\Gamma \quad \forall v \in \mathcal{K},$$
(2)

where $a(u, v) = \int\limits_{\Omega} c_{ijpm}\varepsilon_{pm}(u)\varepsilon_{ij}(v)\,d\Omega$.

The quasi-variational inequality can be rewrite in the following way [7]

$$u \in \arg\min_{v \in \mathcal{K}} \left\{ \frac{1}{2}a(v, v) - \int\limits_{\Omega} f_i v_i \, d\Omega - \int\limits_{\Gamma_P} T_i v_i d\Gamma + \int\limits_{\Gamma_K} \mathcal{F}|\sigma_n(u)| \, |v_t| \right\} d\Gamma.$$

Let us consider the successive approximation method for solving the quasi-variational inequality (2). We specify an initial frictional force $g_0 \in H^{1/2}(\Gamma_K)$ such that $g_0 \geq 0$. At the $(k + 1)$th step, u^{k+1} is defined as the solution to the problem with given friction [1]

$$a(u^{k+1}, v - u^{k+1}) + \int\limits_{\Gamma_K} g_k(|v_t| - |u_t^{k+1}|)d\Gamma \geq$$
$$\geq \int\limits_{\Omega} f_i (v_i - u_i^{k+1}) \, d\Omega + \int\limits_{\Gamma_P} T_i (v_i - u_i^{k+1})d\Gamma \quad \forall v \in \mathcal{K},$$
(3)

where $g_k = \mathcal{F}|\sigma_n(u^k)|$, $k = 1, 2, \ldots$, and u^k is a solution, finding on a previous step k of successive approximations method.

It is easy to show that variational inequality (3) is equivalent to the variational problem

$$\begin{cases} J(v) = \dfrac{1}{2}a(v, v) - \int\limits_{\Omega} f_i v_i \, d\Omega - \int\limits_{\Gamma_P} T_i v_i d\Gamma + \int\limits_{\Gamma_K} g_k|v_t|d\Gamma \to \min, \\ v \in \mathcal{K}. \end{cases}$$
(4)

Moreover, the corresponding boundary value problem for (4) has the same form as the original boundary value problem except that the friction conditions in (1) are replaced with

$$|\sigma_t(u)| \le g_k, \quad \sigma_t(u)\, u_t + g_k |u_t| = 0.$$

The kernel R of the bilinear form $a(v,v)$ is not empty in $[H^1(\Omega)]^2$ and consists of the vector function $\rho = (\rho_1, \rho_2)$, where $\rho_1 = a_1 - b\,x_2$, $\rho_2 = a_2 + b\,x_1$ and a_1, a_2, b are arbitrary fixed scalars.

Define $W = \{v \in [H^1(\Omega)]^2 : v_n = 0 \text{ on } \Gamma_0\}$. The subspace $\tilde{R} = W \cap R$ is a set of virtual rigid displacements (i.e., displacements of Ω as an absolutely rigid body with the strict (two-sides) constructions being preserved). According to Fig. 1 (see [1]), $\tilde{R} = W \cap R$ is a one-dimensional set and looks in the following way $\tilde{R} = \{\rho = (\rho_1, \rho_2) : \rho_1 = a, \rho_2 = 0\}$, where a is an arbitrary constant.

Since Γ_K is not parallel to Γ_0, a unit outward normal vector n on Γ_K satisfies the condition $n_1 \ne 0$. According to Fig. 1, $n_1 > 0$ on Γ_K. Then

$$\tilde{R} \cap \mathcal{K} = \{\rho = (a,0),\ a \le 0\}. \tag{5}$$

Taking into account that $n_1 = \text{const} \ne 0$ on Γ_K, we can conclude (see [1]) that the form

$$\left(\int_\Omega c_{ijpm}\varepsilon_{ij}(v)\varepsilon_{pm}(v)\, d\Omega + \left(\int_{\Gamma_K} v_n\, d\Gamma \right)^2 \right)^{1/2} \tag{6}$$

is the norm in W, which is equivalent to norm in $[H^1(\Omega)]^2$. Assume that

$$\int_\Omega F_1\, d\Omega + \int_{\Gamma_P} T_1\, d\Gamma > 0. \tag{7}$$

An arbitrary $v \in W$ is decomposed into a normal and a tangent component to Γ_K; i.e., $v = (v_n, v_t)$. Let

$$\bar{v} = (\bar{v}_n, 0), \quad \text{where } \bar{v}_n = \frac{1}{\text{mes}\,\Gamma_K} \int_{\Gamma_K} v_n\, d\Gamma,$$

and let $\tilde{v} = v - \bar{v} = (v_n - \bar{v}_n, v_t)$. It is easy to see that

$$\int_{\Gamma_K} \tilde{v}_n\, d\Gamma = \int_{\Gamma_K} (v_n - \bar{v}_n)\, d\Gamma = 0.$$

In this case, norm (6) for $\tilde{v}$ becomes

$$\left(\int_\Omega c_{ijpm}\varepsilon_{ij}(\tilde{v})\varepsilon_{pm}(\tilde{v})\, d\Omega \right)^{1/2} = (a(\tilde{v}, \tilde{v}))^{1/2}. \tag{8}$$

We introduce the linear functional

$$L_1(v) = \int\limits_{\Omega} f_i\, v_i\, d\Omega + \int\limits_{\Gamma_P} T_i\, v_i\, d\Gamma$$

on the set W. Then, for arbitrary $v \in W$, we have

$$J(v) = \tfrac{1}{2}a(\tilde{v},\tilde{v}) - L_1(\tilde{v}) - L_1(\bar{v}) + \int\limits_{\Gamma_K} g_k|v_t|d\Gamma =$$

$$= \tfrac{1}{2}a(\tilde{v},\tilde{v}) - L_1(\tilde{v}) - \bar{v}_n\left(\int\limits_{\Omega} F_1\, d\Omega + \int\limits_{\Gamma_P} T_1\, d\Gamma\right) + \int\limits_{\Gamma_K} g_k|v_t|d\Gamma. \tag{9}$$

Let $v \in \mathcal{K}$. Since $\bar{v}_n \leq 0$ condition (7) implies that $J(v) \to +\infty$ as $\|v\|_W \to \infty$ and $v \in \mathcal{K}$. Thus, we have proved that auxiliary problem (4) is solvable for any $k = 1, 2, \ldots$.

The existence and uniqueness of solutions in the case of minimizing quadratic functionals were studied in detail in [8]. The solvability of quasi-variational inequality (2) was analysed in [1].

Every iteration step in the method of successive approximations involves a solution to the problem (4) with given friction. Apparently, the convergence of the method still remains an open question (see [1]). Nevertheless, the method has been tested in contact elasticity problems with friction. Specifically, in [1], the minimization of auxiliary nondifferentiable functional in (4) was replaced with the problem of finding a saddle point of a differentiable classical Lagrangian functional on the set $\mathcal{K} \times \Lambda$ ($\Lambda = \{\mu \in L_2(\Gamma_K)\colon |\mu| \leq 1 \text{ a.e. on } \Gamma_K\}$), in spite of the fact that semicoercive problem was considered.

Below, the method used to solve the problem (4) is based on a duality scheme that allows us simultaneously to drop constraints of the form "$v_n \leq 0$ on Γ_K" and to smooth the minimizing functionals.

Let us rewrite the problem (4) in equivalent form

$$\begin{cases} \bar{J}(v,w) = a(v,v) - \int\limits_{\Omega} f_i v_i\, d\Omega - \int\limits_{\Gamma_P} T_i v_i\, d\Gamma + \int\limits_{\Gamma_K} g|v\cdot t - w|d\Gamma \to \min, \\ v \in \mathcal{K}, \\ w \in L_2(\Gamma_K),\ w = 0 \text{ on } \Gamma_K, \end{cases} \tag{10}$$

where $t = (-n_2, n_1)$ on Γ_K.

Remark 1. In further consideration we will investigate the problem (10) instead of problem (4). We can construct an iterative method with smooth auxiliary functional for solving the problem (4).

3   Sensitivity Functional and Modified Lagrangian Functional

In this section, we introduce and study the sensitivity functional which plays an important role in further investigations.

We introduce the modified Lagrangian functional [4,5]

$$M(v, w; l_1, l_2) = \bar{J}(v, w) + \frac{1}{2r} \int\limits_{\Gamma_K} \left(\left((l_1 + rv_n)^+ \right)^2 - l_1^2 \right) d\Gamma + \int\limits_{\Gamma_K} \left(l_2 w + \frac{r}{2} w^2 \right) d\Gamma$$

for $(v, w; l_1, l_2) \in (W \times L_2(\Gamma_K)) \times [L_2(\Gamma_K)]^2$, where $(l_1 + rv_n)^+ \equiv \max\{0, l_1 + rv_n\}$.

Let us define the dual functional

$$\underline{M}(l_1, l_2) = \inf_{(v,w) \in W \times L_2(\Gamma_K)} M(v, w; l_1, l_2) \quad \forall (l_1, l_2) \in [L_2(\Gamma_K)]^2.$$

Dual functional has the other representation [4,5]

$$\underline{M}(l_1, l_2) = \inf_{(\mu_1, \mu_2) \in [L_2(\Gamma_K)]^2} \{ \chi(\mu_1, \mu_2) + \\ + \int\limits_{\Gamma_K} (l_1 \mu_1 + l_2 \mu_2) d\Gamma + \frac{r}{2} \int\limits_{\Gamma_K} (\mu_1^2 + \mu_2^2) d\Gamma \}, \tag{11}$$

where

$$\chi(\mu_1, \mu_2) = \inf_{\substack{v_n \le \mu_1 \text{ on } \Gamma_K \\ w = \mu_2 \text{ on } \Gamma_K}} \bar{J}(v, w)$$

is called the sensitivity functional [9].

Denote $K_{\mu_1,\mu_2} = \{(v, w) \in W \times L_2(\Gamma_K) : v_n \le \mu_1, w = \mu_2 \text{ on } \Gamma_K\}$ $\forall (\mu_1, \mu_2) \in [L_2(\Gamma_K)]^2$. Set K_{μ_1,μ_2} may be empty for some $\mu_1 \in L_2(\Gamma)$. For example, if $\mu_1 \in [C(\Gamma_K) \cap L_2(\Gamma_K)] \backslash H^{1/2}(\Gamma_K)$ and μ_1 is unbounded below, then $K_{\mu_1,\mu_2} = \emptyset$ because of $H^1(\Omega) \subset H^{1/2}(\Gamma)$ [10,11]. For all points (μ_1, μ_2) such that $K_{\mu_1,\mu_2} = \emptyset$ we put $\chi(\mu_1, \mu_2) = +\infty$. It is easy to show that under condition (7), $\chi(\mu_1, \mu_2)$ is a proper convex functional in $[L_2(\Gamma_K)]^2$, but its effective domain

$$\text{dom}\,\chi = \{(\mu_1, \mu_2) \in [L_2(\Gamma_K)]^2 : \chi(\mu_1, \mu_2) < +\infty\}$$

doesn't coincide with $[L_2(\Gamma_K)]^2$. Obviously, $\overline{\text{dom}\,\chi} = [L_2(\Gamma_K)]^2$.

Theorem 1. *Let $(\bar{\mu}_1, \bar{\mu}_2) \notin \text{dom}\,\chi$. Then for any sequence $\{(\mu_1^i, \mu_2^i)\}$ in $\text{dom}\,\chi$, such as $\lim\limits_{i \to \infty} \|(\mu_1^i, \mu_2^i) - (\mu_1, \mu_2)\|_{[L_2(\Gamma_K)]^2} = 0$ the formula $\lim\limits_{i \to \infty} \chi(\mu_1^i, \mu_2^i) = +\infty$ is correct.*

Proof. Denote

$$(v^i, w^i) = \arg \min_{(v,w) \in K_{\mu_1^i, \mu_2^i}} \bar{J}(v, w).$$

We show, that $\lim\limits_{i \to \infty} \|(v^i, w^i)\|_{W \times L_2(\Gamma_K)} = +\infty$. Suppose on the contrary, there is a subsequence $\{i_j\}$ and constant $C > 0$, such that $\|(v^{i_j}, w^{i_j})\|_{W \times L_2(\Gamma_K)} \le C$ for any i_j. Since $[H^1(\Omega)]^2 \subset [H^{1/2}(\Gamma)]^2$, we have

$$\|(v^{i_j}, w^{i_j})\|_{[H^{1/2}(\Gamma)]^2 \times L_2(\Gamma_K)} \le C_1,$$

where the constant $C_1 > 0$ does not depend on i_j. Moreover, $\{v^{i_j}\}$ is compact sequence in $[L_2(\Gamma_K)]^2$. Let $(\tilde{v}, \tilde{w}) \in [H^{1/2}(\Gamma)]^2 \times L_2(\Gamma_K)$ be its weak limit point. Without loss of generality we can suppose that (v^{i_j}, w^{i_j}) converges to $(\tilde{v}, \tilde{w})$ weakly. Then $\{v^{i_j}\}$ converges to $\tilde{v}$ strictly in $[L_2(\Gamma_K)]^2$. Since $v_n^{i_j} \le \mu_1^{i_j}$ on Γ_K, we have $\tilde{v}_n \le \mu_1$ on Γ_K, that is $K_{\mu_1,\mu_2} \ne \emptyset$. We obtain contradiction. Therefore $\lim_{i\to\infty} \|(v^i, w^i)\|_{W \times L_2(\Gamma_K)} = +\infty$. It means $\lim_{i\to\infty} \|v^i\|_W = +\infty$. As above (see (9)), we introduce

$$\bar{v}_n^i = \frac{1}{\text{mes } \Gamma_K} \int_{\Gamma_K} v_n^i d\Gamma.$$

Since $v_n^i \le \mu_1^i$ on Γ_K, then

$$\bar{v}_n^i \le \frac{1}{\text{mes } \Gamma_K} \int_{\Gamma_K} \mu_1^i d\Gamma.$$

We have

$$\left| \int_{\Gamma_K} \mu_1^i d\Gamma \right| \le (\text{mes } \Gamma_K)^{1/2} \|\mu_1^i\|_{L_2(\Gamma_K)},$$

$$-(\text{mes } \Gamma_K)^{1/2} \|\mu_1^i\|_{L_2(\Gamma_K)} \le \int_{\Gamma_K} \mu_1^i d\Gamma \le (\text{mes } \Gamma_K)^{1/2} \|\mu_1^i\|_{L_2(\Gamma_K)}.$$

Hence

$$\bar{v}_n^i \le \frac{1}{(\text{mes } \Gamma_K)^{1/2}} \|\mu_1^i\|_{L_2(\Gamma_K)}, \quad i = 1, 2, \ldots.$$

Since $\lim_{i\to\infty} \|\mu_1^i - \mu_1\|_{L_2(\Gamma_K)} = 0$, it follows

$$\|\mu_1^i\|_{L_2(\Gamma_K)} \le \|\mu_1\|_{L_2(\Gamma_K)} + \varepsilon$$

for any $\varepsilon > 0$ and for sufficiently large i. It means

$$\bar{v}_n^i \le \frac{1}{(\text{mes } \Gamma_K)^{1/2}} \left(\|\mu_1\|_{L_2(\Gamma_K)} + \varepsilon \right) \le C_2, \quad C_2 > 0 - \text{const} \tag{12}$$

for sufficiently large i.

As above, we use the representation $v^i = \bar{v}^i + \tilde{v}^i$, where $\bar{v}^i = (\bar{v}_n^i, 0)$, $\tilde{v}^i = (v_n^i - \bar{v}_n^i, v_t^i)$. From $\lim_{i\to\infty} \|v^i\|_W = +\infty$, (9) and (12), it follows $\lim_{i\to\infty} \bar{J}(v^i, w^i) = +\infty$ or

$$\lim_{i\to\infty} \chi(\mu_1^i, \mu_2^i) = +\infty. \tag{13}$$

Theorem 2. *Let $\{(\mu_1^i, \mu_2^i)\} \in dom\chi$ is a sequence such as $\lim_{i\to\infty} \|(\mu_1^i, \mu_2^i) - (\bar{\mu}_1, \bar{\mu}_2)\|_{[L_2(\Gamma_K)]^2} = 0$. Then $\lim_{i\to\infty} \chi(\mu_1^i, \mu_2^i) \ge \chi(\bar{\mu}_1, \bar{\mu}_2)$.*

Proof. We have $\{(\mu_1^i, \mu_2^i)\} \in \text{dom}\,\chi$ and $\lim\limits_{i \to \infty} \|(\mu_1^i, \mu_2^i) - (\bar{\mu}_1, \bar{\mu}_2)\|_{[L_2(\Gamma_K)]^2} = 0$. Let $\{(\mu_1^{i_j}, \mu_2^{i_j})\}$ be a subsequence for with

$$\lim_{j \to \infty} \chi(\mu_1^{i_j}, \mu_2^{i_j}) = \varliminf_{i \to \infty} \chi(\mu_1^i, \mu_2^i)$$

is correct.

We consider the sequence $\{(v^{\mu_1^{i_j}}, w^{\mu_2^{i_j}})\}$, where

$$(v^{\mu_1^{i_j}}, w^{\mu_2^{i_j}}) = \arg \min_{(v,w) \in K_{\mu_1^{i_j}, \mu_2^{i_j}}} \bar{J}(v, w).$$

From formulas (9), (12) it follows that $\{(v^{\mu_1^{i_j}}, w^{\mu_2^{i_j}})\}$ is a bounded sequence in $[H^1(\Omega)]^2 \times L_2(\Gamma_K)$ (otherwise $\lim\limits_{j \to \infty} \chi(\mu_1^{i_j}, \mu_2^{i_j}) = +\infty$ and theorem has been proved). Since $[H^1(\Omega)]^2 \subset [H^{1/2}(\Gamma)]^2$, then $\{(v^{\mu_1^{i_j}}, w^{\mu_2^{i_j}})\}$ is a bounded sequence in $[H^{1/2}(\Gamma)]^2 \times L_2(\Gamma_K)$. Let $(\bar{v}, \bar{w})$ be its weak limit point. Without lose of generality we assume that $\{(v^{\mu_1^{i_j}}, w^{\mu_2^{i_j}})\}$ is a weakly convergent sequence, that is $(\bar{v}, \bar{w})$ is a weak limit of $\{(v^{\mu_1^{i_j}}, w^{\mu_2^{i_j}})\}$ in $[H^{1/2}(\Gamma)]^2 \times L_2(\Gamma_K)$. Since $H^{1/2}(\Gamma)$ is involved to $L_2(\Gamma)$ compactly and $L_2(\Gamma) \subset H^{1/2}(\Gamma)$, then $\{(v^{\mu_1^{i_j}}, w^{\mu_2^{i_j}})\}$ converges to $(\bar{v}, \bar{w})$ in $[L_2(\Gamma)]^2 \times L_2(\Gamma_K)$. We have $(\mu_1^{i_j}, \mu_2^{i_j}) \to (\bar{\mu}_1, \bar{\mu}_2)$ in $[L_2(\Gamma_K)]^2$, $(v_n^{\mu_1^{i_j}}, w^{\mu_2^{i_j}}) \to (\bar{v}_n, \bar{w})$ in $[L_2(\Gamma)]^2 \times L_2(\Gamma_K)$ and $v_n^{\mu_1^{i_j}} \le \mu_1^{i_j}$, $w^{\mu_2^{i_j}} = \mu_2^{i_j}$ on Γ_K (n is a unit outward vector on Γ_K). Then $\bar{v}_n \le \bar{\mu}_1$, $\bar{w} = \bar{\mu}_2$ on Γ_K. Let

$$(\hat{v}, \hat{w}) = \arg \min_{\substack{v_n = \bar{v}_n \\ w = \bar{\mu}_2}} {}_{\text{on } \Gamma_K} \bar{J}(v, w).$$

We have

$$\bar{J}(v^{\mu_1^{i_j}}, w^{\mu_2^{i_j}}) - \bar{J}(\hat{v}, \hat{w}) = a(\hat{v}, v^{\mu_1^{i_j}} - \hat{v}) - \int_\Omega f_k(v_k^{\mu_1^{i_j}} - \hat{v}_k)d\Omega - \int_{\Gamma_P} T_k(v_k^{\mu_1^{i_j}} - \hat{v}_k)d\Gamma +$$
$$+ \int_{\Gamma_K} g(|v^{\mu_1^{i_j}} \cdot t - w^{\mu_2^{i_j}}| - |\hat{v} \cdot t - \hat{w}|)d\Gamma + \frac{1}{2}a(v^{\mu_1^{i_j}} - \hat{v}, v^{\mu_1^{i_j}} - \hat{v}) =$$

$$= <\mu, v^{\mu_1^{i_j}} - \hat{v}> - \int_{\Gamma_P} T_k(v_k^{\mu_1^{i_j}} - \hat{v}_k)d\Gamma +$$
$$+ \int_{\Gamma_K} g(|v^{\mu_1^{i_j}} \cdot t - w^{\mu_2^{i_j}}| - |\hat{v} \cdot t - \hat{w}|)d\Gamma + \frac{1}{2}a(v^{\mu_1^{i_j}} - \hat{v}, v^{\mu_1^{i_j}} - \hat{v}),$$

where

$$<\mu, v> = a(\hat{v}, v) - \int_\Omega f_k v_k \, d\Omega$$

and, moreover, $\mu \in [H^{-1/2}(\Gamma)]^2$ [11,12].

Since $\{v^{\mu_1^{ij}}\}$ weakly converges to $\bar{v}$ in $[H^{1/2}(\Gamma)]^2$, then $\lim\limits_{j\to\infty} <\mu, v^{\mu_1^{ij}} - \hat{v}> = 0$.

Now, because of $\{(v^{\mu_1^{ij}}, w^{\mu_2^{ij}})\}$ converges to $(\bar{v}, \bar{w})$ in $[L_2(\Gamma)]^2 \times L_2(\Gamma)$, we have

$$\lim\limits_{j\to\infty} \bar{J}(v^{\mu_1^{ij}}, w^{\mu_2^{ij}}) \geq \bar{J}(\hat{v}, \hat{w})$$

or

$$\lim\limits_{j\to\infty} \chi(\mu_1^{ij}, \mu_2^{ij}) \geq \chi(\bar{v}_n, \bar{\mu}_2) \geq \chi(\bar{\mu}_1, \bar{\mu}_2).$$

From Theorems 1 and 2 it follows that functional $\chi(\mu_1, \mu_2)$ is lower semicontinuous in $[L_2(\Gamma_K)]^2$. Taking into account that $\chi(\mu_1, \mu_2)$ is convex functional we can conclude that $\chi(\mu_1, \mu_2)$ is weakly lower semicontinuous functional in $[L_2(\Gamma_K)]^2$.

Let us take a sequence $\{(\mu_1^k, \mu_2^k)\} \subset \mathrm{dom}\,\chi$ convergent to $(\bar{\mu}_1, \bar{\mu}_2) \in \mathrm{dom}\,\chi$ in $[L_2(\Gamma_K)]^2$. We select such subsequence $\{(\mu_1^{k_i}, \mu_2^{k_i})\}$, for which

$$\lim\limits_{i\to\infty} \chi(\mu_1^{k_i}, \mu_2^{k_i}) = \underline{\lim\limits_{i\to\infty}} \chi(\mu_1^k, \mu_2^k).$$

From $\{(\mu_1^{k_i}, \mu_2^{k_i})\}$ we will allocate a subsequence convergent to $(\bar{\mu}_1, \bar{\mu}_2)$ almost everywhere. Without loss of generality we assume that sequence $\{(\mu_1^{k_i}, \mu_2^{k_i})\}$ convergent to $(\bar{\mu}_1, \bar{\mu}_2)$ almost everywhere.

Let we suppose that

$$\lim\limits_{i\to\infty} \chi(\mu_1^{k_i}, \mu_2^{k_i}) = \underline{\lim\limits_{i\to\infty}} \chi(\mu_1^k, \mu_2^k) \leq \chi(\bar{\mu}_1, \bar{\mu}_2) = \inf\limits_{(v,w)\in K_{\bar{\mu}_1, \bar{\mu}_2}} \bar{J}(v, w).$$

We consider the sequence $\{(v_n^{\mu_1^{k_i}}, w^{\mu_2^{k_i}})\}$ (n is a unit outward normal vector on Γ_K). It follows from formulas (9) and (12) that $\{(v_n^{\mu_1^{k_i}}, w^{\mu_2^{k_i}})\}$ is a bounded sequence in $H^1(\Omega) \times L_2(\Gamma_K)$. Otherwise $\lim\limits_{i\to\infty} \chi(\mu_1^{k_i}, \mu_2^{k_i}) = +\infty$. Therefore $\{(v_n^{\mu_1^{k_i}}, w^{\mu_2^{k_i}})\}$ is a bounded sequence in $H^{1/2}(\Gamma) \times L_2(\Gamma_K)$. Since the measure of point of Γ_K, where $\{\mu_1^{k_i}\}$ doesn't converge to $(\bar{\mu}_1, \bar{\mu}_2)$ uniformly, tends to zero with increasing i and functions $v_n^{\mu_1^{k_i}}$ are uniformly bounded in $H^{1/2}(\Gamma)$ with respect to i, then taking into account Theorem 2 we can set that $\lim\limits_{i\to\infty} \chi(\mu_1^{k_i}, \mu_2^{k_i}) = \chi(\bar{\mu}_1, \bar{\mu}_2)$. Now from Theorem 1 it follows that $\chi(\mu_1, \mu_2)$ is lower semicontinuous functional in $L_2(\Gamma_K)$. Since $\chi(\mu_1, \mu_2)$ is convex functional then $\chi(\mu_1, \mu_2)$ is weakly lower semicontinuous in $L_2(\Gamma_K)$.

For arbitrary point $(l_1, l_2) \in [L_2(\Gamma_K)]^2$, let us denote

$$F(\mu_1, \mu_2) = \chi(\mu_1, \mu_2) + \int\limits_{\Gamma_K} \left(l_1\mu_1 + \frac{r}{2}(\mu_1)^2\right) d\Gamma + \int\limits_{\Gamma_K} \left(l_2\mu_2 + \frac{r}{2}(\mu_2)^2\right) d\Gamma,$$

$r > 0$—const. From (11), it follows that dual functional $\underline{M}(l_1, l_2)$ has the form

$$\underline{M}(l_1, l_2) = \inf\limits_{(\mu_1, \mu_2)\in [L_2(\Gamma_K)]^2} F(\mu_1, \mu_2).$$

Functional $F(\mu_1, \mu_2)$ is weakly lower semicontinuous on $[L_2(\Gamma_K)]^2$.

Let us consider arbitrary $v \in W$ such as $v_n \leq \mu_1$ on Γ_K. Then

$$\bar{v}_n = \frac{1}{\operatorname{mes} \Gamma_K} \int\limits_{\Gamma_K} v_n d\Gamma \leq \frac{1}{\operatorname{mes} \Gamma_K} \int\limits_{\Gamma_K} \mu_1 d\Gamma \leq \frac{1}{(\operatorname{mes} \Gamma_K)^{1/2}} \|\mu_1\|_{L_2(\Gamma_K)}.$$

From here and (9), it follows that $F(\mu_1, \mu_2)$ is coercive functional in $[L_2(\Gamma_K)]^2$ under arbitrary (l_1, l_2), moreover, $F(\mu_1, \mu_2)$ is strongly convex functional on dom χ, that is

$$F\left((1-\lambda)(\mu_1', \mu_2') + \lambda(\mu_1'', \mu_2'')\right) \leq$$
$$\leq (1-\lambda) F(\mu_1', \mu_2') + \lambda F(\mu_1'', \mu_2'') - \frac{r}{2}\lambda(1-\lambda)\|(\mu_1', \mu_2') - (\mu_1'', \mu_2'')\|^2_{[L_2(\Gamma_K)]^2}$$
$$\forall(\mu_1', \mu_2'), (\mu_1'', \mu_2'') \in \operatorname{dom} \chi, 0 \leq \lambda \leq 1.$$

Therefore we can formulate the following theorem

Theorem 3. *For arbitrary $l = (l_1, l_2) \in [L_2(\Gamma_K)]^2$ there is a unique element*

$$\mu(l) = (\mu_1(l), \mu_2(l)) = \arg \min_{(\mu_1, \mu_2) \in [L_2(\Gamma_K)]^2} F(\mu_1, \mu_2).$$

Theorem 4. *The dual functional $\underline{M}(l_1, l_2)$ is Gateaux differentiable in $[L_2(\Gamma_K)]^2$ and its derivative satisfies the Lipschitz condition with the constant $1/r$; i.e., for any $l^1 = (l_1^1, l_2^1)$, $l^2 = (l_1^2, l_2^2) \in [L_2(\Gamma_K)]^2$*

$$\left\|\nabla \underline{M}(l_1^1, l_2^1)) - \nabla \underline{M}(l_1^2, l_2^2)\right\|_{[L_2(\Gamma_K)]^2} \leq \frac{1}{r} \left\|(l_1^1, l_2^1) - (l_1^2, l_2^2)\right\|_{[L_2(\Gamma_K)]^2}$$

and $\nabla \underline{M}(l_1, l_2) = (\mu_1(l), \mu_2(l))$.

The proof of Theorem 3 is analogous to the proof of Theorem 6 in [16].

Let us consider the problem

$$\begin{cases} \underline{M}(l_1, l_2) - \max, \\ (l_1, l_2) \in [L_2(\Gamma_K)]^2. \end{cases} \tag{14}$$

The problem (14) is called a dual problem for (4). Let $(v^*, w^*; l_1^*, l_2^*)$ be a saddle point of classical Lagrangian functional $L(v, w; l_1, l_2)$. We supposed above that solution u of problem (4) belongs to space $[H^2(\Omega)]^2$ and, moreover, mes$\{x \in \Gamma_K: \sigma_n(u) < 0\} > 0$. Then a saddle point of $L(v, w; l_1, l_2)$ (and modified functional $M(v, w; l_1, l_2)$ too) has the form $(u, 0; -\sigma_n(u), l_2^*)$, and the vector-function $(-\sigma_n(u), l_2^*)$ is a solution of dual problem (14) [4,5].

4   Uzawa Gradient Method

For solving the dual problem (14) we consider the Uzawa iterative method

$$(l_1^{i+1}, l_2^{i+1}) = (l_1^i, l_2^i) + r(\mu_1(l^i), \mu_2(l^i)), \quad i = 0, 1, \dots, \ (l_1^0, l_2^0) \in [L_2(\Gamma_K)]^2,$$

where

$$(\mu_1(l^i), \mu_2(l^i)) = \arg \min_{(\mu_1,\mu_2)\in[L_2(\Gamma_K)]^2} \{\chi(\mu_1,\mu_2)+ \tag{15}$$
$$+ \int_{\Gamma_K} (l_1^i\mu_1 + l_2^i\mu_2)d\Gamma + \frac{r}{2} \int_{\Gamma_K} (\mu_1^2 + \mu_2^2)d\Gamma \},$$

$r > 0$ is a constant.

Let us analyse the mapping $P(l_1, l_2) = (l_1, l_2) + r(\mu_1(l), \mu_2(l))$ for $l = (l_1, l_2) \in [L_2(\Gamma_K)]^2$.

Theorem 5. *Let solution u of problem (4) belongs to space $[H^2(\Omega)]^2$ and $mes\{x \in \Gamma_K: \sigma_n(u) < 0\} > 0$. Then the mapping $P(l_1, l_2)$ satisfies the conditions*
(a) $P(-\sigma_n(u), l_2^*) = (-\sigma_n(u), l_2^*)$;
(b) $\|P(-\sigma_n(u), l_2^*) - P(l_1, l_2)\|_{[L_2(\Gamma_K)]^2} < \|(-\sigma_n(u), l_2^*) - (l_1, l_2)\|_{[L_2(\Gamma_K)]^2}$ for any $(l_1, l_2) \in [L_2(\Gamma_K)]^2$, $l_1 \neq -\sigma_n(u)$.

The proof of Theorem 5 is analogous to the proof of Theorem 4 in [4].

Theorem 6. *The limiting equality*

$$\lim_{i\to\infty} \|(\mu_1(l^i), \mu_2(l^i))\|_{[L_2(\Gamma_K)]^2} = 0$$

takes place for algorithm (15).

Proof. From Theorem 4, it follows that for any $l = (l_1, l_2)$, $h = (h_1, h_2)$ from the space $[L_2(\Gamma_K)]^2$ the following presentation

$$\underline{M}(l + h) - \underline{M}(l) = \int_0^1 (\nabla\underline{M}(l + \tau h), h)_{[L_2(\Gamma_K)]^2} d\tau = \int_0^1 (\mu(l + \tau h), h)_{[L_2(\Gamma_K)]^2} d\tau,$$

where $\mu(l) = (\mu_1(l), \mu_2(l))$ is correct [14].

From this, by analogy with [15, p. 31], the proof of the theorem follows.

Gradient method (15) can be rewritten in the following way [17]:

Step 0. $i := 0$, $(l_1^0, l_2^0) \in [L_2(\Gamma_K)]^2$.

Step 1. Solve the problem: $(v^{i+1}, w^{i+1}) = \arg \min_{(v,w)\in W\times L_2(\Gamma_K)} \{\bar{J}(v, w)+$

$$+\frac{1}{2r} \int_{\Gamma_K} \left(((l_1^i + r\,v_n)^+)^2 - (l_1^i)^2\right) d\Gamma + \int_{\Gamma_K} \left(l_2^i w + \frac{r}{2}w^2\right) d\Gamma \}.$$

Step 2. Set $(l_1^{i+1}, l_2^{i+1}) = (l_1^i, l_2^k) + r\left(\max\left(v_n^{i+1}, -\frac{l_1^i}{r}\right), w^{i+1}\right)$. $\tag{16}$

Step 3. $i := i + 1$, back to **Step 1**.

The functional $\chi(\mu_1, \mu_2)$ is a weakly lower semicontinuous functional, then it can be proved that

$$\lim_{i\to\infty} \bar{J}(v^i, w^i) = \lim_{i\to\infty} \chi(\mu_1(l^i), \mu_2(l^i)) = \chi(0,0) = \bar{J}(u,0) = J(u).$$

Thus the sequence $\{(v^i, w^i)\}$ converges to $(u,0)$ with respect to functional. Now it is easy to show that the sequence $\{(v^i, w^i)\}$ is a bounded in $W \times L_2(\Gamma_K)$.

The following statement takes place.

Theorem 7. *Let the conditions of Theorem 5 be satisfied, friction force g_k of problem (4) belongs to set $L_\infty(\Gamma_K) \cap H^{1/2}(\Gamma_K)$ and points v^i, $i = 1, 2, \ldots$, developed on algorithm (16) satisfies the following conditions:*

(1) $v^i \in [H^2(\Omega)]^2$;
(2) $\|v^i\|_{[H^2(\Omega)]^2} \leq C$;
(3) $\|l_2^i\|_{H^{1/2}(\Gamma_K)} \leq C$, $C > 0$ — const.

Then sequence $\{(v^i, w^i; l_1^i, l_2^i)\}$ converges in $(W \times L_2(\Gamma_K)) \times [L_2(\Gamma_K)]^2$ to a saddle point of modified Lagrangian functional for any starting point $l^0 = (l_1^0, l_2^0) \in [H^{1/2}(\Gamma_K)]^2$ and any fixed parameter $r > 0$.

The proof of Theorem 7 is analogous to the proof of Theorem 4 in [19].

Remark 2. We solve the problem (10) instead of problem (4). It allows us to obtain a problem of minimization of differentiable functional on Step 1 of the method (16). In fact,

$$\min_{v,w} M(v, w; l_1, l_2) = \min_{v,w} \left\{ \frac{1}{2}a(v,v) - \int_\Omega f_i v_i \, d\Omega - \int_{\Gamma_P} T_i v_i \, d\Gamma + \right.$$

$$+ \int_{\Gamma_K} \left(g|v \cdot t - w| + l_2 w + \frac{r}{2}w^2 \right) d\Gamma + \frac{1}{2r} \int_{\Gamma_K} \left(((l_1 + rv_n)^+)^2 - l_1^2 \right) d\Gamma \right\} =$$

$$= \min_v \left\{ \frac{1}{2}a(v,v) - \int_\Omega f_i v_i \, d\Omega - \int_{\Gamma_P} T_i v_i \, d\Gamma + \frac{1}{2r} \int_{\Gamma_K} \left(((l_1 + rv_n)^+)^2 - l_1^2 \right) d\Gamma + \right.$$

$$\left. + \inf_w \int_{\Gamma_K} \left(g|v \cdot t - w| + l_2 w + \frac{r}{2}w^2 \right) d\Gamma \right\} =$$

$$= \min_v \left\{ \frac{1}{2}a(v,v) - \int_\Omega f_i v_i \, d\Omega - \int_{\Gamma_P} T_i v_i \, d\Gamma + \frac{1}{2r} \int_{\Gamma_K} \left(((l_1 + rv_n)^+)^2 - l_1^2 \right) d\Gamma + \right.$$

$$\left. + \int_{\Gamma_K} \inf_w \left(g|v \cdot t - w| + l_2 w + \frac{r}{2}w^2 \right) d\Gamma \right\}.$$

It is known that number function $F(v \cdot t) = \inf_w \left(g|v \cdot t - w| + l_2 w + \frac{r}{2}w^2 \right)$ is continuously differentiable convex function of real argument $(v \cdot t)$ [5].

Therefore, Step 1 of method (16) is reduced to minimization problem of differentiable with respect v functional

$$\frac{1}{2}a(v,v) - \int\limits_{\Omega} f_i\,v_i\,d\Omega - \int\limits_{\Gamma_P} T_i\,v_i\,d\Gamma +$$

$$+\frac{1}{2r}\int\limits_{\Gamma_K}\left(\left((l_1 + rv_n)^+\right)^2 - l_1^2\right)d\Gamma + \int\limits_{\Gamma_K} F(v\cdot t)\,d\Gamma.$$

5 Conclusion

In this paper a new method of solving semicoercive contact elastic problem with friction is constructed and investigated. This method is based on the duality scheme with a modified Lagrangian functional. The modified Lagrange functional allows us simultaneously to drop constraints of the form "$v_n \leq 0$ on Γ_K" in (4) and to reduce the semicoercive nondifferentiable problem to the minimization problem of a differentiable functional.

The authors considered the properties of the modified Lagrangian functional and proved the convergence theorems. Dual functional is used in the presentation (11) with the help of sensitivity functional $\chi(\mu_1,\mu_2)$. The proofs of theorems are presented under the assumption, that functional $\chi(\mu_1,\mu_2)$ is weakly lower semicontinuous functional on dom χ.

References

1. Glaváček, I., Haslinger, J., Nečas, I., Lovišek, J.: Numerical Solution of Variational Inequalities. Springer, Berlin (1988)
2. Kikuchi, N., Oden, T.: Contact Problem in Elasticity: A Study of Variational Inequalities and Finite Element Methods. SIAM, Philadelphia (1988)
3. Kravchuk, A.S.: Variational and Quasi-Variational Inequalities in Mechanics. MGAPI, Moscow (1997). [in Russian]
4. Vikhtenko, E.M., Namm, R.V.: Duality scheme for solving the semicoercive Signorini problem with friction. J. Comp. Math. Math. Phys. **47**, 1938–1951 (2007)
5. Vikhtenko, E.M., Maksimova, N.N., Namm, R.V.: Modified Lagrange functionals to solve the variational and quasivariational inequalities of mechanics. Avtomatika i Telemekhanika. **4**, 3–18 (2012)
6. Namm, R.V., Woo, G., Xie, S., Yi, S.: Solution of semicoercive Signorini problem based on a duality scheme with modified Lagrangian functional. J. Korean Math. Soc. **49**, 843–854 (2012)
7. Antipin, A.S.: Gradient and Extragradient Approaches in Bilinear Equilibrium Programming. Vychisl. Tsentr Ross. Akad, Nauk (2002). [in Russian]
8. Fikera, G.: Existence theorems in elasticity. In: Truesdell, C. (ed.) Linear Theories of Elasticity and Thermoelasticity. Springer, Heidelberg (1976). https://doi.org/10.1007/978-3-662-39776-3_3
9. Antipin, A.S., Golikov, A.I., Khoroshilova, E.M.: Sensitivity function: properties and applications. J. Comp. Math. Math. Phys. **51**, 2126–2143 (2011)
10. Mclean, W.: Strongly Elliptic Systems and Boundary Integral Equations. University Press, Cambridge (2000)

11. Khludnev, A.M.: Elastisy Problems in Nonsmooth Domains. Fizmatlit, Moscow (2010). (in Russian)
12. Temam, R.: Mathematical Problems in Plasticity. Cauthier-Villars, Paris (1985)
13. Ekland, I., Temam, R.: Convex Analysis and Variational Problem. North-Holland Publishing Company, Amsterdam (1976)
14. Kantorovich, L.V., Akilov, G.P.: Functional Analysis. Pergamon Press, Oxford (1982)
15. Polyak, B.T.: Introduction to Optimization. Optimization Software Inc., New-York (1987)
16. Namm, R.V., Vikhtenko, E.M., Woo, G.: Sensitivity functionals in contact problems of elasticity theory. J. Comp. Math. Math. Phys. **54**, 1190–1200 (2014)
17. Namm, R.V., Vikhtenko, E.M.: Modified Lagrangian functional for solving the Signorini problem with friction. In: Advances in Mechanics Research, vol. 1, pp. 435–446. Nova Science Publishers, New-York (2011)
18. Kushniruk, N.N., Namm, R.V.: On a finite element solution of model problem in mechanics with friction based on smoothing lagrange multiplier method. J. Comp. Math. Math. Phys. **52**, 20–30 (2012)
19. Vikhtenko, E.M.: On the method of searching a saddle point of modified Lagrangian functional for elasticity problem with friction. Far Eastern Math. J. **12**(1), 3–11 (2012). (in Russian)

Polynomial Capacity Guarantees PTAS for the Euclidean Capacitated Vehicle Routing Problem Even for Non-uniform Non-splittable Demand

Michael Khachay[1,2,3(✉)] ⓘD and Yuri Ogorodnikov[1,2] ⓘD

[1] Krasovsky Institute of Mathematics and Mechanics, Ekaterinburg, Russia
{mkhachay,yogorodnikov}@imm.uran.ru
[2] Ural Federal University, Ekaterinburg, Russia
[3] Omsk State Technical University, Omsk, Russia

Abstract. The Capacitated Vehicle Routing Problem (CVRP) is the well-known combinatorial optimization problem that has numerous valuable practical applications. It is known, that CVRP is strongly NP-hard even on the Euclidean plane and APX-hard in its metric setting even for any fixed capacity $q \geq 3$. For the Euclidean setting, there are known several approximation schemes. But, to the best of our knowledge, polynomial bounds for their time complexity were proved either for a fixed capacity q or under the restriction $q \ll n$. Moreover, most of these schemes were developed for the simplest case, where each customer has a unit demand, and cannot be extended to the general case of a non-uniform demand (both splittable or not) directly.

In this paper, we are managed to significantly relax the restriction on capacity admitting the existence of PTAS for this problem and propose the first approximation scheme for the CVRP on the Euclidean plane with non-uniform non-splittable demand parameterized by an upper bound for the size of an optimum solution. Time complexity of the proposed scheme is polynomial for any fixed parameter values if $q = poly(n)$.

Keywords: Capacitated Vehicle Routing Problem · Non-uniform splittable demand · Polynomial Time Approximation Scheme

1 Introduction

The Capacitated Vehicle Routing Problem (CVRP) is the classic combinatorial optimization problem [26] making a great impact to the computation complexity theory and having numerous important applications in operations research.

In the simplest case, an instance of the CVRP can be specified by a set of points $\{x_1 \ldots, x_n\} \cup \{y\}$, where $X = \{x_1, \ldots, x_n\}$ consists of *customer locations* (or just *customers*) and y is a dedicated point referred to as *a depot*. Each customer has the unit demand that should be serviced by an unbounded fleet of the identical vehicles having the same integer *capacity* q.

ⓒ Springer Nature Switzerland AG 2020
M. Jaćimović et al. (Eds.): OPTIMA 2019, CCIS 1145, pp. 415–426, 2020.
https://doi.org/10.1007/978-3-030-38603-0_30

The goal is to service all the customer demand by a family of cyclic routes of a minimum total length, such that each of routes has the depot y as its origin and destination point and obeys the capacity constraint.

Introduced by Dantzig and Ramser in their seminal paper [11] as an applied problem that deals with the servicing of a gas station network by a fleet of gasoline trucks, the CVRP became a solid mathematical model for numerous important applications in operations research [2,10] attracting the interest of both practitioners and specialists on algorithm design and analysis.

To date, the significant progress is achieved in solving the CVRP by optimal methods based on reduction to appropriate integer and mixed integer programs (see, e.g. [7,8,26]), in developing fast heuristics and meta-heuristics, among them are local-search [15], genetic and memetic algorithms [6,23,27], ant and bee colonies [24,25], and in design of polynomial time algorithms with theoretically proven accuracy bounds and approximation schemes (see, e.g. [5,19]) dating back to the classic papers by Haimovich and Rinnooy Kan [14], and Arora [3].

Unfortunately, the exact methods are applicable only to rather small instances, since the CVRP is strongly NP-hard both in its general formulation and even in very specific settings. Heuristic techniques, despite their ability to solve some instances stemming from the practice very efficiently, have no theoretical accuracy guarantees and need an additional tuning for any certain problem setting.

For the class of approximation algorithms with performance guarantees, there are a number of promising results, which we overview in Sect. 2. Nevertheless, the conjecture that the Euclidean CVRP admits[1] a Polynomial Time Approximation Scheme (PTAS) in the case of unbounded capacity still remains open. Moreover, almost all the results mentioned were obtained for the simplest case of unit customer demand, which is far from the real-life settings of the problem.

In this paper, we try to bridge these two gaps and propose the first[2] approximation scheme for the CVRP on the Euclidean plane with non-uniform integer non-splittable demand, whose time complexity is polynomial provided the capacity $q = poly(n)$.

The rest of the paper is structured as follows. In Sect. 2 we give a short overview of known approximation results for the CVRP. Then, in Sect. 3, we introduce the mathematical notion of the CVRP with non-uniform Non-Splittable Demand (CVRP-NSD). Further, in Sect. 4, we describe the main idea of our scheme, while Sect. 5 provides a proof sketch of its correctness and time complexity bounds. Finally, in Sect. 6, we summarize the results proposed and discuss some directions of future work.

2   Related Work

Historically, the first approximation scheme for the basic case of the CVRP on the Euclidean plane was introduced by Haimovich and Rinnooy Kan [14] in

---

[1] Like the Euclidean Traveling Salesman Problem.

[2] To the best of our knowledge.

1985. The main idea of their celebrated scheme was as follows. For any given $\varepsilon > 0$, they engaged an accuracy-driven partition of the initial CVRP instance into two subinstances, specified by the *outer* and *inner* customers, respectively. Then, the former subinstance was solved to optimality, whilst the latter one was approximated by the famous Iterated Tour Partition (ITP) heuristic that reduces any CVRP instance to some auxiliary instance of the Euclidean Traveling Salesman Problem (TSP). M. Haimovich and A. Rinnooy Kan showed that their scheme is PTAS in the case of $q = o(\log \log n)$. Despite that this result matched perfectly to instances coming from the practice, where as a rule capacity has large but constant value, further research in the field of approximation of the Euclidean CVRP was directed to weakening this restriction. Thus, in [4], a PTAS was proposed for the case $q = O(\log n / \log \log n)$. Later, these results were extended to the case of Euclidean spaces of an arbitrary fixed dimension [16,22], the case of multiple depots [9,21], and the setting with an additional time windows constraint [17,18,20].

Another approach to obtain an efficient approximation results for the Euclidean CVRP dates back to the famous Arora's PTAS [3] for the Euclidean TSP. Evidently, this scheme can be applied directly to approximation of the CVRP with unit demand in the case when $q \geq n$. Extending this scheme, Das and Mathieu [12,13] introduced Quasi Polynomial Time Approximation Scheme (QPTAS) for the general setting of the CVRP on the Euclidean plane (without any additional constraints on q). For any $\varepsilon > 0$, their scheme provides a $(1 + \varepsilon)$-approximate solution of the problem in time $n^{\log^{O(1/\varepsilon)} n}$.

Combination of these two brilliant approaches admits Adamaszek, Czumaj, and Lingas [1] to propose PTAS for the CVRP on the Euclidean plane within the significantly extended bound on admissible capacity q. Inspired by the former approach, they carried out decomposition of the initial instance into two families of the auxiliary subinstances. Then, for approximation of any subinstance of the first kind they applied the Das-Mathieu's QPTAS, while all remaining substances they approximated by the ITP. As a result, they obtained the PTAS for the CVRP on the Euclidean plane for $q \leq 2^{\log^{\delta} n}$ for some $\delta = \delta(\varepsilon) \ll 1$. Recently [19], their scheme was extended to the CVRP with non-unit splittable demand and time windows with the same restriction on capacity growth.

To the best of our knowledge, to date, there were no approximation results for the CVRP on the plane with non-splittable demand, at least in the class of algorithms with theoretical guarantees. In this paper, we propose the first parameterized approximation scheme for these problem, whose running time is polynomial for any fixed parameter value and $q = poly(n)$.

3   Problem Statement

An instance of CVRP-NSD is given by a complete node- and edge-weighted digraph $G = (X \cup \{y\}, E, d, w)$ and an integer capacity bound q. Here, d is a node-weighting function that assigns to any customer $x \in X$ an appropriate natural-valued demand $d(x)$. For any customer, his (or her) demand $d(x)$ is assumed to be *non-splittable*, i.e. it should be serviced by a single *route*. A route is called *feasible*, if it is a simple cycle $R_j = y, x_{i_1}, \ldots, x_{i_s}, y$ in the graph G and fulfills the capacity constraint, i.e.

$$d(x_{i_1}) + \ldots + d(x_{i_s}) \le q.$$

Further, w is a non-negative edge-weighting function that defines the direct transportation cost for any ordered pair of nodes $(v_1, v_2) \in X \cup \{y\}$. For any feasible route R, we assign its transportation cost $w(R)$ by the following formula

$$w(R) = w(y, x_{i_1}) + w(x_{i_1}, x_{i_2}) + \ldots + w(x_{i_s}, y).$$

Thus, the goal is to find a finite set U of feasible routes of the minimum total transportation cost that satisfies the entire customers demand.

If the function w satisfies the well-known triangle inequality, then the transportation cost for an arbitrary nodes $\{v_1, v_2\} \in X \cup \{y\}$ is called a *distance* (between them) and the given instance of the CVRP is called *metric*. Furthermore, if the set $X \cup \{y\}$ is a point set in some finite-dimensional Euclidean space and $w(v_1, v_2) = \|v_1 - v_2\|_2$, the instance is called *Euclidean* as well.

For the sake of simplicity, in this paper, we only consider so-called *nice* instances [28], satisfying the following conditions

(i) coordinates of all locations (customers and the depot) are integers from $[0, O(n)]$
(ii) for any two distinct locations, the distance between them is at least 4.

Also, we parameterize the CVRP-NSD on Euclidean plane a natural value T assuming that there is an optimum solution with at most T routes.

4   Main Idea

Our scheme extends the famous PTAS proposed by S. Arora for the Euclidean TSP on the plane [3] that consists of the following stages:

(i) accuracy-driven cost-preserving polynomial time reduction of the initial instance to an auxiliary instance specified by a set of integer points from $[0, O(n)]$ such that the distance between any two distinct points is at least four;
(ii) randomized dissection of the box enclosing the auxiliary instance, construction a randomly shifted quadtree rooted at this box, location m equidistant *portals* at each internal side of any sub-box obtained at any level of the dissection, and introduction so-called *portal-respecting* r-*light* tours restricted to cross any side of each sub-box only at portals and at most r times (for some predefined numbers m and r) (see Fig. 1). By the Arora's Structure Theorem, with probability at least $1/2$, randomly shifted quadtree produces in such a tour, which is a $(1 + \varepsilon)$-approximate solution of the initial instance, provided that $m = O(1/\varepsilon \cdot \log n)$ and $r = O(1/\varepsilon)$;

(iii) construction of a minimum cost portal-respecting r-light tour for a randomly shifted quadtree by the dynamic programming;

(iv) derandomization of the algorithm by multi-starting at any integer-valued shift of the initial quadtree.

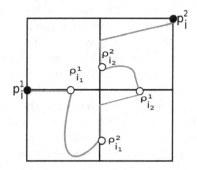

Fig. 1. A portal-respecting r-light tour passing through portals

Unfortunately, this famous scheme ignores the capacity constraint and cannot be applied to efficient approximation of the CVRP directly even in the case of unit demand. Indeed, the scheme is intended to construct a single route of close-to-optimal length that visits all the given locations. As was shown in well-known Structure Theorem (see. e.g. [3]), such a suboptimal route can be found in the class of portal-respecting tours intersecting the boundary of each dissection box at most r times, for $r = O(1/\varepsilon)$. In the case of CVRP, to fulfill the capacity constraint, we should construct a family of routes such a way that boundaries of some boxes[3] can be intersected much more times than it was assumed in the Arora's scheme. Therefore, we cannot employ this scheme for the approximation of CVRP-NSD as a 'black-box'.

Nevertheless, our scheme inherits almost all the aforementioned stages except maybe stages (i) and (iii). Stage (i) is excluded due to our simplifying assumption to consider nice instances of the problem. In the stage (iii), to construct portal-respecting r-light approximate solutions of a nice instance of the CVRP-NSD on the plane (we call them (m, r)-*approximations*), we propose our own dynamic programming algorithm. With these modifications, we obtain our main result.

Theorem 1. *There exists an algorithm that finds an $(1 + O(\varepsilon))$-approximate solution for a nice CVRP-NSD instance for the case of Euclidean plane in time $\left(n^3 \log n \cdot q \cdot T\right)^{O(T/\varepsilon)}$, where T is the given upper bound for the number of routes in an optimum solution of the instance considered.*

Remark 1. For any fixed value of the parameter T, this algorithm is PTAS, if $q = poly(n)$.

---

[3] At least, the boxes containing the depot.

5 Dynamic Programming

First of all, we recall that, during the dissection procedure, current box b is partitioned into four child sub-boxes, if b contains at least two distinct locations. Therefore, any leaf of the constructed quadtree either contains a single location (a customer or the depot) or empty. Such boxes together with not-a-leaf boxes containing no depot, we call *trivial*. On the other hand, each not-a-leaf box of the quadtree that contains the depot (including the root) are called *non-trivial*.

To any box b (trivial or non-trivial), we associate an entry collection of the dynamic programming table (whose is initially empty), each of them is indexed by the corresponding *task* specified by a sequence of quadruples $((p_i^1, p_i^2, q_i, dep_i)$, $i = 1, \ldots, T \cdot J)$, for some number $J = O(r)$. Here, p_i^1 and p_i^2 are portals, where some route segment should enter and leave the box b, respectively, q_i is a demand that should be serviced by this segment, and dep_i is a Boolean flag indicating whether this segment or should not visit the depot inside b. The value J corresponds to the maximum number of segments that can be constructed for an arbitrary route.

For any given task t, the goal is either to show that t is *infeasible* or, for any quadruple, to construct portal-respecting r-light segments connecting the portals p_i^1 and p_i^2, such that

  (i) all the demand inside b is serviced
 (ii) each segment connecting p_i^1 and p_i^2 visits the depot iff $dep_i = 1$ and services q_i units of customer demand exactly
(iii) total length of all segments and additional routes is minimal.

Further, we provide a simple necessarily condition for the feasibility of the task t with respect to some dissection box b with total customer demand D_b.

Lemma 1. *The task t is infeasible if any of he following statements holds*

(i) b is a trivial box and

$$\sum_{i=1}^{I} q_i \neq D_b \ \vee \ (dep_i = 1 \ for \ some \ i)$$

(ii) b is a non-trivial box and $\sum_{i=1}^{I} q_i \geq D_b$.

Notice, that for the non-trivial box case we should construct additional supplementary (m, r)-approximate routes if $\sum_{i=1}^{I} q_i > D_b$. Also, Lemma 1 covers the case of an empty task t_0, which is feasible for any non-trivial box. If b is trivial, t_0 is feasible iff $D_b = 0$.

To solve any task t associated with some quadtree box b

  (i) we check out the necessarily feasibility condition of t by Lemma 1, if t is infeasible, we assign to the appropriate entry the constant FAIL and stop, otherwise
 (ii) if b is a leaf

a. we connect any pair of portals p_i^1 and p_i^2 directly, if b is empty
b. if b contains a customer x, we construct two-legs segment $p_i^1 \to x \to p_i^2$ for any quadruple $(p_i^1, p_i^2, q_i, 0)$, where $q_i = D_b = d(x)$, and connect p_i^1 and p_i^2 directly, otherwise
c. if b contains the depot y, construct the segment $p_i^1 \to y \to p_i^2$ or $p_i^1 \to p_i^2$, if dep_i equals to 1 or 0, respectively,

then, in any case, we return the segments constructed (see Fig. 2)

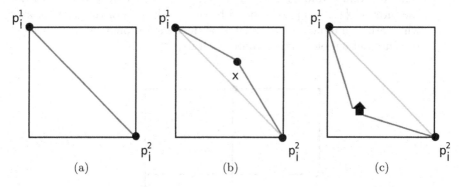

p_i^1 p_i^1 p_i^1

x

p_i^2 p_i^2 p_i^2

(a)   (b)   (c)

Fig. 2. Example: (a) is an empty leaf box; (b) is a leaf box with a customer x; (c) is a leaf with the depot

(iii) if b is a not-a-leaf trivial box, we construct a solution from the best outputs of its child sub-boxes on the sequence of subtasks, each of them is generated as follows

(a) build a sequence

$$((\rho_j^1, \rho_j^2), j = 1, \ldots, I, \text{ where } I = T' \cdot J, \ T' \leq T$$

(here T' is a number of routes needed for covering task's demand q_i, while I is a total number of their segments)

(b) to any pair (ρ_j^1, ρ_j^2), assign a capacity $c_j \in [0, q]$
(c) to any triple $(\rho_j^1, \rho_j^2, c_j)$, assign $dep_j = 0$, since the trivial box b contains no depots
(d) to any tuple $(\rho_j^1, \rho_j^2, c_j, dep_j)$, assign a number $i_j \in \{1, \ldots, I\}$ that associates this quadruple with the quadruple $(p_i^1, p_i^2, q_i, dep_i)$ of the task t to been executed
(e) then, to any tuple $(\rho_j^1, \rho_j^2, c_j, dep_j, i_j)$ assign the child box C_j, which will obtain the corresponding tuple as a part of its own subtask
(f) finally, if $\sum_{i_j=i} c_j = q_i$ for any $i \in \{1, \ldots, I\}$, and $|\{j : i_j = t\}| \leq J$ for any $t \in \{1 \ldots T'\}$, we obtain assumedly correct subtasks for the child boxes specified for any child box C (of current box b) by the subsequence

$$((\rho_j^1, \rho_j^2, c_j, dep_j): j = 1, \ldots, I, \ C_j = C) ; \qquad (1)$$

gather outputs of all child boxes, if all of them are positive, glue the segments associated with quadruples $(p_i^1, p_i^2, q_i, dep_i)$ and update the record

(g) we output the record along with the corresponding glued $p_i^1 \rightarrow p_i^2$ segments, if the record has be updated at least once; otherwise, we return FAIL

(iv) if b is a non-trivial box (including the root), the subtasks generation procedure is almost the same except that
   - at step (c), dep_j can take any value from $\{0,1\}$
   - if the equality $\sum_{i=1}^{I} q_i < D_b$ is true, then the number i_j at step (d) can take value from $\{1,\ldots,I,I+1,\ldots,T\cdot J\}$, where the values from the interval $\{1,\ldots,I\}$ correspond to the quadruples of the initial task t while other numbers are associated with the additional subtasks needed to construct supplementary routes (see Fig. 3)

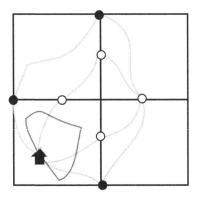

Fig. 3. An example of two routes. The red one is a supplementary route for the bottom left box

   - at step (f), the correctness condition should be replaced as follows

$$\sum_{i_j=i} \begin{bmatrix} c_j \\ dep_j \end{bmatrix} = \begin{cases} \begin{bmatrix} q_i \\ dep_i \end{bmatrix}, \, i \leq I \\ \begin{bmatrix} Q_i \\ 1 \end{bmatrix}, \, i > I, \end{cases} \quad 0 < Q_i < q, \quad \sum_{i=1}^{I} q_i + \sum_{i=I+1}^{T\cdot J} Q_i \leq D_b,$$

$$|\{j : i_j = t\}| \leq J \text{ for any } t \in \{1\ldots T\}$$

As in step (iii), any assumedly correct subtask is assigned to the child boxes. Then, if they output positive answers, we glue the route subsegments obtained, calculate their total length and upgrade the record.

5.1   Proof Sketch

We bound the time complexity in the following lemma.

Lemma 2. *Time complexity of the proposed dynamic programming procedure is*

$$(n \log n \cdot q \cdot T)^{O(T/\varepsilon)}.$$

Proof. Consider an arbitrary box b and determine how many tasks can be assigned to it. Total number of portals located on any side of b is $O(m)$, hence there are $O(m^2)$ ways to construct a portal pair (p_i^1, p_i^2). For every such a pair (p_i^1, p_i^2) we assign an integer capacity $q_i \in [0, q]$ and a depot visiting indicator $dep_i \in \{0, 1\}$ that increases the total number of ways to construct a possible quadruple to $O(m^2 \cdot q)$. As, for any task t, total number of quadruples is bounded by $O(T \cdot J)$, we obtain that the total tasks number for b is $O\left(m^2 \cdot q\right)^{O(T \cdot J)}$.

Consider, how long an arbitrary task can be executed (skipping the trivial case of leaf boxes, for the case of brevity). For an arbitrary task we assign a sequence of subtasks each of them is constructed as follows. At step (a) of the dynamic programming scheme, we should choose a portal pair (ρ_j^1, ρ_j^2) by one of the $O(m^2)$ ways. The capacity value c_j considered at step (b) can be chosen from the interval $[0, q]$ while the depot indicator dep_j at step (c) is set to 0 for any trivial box and can take values $\{0, 1\}$ for each non-trivial one. We set the quadruple number for that the constructing subtask is assigned at step (d), by one of the I ways for trivial box and $T \cdot J$ for non-trivial one. Finally, we need to choose the child box of b at step (e) for that the subtasks is intended, by one of the four ways. So, there are $O\left(m^2 \cdot q \cdot T\right)$ possible subtasks[4] associated with an appropriate quadruple of the given task. In other words, the total number of subtasks for a given task is $O\left(m^2 \cdot q \cdot T\right)^{O(T \cdot J)}$.

For an arbitrary task, checking out its correction (see step (f)) takes time $O(I)$ and $O(T \cdot J)$ for trivial and non-trivial box respectively. Finally, gluing of the subtasks positive outputs is also can be done in time $O(T \cdot J)$. Therefore, the time complexity of processing an arbitrary box b is

$$\left(m^2 \cdot q\right)^{O(T \cdot J)} \cdot \left(m^2 \cdot q \cdot T\right)^{O(T \cdot J)} = \left(m^2 \cdot q \cdot T\right)^{O(T \cdot J)}.$$

Since total number of boxes in any quadtree dissection is $O(n \log n)$ and $m = O(1/\varepsilon \log n)$, for any fixed $\varepsilon > 0$, we obtain the following time complexity upper bound

$$\left(n \log n \cdot m^2 \cdot q \cdot T)\right)^{O(T \cdot J)} = (n \log n \cdot q \cdot T)^{O(T/\varepsilon)}.$$

Lemma is proved.

As we consider all possible ways to construct r-light route segments, we obtain an optimal portal-respecting solution at the root level for the quadtree considered. Our final results relies to the following straight-forward modification of the famous Arora's Structure Theorem (see, e.g. [28])

---

[4] for any fixed $J = O(r) = O(1/\varepsilon)$.

Theorem 2. *Let R be a cycle (not necessarily simple) visiting all the nodes of a complete edge-weighted graph $G = (V, E, w)$, whose nodeset belongs to the plane end enclosed to the box of size L and the weighting function is specified by the Euclidean metric. If integers a and b are picked from $(-L/2, 0]$ uniformly at random, then with probability at least $1/2$, the (a, b)-dissection has an r-light portal-respecting tour of cost at most $(1 + \varepsilon) \cdot w(R)$ for portal parameter $m = O\left(\frac{1}{\varepsilon} \log L\right)$ and $r = \left(\frac{1}{\varepsilon}\right)$.*

Theorem 1, our main result, follows from Lemma 2, Theorem 2, and the simple derandomization procedure, which tries each integer shift of the quadtree in the enclosing box of size $O(n)$.

6   Conclusion

In this paper, we propose an approximation scheme for the Euclidean Capacitated Vechicle Routing Problem on the plane with non-unit non-splittable demand parameterized by the upper bound T of routes in its optimum solution. The scheme has a polynomial running time for any fixed value of T and the capacity $q = poly(n)$.

In the forthcoming paper, we will try to extend this result to the case of an unbounded number of routes and an arbitrary dimension of the Euclidean ground space.

Acknowledgements. This research was funded by the Russian Foundation for Basic Research, grants no. 17-08-01385 and 19-07-01243.

References

1. Adamaszek, A., Czumaj, A., Lingas, A.: PTAS for k-tour cover problem on the plane rof moderately large values of k. Int. J. Found. Comput. Sci. **21**(06), 893–904 (2010)
2. Anbuudayasankar, S.P., Ganesh, K., Mohapatra, S.: Models for Practical Routing Problems in Logistics. Springer, Cham (2014). https://doi.org/10.1007/978-3-319-05035-5
3. Arora, S.: Polynomial time approximation schemes for Euclidean traveling salesman and other geometric problems. J. ACM **45**(5), 753–782 (1998)
4. Asano, T., Katoh, N., Tamaki, H., Tokuyama, T.: Covering points in the plane by k-tours: towards a polynomial time approximation scheme for general k. In: Proceedings of the Twenty-ninth Annual ACM Symposium on Theory of Computing, pp. 275–283. STOC 1997. ACM, New York (1997)
5. Becker, A., Klein, P.N., Saulpic, D.: Polynomial-time approximation schemes for k-center, k-median, and capacitated vehicle routing in bounded highway dimension. In: Azar, Y., Bast, H., Herman, G. (eds.) 26th Annual European Symposium on Algorithms, Helsinki, Finland, ESA 2018, 20–22 August 2018. LIPIcs, vol. 112, pp. 8:1–8:15. Schloss Dagstuhl - Leibniz-Zentrum fuer Informatik (2018). http://www.dagstuhl.de/dagpub/978-3-95977-081-1

6. Blocho, M., Czech, Z.: A parallel memetic algorithm for the vehicle routing problem with time windows. In: 2013 Eighth International Conference on P2P, Parallel, Grid, Cloud and Internet Computing, pp. 144–151 (2013)
7. Borčinova, Z.: Two models of the capacitated vehicle routing problem. Croatian Oper. Res. Rev. **8**, 463–469 (2017)
8. Bula, G.A., Gonzalez, F.A., Prodhon, C., Afsar, H.M., Velasco, N.M.: Mixed integer linear programming model for vehicle routing problem for hazardous materials transportation. IFAC-PapersOnLine 49(12), 538–543 (2016). http://www. sciencedirect.com/science/article/pii/S2405896316309673
9. Cardon, S., Dommers, S., Eksin, C., Sitters, R., Stougie, A., Stougie, L.: A PTAS for the multiple depot vehicle routing problem. Tech. rep., Eindhoven University of Technology, March 2008. http://www.win.tue.nl/bs/spor/2008-03.pdf
10. Cattaruzza, D., Absi, N., Feillet, D., González-Feliu, J.: Vehicle routing problems for city logistics. EURO J. Transp. Logistics **6**(1), 51–79 (2017). https://doi.org/ 10.1007/s13676-014-0074-0
11. Dantzig, G., Ramser, J.: The truck dispatching problem. Manage. Sci. **6**, 80–91 (1959)
12. Das, A., Mathieu, C.: A quasi-polynomial time approximation scheme for Euclidean capacitated vehicle routing. In: Proceedings of the Twenty-First Annual ACM-SIAM Symposium on Discrete Algorithms, pp. 390–403. SODA 2010. Society for Industrial and Applied Mathematics, Philadelphia (2010)
13. Das, A., Mathieu, C.: A quasipolynomial time approximation scheme for Euclidean capacitated vehicle routing. Algorithmica **73**, 115–142 (2015)
14. Haimovich, M., Rinnooy Kan, A.H.G.: Bounds and heuristics for capacitated routing problems. Math. Oper. Res. **10**(4), 527–542 (1985)
15. Hashimoto, H., Yagiura, M.: A path relinking approach with an adaptive mechanism to control parameters for the vehicle routing problem with time windows. In: van Hemert, J., Cotta, C. (eds.) EvoCOP 2008. LNCS, vol. 4972, pp. 254–265. Springer, Heidelberg (2008). https://doi.org/10.1007/978-3-540-78604-7_22
16. Khachai, M.Y., Dubinin, R.D.: Approximability of the vehicle routing problem in finite-dimensional Euclidean spaces. Proc. Steklov Insts Math. **297**(1), 117–128 (2017)
17. Khachay, M., Ogorodnikov, Y.: Improved polynomial time approximation scheme for capacitated vehicle routing problem with time windows. In: Evtushenko, Y., Jaćimović, M., Khachay, M., Kochetov, Y., Malkova, V., Posypkin, M. (eds.) OPTIMA 2018. CCIS, vol. 974, pp. 155–169. Springer, Cham (2019). https:// doi.org/10.1007/978-3-030-10934-9_12
18. Khachay, M., Ogorodnikov, Y.: Efficient PTAS for the Euclidean CVRP with time windows. In: van der Aalst, W.M.P., Batagelj, V., Glavaš, G., Ignatov, D.I., Khachay, M., Kuznetsov, S.O., Koltsova, O., Lomazova, I.A., Loukachevitch, N., Napoli, A., Panchenko, A., Pardalos, P.M., Pelillo, M., Savchenko, A.V. (eds.) AIST 2018. LNCS, vol. 11179, pp. 318–328. Springer, Cham (2018). https://doi. org/10.1007/978-3-030-11027-7_30
19. Khachay, M., Ogorodnikov, Y.: Approximation scheme for the capacitated vehicle routing problem with time windows and non-uniform demand. In: Khachay, M., Kochetov, Y., Pardalos, P. (eds.) MOTOR 2019. LNCS, vol. 11548, pp. 309–327. Springer, Cham (2019). https://doi.org/10.1007/978-3-030-22629-9_22

20. Khachay, M., Ogorodnikov, Y.: Improved polynomial time approximation scheme for capacitated vehicle routing problem with time windows. In: Evtushenko, Y., Jaćimović, M., Khachay, M., Kochetov, Y., Malkova, V., Posypkin, M. (eds.) OPTIMA 2018. CCIS, vol. 974, pp. 155–169. Springer, Cham (2019). https:// doi.org/10.1007/978-3-030-10934-9_12

21. Khachay, M., Dubinin, R.: PTAS for the Euclidean capacitated vehicle routing problem in R^d. In: Kochetov, Y., Khachay, M., Beresnev, V., Nurminski, E., Pardalos, P. (eds.) DOOR 2016. LNCS, vol. 9869, pp. 193–205. Springer, Cham (2016). https://doi.org/10.1007/978-3-319-44914-2_16

22. Khachay, M., Zaytseva, H.: Polynomial time approximation scheme for single-depot Euclidean capacitated vehicle routing problem. In: Lu, Z., Kim, D., Wu, W., Li, W., Du, D.-Z. (eds.) COCOA 2015. LNCS, vol. 9486, pp. 178–190. Springer, Cham (2015). https://doi.org/10.1007/978-3-319-26626-8_14

23. Nalepa, J., Blocho, M.: Adaptive memetic algorithm for minimizing distance in the vehicle routing problem with time windows. Soft Comput. **20**(6), 2309–2327 (2016)

24. Necula, R., Breaban, M., Raschip, M.: Tackling dynamic vehicle routing problem with time windows by means of ant colony system. In: 2017 IEEE Congress on Evolutionary Computation (CEC), pp. 2480–2487 (2017)

25. Ng, K., Lee, C., Zhang, S., Wu, K., Ho, W.: A multiple colonies artificial bee colony algorithm for a capacitated vehicle routing problem and re-routing strategies under time-dependent traffic congestion, vol. 109, pp. 151–168 (2017). http://www.sciencedirect.com/science/article/pii/S0360835217301948

26. Toth, P., Vigo, D.: Vehicle Routing: Problems, Methods, and Applications. MOS-Siam Series on Optimization, 2nd edn. SIAM, Philadelpia (2014)

27. Vidal, T., Crainic, T.G., Gendreau, M., Prins, C.: A hybrid genetic algorithm with adaptive diversity management for a large class of vehicle routing problems with time-windows. Comput. Oper. Res. **40**(1), 475–489 (2013)

28. Williamson, D.P., Shmoys, D.B.: The Design of Approximation Algorithms, 1st edn. Cambridge University Press, New York (2011)

New Version of Mirror Prox for Variational Inequalities with Adaptation to Inexactness

Fedor S. Stonyakin[1,2], Evgeniya A. Vorontsova[3,4(✉)],
and Mohammad S. Alkousa[2]

[1] V. I. Vernadsky Crimean Federal University, Simferopol, Russia
fedyor@mail.ru
[2] Moscow Institute of Physics and Technology (National Research University),
Moscow, Russia
mohammad.alkousa@phystech.edu
[3] Grenoble Alpes University, Grenoble, France
vorontsovaea@gmail.com
[4] Far Eastern Federal University, Vladivostok, Russia

Abstract. Some adaptive analogue of the Mirror Prox method for variational inequalities is proposed. In this work we consider the adaptation not only to the value of the Lipschitz constant, but also to the magnitude of the oracle error. This approach, in particular, allows us to prove a complexity near $O\left(\frac{1}{\varepsilon}\right)$ for variational inequalities for a special class of monotone bounded operators. This estimate is optimal for variational inequalities with monotone Lipschitz-continuous operators. However, there exists some error, which may be insignificant. The results of experiments on the comparison of the proposed approach with some known analogues are presented. Also, we discuss the results of the experiments for matrix games in the case of using non-Euclidean proximal setup.

Keywords: Variational inequality · Mirror Prox · Inexactness ·
Adaptation · Bounded operator · Lipschitz-continuous operator ·
Matrix game

1 Introduction

Variational inequalities (VI) and saddle point problems often arise in a variety of important applications [1]. For solving such problems a lot of algorithmic schemes are known (see e.g. [1–4]). The Mirror Prox method proposed by Nemirovski [4] is currently one of the most popular of such methods. This method goes back to the well-known extragradient method proposed by Korpelevich in [5]. At the same time, unlike the standard extragradient method, Mirror Prox allows to

The research of F. Stonyakin in Sects. 2 and 3 and numerical experiments in Table 1 were supported by Russian Science Foundation (project 18-71-00048). The research of E. Vorontsova was supported by Russian Foundation for Basic Research (project number 18-29-03071 mk and project number 18-31-20005 mol-a-ved).

© Springer Nature Switzerland AG 2020
M. Jaćimović et al. (Eds.): OPTIMA 2019, CCIS 1145, pp. 427–442, 2020.
https://doi.org/10.1007/978-3-030-38603-0_31

effectively solve problems with non-Euclidean norms, as well as with the Hölder-continuous operators:

$$\|g(x) - g(y)\|_* \leqslant L_\nu \|x - y\|^\nu \ \forall x, y \in Q \text{ for some } \nu \in [0; 1], \qquad (1)$$

all notations are explained in Sect. 2 below.

Recently, a universal analogue of the Nemirovski method [4] was proposed in [6,7]. The universality is understood as an adaptive adjustment to the optimal smoothness level ν in (1), as well as the constant value $L_\nu > 0$. Note that universal gradient method for convex optimization problems was proposed by Nesterov [8] (see also Sect. 5 of the textbook [9]). And it is possible to observe the convergence rate of the proposed method, which is typical for the smooth case $\nu = 1$ (Lipschitz-continuous operators), for some problems with bounded operators ($L_0 < +\infty$ and $L_\nu = +\infty$ for all $\nu > 0$).

This paper is devoted to the modification of Mirror Prox method [6,7] for the following analogue of the Lipschitz condition for the operator g with constant $L > 0$

$$\langle g(y) - g(x), \, y - z \rangle \leqslant LV(y, x) + LV(y, z) + \delta \|y - z\| \ \ \forall x, y, z \in Q, \qquad (2)$$

where $\delta > 0$ is a fixed value and $V(y, x)$ is the Bregman divergence (see Sect. 2 below).

We propose an analogue of adaptive Mirror Prox method [6,7]. At the same time, we consider adaptive tuning both for the value of the parameters L and δ. One of the features of the proposed method which are important for applications is the possibility for the value of δ to reflect the inexactness of operator g. In addition, the value of δ can indicate the degree of discontinuity of the operator g. Adaptive tuning to its value can approximate the convergence rate for variational inequalities with bounded operators ($\nu = 0$) to the convergence rate for variational inequalities with Lipshitz-continuous operators ($\nu = 1$). Effects of this approach can be observed for the universal method but without a theoretical justification for the convergence rate $O\left(\frac{1}{\varepsilon}\right)$ for non-smooth operators [6]. This means that the proposed approach in this paper is an alternative to the universal method.

The contribution of the paper can be summarized as follows:

– An analogue of the Mirror Prox method for variational inequalities with a monotone Lipschitz-continuous operator, which allows for adaptive tuning to the value of the Lipschitz constant L, as well as the bounded value of the error δ of the specifying operator g, is proposed.
– The applicability of the proposed method to a certain class of variational inequalities with bounded operators ($\nu = 0$) is discussed. The rate of convergence $O\left(\frac{1}{\varepsilon}\right)$ of this method is proved with some finite error associated with the non-smoothness of the operator. Thus, some alternative to the universal method has been proposed, but with a clearer theoretical rationale for acceleration.

- The results of numerical experiments for finding the equilibrium in a bilinear matrix game (or VI with Lipschitz-continuous operator) with a bounded error in the definition of the operator are given. A comparison of the quality of the calculated solution is given depending on the number of iterations for the adaptive Mirror Prox method from [7] and the method proposed in this paper with an adaptive setting for the magnitude of the error.
- The results of numerical experiments for a variational inequality with a bounded ($\nu = 0$) operator (related to the Fermat-Torricelli-Steiner problem) are presented. These results show that the method due to the proposed adaptation of the non-smoothness error can converge much faster than the optimal lower estimate $O\left(\frac{1}{\varepsilon^2}\right)$ for the corresponding class of problems.

2   Problem Statement and Some Examples

Let $(E, \|\cdot\|)$ be a normed finite-dimensional vector space and E^* be the conjugate space of E with the norm:

$$\|y\|_* = \max_x \{\langle y, x \rangle, \|x\| \leq 1\},$$

where $\langle y, x \rangle$ is the value of the continuous linear functional y at $x \in E$.

Let $Q \subset E$ be a (simple) closed convex set, $d : Q \to \mathbb{R}$ be a distance generating function (d.g.f.), which is continuously differentiable and 1-strongly convex with respect to the norm $\| \cdot \|$ and assume that $\min_{x \in Q} d(x) = d(0)$.

For all $x, y \in Q \subset E$ we consider the corresponding Bregman divergence

$$V(x,y) = d(x) - d(y) - \langle \nabla d(y), x - y \rangle. \tag{3}$$

Let $g : Q \to E^*$ be a continuous operator. We consider the problem of finding a solution to a variational inequality of the form

$$\langle g(x_*), x_* - x \rangle \leqslant 0 \ \forall x \in Q. \tag{4}$$

Under the assumption of the monotony of the operator g, i.e.

$$\langle g(x) - g(y), x - y \rangle \geq 0 \ \forall x, y \in Q,$$

the inequality (4) is equivalent to the following weak variational inequality

$$\langle g(x), x_* - x \rangle \leqslant 0 \ \forall x \in Q. \tag{5}$$

Assume that the operator g satisfies the condition (2). In this section, we show some examples of problems for which a condition of the form (2) naturally arises. First of all, this is due to the inexactness of the oracle for the operator of a variational inequality. But also the value of $\delta\|y - z\|$ can describe the degree of discontinuity of the operator g (i.e. using considered approach, one can propose an approach to the solution of some VI's with bounded operators).

Example 1. Let $g : Q \to \mathbb{R}^n$ be a Lipschitz-continuous operator with constant $L > 0$, i.e.

$$\|g(x) - g(y)\|_* \leqslant L\|x - y\| \quad \forall x, y \in Q.$$

However, suppose that the exact value of the operator g is not available, and only an approximate value of $g(x)$, i.e. $\tilde{g}(x)$, is known:

$$\|\tilde{g}(x) - g(x)\|_* \leqslant \frac{\delta}{2} \quad \forall x \in Q.$$

Then for each x, y, $z \in Q$ we have:

$$|\langle \tilde{g}(y) - \tilde{g}(x), \, y - z \rangle - \langle g(y) - g(x), \, y - z \rangle| = |\langle \tilde{g}(y) - g(y), \, y - z \rangle + \langle g(x) - \tilde{g}(x), \, y - z \rangle| \leqslant$$

$$\leqslant \|\tilde{g}(y) - g(y)\|_* \cdot \|y - z\| + \|\tilde{g}(x) - g(x)\|_* \cdot \|y - z\| \leqslant \left(\frac{\delta}{2} + \frac{\delta}{2} \right) \|y - z\| = \delta \|y - z\|.$$

Therefore,

$$\langle \tilde{g}(y) - \tilde{g}(x), \, y - z \rangle \leqslant \langle g(y) - g(x), \, y - z \rangle + \delta\|y - z\| \leqslant \|g(y) - g(x)\|_* \cdot \|y - z\| + \delta\|y - z\| \leqslant$$

$$\leqslant L\|y - x\| \cdot \|y - z\| + \delta\|y - z\| \leqslant \frac{L}{2}\|y - x\|^2 + \frac{L}{2}\|y - z\|^2 + \delta\|y - z\| \leqslant$$

$$\leqslant LV(y, x) + LV(y, z) + \delta.$$

Example 2. Note that the term $\delta\|y - z\|$ in (2) can describe non-smoothness for the operator g along any fixed vector segment $\{ty + (1 - t)x\}_{0 \leqslant t \leqslant 1}$. In general (if you combine all possible vector segments), on the domain of non-smoothness points there can be an infinite number.

For example, assume that for some subset $Q_0 \subset Q$ the function f is differentiable at all points of $Q \backslash Q_0$ and that for an arbitrary $x \in Q_0$ there exists a finite subdifferential $\partial f(x)$ in the sense of convex analysis.

For fixed $x, y \in Q$ with $t \in [0; 1]$ we denote $y_t := (1 - t)x + ty$.

Definition 1. *([10]) Fix $\delta > 0$ and $L > 0$. We say that the convex function $f : Q \to \mathbb{R}$ ($Q \subset \mathbb{R}^n$) has (δ, L)-Lipschitz subgradient ($f \in C_{L,\delta}^{1,1}(Q)$), if:*

(i) for arbitrary $x, y \in Q$ the function f is differentiable at all points of the set $\{y_t\}_{0 \leqslant t \leqslant 1}$, with the exception of the sequence (possibly finite)

$$\{y_{t_k}\}_{k=1}^{\infty} : \, t_1 < t_2 < t_3 < \ldots \quad \text{and} \quad \lim_{k \to \infty} t_k = 1; \tag{6}$$

(ii) for a sequence of points from (6) there exist finite subdifferentials $\{\partial f(y_{t_k})\}_{k=1}^{\infty}$ and

$$\text{diam} \, \partial f(y_{t_k}) =: \delta_k > 0, \quad \text{where} \quad \sum_{k=1}^{+\infty} \delta_k =: \delta < +\infty, \tag{7}$$

and $\quad \text{diam} \, \partial f(x) = \max\{\|y - z\|_* \mid y, z \in \partial f(x)\};$

(iii) if for $x, y \in Q$ the function f is differentiable at each point y_t, $t \in (0; 1)$, then the following inequality holds:

$$\min_{\substack{\hat{\partial} f(x) \in \partial f(x), \\ \hat{\partial} f(y) \in \partial f(y)}} \|\hat{\partial} f(x) - \hat{\partial} f(y)\|_* \leqslant L\|x - y\|. \tag{8}$$

Indeed, the property of (δ, L)-Lipschitzness for each subgradient $g(x) = \hat{\partial} f(x)$ means that

$$\|g(y) - g(x)\|_* \leqslant L\|y - x\| + \delta. \tag{9}$$

To prove (9), it suffices to split the segment $\{y_t\}_{0 \leqslant t \leqslant 1}$ into the intervals of smoothness and take into account the boundedness of diameters of the subdifferentials at non-smoothness points of f.

The inequality (9) means that

$$\langle g(y) - g(x), \, y - z \rangle \leqslant \|g(y) - g(x)\|_* \cdot \|y - z\| \leqslant L\|y - x\| \cdot \|y - z\| + \delta\|y - z\| \leqslant$$

$$\leqslant \frac{L}{2}(\|y - x\|^2 + \|y - z\|^2) + \delta\|y - z\| \leqslant LV(y, x) + LV(z, y) + \delta\|y - z\|.$$

Let us give a concrete example of a non-smooth functional with a (δ, L)-Lipschitz subgradient with an arbitrarily large Lipschitz constant.

Example 3. We fix some $k > 0$, the value $\delta > 0$ and consider the piecewise linear function $f : [0; 1] \to \mathbb{R}$ (here $Q = [0; 1] \subset \mathbb{R}$) defined as follows

$$f(x) = \begin{cases} kx & ; 0 \leqslant x \leqslant \frac{1}{2}, \\ \left(k + \sum_{i=1}^{n} \frac{\delta}{2^i}\right) x - \sum_{i=1}^{n} \frac{\delta}{2^i}\left(1 - \frac{1}{2^i}\right) & ; 1 - \frac{1}{2^n} < x \leqslant 1 - \frac{1}{2^{n+1}}, \\ \lim_{x \to +1} f(x) & ; x = 1. \end{cases} \tag{10}$$

In this case, $Q_0 = \{1 - \frac{1}{2^n}\}_{n=1}^{\infty}$, $\partial f(q_n) = \left[k + \sum_{i=1}^{n-1} \frac{\delta}{2^i}; k + \sum_{i=1}^{n} \frac{\delta}{2^i}\right]$ with $n > 1$, $\partial f(q_1) = \left[k; k + \frac{\delta}{2}\right]$ (here $q_n = 1 - \frac{1}{2^n}$ with $n = 1, 2, 3, \ldots$). It is clear that $\partial f(q_n) = \frac{\delta}{2^n}$, which is true for the entered value $\delta > 0$. Moreover, on the intervals $(0; q_1)$, $(q_n; q_{n+1})$ the function f has a Lipschitz-continuous gradient with the constant $L = 0$. Therefore, for the function f from (10), we find that $f \in C_{0,\delta}^{1,1}(Q)$.

Any functional with a finite set of non-smooth points along an arbitrary segment will satisfy the proposed Lipschitz condition for the subgradient. Obviously, this condition holds for each objective function with finite points of non-smoothness on each vector segment $[x; y]$. Thus, it is possible to apply this technique to problems of minimization for sum distances to several balls in Hilbert spaces [11]. Such an objective function, obviously, will not be differentiable in the usual sense at the points of the boundaries of the balls of which there are infinitely many. Note that among points of each vector segment $[x; y]$ such an objective function have finite points of non-smoothness. However, the considered Lipschitz condition for a special choice of subgradient holds for some functions with infinitely many points of non-smoothness (e.g. for maximum of linear functions).

3 Adaptive Method for Variational Inequalities with Adaptation to Inexactness

In this section, we introduce a new version of the Mirror Prox method for variational inequalities (see [7]), which we call *Mirror Prox with Adaptation to Inexactness (MPAI)*. In this version, which is listed as Algorithm 1 below, we consider the adaptation not only to the level of operator smoothness, but also to the magnitude of the oracle error, which may allow to receive complexity near $O\left(\frac{1}{\varepsilon}\right)$ for VI with bounded operators, i.e. the optimal complexity for VI with Lipshitz-continuous operators.

We evaluate the solution quality of the problem (4), produced by Algorithm 1, by using the Bregman divergence (3).

Algorithm 1. Mirror Prox with Adaptation to Inexactness (MPAI).

Input: $x^0 = \arg\min_{x \in Q} d(x), L^0, \delta^0$.

1: $N := N + 1$; $L^{N+1} := \frac{L^N}{2}$; $\delta^{N+1} := \frac{\delta^N}{2}$.
2: Calculate
$$y^{N+1} := \arg\min_{x \in Q}\{\langle g(x^N), x - x^N\rangle + L^{N+1}V(x, x^N)\}, \qquad (11)$$

$$x^{N+1} := \arg\min_{x \in Q}\{\langle g(y^{N+1}), x - x^N\rangle + L^{N+1}V(x, x^N)\}. \qquad (12)$$

3: **If**
$$\langle g(y^{N+1}) - g(x^N), y^{N+1} - x^{N+1}\rangle \le L^{N+1}V(y^{N+1}, x^N) + \qquad (13)$$
$$+ L^{N+1}V(x^{N+1}, y^{N+1}) + \delta^{N+1}\left\|y^{N+1} - x^{N+1}\right\|,$$

then go to the next iteration (item 1).
4: **Else** increase L^{N+1} and δ^{N+1} by two times and go to item 2.

Theorem 1. *After N iterations of Algorithm 1, the following estimate holds:*

$$\sum_{k=0}^{N-1} \frac{1}{L^{k+1}}\langle g(y^{k+1}), y^{k+1} - x\rangle \le V(x, x^0) - V(x, x^N) + \sum_{k=0}^{N-1} \frac{\delta^{k+1}}{L^{k+1}}\left\|y^{k+1} - x^{k+1}\right\|.$$

Proof. One can directly check the following inequalities:

$$\left\langle\nabla_x V(x, x^k)\big|_{x=x^{k+1}}, x - x^{k+1}\right\rangle = V(x, x^k) - V(x, x^{k+1}) - V(x^{k+1}, x^k), \quad (14)$$

$$\left\langle\nabla_x V(x, x^k)\big|_{x=y^{k+1}}, x - y^{k+1}\right\rangle = V(x, x^k) - V(x, y^{k+1}) - V(y^{k+1}, x^k). \quad (15)$$

Further, for each $x \in Q$ and $k = \overline{0, N-1}$:

$$\left\langle\nabla_x\left(\langle g(x^k), x - x^k\rangle + L^{k+1}V(x, x^k)\right)\big|_{x=y^{k+1}}, x - y^{k+1}\right\rangle \ge 0,$$

$$\left\langle \nabla_x \left(\langle g(y^{k+1}), x - x^k \rangle + L^{k+1} V(x, x^k) \right) \big|_{x=x^{k+1}}, \ x - x^{k+1} \right\rangle \geqslant 0.$$

Thus,

$$\langle g(y^{k+1}), x^{k+1} - x \rangle \leqslant L^{k+1} V(x, x^k) - L^{k+1} V(x, x^{k+1}) - L^{k+1} V(x^{k+1}, x^k)$$

and

$$\langle g(x^k), y^{k+1} - x \rangle \leqslant L^{k+1} V(x, x^k) - L^{k+1} V(x, y^{k+1}) - L^{k+1} V(y^{k+1}, x^k).$$

Taking into account (13), we have for each $k = \overline{0, \ N-1}$:

$$\langle g(y^{k+1}), y^{k+1} - x \rangle = \langle g(y^{k+1}), x^{k+1} - x \rangle + \langle g(x^k), y^{k+1} - x^{k+1} \rangle$$

$$+ \langle g(y^{k+1}) - g(x^k), y^{k+1} - x^{k+1} \rangle$$

$$\leqslant L^{k+1} V(x, x^k) - L^{k+1} V(x, x^{k+1}) - L^{k+1} V(x^{k+1}, x^k) + L^{k+1} V(x^{k+1}, x^k)$$

$$- L^{k+1} V(x^{k+1}, y^{k+1}) - L^{k+1} V(y^{k+1}, x^k) + L^{k+1} V(y^{k+1}, x^k)$$

$$+ L^{k+1} V(x^{k+1}, y^{k+1}) + \delta^{k+1} \left\| y^{k+1} - x^{k+1} \right\|,$$

i.e.

$$\frac{1}{L^{k+1}} \langle g(y^{k+1}), y^{k+1} - x \rangle \leqslant V(x, x^k) - V(x, x^{k+1}) + \frac{\delta^{k+1}}{L^{k+1}} \left\| y^{k+1} - x^{k+1} \right\|. \tag{16}$$

After summing (16) by $k = \overline{0, \ N-1}$, we have

$$\sum_{k=0}^{N-1} \frac{1}{L^{k+1}} \langle g(y^{k+1}), y^{k+1} - x \rangle \leqslant V(x, x^0) - V(x, x^N) + \sum_{k=0}^{N-1} \frac{\delta^{k+1}}{L^{k+1}} \left\| y^{k+1} - x^{k+1} \right\|.$$

Let us denote

$$S_N = \sum_{k=0}^{N-1} \frac{1}{L^{k+1}}, \ \widetilde{y} = \frac{1}{S_N} \sum_{k=0}^{N-1} \frac{y^{k+1}}{L^{k+1}} \ \text{and} \ R^2 = \max_{x \in Q} V(x, x^0).$$

Theorem 2. *For monotone operator g after N iterations of Algorithm 1, the following estimate holds:*

$$\max_{x \in Q} \langle g(x), \widetilde{y} - x \rangle \leqslant \frac{R^2}{S_N} + \frac{1}{S_N} \sum_{k=0}^{N-1} \frac{\delta^{k+1}}{L^{k+1}} \left\| y^{k+1} - x^{k+1} \right\|. \tag{17}$$

Assume that for fixed ε

$$\sum_{k=0}^{N-1} \frac{1}{L^{k+1}} \geqslant \frac{R^2}{\varepsilon}. \tag{18}$$

Then the following inequality holds:

$$\max_{x \in Q} \langle g(x), \widetilde{y} - x \rangle \leqslant \varepsilon + \frac{1}{S_N} \sum_{k=0}^{N-1} \frac{\delta^{k+1}}{L^{k+1}} \left\| y^{k+1} - x^{k+1} \right\|. \tag{19}$$

If $L^0 \leqslant 2L$, then inequality (18) holds at no more than

$$N = \left\lceil \frac{2LR^2}{\varepsilon} \right\rceil$$

iterations of Algorithm 1.

Proof. By monotony of g we have for each $k = 0, 1, ...$:

$$\langle g(x), y^{k+1} - x \rangle = \langle g(y^{k+1}), y^{k+1} - x \rangle + \langle g(x) - g(y^{k+1}), y^{k+1} - x \rangle \leqslant \langle g(y^{k+1}), y^{k+1} - x \rangle,$$

so the inequality

$$\frac{1}{S_N} \max_{x \in Q} \sum_{k=0}^{N-1} \frac{1}{L^{k+1}} \langle g(y^{k+1}), y^{k+1} - x \rangle$$

$$\leqslant \frac{R^2}{S_N} + \frac{1}{S_N} \sum_{k=0}^{N-1} \frac{\delta^{k+1}}{L^{k+1}} \|y^{k+1} - x^{k+1}\| \leqslant \varepsilon + \frac{1}{S_N} \sum_{k=0}^{N-1} \frac{\delta^{k+1}}{L^{k+1}} \|y^{k+1} - x^{k+1}\|$$

(20)

can be replaced by

$$\max_{x \in Q} \langle g(x), \widetilde{y} - x \rangle \leqslant \frac{R^2}{S_N} + \frac{1}{S_N} \sum_{k=0}^{N-1} \frac{\delta^{k+1}}{L^{k+1}} \|y^{k+1} - x^{k+1}\|$$

(21)

$$\leqslant \varepsilon + \frac{1}{S_N} \sum_{k=0}^{N-1} \frac{\delta^{k+1}}{L^{k+1}} \|y^{k+1} - x^{k+1}\|.$$

Remark 1. Due to adaptive choice of parameters L^{k+1} and δ^{k+1} at each iteration of Algorithm 1 the expression

$$\frac{R^2}{S_N} + \frac{1}{S_N} \sum_{k=0}^{N-1} \frac{\delta^{k+1}}{L^{k+1}} \|y^{k+1} - x^{k+1}\|$$

in (21) may be small enough even in the case of $L = +\infty$ or $\delta = +\infty$ in (2).

Remark 2. Clearly, for each k, we have $\delta_k \leqslant C_L \delta$ ($C_L = \max\left\{1, \frac{2L}{L^0}\right\}$) and:

$$\frac{1}{S_N} \sum_{k=0}^{N-1} \frac{\delta^{k+1}}{L^{k+1}} \|y^{k+1} - x^{k+1}\| \leqslant C_L \delta \max_{k=0,N-1} \|y^{k+1} - x^{k+1}\|.$$

This means that the value associated with the error in the specifying operator g is bounded on the set Q of a finite diameter.

Remark 3. If $g \not\equiv 0$, then the condition $L^0 \leqslant 2L$ can be satisfied by choosing

$$L^0 := \frac{\|g(x) - g(y)\|_*}{\|x - y\|} \text{ at } g(x) \neq g(y).$$

Remark 4. Note that the estimate of the number of iterations $N = \left\lceil \frac{2LR^2}{\varepsilon} \right\rceil$ with accuracy to a numerical factor is optimal for variational inequalities with a Lipschitz-continuous operator [12]. Note that the evaluation of the inexactness of the value of the operator, as we see from the previous remark, is bounded and does not accumulate.

Note that similarly Remark 4 in [13] the total number of attempts to solve (11) and (12) is bounded by $4N + \max \left\{ \log_2 \frac{2L}{L^0}, \ \log_2 \frac{2\delta}{\delta^0} \right\}$.

4   Numerical Experiments for Non-Smooth Optimization Problem: Variational Inequality for Some Analogue of Fermat-Torricelli-Steiner Problem

In this section, to show the advantages of the proposed Algorithm 1, we consider some numerical experiments for the saddle point problem (and the corresponding VI), which corresponds to the convex programming problem for some analogues of the Fermat-Torricelli-Steiner problem with functional constraints. Note that the objective functions are non-smooth and the corresponding operators of the variational inequality of the problem under consideration are bounded ($\nu = 0$). However, experimentally, due to adaptation, we can observe an estimate of the complexity inherent in the case of Lipshitz-continuous operators of VI.

All experiments in this section were implemented in Python 3.4, on a computer equipped with Intel(R) Core(TM) i7-8550U CPU @ 1.80 GHz, 1992 Mhz, 4 Core(s), 8 Logical Processor(s). RAM of the computer is 8 GB.

For a given set of N points $\{A_k = (a_{1k}, a_{2k}, \ldots, a_{nk}); \ k = \overline{1, N}\}$, that represent the centers of the balls ω_k with radii r_k, in the n-dimensional Euclidean space $\mathbb{R}^n$, we need to find such a point $X = (x_1, x_2, \ldots, x_n)$ of the objective function [11] $(XA_k = \sum\limits_{k=1}^{N} \sqrt{(x_1 - a_{1k})^2 + \ldots + (x_n - a_{nk})^2})$

$$f(x) := \sum_{k=1}^{N} d(X, A_k), \tag{22}$$

where

$$d(X, A_k) = \begin{cases} XA_k - r_k, & \text{if } XA_k \geqslant r_k \\ 0, & \text{otherwise,} \end{cases}$$

would take the minimal value on the set Q, which is given by several functional constraints:

$$\begin{aligned} \varphi_1(x) &= \alpha_{11}x_1^2 + \alpha_{12}x_2^2 + \ldots + \alpha_{1n}x_n^2 - 1, \\ \varphi_2(x) &= \alpha_{21}x_1^2 + \alpha_{22}x_2^2 + \ldots + \alpha_{2n}x_n^2 - 1, \\ &\qquad \ldots \\ \varphi_m(x) &= \alpha_{m1}x_1^2 + \alpha_{m2}x_2^2 + \ldots + \alpha_{mn}x_n^2 - 1, \end{aligned} \tag{23}$$

where the coefficients $\alpha_{11}, \alpha_{12}, \ldots, \alpha_{mn}$ are represented by the matrix

$$\begin{pmatrix} \alpha_{11} & \alpha_{13} & \cdots & \alpha_{1n} \\ \alpha_{21} & \alpha_{23} & \cdots & \alpha_{2n} \\ \cdots\cdots\cdots\cdots\cdots \\ \alpha_{m1} & \alpha_{m3} & \cdots & \alpha_{mn} \end{pmatrix},$$

in which one element of each row is an integer belonging to the interval $(1; 10)$, and the remaining elements of the row are equal to 1.

To solve such a problem, we can consider a saddle point problem $\min\limits_x \max\limits_\lambda L(x, \lambda)$, where

$$L(x, \lambda) = f(x) + \sum_{p=1}^{m} \lambda_p \varphi_p(x), \ \overrightarrow{\lambda} = (\lambda_1, \lambda_2, \ldots, \lambda_m).$$

Consider the corresponding variational inequality:

$$\langle G(x_*, \overrightarrow{\lambda}_*), (x_*, \overrightarrow{\lambda}_*) - (x, \overrightarrow{\lambda}) \rangle \leqslant 0 \ \ \forall (x, \overrightarrow{\lambda}) \in B \subset \mathbb{R}^{n+m},$$

where

$$B = \left\{ (x, \overrightarrow{\lambda}) \,\middle|\, \sum_{k=1}^{n} x_k^2 + \sum_{p=1}^{m} \lambda_p^2 \leqslant 1 \right\},$$

$$G(x, \lambda) = \begin{pmatrix} \nabla f(x) + \sum\limits_{p=1}^{m} \lambda_p \nabla \varphi_p(x), \\ -\varphi_1(x), -\varphi_2(x), \ldots, -\varphi_m(x) \end{pmatrix}.$$

We give an example for $n = 100$, $m = 20$, $N = 5$, initial approximation

$$(x^0, \lambda^0) = \left(\frac{1}{\sqrt{m+n}}, \frac{1}{\sqrt{m+n}} \cdots, \frac{1}{\sqrt{m+n}} \right) \in \mathbb{R}^{n+m},$$

and $\delta_0 = \frac{1}{20}$. The coordinates of the points A_k are chosen in such a way that $\|A_k\| \in [1; 2]$. We choose the standard Euclidean proximal setup as a prox-function.

Note that the centers of the balls were chosen with the norm in the interval $[1; 2]$, and the radii of the balls are 1. Therefore, in a single ball with the center at zero there will be points of the boundary of the balls in which the objective function (22) and the operator of the corresponding variational inequality will be bounded. At the same time, it can be shown that the diameter of the subdifferential at such points will be at least 1. It means that theoretically δ can be at least 1. However, experimentally, we can see significantly better solution quality (see the column "General estimate" in Table 1).

The results of the work of Algorithm 1, for objective function (22) are represented in Table 1 below.

Table 1. The results of Algorithm 1 for objective function (22).

Iterations	General estimate	Time (sec.)
17	0.1051	0.264
19	0.0527	0.291
21	0.0266	0.315
22	0.0212	0.342
23	0.0177	0.354
24	0.0133	0.364
25	0.0106	0.380
26	0.0082	0.427
27	0.0063	0.467
28	0.0048	0.423
29	0.0044	0.443

It is known [12,14] that for variational inequalities with bounded operators, the theoretical estimate of the complexity (the convergence rate) $O\left(\frac{1}{\varepsilon^2}\right)$ is theoretically optimal. However, experimentally we see from Table 1 that, for example, an accuracy of 0.1051 is achieved in 17 iterations, and a 10-fold greater accuracy of 0.0106 is achieved in 25 iterations and approximately in the same time. If the method worked without adaptation and strictly according to optimal lower bounds for the specified class of problems, then the increase in costs could be approximately 100 times. Thus, due to the adaptability of the proposed method, we observe a convergence rate close to $O\left(\ln\left(\frac{1}{\varepsilon}\right)\right)$.

Now for a given set of N points $\{A_k = (a_{1k}, a_{2k}, \ldots, a_{nk}); k = \overline{1, N}\}$ in n-dimensional Euclidean space $\mathbb{R}^n$ we need to find such a point $x = (x_1, x_2, \ldots, x_n)$, that the objective function

$$f(x) := \sum_{k=1}^{N} \sqrt{(x_1 - a_{1k})^2 + \ldots + (x_n - a_{nk})^2} = \sum_{k=1}^{N} XA_k \qquad (24)$$

would take the minimal value on the set Q, which is defined by the previous constraints (23). The coordinates of the points A_k for $k = \overline{1, N}$, are chosen as the rows of the matrix $A \in \mathbb{R}^{N \times n}$. The entries of the matrix A are random integers in the closed interval $[-10; 10]$, which are drawn from the discrete uniform distribution.

The results of the work of Algorithm 1, for objective function (24) and for some different values of n, m and N, are presented in Table 2 below. These results demonstrate the number of iterations produced by Algorithm 1 to reach the solution of the problem, the quality of the solution "General estimate", which is the right side of inequality (17), and the running time of the algorithm in seconds.

Table 2. The results of Algorithm 1 for objective function (24).

$n = 600, m = 400, N = 25$			$n = 1000, m = 500, N = 50$		
Iteration	General estimate	Time (sec.)	Iteration	General estimate	Time (sec.)
22	0.122	67.955	19	0.1343	252.151
23	0.061	70.587	20	0.0672	252.673
24	0.0305	75.107	21	0.0336	266.636
25	0.0153	72.917	22	0.0168	279.883
26	0.0076	76.686	23	0.0084	293.866

As we see from Table 2 we also observe a convergence rate close to $O\left(\ln\left(\frac{1}{\varepsilon}\right)\right)$, due to the adaptability of the proposed method.

Remark 5. Now we take all previous parameters but with points $A_k (k = \overline{1, N})$ in the unit ball. The results of Algorithm 1, for objective function (24) and for some different values of n, m and N, with constraints (23), are presented in Table 3 below.

Table 3. The results of Algorithm 1 for objective function (24), with points A_k in the unit ball.

$n = 100, m = 50, N = 25$			$n = 200, m = 100, N = 50$		
Iteration	General estimate	Time (sec.)	Iteration	General estimate	Time (sec.)
318	0.2539	35.805	684	0.2522	441.744
468	0.1702	52.809	1016	0.1688	645.082
618	0.1279	71.484	1350	0.1267	857.185
768	0.1026	87.851	1682	0.1015	1049.885
918	0.0857	103.686	2016	0.0846	1305.006
1068	0.0736	123.056	2349	0.0727	1534.333
1218	0.0645	141.044	2683	0.0637	1753.489
1368	0.0575	153.877	3014	0.0567	2026.402
1518	0.0518	173.538	3348	0.0511	2210.362
1668	0.0472	190.714	3680	0.0465	2301.327
1818	0.0434	208.966	4014	0.0427	2611.942
1968	0.0401	228.198	4346	0.0394	2970.324
2118	0.0373	243.970	4678	0.0367	3153.905
2268	0.0349	253.112	5012	0.0343	3387.451
2426	0.0323	266.583	5346	0.0322	3619.831

In this case, since the points A_k are chosen in the unit ball, the operator of the variational inequality is bounded. The results of experiments in Table 3 show that the rate of convergence of the proposed method is close to $O\left(\frac{1}{\varepsilon}\right)$, which is significantly better than the optimal one, which is $O\left(\frac{1}{\varepsilon^2}\right)$, for non-smooth convex optimization problems and bounded variational inequalities [14].

5   Numerical Experiments for Matrix Games with Inexactness

We continue our experiments with computing a Nash equilibrium of a matrix game. For that purpose one should solve the saddle point problem

$$\min_{x \in \Delta_n} \max_{y \in \Delta_m} x^T A y, \tag{25}$$

where $x = (x_1, x_2, \ldots, x_n) \in \mathbb{R}^n$, $y = (y_1, y_2, \ldots, y_n) \in \mathbb{R}^m$, Δ_n is a standard simplex in $\mathbb{R}^n$, i.e. $\Delta_n = \{x \in \mathbb{R}^n \mid x \geq 0, \sum_{i=1}^{n} x_i = 1\}$, Δ_m is a standard simplex in $\mathbb{R}^m$, A is the payoff matrix for the y player. In all experiments we use payoff distributions centered at zero. Consider the following operator

$$g(u) = \begin{pmatrix} \nabla_x(x^T A y) \\ -\nabla_y(x^T A y) \end{pmatrix} = \begin{pmatrix} A^T y \\ -A x \end{pmatrix}, \ u = (x, y) \in Q \equiv \Delta_n \times \Delta_m. \tag{26}$$

The operator $g(u)$ from (26) is monotone on Q, and with this operator the VI (5) has the same solution as the saddle point problem (25). So, Mirror Prox methods could be used for solving it.

In all experiments with matrix games we use the entropy prox-function $d(x) = \sum_{i=1}^{n} x_i \ln x_i$ in Bregman divergence (3). Entropy prox for matrix games on simplex is the best option (see [15]).

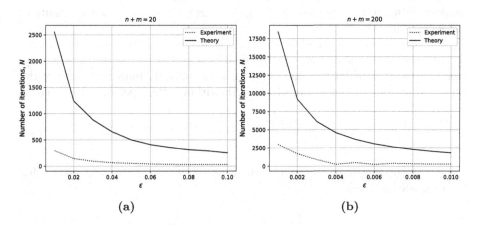

(a)   (b)

Fig. 1. The first kind of experiments with matrix games. Dependence between experimental and theoretical number of iterations for different ε. (a) 10×10 matrix A from (25), (b) 100×100 matrix A from (25).

First of all, we calculate experimental numbers of iterations for adaptive proximal method for VI [7] and compare these numbers with theoretical estimation
$N = \frac{C}{\varepsilon}$, for some $C > 0$ [4]. For that kind of experiments we run simulations on
two classes of random matrix games: 10×10 and 100×100 normally distributed
payoff matrices. For the first setting we create 50 games at random and calculate average experimental number of iterations over all games. Figure 1 shows
the results for different ε. The experimental results are better than theoretical
estimation in all cases.

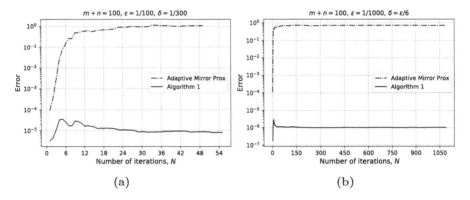

Fig. 2. The second kind of experiments with matrix games. The logarithmic scale on
the Error-axis. (a) $\varepsilon = 1/100$, $\delta = 1/300$, (b) $\varepsilon = 1/1000$, $\delta = 1/6000$.

In the second part of experiments with matrix games we compare the proposed Algorithm MPAI with adaptive Mirror Prox method from [7]. In this part
of experiments we modified problem (25) by adding inexactness (a bounded by δ
random noise) to the operator g of VI (5). Figures 2 and 3 show the results of
these experiments.

For comparison we calculate the specific values that determine the degree
of influence of the inexactness on the final estimate of the decisions' accuracy.
Not the whole error estimations are compared, but only improved parts of the
error estimations by our approach. For Algorithm 1 this specific value is equal
to (see (19)):

$$\frac{1}{S_N} \sum_{k=0}^{N-1} \frac{\delta^{k+1}}{L^{k+1}} \left\| y^{k+1} - x^{k+1} \right\| \tag{27}$$

and for adaptive Mirror Prox [7] method we can estimate the analogous value
in the following way:

$$\frac{1}{S_N} \sum_{k=0}^{N-1} \frac{\delta}{L^{k+1}} \left\| y^{k+1} - x^{k+1} \right\|.$$

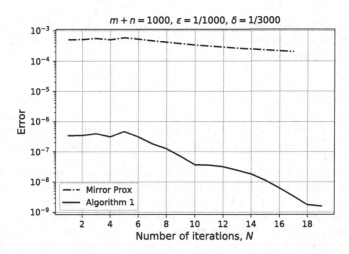

Fig. 3. The second kind of experiments with huge-scale matrix games. The logarithmic scale on the Error-axis. 1000×1000 matrix A from (25).

According to the scheme of the proofs of Theorems 1 and 2, we can obtain the analogous value for non-adaptive Mirror Prox with constant step $1/L$:

$$\frac{1}{N} \sum_{k=0}^{N-1} \delta \left\| y^{k+1} - x^{k+1} \right\|. \tag{28}$$

It is worth mentioning that estimate (27) should be less than (28) because of adaptive reduction $\delta^{k+1} < \delta$. We may see that the accumulated error for proposed MPAI method is smaller than the error for adaptive Mirror Prox method from [7].

6 Conclusion

The paper proposes an analogue of the Mirror Prox method for variational inequalities with adaptive tuning not only for the constant L_ν, but also for the magnitude of the operator's error. Moreover, such an error can set the degree of discontinuity of the operator. It is proved that the proposed method converges with the optimal estimate of complexity $O\left(\frac{1}{\varepsilon}\right)$ for the Lipschitz-continuous operator and the magnitude of the error is limited. It is important that the result applies to a certain class of variational inequalities with bounded (generally, discontinuous) operators ($\nu = 0$). The paper also presents the results of experiments that demonstrate the ability to work with the estimate of the complexity close to $O\left(\ln\left(\frac{1}{\varepsilon}\right)\right)$ even for a problem with bounded operator ($\nu = 0$) and the experimental comparison between adaptive Mirror Prox and the proposed algorithm for a special application (matrix games with inexactness). We also note that the considered method in this paper is applicable for VI with relatively smooth operators. More precisely, the prox-function and Bregman divergence in (2) may not be strongly convex (for convex optimization problems this situation was studied e.g. in [16,17]).

References

1. Facchinei, F., Pang, J.S.: Finite-Dimensional Variational Inequality and Complementarity Problems, 1st edn. Springer-Verlag, New York (2003). https://doi.org/10.1007/b97543
2. Antipin, A.S., Jaćimović, V., Jaćimović, M.: Dynamics and variational inequalities. Comput. Math. Math. Phys. **57**(5), 784–801 (2017)
3. Chambolle, A., Pock, T.: A first-order primal-dual algorithm for convex problems with applications to imaging. J. Math. Imaging Vis. **40**(1), 120–145 (2011)
4. Nemirovski, A.: Prox-method with rate of convergence $O(1/T)$ for variational inequalities with Lipschitz continuous monotone operators and smooth convex-concave saddle point problems. SIAM J. Optim. **15**(1), 229–251 (2004)
5. Korpelevich, G.M.: Extragradient method for finding saddle points and other problems. Ekon. Mat. Metody **12**(4), 747–756 (1976)
6. Stonyakin, F., Gasnikov, A., Dvurechensky, P., Alkousa, M., Titov, A.: Generalized Mirror Prox for Monotone Variational Inequalities: Universality and Inexact Oracle (2019). https://arxiv.org/pdf/1806.05140.pdf
7. Gasnikov, A.V., Dvurechensky, P.E., Stonyakin, F.S., Titov, A.A.: Adaptive proximal method for variational inequalities. Comput. Math. Math. Phys. **59**(5), 836–841 (2019)
8. Nesterov, Y.: Universal gradient methods for convex optimization problems. Math. Program **152**(1–2), 381–404 (2015)
9. Gasnikov, A.V.: Modern Numerical Optimization Methods, Universal Gradient Descent. MIPT, Moscow (2018). https://arxiv.org/abs/1711.00394
10. Stonyakin, F.S.: Some analogue of quadratic interpolation for a special class of nonsmooth functionals and its application to the adaptive mirror descent method. Dyn. Syst. **9**(37), 3–16 (2019)
11. Mordukhovich, B.S., Nam, N.M.: Applications of variational analysis to a generalized Fermat-Torricelli problem. J. Optim. Theory Appl. **148**(3), 431–454 (2011)
12. Nemirovsky, A.S.: Information-based complexity of linear operator equations. J. Complexity **8**(2), 153–175 (1992)
13. Stonyakin, F.S.: An adaptive analog of Nesterov's method for variational inequalities with a strongly monotone operator. Numer. Anal. Appl. **2**(12), 166–175 (2019)
14. Nemirovsky, A., Yudin, D.: Problem Complexity and Method Efficiency in Optimization. Wiley, New York (1983)
15. Nesterov, Y.: Smooth minimization of non-smooth functions. Math. Program **103**(1), 127–152 (2005)
16. Haihao, L., Freund, R.M., Nesterov, Y.: Relatively-smooth convex optimization by first-order methods, and applications. SIAM J. Optim. **28**(1), 333–354 (2018)
17. Stonyakin, F.S., et al.: Gradient methods for problems with inexact model of the objective. In: Khachay, M., Kochetov, Y., Pardalos, P. (eds.) MOTOR 2019. LNCS, vol. 11548, pp. 97–114. Springer, Cham (2019). https://doi.org/10.1007/978-3-030-22629-9_8

Computational Experience and Challenges with the Conjugate Epi-Projection Algorithms for Non-smooth Optimization

Evgeni A.Nurminski[1] and Natalia B. Shamray[2]([✉])

[1] School of Natural Sciences, Far Eastern Federal University,
Ajax St., Vladivostok, Russky Island, Russia
`nurminskiy.ea@dvfu.ru`
[2] Institute for Automation and Control Processes, ul. Radio, 5, Vladivostok, Russia
`shamray@dvo.ru`

Abstract. This paper considers implementable versions of a conceptual convex optimization algorithm which provides a high-speed (super-linear, quadratic and finite) convergence for broad classes of convex optimization problems. The algorithm can be best viewed in the space of conjugate variables and as such it implicitly solves optimality conditions by sequential projection on the epigraph of conjugate function. The implementable version of this algorithm tries to solve projection problems approximately by construction of the inner approximations of the epigraph up to sufficient accuracy.

This paper suggests also a version of the algorithm with additional linear cuts imposed on the epigraph which requires solution of an nontraditional auxiliary one-dimensional optimization problem. We derive an explicit form of this subproblem and provide convergence theorem for the resulting algorithm.

Keywords: Convex optimization · Conjugate function · Approximate sub-differential · Super-linear convergence · Quadratic convergence · Finite convergence · Projection · Epigraph

Introduction

This work considers some computational ideas related to numerical solution of convex optimization problems

$$\min_x f(x) \tag{1}$$

where the objective function f does not need to be a differentiable in a classical sense.

This work is supported by RF Ministry of Education and Science, project 1.7658.2017/6.7 and RFBR grant 18-29-03071.

The main idea is to consider the equivalent problem in the conjugate (sub-gradient) space of computing the value and subgradient of a conjugate function at zero. Convexity allows to guarantee a number of attractive features of such approach [1]: uniform treatment of conditional and unconditional optimization problems, development of projection-type algorithms with super-linear convergence in the general case, quadratic rate of convergence in sub-quadratic case and finite convergence in the case of sharp minima. In any case these algorithms are globally convergent and do not need favorable initial points.

It was further suggested in [2,3] to impose certain additional cuts to improve the relaxation properties of the algorithm. Convergence of the resulting algorithms was proved under very general conditions however the computational efficiency of these algorithms remained under the question. Here we intend to study it at least experimentally starting with the single linear cut for which can be made a couple of simple natural choices.

1   Notations and Preliminaries

Throughout the paper we use the following notations: E is a finite dimensional euclidean space of primal variables of any dimensionality. The inner product of vectors x, y from E is denoted as xy. The cone of non-negative vectors of E is denoted as E_+. The set of real numbers is denoted as $\mathbb{R}$ and $\mathbb{R}_\infty = \mathbb{R} \cup \{\infty\}$.

The norm in E is defined in a standard way: $\|x\| = \sqrt{xx}$ and for $X \subset E$ $\|X\| = \sup_{x \in X} \|x\|$. This norm defines of course the standard topology on E with the common definitions of open and closed sets and closure and interior of subsets of E. The interior of a set X is denoted as $\mathrm{int}(X)$.

The unit ball in E is denoted as $B = \{x : \|x\| \leq 1\}$. The support function of a set $Z \subset E$ is denoted and defined as $(Z)_x = \sup_{z \in Z} xz$.

A vector of ones of a suitable dimensionality is denoted by $e = (1, 1, \ldots, 1)$. A standard simplex $\{x : x \geq 0, xe = 1\}$ with $x \in E, \dim(E) = n$ is denoted by Δ_E.

We use the standard definitions of convex analysis (see f.i. [6]) related mainly to functions $f : E \to \mathbb{R}_\infty$: the domain $\mathrm{dom}\, f$ of a function f is the set $\mathrm{dom}\, f = \{x : f(x) < \infty\}$, the epigraph $\mathrm{epi}\, f$ of a function f is a set $\mathrm{epi}\, f = \{(\mu, x) : \mu \geq f(x)\} \subset \mathbb{R}_\infty \times E$.

Further on all functions are closed convex in a sense that their epigraphs are *closed convex* subsets of $\mathbb{R}_\infty \times E$.

Definition 1. *For a convex function $f : E \to \mathbb{R}$ and fixed $x \in E$ the set $\partial f(x) = \{g : f(y) - f(x) \geq g(y - x)$ for all $y \in \mathrm{dom}\, f\}$ is called a sub-differential of f at the point x.*

The sub-differential $\partial f(x)$ of f at point x is well-defined and is a closed bounded convex set for all $x \in \mathrm{int}(\mathrm{dom}\, f)$. At the boundary of $\mathrm{dom}\, f$ it may or may not exists. The sub-differential $\partial f(x)$ is also upper semi-continuous as a multi-function of x when exists.

Definition 2. *The directional derivative of a finite convex function f at point x in direction d is denoted and defined as $\partial f(x; d) = \lim_{\delta \to +0} (f(x + \delta d) - f(x))/\delta$.*

It is well-known from convex analysis that $\partial f(x; d) = \sup_{g \in \partial f(x)} gd = (\partial f(x))_d$.

Definition 3. *For a convex function $f : E \to \mathbb{R}_\infty$ the function*

$$f^*(g) = \sup_x \{gx - f(x)\} = (\text{epi } f)_{\bar{g}}, \text{ where } \bar{g} = (-1, g) \in \mathbb{R}_\infty \times E \qquad (2)$$

is called a conjugate function of f.

The key result of convex analysis is that for a closed convex function f

$$\sup_g \{gx - f^*(g)\} = (\text{epi } f^*)_{\bar{x}} = f(x), \qquad (3)$$

where $\bar{x} = (-1, x) \in \mathbb{R}_\infty \times E$.

It is also easy to see that if $(\text{epi } f^*)_{\bar{x}} = g_x x - f^*(g_x)$ then $g_x \in \partial f(x)$ and the other way around: for $\bar{g} = (-1, g)$ if $(\text{epi } f)_{\bar{g}} = gx_g - f(x_g)$ then $x_g \in \partial f^*(g)$.

The trivial consequence of the Definition 3 is that $f^*(0) = -\inf_x f(x)$ which is the key correspondence used by the conjugate epi-projection algorithm, considered further on. As the conjugate epi-projection algorithm operates in the conjugate space its convergence properties depend upon the properties of the conjugate function of the objective. Therefore we introduce some additional classes of primal functions to ensure the desired behavior of the conjugates.

Definition 4. *Convex function f is called sup-quadratic with respect to a point $x \in \text{int}(\text{dom } f)$ if there exists a constant $\tau > 0$ such that*

$$f(y) - f(x) \geq g(y - x) + \frac{1}{2}\tau \|y - x\|^2 \qquad (4)$$

for any $g \in \partial f(x)$ and any y.

We will call τ the sup-quadratic characteristic of f at x. Notice that strongly convex functions are sup-quadratic at any x from their domains, however a function f, sup-quadratic at some x, need not to be strongly convex.

A symmetric definition can be given for *sub-quadratic* functions.

Definition 5. *Convex function f is called sub-quadratic with respect to a point $x \in \text{int}(\text{dom } f)$ if there exists a constant $\tau > 0$ such that*

$$f(y) - f(x) \leq g(y - x) + \frac{1}{2}\tau^{-1} \|y - x\|^2 \qquad (5)$$

for any $y \in \text{dom } f$ and some $g \in \partial f(x)$.

Notice that it follows from this definition that the function f, sub-quadratic at point x is in fact differentiable at this point. Of course not all functions differentiable at x are sub-quadratic.

From the point of view of non-smooth optimization namely sup-quadratic functions are of particular interest, as the class of such functions contains, for instance, the common case of a maximum of a finite set of quadratic functions. The Definitions 4 and 5 establish important properties of conjugates functions for sup-quadratic primal functions.

Lemma 1. *Let $f : E \to \mathbb{R}$ attains its minimum value $f_\star$ at the point $x^\star$ and f is sup-quadratic at point $x^\star$ with the positive sup-quadratic characteristic τ. Then $f^\star(g)$ is sub-quadratic at $g = 0$ with the corresponding sub-quadratic characteristic not lower then τ^{-1}.*

Proof. By definition for any x

$$\frac{1}{2}\tau\|x^\star - x\|^2 \leq f(x) - f_\star = f(x) + f^\star(0) \tag{6}$$

and hence

$$f^\star(g) - f^\star(0) = x_g g - (f(x_g) + f^\star(0)) \leq x_g g - \frac{1}{2}\tau\|x^\star - x_g\|^2 \tag{7}$$

for any $x_g \in \partial f^\star(g)$. Hence

$$\begin{aligned} f^\star(g) - f^\star(0) &\leq x^\star g + (x_g - x^\star)g - \frac{1}{2}\tau\|x^\star - x_g\|^2 \leq \\ x^\star g + \sup_z\{zg - \frac{1}{2}\tau\|z\|^2\} &= x^\star g + \frac{1}{2}\tau^{-1}\|g\|^2. \end{aligned} \tag{8}$$

■

Another interesting subclass of convex functions are those which have zero in the interior of the subdifferential at the solution $x^\star$ of (1), that is $0 \in \text{int}(\partial f(x^\star))$. This condition is also known as "sharp minimum" and extended further on in [7] and others. The special attraction of this case is that the well-known proximal method has then a finite termination [8] for such problems. The conjugate epi-projection optimization algorithm has the same property which is based on the fact that the conjugate functions for the primal functions with sharp minimum have very simple behavior in the vicinity of zero.

Lemma 2. *If solution $x^\star$ of (1) is such that $0 \in \text{int}(\partial f(x^\star))$ then there is $\rho > 0$ such that $f^\star(g) = gx^\star - f(x^\star)$ for $\|g\| < \rho$.*

Proof. If ρ is small enough then sharp minimum condition implies $0 \in \partial(f(x^\star) - gx^\star) = \partial f(x^\star) - g$ for any $g \in \rho B$ and therefore

$$f^\star(g) = \sup_x\{gx - f(x)\} = gx^\star - f(x^\star)$$

is a linear function of g. ■

Namely this property guarantees the finite termination of the conjugate epi-projection optimization algorithm.

For additional results on connections between sharp minimum and properties of conjugate functions see also [9].

2   Conjugate Epi-Projection Algorithm

As it was already mentioned the basic idea of the conjugate epi-projection algorithms consists in considering the convex problem (1) as the problem of computing the conjugate function of the objective at the origin:

$$f^\star(0) = -\min_x f(x) = -f_\star = \inf_{(0,\mu)\in\text{epi } f^\star} \mu.$$

We suggest to use for computing $f^\star(0)$ the algorithms based on projection onto the epigraph epi $f^\star$. This idea demonstrates some promises for effective solution of (1) and suggests some new computational ideas.

The algorithms considered here consist in execution of an infinite sequence of iterations, which generates the corresponding sequence of points $\{(\xi_k, 0) \in \mathbb{R} \times E, k = 0, 1, \dots\}$ with $\xi_k \to f^\star(0)$ when $k \to \infty$. For each of these iterations they call a subgradient oracle which for any $x \in E$ computes $f(x)$ and some arbitrary $g \in \partial f(x)$. Also they require solution of nonlinear projection problems which make the algorithms, strictly speaking, un-implementable. However the analysis of the algorithm demonstrate its potential and can show the ways to its practical implementations.

We give here first the original version of a conceptual conjugate epi-projection algorithm and cite here the key results about its convergence. This is followed by a few simple numerical experiments just to provide a reference point for further modifications and to indicate some numerical problems which can arise in its straightforward implementation.

2.1   Basic Computational Scheme

The principal details of the iteration of the conjugate epi-projection algorithm are given on the figure Algorithm 1. Convergence of the Algorithm 1 is confirmed by the following theorem.

Theorem 1. *Let f be a finite convex function with the finite minimum $f_\star = \min_x f(x) = -f^\star(0)$ and $\xi_k, k = 1, 2, \dots$ are defined by the Algorithm 1 with $\xi_0 < f^\star(0)$. Then*

$$\lim_{k\to\infty} \xi_k = f^\star(0) = -f_\star$$

and

$$f^\star(0) - \xi_{k+1} \le \lambda_k(f^\star(0) - \xi_k)$$

with $\lambda_k \to 0$ when $k \to \infty$.

It means that Algorithm 1 in general case has at least super-linear rate of convergence.

Next we consider the problem (1) with sup-quadratic objective function f where we can claim global convergence of the conceptual conjugate epi-projection algorithm and asymptotic quadratic rate of convergence.

Data: The convex function $f : E \to \mathbb{R}$, the epigraph epi f^*, the current
iteration number k and the current approximation $\xi_k \leq f^*(0)$.
Result: The next approximation ξ_{k+1} such that $\xi_k \leq \xi_{k+1} \leq f^*(0)$
Each iteration consists of two basic operations: **Project** and **Support-Update**
Project. Solve the projection problem of the point $(\xi_k, 0)$ onto epi f^*:

$$\min_{(\xi,g)\in\text{epi } f^*}\{(\xi - \xi_k)^2 + \|g\|^2\} = (\xi_k^p - \xi_k)^2 + \|g_p^k\|^2 \tag{9}$$

with the corresponding solution $(\xi_k^p, g_p^k) = (f^*(g_p^k), g_p^k) \in \text{epi } f^*$. We demonstrate
in the analysis of the algorithm convergence that $f^*(0) \geq \xi_k^p > \xi_k$ if $\xi_k < f^*(0)$.
Support-Update Compute support function of epi f^* with the support vector
$z^k = -(\xi_k^p - \xi_k, g_p^k) \in \mathbb{R} \times E$

$$(\text{epi } f^*)_{z^k} = \sup_{(\mu,g)\in\text{epi } f^*}\{-(\xi_k^p - \xi_k)\mu + g_p^k g)\} =$$

$$(\xi_k^p - \xi_k)\sup_{(\mu,g)\in\text{epi } f^*}\{-\mu + \frac{g_p^k}{(\xi_k^p - \xi_k)}g\} = (\xi_k^p - \xi_k)\sup_{(\mu,g)\in\text{epi } f^*}\{-\mu + x_p^k g\} =$$

$$(\xi_k^p - \xi_k)(x_p^k g_p^k - f^*(g_p^k))\} = (\xi_k^p - \xi_k)f(x_p^k),$$

where $x_p^k = g_p^k/(\xi_k^p - \xi_k)$. Notice that as f is assumed to be a finite function this
operation is well-defined.
Finally we update the approximate solution with ξ_{k+1} using the relationship

$$\bar{\xi}_{k+1}z^k = (\text{epi } f^*)_{z^k}, \text{ where } \bar{\xi}_{k+1} = (\xi_{k+1}, 0) \in \mathbb{R} \times E,$$

which actually amounts to $\xi_{k+1} = -f(x_p^k)$, increment iteration counter
$k \to k + 1$, etc.

Algorithm 1: The basic iteration of the conceptual conjugate epi-
projection algorithm

Theorem 2. *Let objective function f in problem (1) is locally sup-quadratic with
sup-quadratic characteristic τ and $\xi_k, k = 1, 2, \ldots$ are defined by the Algorithm 1
with $\xi_0 < -f_\star$. Then $\lim_{k\to\infty} \xi_k = f^*(0)$ (Algorithm 1 converges) and for k large
enough $f^*(0) - \xi_{k+1} \leq \tau^{-1}(f^*(0) - \xi_k)^2$ (that is convergence is quadratic).*

Finite convergence of this algorithm for sharp minimum is established by the
following theorem.

Theorem 3. *Let the objective function of (1) has a sharp minimum at solution
point $x^\star$, all assumptions of the Theorem 1 are satisfied and $\xi_k, k = 1, 2, \ldots$
are defined by the Algorithm 1 with $\xi_0 < -f_\star$. Then there exists $k^\star$ such that
$\xi_{k^\star} = f^*(0) = -f_\star$.*

Notice that in all cases convergence is global and does not require any additional
assumptions.

2.2   Implementation Issues

The critical part in implementation of Algorithm 1 is the projection step (9),
where the point $(\xi_k, 0)$ is projected onto epi f^*. The set epi f^* is given implicitly

only, however due to Fenchel-Morou duality we can easily compute the supremum on it of any linear function $\bar{p}\bar{z}$ where $\bar{z} = (\mu, z), \mu \geq f^\star(z)$, $\bar{p} = (\pi, p)$. This supremum is finite when $\pi < 0$ and then

$$(\sup_{\bar{z} \in \text{epi}\, f^\star} \{\pi\mu + pz\} = |\pi| \sup_{z,\mu \geq f^\star(z)} \{pz/|\pi| - \mu\} = |\pi| \sup_z \{pz/|\pi| - f^\star(z)\} = |\pi| f(pz/|\pi|).$$

It gives a chance to suggest simple iteration-like algorithms, using implementable projection onto inner approximation P_k of epi $f^\star$ which is represented on Algorithm 2. This algorithm in practice is interrupted when desirable accuracy is achieved. The quadratic optimization problem (10) can be solved by many off-the-shelf quadratic solvers, however our experience is that the specialized algorithms like [4] outperforms them. One can find the OCTAVE-version of the code as DOI: 10.13140/RG.2.2.21281.86882 at [5].

Data: The epigraph epi $f^\star$, its polyhedral approximation, the point
$\bar{q} = (\xi, 0) \notin \text{epi}\, f^\star$
Result: The sequence $\{\bar{g}^k = (\xi_k, g^k) \in \text{epi}\, f^\star, k = 1, 2, \dots\}$ such that
$\bar{g}^k \to \bar{g}^\star \in \text{epi}\, f^\star$ and $\|\bar{g}^\star - \bar{q}\| = \min_{\bar{g} \in \text{epi}\, f^\star} \|\bar{g} - \bar{q}\|$
Initialize; $P_0 = g^0, k = 0$
While;
Solve quadratic optimization problem:

$$\min_{g \in P_k} \|g - \bar{q}\|^2 = \|g^k - \bar{q}\|^2 \qquad (10)$$

Upgrade:

$$P_{k+1} = \text{co}\{P_k, g^k\}, \quad k \to k + 1$$

end while;

Algorithm 2: Iterative algorithm for projection on epi $f^\star$

2.3 Numerical Example

The most interesting and difficult tests of non-smooth optimization consist in minimization of piece-wise quadratic problems which are constructed as finite maximum of convex quadratics. We demonstrate performance of the implementable version of Algorithm 1 with iterative Algorithm 2 for approximate solution of the auxiliary projection problem (9) on a simple problem (1) with $f(x) = \max_{i=1,2}(x - a^i)A_i(x - a^i)$ with $a^1 = (0,0,0), a^2 = (2,3,9)$ and A_i are diagonal matrices: $A_1 = \text{diag}(9,4,1), A_2 = \text{diag}(1,4,9)$. The optimization solver CONDOR 1.06, running on NEOS optimization solver [10] reported successful completion after 63 function evaluations with the objective value of 0.4348696068. Our solver attained slightly worse 0.43673 with 27 function evaluations.

The loss in the value of objective function can be probably explained by the numerical instability of projection problems (9) at the final iterations of optimization process. The Fig. 1 demonstrates the peculiar features of SU-step during solution of minimization problem. It shows convergence of the simple projection Algorithm 2 in solution of the projection problem (9) in terms of optimality condition $\delta_k = \|z^k\|^2 - \inf_{z \in \text{epi } f^\star} zz^k$ where $z^k \in$ epi $f^\star$—an approximate solution of (2) obtained on k-th iteration of this algorithm. For any k the value of δ_k is non-negative and if $\delta_k = 0$ then z^k is the solution of (2).

It can be seen from the Fig. 1 that in all cases the auxiliary projection problem was solved sufficiently quickly with at least the linear rate of convergence. However, it also can be seen that the projection Algorithm 2 slows down when projected point approaches the epigraph epi $f^\star$. This was expected behavior of the algorithm and there are known technics to improve solution of 2 in this case, but this issue requires additional investigation.

3   Conjugate Epi-Projection Algorithm with a Skew Cut

One of the other possible ways to improve computational behavior of the conjugate epi-projection algorithm is to introduce additional constraints in **Support-Update** (SU) step of this algorithm. Namely, if we assume that there is an additional condition $(\mu, g) \in Q \subset E \times \mathbb{R}$ with $(f^\star(0), 0) \in Q$ then

$$\omega_x = \sup_{(\mu, g) \in \text{epi } f^\star \cap Q} \{xg - \mu\} = xg_x - f^\star(g_x) \le xg_x - xg_x + f(x) = f(x)$$

so ω_x will provide better (lower) upper estimates for $\min_x f(x) = -f^\star(0)$. Of course it will be necessary to ensure that an additional constraint $(\mu, g) \in Q$ does not cut off the solution $(f^\star(0), 0)$. It implies that $(f^\star(0), 0) \in Q$ which can be ensured in different ways.

The corresponding modification of SU-step is shown as Algorithm 3.

Convergence of the Algorithm 3 is confirmed by the following theorem.

Theorem 4. *Let f be a finite convex function with the finite minimum $f_\star = \min_x f(x) = -f^\star(0)$ and $\xi_k, k = 1, 2, \ldots$ are defined by the Algorithm 3 with $\xi_0 < f^\star(0)$. Then $\lim_{k \to \infty} \xi_k = f^\star(0) = -f_\star$, that is the algorithm converges;*

Proof. Assume that on k-th iteration we have $\xi_k < f^\star(0)$ as the approximation of $f^\star(0)$. According to Algorithm 3 to construct the next (k+1-th) approximation ξ_{k+1} the point $(\xi_k, 0) \in \mathbb{R} \times E$ is to be projected onto epi $f^\star \cap Q_k$ first:

$$\min_{(\xi, g) \in \text{epi } f^\star \cap Q_k} \{(\xi - \xi_k)^2 + \|g\|^2\} = (\xi_k^p - \xi_k)^2 + \|g_p^k\|^2 \qquad (11)$$

The solution $(\xi_k^p, g_p^k) = (f^\star(g_p^k), g_p^k) \in$ epi $f^\star$ of this problem satisfies optimality conditions

$$(f^\star(g_p^k) - \xi_k)(\xi - \xi_k^p) + g_p^k(g - g_p^k) \ge 0 \qquad (12)$$

for any $(\xi, g) \in$ epi $f^\star \cap Q_k$.

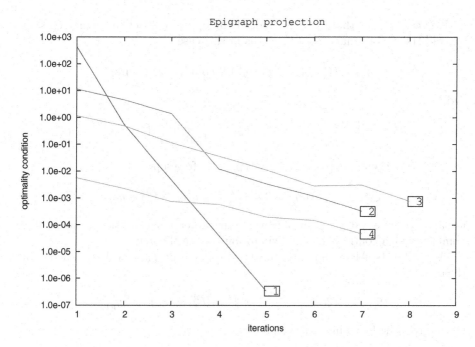

Fig. 1. Projection operation on epi f^* on different iterations of Algorithm 1. Boxed numbers on the Figure denote the major iterations of Algorithm 1

Data: The epigraph epi f^*, the current iteration number k, the current approximation $\xi_k \leq f^*(0)$, and projection vector z^k obtained from **Project** step.

Result: The next approximation ξ_{k+1} such that $\xi_k \leq \xi_{k+1} \leq f^*(0)$.

Modified Support-Update

Compute support function of $G_k = $ epi $f^* \cap Q_k$ with the support vector $z^k = -(\xi_k^p - \xi_k, g_p^k) \in \mathbb{R} \times E$

$$(G_k)_{z^k} = \sup_{(\mu,g)\in G_k}\{-((\xi_k^p - \xi_k)\mu + g_p^k g)\} =$$

$$(\xi_k^p - \xi_k)\sup_{(\mu,g)\in G_k}\left\{-\mu + \frac{g_p^k}{(\xi_k^p - \xi_k)}g\right\} = (\xi_k^p - \xi_k)\sup_{(\mu,g)\in G_k}\{-\mu + x_p^k g\} =$$

$$(\xi_k^p - \xi_k)(x_p^k \tilde{g}_p^k - f^*(g_p^k)).$$

where $x_p^k = g_p^k/(\xi_k^p - \xi_k)$.

Notice that now $g_p^k \notin \partial f(x_p^k)$ and we need an additional operation to recover the support vector to epi f^* at the point $(f^*(g_p^k), g_p^k)$.

Algorithm 3: Modified **Support-Update** (MSU) step

It is easy to see that $\xi_k^p > \xi_k$. Indeed the opposite strict inequality $\xi_k^p < \xi_k$ contradicts the optimality of (ξ_k^p, g_p^k) as in this case

$$(\xi_k, g_p^k) = (\xi_k^p + (\xi_k - \xi_k^p), g_p^k) \in \operatorname{epi} f^\star \cap Q_k \subset \operatorname{epi} f^\star,$$

and

$$(\xi_k - \xi_k)^2 + \|g_p^k\|^2 < (\xi_k - \xi_k^p)^2 + \|g_p^k\|^2 = \min_{(\xi,g)\in\operatorname{epi} f^\star} \{(\xi_k - \xi)^2 + \|g\|^2\}.$$

If $\xi_k^p = \xi_k$ then $\mathbb{R} \times \{0\}$ is strictly separable from $\operatorname{epi} f^\star$:

$$\xi(\xi_k - \xi_k^p) + 0 g_p^k = 0 < \|g_p^k\|^2 \le \mu(\xi_k - \xi_k^p) + g g_p^k$$

for any $(\mu, g) \in \operatorname{epi} f^\star$ as it follows from projection conditions. Hence $0 \notin \operatorname{dom}(f^\star)$ which contradicts the assumptions of the theorem.

According to Algorithm 3 the next approximation ξ_{k+1} is determined from the equality

$$(\xi_k^p - \xi_k)(\xi_{k+1} - \xi_k) - \|g_p^k\|^2 = (\xi_k^p) - (\xi_k)^2 + \|g_p^k\|^2$$

which gives the following expression for ξ_{k+1}:

$$\xi_{k+1} = \xi_k + \|g_p^k\|^2/(\xi_k^p - \xi_k) \ge \xi_k,$$

and $\xi_{k+1} = \xi_k$ if and only if $g_p^k = 0$ which means that we already obtained the solution.

Repeating this operation we obtain the monotone sequence $\xi_k, k = 0, 1, \ldots$ such that

$$\xi_k \le \xi_{k+1} \le f^\star(0), k = 0, 1, \ldots$$

where inequalities turn into equalities only if either $\xi_k = f^\star(0)$ or $\xi_{k+1} = f^\star(0)$ which of course makes no difference. Under these conditions $\lim_{k\to\infty} \xi_k = f^\star(0)$ which proves the convergence of the Algorithm 1.

■

3.1   Projection in Modified Support-Update Step

The key step in MSU step of Algorithm 3 is the computation of projection of a given point, say z, on $G_k = \operatorname{epi} f^\star \cap Q_k$ where Q_k is a cutting set. It can be approximately solved by the iterative Algorithm 2 which in turn requires computing of the support function $(G_k)_{z^k}$ of the set $G_k = \operatorname{epi} f^\star \cap Q_k$ with the given support vector $z^k \in \mathbb{R} \times E$. By dropping for simplicity the iteration index k we face the following problem

$$(G)_z = \sup_{\substack{g \in \operatorname{epi} f^\star \\ g \in Q}} zg = zg^\star$$

the computational difficulty of which critically depends on cutting set Q. To begin with something constructive we consider here the simplest choice of $Q = H_{p,\beta}$ where $H_{p,\beta}$ is the half-space, described by linear inequality $H_{p,\beta} = \{(\mu, g) : pg - \mu \geq \beta\}$, where $p \in E$ and $\beta \geq f^\star(0)$ to guarantee that $(f^\star(0), 0) \in H_{p,\beta}$. Such β is easy to obtain from the inner approximation D of epi $f^\star$ if available. If the vertical line $\mathbb{R} \times \{0\}$ intersects D at some point $(-\beta, 0)$. Then $-\beta \geq f^\star(0)$ and therefore $(f^\star(0), 0) \in H_{p,\beta}$.

For the choice of vector p we have almost unlimited freedom and choice of the best p might be an interesting subject for further consideration.

Then

$$\sup_{\substack{\mu \geq f^\star(g) \\ pg - \mu \geq \beta}} \{xg - \mu\} = \inf_{\theta \geq 0} \sup_{\mu \geq f^\star(g)} \{xg - \mu + \theta(pg - \mu - \beta)\} =$$

$$\inf_{\theta \geq 0} \sup_{\mu \geq f^\star(g)} \{g(x + \theta p) - \mu(1 + \theta)\} - \beta\theta =$$

$$\inf_{\theta \geq 0} \{-\beta\theta + (1 + \theta) \sup_{\mu \geq f^\star(g)} \{g\frac{x + \theta p}{1 + \theta} - \mu\}\} =$$

$$\inf_{\theta \geq 0} \{-\beta\theta + (1 + \theta) f(\frac{x + \theta p}{1 + \theta})\} = \inf_{\theta \geq 0} \{-\beta\theta + (1 + \theta) f(x + \frac{\theta}{1 + \theta}(p - x))\}.$$

By introduction of new variable $\gamma = \dfrac{\theta}{1 + \theta}$ the last expression can be transformed in

$$\inf_{\gamma \in [0,1)} \{-\frac{\gamma}{1 - \gamma}\beta + \frac{1}{1 - \gamma} f(x + \gamma(p - x))\} = \inf_{\gamma \in [0,1)} (1 - \gamma)^{-1}\{f(x_\gamma) - \gamma\beta\} = \psi(\gamma),$$

where $x_\gamma = x + \gamma(p - x)$ and so the support problem is reduced to one-dimensional minimization.

Conclusion

It is shown that an implementable version of the conceptual the dual epi-projection algorithm of convex optimization demonstrate competitive efficiency even for rather simple-minded approximation of projection operator, however the resulting computational complexity requires further investigations. The important question is numerical stability of the projection operation especially on the final iterations when projected points are becoming close to the surface of the epigraph. The interesting possibility to improve convergence properties of the epi-projection algorithm is insertion of additional cuts into the projection problem. This can be easily done for the linear cuts in the form of an auxiliary one-dimensional optimization. The convergence of such algorithms is proved under very general conditions, but here again we have an open question about the best choice for such cuts.

References

1. Nurminski, E.A.: A conceptual conjugate epi-projection algorithm of convex optimization: superlinear, quadratic and finite convergence. Optim. Lett. **13**, 23–34 (2019). https://doi.org/10.1007/s11590-018-1269-3
2. Vorontsova, E.A., Nurminski, E.A.: Synthesis of cutting and separating planes in a nonsmooth optimization method. Cybern. Syst. Anal. **51**(4), 619–631 (2015)
3. Nurminski, E.: Multiple cuts in separating plane algorithms. In: Kochetov, Y., Khachay, M., Beresnev, V., Nurminski, E., Pardalos, P. (eds.) DOOR 2016. LNCS, vol. 9869, pp. 430–440. Springer, Cham (2016). https://doi.org/10.1007/978-3-319-44914-2_34
4. Nurminski, E.A.: Convergence of the suitable affine subspace method for finding the least distance to a simplex. Comput. Math. Math. Phys. **45**(11), 1915–1922 (2005)
5. Nurminski, E.A.: Orthogonal projection on convex hull of a finite set of points of a finite-dimensional Euclidean space. Version 1.6, updates 1.5. https://doi.org/10.13140/RG.2.2.21281.86882
6. Hirriart-Uruty, J.-B., Lemarechal, C.: Convex Analysis and Minimization Algorithms II: Advanced Theory and Bundle Methods. A Series of Comprehensive Studies in Mathematics, vol. 306. Springer, Heidelberg (1993). https://doi.org/10.1007/978-3-662-06409-2
7. Ferris, M.C.: Weak sharp minima and penalty functions in mathematical programming. Ph.D. dissertation, University of Cambridge, Cambridge, UK (1988)
8. Ferris, M.C.: Finite termination of the proximal point algorithm. Math. Program. **50**, 359–366 (1991)
9. Zhou, J., Wang, C.: New characterizations of weak sharp minima. Optim. Lett. **6**, 1773 (2012). https://doi.org/10.1007/s11590-011-0369-0
10. NEOS Server: State-of-the-Art Solvers for Numerical Optimization. https://neos-server.org/neos/

A Criterion of Optimality of Some Parallelization Scheme for Backtrack Search Problem in Binary Trees

Roman Kolpakov[1,2(✉)] and Mikhail Posypkin[2,3,4]

[1] Moscow State University, Moscow, Russia
foroman@mail.ru
[2] Dorodnicyn Computing Centre of Federal Research Center
"Computer Science and Control", Moscow, Russia
mposypkin@gmail.com
[3] Moscow Institute of Physics and Technology, Moscow, Russia
[4] Institute for Information Transmission Problems of the Russian Academy of
Sciences, Moscow, Russia

Abstract. The backtracking is a basic combinatorial search algorithm. As many other deterministic methods it suffers from the high complexity. Fortunately the high performance computing can efficiently cope with this issue. It was observed that the structure of the search tree can dramatically affect the efficiency of a parallel search. We study the complexity of a frontal autonomous scheme for the backtrack search parallelization. In this approach several independent cores perform a number of first steps of the backtrack search. After that each core takes one subproblem and solves it completely. Then the results are merged to select the best solution. To study the impact of the tree structure on the performance of the frontal autonomous scheme we formalize the notion of a perfectly scalable problem and prove a criterion for a search tree to fit to this class.

Keywords: Backtrack search · Parallel tree search · Complexity analysis

1 Introduction

The *backtrack search* [6] is a basic pattern for many algorithms locating a solution to an equation or an extreme for a function. The backtracking consists in a systematic exploration of a search tree combined with discarding of redundant branches. A branch is discarded if the corresponding sub-problem is infeasible or has a trivially obtainable solution. Though the discarding can significantly reduce

Partially supported by Russian fund for basic research, project 18-07-00566, and by Program 2 of Presidium of RAS "Mechanisms for ensuring fault tolerance in modern high-performance and highly reliable computing".

© Springer Nature Switzerland AG 2020
M. Jaćimović et al. (Eds.): OPTIMA 2019, CCIS 1145, pp. 455–464, 2020.
https://doi.org/10.1007/978-3-030-38603-0_33

the size of a search tree in many practical cases it is insufficient to solve a problem in a reasonable time. Fortunately the backtrack search can take an advantage of many-core parallel architectures to reduce the running time dramatically.

Last decades many efforts has been invested to the development and efficient implementation of parallel tree search algorithms [3, 11, 12]. It was observed that the structure of a search tree has a tremendous impact to the performance of a parallel algorithm. Figure 1 depicts two trees. The tree (a) is well-balanced and the tree (b) has a one-sided narrow structure. Obviously the tree (a) is much better for a parallel execution than the tree (b). The latter can't benefit from large number of cores due to data dependencies preventing concurrent processing of more than two sub-problems simultaneously. Thus the potential speedup is limited to two.

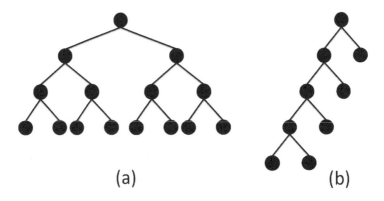

(a)   (b)

Fig. 1. "Good" (a) and "bad" (b) trees from the parallel execution point of view

In this paper we consider a *frontal autonomous scheme of parallelization (FAP-scheme)* for the backtrack search parallelization. In the FAP-scheme several independent cores perform a number of first steps of the backtrack search. After that each core takes one sub-problem and solves it completely. Then the results are merged to select the best solution. This scheme meets the requirements of loosely-coupled distributed systems or GPU-accelerators where the communication among nodes is problematic or very expensive. In this paper we study the impact of the tree structure on the performance of the FAP-scheme. We formalize the notion of a perfectly scalable problem and prove a criterion for a search tree to fit to this class.

The following notations are used through out the paper. Let X be an arbitrary set. Then

$f(x) = O(g(x))$ if $f(x) \leq Cg(x)$ for some $C > 0$ and any $x \in X$;

$f(x) = \Omega(g(x))$ if $f(x) \geq Cg(x)$ for some $C > 0$ and any $x \in X$;

$f(x) = \Theta(g(x))$ if $C'g(x) \leq f(x) \leq C''g(x)$ for some $C', C'' > 0$ and any $x \in X$.

2   Problem Statement

The backtrack search problem for binary trees can be stated as follows. Let D be a rooted binary tree. For the sake of convenience by a rooted binary tree we will mean a full tree (i.e. a tree in which every internal node has exactly two children) such that to each leaf v of D some cost $c(v)$ be assigned. It is assumed that from each node v of D any node adjacent to v (the parent or the children of v) can be accessed in constant time t_s. The backtrack search problem for D is to find the minimum cost of leaves of D, proceeding from the root of D. We will denote this problem by $P(D)$.

The serial code for finding the solution of $P(D)$ is as follows:

Procedure BackTrack(v)
Input:
   v — the tree root
1: $(v_l, v_r) := children(D)$
2: **if** $(v_l, v_r) = \emptyset$ **then**
3:    **return** $c(v)$
4: **else**
5:    $c_l = BackTrack(v_l)$
6:    $c_r = BackTrack(v_r)$
7:    **return** $best(c_l, c_r)$
8: **end if**

Initially the pair of children of the node v is obtained (line 1). If the pair is empty (v is a terminal node) then the cost of a node is returned (line 3). Otherwise left and right descendants are recursively processed and the best result is returned (lines 5–7).

Let v be a node of a rooted binary tree D. By the depth $d(v)$ of the node v we mean the length of the path from the root of D to v^1. By $d(D)$ we denote the depth of the tree D, i.e. the maximum length of paths from the root of D to leaves of D. Thus the depth $d(D)$ of a tree D is a maximal depth of its nodes.

Let $L(D)$ denote the number of nodes in a rooted binary tree D. By $T(D)$ we denote the time of sequential solving of the problem $P(D)$ by one processor. If one step of the BackTrack algorithm consumes time t_0, then the time for solving the problem $P(D)$ is $L(D) \cdot t_0$. Thus, $T(D) = \Theta(L(D))$.

By $D[v]$ we denote the subtree of D rooted in v. Two subtrees $D[v']$ and $D[v'']$ of D are called *sibling subtrees* if the nodes v' and v'' have the same parent node. For any node v of D the problem $P(D[v])$ will be called a *subproblem* of the problem $P(D)$. By the *level* of a subproblem $P(D[v])$ we will mean the depth $d(v)$ of the node v (the initial problem $P(D)$ is assumed to have level 0). By $ST^l(D)$ we denote the set of all subtrees of D rooted in nodes of depth l, and by $SP^l(D)$ we denote the set of all subproblems of $P(D)$ which have the level l, i.e.

$$SP^l(D) = \{P(D') : D' \in ST^l(D)\}.$$

---

[1] By the length of a path we mean the number of edges contained in the path.

Let v', v'' be the children of the root of the tree D. Note that the solution of the problem $P(D)$ is the minimum of the solutions of the subproblems $P(D[v'])$ and $P(D[v''])$, so the solving of $P(D[v])$ is reduced to the solving of the subproblems $P(D[v'])$ and $P(D[v''])$. We call this operation *decomposition* of the problem $P(D)$ into subproblems $P(D[v'])$ and $P(D[v''])$. The subproblems $P(D[v'])$ and $P(D[v''])$ will be called *children* of $P(D)$, and $P(D)$ will be called the *parent* of $P(D[v'])$ and $P(D[v''])$.

We study a parallel solving of backtrack search problem for binary trees by distributed memory many-core parallel systems. Independent processes comprising a parallel application communicate via network. We use a simple but adequate performance model. The time for transferring of n bytes of data is computed as $\alpha + \beta n$, where α and β are the network's latency and the bandwidth respectively. The node's network interface is assumed to be single ported: at most one message can be sent and one message can be received simultaneously. Since the algorithm considered in the paper uses only constant size data transfers, further we assume that all data transfers are performed in constant time.

The FAP-scheme uses the following algorithm of parallel solving of backtrack search problem for binary trees. Below we assume that the number of processors m is a power of 2: $m = 2^s$, where $s \leq d(D)/2$. It is a common choice in practice and it perfectly fits the needs of an asymptotic analysis. The second assumption is that the problem data is available on all processors prior to computations. Though not always true in practice, we can take this assumption because the distribution of the problem initial data is not usually a resource-consuming operation.

The FAP-scheme has three stages. At the first stage the initial problem $P(D)$ is decomposed into the subproblems at the level s. This stage is performed in parallel by each processor. After the first stage the problem $P(D)$ is decomposed into all the subproblems from the set $SP^s(D)$ and each processor has no more than one subproblem from the set $SP^s(D)$. At the Fig. 2 the first stage is illustrated for the case of $s = 2$ and $m = 4$ processors. Black circles depict subproblems selected for further processing (decomposition or sending). White circles—discarded subproblems. This first stage requires s parallel steps. Since each of the steps of the first stage can be performed in constant time, the first stage can be performed in $\Theta(s)$ time.

At the second stage all the subproblems from the set $SP^s(D)$ are simultaneously solved by the assigned processors. The time of this stage is the maximum time needed for solving of subproblems from the set $SP^s(D)$ by a serial algorithm. Thus, this time is $\Theta(L^s(D))$, where $L^s(D)$ is the maximum number of nodes in a subtree from the set $ST^s(D)$.

The third stage consists in joining the results obtained by processors. This operation is a reduction pattern widely used in parallel programming. By using binary tree reduction algorithm [1] this stage can be performed in a time $O(s)$.

Denote by $T_{FAP}^s(D)$ the time of solving of the problem $P(D)$ by FAP-scheme with the parallelization depth s. Note that this time is the total time of all the

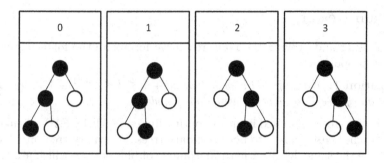

Fig. 2. The illustration of two parallel steps of the FAP 1-st stage

stages of this scheme, so from the above reasoning we conclude that $T^s_{FAP}(D) = \Theta(s + L^s(D))$.

Let $\mathcal{D}$ be some class of rooted binary trees D such that

$$T^s_{FAP}(D) = O\left(\frac{L(D)}{2^s} + d(D)\right)$$

where $s \leq d(D)/2$. Note that in this case FAP-scheme, independently on the parallelization depth, is time optimal algorithm for parallel solving of all problems $P(D)$ such that $D \in \mathcal{D}$. In this case the class $\mathcal{D}$ is called *optimally resolvable by FAP-scheme*. In this work we establish characteristic properties of classes optimally resolvable by FAP-scheme.

3   Related Work

The parallel complexity of a tree search algorithms was addressed in numerous papers. In [7–9] various parallel execution schemes for a branch-and-bound method for subset sum problem were studied. In this paper we address a parallelization of an arbitrary tree search.

The running time of a parallel algorithm is bounded by the execution time of a critical path in the dependency graph. Thus any algorithm for a parallel backtrack solving a problem $P(D)$ by m processors in time $O(L(D)/m + d(D))$ is time optimal [5].

In [13] the authors proposed an algorithm for parallel solving of the problem $P(D)$ in a distributed memory model in time $O(T(D)/m + md(D))$ where m is the number of processors. This approach was further developed in [2]. A randomized algorithm for parallel solving of the problem $P(D)$ in a distributed memory model in optimal time $O(T(D)/m + d(D))$ with a high probability was proposed in [5]. Further improvements in parallel solving of the backtrack search problem were made for shared memory model. For example, a shared memory parallel algorithm for the problem $P(D)$ in time $O((T(D)/m + d(D))(\log \log \log m)^2)$ is proposed in [4]. A space-efficient algorithm for parallel solving of $P(D)$ in a shared memory model in time $O((T(D)/m + d(D)) \log m)$ was proposed in [10].

4   Main Result

Before stating and proving the main result we formulate the following evident auxiliary proposition.

Proposition 1. *For any odd l such that $2d + 1 \leq l \leq 2^{d+1} - 1$ there exist a rooted binary tree D of depth d which has l nodes.*

Note that trees contained in classes optimally resolvable by FAP-scheme have to be in some sense "close" to perfect binary trees, i.e. binary trees in which all leaves have the same depth. A natural example of binary trees which are "close" to perfect trees is binary trees satisfying a balance property, i.e. rooted binary trees in which any two sibling subtrees have approximately the same size.

We define the balance property in the following way. Let $\delta : \mathbb{N} \to \mathbb{N}$ be a non-decreasing function, i.e. $\delta(n) \leq \delta(n+1)$ for any $n \in \mathbb{N}$. We will call a rooted binary tree D δ-*balanced* if for any sibling subtrees $D[v']$ and $D[v'']$ of D such that the depth of nodes v' and v'' is not greater than $d(D)/2$ the difference between the numbers of nodes of $D[v']$ and $D[v'']$ is not greater than $\delta(d)$ where d is the depth of the subtree $D[v]$ rooted in the parent node v of nodes v' and v''. Denote by $\mathcal{D}^{(\delta)}$ the class of all δ-balanced rooted binary trees. In this work we give the following criterion for a given class $\mathcal{D}^{(\delta)}$ to be optimally resolvable by FAP-scheme.

Theorem 1. *A class $\mathcal{D}^{(\delta)}$ is optimally resolvable by FAP-scheme if and only if*

$$\sum_{i=1}^{s} \frac{\delta(d + i)}{2^i} = O(s + d)$$

where $s, d \in \mathbb{N}$.

Proof. Denote $\Delta(s, d) = \sum_{i=1}^{s} \frac{\delta(d + i)}{2^i}$. Let $\Delta(s, d) = O(s + d)$. Then we prove that $\mathcal{D}^{(\delta)}$ is optimally resolvable by FAP-scheme in the following way. Let D be a tree from $\mathcal{D}^{(\delta)}$, and $s \leq d(D)/2$ be a parallelization depth of FAP-scheme for solving of $P(D)$ by 2^s processors. Consider a subtree $D[v_0]$ from $ST^s(D)$ with the maximal number of nodes, i.e. $L(D[v_0]) = L^s(D)$. Let the path from v_0 to the root of D traverse nodes $v_0, v_1, \ldots, v_s$, i.e. v_i be the parent of v_{i-1} for $i = 1, 2, \ldots, s$ and v_s be the root of D. Denote by d_0 the maximal depth of subtrees from $ST^s(D)$. Note that $d(D[v_i]) \leq d_0 + i$ for any $i = 1, 2, \ldots, s$ and $d(D) = d_0 + s$. By induction for $n = 1, 2, \ldots, s$ we prove that

$$L(D[v_n]) \geq 2^n L(D[v_0]) + (2^n - 1) - \sum_{i=1}^{n} 2^{n-i} \delta(d_0 + i). \tag{1}$$

Let $n = 1$. Consider the sibling subtree $D[v_0']$ for the subtree $D[v_0]$. Since $D \in \mathcal{D}^{(\delta)}$, we have that the difference between the numbers of nodes of $D[v_0]$ and $D[v_0']$ is not greater than $\delta(d(D[v_1]))$, i.e.

$$L(D[v_0']) \geq L(D[v_0]) - \delta(d(D[v_1])).$$

Since $d(D[v_1]) \le d_0 + 1$, we have also $\delta(d(D[v_1])) \le \delta(d_0 + 1)$, so

$$L(D[v_0']) \ge L(D[v_0]) - \delta(d_0 + 1).$$

Thus,

$$L(D[v_1]) = L(D[v_0]) + L(D[v_0']) + 1 \ge 2L(D[v_0]) - \delta(d_0 + 1) + 1$$
$$= 2L(D[v_0]) + (2 - 1) - \sum_{i=1}^{1} 2^{1-i}\delta(d_0 + i).$$

Assume now that $n > 1$ and

$$L(D[v_{n-1}]) \ge 2^{n-1}L(D[v_0]) + (2^{n-1} - 1) - \sum_{i=1}^{n-1} 2^{n-1-i}\delta(d_0 + i). \qquad (2)$$

Consider the sibling subtree $D[v_{n-1}']$ for the subtree $D[v_{n-1}]$. Since $D \in \mathcal{D}^{(\delta)}$, we have that

$$L(D[v_{n-1}']) \ge L(D[v_{n-1}]) - \delta(d(D[v_n])).$$

Since $d(D[v_n]) \le d_0 + n$, we have also $\delta(d(D[v_n])) \le \delta(d_0 + n)$, so

$$L(D[v_{n-1}']) \ge L(D[v_{n-1}]) - \delta(d_0 + n).$$

Thus,

$$L(D[v_n]) = L(D[v_{n-1}]) + L(D[v_{n-1}']) + 1 \ge 2L(D[v_{n-1}]) - \delta(d_0 + n) + 1.$$

Therefore, applying inequality (2), we obtain that

$$L(D[v_n]) \ge 2\left(2^{n-1}L(D[v_0]) + (2^{n-1} - 1) - \sum_{i=1}^{n-1} 2^{n-1-i}\delta(d_0 + i)\right)$$
$$- \delta(d_0 + n) + 1 = 2^n L(D[v_0]) + (2^n - 1) - \sum_{i=1}^{n} 2^{n-i}\delta(d_0 + i).$$

Thus, taking into account that $D[v_s] = D$, for $n = s$ we have

$$L(D) = L(D[v_s]) \ge 2^s L(D[v_0]) + (2^s - 1) - \sum_{i=1}^{s} 2^{s-i}\delta(d_0 + i)$$
$$> 2^s L(D[v_0]) - \sum_{i=1}^{s} 2^{s-i}\delta(d_0 + i).$$

Dividing the last inequality by 2^s, we obtain that

$$\frac{L(D)}{2^s} > L(D[v_0]) - \sum_{i=1}^{s} \frac{\delta(d_0 + i)}{2^i} = L(D[v_0]) - \Delta(s, d_0).$$

Therefore,

$$L^s(D) = L(D[v_0]) < \frac{L(D)}{2^s} + \Delta(s, d_0) = \frac{L(D)}{2^s} + O(s + d_0) = \frac{L(D)}{2^s} + O(d(D)).$$

Thus,

$$T_{FAP}^s(D) = O(s + L^s(D)) = O(d(D) + L^s(D)) = O(\frac{L(D)}{2^s} + d(D)),$$

i.e. $\mathcal{D}^{(\delta)}$ is optimally resolvable by FAP-scheme.

Now let $\Delta(s, d)$ be not $O(s + d)$. First we prove that in this case $\delta(d)$ is not $O(d)$. Suppose by contradiction that $\delta(d) \leq Cd$ for some constant C. Then

$$\Delta(s, d) \leq \sum_{i=1}^{s} \frac{C(d+i)}{2^i} = C(\sum_{i=1}^{s} \frac{d}{2^i} + \sum_{i=1}^{s} \frac{i}{2^i})$$

$$< C(\sum_{i=1}^{\infty} \frac{d}{2^i} + \sum_{i=1}^{\infty} \frac{i}{2^i}) = C(d+2) = O(d)$$

which contradicts that $\Delta(s, d)$ is not $O(s + d)$. Thus, for any $C \geq 2$ we can consider a great enough d such that $\min(\delta(d+1), 2^{d+1} - 2) \geq C(d+1)$. Denote by D' the rooted binary tree defined in the following way. If $\delta(d+1) \geq 2^{d+1} - 2$ then D' is the perfect binary tree of depth d. Otherwise, consider the greatest odd number l not exceeding $\delta(d+1) + 1$. Note that

$$l > \delta(d+1) - 1 \geq C(d+1) - 1 \geq Cd + 1 \geq 2d + 1. \tag{3}$$

and $l \leq \delta(d+1) + 1 < (2^{d+1} - 2) + 1 = 2^{d+1} - 1$. Thus, by Proposition 1 there exists a rooted binary tree of depth d which has l nodes and is defined as D'. Consider a rooted binary tree obtained by attaching of the tree D' to a perfect binary tree as a subtree rooted in some leaf of this perfect tree. Denote the obtained tree by D''. Note that $d(D'') = 2d$ and

$$L(D'') = 2^{d+1} - 2 + L(D') \leq 2^{d+1} - 2 + 2^{d+1} - 1 < 2^{d+2},$$

so $L(D'')/2^d + d(D'') < 4 + 2d = O(d)$. Note also that D' is the maximal subtree from $ST^d(D'')$, so $L^d(D'') = L(D')$. Hence, by (3), $L^d(D'') > Cd$. Thus,

$$T_{FAP}^d(D'') = \Theta(d + L^d(D'')) = \Omega(L^d(D'')) = \Omega(Cd).$$

As C can be arbitrarily large, from $L(D'')/2^d + d(D'') = O(d)$ and $T_{FAP}^d(D'') = \Omega(Cd)$ we conclude that $T_{FAP}^s(D)$ is not $O(\frac{L(D)}{2^s} + d(D))$ for any tree D from $\mathcal{D}^{(\delta)}$ and any parallelization depth $s \leq d(D)/2$, i.e. $\mathcal{D}^{(\delta)}$ is not optimally resolvable by FAP-scheme.

5   Conclusions

In this paper we proved an optimality criterion of a basic parallel backtracking algorithm (FAP-scheme) for binary trees. The obtained criterion could be applied in practice in the case when the considered binary trees can be completely defined. The experimental evaluation of the obtained criterion is a subject of further research. Recall also that the obtained criterion is stated under the condition $s \leq d(D)/2$ (this constraint is based on an assumption that the number of processors assigned for solving of the problem $P(D)$ have to be significantly smaller than the number of nodes of D). A generalization of this criterion to the case of trees with arbitrary degree of branching of internal nodes or to the case when the parallelization depth s can be greater than $d(D)/2$ is one of the interesting directions of further research. Note that the obtained conditions of optimality of FAP-scheme are quite strict since FAP-scheme is quite simple algorithm of parallel solving of the backtrack search problem. Further we intend to study more complicated parallel schemes for backtrack search problems.

References

1. Barnett, M., Shuler, L., van De Geijn, R., Gupta, S., Payne, D.G., Watts, J.: Interprocessor collective communication library (intercom). In: Proceedings of IEEE Scalable High Performance Computing Conference, pp. 357–364. IEEE (1994)
2. Bhatt, S., Greenberg, D., Leighton, T., Liu, P.: Tight bounds for on-line tree embeddings. SIAM J. Comput. **29**(2), 474–491 (1999)
3. Evtushenko, Y., Posypkin, M., Sigal, I.: A framework for parallel large-scale global optimization. Comput. Sci. Res. Dev. **23**(3–4), 211–215 (2009)
4. Herley, K.T., Pietracaprina, A., Pucci, G.: Deterministic parallel backtrack search. Theoret. Comput. Sci. **270**(1–2), 309–324 (2002)
5. Karp, R.M., Zhang, Y.: Randomized parallel algorithms for backtrack search and branch-and-bound computation. J. ACM (JACM) **40**(3), 765–789 (1993)
6. Knuth, D.E.: Art of Computer Programming, volume 2: Seminumerical Algorithms. Addison-Wesley Professional, Boston (2014)
7. Kolpakov, R., Posypkin, M.: Estimating the computational complexity of one variant of parallel realization of the branch-and-bound method for the knapsack problem. J. Comput. Syst. Sci. Int. **50**(5), 756 (2011)
8. Kolpakov, R., Posypkin, M.: The lower bound on complexity of parallel branch-and-bound algorithm for subset sum problem. In: AIP Conference Proceedings, vol. 1776, p. 050008. AIP Publishing (2016)
9. Kolpakov, R.M., Posypkin, M.A., Sigal, I.K.: On a lower bound on the computational complexity of a parallel implementation of the branch-and-bound method. Autom. Remote Control **71**(10), 2152–2161 (2010)
10. Pietracaprina, A., Pucci, G., Silvestri, F., Vandin, F.: Space-efficient parallel algorithms for combinatorial search problems. J. Parallel Distrib. Comput. **76**, 58–65 (2015)
11. Sergeyev, Y., Grishagin, V.: Parallel asynchronous global search and the nested optimization scheme. J. Comput. Anal. Appl. **3**(2), 123–145 (2001)

12. Strongin, R., Sergeyev, Y.: Global multidimensional optimization on parallel computer. Parallel Comput. **18**(11), 1259–1273 (1992)
13. Wu, I.C., Kung, H.T.: Communication complexity for parallel divide-and-conquer. In: 32nd Annual Symposium on Foundations of Computer Science, 1991 Proceedings, pp. 151–162. IEEE (1991)

Well Posedness of the Nearest Points Problem for Two Sets in Asymmetric Seminormed Spaces

Mariana S. Lopushanski[✉][iD]

Moscow Institute of Physics and Technology,
9 Institutskiy per., Dolgoprudny, Moscow Region 141700, Russian Federation
masha.alexandra@gmail.com

Abstract. In this paper we consider spaces with an asymmetric seminorm and continue to study weakly convex sets. If we consider the Minkowski functional of the epigraph of some convex function as a seminorm, then the results obtained for weakly convex sets can be applied to weakly convex functions whose epigraphs are weakly convex sets with respect to this seminorm. We consider two sets in an asymmetric seminormed space, one of which is weakly convex with respect to an asymmetric seminorm, and the other one is strongly convex with respect to the asymmetric seminorm. We study the nearest points (in the sense of seminorm) problem and prove that this problem is well posed. Well posedness is an important property in the optimization theory. If a minimization problem is well posed, then one can build stable numerical algorithms, used for finding the solution of the problem.

Keywords: Asymmetric seminorm · Well posedness · Minimization · Convex projection · Weakly convex sets · Strongly convex sets

1 Introduction

Convex analysis is an important instrument of optimization and approximation. To obtain the necessary estimates in optimization we often need some quantitative characteristics of convexity. The parametrical convex analysis considers these characteristics of sets, providing the necessary information on the geometric and approximative properties of sets and functions. Thus, parametrically convex analysis deals with weakly convex and strongly convex sets.

Weakly convex sets were first introduced in [6] under the name of sets with positive reach in $\mathbb{R}^n$. [2] studied proximally smooth sets in Hilbert space – sets with continuously differentiable distance function in some neighbourhood of the set. Later [24] considered prox-regular sets in Hilbert spaces and proved that in a Hilbert space the class of uniformly prox-regular sets coincides with the classes of proximally smooth and weakly convex sets.

Supported by the Russian Foundation for Basic Research, grant 18-01-00209.

Weakly convex and strongly convex sets are often used in differential games for the construction of more effective algorithms. In [7] the alternative theorem for a linear differential game with strongly convex admissible control sets and smooth target set was obtained. In [8] sufficient conditions of smoothness of reachable sets for nonlinear differential games were considered.

We strive to develop a unified approach for weakly convex sets and functions. So we study weakly convex sets in asymmetric seminormed spaces. Then the results obtained for weakly convex sets can be used for solving optimization problems for weakly convex functions, working with their epigraphs. Asymmetric normed and seminormed spaces were studied in [4,5,21]. Approximation problems for weakly convex sets in such spaces were considered in [1,11–16,22,23]. In [13] weakly convex function were characterized as functions with weakly convex epigraphs in a seminormed space.

An important notion in optimization is the notion of "well posed problem". A minimization problem is Tikhonov well posed (further - well posed) if it admits a unique minimizer and any minimizing sequence converges to this minimizer. Well posedness property of a minimization problem is needed for stability of numerical algorithms, used for finding the solution of the problem.

In this paper we consider the nearest points problem for two sets, one of which is weakly convex, and the other one is strongly convex (in the sense of the seminorm). We prove that this problem is well posed. This paper generalizes the results obtained in [10] for Banach spaces with a quasiball. In [10] a Banach space was considered, in which we then defined a quasiball M as a closed convex set such that $0 \in \text{int } M$. However, the topology of this space was induced by the norm of the Banach space, and was not connected to the quasiball introduced. Also the modulus of bounded uniform convexity (a generalization of uniform convexity for unbounded sets) was defined using both the initial norm and the quasiball. In this paper we deal directly with asymmetric normed spaces without any predefined norm.

2   Definitions and Notation

Let E be a real vector space. Next, we provide some necessary definitions and notation, used throughout this paper.

Definition 1 (See [3]). *A function* $\mu : E \to \mathbb{R}$ *is called* sublinear *if it is positively homogeneous:*

$$\mu(\lambda x) = \lambda \mu(x) \qquad \forall x \in E \quad \forall \lambda \geq 0$$

and subaditive*:*

$$\mu(x + y) \leq \mu(x) + \mu(y) \qquad \forall x, y \in E.$$

Definition 2 (See [3]). *A sublinear function* $\mu : E \to [0, +\infty)$ *such that*

$$\max\{\mu(x), \mu(-x)\} > 0 \qquad \forall x \in E \setminus \{0\} \tag{1}$$

is called asymmetric seminorm*. The pair* (E, μ) *is called* asymmetric seminormed space*.*

Definition 3. *If asymmetric seminorm μ satisfies additional condition*

$$\mu(x) > 0 \qquad \forall x \in E \setminus \{0\},$$

then it is called asymmetric norm. *In such a case the pair (E, μ) is called* asymmetric normed space.

In some literature (e.g. [5]) a bit different terminology is used: an asymmetric seminorm is called an asymmetric norm and a sublinear function $\mu : E \to [0, +\infty)$ (without axiom (1)) is called an asymmetric seminorm.

Definition 4. *For $\varepsilon > 0$ and $x \in E$ use $U_\varepsilon(x)$ to denote ε-neighbourhood of x:*

$$U_\varepsilon(x) = \{y \in E : \ \mu(x - y) < \varepsilon\}.$$

A set $X \subset E$ is called μ-open if for any $x \in X$ there exists $\varepsilon > 0$ such that $U_\varepsilon(x) \subset X$. We shall use τ_μ to denote the family of all μ-open subsets $X \subset E$.

Remark 1. (E, τ_μ) is a topological space.

Remark 2. If μ is an asymmetric seminorm, then (E, τ_μ) is not a Hausdorff space in general. Consider, for example, an asymmetric seminormed space (E, μ) with $E = \mathbb{R}$ and

$$\mu(x) = \begin{cases} x, \ x \geq 0, \\ 0, \ x < 0. \end{cases}$$

Then for any $x \in \mathbb{R}$, $\varepsilon > 0$ we have $U_\varepsilon(x) = (x - \varepsilon, +\infty)$ and (E, τ_μ) is not a Hausdorff space.

Definition 5 (See [5]). *Let X be an arbitrary set. A function $\varrho : X \times X \to [0, +\infty)$ is called a* quasi-metric *if*

$$\varrho(x, x) = 0 \quad \forall x \in X,$$

$$\varrho(x, z) \leq \varrho(x, y) + \varrho(y, z) \qquad \forall x, y, z \in X,$$

$$\varrho(x, y) = \varrho(y, x) = 0 \quad \Rightarrow \quad x = y \qquad \forall x, y \in X.$$

Remark 3. Any asymmetric seminormed space (E, μ) possesses a quasi-metric

$$\varrho_\mu(x, y) = \mu(x - y), \qquad x, y \in E,$$

a metric

$$\varrho_\mu^s(x, y) = \max\{\mu(x - y), \mu(y - x)\}, \qquad x, y \in E,$$

and a norm

$$\|x\|_\mu = \max\{\mu(x), \mu(-x)\}, \qquad x \in E.$$

We shall use $\mathrm{cl}X$ to denote the closure of the subset $X \subset E$ with respect the topology induced by metric ϱ_μ^s. Let E^* be the space of linear functionals $p : E \to \mathbb{R}$, which are continuous with respect to metric ϱ_μ^s.

Definition 6. *The* diameter *of a set $X \subset E$ is defined as*

$$\operatorname{diam} X = \sup_{x,y \in X} \mu(x - y).$$

We generalize the definition given in [5] for an asymmetric normed space as follows.

Definition 7. *A sequence $\{x_k\}$ in an asymmetric seminormed space (E, μ) is called*

– ϱ_μ^s-Cauchy *if*

$$\forall \varepsilon > 0 \; \exists N_\varepsilon : \; \forall n, k \geq N_\varepsilon \qquad \varrho_\mu^s(x_n, x_k) < \varepsilon,$$

– convergent to $x \in E$ *(we write $x_k \to x$) if $\varrho_\mu^s(x_k, x) \to 0$ as $k \to \infty$.*

Definition 8. *An asymmetric seminormed space (E, μ) is called* biBanach space *if any ϱ_μ^s-Cauchy sequence $\{x_k\}$ converges to some $x \in E$.*

Remark 4. If (E, μ) is a biBanach space, then $(E, \|\cdot\|_\mu)$ is a Banach space.

Definition 9. *Let (E, μ) be an asymmetric seminormed space.*
The μ-distance from a point $x \in E$ to a set $A \subset E$ is

$$\varrho_\mu(x, A) = \inf_{a \in A} \mu(x - a).$$

The μ-distance from a set $C \subset E$ to a set $A \subset E$ is

$$\varrho_\mu(C, A) = \inf_{\substack{c \in C \\ a \in A}} \mu(c - a).$$

The μ-projection (or μ-nearest point) for a point $x \in E$ on a set $A \subset E$ is

$$P_\mu(x, A) = \{a \in A : \mu(x - a) = \varrho_\mu(x, A)\}.$$

Given $\varepsilon > 0$, the ε-μ-projection of a point $x \in E$ on a set $A \subset E$ is defined as

$$P_\mu^\varepsilon(x, A) = \{a \in A : \mu(x - a) \leq \varrho_\mu(x, A) + \varepsilon\}.$$

The cone of proximal normals to a set $A \subset E$ at a point $a \in A$ is

$$N_\mu(a, A) = \{z \in E \mid \exists t > 0 : \; a \in P_\mu(a + tz, A)\}.$$

Definition 10. *A set $A \subset E$ is called μ-weakly convex if for any $a \in A$ and $z \in N_\mu(a, A)$ with $\mu(z) = 1$ one has $a \in P_\mu(a + z, A)$.*

Ivanov in [9] suggested the following definition, which generalized the notion of parabolic sets, also introduced by him (see, for example, [10]).

Definition 11. *An asymmetric seminormed space* (E, μ) *is called* parabolic *if for any* $b \in E$

$$\sup_{\substack{x \in E:\ \mu(x) \leq 1,\\ \mu(x+b) > 2}} \mu(-x) < +\infty.$$

Definition 12. *An asymmetric seminormed space* (E, μ) *is called* uniformly convex *if* $\delta_\mu(\varepsilon, R) > 0$ *for any positive* ε *and* R, *where*

$$\delta_\mu(\varepsilon, R) = \inf \left\{ 1 - \mu\left(\tfrac{x_1 + x_2}{2}\right) \middle| \begin{array}{l} x_1, x_2 \in E : \\ \mu(x_1) \leq 1, \ \mu(x_2) \leq 1, \\ \mu(-x_1) \leq R, \ \mu(-x_2) \leq R, \\ \mu(x_1 - x_2) \geq \varepsilon \end{array} \right\}. \tag{2}$$

Remark 5. If (E, μ) is a uniformly convex asymmetric seminormed space, then $(E, \|\cdot\|_\mu)$ is a uniformly convex (and consequently reflexive) normed space.

Remark 6. An asymmetric seminormed space (E, μ) is *uniformly convex*, if $\lim_{t \to +0} \psi_\mu(t, R) = 0$ for any $R > 0$, where $\psi_\mu(\cdot, \cdot)$ is

$$\psi_\mu(t, R) = \sup \left\{ \mu(x_1 - x_2) \middle| \begin{array}{l} x_1, x_2 \in E : \\ \mu(x_1) \leq 1, \ \mu(x_2) \leq 1, \\ \mu(-x_1) \leq R, \ \mu(-x_2) \leq R, \\ 1 - \mu\left(\tfrac{x_1 + x_2}{2}\right) \leq t, \end{array} \right\}, \quad R > 0, \ t \geq 0. \tag{3}$$

Note that $\psi_\mu(t, R)$ is increasing with respect to t.

3   Motivation of the Problem

In this section we will discuss the application to optimization problems of the main result of this paper. But first, let us give the necessary definitions.

Given a function $f : A \to \mathbb{R} \cup \{+\infty\}$, let us consider the problem

$$\min_{x \in E} f(x). \tag{4}$$

A sequence $\{x_k\} \subset E$ is said to be a *minimizing sequence* if

$$\lim_{k \to \infty} f(x_k) = \inf_{x \in E} f(x).$$

Definition 13. *The problem (4) is called* well posed, *if every minimizing sequence of this problem converges.*

Remark 7. If the problem (4) is well posed and function $f(\cdot)$ is lower-semicontinuous, then the minimizer is unique.

Recall that the *epigraph* of function $f : E \to \mathbb{R} \cup \{-\infty, +\infty\}$ is defined by

$$\text{epi } f = \{(x, y) \in E \times \mathbb{R} \mid x \in E,\ y \geq f(x)\};$$

the *effective domain* of f is

$$\text{dom } f = \{x \in E \mid f(x) \in \mathbb{R}\}.$$

The *infimal convolution* of the functions $f : E \to \mathbb{R} \cup \{+\infty\}$ and $g : E \to \mathbb{R} \cup \{+\infty\}$ is the function $f \boxplus g$, defined by

$$(f \boxplus g)(x) = \inf_{u \in E} \Big(f(x - u) + g(u) \Big), \qquad x \in E.$$

Remark 8. For any functions $f, g : E \to \mathbb{R} \cup \{+\infty\}$ the following inclusions hold

$$\text{epi } f + \text{epi } g \subset \text{epi } (f \boxplus g) \subset \text{clepi } f + \text{epi } g.$$

So, the infimal convolution of functions f and g is a function whose epigraph coincides up to closure with the Minkowski sum of the epigraphs of the functions f and g.

In [9] the nearest point problem was considered. Let there be given functions $f, g : E \to \mathbb{R} \cup \{+\infty\}$ and the set $A = \text{epi } f$. Put $M = \text{epi } g$. Let $\mu(\cdot)$ be the Minkowski functional of M and $c = (x_0, y_0) \in E \times \mathbb{R}$ (here $x \in E$ and $y \in \mathbb{R}$) be such that $\rho_\mu(c, A) = 1$. The nearest point problem can be rewritten as follows

$$\min_{a \in A} \mu(c - a). \tag{5}$$

Let us show that we can reformulate this problem using the notion of infimal convolution.

Consider $\inf_{u \in E} (f(x - u) + g(u)) = \lambda$. Suppose that the minimum is attained. Then there exists $u_0 \in E$ such that

$$g(x_0 - u_0) + f(u_0) = \min_{u \in E}(f(u) + g(x_0 - u)).$$

Consider $a = (x, y) \in A$, where $x \in E$ and $y \in \mathbb{R}$. Problem (5) can be rewritten as follows

$$\min_{\substack{x \in E,\, y \geq f(x), \\ (x_0, y_0) - (x, y) \in t M}} t.$$

As $\rho_\mu(c, A) = 1$, we have that $\inf\limits_{\substack{c - a \in t M, \\ a \in A}} t = 1$. The inclusion $c - a \in M$ can be rewritten as $(x_0 - x, y_0 - y) \in \text{epi } g$. Thus, $y_0 - y \geq g(x_0 - x)$. As we search the minimum on condition that $y \geq f(x)$, we obtain that

$$f(x) \leq y_0 - g(x_0 - x).$$

Hence,

$$f(x) + g(x_0 - x) \leq y_0.$$

If the minimum by t is obtained and unique, then there exists $u_0 \in E$ such that $f(u_0) + g(x_0 - u_0) = y_0$. Thus for any $u \in E \setminus \{u_0\}$ we have that $f(u) + g(x_0 - u) > y_0$. Thus the well-posedness of nearest points problem (5) for sets that can be described as the epigraph of some function can be reduced to the problem of well-posedness of infimal convolution (see Theorem 3.6 [13]).

The infimal convolution problem is very important in optimization. Applications of the infimal convolution to optimal control are considered e.g. in [17–19]. In [20] it is shown that well-posedness properties of optimization problems are very important in subdifferential calculus.

4   Auxiliary Results

In this section, we provide some auxiliary results and also recall some lemmas, proved in previous works. In this section we will prove that the ε-μ-projection of a r-μ-strongly convex set (the definition will be given later) on a μ-weakly convex set is bounded. In the main result section we will prove that μ-projection of a r-μ-strongly convex set (the definition will be given later) on a μ-weakly convex set consists of a single point.

Lemma 1. *Let* (E, μ) *be a uniformly convex biBanach asymmetric seminormed space and let the vectors* $x, y \in E$ *and the number* $R > 0$ *satisfy the inequalities*
$$0 < \mu(-x) \leq R \cdot \mu(x), 0 < \mu(-y) \leq R \cdot \mu(y). \text{ Then}$$
$$\mu\left(\frac{x}{\mu(x)} - \frac{y}{\mu(y)}\right) \leq \psi_\mu\left(\frac{\mu(x) + \mu(y) - \mu(x + y)}{2 \min\{\mu(x), \mu(y)\}}, R\right),$$
where $\psi_\mu(\cdot, \cdot)$ *is defined by formula (3).*

Proof. Let us denote
$$a = \frac{x}{\mu(x)}, \qquad b = \frac{y}{\mu(y)},$$
$$c = \frac{a + b}{2}, \qquad \mu_1 = \mu(x), \qquad \mu_2 = \mu(y),$$
$$t = \frac{\mu(x) + \mu(y) - \mu(x + y)}{\min\{\mu(x), \mu(y)\}}.$$
Without loss of generality we assume that $\mu_2 \leq \mu_1$. Using the sublinearity of function μ and taking into account that $\mu(a) = 1$, we have that
$$\mu(x + y) = \mu((\mu_1 - \mu_2)a + \mu_2(a + b))$$
$$\leq (\mu_1 - \mu_2) \cdot \mu(a) + \mu_2 \cdot \mu(a + b) = \mu_1 + \mu_2 - 2\mu_2 \cdot (1 - \mu(c)).$$
Thus
$$\mu(c) \geq 1 - \frac{t}{2}. \tag{6}$$
Inequality (6) and the facts that $\mu(a) = \mu(b) = 1$ and $\mu(-a) \leq R$, $\mu(-b) \leq R$, according to the equality (3), imply that $\mu(a - b) \leq \psi_\mu(t, R)$.

Lemma 2. *Let* (E, μ) *be a biBanach asymmetric seminormed space and let the set* $A \subset E$ *be* μ*-weakly convex, and let the point* $x_0 \in E$ *and the number* $\varepsilon > 0$ *be such that* $0 < \varrho_\mu(x_0, A) = \varrho < 1 - \varepsilon$. *Let* $a_0 \in P_\mu(x_0, A)$, $a \in P_\mu^\varepsilon(x_0, A)$ *and* $R = \max\left\{ \frac{\|x_0 - a_0\|_\mu}{\mu(x_0 - a_0)}, \frac{\|x_0 - a\|_\mu}{\mu(x_0 - a)} \right\}$. *Then*

$$\|a - a_0\|_\mu \leq \varepsilon R + \varrho \psi_\mu \left(\frac{\varepsilon}{2 \min\{\varrho, 1 - \varrho\}}, R \right).$$

Proof. Let us denote $z = \frac{x_0 - a_0}{\varrho}$, $x = x_0 - a$, $y = (1 - \varrho)z$. As the set A is μ-weakly convex and $z \in N_\mu(a_0, A)$, $\mu(z) = 1$, then $\varrho_\mu(a_0 + z, A) = 1$. Thus, $\varrho \leq \mu(x) \leq \varrho + \varepsilon$, $\mu(x + y) = \mu(a_0 + z - a) \geq \varrho_\mu(a_0 + z, A) = 1 \geq \mu(x) + \mu(y) - \varepsilon$. So, according to Lemma 1 and Remark 3 we have that

$$\left\| \frac{x}{\mu(x)} - \frac{y}{\mu(y)} \right\|_\mu \leq \psi_\mu \left(\frac{\varepsilon}{2 \min\{\varrho, 1 - \varrho\}}, R \right).$$

The relations $\left\| \frac{x}{\mu(x)} - \frac{x}{\varrho} \right\|_\mu = \frac{\|x\|_\mu}{\mu(x)} \frac{|\mu(x) - \varrho|}{\varrho} \leq \frac{\varepsilon R}{\varrho}$, $\left\| \frac{y}{\mu(y)} - \frac{x}{\varrho} \right\|_\mu = \frac{\|a - a_0\|_\mu}{\varrho}$ imply that the inequality

$$\|a - a_0\|_\mu \leq \varepsilon R + \varrho \psi_\mu \left(\frac{\varepsilon}{2 \min\{\varrho, 1 - \varrho\}}, R \right)$$

holds.

Let (E, μ) be a biBanach asymmetric seminormed and two sets $A, C \subset E$. Consider the problem

$$\min_{a \in A, \ c \in C} \mu(c - a). \tag{7}$$

According to the Definition 13 the problem (7) is well posed if any two sequences $\{a_k\} \subset A$ and $\{c_k\} \subset C$ such that $\lim_{k \to \infty} \mu(c_k - a_k) = \varrho_\mu(C, A)$ converge to some points $\hat{a} \in A$ and $\hat{c} \in C$ respectively. In addition, due to the continuity of the Minkowski functional, $\mu(\hat{c} - \hat{a}) = \varrho_\mu(C, A)$, i.e. the pair of points $(\hat{a}, \hat{c})$ is the solution of the problem (7).

Let (E, μ) be a biBanach asymmetric seminormed. We will say that

$$P_\mu^\varepsilon(C, A) = \{a \in A \mid \exists c \in C : \mu(c - a) \leq \varrho_\mu(C, A) + \varepsilon\}$$

.

By μ^- we denote the function $\mu^- : E \to \mathbb{R}$ such that $\mu^-(x) = \mu(-x)$.

Remark 9. Let E be a Banach space, $M \subset E$ a quasiball and let sets $A \subset E$ and $C \subset E$ be closed. The following statements are equivalent:

(1) $\lim_{\varepsilon \to +0} \operatorname{diam} P_\mu^\varepsilon(C, A) = 0$ and
   $\lim_{\varepsilon \to +0} \operatorname{diam} P_{\mu^-}^\varepsilon(A, C) = 0$;
(2) the problem (7) is well posed.

Proposition 1. *[Lemma 4, [9]] Let (E, μ) be a parabolic asymmetric semi-normed space. Then for any $\lambda_1 > 0$, $\lambda_2 > \lambda_1$ and $x_1, x_2 \subset E$ one has*

$$\sup_{\substack{x \in E: \ \mu(x - x_1) \leq \lambda_1, \\ \mu(x - x_2) > \lambda_2}} \mu(-x) < +\infty.$$

Remark 10. In Proposition 1 the subadditivity of function $\mu(\cdot)$ implies that $\mu(x) \leq \lambda_1 + \mu(x_1)$. Thus

$$\sup_{\substack{x \in E: \ \mu(x - x_1) \leq \lambda_1, \\ \mu(x - x_2) > \lambda_2}} \|x\|_\mu < +\infty.$$

The *Minkowski sum* of sets $A, B \subset E$ is

$$A + B = \{a + b \mid a \in A, \ b \in B\}.$$

Definition 14. *Let there be given a set $M \subset E$ such that $M = \{x \in E : \mu(x) \leq 1\}$. We will say that a set $C \subset E$ is r-μ-strongly convex if it is convex, μ-closed (that it, the set $E \setminus C$ is μ-open) and there exists a set $C_1 \subset E$ such that $C + C_1 = -rM$, $r > 0$.*

Theorems 1 and 2 [9] imply the following Proposition.

Proposition 2. *Let (E, μ) be a uniformly convex parabolic biBanach asymmetric seminormed space and let $A \subset E$ be a μ-closed set. Then the following statements are equivalent:*

(1) the set A in μ-weakly convex;
(2) for any point $x_0 \in E$ such that $0 < \varrho_\mu(x_0, A) < 1$ the problem

$$\min_{a \in A} \mu(x_0 - a)$$

is well posed;

Lemma 3. *Let (E, μ) be a uniformly convex parabolic biBanach asymmetric seminormed space. Let the set $A \subset E$ be μ-closed and μ-weakly convex. Let the set $C \subset E$ be r-μ-strongly convex, $r \in (0, 1)$, $\varepsilon > 0$, $0 < \varrho_\mu(C, A) < 1 - r - \varepsilon$. Then $\operatorname{diam} P_\mu^\varepsilon(C, A) < +\infty$, $\operatorname{diam} P_{\mu^-}^\varepsilon(A, C) < +\infty$.*

Proof. Consider an arbitrary point $x \in E \setminus A$ such that $\varrho_\mu(x, A) \geq 1$. The definition of r-μ^--strong convexity implies that there exists a set $C_1 \subset E$ such that $C + C_1 = -rM$, where $M = \{x \in E : \mu(x) \leq 1\}$. Consider an arbitrary point $c_1 \in C_1$. Then $c_1 + C \subset -rM$. Consider an arbitrary point $y \in P_\mu^\varepsilon(C, A)$ and $y' \in P_{\mu^-}^\varepsilon(A, C)$. Then

$$\mu(y) > \mu(x - z), \quad \mu(y') > \mu(x - (1 - \varrho_\mu(C, A) - \varepsilon)z)$$

and

$$\mu(y) \leq \mu(-c_1 + (\varrho_\mu(C, A) + r + \varepsilon)z),$$
$$\mu(y') \leq \mu(-c_1 - rz), \quad \forall z : \mu(z) \leq 1.$$

Thus, according to Proposition 1, the assertion of the lemma is true.

We say that a set $A \subset E$ is μ^--weakly convex if for any $a \in A$ and $z \in N_{\mu^-}(a, A)$ with $\mu(-z) = 1$ one has $a \in P_{\mu^-}(a + z, A)$, where

$$P_{\mu^-}(x, A) = \{a \in A : \mu(a - x) \leq \inf_{a' \in A} \mu(a' - x)\}.$$

and

$$N_{\mu^-}(a, A) = \{z \in E \mid \exists t > 0 : a \in P_{\mu^-}(a + tz, A)\}.$$

Lemma 3.1 [10] implies the following proposition.

Proposition 3. *Let (E, μ) be a uniformly convex biBanach asymmetric seminormed space. Let set $C \subset E$ be a summand of the set $M = \{x \in E : \mu(x) = 1\}$. Then*

$$C - c_0 \subset M - z \qquad \forall c_0 \in C \quad \forall z \in N_\mu(c_0, C) \bigcap \partial M.$$

5   Main Result

Theorem 1. *Let (E, μ) be a uniformly convex parabolic biBanach asymmetric seminormed space. Let $A \subset E$ be μ-closed and μ-weakly convex. Let $C \subset E$ be r-μ-strongly convex, $r \in (0, 1)$. Let $0 < \varrho_\mu(C, A) < 1 - r$. Then problem (7) is well posed.*

Proof. We denote

$$\varrho_0 = \varrho_\mu(C, A), \qquad A_\varepsilon = P_\mu^\varepsilon(C, A), \qquad C_\varepsilon = P_{\mu^-}^\varepsilon(A, C).$$

Let us fix a number $\varepsilon_0 \in (0, 1 - r - \varrho_0)$. According to Lemma 3 the sets A_{ε_0} and C_{ε_0} are bounded in the sense of $\|\cdot\|_\mu$. This and the inequality $\mu(c - a) \geq \varrho_0 > 0$, which holds for any $a \in A$, $c \in C$, imply that

$$R := \sup_{a \in A_{\varepsilon_0}, \, c \in C_{\varepsilon_0}} \frac{\|c - a\|_\mu}{\mu(c - a)} < +\infty. \tag{8}$$

We fix a number $\varepsilon \in (0, \varepsilon_0]$ and a point $a_\varepsilon \in A_\varepsilon$. As the set C is convex, it is μ^--weakly convex (see Lemma 2 [9]). This and Proposition 2 imply that there exists a point $c_\varepsilon \subset E$ such that $c_\varepsilon \in P_{\mu^-}(a_\varepsilon, C)$. As the vector $z_\varepsilon = \frac{c_\varepsilon - a_\varepsilon}{\mu(c_\varepsilon - a_\varepsilon)}$ satisfies the inclusion $z_\varepsilon \in -N_{\mu^-}(c_\varepsilon, C)$ and $\mu(z_\varepsilon) = 1$ then, according to Proposition 3, we have

$$C - c_\varepsilon \subset r(z_\varepsilon - M), \tag{9}$$

where $M = \{x \in E : \mu(x) \leq 1\}$. Using the inclusion $a_\varepsilon \in A_\varepsilon$, we get $\varrho_{-\mu}(a_\varepsilon, C) = \inf_{c \in C} \mu(c - a_\varepsilon) \leq \varrho_0 + \varepsilon$. Therefore

$$\varrho_0 \leq \varrho(c_\varepsilon, A) \leq \mu(c_\varepsilon - a_\varepsilon)$$
$$= \mu(-a_\varepsilon + c_\varepsilon) = \varrho_{\mu^-}(a_\varepsilon, C) \leq \varrho_0 + \varepsilon < 1 - r. \tag{10}$$

Proposition 2 implies that there exists a point

$$a'_\varepsilon \in P_\mu(c_\varepsilon, A). \tag{11}$$

Then the vector $z'_\varepsilon = \frac{c_\varepsilon - a'_\varepsilon}{\mu(c_\varepsilon - a'_\varepsilon)}$ satisfies the inclusion $z'_\varepsilon \in N_\mu(a'_\varepsilon, A)$ and the equality $\mu(z) = 1$. As the set A is μ-weakly convex, we have

$$a'_\varepsilon \in P_\mu(a'_\varepsilon + z'_\varepsilon, A). \tag{12}$$

According to the relation (10), we have $a_\varepsilon \in P^\varepsilon_\mu(c_\varepsilon, A)$. The equality (8) and the inclusions $a_\varepsilon \in A_\varepsilon$, $a'_\varepsilon \in A_\varepsilon$, $c_\varepsilon \in C_\varepsilon$ imply the inequalities

$$\|z_\varepsilon\|_\mu \le R, \qquad \|z'_\varepsilon\|_\mu \le R. \tag{13}$$

Lemma 2 and the relations (10), (11) imply that

$$\|a_\varepsilon - a'_\varepsilon\| \le \Delta(\varepsilon) := \varepsilon R + \psi_\mu \left(\frac{\varepsilon}{\min\{\varrho_0, r\}}, R \right). \tag{14}$$

As

$$
\begin{aligned}
z_\varepsilon - z'_\varepsilon &= \frac{c_\varepsilon - a_\varepsilon}{\mu(c_\varepsilon - a_\varepsilon)} - \frac{c_\varepsilon - a'_\varepsilon}{\mu(c_\varepsilon - a'_\varepsilon)} \\
&= \frac{a'_\varepsilon - a_\varepsilon}{\mu(c_\varepsilon - a_\varepsilon)} + \left(\frac{1}{\mu(c_\varepsilon - a_\varepsilon)} - \frac{1}{\mu(c_\varepsilon - a'_\varepsilon)} \right)(c_\varepsilon - a'_\varepsilon),
\end{aligned}
$$

we have

$$\|z_\varepsilon - z'_\varepsilon\|_\mu \le \frac{\|a_\varepsilon - a'_\varepsilon\|_\mu}{\varrho_0} + R \frac{|\mu(c_\varepsilon - a'_\varepsilon) - \mu(c_\varepsilon - a_\varepsilon)|}{\varrho_0}.$$

Remark 3 implies that function $\mu(\cdot)$ is Lipschitz with constant 1. Therefore

$$\|z_\varepsilon - z'_\varepsilon\|_\mu \le \frac{\|a_\varepsilon - a'_\varepsilon\|_\mu}{\varrho_0}(1 + R). \tag{15}$$

Since

$$\mu(c_\varepsilon - a'_\varepsilon) = \varrho_\mu(c_\varepsilon, A) \le \mu(c_\varepsilon - a_\varepsilon) \le \varrho_0 + \varepsilon < 1 - r$$

and

$$c_\varepsilon + r z'_\varepsilon - a'_\varepsilon = \left(\mu(c_\varepsilon - a'_\varepsilon) + r \right) z'_\varepsilon, \tag{16}$$

we have $c_\varepsilon + r z'_\varepsilon \in [a'_\varepsilon, a'_\varepsilon + z'_\varepsilon]$. This and the inclusion (12) imply that

$$a'_\varepsilon \in P_\mu(c_\varepsilon + r z'_\varepsilon, A). \tag{17}$$

We denote

$$\Delta_1(\varepsilon) = \varepsilon + \frac{r(1 + R)}{\varrho_0} \Delta(\varepsilon). \tag{18}$$

Let us prove that

$$A_\varepsilon \subset P_\mu^{\Delta_1(\varepsilon)}(c_\varepsilon + r z'_\varepsilon, A). \tag{19}$$

Indeed, let $a \in A_\varepsilon$. Using the inclusion (9), we obtain that $a \in C - (\varrho_0 + \varepsilon)M \subset c_\varepsilon + r z_\varepsilon - (r + \varrho_0 + \varepsilon)M$, i.e. $\mu(c_\varepsilon + r z_\varepsilon - a) \le r + \varrho_0 + \varepsilon$. Therefore the relations (14), (15), (18) imply that

$$
\begin{aligned}
\mu(c_\varepsilon + r z'_\varepsilon - a) &\le r + \varrho_0 + \varepsilon + r\mu(z'_\varepsilon - z_\varepsilon) \\
&\le r + \varrho_0 + \varepsilon + r\|z'_\varepsilon - z_\varepsilon\|_\mu \le r + \varrho_0 + \Delta_1(\varepsilon).
\end{aligned}
\tag{20}
$$

The relations (16), (17) imply that

$$\varrho_\mu(c_\varepsilon + rz'_\varepsilon, A) = \mu(c_\varepsilon + rz'_\varepsilon - a'_\varepsilon)$$
$$= \mu(c_\varepsilon - a'_\varepsilon) + r \geq \varrho_0 + r. \tag{21}$$

Combining the inequalities (20), (21), we obtain that $\mu(c_\varepsilon + rz'_\varepsilon - a) \leq \varrho(c_\varepsilon + rz'_\varepsilon, A) + \Delta_1(\varepsilon)$, which proves the inclusion (19).

The inequalities (13), (21), the inclusions $a'_\varepsilon \in A_\varepsilon \subset A_{\varepsilon_0}$, $c_\varepsilon \in C_\varepsilon \subset C_{\varepsilon_0}$, and the boundedness of the set A_{ε_0}, C_{ε_0} imply that

$$\sup_{\varepsilon \in (0,\varepsilon_0]} \sup_{a \in A_\varepsilon} \frac{\|c_\varepsilon + rz'_\varepsilon - a\|_\mu}{\mu(c_\varepsilon + rz'_\varepsilon - a)} < +\infty.$$

Then, according to Lemma 2 and the relations (17)–(21), we obtain that diam $A_\varepsilon \to 0$ as $\varepsilon \to +0$. As the sets A_ε are included in each other and closed, there exists a point $\hat{a} \in \bigcap_{\varepsilon>0} A_\varepsilon$.

Let us prove that

$$C_\varepsilon \subset P_{\mu^-}^{\varepsilon + \text{diam } A_\varepsilon}(\hat{a}, C) \qquad \forall \varepsilon \in (0, \varepsilon_0]. \tag{22}$$

Let $\varepsilon \in (0, \varepsilon_0]$, $c \in C_\varepsilon$. According to Proposition 2 there exists a point $a \in P_\mu(c, A)$. Then $\varrho_{-\mu}(a, C) \leq \mu(c - a) = \varrho_\mu(c, A) \leq \varrho_0 + \varepsilon$. Therefore $a \in A_\varepsilon$ and $\mu(c - \hat{a}) \leq \mu(c - a) + \mu(a - \hat{a}) \leq \varrho_0 + \varepsilon + \text{diam } A_\varepsilon \leq \varrho_{-\mu}(\hat{a}, C) + \varepsilon + \text{diam } A_\varepsilon$. This implies inclusion 22. Inclusion 22 and Lemma 2 imply that diam $C_\varepsilon \to 0$ as $\varepsilon \to +0$. According to Remark 9 this completes the proof.

Example 1 (proposed by professor G.E. Ivanov). Let us consider a two-dimensional real vector space E and the function $g : \mathbb{R} \to \mathbb{R}$, $g(x) = x^2 - 1$. Put $M = \text{epi } g$, and let $\mu : E \to \mathbb{R}$ be the Minkowski function of M. The pair (E, μ) is an asymmetric seminormed space.

Consider function $a > 0$ and $f : \mathbb{R} \to \mathbb{R} \cup \{+\infty\}$,

$$f = \begin{cases} -ax^2, & |x| \leq 1, \\ +\infty, & |x| > 1. \end{cases}$$

Put $A = \text{epi } f$ and $C = \{(x_0, -0.5)\}$.

Consider the infimal convolution of functions f and g

$$f \boxplus g(x) = \inf_{u \in \mathbb{R}} (f(u) + g(x - u)).$$

Let us prove that for $a \in (0, 1)$ the problem is well-posed, and for $a = 1$ the problem is ill-posed. Consider $a \in (0, 1)$. Then

$$f \boxplus g(x_0) = \inf_{u \in [-1,1]} (-au^2 + (x_0 - u)^2 - 1)$$

$$= \min_{u \in [-1,1]} ((1 - a)u^2 + x_0^2 - 2x_0 u - 1) = -1 - \frac{ax_0^2}{1 - a}.$$

Thus $u_min = -\frac{x_0}{1-a}$, and the problem is well-posed for any $x_0 \in \mathbb{R}$ (see also [13]).

Consider now $a = 1$. Then

$$f \boxplus g(x_0) = \inf_{u \in [-1,1]} (-u^2 + (x_0 - u)^2 - 1) = \inf_{u \in [-1,1]} (x_0^2 - 2x_0 u - 1).$$

We obtain that if $x_0 = 0$, then the minimum is attained on $[-1, 1]$. If we consider $x_0 \neq 0$, then the minimum is attained at $u = \text{sign } x_0$. Thus the argmin $(\cdot)$ function of $-u^2 + (x_0 - u)^2 - 1$ is not continuous, and the problem is ill-posed.

6   Conclusion

The well posedness of the nearest points problems is closely related to the well posedness of the infimal convolution, which is a very useful instrument in optimization problems.

The author thanks G.E. Ivanov for fruitful discussions and help.

References

1. Borodin, P.: On the convexity of n-chebyshev sets. Izv. RAN. Ser. Mat. **75**(5), 19–46 (2011)
2. Clarke, F.H., Stern, R.J., Wolenski, P.R.: Proximal smoothness and lower-c^2 property. J. Convex Anal. **2**(1–2), 117–144 (1995)
3. Cobzas, S.: Separation of convex sets and best approximation in spaces with asymmetric norm. Quaestiones Math. **27**(3), 275–296 (2004)
4. Cobzas, S.: Ekeland variational principle in asymmetric locally convex spaces. Topol. Appl. **159**(10–11), 2558–2569 (2012)
5. Cobzas, S.: Functional Analysis in Asymmetric Normed Spaces. Birkhauser, Basel (2013)
6. Federer, H.: Curvature measures. Trans. Am. Math. Soc. **93**, 418–491 (1959)
7. Ivanov, G.E., Golubev, M.O.: Alternative theorem for differential games with strongly convex admissible control sets. In: Evtushenko, Y., Jaćimović, M., Khachay, M., Kochetov, Y., Malkova, V., Posypkin, M. (eds.) OPTIMA 2018. CCIS, vol. 974, pp. 321–335. Springer, Cham (2019). https://doi.org/10.1007/978-3-030-10934-9_23
8. Ivanov, G., Golubev, M.: Strong and weak convexity in nonlinear differential games. IFAC-PapersOnLine **51**, 13–18 (2018)
9. Ivanov, G.E., Lopushanski, M.S., Golubev, M.O.: The nearest point theorem for weakly convex sets in asymmetric seminormed spaces. In: Evtushenko, Y., Jaćimović, M., Khachay, M., Kochetov, Y., Malkova, V., Posypkin, M. (eds.) OPTIMA 2018. CCIS, vol. 974, pp. 21–34. Springer, Cham (2019). https://doi.org/10.1007/978-3-030-10934-9_2
10. Ivanov, G., Lopushanski, M.: Well-posedness of approximation and optimization problems for weakly convex sets and functions. J. Math. Sci. (United States) **209**, 66–87 (2015)
11. Ivanov, G.E.: On well posed best approximation problems for a nonsymmetric seminorm. J. Convex Anal. **20**(2), 501–529 (2013)

12. Ivanov, G.E.: Continuity and selections of the intersection operator applied to nonconvex sets. J. Convex Anal. **22**(4), 939–962 (2015)
13. Ivanov, G.E.: Weak convexity of sets and functions in a banach space. J. Convex Anal. **22**(2), 365–398 (2015)
14. Ivanov, G.E., Lopushanski, M.S.: Separation theorem for nonconvex sets and its applications. Fundam. Appl. Math. **21**(4), 23–65 (2016)
15. Ivanov, G.E., Lopushanski, M.S.: Separation theorems for nonconvex sets in spaces with non-symmetric seminorm. J. Math. Inequalities Appl. **20**(3), 737–754 (2017)
16. Ivanov, G.: Weak convexity of functions and the infimal convolution. J. Convex Anal. **23**, 719–732 (2016)
17. Ivanov, G., Thibault, L.: Infimal convolution and optimal time control problem ii: limiting subdifferential. Set-Valued Variational Anal. **25**(3), 517–542 (2017)
18. Ivanov, G., Thibault, L.: Infimal convolution and optimal time control problem i: Fréchet and proximal subdifferentials. Set-Valued Variational Anal. **26**, 581–606 (2018)
19. Ivanov, G., Thibault, L.: Infimal convolution and optimal time control problem iii: minimal time projection set. SIAM J. Optim. **28**(1), 30–44 (2018)
20. Ivanov, G., Thibault, L.: Well-posedness and subdifferentials of optimal value and infimal convolution. Set-Valued Variational Anal. **27**, 841–861 (2019)
21. Jordan-Pérez, N., Sánchez-Perez, E.: Extreme points and geometric aspects of compact convex sets in asymmetric normed spaces. Topol. Appl. **203**, 15–21 (2016)
22. Lopushanski, M.: Weakly convex sets in asymmetric normed spaces. In: Abstracts of the International Conference "Constructive Nonsmooth Analysis and Related Topics" Dedicated to the Memory of Professor V. F. Demyanov, pp. 34–38. Sankt-Petersburg (2017)
23. Lopushanski, M.S.: Normal regularity of weakly convex sets in asymmetric normed spaces. J. Convex Anal. **25**(3), 737–758 (2018)
24. Poliquin, R.A., Rockafellar, R.T., Thibault, L.: Local differentiability of distance functions. Trans. Am. Math. Soc. **352**, 5231–5249 (2000)

Two Optimization Problems
for a Material Point Moving Along
a Straight Line in the Presence of Friction
and Limitation on the Velocity

Nikolai Osmolovskii$^{1(\boxtimes)}$ ⓘ, Adam Figura^{2}ⓘ, and Marian Kośka^{2}ⓘ

^{1} Systems Research Institute, Polish Academy of Sciences,
ul. Newelska 6, 01-447 Warsaw, Poland
Nikolai.Osmolovskii@ibspan.waw.pl
^{2} University of Technology and Humanities,
ul. Malczewskiego 20A, 26-600 Radom, Poland
{dziekan.wim,km}@uthrad.pl

Abstract. We discuss two optimization problems: minimization of time and minimization of energy – for a material point, controlled by a limited force, moving along a straight line in the presence of friction and under limitation on the velocity. In the second problem, the time interval is fixed, and the recuperation of energy is taken into account. We describe extremals of these problems, satisfying the maximum principle for problems with state constraints in the Dubovitskii – Milyutin form.

Keywords: Pontryagin's maximum principle · State constraint · Optimal control · Singular arc · Energy consumption

1 Introduction

In this paper, we consider a train (a tram, a bus, a trolley, etc.) as a material point of the mass equal to one moving along a given segment of a horizontal straight line, under the influence of traction or breaking force, in the presence of friction and subject to limitation on the velocity. The initial position and the initial velocity, the final position and the final velocity of the point are fixed. The traction and breaking forces are limited and considered as a control. We consider two optimality criteria for this problem: minimization of time and minimization of energy. In the second case, the time interval is fixed and the system admits a recuperation of a part of the energy, i.e. returning it to the system during deceleration. We study these two problems using the maximum principle (MP) for problems with state constraints in the Dubovitskii-Milyutin form.

There is an important literature on the subject, see, e.g., [1–4,7] and the literature therein. This article is a presentation of the ideas, developed in [7] and [2].

© Springer Nature Switzerland AG 2020
M. Jaćimović et al. (Eds.): OPTIMA 2019, CCIS 1145, pp. 479–493, 2020.
https://doi.org/10.1007/978-3-030-38603-0_35

The paper is organized as follows. In Sect. 2, we give a formulation of the time optimal control problem for a material point, which is briefly called Problem A. We note that the solution exists in this problem. In Sect. 3, for the reader's convenience, we formulate the Dubovitskii-Milyutin maximum principle, first for a class of problems with state constraint, containing Problem A, and then for this particular problem. In Sect. 4, we describe all possible types (A)-(E) of extremals of Problem A, and note that the extremals of the type (E) does not satisfy the second order necessary optimality condition. Therefore one should find an optimal solution among extremals of the types (A)-(D). At the end of Sect. 4 we give a numerical example proposed by H. Maurer, where the solution of Problem A was found for given data. The energy optimal control problem, briefly called Problem B, is formulated in Sect. 5. In Sect. 6, we recall the formulation of the Dubovitskii-Milyutin maximum principle for the class of problems with state constraints, containing Problem B, and then we formulate it for this problem. In Sect. 7, we describe all possible types of extremals of Problem B. In Sect. 8, we summarize the results of the paper.

2   Time Optimal Control Problem

Let us give a formal description of the problem. Let the motion occur on an interval $[0, T]$, $T > 0$. Denote by $x(t) \in \mathbb{R}$ and by $y(t) \in \mathbb{R}$ the *position* and the *velocity* of the point at the time $t \in [0, T]$, respectively. Then $\dot{x}(t) = y(t)$ for all $t \in [0, T]$, where $\dot{x} = dx/dt$ is the derivative of x with respect to the time t. Denote by $u(t) \in \mathbb{R}$ the value of traction or breaking force acting on the material point at time t. The inequality $u > 0$ corresponds to the real *traction*, while the inequality $u < 0$ corresponds to the real *braking*. The function $u : [0, T] \to \mathbb{R}$ plays the role of the *control*, which belongs to an interval $[a, b]$ with $a < 0$ and $b > 0$. We consider $u : [0, T] \to \mathbb{R}$ as a measurable and essentially bounded function, while $x : [0, T] \to \mathbb{R}$ and $y : [0, T] \to \mathbb{R}$ are assumed to be Lipschitz continuous functions. We call $x(\cdot)$ and $y(\cdot)$ the *state* variables.

According to Newton's second law, the dynamics of the point satisfies the equation

$$\ddot{x}(t) = -w(y(t)) + u(t) \quad \text{a.e. in} \quad [0, T],$$

where $w(y)$ is the *force of resistance to the motion (friction)*, depending on the velocity y.

Assumption 1. The function $w : \mathbb{R} \to \mathbb{R}$ is odd, continuous, twice continuously differentiable on the half-line $(0, \infty)$, and satisfies the conditions

$$w'(y) > 0 \quad \text{and} \quad w''(y) \geq 0 \quad \text{for all} \quad y > 0.$$

Note, that Assumption 1 implies: $w(0) = 0$ and $w(y) > 0$ for all $y > 0$.

For example, let $k_1 > 0$ and $k_2 \geq 0$. Then the function, determined for $y > 0$ by $w(y) = k_1 y + k_2 y^2$, prolonged oddly on the entire axis, satisfies conditions of Assumption 1.

Thus we come to the control system

$$\dot{x}(t) = y(t) \quad \text{a.e. in} \quad [0, T], \tag{1}$$

$$\dot{y}(t) = -w(y(t)) + u(t) \quad \text{a.e. in} \quad [0, T], \tag{2}$$

$$u(t) \in [a, b] \quad \text{a.e. in} \quad [0, T]. \tag{3}$$

As was said, the initial position, the initial velocity, the terminal position, and the terminal velocity are fixed, that is

$$x(0) = x_0, \quad y(0) = y_0, \quad x(T) = x_T, \quad y(T) = y_T, \tag{4}$$

where x_0, x_T, y_0, y_T are prescribed values, $x_0 < x_T$.

There is a state constraint of the form

$$y(t) \leq V \quad \text{for all} \quad t \in [0, T], \tag{5}$$

where $V > 0$ is the maximal possible speed. To avoid the trivial maximum principle, we make the following assumption.

Assumption 2. $0 \leq y_0 < V$ and $0 \leq y_T < V$.

Also we will need the following important assumption.

Assumption 3. $w(V) < b$.

This implies that if $0 \leq y \leq V$, then $0 \leq w(y) \leq w(V)$, and hence $-w(y) + b \geq -w(V) + b > 0$, while $-w(y) + a < 0$. Consequently, the case $u = b$ corresponds to acceleration, and the case $u = a$ corresponds to deceleration.

Our goal now is to minimize the time of movement:

$$\text{minimize} \quad T. \tag{6}$$

For brevity, problem (1)–(6) is called the *Problem A*. Obviously, an optimal process $(x(\cdot), y(\cdot), u(\cdot), T)$ satisfies

Assumption 4. $y(t) \geq 0$ for all $t \in [0, T]$, and there is no interval $[t_1, t_2] \subset [0, T]$ of positive measure such that $y(t) = 0$ for all $t \in [t_1, t_2]$.

Otherwise the time T can be reduced. We consider only processes satisfying this assumption.

In [7], it was shown that the optimal trajectory in this problem exists, and hence it should be chosen among extremals satisfying the maximum principle.

3   The Dubovitskii-Milyutin Maximum Principle for Time Optimal Control Problem

3.1   Maximum Principle for a General Problem of the Type A

For convenience of the reader, we first formulate the maximum principle obtained by Dubovitskii and Milyutin for a class of problems, containing Problem A, see [3] or [5]. Consider the following optimal control problem:

$$\text{minimize} \quad T \tag{7}$$

subject to

$$\dot{x}(t) = f(x(t), u(t)), \quad u(t) \in U \quad \text{for a.a.} \quad t \in [0, T], \tag{8}$$

$$x(0) = x_0, \quad x(T) = x_T, \tag{9}$$

$$\varphi(x(t)) \leq 0 \quad \text{for all} \quad t \in [0, T]. \tag{10}$$

Here the *state variable* $x : [0, T] \to \mathbb{R}^n$ is a Lipschitz continuous function, the *control variable* $u : [0, T] \to \mathbb{R}^m$ is a measurable and essentially bounded function, the mapping $f : \mathbb{R}^{n+m} \mapsto \mathbb{R}^n$ is assumed to be continuous together with its partial derivative f_x, the mapping $\varphi : \mathbb{R}^n \mapsto \mathbb{R}$ is continuously differentiable, $x_0, x_T \in \mathbb{R}^n$ are given vectors, and $U \subset \mathbb{R}^m$ is an arbitrary set.

A pair of functions $(x(\cdot), u(\cdot))$ together with their domain of definition $[0, T]$ is called the *process* of the problem. A process $(x(\cdot), u(\cdot), T)$ is called *admissible* if it satisfies all constraints of the problem. An admissible process $(\hat{x}(\cdot), \hat{u}(\cdot), \hat{T})$ is called a *strong local minimum* if there is an $\varepsilon > 0$ such that $T \geq \hat{T}$ for all admissible processes satisfying $|T - \hat{T}| < \varepsilon$ and $|x(t) - \hat{x}(t)| < \varepsilon$ for all $t \in [0, T] \cap [0, \hat{T}]$.

In order to formulate necessary conditions for a strong local minimum for a process $(\hat{x}(\cdot), \hat{u}(\cdot), \hat{T})$, we introduce the *Pontryagin function* (or *pre-Hamiltonian*)

$$H(x, u, p) := p\, f(x, u), \tag{11}$$

where p is a row vector of the dimension n. The function

$$\mathcal{H}(x, p) := \sup_{u \in U} H(x, u, p) \tag{12}$$

is called the *Hamiltonian*.

We say that an admissible process $(\hat{x}(\cdot), \hat{u}(\cdot), \hat{T})$ satisfies conditions of the *maximum principle* if there exist left-continuous functions of bounded variation $p : [0, T] \to \mathbb{R}^n$ and $\mu : [0, T] \to \mathbb{R}$, defining measures $\mathrm{d}p$ and $\mathrm{d}\mu$, respectively, such that

$$\mathrm{d}\mu \geq 0, \quad \varphi(\hat{x}(\cdot))\, \mathrm{d}\mu = 0, \tag{13}$$

$$\mathrm{Var}(p) + \int_{[0, T]} \mathrm{d}\mu > 0, \tag{14}$$

$$- \, \mathrm{d}p = H_x(\hat{x}(\cdot), \hat{u}(\cdot), p(\cdot)) \, \mathrm{d}t - \varphi'(x(\cdot)) \, \mathrm{d}\mu, \tag{15}$$

$$\max_{u \in U} H(\hat{x}(t), u, p(t)) = H(\hat{x}(t), \hat{u}(t), p(t)) \quad \text{for a.a. } t \in [0, T], \tag{16}$$

$$H(\hat{x}(t), \hat{u}(t), p(t)) = \text{const} =: \alpha_0 \geq 0 \quad \text{for a.a. } t \in [0, T]. \tag{17}$$

Here, conditions (13) are called the *nonnegativeness of the measure* and *complementarity condition*, respectively, inequality (14) is called the *nontriviality condition*, (15) is the *adjoint equation*, (16) is the *maximum condition for the Pontryagin function*, and (17) is the condition of the *constancy and nonnegativeness of the Hamiltonian*.

Theorem 1. *If a process $(\hat{x}(\cdot), \hat{u}(\cdot), \hat{T})$ is a strong local minimum in problem (7)–(10), then it satisfies conditions of the maximum principle.*

3.2 Maximum Principle for Problem A

Observe, that for Problem A we have:

$$m = 1, \ n = 2, \quad U = [a, b],$$

$$f_1(x, y) = y, \ f_2(x, y) = -w(y) + u, \quad \varphi(x, y) = y - V,$$

and obviously, all assumptions of problem (7)–(10) are fulfilled. According to (11), the Pontryagin function for Problem A has the form:

$$H = p_1 y + p_2(-w(y) + u). \tag{18}$$

Let a process $(x(\cdot), y(\cdot), u(\cdot), T)$ be *admissible* in Problem A, and $y(t) \geq 0$ for all $t \in [0, T]$. The maximum principle for this process consists in the following: there exist left-continuous functions of bounded variation $p_1 : [0, T] \to \mathbb{R}$, $p_2 : [0, T] \to \mathbb{R}$, and $\mu : [0, T] \to \mathbb{R}$ such that

$$\mathrm{d}\mu \geq 0, \quad (y(\cdot) - V) \, \mathrm{d}\mu = 0, \tag{19}$$

$$\text{Var}(p_1) + \text{Var}(p_2) + \int_{[0, T]} \mathrm{d}\mu > 0, \tag{20}$$

$$- \, \mathrm{d}p_1 = 0, \quad - \, \mathrm{d}p_2 = (p_1(\cdot) - p_2(\cdot) w'(y(\cdot))) \, \mathrm{d}t - \mathrm{d}\mu, \tag{21}$$

$$\max_{u \in [a, b]} p_2(t) u = p_2(t) u(t) \quad \text{a.e. in} \quad [0, T], \tag{22}$$

$$p_1(t) y(t) + p_2(t)(-w(y(t)) + u(t)) = \text{const} =: \alpha_0 \geq 0 \quad \text{a.e. in} \quad [0, T]. \tag{23}$$

Let a triple of multipliers $(p_1(\cdot), p_2(\cdot), \mathrm{d}\mu)$ satisfy conditions (19)–(23) of the maximum principle. According to (21), $\mathrm{d}p_1 = 0$, consequently $p_1 = const$. Set

$$\beta = -p_1, \quad p = p_2.$$

Then the second equation in (21) takes the form

$$\mathrm{d}p = (p(\cdot)w'(y(\cdot)) + \beta)\,\mathrm{d}t + \mathrm{d}\mu, \qquad (24)$$

and from the maximum condition (22) it follows that

$$
\begin{array}{llll}
\text{if} & p(t) < 0, & \text{then} & u(t) = a, & (25) \\
\text{if} & p(t) = 0, & \text{then} & u(t) \in [a, b], & (26) \\
\text{if} & p(t) > 0, & \text{then} & u(t) = b. & (27)
\end{array}
$$

As was said, the first case corresponds to *deceleration*, the third one is called *acceleration*. the second one is said to be a *singular regime*.

It was proved in [7] that (due to the fact that the state constraint has the order one, that is, control appears after the first differentiation of the state constraint) the measure $\mathrm{d}\mu$ is absolutely continuous, and hence $\mathrm{d}\mu = \dot{\mu}\,\mathrm{d}t$, where the density $\dot{\mu}$ is an integrable function. Then the same is true for the measure $\mathrm{d}p$, and moreover, the densities $\dot{\mu}$ and $\dot{p}$ are measurable and essentially bounded functions. This implies that the adjoint equation can be written in the form

$$\dot{p}(t) = p(t)w'(y(t))) + \beta + \dot{\mu}(t).$$

This fact considerably simplifies the analysis of the maximum principle, see [7].

4   Extremals of Problem A

It was shown in [7] that there are only five possible cases for the adjoint variable $p(\cdot)$, defining five types of extremals in Problem A (depending on initial and final conditions for $x(\cdot)$ and $y(\cdot)$):

(A) $p(t) > 0$ for all $t \in (0, T)$. In this case

$$u(t) = b \quad \text{a.e. in} \quad [0, T].$$

This corresponds to *acceleration* on the whole segment $[0, T]$.

(B) $p(t) < 0$ for all $t \in (0, T)$. In this case

$$u(t) = a \quad \text{a.e. in} \quad [0, T].$$

This corresponds to *deceleration* on the whole segment $[0, T]$.

(C) There is a point $t_1 \in (0, T)$ such that $p(t) > 0$ for all $t \in (0, t_1)$ and $p(t) < 0$ for all $t \in (t_1, T)$. In this case

$$u(t) = b \quad \text{a.e. in} \quad (0, t_1) \quad \text{and} \quad u(t) = a \quad \text{a.e. in} \quad (t_1, T).$$

This corresponds to "acceleration – deceleration" mode with one switching at the point $t_1 \in (0, T)$.

(D) Main case: there are two points $t_1, t_2 \in (0, T)$, $t_1 < t_2$, such that $p(t) > 0$ for all $t \in (0, t_1)$, $p(t) = 0$ for all $t \in [t_1, t_2]$, and $p(t) < 0$ for all $t \in (t_2, T)$. In this case

$$u(t) = b \quad \text{a.e. in} \quad (0, t_1)$$
$$u(t) = a \quad \text{a.e. in} \quad (t_2, T),$$
$$u(t) = w(V) \quad \text{a.e. in} \quad (t_1, t_2), \quad y(t) = V \quad \text{for all} \quad t \in [t_1, t_2].$$

This corresponds to "acceleration – singular – deceleration" mode with two switchings at the points $t_1, t_2 \in (0, T)$. Moreover, the domain $[t_1, t_2]$ of the *singular arc*, where $p(t) = 0$, coincide with the domain of the *boundary arc*, where $y(t) = V$ (see Fig. 1).

(E) There is a point $t_1 \in (0, T)$ such that $p(t) < 0$ for all $t \in (0, t_1)$, $p(t_1) = 0$, and $p(t) > 0$ for all $t \in (t_1, T)$. In this case

$$u(t) = a \quad \text{a.e. in} \quad (0, t_1) \quad \text{and} \quad u(t) = b \quad \text{a.e. in} \quad (t_1, T).$$

Moreover, $y(t_1) = 0$.

This corresponds to "deceleration – acceleration" mode with one switching at the point $t_1 \in (0, T)$ (see Fig. 2).

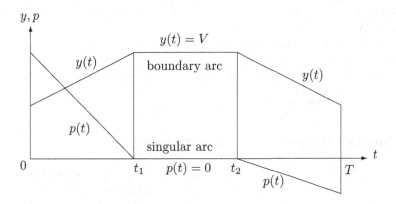

Fig. 1. Extremal D of Problem A

The decomposition in cases (A)–(E) is complete. In [7], it was shown that in case (E) the extremal $(x(\cdot), y(\cdot), u(\cdot), T)$ does not satisfy the second-order necessary conditions for a strong minimum, obtained by Osmolovskii in [6], and hence this extremal is not a strong local minimum in Problem A. Consequently, for each collection a_0, b_0, a_T, b_T of initial and final values of x and y, we must choose the optimal solution among extremals corresponding to the cases (A)–(D).

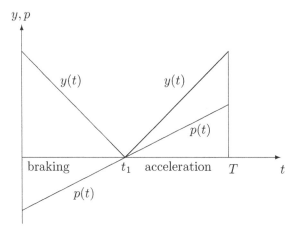

Fig. 2. Extremal E of Problem A

Example 1. The following numerical example belongs to H. Maurer. We choose the following data:

$$w(y) = 0.1\,y^2, \quad x(0) = y(0) = 0, \ x(T) = 10, \ y(T) = 0, \quad a = -1, \ b = 1.$$

Omitting the state $y(t) \leq V$ we see that the optimal control is bang-bang switching from $u(t) = 1$ to $u(t) = -1$ at $t_1 = 4.2507$. The minimal terminal time is $T = 6.51$. We find $\max y(t) = 2.7597$.

Hence, let us choose the state constraint $y(t) \leq V = 1.5$. We get the numerical results

$$T = 8.171755, \ p_1 = 2/3, \ p_2(0) = 1,$$

and the boundary arc in $[1.631, 6.771]$.

5   Energy Optimal Control Problem

Now we assume that the time interval $[0, T]$ is fixed. On this interval, we consider the same control system (1)–(3) with boundary conditions (4) and state constraint (5):

$$\dot{x}(t) = y(t) \quad \text{a.e. in} \quad [0, T],$$
$$\dot{y}(t) = -w(y(t)) + u(t) \quad \text{a.e. in} \quad [0, T],$$
$$u(t) \in [a, b] \quad \text{a.e. in} \quad [0, T],$$
$$x(0) = x_0, \quad y(0) = y_0, \quad x(T) = x_T, \quad y(T) = y_T,$$
$$y(t) \leq V \quad \text{for all} \quad t \in [0, T],$$

where again $x_T > x_0$ and Assumptions 1–3 are fulfilled.

Set $S = x_T - x_0$. Then S is a *specified route*, which can be calculated as

$$\int_0^T y(t)\, dt = S.$$

The *energy consumption* on the interval $[0, T]$ is equal to the integral

$$J = \int_0^T y(t)(u^+(t) - \varepsilon u^-(t))\, dt,$$

where $u^+ = \max\{u, 0\}$, $u^- = \max\{-u, 0\}$, and $0 \le \varepsilon < 1$ is the *recuperation coefficient*, i.e. the portion of energy returned to the system during deceleration; the value $\varepsilon = 0$ corresponds to the case when the energy does not return. Let us explain this formula. If $u(t) \ge 0$ on some interval $[t_1, t_2]$, i.e., there is no deceleration on this interval, then

$$\int_{t_1}^{t_2} y(t)(u^+(t) - \varepsilon u^-(t))\, dt$$

$$= \int_{t_1}^{t_2} y(t)u(t)\, dt = \int_{t_1}^{t_2} y(t)\dot{y}(t)\, dt = \frac{1}{2}(y(t_2)^2 - y(t_1)^2).$$

As is known, this difference is the increment of the kinetic energy of the point of mass $m = 1$ on the interval $[t_1, t_2]$. If $u(t) \le 0$ on some interval $[t_1, t_2]$, i.e., there is no acceleration on this interval, then $u^+(t) = 0$, $u^-(t) = -u(t)$ on $[t_1, t_2]$ and

$$\int_{t_1}^{t_2} y(t)(u^+(t) - \varepsilon u^-(t))\, dt = \varepsilon \int_{t_1}^{t_2} y(t)u(t)\, dt = \frac{\varepsilon}{2}(y(t_2)^2 - y(t_1)^2).$$

The modulus of this value is the portion of kinetic energy returning to the system on the interval $[t_1, t_2]$.

It is obvious, that the variable x may be removed. Indeed, if we know the pair $(y(\cdot), u(\cdot))$ such that $\int_0^T y(t)\, dt = S$, then the function $x(t) = x_0 + \int_0^t x(\tau)\, d\tau$ satisfies $x(0) = x_0$, $x(T) = x_T$, $\dot{x} = y$.

Moreover, without loss of generality, we can take $a = -1$ (passing, if necessary, to the values of the control in different conventional units).

Thus we come to the problem:

$$\text{minimize} \quad J = \int_0^T y(t)(u^+(t) - \varepsilon u^-(t))\, dt, \tag{28}$$

subject to

$$\dot{y}(t) = -w(y(t)) + u(t), \quad u(t) \in [-1, b] \quad \text{a.e. in} \quad [0, T], \tag{29}$$

$$y(0) = y_0, \quad y(T) = y_T, \quad \int_0^T y(t)\, dt = S, \tag{30}$$

$$y(t) \le V \quad \text{for all} \quad t \in [0, T]. \tag{31}$$

For brevity, problem (28)–(31) will be called *Problem B*. Note that the optimal solution in this problem exists, see [2], and then it contains among extremals of this problem, satisfying the maximum principle.

6   The Dubovitskii-Milyutin Maximum Principle for Energy Optimal Control Problem

6.1   Maximum Principle for a General Problem of the Type B

Again, for convenience of the reader, we first formulate the Dubovitskii-Milyutin maximum principle, see [3] or [5], obtained for a class of problems, containing Problem B. Consider the following optimal control problem on a fixed time interval $[0, T]$:

$$\text{minimize} \quad J(y, u) := \int_0^T F(y(t), u(t)) \, dt \tag{32}$$

subject to

$$\dot{y}(t) = f(y(t), u(t)), \quad u(t) \in U, \quad \text{for a.a.} \quad t \in [0, T], \tag{33}$$

$$y(0) = y_0, \quad y(T) = y_T, \tag{34}$$

$$\int_0^T G(y(t), u(t)) \, dt = S, \tag{35}$$

$$\varphi(x(t)) \leq 0 \quad \text{for all} \quad t \in [0, T]. \tag{36}$$

Here the *state variable* $y : [0, T] \to \mathbb{R}^n$ is a Lipschitz continuous function, the *control variable* $u : [0, T] \to \mathbb{R}^m$ is a measurable and essentially bounded function, the mappings $f : \mathbb{R}^{n+m} \mapsto \mathbb{R}^n$, $F : \mathbb{R}^{n+m} \mapsto \mathbb{R}$, $G : \mathbb{R}^{n+m} \mapsto \mathbb{R}$, $\varphi : \mathbb{R}^n \to \mathbb{R}$ are assumed to be continuous together with their partial derivatives f_y, F_y, G_y, and φ_y; the vectors $y_0, y_T \in \mathbb{R}^n$ and the real $S \in \mathbb{R}$ are fixed, and $U \subset \mathbb{R}^n$ is an arbitrary set.

A pair of functions $(y(\cdot), u(\cdot))$ is called the *process* of the problem. A process $(y(\cdot), u(\cdot))$ is called *admissible* if it satisfies all constraints of the problem. An admissible process $(\hat{y}(\cdot), \hat{u}(\cdot))$ is called a *strong local minimum* if there is an $\varepsilon > 0$ such that $J(y, u) \geq J(\hat{y}, \hat{u})$ for all admissible processes satisfying $|y(t) - \hat{y}(t)| < \varepsilon$ for all $t \in [0, T]$.

In order to formulate necessary conditions for a strong local minimum for a process $(\hat{y}(\cdot), \hat{u}(\cdot))$, we introduce the *Pontryagin function*

$$H(y, u, p, \alpha, \beta) = p \, f(y, u) - \alpha \, F(y, u) - \beta \, G(y, u), \tag{37}$$

where p is a row vector of the dimension n, and α and β are real numbers.

We say that a process $(\hat{y}(\cdot), \hat{u}(\cdot))$ satisfies the conditions of the *Dubovitskii-Milyutin maximum principle* if there are real numbers α, β, left-continuous functions of bounded variation $p : [0, T] \to \mathbb{R}^n$ and $\mu : [0, T] \to \mathbb{R}$, defining measures dp and $d\mu$, respectively, such that

$$\alpha \geq 0, \quad d\mu \geq 0, \quad \varphi(\hat{y}(\cdot)) \, d\mu = 0, \tag{38}$$

$$\alpha + |\beta| + \text{Var}(p) + \int_{[0, T]} d\mu > 0, \tag{39}$$

$$- \, dp = H_y(\hat{y}(\cdot), \hat{u}(\cdot), p(\cdot)) \, dt - \varphi'(y(\cdot)) \, d\mu, \tag{40}$$

$$\max_{u \in U} H(\hat{y}(t), u, p(t)) = H(\hat{y}(t), \hat{u}(t), p(t)) \quad \text{for a.a. } t \in [0, T], \tag{41}$$

$$H(\hat{y}(t), \hat{u}(t), p(t)) = \text{const} \quad \text{for a.a. } t \in [0, T]. \tag{42}$$

Here, inequality (39) is called the *nontriviality condition*, (40) is the *adjoint equation*, (41) is the *maximum condition for the Pontryagin function*, and (42) is the condition of the *constancy of the Hamiltonian*.

Theorem 2. *If a process $(\hat{y}(\cdot), \hat{u}(\cdot))$ is a strong local minimum in problem (32)–(36), then it satisfies conditions of the Dubovitskii-Milyutin maximum principle.*

6.2   Maximum Principle for Problem B

For Problem B we have:

$$m = n = 1, \quad U = [-1, b], \quad f = -w(y) + u,$$

$$F = y(u^+ - \varepsilon u^-), \quad G = y, \quad \varphi = y - V,$$

and obviously, all assumptions of problem (32)–(36) are fulfilled. According to (37), the Pontryagin function for Problem B has the form:

$$H = p(-w(y) + u) - \alpha y(u^+ - \varepsilon u^-) - \beta y. \tag{43}$$

Let a pair $(y(\cdot), u(\cdot))$ be a strong minimum in Problem B. It is natural to assume that $y(t) \geq 0$ for all $t \in [0, T]$, i.e. there is no movement in the reverse direction.

The maximum principle for problem B at the point $(y(\cdot), u(\cdot))$ consists in the following: there exist real numbers $\alpha \geq 0$, β and left-continuous functions of bounded variation $p : [0, T] \to \mathbb{R}$ and $\mu : [0, T] \to \mathbb{R}$ such that

$$\alpha + |\beta| + \text{Var}(p) + \int_{[0,T]} d\mu > 0, \tag{44}$$

$$dp = (p(\cdot)w'(y(\cdot)) + \alpha(u^+(\cdot) - \varepsilon u^-(\cdot)) + \beta) \, dt + d\mu \quad \text{a.e. in} \quad [0, T], \tag{45}$$

$$\max_{u \in [-1, b]} (p(t)u - \alpha y(t)(u^+ - \varepsilon u^-))$$

$$= p(t)u(t) - \alpha y(t)(u^+(t) - \varepsilon u^-(t)) \quad \text{a.e. in} \quad [0, T], \tag{46}$$
$$p(t)(-w(y(t)) + u(t)) - \alpha y(t)(u^+(t) - \varepsilon u^-(t)) - \beta y(t)$$

$$= \text{const} \quad \text{a.e. in} \quad [0, T]. \tag{47}$$

If the cost multiplier α is equal to zero (the degenerate case), then, as is easily seen, we obtain the same extremals (y, u) as in the time optimal control problem - Problem A. Therefore, consider the case $\alpha > 0$, and then we can put $\alpha = 1$.

It was proved in [2] that (due to the fact that the state constraint has the order one) the measure $d\mu$ is absolutely continuous, and hence $d\mu = \dot{\mu}\,dt$, where the density $\dot{\mu}$ is an integrable function. Then the same is true for the measure dp, and moreover, the densities $\dot{\mu}$ and $\dot{p}$ are measurable and essentially bounded functions. This implies that the adjoint equations (45) with $\alpha = 1$ can be written in the form

$$\dot{p}(t) = p(t)w'(y(t)) + (u^+(t) - \varepsilon u^-(t)) + \beta + \dot{\mu},$$

and the absolute continuity of the measures considerably simplifies the analysis of the maximum principle, see [2].

It easily follows from the maximum condition (46) that

$$\text{if} \qquad p(t) < \varepsilon y(t), \qquad \text{then} \quad u(t) = -1, \tag{48}$$
$$\text{if} \qquad p(t) = \varepsilon y(t), \qquad \text{then} \quad u(t) \in [-1,0], \tag{49}$$
$$\text{if} \quad \varepsilon y(t) < p(t) < y(t), \quad \text{then} \quad u(t) = 0, \tag{50}$$
$$\text{if} \qquad p(t) = y(t), \qquad \text{then} \quad u(t) \in [0,b], \tag{51}$$
$$\text{if} \qquad p(t) > y(t), \qquad \text{then} \quad u(t) = b. \tag{52}$$

Let us show this. Indeed, for any $t \in [0, T]$, we have to maximize the function

$$h(t, u) := p(t)u - y(t)(u^+ - \varepsilon u^-), \quad u \in [-1, 0].$$

Observe that

$$u \geq 0 \quad \Rightarrow \quad h(t, u) = (p(t) - y(t))u, \tag{53}$$
$$u \leq 0 \quad \Rightarrow \quad h(t, u) = (p(t) - \varepsilon y(t))u. \tag{54}$$

(1) Suppose that $p(t) < \varepsilon y(t)$. Then $p(t) < y(t)$, since $y(t) \geq 0$, $0 < \varepsilon < 1$. If $u \geq 0$, then by (53), $h(t, u) \leq 0$. If $u < 0$, then by (54), $h(t, u) > 0$, and the maximum is attained for $u = -1$.

(2) Suppose that $p(t) = \varepsilon y(t)$. Then $p(t) \leq y(t)$. If $u > 0$, then by (53), $h(t, u) \leq 0$. If $u \leq 0$, then by (54), $h(t, u) = 0$. Hence every $u \in [-1, 0]$ is a point of maximum of $h(t, u)$.

Note that conditions (49) and (51) define *singular arcs*, while conditions (48), (50), and (52) correspond to *braking*, *overshoot*, and *acceleration*, respectively. Further analysis of the case $\alpha > 0$ see in [2].

7   Extremals of Problem B

In the non-degenerate case we have the following types of extremals (see [2]), which correspond to certain types of control $u(t)$. Below we will write $u = (u_1, \ldots, u_k)$ if for some division $0 < t_1 < \ldots < t_k < T$ of the interval $(0, T)$ into intervals (t_{i-1}, t_i), $i = 1, \ldots, k+1$ with $t_0 = 0$ and $t_{k+1} = T$ we have $u(t) = u_i$ a.e. in (t_{i-1}, t_i), $i = 1, \ldots, k+1$.

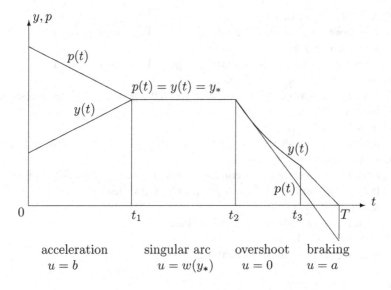

Fig. 3. Extremal A (basic) of Problem B

Case A (main): $u = (b, w(y_*), 0, -1)$, which corresponds to acceleration $u(t) = b$ in the interval $(0, t_1)$, movement with constant velocity $y(t) = y_* \leq V$ in the interval (t_1, t_2), overshoot $u(t) = 0$ in the interval (t_2, t_3), and deceleration (braking) $u(t) = -1$ in the interval (t_3, T), or briefly: acceleration $\to$ constant velocity $y = y_* \leq V \to$ overshoot $\to$ braking (see Fig. 3). In this case $p(t) > y(t)$ in the interval $(0, t_1)$, $p(t) = y(t)$ (singular arc) in the interval (t_1, t_2), $\varepsilon y(t) < p(t) < y(t)$ in the interval (t_2, t_3), and $p(t) < \varepsilon y(t)$ in the interval (t_3, T). At the point t_3 we have $p(t_3) = \varepsilon y(t_3)$.

When $t_1 = t_2$, i.e. when $p(t) = y(t)$ only in one point t_1, then the control has the type $u = (b, 0, -1)$, that means acceleration, overshoot, and braking.

Also from the Case A let us separate an important

Case AV: $u = (b, w(V), 0, -1)$, i.e.

  acceleration $\to$ constant maximal velocity $y = V \to$ overshoot $\to$ braking.

  The remaining cases, depending on the initial and final conditions for x and y, correspond to the following types of control.

Case B: $u = (b, w(y_*), 0)$, i.e.

  acceleration $\to$ constant velocity $y = y_* \leq V \to$ overshoot.

Case C: $u = (b, w(y_*), b)$, i.e.

  acceleration $\to$ constant velocity $y = y_* < V \to$ acceleration.

Case D: $u = (0, w(y_*), 0, -1)$, i.e.

  overshoot $\to$ constant velocity $y = y_* < V \to$ overshoot $\to$ braking.

Case E: $u = (0, w(y_*), 0)$, i.e.
    overshoot → constant velocity $y = y_* < V$ → overshoot.
Case F: $u = (0, w(y_*), b)$, i.e.
    overshoot → constant velocity $y = y_* < V$ → acceleration.
Case G: $u = (-1, 0, w(y_*), 0, -1)$, i.e.
    braking → overshoot → constant velocity $y = y_* < V$ → overshoot →
    braking.
Case H: $u = (-1, 0, w(y_*), 0)$, i.e.
    braking → overshoot → constant velocity $y = y_* < V$ → overshoot.
Case I: $u = (-1, 0, w(y_*), b)$, i.e.
    braking → overshoot → constant velocity $y = y_* < V$ → acceleration.
    If $t_2 = t_3$, from Case I we separate the following
Case J: $u = (-1, 0, b)$, i.e.
    braking → overshoot → acceleration.
Case K: $u = (-1, 0, -1)$, i.e.
    braking → overshoot → braking.

The remaining trajectories

$$u \equiv b, \quad u \equiv 0, \quad u \equiv -1, \quad u = (0, -1), \quad u = (-1, 0)$$

are special cases of the types given above; moreover, they are in some sense degenerate.

Thus all the trajectories which satisfy the MP with $\alpha > 0$ have one of the types A-K, and the optimal trajectory, for given boundary values for x and y, belongs to one of these types.

8   Conclusions

This paper presents all possible types of extremals for two optimal control problems: minimization of time and minimization of energy for a material point, controlled by a limited force, moving along a straight line in the presence of friction and under limitation on the velocity. An optimal solution exists in each of these problems, therefore it contains among extremals. One type of extremals in the time optimal control problem does not satisfy second order necessary optimality conditions and therefore cannot be optimal. All extremals of the time optimal control problem correspond to extremals of the energy optimal control problem with a cost multiplier equal to zero. A detailed study and description of the extremals of these problems, based on the Dubovitskii-Miltyutin maximum principle for problems with state constraints, can be found in [7] and [2].

References

1. Albrecht, A., Howlett, P., Pudney, P., Vu, X., Zhou, P.: Energy-efficient train control: the two-train separation problem on level track. J. Rail Transp. Plan. Manag. 5(3), 163–182 (2015)
2. Asnis, I.A., Dmitruk, A.V., Osmolovskii, N.P.: Solution of the problem of the energetically optimal control of the motion of a train by the maximum principle. USSR Comput. Math. Math. Phys. 25(6), 37–44 (1985)
3. Dubovitskii, A.Y., Milyutin, A.A.: Extremum problems in the presence of restrictions. USSR Comput. Math. Math. Phys. 5(3), 1–80 (1965)
4. Ichikawa, K.: Application of optimization theory for bounded stategy variable problems to the operation of train. Bull. Japans. Soc. Math. and Engng 11(47), 857–865 (1968)
5. Milyutin, A.A., Dmitruk, A.V., Osmolovskii, N.P.: Maximum principle in optimal control. Faculty of Mechanics and Mathematics of Moscow State University, Moscow (2004). (in Russian)
6. Milyutin, A.A., Osmolovskii, N.P.: Calculus of Variations and Optimal Control, vol. 180. American Mathematical Society, Providence (1998)
7. Osmolovskii, N.P., Figura, A., Kośka, M., Wójtowicz, M.: Extremals of the time optimal control problem for a material point moving along a straight line in the presence of friction and limitation on the velocity. Control Cybern. 46(4), 305–324 (2017)

Author Index

Printed in the United States
By Bookmasters